Handbook of Research on Resident and Tourist Perspectives on Travel Destinations

Patrícia Pinto
Faculty of Economics and Research Centre for Tourism, Sustainability, and Well-Being (CinTurs), University of Algarve, Portugal

Manuela Guerreiro
Faculty of Economics and Research Centre for Tourism, Sustainability, and Well-Being (CinTurs), University of Algarve, Portugal

IGI Global
PUBLISHER of TIMELY KNOWLEDGE

A volume in the Advances in Hospitality, Tourism, and the Services Industry (AHTSI) Book Series

Published in the United States of America by
IGI Global
Business Science Reference (an imprint of IGI Global)
701 E. Chocolate Avenue
Hershey PA, USA 17033
Tel: 717-533-8845
Fax: 717-533-8661
E-mail: cust@igi-global.com
Web site: http://www.igi-global.com

Library of Congress Cataloging-in-Publication Data

Names: Pinto, Patrícia, 1970- editor. | Guerreiro, Manuela, 1970- editor.
Title: Handbook of research on resident and tourist perspectives on travel destinations / Patrícia Pinto and Manuela Guerreiro, editors.
Description: Hershey, PA : Business Science Reference, 2020. | Includes bibliographical references and index. | Summary: "This book explores resident and tourist perspectives to understand the composite reality of tourism destinations and to support the development of marketing strategies for these destinations"-- Provided by publisher.
Identifiers: LCCN 2019048549 (print) | LCCN 2019048550 (ebook) | ISBN 9781799831563 (hardcover) | ISBN 9781799831570 (ebook)
Subjects: LCSH: Tourism--Public opinion. | Tourism--Marketing. | Tourists--Attitudes.
Classification: LCC G155.A1 H3527 2020 (print) | LCC G155.A1 (ebook) | DDC 338.4/791--dc23
LC record available at https://lccn.loc.gov/2019048549
LC ebook record available at https://lccn.loc.gov/2019048550

This book is published in the IGI Global book series Advances in Hospitality, Tourism, and the Services Industry (AHTSI) (ISSN: 2475-6547; eISSN: 2475-6555)

British Cataloguing in Publication Data
A Cataloguing in Publication record for this book is available from the British Library.

For electronic access to this publication, please contact: eresources@igi-global.com.

Advances in Hospitality, Tourism, and the Services Industry (AHTSI) Book Series

Maximiliano Korstanje
University of Palermo, Argentina

ISSN:2475-6547
EISSN:2475-6555

Mission

Globally, the hospitality, travel, tourism, and services industries generate a significant percentage of revenue and represent a large portion of the business world. Even in tough economic times, these industries thrive as individuals continue to spend on leisure and recreation activities as well as services.

The Advances in Hospitality, Tourism, and the Services Industry (AHTSI) book series offers diverse publications relating to the management, promotion, and profitability of the leisure, recreation, and services industries. Highlighting current research pertaining to various topics within the realm of hospitality, travel, tourism, and services management, the titles found within the AHTSI book series are pertinent to the research and professional needs of managers, business practitioners, researchers, and upper-level students studying in the field.

Coverage

- Service Training
- International Tourism
- Tourism and the Environment
- Destination Marketing and Management
- Cruise Marketing and Sales
- Health and Wellness Tourism
- Service Management
- Leisure & Business Travel
- Travel Agency Management
- Hotel Management

IGI Global is currently accepting manuscripts for publication within this series. To submit a proposal for a volume in this series, please contact our Acquisition Editors at Acquisitions@igi-global.com or visit: http://www.igi-global.com/publish/.

The Advances in Hospitality, Tourism, and the Services Industry (AHTSI) Book Series (ISSN 2475-6547) is published by IGI Global, 701 E. Chocolate Avenue, Hershey, PA 17033-1240, USA, www.igi-global.com. This series is composed of titles available for purchase individually; each title is edited to be contextually exclusive from any other title within the series. For pricing and ordering information please visit http://www.igi-global.com/book-series/advances-hospitality-tourism-services-industry/121014. Postmaster: Send all address changes to above address.

Titles in this Series

For a list of additional titles in this series, please visit:
http://www.igi-global.com/book-series/advances-hospitality-tourism-services-industry/121014

Accelerating Knowledge Sharing, Creativity, and Innovation Through Business Tourism
Muhammad Waseem Bari (Government College University, Faisalabad, Pakistan) Sadia Shaheen (Government College University, Faisalabad, Pakistan) and Meng Fanchen (Beijing Institute of Technology, China)
Business Science Reference • © 2020 • 300pp • H/C (ISBN: 9781799831426) • US $225.00

Contemporary Management Approaches to the Global Hospitality and Tourism Industry
Abraham Pius (Arden University, UK) Husam H. Alharahsheh (University of Wales Trinity Saint David, UK) and Adenike A. Adesanmi (University of East London, UK)
Business Science Reference • © 2020 • 290pp • H/C (ISBN: 9781799822042) • US $225.00

Global Entrepreneurial Trends in the Tourism and Hospitality Industry
Debasish Batabyal (Amity University, Kolkata, India)
Business Science Reference • © 2020 • 300pp • H/C (ISBN: 9781799826033) • US $225.00

Handbook of Research on the Impacts, Challenges, and Policy Responses to Overtourism
Cláudia Ribeiro de Almeida (University of Algarve, Portugal) Alfred Quintano (Independent Researcher, Malta) Moisés Simancas (University of La Laguna, Spain) Raquel Huete (University of Alicante, Spain) and Zélia Breda (University of Aveiro, Portugal)
Business Science Reference • © 2020 • 447pp • H/C (ISBN: 9781799822240) • US $285.00

Economic, Educational, and Touristic Development in Asia
Bryan Christiansen (Global Training Group, Ltd, UK) and Hakan Sezerel (Anadolu University, Turkey)
Business Science Reference • © 2020 • 241pp • H/C (ISBN: 9781799822394) • US $215.00

Industrial and Managerial Solutions for Tourism Enterprises
Atilla Akbaba (Izmir Katip Çelebi University, Turkey) and Volkan Altıntaş (Izmir Katip Celebi University, Turkey)
Business Science Reference • © 2020 • 336pp • H/C (ISBN: 9781799830306) • US $225.00

Handbook of Research on Smart Technology Applications in the Tourism Industry
Evrim Çeltek (Gaziosmanpasa University, Turkey)
Business Science Reference • © 2020 • 569pp • H/C (ISBN: 9781799819899) • US $295.00

IGI Global
PUBLISHER of TIMELY KNOWLEDGE

701 East Chocolate Avenue, Hershey, PA 17033, USA
Tel: 717-533-8845 x100 • Fax: 717-533-8661
E-Mail: cust@igi-global.com • www.igi-global.com

List of Contributors

Table of Contents

Detailed Table of Contents

The main aim of this chapter is to propose and test a model to reflect on the factors that, associated with the tourist development of a destination, are perceived by the local community to be influential at the personal and collective level, as well as to what extent they impact people's satisfaction with the community they belong to. An enclave with a fishing tradition in the province of Huelva (Spain) was selected for primary data collection. PLS technique was used for hypotheses testing. It is concluded the contingent nature of the categorisation of the impacts with an influence on the local community's perception of tourism development, in addition to the inexistence of a direct relationship between the perception of the impacts on a personal level and residents' satisfaction with their community, although an indirect relationship exists through the perception of the effects on the town as a whole.

This study aims to identify the factors that influence subjective well-being and their support for tourism development. The questionnaire was first designed and distributed online to residents in Alxa. Data were analysed, and the results indicated that positive economic impacts, positive socio-cultural impacts, and positive environmental impacts influence subjective well-being. Residents' perceived negative economic impacts and negative environmental impacts influence negatively their subjective well-being. Also, positive environmental impacts of the tourism development are the most significant factor that affects residents' subjective well-being. Furthermore, subjective well-being was also found to positively influence residents' support for the tourism development in Alxa. Ultimately, based on the findings, this study proposed and discuss both theoretical and managerial implications, as well as the limitations and directions of further studies.

The present study aims to examine residents' support for sustainable tourism. Karuizawa town was selected due to the increasing number in both domestic and foreign visitors, as well as the existence of seasonal residents. Residents' participation in tourism and their support are vital for sustainable tourism. In doing so, this study primarily used the theory of reasoned action, social exchange theory, and emotional solidarity to examine residents' support for sustainable tourism. Data were collected using qualitative approach, and semi-structured interviews were adopted to explore the relationship between theoretical frameworks. Twenty participants were selected by non-probability sampling to examine differences among residents in tourism-related businesses and those who are not, and among permanent and seasonal residents. Deductive thematic analysis was chosen because of the ability to produce themes from theoretical frameworks. Lastly, an in-depth discussion, as well as theoretical and managerial implications are offered along with the limitations and avenues for future studies.

This chapter explores how local community involvement contributed to reviving cultural and natural heritage as assets for sustaining tourism on a small offshore island in Hong Kong. Yim Tin Tsai is a village established by the Hakka Chan clan and the main economic activity was salt production in the past. After the village became deserted due to villagers' migration, the existing salt fields were revived by the community. This process has increased the interest of visitors to travel to the village to experience the cultural, religious, and natural heritage of the destination. This case study combines in loco observations, interviews with key stakeholders on the island, and consultation of secondary sources. Opportunities and challenges regarding community involvement in tourism and its contribution to the conservation of tangible and intangible heritage assets of this offshore island are discussed.

The purpose of this study is to examine residents' support for tourism development in South Korea by the lens of social exchange theory and theory of reasoned. Random sampling strategy was used to collect South Koreans residents' opinions about the perceived impacts of tourism and their level of support for tourism development. Data were collected online using a questionnaire a link was sent to residents in South Korean through several social media platforms, and 693 respondents answered the questionnaire. By using regression analysis, the findings showed that residents who perceive positive impacts of tourism more than negative impacts are more likely to support tourism, while those who perceive more negative impacts of tourism are less likely to support tourism. Both personal economic benefits and community attachment were found to be important determinants of residents' perception and support toward tourism. Both theoretical and managerial implications are discussed, as well as the limitations and recommendations for future studies.

This chapter offers an in-depth analysis of cultural tourism and namely of tourists' awareness of Portuguese World Heritage (WH). In accordance, firstly, there is the discussion of a theoretical framework associated to a cultural role and cultural identity in the sustainable development of tourism is presented. Secondly, it follows a case study approach based on the evaluation of Algarvian tourists' awareness of Portuguese WH. A questionnaire was applied in the Algarve region with the aim to determine the importance of cultural motivation and the level of knowledge regarding Portuguese culture by tourists. The results of this exploratory analysis underline the growing importance of cultural motivation and the recognition of culture's importance in the sustainable development of destinations. However, it also shows the low level of cultural motivation by Algarvian tourists as well as the low level of Portuguese WH knowledge, namely by foreign tourists. Nationality, age, academic background are key variables in tourism knowledge about Portuguese culture.

Creative tourism is characterized by the opportunity that gives visitors to develop their creativity through active participation in learning experiences that are characteristic of their destination. That implies a relationship between visitors and residents in which the exchange of knowledge is central. Being an interactive process, the learning and doing of a creative tourism activity is an exercise of self-affirmation that expresses the creative potential of each participant. That is to say, creative tourism experiences are co-created by tourists and their hosts. Tourists and residents are co-designers in creative tourism, as the CREATOUR experiences prove. In this chapter, the authors intend to present the project activities in the Algarve. After contextualizing co-creation within the creative processes, their research is centred on the creative experiences and discusses the opportunities resulting from co-created activities. They also look at the survey applied to the participants in the experiences to better understand their feelings towards co-creation and creative processes.

This chapter aimed to analyse the profile and patterns of consumption of non-resident spectators of a sporting event that took place in an insular tourist city: Funchal (Madeira Island, Portugal). Data collection was performed at the entrance of the football stadium where 108 (n) spectators were randomly surveyed. Data were analysed based on descriptive and inferential statistics. Results showed that on average, each tourist spent 231 euros in the city; the highest cost was due to accommodation and secondly to tickets; 'arrival time at the stadium' was the variable that most positively influenced the total expenditure; individuals with higher education/wage levels had a greater predisposition to spend more money on food/drink and accommodation. As spectators' opinion on the city/stadium showed high levels of satisfaction, the event improved the visitors' perception and image of the city. This study highlights how sporting events can increase the insular tourism. Future research should replicate this study in similar events hosted in comparable destinations.

The relationship between tourism and sex has been intensively explored over the past decades. The great majority of the studies examining this relationship focus disproportionately on commercial sex tourism as if all the sexual encounters in tourism context are related to prostitution, commercialization, trafficking, inequalities, and disrespect for the human rights. Despite these contributions, the purpose of this chapter is to show that most of the romantic and sexual encounters that occur in a tourism setting are positive, gratifying, and nonpecuniary. Therefore, these relationships should be analyzed outside "sex tourism paradigm." Results from 29 in-depth interviews with tourists in the Algarve (South Portugal) confirm the strong relationship between holidays and romantic and sexual behaviors, due to the liminal nature of tourism. In addition, results show different perspectives and behaviors according to gender, due to the continuous sexual double standard.

According to the World Tourism Organization (UNWTO), in 2017 tourism had the greatest international growth in seven years, and in 2018, international tourism grew 5% reaching the mark of 1.4 billion, a figure reached two years earlier than predicted. At the same time, in the last 40 years, the video game industry has grown steadily, with games beginning to be seen as one of the primary sources of entertainment. However, there are still few studies analyzing the impact of advertising tourist destinations on digital platforms such as video games. The use of video games in the tourist context may be an inspirational tool, supporting the development of new advertising strategies for tourism marketing. Although the connection between tourism and cinema is widely documented, little research has demonstrated a credible correlation between video games and tourists' attitude towards destinations.

Preface

The tourism destination corresponds to a geographical area with a coherent structure of offer of tourism products and services, marketed in an integrated manner and under a distinctive brand. The definition of the tourism destination in spatial terms may have different scales, so it is possible to apply the concept to countries, regions or cities.

An amalgam of specific services and products (accommodation, transport, catering, entertainment, among others), from the private or public initiative, and the local attributes (landscapes, sea, beach, heritage, among others), create a global and composite offer. Tourism destinations must be managed to provide satisfaction to tourists who increasingly look for unique and memorable experiences in the visited places. Indeed, tourism is an economic and social phenomenon centred on the tourists' experience and going beyond the ordinary (place of residence) to the extraordinary (visited destination). In this gaze, tourists' personal drivers not always enable the desired outcomes of maximum satisfaction and intention to return, since these are mediated by the destination image (before and after the trip) and by the experience (during the trip). These elements are critical elements for the quality of the whole experience in a sector (tourism) so dependent on the experiences co-created and provided. So, the focal point of any tourism destination manager and stakeholders is to know what tourists perceive as engaging, intense and memorable.

Moreover, tourist activities should contribute to community residents' well-being and their quality of life. Currently, there is a growing awareness of the need to preserve and involve local populations and entities in preparing the place where they reside to receive tourists in search for opportunities to escape their daily routine, enjoying authentic experiences related to local culture and the identity of places. So, besides looking at the visitors of tourism destinations, destinations management organizations should also give attention to the communities living in the visited places. Indeed, the current paradigm in tourism development relies on the concept of sustainable tourism which implies that tourism planning and development should have the wellbeing of local communities as a central concern. Residents' perceptions about tourism impacts affect their attitudes towards tourism development that, in turn, is an indicator of the extent local communities accept tourism. This is a fundamental aspect of tourism development since host communities that support tourism development tend to be more hospitable with tourists, which influences their satisfaction and loyalty. Furthermore, the interaction with residents of host communities is a crucial component of the quality of tourists' experience, contributing, therefore, to the long-term success and sustainability of destinations.

The *Handbook of Research on Resident and Tourist Perspectives on Travel Destinations* aims to provide the reader with a glimpse of a place which, in addition to its central function as a place of residence, increasingly defines itself as a place of passage/visit in which its tourism functions assume a growing role. Therefore, this book intends to bring insights from residents and tourists' perspectives about the destination, able to inform the decision-making processes towards more competitive destinations, more attractive to visitors and, at the same time, contributing to residents' quality of life and happiness. In addition, it will present valuable insights that will be helpful to understand the composite reality of the tourism destination. In this sense, it can be an inspiring tool to support the development of marketing strategies within the places that are or are willing to become tourism destinations.

The book is organized in two sections with 19 chapters devoted to the theoretical and methodological approaches considering different aspects related to the residents' and tourists' relationships with travel destinations. More theoretical articles coexist with works of a practical and applied nature, with the purpose of supporting the transfer of knowledge between theory and its practical application. The book includes different geographical regions and realities in order to embrace a more diverse and inclusive perspective on travel destinations.

Section 1, "Residents' Perspectives on Travel Destinations," includes eight chapters.

Chapter 1, "Gentrifiers Against Gentrification: Tourism Gentrification in Algarve, South Portugal", approaches the topic of tourism gentrification, by presenting a case study of in a fishing town in Algarve, Olhão. The emphasis is put on the displacement of the original residents of the historical centre and on the transformations occurring in that part of the city.

Chapter 2, "Climate Friendly Hotels and the Promotion of Sustainability in Tourist Destinations: A Study in Algarve from Hotel Management Perspectives", brings to light the potential environmental problems caused by tourism and how they are being perceived and reflected in pro-environmental behaviours by those living in the region but with responsibility in hotels' management.

Chapter 3, "Impacts of Tourism Development and Residents' Satisfaction: An Explanatory Model – The Case of Huelva Spain", proposes and tests a model to understand the factors that residents perceive as affecting tourism development at the personal and collective level, and thus people's satisfaction with their community

Chapter 4, "Residents' Subjective Well-Being and Support for Tourism Development: The Case of Alxa, China", proposes a model to assess the effect of economic, socio-cultural and environmental factors on residents' subjective well-being and their support for tourism development in Alxa, a city in the north of China.

Chapter 5, "Residents' Support for Sustainable Tourism Development in Rural Areas: The Case of Karuizawa, Japan", combines three theoretical frameworks, the theory of reasoned action, the social exchange theory, and emotional solidarity, to examine residents' support for sustainable tourism in Karuizawa, a town that, in the recent years, reported an increasing number in both domestic and foreign visitors, as well as the existence of seasonal residents.

Chapter 6, "Perspectives on Community Involvement in the Revitalisation of Cultural and Natural Heritage for Sustainable Tourism: The Case of an Off-Shore Island in Hong Kong", addresses how local community involvement can contribute to the conservation of tangible and intangible heritage, as assets for sustaining tourism, on a small offshore island in Hong Kong, the Yim Tin Tsai village.

Chapter 7, "Residents' Perception and Their Support for Tourism Development: The Case of South Korea", aims to analyse residents' support for tourism development in South Korea by considering the social exchange theory and the theory of reasoned action. Based on these theories, the influence of perceptions about environmental sustainability, personal economic benefits and community attachment on support for tourism development is proposed and tested.

Chapter 8, "Different Type of Residents, Different Type of Attitudes? The Case of Tourism Development in East-Timor", uses the sustainable tourism attitude scale (SUS-TAS) to assess residents' attitudes towards tourism development in Timor-Leste, a new country that is also in the initial stages of tourism development. In this study, the perspectives of three groups of residents, with different roles and responsibilities, are compared: non-leading residents and leading residents in the public and private sectors.

Section 2, "Tourists' Perspectives on Travel Destinations," includes 11 chapters.

Chapter 9, "Experience Economy Domains in Tourism: Extending the Knowledge and Understanding of Experiences", aims to contribute to set a common ground and understanding of experiences and their associated approaches related to the experience economy in tourism through a systematic literature review.

Chapter 10, "A Conceptualization of the Senses, Emotions, and Memories in Memorable Tourism Experience", brings to light the role of senses, emotions and memories in memorable tourism experiences context. Memorable tourism experiences are the result of a multisensory stimulus and perception with connection with a positive emotional activity.

Chapter 11, "Measuring Attendees' Cultural Event Experiences: Insights From 365 Algarve", explores the concept of Cultural Event Experience, as the interaction between an individual and the cultural event environment, altered by the level of engagement with multiple experiential elements and outputs, measured in four dimensions: experience novelty, affective engagement, cognitive engagement and conative engagement. The Event Experience Scale (EES) emerges as a valid measurement instrument for cultural event experiences, producing valuable insights for stakeholders and marketing management teams in destination contexts.

Chapter 12, "Tourists Awareness of World Heritage: The Case of Tourists Visiting the Algarve (Portugal)", aims at investigating the awareness of Portuguese World Heritage (WH) by tourists visiting the Algarve, the main travel destination in Portugal. It intends to underline the multiple interactions between tourism and culture associated with Portugal and its history/identity, based on WH.

Chapter 13, "Creative Tourism as an Inductor of Co-Creation Experiences: The Creatour Project in the Algarve", describes the CREATOUR project which is focused on the development of co-created experiences in the Algarve region, a leading travel destination in Portugal. Motivated by the cultural and original nature of the region and travelling on their own, visitors go to the inland territories searching for creative experiences.

Chapter 14, "Operant Resources of Young Independent Travelers", aims to understand the operant resources involved in the young travellers' travel experience, when this experience takes place independently of the mass tourism supply chain and the tourist is this young independent traveller, who prefers to venture in the trip neglecting traditional, organized circuits and tours.

Chapter 15, "The Unfolding Theories on Destination Image, 1990-2020: A Content Analysis Approach", deals with the tourism Destination Image from a conceptual perspective and it sets the most relevant theoretical constructs that cutting-edge tourism and management studies have used since 1990 using the systematic literature review method. Recognized as critical in the tourism destination management and marketing strategies, it has been defined from different approaches. Mainly considered mental or holistic impressions the destination image refers to the perceptions held by tourists about a destination.

Chapter 16, "Insular Tourism: Profile and Consumption Patterns of an International Football Event Held in Madeira Island", analyses the profile and patterns of consumption of non-residents spectators of a sports event that took place in an eminently insular touristic city, Funchal (Madeira Island, Portugal). In this event, the national football team of Portugal confronted the Sweden national football team and this game placed the city in the spotlight of the media arena worldwide.

Chapter 17, "Tourism and Sex Outside the Sex Tourism Paradigm: Tourists' Perspectives and Practices in a Sun and Beach Destination", explores tourists' perspectives and practices related to sex and romance in one of the most recognized 'sun and beach' tourism destinations in Europe - the Algarve (south Portugal). The Algarve is a destination for families, couples, and friends on vacation, where sex plays a central role, far away from home and the daily life constraints.

Chapter 18, "Electronic Games as a Touristic Promotional Tool", intends to present a global vision on the theme of video games and tourist destinations, emphasizing the characteristics of video games, such as image and sound. The video game industry has grown steadily in value and in public, with games beginning to be one of the primary sources of entertainment for families. This may be of great interest for destinations targeted to families or groups of friends looking for a place of gathering and leisure to enjoy their holidays.

Chapter 19, "A Content Analysis on Child-Friendly Hotels as an Emerging Concept in Tourism Marketing", approaches the child-friendly hotel concept, which is defined as a hotel that offers additional services intended for children's and parents' special needs. It recognized as an emerging concept for attracting families with children in the tourism market. The purpose of this chapter is to identify what kinds of services child-friendly hotels are being provided to families with children.

Throughout the mixed methodological approaches and cases studies presented throughout the book chapters, the book' editors believe this handbook will help professionals, students and scholars to better comprehend the complexity of tourism destinations

Patrícia Pinto
Faculty of Economics and Research Centre for Tourism, Sustainability, and Well-Being (CinTurs), University of Algarve, Portugal

Manuela Guerreiro
Faculty of Economics and Research Centre for Tourism, Sustainability, and Well-Being (CinTurs), University of Algarve, Portugal

Acknowledgment

The editors would like to express their gratitude to all those who contributed to make this project a reality.

Specially, we would like to acknowledge the authors who accepted to send us their contributions and, therefore, share their knowledge and expertise in this joint publication.

We are also very grateful to the colleagues who were involved in the reviewing process. They had a fundamental role in improving the quality of the manuscripts, as well as the contents' presentation and organization.

Our sincere acknowledge goes also to the IGI global team who were available to help us with all the technical support we needed to make this book possible.

The editors would also like to acknowledge to the Research Centre for Tourism, Sustainability and Well-being (CinTurs) (UID/SOC/04020/2019).

The editors would also to acknowledge to the Portuguese Foundation for Science and Technology (FCT), through project UIDB/04020/2020, which funds their research work, and also to the European Regional Development Fund by CRESC Algarve 2020, through the projects SAICT-ALG/39584/2018 (RESTUR - Residents' attitudes and Behaviours towards Sustainable Tourism Development in the Algarve) and SAICT-ALG/39588/2018 (TurExperience - Tourist experiences' impacts on the destination image: searching for new opportunities to the Algarve).

Patrícia Pinto
Faculty of Economics and Research Centre for Tourism, Sustainability, and Well-Being (CinTurs), University of Algarve, Portugal

Manuela Guerreiro
Faculty of Economics and Research Centre for Tourism, Sustainability, and Well-Being (CinTurs), University of Algarve, Portugal

Section 1

Chapter 1
Gentrifiers Against Gentrification:
Tourism Gentrification in Algarve, South Portugal

Jorge André Guerreiro
https://orcid.org/0000-0003-4488-2343
Centre for Social Studies, University of Coimbra, Portugal

João Filipe Marques
Faculty of Economics and Research Centre for Tourism, Sustainability, and Well-Being (CinTurs), University of Algarve, Portugal

ABSTRACT

This chapter presents a case-study of tourism gentrification in a fishing town in Algarve, South Portugal. Olhão is a former industrial city that saw much of its fishing industry disappear since the 1980s. Over the last few years, hundreds of foreigners have moved into its historic centre. This rapid influx of citizens to derelict neighbours mostly comprised of old retirees and few active fishermen prompted a gentrification process. Olhão now faces the threat of mass displacements of its older and most vulnerable citizens, a fact that is worrying the foreigners that criticize the touristification of the city, with rents at historical highs and landlords forcing out the Portuguese residents in order to promote short-term rentals to tourists. The chapter presents the results obtained from an extensive mixed-methods research, ending with some notes about the future of the city and the implications that can be taken from this case.

DOI: 10.4018/978-1-7998-3156-3.ch001

The initial stage of an invasion has to do with the point of entry, the resistance or inducement offered the invader by the prior inhabitants of the area, the effect upon land values and rentals. The invasion, of course, may be into an unoccupied territory or into territory with various degrees of occupancy. The resistance to invasion depends upon the type of the invader together with the degree of solidarity of the present occupants. The undesirable invader, whether in population type or in use form, usually makes entry (that is, within an area already completely occupied) at the point of greatest mobility.

Park, Burges and McKenzie (1967[1925], p. 75)

INTRODUCTION

At the end of the XX century, the renowned British sociologist Anthony Giddens (2000) wrote that modernity advanced like a runway train, a juggernaut that spared nothing and no one. Arguing that globalization has brought forward a series of transformations that resulted in what Ritzer described as a "*set of processes involving increasing liquidity and the growing multi-directional flows of people, objects, places and information as well as the structures they encounter and create that are barriers to, or expedite, those flows* (2011, p. 11)."

The terms *here* and *now* have thus gained a different understanding, just as our understanding of time and space has changed as a result of these transformations. While the physical world remains much the same, the social world and the meaning attached to it have greatly changed over the last decades. This space/time compression that globalization represents has affected the lifeworld and all of its dimensions and cities are no exception (Bauman, 1998).

The study of modernity has since long focused on cities as the unit of measure to grasp social change, transformations, progress and conflict. One needs only to consider the streets of Paris in the XIXth century (Benjamin, 1969); the British industrial towns and advent of unions; the death of the Romanovs or the transformations of Chicago during the 20s and 30s. If history is told from the point of view of nations, then cities are the playing fields where it actually happens.

Understanding cities nowadays is no longer solely a spatially bound exercise, since much of what happens in a specific territory, happens for reasons often extrinsic to that territory. Transnational capital, as much as the flows of people, potentiates change and affects everything. And while butterflies batting their wings cannot cause a hurricane, a piece of legislation created in China or in the United States can greatly affect cities on the other side of the globe. If anything, this is a testament to the global financial and economic markets and how integrated the world economy has become. But while such transformations are often looked at a nation-state level, they happen at city levels as well (Wallerstein, 1999).

Gentrification as a social phenomenon perfectly encapsulates these ideas. Once a very specific process characteristic of industrial towns, it evolved as much as the cities themselves. As gentrification became more complex and began to occur in global cities as much as industrial towns, its theory expanded to accommodate these novel forms of spatial change (Sassen, 2002). Today, gentrification is mostly found in post-industrial settings and is driven by other factors, while also becoming highly imbricated in the processes of social and spatial change at local levels that themselves result from complex dynamics of mobility and transnational movements of people and capitals.

This chapter focuses on a case-study of Olhão, a fishing town that was once part of the industrial centre of Algarve, South Portugal, to argue that novel forms of gentrification, such as tourism gentrification and transnational gentrification, are tightly linked to contemporary forms of mobilities. Having recently experienced a gentrification process, the transformations in Olhão are a microcosmos of what spaces of hypermobility can lead to, as well as of the lifestyle mobilities that prompt such processes as much as public policies or state legislations.

The chapter analyses data collected from interviews, focus-groups and ethnography, using a phenomenological approach to recount how the gentrification process began, how it evolved and ends with some guesses to where it might go in the near future. Emphasis was put on the displacement of the original residents of the historical centre and how the gentrifiers see these process and the transformations occurring in that part of the city.

We begin by reviewing some of the key concepts used in this chapter and the relevant literature on the subject, before proceeding to description of the methodology used to obtain the data presented and how it as processed. The chapter then analyses the process of gentrification that Olhão experienced since its roots to the present day, ending with some conclusions and future research directions for similar lines of inquiry on this field.

BACKGROUND

Understanding Tourism Gentrification

The term gentrification was originally coined by the British sociologist Rush Glass, in 1964, to describe "*the reinvestment of real estate capital into declining, inner city neighbourhoods to create a new residential infrastructure for middle and high income inhabitants*" (Patch & Brenner, 2007, p. 1917). The influx of affluent gentry to otherwise lower-class neighbourhoods leads to the displacement of those who could no longer support the raising living costs of gentrified areas, with Glass stressing the underlying class relations of the phenomena.

Since then, gentrification changed as much as the cities themselves. Once a process mostly identified in great metropoles and industrial cities, or, more recently, in global cities (Atkinson & Bridges, 2010), it has, become something that can be witnessed in urban fringes of rural territories and even in small towns all over the world (Clark, 2005).

Even if new forms of gentrification have become more prevalent, the same mass displacements and ensuing segregation of lower-class individuals that can no longer afford to live in areas where living expenses rapidly increase remain at its core. Gentrification as a process has thus mutated, with recent takes on the concept attempting to capture these novel faces (Lopes, Rodrigues & Vera-Cruz, 2019). Tourism gentrification (Foulds, 2004) is but one of such takes, amongst several others that have been proposed since the turn of the century (Janoschka, Sequera & Salinas, 2014; Sigler & Wachsmuth, 2016).

Tourism gentrification was introduced by Gotham (2005, p. 1099), who defined it as "a transformation of a middle-class neighbourhood into a relatively affluent and exclusive enclave marked by a proliferation of corporate entertainment and tourism venues." On the production-side, it is understood as a by-product of tourism development and can occur both in large scale settings and destinations permeated by mass tourism, as well as in smaller locations that have recently become touristic destinations (Durr & Jaffe, 2012; Herrera, Smith & Vera, 2013).

On the other hand, tourism gentrification can also be demand-driven, especially in settings where tourism is still in an early stage of development. The concept thus bridges the production-side and demand-side explanations of gentrification that compete since the 80s, proving itself as a more heuristic concept than that of classic gentrification (Gotham, 2005).

While influential, Gotham's definition is somewhat limited in scope. For instance: tourism gentrification can occur in lower class neighbourhoods (Durr & Jaffe, 2012). One can also argue that corporate entertainment is not a *sine qua non* condition of tourism gentrification, especially in villages, small cities and other small-scale settings, as well as in urban fringes or rural territories.

Production-side tourism gentrification often sees political institutions planning and implementing policies that promote tourism development. State-led gentrification is understood as "*the means and rationales by which policy-makers and administrators justify and implement the class-based production of urban space*." (Gent & Boterman, 2018, p. 35). According to this definition, the role of the state is to make policies that regulate or affect urban space, but it could as well include incentives for the tourism industry or legislation that facilitates the investment of transnational capital.

The role of public policies in the promotion (or containment) of gentrification has since long been acknowledged by scholars (Hackworth & Smith, 2001; Rérat, Soderstrom & Piguet, 2009). In fact, the role of the state in promoting social segregation and the reproduction of spatial inequalities was discussed even before the term gentrification had been coined (Smith, 2010) and it became a recurring subject in gentrification literature from the 80s, given the new neoliberal orientations that housing markets assumed from then on, especially in Anglo-Saxon countries, replacing the Keynesian-managerialism that dominated the post-depression housing policies (Hackworth, 2018).

The latter advances on the knowledge of state-led gentrification have, however, pulled away from the macro level understanding of the state role, to a more meso level, with specific policies or local regulations receiving the attention, instead of the ideological tenets that govern the housing or financial markets at a national or international level. The relation between class, place and state has also become a major focus of the state-led approach to the study of gentrification, instead of the state being treated as detached from social and spatial relations (Gent & Boterman, 2018).

This reconfiguration allowed the study of gentrification through the lens of local governance, conceiving it as a spatially located phenomena that manifests itself differently based both on the material and symbolic contexts, even if influenced by large-scale processes. Tourism gentrification is one of such forms, with public institutions often welcoming and cooperating with the private sector's investment in rundown or derelict neighbourhood and promotion new forms of consumption that often result in "trendy" areas in cities that attract more investment and prompt the development of contiguous areas (Gonzélez-Pérez, 2019). This is also the case of transnational gentrification, which can be defined as a

Gentrification phenomenon that connects redevelopment capital to housing demand not within a single city-region but transnationally, and thus creates new possibilities for profitable housing reinvestment – in markets where such possibilities would not have existed on the basis of local demand alone. (Sigler & Wachsmuth, 2016, p. 2).

Much like tourism gentrification, transnational gentrification has only recently become a major subject in social research, even if it has been manifesting itself for quite a few decades now, in one form or another. One could even argue that migrations, as transnational flows, led to displacements and social segregation before gentrification had been defined by Ruth Glass, by pointing to the works of Robert Park and Ernest Burgess (Smith, 2010).

Gentrification has been mostly studied as a locally bound process and much of its theory either comes from case studies of particular cities or from the theorization of macro-processes that lead to the political and economic conditions that allowed for those cases to happen in the first place. Yearly this century, Smith (2002) argued that gentrification could and, indeed, should be seen rather as a global urban strategy. This suggestion was followed upon by many researchers, who began looking at gentrification in the awake of globalization and the mobilities paradigm (Sheller e Urry, 2006).

With that said, gentrification, in all forms, is still a locally bound phenomenon, studied at neighbourhood, city or regional level. But capital, as Marx put it, has no country and flows of people and goods have long surpassed national borders. Therefore, understanding gentrification in this globalized world, especially in hypermobile territories like the European Union, requires an understanding of the world itself and the underlying processes of mobility.

Even if this proposition is not, by any means, revolutionary, some still stress the necessity of gentrification theory to detach itself from the northern European and anglophone experiences, and pay more attention to the new forms of gentrification that are occurring all over the world. As Sigler and Wachsmuth (2016, p. 20) said, there's

A need to disassociate the process from deindustrialisation and working-class displacement – a near-ubiquitous attribute of the late 20th century North Atlantic cases – in favour of a focus on macro-scale class-based change framed by global flows and local urban restructuring.

Tourism and transnational gentrification manifest these transformations both on a macro-level, given they very much are a consequence of the new mobilities paradigm, and on a meso-level, considering the implications that the increasing flows of people and capital have had in many territories world-wide, culminating in gentrification nowadays being a very different phenomena than what it was, hence several authors stressing the need to distance the field from distant-past experiences and to rethink its object in today's world, while retaining goal of "understanding the changing relationship between people and where they live" (Butler, 2007, p. 163).

It is no secret that tourism and mobilities in general play a major role in the social, cultural, spatial and economic transformation of territories. Tourism has become a staple of southern European countries, given its contribution to the economic rebound in the financial crisis aftermath. It is also one of the few industries that continues to grow, despite recent economic turmoil (Baptista, Nofre & Jorge, 2018).

Some might think that, because of this, tourism gentrification tends to occur in sun and sea touristic destinations, especially in the context of southern European countries, but, if anything, this particular form of gentrification has, so far, shown itself more ubiquitous than the original concept of gentrification. Tourism gentrification is not bound by the setting's dimension, exclusive to urban areas nor dependant on specific types of tourism. And over the decade, many case studies have been published focusing on tourism gentrification all over the world (Baptista, Nofre & Jorge, 2018; Foulds, 2014; González-Pérez, 2019; Herrera, Smith & Vera, 2013; Hughes, 2018; Lopes, Rodrigues & Vera-Cruz, 2019; Mendes, 2017; Miró, 2011; Pinkster & Boterman, 2017; Prytherch & Maiques, 2009; Sarafa & Brito-Henriques, 2017; Skoll & Korstanje, 2014).

One particularity of tourism gentrification is that it tends to lead to more malign consequences than "ordinary" gentrification does. Indeed, most studies on tourism gentrification point to the raise of anti-tourist sentiments developing and the growth of tensions between locals and tourists. The images of graffiti proudly displaying the words "tourists go home" in several streets in Barcelona have become a staple of tourism saturation (Hughes, 2018). The same goes for tourism gentrification.

It should be noted, however, that tourism gentrification is but one of many consequences that mass tourism can lead to, while also having a dialectical relation with tourism, in the sense that many of the transformations that touristic settings experience occur in order to attract more tourists and investment, resulting in a cycle in which tourism leads to gentrification, that happens for the sake of attracting more tourists.

While the urban regeneration that follows gentrification is often viewed as positive, especially when it mostly affects derelict districts or rundown neighbourhoods, tourism gentrification tends to happen in areas that are not as deteriorated, which has graver human consequences, given the displacement that it often prompts. The touristification of specific zones that precedes tourism gentrification also results in higher costs for housing, goods and services that can be felt on a city-wide scale, affecting not only those who live in gentrified neighbourhoods, but the remaining population as well. And even if this is not directly caused by tourism gentrification, it is aggravated further by it.

Mobilities and Socio-Spatial Transformations

Gentrification is a very different phenomena today from what it was in the 60s or 70s. Globalization has played a major role as the driving force behind new market tendencies and the way in which communication and flows occur (Smith, 2002). This is specially the case of mobilities, which have become one of the defining characteristics of the developed world (Sheller, 2014). And given that tourism is a manifestation of mobilities, one could hardly debate tourism gentrification without addressing it.

Hannam, Sheller and Urry (2006, p. 1) state that mobilities

Encompasses both the large-scale movements of people, objects, capital and information across the world, as well as the more local processes of daily transportation, movement through public space and the travel of material things within everyday life.

The term mobilities is often used interchangeably with that of movement. One would say that going from point A to point B is a form of mobility, which would not be wrong, but would remiss all that differs between movement and mobility: mobility is more than just movement, as it implies strategies, relational moments and social contexts, while movement by itself implies none of these things (Cresswell, 2006). To better understand this difference, one needs only to point to animals: they can move, but that movement would not constitute a form of mobility, as it lacks social meaning, symbolic value or strategy.

Human mobility, on the other hand, has meaning attributed to it, provided by the self as much as it is given by others. It is negotiated with other social actors and produced in everyday life, while manifesting choices, preferences or states of mind. What this means is that mobility is socially produced and not merely a simple act of going from one point to another. And after such elements are factored into movement, "*getting from A to B becomes synonymous with freedom, with transgression, with creativity, with life itself. […] mobility is practiced, it is experienced, it is embodied. Mobility is a way of being in the world.*" (Cresswell, 2006, p. 3).

But mobilities is not solely based on human movements – it also encompasses the movement of objects and information, factoring in their relational dynamics, as well as immobilities. Mobilities combines concerns from social (social stratification, inequalities, power relations, hierarchies, etc), spatial (territories, borders, space appropriations) and cultural (discourses, ideologies, social representations) dimensions, resulting in "*a relational ontology of the constitution of subjects, spaces and meaning*" (Sheller, 2011, p. 2). The main argument behind the idea of a mobilities paradigm is that the study of the aforementioned dimensions uses contributes across social sciences to make sense of the social world in a relational and dynamic approach, instead of a static point of view.

The intersection of mobilities and gentrification is a fortuitous one. Tourism and transnational gentrifications have contributed to a more fluid vision of gentrification theory, instead of a static, immutable one (Latour, 1987). Gentrification and social sciences in general have spent most of their history focusing on singularities: phenomena bound to a specific time and place. This would often manifest as a tendency to ignore distance, change, transformation and placelessness as dimensions to factor in social research, assuming a synchronic ontology (Sheller & Urry, 2006).

If anything, mobility studies have proven that social relations are increasingly more fluid and dynamic, having attained a quasi-status of *liquidity* or *weightlessness* as many of the globalization theorist have argued (Bauman, 2000; Beck, 2008; Ritzer, 2011). The sedentarism that characterized social theory has been replaced by a negotiable approach to the social inquire of reality, in the sense the phenomena can – and indeed do – happen across time and place. This means that even if a specific scientific object is bound to a particular place, the understanding of the phenomenon can only can achieved by situating it on the larger plane of social relations and macro processes that influence it. Hence the idea of a mobilities paradigm that "emphasises that all places are tied into at least thin networks of connections that stretch beyond *each such place and mean that nowhere can be an 'island'*" (Sheller & Urry, 2006, p. 209).

The key idea about mobilities that one should retain when discussing transformations at a local scale is the fact that being on the move has become much more than a state that individuals find themselves when physically dislocating, but rather a way of life, a question of identity – a lifestyle (Urry, 2002). Mobilities have become such a defining characteristic of today that much of our daily lives are predicated on it, even without most people realizing it.

This argument nears the reason why tourism gentrification and transnational gentrification are addressed together in this chapter, even if the main focus is on tourism gentrification: both are products of contemporary mobilities. Without today's novel forms of mobility, transnationalism or tourism gentrification would hardly exist on such a large scale. And understanding transformations in territories requires that the larger context is addressed as well, as to make sense of the material and symbolic flows that contribute to social change of social actors as well as social spaces.

While both transnational and tourism gentrification can manifest separately, they can also be closely intertwined, in the sense that zones that experience high flows of tourism can often experience the immigration of previous tourists, especially during retirement or later stages of the life-course, as international retirement migration or lifestyle migration studies have showed (Benson & O'Reilly, 2009; Gustafson, 2009; McIntyre, 2013; Torkington, 2012). Such forms of gentrification can only be understood in the context of the mobilities paradigm (Sheller & Urry, 2006), since being "on the move" has become a lifestyle in hypermobile territories like the European Union and that has given birth to a new transnational elite characterize by their high motility (Aalbers, 2019).

As the research presented here demonstrates, mobilities and the novel forms of gentrification like tourism and transnational gentrification are strictly related and can be combined in order to create a framework suitable for the study of local scale, while factoring transcendental dimensions that influence the everyday lives of residents. Lifestyle mobilities, in particular, can greatly contribute to this cause.

Where an individual resides is the perfect example of a static, place-bound notion. Statistics capture such data by counting who resides at a specific place. If the individual has multiple households or divides his or her time between different locations, then the plot thickens and statistics begin to waiver. In such cases, time is factored in: how much time does the person spend in which house or location? And between two objective criteria, place and duration, a classification is achieved: resident, tourist, immigrant, etc. Yet, this approach based on the quantification of time neglects one key dimension: the relations that the individual forms with the place or the meaning of his presence in that particular locale. This attachment or belonging is instrumental for mobilities, as transcends both time and place and can only be understood by looking at the larger context at hand.

The term Lifestyle Mobilities was introduced by McIntyre (2013: 194) to describe "the movements of people, capital, information and objects associated with the process of voluntary relocation to places that are perceived as providing an enhanced or, at least, different lifestyle." It was latter conceptualized to use *attachment* and *belonging* as means to better distinguish and understand the forms of mobilities that situated between tourism and migration (Cohen, Duncan & Thulemark, 2015), making it a suitable concept for the study of complex forms of mobility.

The sociological use of the concept of lifestyle follows upon the works of Weber (1963) (through his use of the concept of *stutus*), Bourdieu (1996), Giddens (1991), Featherstone (1987) and Baudrillard (1981), who argued that consumption, in the late capitalism, had changed from use-value to sign value, meaning that consumption is not only about fulfilling human needs, but have a symbolic value to them, i.e. they are ideological and actively reflect a lifestyle (what brands to wear, what kind of car to drive, where to live, etc).

In fact, the concept of 'lifestyle' originally referred to the consumption patterns in modern societies, dating back to Marx (1982), but recently its use has been shifted to also describe the identity consequences of social practices (Veal, 1993). According to Stebbins, "a lifestyle is a distinctive set of shared patterns of tangible behaviour that is organized around a set of coherent interests or social conditions or both, that is explained and justified by a set of related values, attitudes, and orientations and that, under certain conditions, becomes the basis for a separate common social identity for its participants" (Stebbins, 1997, p. 350).

Its pairing with mobilities has been justified with the need to consider the mobilities reflect ongoing lifestyle choices, given how they (Cohen, Duncan & Thulemark, 2015: 156) "(1) blur the boundaries between travel, leisure and migration; (2) are exemplary of how a binary divide between work and leisure may be collapsed; (3) destabilise dichotomies of 'home' and 'away'; and (4) illustrate complexities of belonging and identity associated with sustained mobility."

Methodology, Universe and Sampling

This section recounts the gentrification of Olhão, from its origins to the current days, while addressing the pioneers' feelings about the new gentrifiers that are moving in, and the touristification of the city. These excerpts were collected by interviews and focus groups, while the remaining information was obtained from ethnographical observation over a period of two years, carried at diverse moments (both week days and weekend days, mornings and afternoons, in all seasons), informal conversations and self-administered questionnaires. Most of these were conducted under strict guarantee of anonymity and confidentiality, hence all names and personal identification being omitted.

The study encompassed documental analysis, two focus groups, 19 semi-structured interviews, roughly 80 hours of ethnography. Non-structured interviews and conversations were held with approximately 30 residents. While all these data sources contributed to the overall conclusions, details will only be provided about the data sources that directly contributed to this chapter, namely the semi-structured interviews, the non-structured interviews, one focus group and some items on the survey.

The data collected from the interviews and focus group was interpreted according to phenomenological and ethnographical approaches to qualitative research (Creswell, 1998). The study attempted to describe the evolution of the case study at hand, while stressing the transformations that gradually occurred and how the subjects – i.e., the foreign residents –, saw these transformations. Emphasis was put on the interviewed foreigner's subjective point of view, paying particular attention to what happened and how it happened, based on their personal experience. The data is represented in a narrative manner, with excerpts of the transcriptions used to enrich the section. The section also factors in some insights from other data sources to a lesser extent.

Interviews, focus group and conversations were around three main subjects: (1) why did the foreign residents choose the historical centre of Olhão to reside, (2) what is their sense of belonging in regards to the historical centre of Olhão and (3) what do they think about the transformations happening in Olhão. Several follow-up questions were made to further delve into some subjects as the interviewees talked, but there was a deliberate attempt to let them talk freely and choose what subjects and issues were more important to them. Specific questions were only introduced at the final stage of the interviews to find how they felt and what they thought about specific subjects that were left out of their previous answers.

We acknowledge that the French residents were underrepresented, as they, alongside the British, constitute the largest nationality. Since the size of the universe is unknown, it was estimated that between 300 and 400 foreigners live in the historical centre.

All participants were chosen by convenience and referred to by other residents. Most of the foreign citizens contacted were reluctant in participating, far more so that it is usually expected in such studies. These transformations have garnered some political interest and given that several foreigners are civically active, there were many reservations about publicly discussing the gentrification process. The upmost care was taken when collecting data, as to ensure that no personal information could be unintentionally included. All excerpts were validated by the people who enunciate them, and their transcription and reproduction were previously authorized. This is an extremely important aspect of this case-study and should be considered a deontological imperative of scientific research. Some participants were unwilling to disclose their age or any personal information besides country of origin and also refused the recording of the interview and, in some cases, that notes were made during the interview, becoming more or less an informal conversation on their own terms. This reluctance in sharing their views on the matter greatly difficulted the data gathering process, but were, themselves, a rather curious fact.

Olhão: A Former Industrial, Fishing Town in Algarve

Portugal first started to experience tourism in the first decades of the XX century, not unlike many other South European countries. Portugal's dictatorial government that seated in power from 1928 to 1974, however, did not take kindly upon the idea of mass tourism, especially given the secluded nature of its policies that placed heavy restrictions on the flow of people and goods in or out of the country. By the 1940s, Salazar's government realised the economic potential of tourism and began investing in infrastructures and promoting Portugal overseas, making tourism the countries principal export after a short time (Agarez, 2013).

At this time, Algarve was an isolated region, even by Portuguese standards. Located on the south-western tip of the Iberian Peninsula, the region was separated from the rest of the country by several mountain ranges that stand but a few kilometres from the sea and the semi-deserted Alentejo. There were hardly any connections to Lisbon and traveling the 300 kilometres that separated the region from the capital took an entire day. This geography motivated its inhabitants to turn to the sea since immemorial times and continued to be at the heart of its history and culture during the last century (Brandão, 2012).

Aware of Algarve's appeal and potential as a touristic destination, with its white sand beaches, mountain ranges over the sea and old century old villages and towns, the Portuguese government decided Algarve would be the countries' main touristic destination. As such, the national propaganda office, under António Ferro's tutelage, used depictions from Olhão and of its singular architecture to represent the region, both abroad or at home (Agarez, 2013), and thus all of the Algarve became known for its white houses and square rooftops, even if such houses were only found in Olhão (Antunes & Azevedo, 1988).

Olhão's townscape was a fixture in representations of Algarve from the 1920s, seducing adepts of the picturesque and modernists alike. With overlapping whitewashed volumes, terrace roofs and box-like turrets, it was repeatedly called Portugal's 'cubist village', a title first used in 1922 by António Ferro. [...] Ferro deployed the Olhão 'cubist' house as a surrogate for the Algarve as a whole, whenever the region was to be represented, at home and abroad. Agarez (2013, p. 726-727)

Olhão distinguished itself as a fishing town since the early XVIII century, quickly becoming one of the most important ports in the southern half of Portugal. In the 1880s, Olhão became a prime location for fish factories that were popular in countries like Italy, France or Spain (Nobre, 2008). This led to an industrial development that defined the town until the 1970s, when deindustrialization became a common thread all over Portugal. By the 1990s, very few factories remained of the several dozen that existed few decades ago (Bonança, 2014).

This did not change Olhão's relation to the sea, as it remained the town's main source of income until the 90s, when the European Economic Community fishing policies caused the sector to nearly collapse in the region. Since then, fishing maintains its importance for the lower classes and older inhabitants that still rely on it to survive. Most of these fishermen reside in the Fisherman's Neighbourhood, built in 1955 by the Portuguese dictatorial government, and in Barreta, the birthplace of Olhão, that once stood by what was Olhão's beach.

Given its industrial history and social problems that arose from deindustrialization, Olhão was never a touristic destination. Its beach was destroyed to build a harbour decades ago and its islands are only reachable by boat. Its deteriorated urban landscape contrasted with the high story buildings that began being built all over the city, since the late 80s. The historical zone near Barreta remained mostly untouched, as it had no real estate value and the younger generations did not want to live in the old neighbourhoods. As time passed, the ruins multiplied as the inhabitants dwindled.

The residents that remained, mostly retirees and old fishermen, had no means of restoring their old houses and when they passed, their descendants had no reason to invest into houses that had no value nor buyers. Many streets felt abandoned, as dozens of ruins stood empty and several buildings that were once warehouses for factories or artisan's workshops were now forgotten and boarded shut.

Today, the city of Olhão has around 28 000 inhabitants and is the centre of the municipality with the same name, with a population of over 45 000, which makes it the fourth highest populated municipality of Algarve. The historical centre is located on the parish of Olhão, at the heart of the city, by the Ria Formosa lagoon, that has a population close to 15 000 (INE, 2015).

The Gentrification a Fishing Town

The first foreigners moved to Olhão in the late 80s. Their number slowly raised until 2012, when it became obvious that a considerable number of foreigners had moved into the historic centre, with Barreta neighbourhood at its heart.

Olhão was so unique. Here were fishermen and old ladies, every day, going to the fish market right by the sea. Every morning, in front of my house were seated 15 or 20 old ladies and there was discussion about everything. I don't know what they were discussing, but it was a special feeling. Every morning, at 7 o'clock they were sitting there [I1, 64 years old German male].

[In the 80s] I saw that a lot of houses were in very bad condition and I said: I got a feeling that young people don't care about houses, just the old people, but they don't have money. Since then, much has changed. Now foreigners come and make houses, but I'm not so sure I agree with this, I think we have to do this with the Portuguese together, instead of pushing them aside and letting foreigners do it by themselves [I1, 64 years old German male].

After 2012, their number exponentially grew and by 2015, Olhão was one of the most popular destinations for foreigners looking to live in the Algarve. The majority of these new residents either acquired old buildings and restore them or bought previously restored buildings. As this phenomenon grew in scale, the number of available buildings diminished. A most curious fact was that these foreigners were mostly interested in traditional houses, even if they were a century old. This resulting in the historical centre becoming a convergence point for northern and centre Europeans who wanted to reside in Olhão, as buildings such as these could seldom be found elsewhere and the foreigners did not want apartments nor properties in the countryside.

I feel like the buildings are a living thing here. You never know what you will find. Anne found a sardine can with almost 100 years. I found an old pair of shoes that were handmade and that no one knew how to make any more [...] When I cleaned up my house, I felt like an archaeologist, finding artefacts of a bygone era. You can't get that with a new house, there's no history there [I2, 56 years old British female].

During the years that followed the subprime crisis of 2008, Olhão had some of the lowest value urban properties in Portugal. Its historic centre had even lower prices, as few wanted to reside there. But after 2015 the pattern started to change: soon the real estate values began to raise and ruins that once had no value whatsoever competed with brand-new apartment buildings. Since then, this pattern only became more noticeable and in 2019 very few available buildings remain in this part of the city.

When I see old houses fall apart and be replaced by something with more than three floors, I'm crying on the inside. These old buildings are the soul of Olhão. It is unique because nothing has changed and the real treasure of Olhão are these small, white houses, true people, nice people always here. But that is changing. Rentals, condos, gated communities, that's not Olhão. The day that everything here is just like a hostel, it will have no interest at all. That's the problem, so do not modernize Olhão to keep it authentic [I3, 53 years old French female].

It is going very fast. I think it is too fast. Everything is changing. A lot of French people are coming. They have to pay too much taxes and in Côte D'azur, they sell their houses, because it's cheaper in Portugal, so they come here. Every other house is a French house now [I1, 64 years old German male].

In the last couple of years, real estate agencies began aggressively inquiring the old inhabitants of the historic centre, enticing them with offers for their old houses, pointing how they could afford new flats in the suburbs with the profits from the sale of their old houses. Some accepted the deal, trading old homes with barely running water for new apartments houses with all the comforts they had never experienced during their lives.

There are so many widows around us, several of them now feel scared because many of their neighbours have left or died and they have no family left. The real estate agents are knocking at their doors asking – "Why don't you sell your house and move to a brand-new apartment building without infiltrations or bad pluming?" And that is why I am really concerned about it, because the more we lose that cultural integrity that is knitted so tightly into our community base, the more homogenous these villages will become, because they'll just become these white, tourist houses, that are Airbnb ghost towns, that have people walking down with their phones, not even looking at the environment, just trying to find where they are supposed to go, with carts and coming in and out, not living in our communities, so lots of white, new, renovated buildings with no life or families inside [I4, 58 years old female American].

The influx of foreigners from northern and centre European countries came as a surprise to everyone. Olhão had no tourism industry, no promotion overseas nor had been known as a touristic attraction. Located 10 kilometres east of Faro, the region's capital, it functioned as a dormitory for families or recently graduates from the university of Algarve (located in Faro) that were starting their lives. Olhão was merely a passing point to more popular touristic towns, like Tavira or Monte Gordo.

Olhão did not have a good reputation. We wouldn't have considered it. We didn't know it. But when we finally sold the house in Algoz, a number of people said – "now you sold the house, where are you going go to?" – By that time, we were considering Olhão. We told people "well, we are thinking about going to Olhão" and people just stared at me... And I knew what they were thinking, because I shared that same hesitation. But then we came and we saw it and thought ok, let's try [I5, 73 years old American female & I6, 60 years old British male].

These foreigners were mostly from northern and centre European countries and had become disenchanted with life back home. Many of the pioneers are artists and intellectuals that found the life of a small town by the see call-back to what many experienced during their childhoods. It is no coincidence that they fell in love in the architecture of the city and started this trend of restoring old buildings.

I first came with my daughter; we came for a holiday to stay with a friend that has been here for 25 years. We liked it very much. And I had been thinking of having a holiday place, so I bought one of his houses and I found I was spending more and more time here. That served me well, since as a painter I can work anywhere [I7, 74 years old British female].

I was considering getting a vacation home. I thought of Ibiza, but that was too expensive… It was a pleasant surprise when I realized that Olhão was very cheap. It fit the criteria, because it had all the benefits you look for, if you want a house somewhere without it being touristic, which goes down do to what I was saying, I prefer living in the place.

I chose an older house because it has more character. Because the house I have here has a very high ceilings, 4-5-meter-high ceilings and all moulded decorated. You don't get that in England. You don't get many of these houses and if you do, you pay a lot of money for them. Here I can live in a much grander style, you know, more luxurious. The doors are tall, the sealing is high, the rooms are big and there's so much natural light, which makes such a difference here in the south. It is such a beautiful style.

Algarve has had tourists turn into residents since the 30s (Williams & Patterson, 1998), but there has been a tendency for these new residents to settle in rural areas or urban fringes of rural territories, often close to beaches, resorts, golf courses or touristic urbanizations, rather than in urban centres. That is why the case of Olhão is a surprising one.

Why am I living here, in the city? Well, maybe because I'm single. If I was married or had a partner I might not. As a single person, there's a lot of people here. Most of my friends are single, so to be 5 or 10 minutes from each other and everything we want is good and not having to drive a car at night if you drink makes such difference. I think it is much easier this way. Almost every night I see one or another friend, so it's easy just to walk from one house to the other. If you are in the country you go: "I can't be bothered to go out in the evening", you feel more alone in the countryside, it's not safe. I wouldn't feel safe on my own in the countryside [I7, 74 years old British female].

However, many of these foreign citizens are now complaining that Olhão is changing too fast and they fear that the lifestyle they fell in love with might be threatened by tourism development. They stress that the sense of community is what makes Olhão unique: the old people having coffee in the morning and the afternoon in the terraces, the fishermen coming and going, people talking loudly in the streets and drinking by the sea in large groups. They worry all this could be replaced by hordes of tourists with smartphones, walking sticks, cameras and tour maps, like elsewhere in Algarve.

It is wonderful on Saturday to see people come from everywhere to sell the carrots that they planted a few kilometres from here. And I think this is life, this is what it means to live in a small community and this is what we have totally lost in France, with all of these big supermarkets and we don't need supermarkets, we need people. That is what makes Olhão unique, because nothing has changed and the real treasure of Olhão are these small, white houses, true people, nice people always here. And I fear that it might be changing. The day that everything here is just like a hostel, Olhão will have no interest at all. That's the problem, so I say don't modernize Olhão, so it can be kept authentic [I3, 53 years old French female].

It is a pleasure for me to say "- hello, how are you?". We can see people here. We can see people work, we can see people alone, just like my neighbour, that I like very much. I know that it is cold in his house, so now, when I see him on the street, with the light of the sun, I know it's just because it is so cold inside, you know? And it is good to know the people that live around you. In France, I was in the suburbs, with a very nice house and it was hard for me to meet my neighbours [...]and we're never together in a relationship.

Over the last couple of years, a different type of foreigners has been purchasing houses in Olhão. These foreigners can be described as members of the transnational elites (Aalbers, 2019; Rofe, 2003) and possess a larger economic capital than that of the original foreigners, which was already far above that of local residents. In such cases, the result is a process of super-gentrification, that consists in a new gentrification of previously gentrified areas, with social elites moving in and changing the social dynamics and relations of those that came before (Butler & Lees, 2006). These super-gentrifiers differ from the original gentrifiers, the pioneers, by their social status and high economic capital. They are more interested in consuming places than living in them, dividing their times between several different destinations across the world, always seeking what is trendy and fashionable this season.

This worries me, since the people who [used to] come here are not rich people. Here is not for rich people and that's why it's good. Here is for people who want to have an authentic relationship with the people that live here. If you don't want to learn Portuguese don't come here, go to Vilamoura, go somewhere with English, French and Germans... but when you are here, you are here to be involved in the community. But some of the new people moving in, they don't care about that. They come here for a few days, throw their parties, drive Porsche, Mercedes-Benz, Aston Martins, rent the big boats at the marina and then they leave their houses closed for the next months [I3, 53 years old French female].

We lived in the centre, but in the last two, three years, so much northern money has come in... so many "special" people living here... there are so many now that it's not comfortable. In the summer, for example, people are walking like they are in Disneyworld, I don't know how else to say it... like it's a tourist attraction rather than a living place. And because of the money they have, there's a false importance... and they tend to rather than to accept where they are, that you are visiting, the privilege of another nationality and country, that you should feel lucky... they prefer to come in and say "I'm from Holland, I'm from England or I'm from Germany, and I am very important and only poor people live here..." we know tourism in the government goes its way, be we were kind of hoping Olhão would be left alone [I8, 66 years old American male].

The pioneers have been calling for measures from the municipal government and while a controlled rent programme was instituted recently, aimed at the local Portuguese youth with the intention to avoid more locals from abandoning the historical centre, it had very few people interested in it. This might be due to how high the rent was, even with the municipality support, that covered up to a specific value. On the other hand, downtown Olhão was never appealing to younger inhabitants, who prefer other zones of the city or in the outskirts.

When interviewed as part of this investigation, the town mayor, António Pina, stated that he was indeed worried with the displacements and the prospect of only foreigners living in the downtown, but also stressed that for decades, no one wanted to move into the historical centre and that it was in ruins because of this lack of interest. While the prospection of the historic centre losing its identity is troubling, the mayor feels like there's not much that can be done to avoid it in the long run. Olhão has lost most of its fishing culture, with only the older inhabitants and some of their descendants still retaining such traditions. It might as well be impossible to prevent that culture from disappearing altogether, even if no foreigners were moving into the city.

FUTURE RESEARCH DIRECTIONS

Portugal is currently facing a shortage of long-term housing at accessible prices. This is happening mostly in the coast line territories, but several interior cities are beginning to experience the same shortages. In all cases, there's a relation between tourism and rent increases. The increase of short-term rentals and platform tourism has grown exponentially over the last few years and even middle-class familiars are having trouble finding affordable housing. This has also cause severe issues to the student population of cities that have universities, as students cannot support the increasing housing and living expenses.

Tourism is being blamed for all these issues and there is an anti-tourism sentiment growing, as people are claiming that the benefits of mass tourism are lesser than its consequences. While Portugal has experienced mass tourism for decades, the new forms of unplanned tourism, like platform tourism, are straining the already fragile balance of touristic regions that have more touristic visits than they were designed for, putting added pressure on the already deficient public services, like garbage disposal, hospitals and medical facilities, public transportation or regional highways.

These new forms of tourism accommodation along with the tourism gentrification and the cultural relations between 'hosts and guests' are still largely unresearched in Portugal and require understanding, as do the consequences described above. While most Portugal is still far from the outcries in Barcelona, Lisbon has experienced tremendous touristic pressure over the last years. In turn, it is also the city that has more organized social movements against tourism, gentrification and the proliferation of short-term rentals.

Such processes will define the urban landscapes in Portugal for the next decades and need to be closely studied and followed upon. Portuguese economy relies on tourism as its largest export and finding a balance or alternatives for the current paradigm is essential to avoid oversaturation and an escalation of the anti-tourism sentiment.

CONCLUSION

All forms of gentrification have one thing in common: they address the social transformations that result from the inflow of new residents to neighbourhoods. In the case of this chapter, the original process of gentrification and the subsequent process of super-gentrification, occurred as a result of tourism development in the wake of Olhão becoming a popular tourist attraction. Even if the first wave of foreigners were responsible for this process, they now find themselves denouncing the touristification of the city and the displacement of the original residents from the neighbourhood. As Lees, Slater & Wyly (2008, p. 118) state,

One of the most commonly noted trends in the process of gentrification is that places and people once deemed hip, authentic, trendy, and subversive quickly become appropriated, manufactured, and mass-produced kitsch for higher-earning groups. Thus, if we speak of a gentrification aesthetic, we must remember that this aesthetic is far from frozen, and leads to enormous profits as cultural capital becomes economic capital.

In the case of Olhão, the tourism gentrification is hard to define. While on the one hand, it cannot be described as strictly created by the tourism industry, since foreigners were moving in even before there was a touristic industry in the city, on the other hand, it was the growth of a tourism industry in the city that saw the displacement of the local inhabitants, as well as the emergence of some elements of a super-gentrification process, such as the construction of luxury buildings for leisure purposes, high end private condos with panoramic swimming pools on the rooftops, high profile cars, a marina, boat tours industry, cosmopolitan gourmet and ethnic restaurants, not unlike what can be seen in Vilamoura and Vale do Lobo, two other prominent luxurious touristic destinations in South Portugal.

While it is still too early to fully grasp the implications of these transformation, the fact is that dozens of original residents are being harassed to sell their houses or evicted by land lords, who much prefer to engage in short-term rentals. The rapidly increasing living-costs are affecting the locals who find it difficult to maintain their lifestyle, since restaurants, bars, stores and retail have catered to the new foreign residents and the tourists that visit Olhão. Even being aware of the general negative social and cultural impacts of the gentrification processes, one may ask what would be the destiny of most of the traditional buildings and the labyrinth shaped streets of the historical neighbourhoods of Olhão if this phenomenon did not happen. At least, the first generation of gentrifiers – the pioneers, seemed truly concerned about the local inhabitant's future. The same cannot be said about the new vague, who sees Olhão merely as a trendy location they eagerly consume as another touristic attraction.

The pioneers of the gentrification worry about the older residents that are being pushed out of the historic centre and fear that Olhão might lose what makes it authentic to their eyes: its architecture and its people, becoming an Airbnb ghost town or a Disneyland, devoid of identity, embodied in the dramatization of everyday life as a staged touristic attraction (Guerreiro & Marques, 2017; MacCannell, 1973), a *non lieu* (Augé, 1995).

While cases of gentrifiers standing against gentrification is not something unheard of, Olhão is a case where such has happened in record time span. In previous documented cases, the pioneers were also artists, intellectuals or liberal professionals that criticized newcomers' decades after they had arrived. But in such cases, the pioneers weren't all that sympathetic towards the original inhabitants of the neighbourhoods. In one case (Tissot, 2015), these pioneers actively protested and caused an embargo over the construction of social housing intended for African Americans and Hispanics during the 70s, in Boston's South End, while now criticizing the yuppies and social elites that prompted a super-gentrification process.

One can also gain a deeper degree of understanding of the original gentrifiers' motivations for condemning the touristification of Olhão by looking at the process through the lenses of Elias' and Scotson's (1994) work *The Established and the Outsiders*, in which the authors study how a small neighbourhood community constituted of two distinguishable social groups coexist and how the "old families" see the newcomers and threat them as *outsiders*. In Olhão's case, one might consider as well that the first gentrifiers fear that the new wave of gentrification induced by mass tourism will change the town they know and enjoy.

Olhão arrived late to the touristification of the Algarve. Most fishing towns that once had hundreds of fishing boats and ships were replaced with leisure and recreational boats. Fishing is slowly declining and the fishing culture of the region is all but gone. Few places retain that culture, still impervious to the mass tourism transformations that have occurred all over Algarve. Cities like Quarteira, Albufeira, Portimão, Lagoa, Monte Gordo and Lagos, once great fishing ports or towns, are now composed of several story high apartment blocks by the sea, towering over the towns that once saw blue from their windows. That is what Olhão's pioneers fear above all: that Olhão might become another Quarteira or Albufeira: devoid of identity, culture and little more than a vacation place for the transnational elites.

At the same time, the Portuguese residents of the historical zone can do little to stem the tide. While they appreciate the recuperation of the derelict buildings, they fear that their town is being taken from them, as more families are evicted by landlords and older inhabitants sell their houses to foreigners.

ACKNOWLEDGMENT

The second author's contribution in this chapter is financed by National Funds provided by FCT - Foundation for Science and Technology, through project UIDB/04020/2020.

REFERENCES

Aalbers, M. B. (2019). Introduction To The Forum: From Third To Fifth-Wave Gentrification. *Tijdschrift voor Economische en Sociale Geografie*, *110*(1), 1–11. doi:10.1111/tesg.12332

Agarez, R. (2013). Regional identity for the leisure of travellers: Early tourism infrastructure in the Algarve (Portugal), 1940-1965. *Journal of Architecture (London)*, *18*(5), 721–743. doi:10.1080/13602365.2013.838982

Antunes, A.M., & Azevedo, A. (1988). *Arquitectura Popular em Portugal, 3rd Volume*. Lisbon: Associação dos Arquitectos Portugueses.

Atkinson, R., & Bridge, G. (2010). Globalisation and the New Urban Colonialism. In J. Brown-Saracino (Ed.), *The Gentrification Debates* (pp. 51–61). Routledge.

Augé, M. (1995). *Non-Places: Introduction to an Anthropology of Supermodernity*. Verso.

Baptista, L.V., Nofre, J., & Jorge, M.R. (2018). Mobilidade, Cidade e Turismo: pistas para analisar as transformações em curso no centro histórico de Lisboa. *Revista de Sociologia da Faculdade de Letras da Universidade do Porto*, 14-32.

Baudrillard, J. (2001). Simulacra and Simulations. In M. Porter (Ed.), *Jean Baudrillard – Selected Writings* (pp. 166–184). Polity Press.

Bauman, Z. (1998). *Globalization: The human consequences*. Polity Press.

Bauman, Z. (2000). *Liquid Modernity*. Polity Press.

Beck, U. (2008). *Qué es la Globalización?* Barcelona: Paidós.

Benjamin, W. (1969). Paris: Capital of the Nineteenth Century. *Perspecta*, *12*, 163–172.

Benson, M., & O'Reilly, K. (2009). Migration and the search for a better way of life: A critical exploration of lifestyle migration. *The Sociological Review*, *57*(4), 608–625. doi:10.1111/j.1467-954X.2009.01864.x

Bonança, J. (2014). *O Milagre do 25 de Abril: Transformação de Olhão*. Arandis.

Bourdieu, P. (1996). *Distinction: A Social Critique of the Judgement of Taste* (8th ed.). Harvard University Press.

Brandão, V. G. J. (2012). *O Compromisso Marítimo de Olhão (1765-1873)* (Unpublished master's dissertation). University of Algarve.

Butler, T. (2007). For gentrification? *Environment & Planning*, *39*(1), 162–181. doi:10.1068/a38472

Butler, T., & Lees, L. (2006). Super-Gentrification in Barnsbury, London: Globalization and Gentrifying Global Elites at the Neighbourhood Level. *Transactions of the Institute of British Geographers*, *31*(4), 467–487. doi:10.1111/j.1475-5661.2006.00220.x

Clark, E. (2005). The order and simplicity of gentrification – a political challenge. In R. Atkinson & G. Bridge (Eds.), *Gentrification in a Global Context: The new urban colonialism* (pp. 261–269). Routledge. doi:10.4324/9780203392089_chapter_16

Cohen, S. A., Duncan, T., & Thulemark, M. (2015). Lifestyle Mobilities: The Crossroads of Travel, Leisure and Migration. *Mobilities*, *10*(1), 155–172. doi:10.1080/17450101.2013.826481

Cresswell, T. (2006). *On The Move – Mobility in the Modern Western World*. Routledge.

Creswell, J. W. (1998). *Qualitative Inquiry and Research Design*. Sage Publications.

Durr, E., & Jaffe, R. (2012). Theorizing Slum Tourism: Performing, Negotiating and Transforming Inequality. *Revista Europea de Estudios Latinoamericanos y del Caribe*, *93*(93), 113–123. doi:10.18352/erlacs.8367

Elias, N., & Scotson, J. L. (1994). *The Established and the Outsiders: A Sociological Enquiry into Community Problems*. Sage Publications.

Featherstone, M. (1987). Lifestyle and Consumer Culture. *Theory, Culture & Society*, *4*(1), 55–70. doi:10.1177/026327687004001003

Foulds, A. (2014). *Buying a colonial dream: the role of lifestyle migrants in the gentrification of the historic center of Granada, Nicaragua* (Unpublished master's dissertation). University of Kentucky.

Gent, W. V., & Boterman, W. (2018). Gentrification of the Changing State. *Tijdschrift voor Economische en Sociale Geografie*, *110*(1), 35–46. doi:10.1111/tesg.12331

Giddens, A. (1991). *Modernity and Self-Identity: Self and Society in the Late Modern Age*. Polity Press.

Giddens, A. (2000). *As Consequências da Modernidade*. Celta.

González-Pérez, J. M. (2019). The dispute over touristic cities. Tourism gentrification in the historic Centre of Palma (Majorca, Spain). *Tourism Geographies*. Advance online publication. doi:10.1080/14616688.2019.1586986

Gotham, K. F. (2005). Tourism Gentrification: The Case of New Orlean's Vieux Carre (French Quarter). *Urban Studies (Edinburgh, Scotland)*, *42*(7), 1099–1121. doi:10.1080/00420980500120881

Guerreiro, J. A., & Marques, J. F. (2017). Visita Guiada à Fábrica de Antiguidades Sociologia, Turismo e Autenticidade. *Anais Brasileiros de Estudos Turísticos*, *1*(7), 8–22. doi:10.34019/2238-2925.2017.v7.3160

Gustafson, P. (2009). Your Home in Spain: Residential Strategies in International Retirement Migration. In M. Benson & K. O'Reilly (Eds.), *Lifestyle Migrations: Expectations, Aspirations and Experiences* (pp. 69–86). Ashgate.

Hackworth, J. (2018). Gentrification as a Politico-economic Window: Reflections on the Changing State of Gentrification. *Tijdschrift voor Economische en Sociale Geografie*, *110*(1), 47–53. doi:10.1111/tesg.12330

Hackworth, J., & Smith, N. (2001). The Changing State of Gentrification. *Tijdschrift voor Economische en Sociale Geografie*, *92*(4), 464–477. doi:10.1111/1467-9663.00172

Herrera, L. M., Smith, N., & Vera, M. A. M. (2013). Gentrification, Displacement, and Tourism in Santa Cruz De Tenerife. *Urban Geography*, *28*(3), 276–298. doi:10.2747/0272-3638.28.3.276

Hughes, N. (2018). 'Tourists go home': Anti-tourism industry protest in Barcelona. *Social Movement Studies*, *17*(4), 471–477. doi:10.1080/14742837.2018.1468244

INE. (2015). *Anuário Estatístico da Região do Algarve 2014*. Instituto Nacional de Estatística.

Janoschka, M., Sequera, J., & Salinas, L. (2014). Gentrification in Spain and Latin America: A Critical Dialogue. *International Journal of Urban and Regional Research*, *38*(4), 1234–1265. doi:10.1111/1468-2427.12030

Latour, B. (1987). *Science in Action: How to follow scientists and engineers through society*. Harvard University Press.

Lees, L., Slater, T., & Wyly, E. (2008). *Gentrification*. Routledge.

Lopes, A. C., Rodrigues, E. B., & Vera-Cruz, R. (2019). Tourism Gentrification. *IOP Conf. Series: Materials Science and Engineering*, 471.

MacCannell, D. (1973). Staged Authenticity: Arrangements of Social Space in Tourist Settings. *American Journal of Sociology*, *79*(3), 589–603. doi:10.1086/225585

Marx, K. (1982). *Capital* (3rd ed., Vol. 1). Penguin Books.

McIntyre, N. (2013). Mobilities, Lifestyles and Imagined Worlds: Exploring the Terrain of Lifestyle Migration. In T. Duncan, S. A. Cohen, & M. Thulemark (Eds.), *Lifestyle Mobilities: Intersections of Travel, Leisure and Migration* (pp. 193–207). Ashgate.

Mendes, L. M. (2017). Gentrificação turística em Lisboa: Neoliberalismo, financeirização e urbanismo austeritário em tempos de pós-crise capitalista 2008-2009. *Cadernos Metrópole*, *19*(39), 479–512. doi:10.1590/2236-9996.2017-3906

Miró, S. V. (2011). Producing a "Successful City": Neoliberal Urbanism and Gentrification in the Tourist City – The Case of Palma (Majorca). *Urban Studies Research*, *2011*. Advance online publication. doi:10.1155/2011/989676

Nobre, A. (2008). *História Breve da Vila de Olhão da Restauração*. Olhão: APOS.

Park, R., Burgess, E., & McKenzie, R. D. (1967). *The City. Suggestions for Investigation of Human Behavior in the Urban Environment*. The University of Chicago Press. (Original work published 1925)

Patch, J., & Brenner, N. (2007). Gentrification. In G. Ritzer (Ed.), *The Blackwell Encyclopedia of Sociology* (pp. 1917–1920). Blackwell Publishing. doi:10.1002/9781405165518.wbeosg035

Pinkster, F. M., & Boterman, W. R. (2017). When the spell is broken: Gentrification, urban tourism and privileged discontent in the Amsterdam canal district. *Cultural Geographies*, *24*(3), 457–472. doi:10.1177/1474474017706176 PMID:29278248

Prytherch, D. L., & Maiques, J. V. B. (2009). City profile: Valencia. *Cities (London, England)*, *26*(2), 103–115. doi:10.1016/j.cities.2008.11.004

Rérat, P., Söderström, O., & Piguet, E. (2009). New Forms of Gentrification: Issues and debates. *Population Space and Place*, *16*(5), 335–343. doi:10.1002/psp.585

Ritzer, G. (2011). *Globalization: The essentials*. Wiley-Blackwell.

Rofe, M. W. (2003). 'I Want to be Global': Theorising the Gentrifying Class as an Emergent Élite Global Community. *Urban Studies (Edinburgh, Scotland)*, *40*(12), 2511–2526. doi:10.1080/0042098032000136183

Sarafa, J., & Brito-Henriques, E. (2017). O Hotel Santiago de Alfama Como Alegoria da Política Recente de Regeneração Urbana. *Finisterra*, *52*(106), 65–84.

Sassen, S. (2007). *A Sociology of Globalization*. W. W. Norton & Company.

Sheller, M. (2011). Mobility. *Sociopedia.isa*. Doi:10.1177/205684601163

Sheller, M. (2014). The new mobilities paradigm for a live sociology. *Current Sociology Review*, *62*(6), 1–23. doi:10.1177/0011392114533211

Sheller, M., & Urry, J. (2006). The New Mobilities Paradigm. *Environment & Planning A*, *38*(2), 207–226. doi:10.1068/a37268

Sigler, T., & Wachsmuth, D. (2016). Transnational gentrification: Globalisation and neighbourhood change in Panama's Casco Antiguo. *Urban Studies (Edinburgh, Scotland)*, *53*(4), 705–722. doi:10.1177/0042098014568070

Skoll, G. R., & Korstanje, M. (2014). Urban heritage, gentrification, and tourism in Reverwest and El Abasto. *Journal of Heritage Tourism*, *9*(4), 349–359. doi:10.1080/1743873X.2014.890624

Smith, N. (2002). New Globalism, New Urbanism: Gentrification as Global Urban Strategy. *Antipode, 34*(3), 434–457. doi:10.1111/1467-8330.00249

Smith, N. (2010). A Short History of Gentrification. In J. Brown-Saracino (Ed.), *The Gentrification Debates* (pp. 31–36). Routledge.

Stebbins, R. (1997). Lifestyle as a Generic Concept in Ethnographic Research. *Quality & Quantity, 31*(4), 347–360. doi:10.1023/A:1004285831689

Tissot, S. (2015). *Good Neighbors: Gentrifying Diversity in Boston's South End.* Verso.

Torkington, K. (2012). Place and Lifestyle Migration: The Discursive Construction of 'Glocal' Place-Identity. *Mobilities, 7*(1), 71–92. doi:10.1080/17450101.2012.631812

Urry, J. (2002). Mobility and Proximity. *Sociology, 36*(2), 255–274. doi:10.1177/0038038502036002002

Veal, A. J. (1993). The concept of lifestyle: A review. *Leisure Studies, 12*(4), 233–252. doi:10.1080/02614369300390231

Wallerstein, I. (1999). Análise dos Sistemas Mundiais. In A. Giddens & J. Turner (Eds.), *Teoria Social Hoje* (pp. 447–470). UNESP.

Weber, M. (1963). *Economy and Society: An outline of interpretive Sociology.* University of California Press.

Williams, A. M., & Patterson, G. (1998). 'An Empire Lost but a Province Gained': A Cohort Analysis of British International Retirement in the Algarve. *International Journal of Population Geography, 4*(2), 135–155. doi:10.1002/(SICI)1099-1220(199806)4:2<135::AID-IJPG99>3.0.CO;2-6 PMID:12348626

ADDITIONAL READING

Adey, P., Bissell, D., Hannam, K., Merriman, P., & Sheller, M. (Eds.). (2014). *The Routledge Handbook of Mobilities* (pp. 45–54). Routledge. doi:10.4324/9781315857572

Atkinson, R., & Bridge, G. (Eds.). (2005). *Gentrification in a Global Context: The new urban colonialism.* Routledge.

Benson, M., & O'Reilly, K. (Eds.). (2009). *Lifestyle Migrations: Expectations, Aspirations and Experiences.* Ashgate.

Brown-Saracino, J. (Ed.). (2010). *The Gentrification Debates.* Routledge.

Cohen, S. A., Duncan, T., & Thulemark, M. (2015). Lifestyle Mobilities: The Crossroads of Travel, Leisure and Migration. *Mobilities, 10*(1), 155–172. doi:10.1080/17450101.2013.826481

Cresswell, T. (2006). *On The Move – Mobility in the Modern Western World.* Routledge.

Duncan, T., Cohen, S. A., & Thulemark, M. (Eds.). (2013). *Lifestyle Mobilities: Intersections of Travel, Leisure and Migration*. Ashgate.

Hall, C. M., & Müller, D. K. (Eds.). (2004). *Tourism, Mobility and Second Homes: Between Elite Landscape and Common Ground* (pp. 273–278). Cromwell Press. doi:10.21832/9781873150825

Merriman, P. (2012). *Mobility, Space and Culture*. Routledge. doi:10.4324/9780203842102

Montezuma, J., & McGarrigle, J. (2018). What motivates international homebuyers? Investor to lifestyle 'migrants' in a tourist city. *Tourism Geographies*, *21*(2), 214–234. doi:10.1080/14616688.2018.1470196

KEY TERMS AND DEFINITIONS

Anti-Tourism: Negative or even hostile attitudes towards tourism, tourists and the touristic industry. Often begins as sentiments and can develop into social movements when tourism begins to have a perceived negative impact in the resident's lives. Often closely related to touristification, overtourism, and tourism gentrification.

Gentrification: A transformation process caused by the influx of middle and high income individuals to urban areas previously inhabited by the working classes with lower income, resulting in higher living and housing costs, which results in the dislocation of the original inhabitants from the affected area.

Lifestyle Mobilities: Consists in the voluntary relocation of individuals to places they perceive as allowing them to adopt or manifest a different lifestyle. Can also encapsulate frequent movements (circular residential patterns) or lifestyles based on frequent residence changes (such as backpackers or neo-nomads).

Mobilities: Mobilities (not to be confused with social mobility) refers to the movement of people, objects, capital and information, and to the social meaning, implications and contexts of such movements. It distinguishes itself from movement because of this aspect: movement does not require meaning, while mobilities have meaning attached to them (what does it mean to go to one place instead of the other, what some places represent to some people and how these meanings change according to the individual, etc.).

Platform Tourism: Tourism goods or services offered by private individuals to visitors through digital platforms, such as Airbnb. It is also referred to as Platform Tourism Services.

Super Gentrification: Process that occurs when previously gentrified areas begin experiencing a new wave of gentrification that affects not only the original inhabitants, but also the previous gentrifiers. It is often caused by individuals with even higher economic capital (often transnational elites) move to gentrified areas.

Tourism Gentrification: Gentrification process not directly caused by the influx of new residents with higher income, but rather by the touristification of areas and the ensuing escalating housing prices inflation. Lately it has become closely tight with platform tourism.

Touristification: An increase of the number of tourists and tourism businesses in a given territory the impact created by it. The term has lately been used in conjunction (or interchangeably) with overtourism to designate the overcrowding and excess of tourism, as well as its impacts to the local population and to the tourism industry itself.

Transnational Gentrification: Gentrification process caused by the influx of individuals from another country, with higher income than that of the original residents of the area. It is closely related to mobilities and differs from tourism gentrification in the sense that these are either new permanent inhabitants or individuals that acquire housing for temporary stays. Both forms lead to increasing housing and cost of living costs, thus leading to the displacement of the original inhabitants.

Chapter 2
Climate-Friendly Hotels and the Promotion of Sustainability in Tourist Destinations:
A Study in Algarve From Hotel Management Perspectives

Margarida Arraes Viegas
https://orcid.org/0000-0002-0368-7157
School of Management, Hospitality and Tourism, University of Algarve, Portugal

ABSTRACT

Tourism promotes the interaction of visitors with destination places and their communities, resulting in a set of impacts for the local population. Although traditionally no major environmental damage due to tourism in general, and hospitality, in particular, has been recognized, as the concept of sustainability develops, its negative impacts are becoming evident. As the hotel sector is one of the most important components of the tourism offer, minimizing its environmental impacts plays a fundamental role in the sustainability of tourist destinations. Since Algarve is one of the top 20 travel destinations worldwide, the present study aims to identify the environmental behaviour of the hotels in the region and thus evaluate their contribution to the sustainability of this tourist destination. In the empirical analysis, statistical techniques such as principal components and cluster analysis were used, and the results allowed the identification of three groups with different environmental behaviours.

INTRODUCTION

The quest for a more sustainable society has become, over the past few decades, a major concern of the international community. Throughout the second half of the twentieth century, the emergence of global socio-environmental problems triggered an active search for a new model of development that could reconcile economic, social and environmental aspects. This new paradigm, known as sustainable

DOI: 10.4018/978-1-7998-3156-3.ch002

development, emerged in the early seventies and brought with it a humanization of the concept of development, including the concern with the preservation of resources for future generations. Corporations, the basic units of any economic organization, play a crucial role in addressing the challenges posed by the paradigm of sustainability, from which a set of influences and practices stems, such as Corporate Social Responsibility (CSR) and Corporate Environmental Responsibility (CER), to achieve a socially and environmentally responsible performance. In this context, and according to Boiral (2006), the analysis of the corporations' environmental management strategies, is a way to assess their contribution to sustainable development. As the Algarve is the most popular tourist region in Portugal and tourism its main economic activity, this research focuses on the environmental strategies of its hotel industry, to assess their contribution to the sustainable development of the region. In this sense, the research question underlying this study can be formulated as follow: do Algarve hotels, through their environmental management strategies, contribute to the sustainable tourism development of the region?

The empirical analysis starts with the classification of the hotels according to their environmental strategy, which is defined according to the level of implementation of a set of organizational and operational environmental management actions. The identified groups will then be differentiated, through a series of hypothesis tests, regarding contextual factors and environmental performance since, according to Carmona et al. (2004, p.108): "environmental strategy is affected by certain contextual factors and in turn affects environmental performance". In the first case, the variables used concern both externally observable aspects (explicit factors) and aspects that relate to the internal dynamics of the company (implicit factors). Environmental performance is measured by the level of implementation of 9 environmental action areas, adapted from the priority environmental action areas defined in "Agenda21 for the Travel and Tourism Industry". The analysis process is outlined in Figure 1:

Figure 1. Analysis process

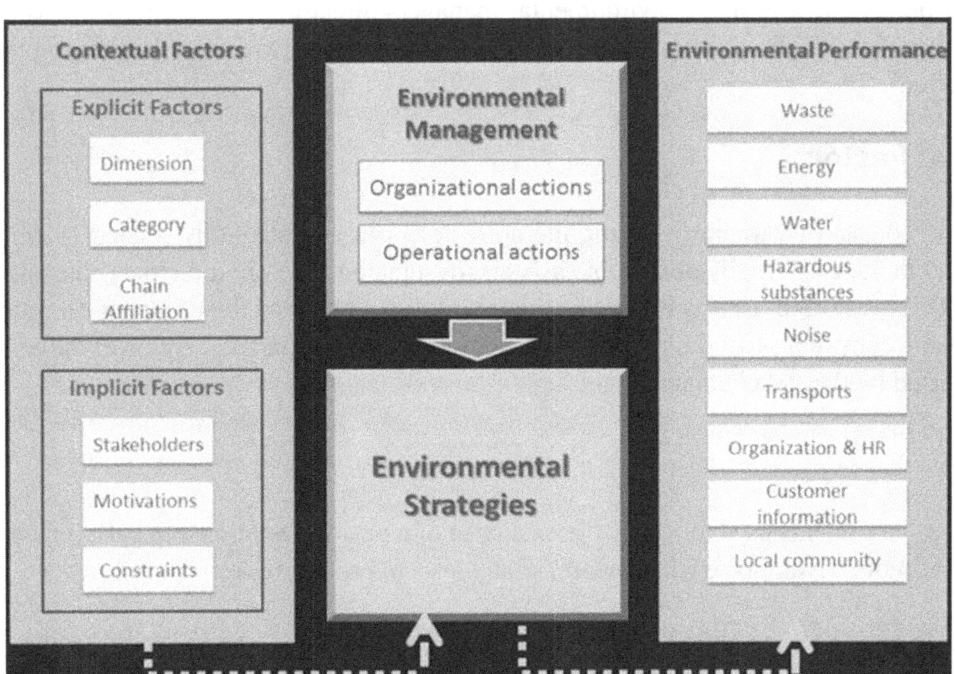

CONCEPTS, VARIABLES AND HYPOTHESES

Concepts are the basic pillars of scientific research. According to Davis and Cosenza (1994), concepts can be defined constitutively (using other concepts) or operationally (specifying the procedures for their quantification). Operationally defined concepts become variables and the concrete description of the researcher's expectation about their relationships is called hypotheses. Hypotheses thus allow the structuring of the problem under study in terms of relationships between variables and will guide the process of data analysis (in the case of this study, unless otherwise indicated, the hypotheses will be tested to a significance level of 5%).

Environmental Strategy

According to Hoffman (1999), the environmental strategy can be seen as a management tool that allows organizations to plan and develop a consistent response to the environmental challenge and thus contribute to strengthening their business strategies. Henriques and Sadorsky (1999) state that this concept can be operationally defined through the level of implementation of a set of organizational and operational/technical environmental management actions. Several authors, such as Hunt and Auster (1990), Roome (1992), Vastag, Kerekes, and Rondinelli (1996), Henriques and Sadorsky (1999), contributed to the consolidation of a theoretical basis that allows the classification of environmental strategies according to their degree of proactivity, which is defined as the level reached in the adoption of these environmental management actions. Despite the differences between the various typologies proposed, all of them are based on a continuum of evolution ranging from the most reactive to the most proactive (or anticipatory) strategy, with different intermediate strategies. For the present study, 19 of these actions were selected to cover the most important aspects of environmental management, according to the classifications mentioned in the literature (Aragón,1998; Henriques & Sadorsky, 1999; Céspedes & de Burgos, 2004). In the questionnaire sent to hotel environmental managers/directors, they were asked to evaluate the implementation level of these actions using a 5-point Likert-type scale (1-null implementation; 5-full implementation).

Contextual Factors

These factors consist of variables that the literature refers to as potentially related to environmental strategies and include externally observable aspects, designated as explicit factors (hotel size, category and chain affiliation), and aspects related to the internal dynamics of the company (implicit factors): motivations and constraints to the adoption of environmental actions and perception of the environmental pressure exerted by the stakeholders.

Hotel size

The theoretical arguments used to justify the existence of a positive relationship between company size and the use of more developed environmental strategies can be summarized as follows:

i. Larger companies have a greater environmental impact in absolute terms and are therefore under greater pressure (Henriques and Sadorsky, 1996);

ii. Larger companies have more resources to invest in environmental protection (Aragón,1998);
iii. Larger companies adopt a more formal management model, which applies to environmental management (Merritt,1998). These arguments legitimize the following hypothesis:
Hypothesis 1: Larger hotels adopt more proactive environmental strategies.

Category

The reasons for the inclusion of this variable are related to those mentioned in the previous one, considering that the theoretical arguments previously expressed could be extended to the hotel category. Also, higher-class hotels have greater visibility to the general public, which may act as an incentive for more proactive environmental practices. The hypothesis thus stands:

Hypothesis 2: Higher category hotels adopt more proactive environmental strategies.

Chain Affiliation

In addition to aspects related to hotel size, and according to Darr et al. (as cited in Céspedes & de Burgos, 2004, p.12), the fact that the hotel is chain affiliated allows it to take advantage of the knowledge and experience accumulated by the other members so, in this situation, it is expected to implement more proactive environmental strategies. This leads to the hypothesis:

Hypothesis 3: Chain affiliated hotels have more proactive environmental strategies.

Motivations

Similar to the works of authors such as Bansal and Roth (2000) and González and González (2005), motivations are understood as the expectations of the organization about potentials results of environmental proactivity, that is, the expectations regarding the competitive advantages that can be obtained. According to Murillo et al. (2008), the level of proactivity of the environmental strategy is influenced by specific motivations, therefore, by including this variable, it is intended to differentiate the strategic orientations of the hotels according to the expectations arising from environmental protection measures. To identify the motivations most valued by the Algarve hotels, 11 variables measured on a 5-point Likert-type scale were used, according to which the hotel environmental managers should indicate the importance of each motivation for the implementation of environmental management actions. The variables were selected according to the literature on the subject and to contemplate the categories indicated by Bansal and Roth (2000) and González and González (2005), which suggest the existence of three types of motivations for a more proactive strategic positioning: competitiveness (productive and commercial), legitimacy (or relational) and environmental responsibility (or ethical). The hypothesis (H4) under analysis is that hotels with more proactive strategies value more the motivations:

Hypothesis 4a: of competitiveness
Hypothesis 4b: of legitimacy
Hypothesis 4c: of environmental responsibility.

Constraints

According to Aragón et al. (2008), the environmental strategic behaviour of companies is determined by the relative importance of the factors that motivate it and the factors that prevent or hinder its adoption. The latter can be classified into external (or of the industry) and internal (or organizational) constraints. According to this classification, 20 variables were used (10 external and 10 internal), evaluated in a Likert-type scale with 5 levels of importance, to identify the constraints perceived by the Algarve hotels to their environmental effort. Internal constraints are considered by some authors (Post & Altman, 1994; Hillary, 2003) as the main obstacle to more proactive environmental behaviours because companies will only be free to devote time and resources to external constraints after overcoming internal ones. The above justifies the formulation of the hypothesis (H5) that in hotels with a more proactive environmental stance there will be a lower valuation of internal constraints and a higher valuation of external ones:

Hypothesis 5a: In hotels with more proactive strategies, internal constraints are less valued
Hypothesis 5b: In hotels with more proactive strategies, external constraints are most valued.

Stakeholders

As noted by Ayuso (2003), companies can adopt environmental strategies in response to the expectations and pressures of their interest groups or stakeholders. Although Ortiz, Martínez, Rogel, and Rochel (2016) reported a lack of research on hotel industry stakeholders, according to Alvarez et al. (2001a), the groups to consider for hotels are shareholders, public administration, travel agencies and tour operators, association or chain to which they belong, clients, competitors, hotel director, employees, suppliers and NGOs. Still, according to these authors, stakeholder pressure can be measured by the geometric mean between their power of influence and their interest in the hotel's environmental actions, attributes that respondents should assess on a 5-point Likert-type scale. The resulting indicator will allow ascertaining whether the perception of environmental pressure from stakeholders is, as stated by Sharma and Starik (2004), a driving force of environmental proactivity. The hypothesis thus stands:

Hypothesis 6: Hotels with more proactive strategies perceive greater environmental pressure from their stakeholders.

Environmental Performance

Environmental performance can be defined as "the minimization of negative environmental impacts arising from the company's activity" (de Burgos & Céspedes 2001, p.1561). As the authors point out, this definition is difficult to operationalize in the context of hotels because there is a great difficulty, on the part of hotel managers, to provide quantitative information on environmental impacts, information that they do not normally have. Therefore, respondents are asked to indicate (on a 5-point Likert-type scale) to what extent are the various environmental action areas implemented, assuming that a better implementation level implies greater minimization of environmental impact, and therefore better performance, which leads to the hypothesis:

Hypothesis 7: Hotels with more proactive strategies have better implementation levels in all environmental action areas.

Based on the results reported by Viegas (2011), indicating that the lowest levels of implementation were in environmental areas without a specific legal framework and where cost reductions are less visible (organization and human resources, customer information and engagement with the local community), it is also hypothesized that:

Hypothesis 8: Hotels with a more proactive strategy have better levels of implementation in these environmental action areas.

QUESTIONNAIRE DESIGN AND DATA COLLECTION

Although there are no universal principles for questionnaire design, there are general guidelines, aimed at maximizing the response rate, that was taken into account in this study. The questionnaire was therefore drawn up following the main orientations related to the order, clarity, intelligibility and unambiguity of the questions asked. In general, the questions were previously validated in studies of a similar nature, but there was a need to adapt them to the particular conditions of this investigation, both in terms of its presentation and its proper Portuguese translation. Given the characteristics of the theme under study, most of them are qualitative and, in terms of form, almost all of them are closed and follow a five-category rating scale, known as Likert-type scale, inducing what can be considered quantitative responses of interval measurement level. The initial questionnaire was pre-tested with potential respondents, questionnaire design specialists and hotel managers, which led to a change in the wording and sequence of some of the questions. The final version was drawn on the Surveygyzmo platform and the link emailed to the environmental managers/directors of the 136 hotels in the sampling base, which was the National Register of Tourism Enterprises (RNET), available on the Turismo de Portugal portal.

The universe or target population of the study is defined as the two, three, four and five-star hotels in the Algarve. According to Decree-Law no. 39/2008 of March 7, hotels fall into the category of hotel establishments, which also includes apartment-hotels and guesthouses. However, the latter categories, as well as the one-star hotels, are not included in the target population, as they are considered to have particularities (size, environmental dynamism) that differentiate them from other hotel establishments. Given the finite character and relatively small size of the universe (136 hotels), no sampling was carried out and the questionnaire was sent to all members of the target population.

RESULTS AND DISCUSSION

To identify and characterize the environmental behaviour of the Algarve hotels, it was first analyzed the existence of different patterns in their environmental strategies to define a typology of hotel groups such that, in each group, they are homogeneous concerning environmental management actions implemented by them.

Identification of Environmental Strategies

For the study of different typologies/taxonomies, the statistical technique more commonly used is Cluster Analysis (CA), which was adopted in the present study. The environmental management actions carried out by the hotels were considered as segmentation variables. However, the original variables were not used but the dimensions resulting from subjecting these variables to a Principal Component Analysis (PCA). The use of these dimensions removes possible interdependencies between variables, which could affect the result of cluster analysis, allowing the data to be represented by a set of independent factors. The results indicate that the 19 variables chosen to measure the hotels' environmental management can be grouped into two dimensions results coinciding with those obtained by several authors (Aragón,1998; Alvarez, Burgos, & Céspedes 2001a; Carmona et al., 2004).

As previously mentioned, these dimensions were used as segmentation variables in the cluster analysis, performed by the Ward method (an agglomerative hierarchical method which, due to the minimization of internal variation, tends to result in clusters with greater internal homogeneity and similar dimensions) that led to the identification of 3 segments. The internal validation of this solution was performed using the Kruskal-Wallis test since the normality and homoscedasticity assumptions for ANOVA application are not met.

The results show that group 1 has a negative standardized value (below average 0) in both dimensions, meaning that this group has a low level of implementation of both types of environmental actions. The same is true for the organizational initiatives of the second group, which, however, already has a positive value in operational actions, being the third segment the only one to present positive values in both types of environmental procedures. To analyze between which groups there are statistically significant differences, Fisher's LSD test was applied, which indicated that, concerning organizational initiatives, group 3 is positively distinguished from the others and that, in operational actions, all the groups present statistically significant differences among themselves. However, this test does not allow the interpretation of the behaviour of the different variables in each of the groups, so it is necessary to complement this information with the average values of the variables in each one of them.

The results lead, similar to the typology suggested by Claver et al. (2007), to the identification of the following environmental strategies:

Reactive Strategy (Group 1): It is the group with the poorest environmental balance, both in terms of operational and organizational environmental initiatives. The values are consistently lower than the sample average and, in some procedures, almost nonexistent (close to 1 on a scale of 1 to 5). This is the case of the implementation of an environmental management system, the existence of a code of good practice and external communication of environmental performance. The highest values are those that do not involve a specific effort or investment in environmental protection (encouraging guest collaboration) or whose main objective is immediate cost savings (water and energy-saving measures).

Intermediate Strategy (Group 2): Hotels in this group appear to be moderately aware of environmental issues. They differ significantly from the previous cluster in the dimension related to operational initiatives which, when analyzed individually, register all values above the global average and, although this differentiation is not registered in the organizational dimension, already shows some commitment to environmental aspects by top management (average equals global average). Nevertheless, its environmental strategy maintains a focus on cost reduction and the environmental issue is not yet assimilated by its organizational structure.

Proactive Strategy (Group 3): This group of hotels presents a more developed environmental strategy, differing significantly from previous clusters in operational and organizational initiatives, and presenting for all of them higher values than the other groups and the global average. Its response to environmental issues is quite homogeneous as all average values are higher than 4 (on a scale of 1 to 5), thus revealing an environmental strategy that encompasses both operational and organizational aspects.

Based on the above, we can state that the two extreme groups of this taxonomy of environmental strategies are perfectly characterized. The reactive group for its indifference to environmental issues, the proactive group for implementing environmental measures with broader objectives than mere cost savings and revealing a more environmentally responsible stance. Between the two we find the intermediate group. This group already reveals some environmental awareness, namely in terms of the operational measures, that register an implementation level close to that in the proactive segment. This is not the case for organizational initiatives, that are mostly weakly implemented.

Characterization of Environmental Strategies

Having identified the typology described above, it is now important to perform its external validation through significance tests that allow the comparison of the three groups in the variables not used to generate the solution. The execution of these tests will lead to the confirmation, or rejection, of the initially formulated hypotheses and thus deepen the characterization of the identified segments. As previously explained, the variables used are hotel size, category, chain affiliation, motivations, constraints, stakeholders (contextual factors) and environmental performance.

Hotel Size

Since the hotel size was measured by the number of rooms, ANOVA was used to determine the differences between the average number of rooms registered in each cluster.

After verifying the assumptions for the application of this technique and concluding by rejecting its null hypothesis (which indicates the existence of statistically significant differences between the means of at least two groups), the Scheffé multiple comparisons test was performed, indicating the existence of a statistically significant difference between the third group and the others. This confirms the first hypothesis (H1) that larger hotels adopt more proactive environmental strategies, a result identical to that reported by Alvarez et al. (2001b).

Category and Chain Affiliation

The application of the $\chi 2$ test leads to the conclusion of independence between the hotel category and its environmental strategy. However, there is a loss of rigour in the application of this test because, although the four initial categories were grouped into only two (2 and 3 stars/4 and 5 stars), it was not possible to comply with the rule on a number of cells with an expected value less than 5 (33.3%), which should not exceed 20%. The adjusted standardized residuals were therefore analyzed and did not indicate in any case, any significant departure from the independence hypothesis. Thus, it is not possible to confirm the hypothesis (H2) that higher category hotels adopt more proactive environmental strategies.

As for the chain affiliation, it appears that the reactive group is mostly made up of hotels not affiliated to any hotel chain, while in the proactive group the opposite is true. In the middle group, there is an almost equal distribution of both cases. The application of the $\chi2$ test leads to the rejection of the null hypothesis, that is, the variables are not independent. Although the association between them is not strong, as indicated by the value of 0.39 of Crámer's V coefficient, it can be concluded, as formulated in hypothesis H3, that environmental proactivity is higher in hotels that are chain affiliated.

Motivations

Regarding the expectations of achieving competitive advantages, through the adoption of environmental management measures, the average values indicate that both overall and in each strategy, environmental management is associated with the possibility of obtaining each of the advantages (all values above 3, on a scale of 1 to 5), but cost reduction is the motivation most valued by all segments. For the first two groups follow the motivations regarding the improvement of profitability and resource utilization, while in the proactive group, compliance with legislation and improvement of the organizational image emerge in the second and third place. Following the classification of Bansal and Roth (2000) and González and González (2005), it appears that, in general, productive competitiveness motivations are more valued and those of commercial competitiveness less valued.

The results do not indicate significant differences regarding any of the productive competitiveness advantages (all highly valued in all three groups, with all values> 4) and also in the commercial advantage "access new markets" (the least valued in all groups), which means that the reasons for the different degrees of proactivity do not rely on these aspects. This is not the case for the other commercial motivations as well as for the environmental responsibility and legitimacy motivations. In almost all of them, the proactive group differs from the others, indicating that these are decisive motivations in the evolution from the reactive and intermediate strategies to the proactive environmental strategy. The only exception is "meeting stakeholder requirements" motivation. In this case, all groups present statistically significant differences among themselves, which indicates that this is an important motivation to evolve along with the three identified strategic levels. Overall, the results indicate that the lower strategic levels present a more similar pattern as regards their expectations for competitive advantage, being commercial (except for "access to new markets"), ethical and relational motivations decisive in the evolution from less advanced environmental strategies to proactive environmental strategy, which is differentiated by the superior importance of this kind of motivations. These results thus allow to globally confirm hypothesis H4, except partial confirmation of H4a as no empirical evidence was found that the productive competitiveness motivations exert influence on the proactivity level of environmental strategies.

Constraints

In general terms, it can be seen that respondents' value external constraints more than the internal ones. In the first case, the highest value goes to investment issues and bureaucratic complexity and in the second, to the scarcity of resources and the difficulty of strategic adaptation. Analyzing the evolution of the average values in the three environmental strategies, different trends are however detected in the two types of constraints: in the external ones, most values increase with the level of proactivity, while in the internal the opposite is true. These results seem to indicate that more proactive environmental behaviours entail overcoming internal constraints and greater valuing of the external ones, which is consistent with the argument of Post and Altman (as cited in Murillo, Garcés, & Rivera, 2007, p.37), that the companies will only devote time and resources to external obstacles once internal ones are overcome.

To confirm the previous interpretation, and as done by Murillo et al. (2007), two principal component analyzes were carried out, that is, this technique was applied separately to the set of external constraints and the set of internal ones.

According to the criterion of cumulative percentage of variance, and valuing the possible interpretation of the factors to be extracted, a three-component solution was chosen, although Kaiser's method suggests the extraction of only two.

To analyze the relationship between the two types of constraints to environmental adaptation and the proactivity of environmental strategies, the Kruskal-Wallis test was applied, since the assumptions for the application of ANOVA were not met. According to the authors mentioned above, five new variables were created, representing the principal components and resulting from the arithmetic mean of the values of the variables associated with each one. Only the component related to financial and organizational constraints presents statistically significant differences, with the proactive group standing out due to the lower appreciation of these type of constraints (either don't have them or already overcome them).

Regarding external constraints, it was not possible to confirm hypothesis H5b: the results are not consistent with those of Murillo et al. (2007), as no relationship was found between the proactivity level of the environmental strategy and the valuation of these barriers, which may be explained by the fact that the authors' work is applied to industrial companies, where such constraints will have a greater impact than in the hotel sector. Regarding internal constraints, the results are similar to those of the authors, indicating that more proactive environmental behaviour involves overcoming internal barriers related to lack of financial, organizational and strategic capacity (financial and organizational constraints). However, hypothesis H5a is not fully confirmed as this finding did not cover internal barriers related to human resources.

Stakeholders

To determine whether the strategic groups differentiate among them according to the environmental pressure exerted by their stakeholders, the procedure adopted by Álvarez et al. (2001a) was applied. Thus, a new variable was created to translate the environmental pressure of each of the stakeholders, resulting from the combination of their influence with their interest in the environmental management actions carried out by the hotel units. This variable resulted from the weighting (through the geometric mean) of these two attributes and revealed high internal consistency, with a Cronbach's α of 0.879. Since the assumptions for the application of ANOVA were not met, the nonparametric Kruskal-Wallis test was used. The stakeholder "director" is the one who is perceived to have the greatest environmental pressure. However, when comparing the values between the three strategic groups, it appears that the reactive group is differentiated by the lower pressure it recognizes this social agent. The same applies regarding the pressure it perceives on competitors, customers, employees and suppliers, which indicates that the environmental pressure of these stakeholders is fundamental for the hotels to surpass a reactive environmental strategy. To achieve a truly proactive strategy, however, it is the stakeholder "chain affiliation" that plays a key role, as it is the only one that differentiates this strategy from the previous two (reactive and intermediate). Hypothesis H6 is therefore partially confirmed since empirical evidence has not been found for all stakeholders.

Environmental Performance

Overall, the environmental action areas have an average implementation level of 3.61, indicating that the surveyed hotels are making some effort to minimize their environmental impacts. However, the areas with the highest values are those that include measures that can more easily provide cost savings, while actions relating to the engagement with the local community have the lowest level of implementation.

Regarding the different strategies, there is a growing level of performance for all environmental areas as their degree of proactivity evolves. The proactive group differs significantly in most environmental action areas, except for waste, energy and water management, where its average level of performance is statistically identical to that of the intermediate group, so hypothesis H7 is confirmed only partially. Hypothesis H8 is confirmed because proactive hotels have a higher level of performance in areas where cost reduction is less visible, such as organization and human resources, customer information and awareness and involvement with the local community. These results are consistent with those obtained for environmental management procedures in which the proactive group differed by the greater implementation of organizational measures, precisely the type of measures with the greatest impact in these areas.

FUTURE RESEARCH DIRECTIONS

The topic under study is a relatively recent subject in the academic literature and there are no solid definitions for the concepts involved or globally validated measurement scales for the constructs (environmental management, stakeholder pressure, environmental performance) that, by their very nature, are neither easy nor completely measurable. Although the authors sought to define the constructs as precisely as possible and select their indicators based on their presence in the theoretical and empirical literature, one cannot fail to mention the exploratory character of some scales used. Assuming that future development of environmental management practices will facilitate access to quantitative information (eg emissions volume, resource savings), attention should be given to the use of these objective indicators in further research. Also, since the data obtained for each hotel were provided by a single respondent, one cannot rule out the possibility that their perceptions do not fully reflect the actual situation. For this reason, in future investigations, it would be important to take into account the opinion of other social agents. Since the use of cross-sectional data makes it impossible to analyze the evolution of the variables, would be equally important further research from a longitudinal perspective, to definitively confirm the established relationships. The application of other research methodologies, such as case studies, should also be considered, as it would provide complementary information of qualitative character, relevant to deepen the analysis of these relationships. Finally, given that the present investigation focused on a specific subsector (the hotelier) and a specific region (the Algarve), the results should not be extrapolated outside this specific scope. Future application of the research to other tourist developments (e.g. resorts, apartment-hotels), other activities (e.g. golf) or other destinations (e.g. Andalusia), would make it possible to determine its applicability to other tourist service companies and to contribute to a better knowledge of the environmental positioning of the sector and the Algarve travel destination.

CONCLUSION

Throughout the 1990s, the sustainable development paradigm brings the corporate world a new way of doing business, with socio and environmental concerns exerting a growing influence on its decisions. The concept that best reflects this new positioning of companies, concerning society and the environment, is the corporate social responsibility that, despite not having a consensual definition, can be understood as a form of management that involves the organizations' commitment to ethically conduct their activity, contributing to the socio-economic development of society and the preservation of environmental resources for future generations. Nowadays, the environment is seen as the most general and encompassing dimension of social life, since environmental degradation involves, to a greater or lesser extent, the entire planet. Responsible environmental behaviour is thus a necessary condition for the exercise of corporate social responsibility and, therefore, one of the ways how companies can contribute to the construction of sustainable development. In this context, this research was developed to achieve, through the analysis of the environmental behaviour of the Algarve hotels, an evaluation of the contribution of this sector to the sustainable development of the region. Companies' environmental behaviour is inevitably associated with their image and role in contemporary society, which makes concern about climate issues a more or less relevant variable in the definition of more or less proactive environmental management strategies. In the case of Algarve hotels, the present investigation allowed their classification according to three distinct strategic positions, a result similar to that reported by Claver et al. (2007), for the hotels of Alicante:

Reactive Strategy - Group consisting of smaller hotels, mostly (61%) not chain affiliated and presenting a poorer environmental balance, both in terms of operational and organizational environmental actions. The values are consistently lower than the sample average and, in some cases, almost nonexistent (close to 1 on a scale of 1 to 5). This is the case of the implementation of an environmental management system, the existence of a code of good practices and the external reporting of the environmental performance. The highest values are registered in actions that do not involve a specific investment in environmental protection (encouraging guest collaboration) or whose main objective is an immediate cost reduction (water and energy-saving measures);

Intermediate Strategy - Hotels in this group are of an intermediate dimension, almost evenly split between chain affiliated (52%) and independent units (48%) and appear to be moderately aware of environmental issues. This group differs significantly from the previous one in the operational dimension and, although there is no such difference in the organizational dimension, already reveals some management environmental commitment (average equals global average). Nevertheless, the focus on cost reduction remains and the environmental issue is not yet quite assimilated by the organizational structures;

Proactive Strategy - This group of hotels has a significantly larger average dimension and the vast majority is hotel chain affiliated (87%). It presents a more developed environmental strategy, differing significantly from the previous groups in the operational and organizational actions and presenting, in all of them, values higher than the other groups and the global average. Its response to environmental issues is quite homogeneous as all average values are higher than 4 (on a scale of 1 to 5), revealing an environmental strategy that encompasses both operational and organizational aspects.

For all groups, the main motivation for implementing environmental management actions is cost reduction. Although there is nodifferentiation in terms of productive competitiveness motivations, practically all other motivations positively differentiate the proactive group. Increasing market share, improving organizational image, differentiating from the competition (commercial competitiveness), complying with legislation (legitimacy), preserving the environmental quality of the region and contributing to its sustainable development (ethics) are the motivations that drive forward hotels for the proactive environmental strategy. The importance given to meeting the stakeholders' demands is the only motivation to significantly differentiate between all strategic groups, which indicates that stakeholder management (understood as the integration of their expectations/requirements into the company's environmental management strategy), it's a key factor in the evolution along with the three identified strategic levels. The stakeholder director is the one who is perceived to have the strongest environmental pressure in all strategic groups. However, the reactive group is distinguished by the lower pressure it recognizes, both on this stakeholder and on all other primary social agents (competitors, customers, employees and suppliers). The pressure from primary interest groups is therefore crucial for hotels to exceed the merely reactive level of their environmental strategies, however, to achieve a truly proactive strategy, the "hotel chain" stakeholder has a key influence, as it alone differentiates the proactive strategy from the others.

According to authors such as Post and Altman (1994) and Zilahy (2004), the process of implementing an environmental strategy is hampered by external (related to the type of business activity) and internal (inherent to the organization) constraints, the latter being the main obstacle to the adoption of proactive environmental actions. Although the surveyed hotels, as a whole, attach greater importance to external obstacles, particularly those related to investment and bureaucratic complexity, it appears that the strategic groups do not differ based on any of these constraints but on constraints intrinsic to the organization itself. The group that stands out for the lower valuation of this kind of constraints is the proactive one, which suggests that the implementation of this strategy involves overcoming internal obstacles related to lack of financial, organizational and strategic capacity.

Overall, the environmental action areas have an average level of implementation of 3.61, indicating that Algarve hotels are concerned about minimizing their environmental impacts. However, the areas with the highest values are those that include measures that can most easily provide cost savings (energy, water and waste). The reactive strategy is characterized by its inferior performance, which distinguishes it from the other strategies in the above areas, while the proactive group differs positively in environmental areas that require more complex environmental actions and where cost reduction is less visible, such as organization and human resources, customer awareness and involvement with the local community.

The different strategic positions identified for the Algarve hotels reflect the existence of different reactions to the environmental problem, which means that it is recognized by these establishments. Thus, climate concerns are present in the sector, although the vast majority of the hotels surveyed (69%) do not express a proactive stance: environmental management actions remain cost-centred, their implementation is mainly driven by expectations of achieving competitive advantages at the level of efficiency and are conditioned by characteristics intrinsic to the hotels themselves, such as inadequate organizational structure, unfavourable organizational culture and difficulty of strategic adaptation.

Given these results, it can be concluded that the environmental behaviour of Algarve hotels is still far from ideal and that the answer to the research question must be negative. However, there are positive signs: climate concerns are present in the sector, practically one-third of the hotels show a proactive environmental stance and evolution in this direction proves to be sensitive to demands of social agents. As these demands are nothing more than a reflection of the concerns of society as a whole, the response of Algarve hotels is expected to be increasingly climate-friendly and committed to promoting the sustainability of this tourist destination.

REFERENCES

Alvarez, M., de Burgos, J., & Céspedes, J. (2001a). Grupos de interés y estratégias medioambientales de los hoteles españoles. *Tribuna de Economía, 790*, 163–181.

Alvarez, M., de Burgos, J., & Céspedes, J. (2001b). An analysis of environmental management, organizational context and performance of Spanish hotels. *Omega, Elsevier, 29*(6), 457–471. doi:10.1016/S0305-0483(01)00033-0

Aragón, J. (1998). Strategic proactivity and firm approach to the natural environment. *Academy of Management Journal, 41*(5), 558–567. doi:10.2307/256942

Aragón, J., Hurtado, N., Sharma, S., & García, V. (2008). Environmental Strategy and Performance in Small Firms: A Resource-Based Perspective. *Journal of Environmental Management, 86*(1), 88–103. doi:10.1016/j.jenvman.2006.11.022 PMID:17239519

Ayuso, S. (2003). *Gestión sostenible en la industria turitica: retórica y práctica en el sector hotelero español* (Unpublished doctoral dissertation). Universitat Autònoma de Barcelona, Barcelona, Spain.

Bansal, P., & Roth, K. (2000). Why companies go green: A model of ecological responsiveness. *Academy of Management Journal, 43*, 717–736. doi:10.2307/1556363

Boiral, O. (2006). Global Warming: Should companies adopt proactive strategy? *Long Range Planning, 39*(3), 315–330. doi:10.1016/j.lrp.2006.07.002

Carmona, E., Céspedes, J., & de Burgos, J. (2004). Environmental Strategies in Spanish Hotels: Contextual factors and performance. *Service Industries Journal, 24*(3), 101–130. doi:10.1080/0264206042000247786

Céspedes, J., & de Burgos, J. (2004). Un análisis de las dimensiones de la gestión ambiental en los servicios hoteleros. *Dirección y Organización, 30*, 5–15.

Claver, E., Molina, J., Pereira, J., & López, M. (2007). Environmental strategies and their impact in hotel performance. *Journal of Sustainable Tourism, 15*(6), 663–679. doi:10.2167/jost640.0

Davis, D. L., & Cosenza, R. M. (1994). *Business Research for Decision Making* (3rd ed.). Wadsworth.

de Burgos, J., & Céspedes, J. (2001). Environmental Performance as an Operations Objective. *International Journal of Operations & Production Management, 21*(12), 1553–1572. doi:10.1108/01443570110410900

González, J., & González, O. (2005). An Analysis of the relationship between environmental motivations and ISO 14001 certification. *British Journal of Management, 16*(2), 133–148. doi:10.1111/j.1467-8551.2005.00436.x

Henriques, I., & Sadorsky, P. (1996). The determinants of an environmental responsive firm: An empirical approach. *Journal of Environmental Economics and Management, 30*(3), 381–395. doi:10.1006/jeem.1996.0026

Henriques, I., & Sadorsky, P. (1999). The relationship between environmental commitment and managerial perceptions of stakeholder importance. *Academy of Management Journal, 42*, 87–99. doi:10.2307/256876

Hillary, R. (2003). Environmental management systems and the smaller enterprise. *Journal of Cleaner Production, 12*(6), 561–569. doi:10.1016/j.jclepro.2003.08.006

Hoffman, A. (1999). Institutional Evolution and Change: Environmentalism and the U.S. Chemical Industry. *Academy of Management Journal, 42*(4), 351–372. doi:10.2307/257008

Hunt, C., & Auster, E. (1990). Proactive environmental management: Avoiding the toxic trap. *Sloan Management Review, 31*(2), 7–18.

Merritt, J. (1998). EM into SME won't go? Attitudes, awareness and practices in the London Borough of Croydon. *Business Strategy and the Environment, 7*(2), 90–100. doi:10.1002/(SICI)1099-0836(199805)7:2<90::AID-BSE141>3.0.CO;2-F

Murillo, J., Garcés, C., & Rivera, P. (2007). What prevents firms from advancing in their environmental strategy? *International Advances in Economic Research, 13*(1), 35–46. doi:10.100711294-006-9059-6

Murillo, J., Garcés, C., & Rivera, P. (2008). Estrategia medioambiental y expectativas de ventajas competitivas. *Cuadernos de Estudios Empresariales, 18*, 9–31.

Ortiz, Y., Martínez, E., Rogel, R., & Rochel, M. (2016). Los stakeholders de la industria hotelera: Una clasificación a partir de sus intereses ambientales. *Universidad & Empresa, 18*(30), 97–120. doi:10.12804/rev.univ.empresa.30.2016.05

Post, J., & Altman, B. (1994). Managing the Environmental Change Process: Barriers and Opportunities. *Journal of Organizational Change Management, 7*(4), 64–81. doi:10.1108/09534819410061388

Roome, N. (1992). Developing environmental management strategies. *Business Strategy and the Environment, 1*(1), 11–24. doi:10.1002/bse.3280010104

Vastag, G., Kerekes, S., & Rondinelli, D. (1996). Evaluation of Corporate Environmental Management Approaches: A Framework and Application. *International Journal of Production Economics, 43*(2-3), 193–211. doi:10.1016/0925-5273(96)00040-0

Vicente, J., Molina, V., & Ruiz, I. (2008). *La Gestión Medioambiental en las Empresas de Economía Social*. Paper presented at VII Colóquio Ibérico de Estudos Rurais, Coimbra, Portugal.

Viegas, M. (2011). Good Environmental Practices in the Hospitality Sector. *Journal of Tourism and Sustainability, I*(2), 5–13. Retrieved December 18, 2019, from https://sapientia.ualg.pt/bitstream/10400.1/13382/3/goodpractices.pdf

World Travel & Tourism Council. (1997). *Agenda 21 for the travel and tourism industry: towards environmentally sustainable development*. World Travel & Tourism Council.

Zilahy, G. (2004). Organizational factors determining the implementation of cleaner production measures in the corporate sector. *Journal of Cleaner Production, 12*(4), 311–319. doi:10.1016/S0959-6526(03)00016-7

ADDITIONAL READING

Baker, M., Davis, E., & Weaver, P. (2014). Eco-friendly attitudes, barriers to participation, and differences in behavior at green hotels. *Cornell Hospitality Quarterly*. 55, 89–99. doi: 10.1177%2F1938965513504483

de Grosbois, D. (2012). Corporate social responsibility reporting by the global hotel industry: Commitment, initiatives and performance. *International Journal of Hospitality Management*, *31*(3), 896–905. doi:10.1016/j.ijhm.2011.10.008

El Dief, M., & Font, X. (2011). Determinants of environmental management in the Red Sea Hotels: Personal and organizational values and contextual variables. *Journal of Hospitality & Tourism Research (Washington, D.C.)*, *36*(1), 115–137. doi:10.1177/1096348010388657

Hall, M. (2019). Constructing sustainable tourism development: The 2030 agenda and the managerial ecology of sustainable tourism. *Journal of Sustainable Tourism*, *27*(7), 1044–1060. doi:10.1080/09669582.2018.1560456

Horng, J., Hsu, H., & Tsai, C. (2018). An assessment model of corporate social responsibility practice in the tourism industry. *Journal of Sustainable Tourism*, *26*(7), 1085–1104. doi:10.1080/09669582.2017.1388384

Maciej, A., & Jarząbek, J. (2016). Corporate social responsibility in hotel industry -environmental implications. *Scientific Journal of the European Ecocycles Society*, *2*(1), 9–16. doi:10.19040/ecocycles.v2i2.51

Theodoulidis, B., Diaz, D., Crotto, F., & Rancati, E. (2017). Exploring corporate social responsibility and financial performance through stakeholder theory in the tourism industries. *Tourism Management*, *62*, 173–188. doi:10.1016/j.tourman.2017.03.018

UNWTO & UNDP. (2017). *Tourism and the sustainable development goals – Journey to 2030*. UNWTO.

Xavier, F., & Jennifer, L. (2018). Corporate social responsibility in tourism and hospitality. *Journal of Sustainable Tourism*, *26*(7), 1027–1042. doi:10.1080/09669582.2018.1488856

KEY TERMS AND DEFINITIONS

Algarve: Portugal's most important tourist region. Located in the extreme south of the country, it is characterized by a temperate Mediterranean climate and known for its beaches and golf resorts.

Corporate Environmental Responsibility (CER): Company's duties to incorporate environmental issues into their operations to maximize the efficiency and productivity of their resources without damaging natural environments.

Corporate Social Responsibility (CSR): Self-regulating business model that helps a company to act for the benefit of society at large, achieving a balance of economic, environmental and social imperatives.

Stakeholder: Any person, organization, social group, or society at large, that has an interest in a company and can either affect or be affected by its business.

Sustainable Tourism Development: One that establishes a suitable balance between the environmental, economic and socio-cultural aspects of tourism development. It attempts to minimize its impact on the environment and local culture so that it will be available for future generations while contributing to generate income, employment, and the conservation of local ecosystems.

Taxonomies: Empirically derived configurations.

Typologies: Conceptually developed configurations.

Chapter 3
Impacts of Tourism Development and Residents' Satisfaction:
An Explanatory Model for the Case of Huelva, Spain

Alfonso Vargas-Sánchez
https://orcid.org/0000-0003-0588-8654
University of Huelva, Spain

Francisco J. Riquel-Ligero
University of Huelva, Spain & Isabel I University, Spain

ABSTRACT

The main aim of this chapter is to propose and test a model to reflect on the factors that, associated with the tourist development of a destination, are perceived by the local community to be influential at the personal and collective level, as well as to what extent they impact people's satisfaction with the community they belong to. An enclave with a fishing tradition in the province of Huelva (Spain) was selected for primary data collection. PLS technique was used for hypotheses testing. It is concluded the contingent nature of the categorisation of the impacts with an influence on the local community's perception of tourism development, in addition to the inexistence of a direct relationship between the perception of the impacts on a personal level and residents' satisfaction with their community, although an indirect relationship exists through the perception of the effects on the town as a whole.

DOI: 10.4018/978-1-7998-3156-3.ch003

INTRODUCTION

It is clear that tourism, as we consider it today, is a system where a large number of agents interact, both public and private, which includes citizens, who have frequently been excluded in the traditional decision-making models when planning the tourism development of a destination or territory (Balbuena, 2014). However, for authors such as Vargas, Plaza, and Porras (2007), the local community is established as a stakeholder that must be considered by tourism destination managers when making strategic decisions on it, even though empirical evidence shows that in practice the local community tends to be overlooked in the tourism planning processes, as well as in the management of the projects resulting from it. This affirmation becomes even more evident when authors like Teye, Sirakaya, and Sönmez (2002) show us that lack of participation occurs completely independently of the destination's socio-economic development.

This paper aims to present a model, with sufficient explanatory capacity, which allows us to reflect on the factors, associated with the tourism development of the territory under study, perceived by the residents to be influential on a personal and collective level at the destination, as well as to what extent they impact the level of satisfaction with the community. It is focused, in short, on perceptions of tourism impacts and satisfaction with the community.

To do so, we have started with the traditional and widely used models to explain the perceptions of the residents towards the effects of tourism (Lui & Var, 1987: Akis et al., 1996; Santana, 1997; Gursoy, Jurowski, & Uysal, 2002, Vargas et al., 2007; Martín de la Rosa, 2009; Monterrubio, Cordero & García, 2011; Balbuena, 2014). These works determine the impacts of tourism development on the local community can be clustered into three large categories: economic, socio-cultural and environmental.

For contextualising this piece of research, we have selected the enclave of El Rompido, a town in the municipality of Cartaya in the province of Huelva (Spain). This town -with a fishing tradition- has the Río Piedras Natural Reserve (Paraje Natural Marismas del Río Piedras) as one of its main tourism resources and presents itself as a typical sun and beach destination, that went through tourism development in the nineteen-eighties, although it was at the start of this century when a commitment (by numerous municipal governments) was made to residential tourism development linked to building golf courses as a complementary offering and a comparative advantage for attracting tourists. This fact, along with the construction of four high capacity hotels, has led us to consider it is a good unit of analysis for the objective pursued in this study.

After this introduction, the structure of the chapter includes the following sections: theoretical framework, methodology, results, discussion and conclusions.

THEORETICAL FRAMEWORK

It is well known that tourism can have a positive or negative impact on local communities (López and Palomino, 2008; Suess, 2014), taking into account that, as Vela and Molina (2009) state, having a hospitable local population is a key factor for a tourism destination's success. In this line, it is worth noting the works authored by Ramón-Cardona (2014, 2015), in which the importance of the interaction between residents and visitors is highlighted. Undoubtedly, therefore, it is necessary to know the local populations' perceptions of the tourism development model implemented by decision-makers on the matter in a given territory. An interest that has been expressed in an ample number of scientific papers that try to delimit the factors that influence those perceptions (Belisle & Hoy, 1980; Liu & Var, 1987; Long, Perdue,

& Allen, 1990; Lankford & Howard, 1994; Akis, Peristianis, & Warner, 1996; Chen, 2000; Jurowski & Gursoy, 2004; Gutiérrez, 2010; Vargas, Plaza, & Porras, 2009, 2011; Nunkoo & Gursoy, 2012; Nunkoo & So, 2016; Masa'deh et. al, 2017; among others).

In this regard, Royo and Ruíz (2009) affirm that a local community which is apathetic about, displeased with or distrustful of tourism activity can give tourists a negative perception of a given destination. It is with good reason that authors such as Ko and Steward (2002) and Vargas et al. (2007) defend the need for further efforts that analyse residents' perception of tourism impacts, including their satisfaction with their local community (Mendoza et al, 2012). The contribution of Sinclair-Maragah (2017) confirms these studies, by relating the degree of acceptance/support of the tourist activity with its influence on the vital aspects of the local population.

Díaz and Gutiérrez (2010) defend that knowing the perceptions and attitudes of the local communities toward tourism is a competitive advantage, facilitating the creation of action plans and helping to reduce the negative impacts perceived by them. Consequently, studying the perceptions of the local community toward tourism development implies having identified the factors that can influence them. In this same line of enquiry, Sharpley (2014) studied the possible variables with an influence on the local community, concluding that there are a set of extrinsic variables that have an impact on the residents' perception. For Egly and Chaiken (1993) the attitude is a psychological trend that is expressed through a gradual assessment of a specific object as favourable or unfavourable. We are referring, therefore, to an internal state that is going to mediate and intervene between the external aspects (factors) and peoples' reactions (assessment response or perceptions). That favourable or unfavourable assessment score on tourism development is what we attempt to delimit in our unit of analysis.

Therefore, it is necessary to determine the factors that could influence perceptions or assessment responses of the local community toward tourism development.

Determining Factors of the Local Community's Perceptions

Balbuena (2014) affirms that there is a wide consensus in scientific studies on the impacts of tourism in the sense that they can be grouped into three categories: economic, social and environmental factors and impacts (Belisle & Hoy, 1980; Liu & Var, 1986; Long, Perdue, & Allen,1990; Lankford, 1994; Akis et al., 1996; Chen, 2000; Jurowski & Gursoy, 2004; Gutiérrez, 2010; Vargas, et al., 2009, 2011; Nunkoo & Gursoy, 2012; Brida, Riaño, & Zapata, 2012). However, Page and Hall (2014) criticize this approach as simplistic, since it ignores the diversity of relationships within this activity and the potential of tourism to impact different subgroups of residents in different ways.

The economic factors, which are the ones that usually have a greater impact on the local community (Liu & Var. 1986; Akis et al. 1996), can generate positive effects, coming from an increase in tourists arriving and the creation of greater investment in services, increasing the quality and the quantity of them, as well as entrepreneurial activity and, therefore, employment (Vargas, Porras, Plaza, & Riquel, 2008). But they can also be the bearers of negative impacts, coming from an increased need to consume resources and alternative uses of those resources, which are different from the traditional ones (Ponce, 2004, Monterrubio, 2018 and 2008; Monterrubio & Mendoza, 2011).

Andereck, Valentine, Knopff, and Vogot (2005) and García, Vásquez, and Macías (2015) are based on the proposal by Dogan (1989) to establish a correlation between tourism development at a given destination and the evolution of its sociocultural characteristics, which can be perceived by the local community in a positive (Sirakaya, Teye, & Sönmez, 2002; Besculides, Lee, & McCormick, 2002) or negative way

(Andriotis, 2005). However, González-García, Parra, Añó, and Calabuig (2016) state that there is little scientific evidence in this area. Be that as it may, the assessment of these impacts has an important subjective side (Santana, 2011), which increases the complexity when making it, that will have to take the interaction between the tourists and the destination, including its resident population, into account in a broad sense. In this regard, the effects of tourism development are going to have a personal assessment component that authors such as Royo and Ruíz (2009) and Monterrubio and Mendoza (2011) define as in the form of attitude, opinion and debates. That assessment of tourism's impacts on a personal level can differ from the perception that the same individual has of the impacts of those same factors in the whole community (Zhnag & Ibakaran, 2006), producing a degree of satisfaction or dissatisfaction in the subject with it (Mason, 2003).

Another factor with a broad scientific consensus regarding its effect on the perceptions that the local community has on tourism is the environmental one. Authors such as Liu and Var (1986) underscore how that factor affects the attitude of the residents toward the development of tourism activity. However, more recent publications, such as the paper by Balbuena, Barrera, and Hernández (2017), refer to that, among the factors with an impact on the perceptions and attitudes of the local community toward tourism, the environment is the one that has the least negative influence, since the residents prefer to support tourism development of a destination, prioritising its advantages over potential environmental harm that it could lead to.

Despite the widespread acceptance that this classification of the factors with an impact on the perceptions and attitudes of the local community (economic, social and environmental), there are works, such as the one by Figuerola, Fernández, Marzal, Gómez, and Martín (2015), that independently consider the impacts on jobs and the distinction between positive and negative economic impacts, following the approaches of authors such as García (1999), that make a distinction between the positive and negative socio-cultural impacts (Martín-Duque, 2017, Andereck & Nyupane, 2011).

Hence, in this chapter, for greater methodological rigour, the technique of factor analysis was applied first to verify the grouping of the variables under study into the blocks suggested above, or for establishing the groups of variables that relate specifically to the unit of analysis used here.

Types of Perceptions and Their Consequences

For Lindsay and Norman (1977), perception is the process by which sensations are interpreted and organised to explain and assess our experiences. Perception has been widely studied in the field of psychology and in particular by social psychology, where the construction of reality based on the stimuli that the perceiver receives has been made clear, which depends to a large extent on their internal factors (personality traits, motivations, goals, previous experiences, etc.) and cultural factors determined by the context they are immersed in (González-García, 2016; Rasoolimanesh et. al, 2017).

As has been mentioned, tourism is an economic activity that has impacts on the territory, which are perceived by the resident population. This community categorises them and creates states of opinion on those stimuli according to widely accepted principles of psychological models. In this respect, Andriotis and Vaughan (2003) highlight two large theoretical streams: "Social Representations Theory" and "Social Exchange Theory", although the latter has been dominant in the tourism literature (Vargas, Valle, Mendes, & Silva, 2015).

If the first (with social psychologist Serge Moscovici at the top) focuses on the construction of collective thought based on social, cultural or symbolic objects (Hoijer, 2011), the second places the emphasis on resource exchanges between groups or individuals in terms of generating a tourism development assessment through cost-benefit analysis, both from an individual as well as a group perspective (Ap, 1992; Madrigal, 1995; Presenza, Del Chiappa, & Sheehan, 2013). With their evolution and changes, when such judgements produce feelings (both collectively and as individuals) on tourism development, they affect the perception that the resident population has on the level of satisfaction with the place in question (Covarrubias, Vargas, & Rodríguez, 2010). If the effect is negative, it will make the destination less competitive in the long term, with future economic implications (Yüruk, Akyol, & Şimşek, 2017). Thus, according to the rationale of the Social Exchange Theory, the perceptions of the impacts caused by tourism development (with the corresponding positive-negative analysis) and the level of satisfaction are intimately linked, as displayed next in our research model.

Initial Research Model

If we start from the foundations of our theoretical reference framework and based on the approaches of several authors (Belisle & Hoy, 1980; Liu & Var, 1986; Long et al., 1990; Lankford, 1994; Akis et al., 1996; Chen, 2000; Jurowski & Gursoy, 2004; Gutiérrez, 2010; Vargas, et al., 2009 and 2011; Nunkoo & Gursoy, 2012; Martín-Duque, 2017), the initial research model will be the one shown in figure 1. The arrows indicate the relationships between the constructs (variables of perception), that constitute the hypothesis to be tested.

Figure 1. Starting Research Model

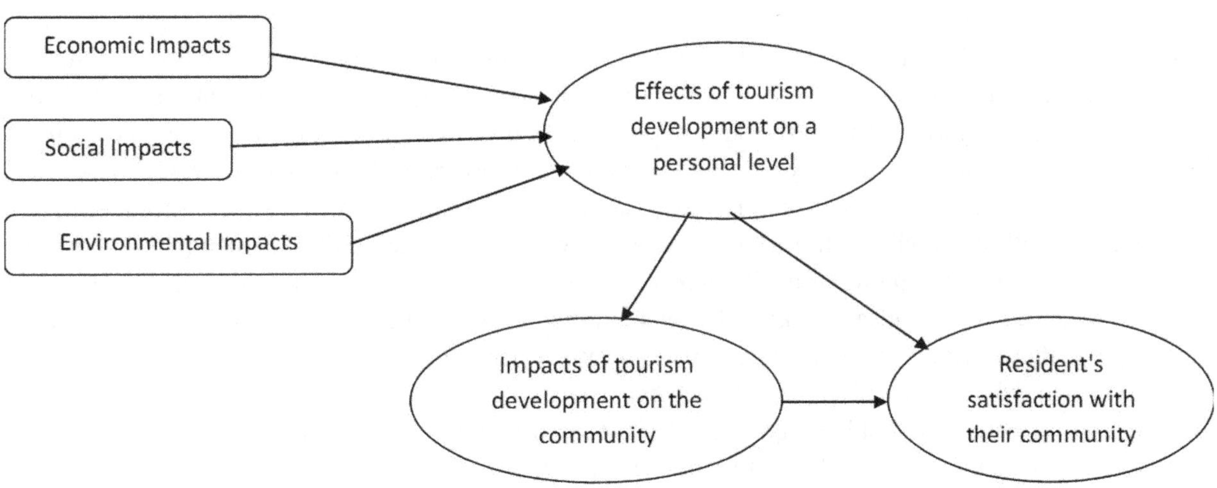

METHODOLOGY

Unit of Analysis: El Rompido

Our unit of analysis is going to focus on the urban core of El Rompido, belonging to the geographical area of the municipality of Cartaya, in the province of Huelva (Spain). Historically, we can find that it began as a settlement undertaking fishing activity on the mouth of the Piedras River.

The tourism development of that enclave has been ongoing since the 1960s of the twentieth century, although the greatest boom began in the 1980s with tourism becoming the main activity in the municipality. This meant that at the end of the 1980s hotel and low-density residential development began. From the accommodation perspective, it is currently characterised mainly by an offering of 4 and 5-star establishments that carry out a strategy specialising in the sun and beach, water sport and golf tourism sectors.

About its offer, the number of hotel beds, according to the Andalusian Tourism Registry, reaches a total of 2,353 (81% of 4-stars category, 16.83% of 5-stars and the remaining 2.17% of 2-stars). Over a total population of 1,779 inhabitants, the number of hotel beds per resident (as an indicator of the territorial pressure of tourism) accounts for 1.32, well above the average of the municipality which this enclave belongs to (0.16) and the whole coastline of the province of Huelva, known as "Costa de la Luz"[1] (0.14). All these figures are referred to the same year that those of the sample detailed below, with no relevant changes since then.

Concerning tourism demand, a high level of seasonality is a relevant feature. Although there are no statistics with the necessary level of disaggregation to present data from El Rompido, the last available figures for the "Costa de la Luz", provided by the National Statistics Institute (INE) and corresponding to the year 2018, show that the seasonal amplitude ratio (that is, the ratio between the number of overnight stays of the peak month and that of the valley month) amounts to 26.86 (specifically, the number of overnight stays in August is almost 27 times those of December). Another relevant indicator is the seasonality rate (the ratio between the sum of overnight stays in the three months of greatest influx and the total overnight stays of the year), whose value was 0.50 (half of the annual overnight stays are concentrated in the three summer months: July, August and September). Nevertheless, because of the existence of a golf resort with 36 holes and the countercyclical character of this segment, it is expected a lower level of seasonality in this case, although still pretty high. The total overnights for that year can be estimated at around 400 thousand, taking the total number of overnights (3,648,944) and the percentage of hotel beds that El Rompido represents in it (11% approximately).

The overlap, when not displacement, which has occurred in this small coastal hub between its traditional economic activities (fishing, agriculture) and tourism, makes it a good testing ground to find out how the local community perceives these types of transformations where they live.

Questionnaire

In the widest research framework, the prepared questionnaire integrates a series of indicators for measuring the constructs that shape the initial research model, expressed in figure 1. Following the approaches referred to in the literature review, table 1 gives details on the items used. The scale used in all cases for their assessment has been between 0 and 10. Because of the process to implement the questionnaire, explained in the next section, this scale was considered more intuitive and, therefore, with the ability to ease the responses.

Table 1. Classification of items in the initial model

Construct (no. of items)	Items	Definition
Economic impacts (8)	P10	The arrival of more tourists.
	P13	Economic activity too limited to summer.
	P25	Increase in tourists during winter.
	P22	Increase in housing costs.
	P24	Loss of land for agriculture.
	P17	Increase in wealth in the population.
	P19	Increase in businesses by people in El Rompido.
	P08	Infrastructure improvement.
Social impacts (7)	P07	Increase in the size of El Rompido.
	P12	More services for tourists.
	P15	Improve how much people value things about El Rompido.
	P16	Increase in the population living in El Rompido.
	P18	Improve the quality of the services for tourists.
	P21	People now cannot go to places that were previously open to the public.
	P23	Loss of cultural identity.
Environmental Impacts (3)	P11	Less water is available for non-tourism uses.
	P14	River and groundwater contamination.
	P20	Damage to the environment.
Effects of tourism development on a personal level (1)	P02B	How much they have benefited (*).
Effects of tourism development on the community (1)	P02	In general, the changes that tourism projects (golf, hotels and developments) in El Rompido have led to are... (from very negative to very positive).
Resident satisfaction with their community (1)	P26	How satisfied are you by living in El Rompido?

(*) For control purposes, "how much has it adversely affected them" was also asked (indicator P02A).
Source: Authors.

Additionally, socio-demographic variables were included such as gender, age, marital status, level of education completed, income level, employment situation, etc.

Sampling

Due to cost and access to the target population, the surveys were done over the phone using a CATI system (*Computer Assisted Telephone Interviewing*), with a structured questionnaire and with the alternative of doing it using CAWI (*Computer Assisted Web Interviewing*) on the website elrompido.org. The characteristics of the sampling process are detailed below:

- Official population data: 1,433 residents in the town of El Rompido (belonging to the municipality of Cartaya) 18 years of age or older.
- Sampling frame: residents in El Rompido with a landline telephone.

- Coverage bias: the sampling frame has covered 86.9%2 of the theoretical framework. It is an acceptable coverage for dealing with a probabilistic sampling, with the target of error being below 7%.
- Weighting: given the reduced number of lines available in the telephone framework (less than a thousand), it is not viable to establish quotas for sex and age when faced with the risk of ending up without a sample. For this reason, the decision is made to not establish quotas a priori and to balance the sample obtained with the target population being studied through the weighting expressed in table 2.

Table 2. Weightings

		Real Population (*)		Sample		Weightings
Man	18 to 29	129	9%	11	5%	1.6741885
Man	30 to 44	262	18%	32	16%	1.1652365
Man	45 to 64	254	18%	32	16%	1.1296568
Man	65 and older	103	7%	18	9%	0.8143807
Woman	18 to 29	100	7%	6	3%	2.3719827
Woman	30 to 44	249	17%	33	16%	1.0738613
Woman	45 to 64	237	17%	55	27%	0.6132653
Woman	65 and older	99	7%	17	8%	0.8287987
	Total	**1,433**		**204**		

(*) Source: National Statistics Institute (INE), Register of Inhabitants 01/01/2014.

- Sampling: for the sampling process an RDD (Random Digit Dialling) technique was used, i.e. randomly dialling all possible telephone ranges, this is equivalent to Simple Random Sampling on the sampling framework.
- Fieldwork: 1,109 different numbers were dialled, with an average of 2.8 redials per telephones not contacted, contacting 336 numbers, in the end, belonging to the study target population, with a participation rate of 60.7% of those contacted in the framework. The protocol established for the start of a call, short and neutral, minimised possible rejection due to a bias of opinion regarding the subject of the study.
- Average interview duration: 10 minutes.
- Date performed: 18/11/2015 to 10/01/2016.
- Sampling error: Considering a simple random sampling, with the estimation of proportions supposed p=q=50%, the maximum error for a confidence level of 95% is 6.36%.

Final Model Delineation: Factor Analysis

Based on the initial model, exploratory factor analysis was performed, that is, a principal components analysis with Varimax rotation for either corroborating or not if the items on the questionnaire are classified in this case in agreement with the impact factors (economic, social and environmental) proposed, i.e. following what is indicated in the theoretical framework.

To do so, starting with the data matrix, the matrix of correlations between the independent variables (or indicators) was obtained, since among the requirements that the matrix must meet, there is the one that the variables have to be highly correlated, and to do so, the determinant of the correlation matrix has to be taken into account (Vargas & Riquel, 2012; Montoya, 2007). In this case, the determinant has a value very close to zero (0.002); likewise, the significance level in Bartlett's sphericity test is 0.000 and the sampling adequacy measurement of Kaiser-Meyer-Olkin is 0.818. All these parameters show the suitability of our data matrix for performing a factory analysis.

In the analysis of the principal components identified, four factors came out with an eigenvalue greater than 1, with a total explained variance of 57.04%. Varimax with Kaiser Normalisation was used as a rotation method to facilitate the interpretation of the factors through the observation of where each of the eighteen considered impacts loads. The results were as follows (table 3).

This exploratory analysis was completed with another of a confirmatory nature performed with Warp-PLS 6.0, proving that, indeed, the items considered load in the same factors (from 1 to 4). The results of the confirmatory factor analysis can be consulted in the appendix.

In sum, we can conclude that the impacts that have an impact on the perception of the local community that is the target of this study are grouped as reflected in table 3. Besides, considering, as was expressed

Table 3. Exploratory factor analysis (results)

Constructs (Cronbach's Alpha)	Indicators	Questionnaire items
F1: Scale factor (0.795)	Increase in the size of El Rompido.	P07
	The arrival of more tourists.	P10
	More services for tourists.	P12
	Economic activity too limited to summer.	P13
	Improve how much people value things about El Rompido.	P15
	Increase in the population living in El Rompido.	P16
	Improve the quality of the services for tourists.	P18
	Increase in tourists during winter.	P25
F2: Negative effects (0.794)	Less water is available for non-tourism uses.	P11
	River and groundwater contamination.	P14
	Damage to the environment.	P20
	People now cannot go to places that were previously open to the public.	P21
	Increase in housing costs.	P22
	Loss of cultural identity.	P23
	Loss of land for agriculture.	P24
F3: Positive effects (0.603)	Increase in wealth in the population.	P17
	Increase in businesses by people in El Rompido.	P19
F4: Infrastructure	Infrastructure improvement.	P08

Source: Authors.

at the beginning, the three constructs corresponding to the effects (impacts of tourism development on a personal level, impacts of tourism development in the community and the satisfaction of residents with their community), the model to be tested (or final research model) is shown graphically in figure 2, which will form the basis for our empirical work.

Therefore, in line with the results of the factor analysis that has been performed and from the theoretical foundation of this study, we can state the following hypothesis, as denoted in figure 2:

Figure 2. Final Research Model

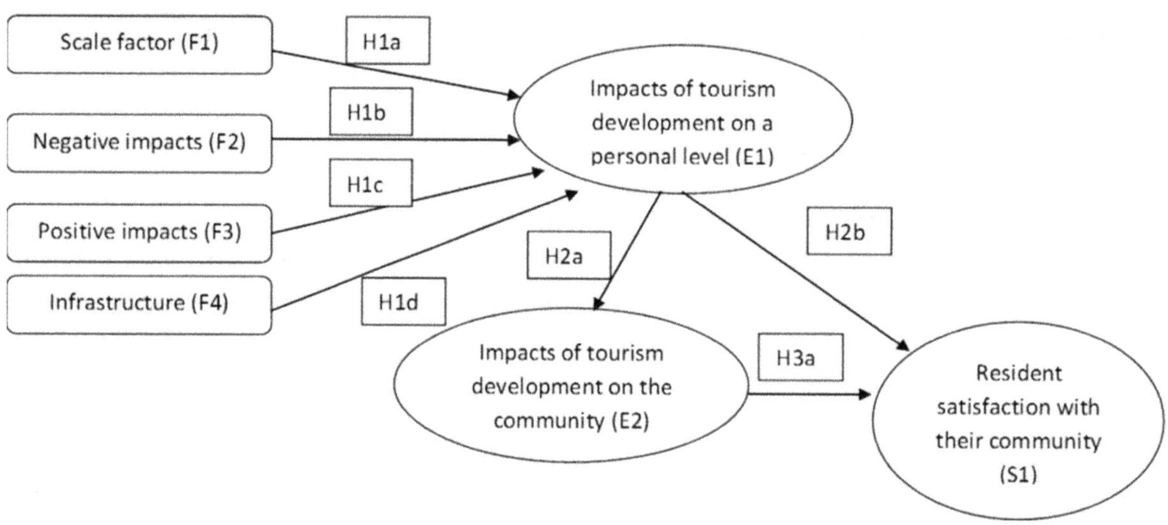

Hypothesis 1a: The perception of the scale of tourism development positively influences perception on the impact of that development on a personal level.

Hypothesis 1b: The perception of the negative effects of tourism development negatively influences the perception of the impact of that development on a personal level.

Hypothesis 1c: The perception of the positive effects of tourism development positively influences the perception of the impact of that development on a personal level.

Hypothesis 1d: The perception of the development of infrastructure linked to tourism positively influences the perception of the impact of that development on a personal level.

Hypothesis 2a: The perception of the impact of tourism development on a personal level positively influences the perception of how it has affected the destination itself.

Hypothesis 2b: The perception of the impact of tourism development on a personal level positively influences the perception of resident satisfaction with their community.

Hypothesis 3a: The perception of the impact of tourism development on the destination itself positively influences the perception of the satisfaction of the resident with their community.

Hypotheses Testing

The data processing for testing the hypotheses displayed in figure 2 (final model) has been performed using structural equations models based on the *Partial Lease Square* (PLS) technique. This is a widely accepted method, particularly in the last decade in the field of social sciences, since it provides a more flexible approach than the traditional one (Martínez & Fierro, 2018). It is a method which is especially recommended for increasing the explanatory power of the empirical analysis of a theory (Roldán & Sánchez, 2012), as is the case we are concerned with, although it can also be used to increase theoretical knowledge in situations where this is limited.

This technique is comprised of the analysis of the structural model and the analysis of the measurement model. In the first of them, the dependency relationships are shown between independent and dependent variables. In the measurement model, the relationships between the latent variables (constructs) and the indicators (observable variables) are analysed, being able to determine the contribution of each item to the measurement scale reflecting which indicators define each construct, as well as evaluating their reliability and validity.

As a methodological approach in the application of the PLS technique, we follow the systematic process proposed by authors such as Martinez and Fierro (2018), Hair, Hult, Ringle, and Sarstedt (2017) and Ramírez, Mariano, A., and Salazar (2014). Therefore, the phases implemented in this research were as follows: (1) Design and description of the research model. (2) Validity of the measurement model. (3) Reliability of the measurement model. (4) Assessment of the structural model. (5) Bootstrapping. (6) Evaluation of the structural model. (7) Interpretation of the results.

RESULTS

Sociodemographic Profile

The profile of the sample can be outlined as follows: the majority are men (52.2%), married (57.3%), with a university education (50.7%), born in the province of Huelva (64.5%) but not in the target town (only 7.4%), owning homes (85.5%) and employed workers (40.0%).

Analysis of the Measurement Model

We start this phase analysing the individual reliability of each item according to the criteria established by Falk and Miller (1992), who postulate that the loadings for the items must reach a value of at least 0.55, as indicated by Chin (1998) and Ramírez et al. (2014) for initial stages of development of research models, thus relaxing the criteria proposed by Carmines and Zeller (1979), which requires values above 0.707. In the appendix (see "Individual reliability of each indicator"), the loadings for each item are shown after the respecification of the proposed model, which requires the elimination of the indicators P13, P15, P18 and P11 due to insufficient factor loadings.

Table 4 measures the internal consistency of the latent variables through the composite reliability (pc) of the respective constructs, as well as their convergent validity, through the Average Variance Extracted (AVE). To do so we have taken the criteria defined by Numally (1978) and Fornell and Larcker (1981) into consideration.

Table 4. Internal Consistency and Convergent Validity

Constructs	Composite Reliability (pc) ≥ 0.7 (Numally, 1978)	AVE ≥ 0.5 (Fornell & Larcker, 1981)
F1: Scale factor	0.843(*)	0.517
F2: Negative impacts	0.859 (*)	0.506
F3: Positive impacts	0.834 (*)	0.715
F4: Infrastructure	1.000	1.000
E1: Impacts on the personal level	1.000	1.000
E2: Impacts on the community level	1.000	1.000
S1: Community satisfaction	1.000	1.000

(*) The corresponding values of Cronbach's Alpha are: F1 = 0.766; F2 = 0.803; F3 = 0.602.
Source: Authors.

Finally, we will analyse the discriminant validity of our model's constructs. The table that, in this regard, can be checked in the appendix shows this property's compliance, by all of the square roots of the AVES (in the diagonal) being above the correlations with the rest of the constructs in the corresponding column. In addition, in order to establish that validity, the heterotrait-monotrait (HTMT) ratio has been calculated, with the following values, all of them below the thresholds of 0.90 and 0.85 recommended by Henseler, Ringle, and Sarstedt (2015) and Kline (2011), respectively: HTMT F1-F2 = 0.260; HTMT F1-F3 = 0.775; HTMT F2-F3 = -0.095.

Figure 3. Analysis of structural model

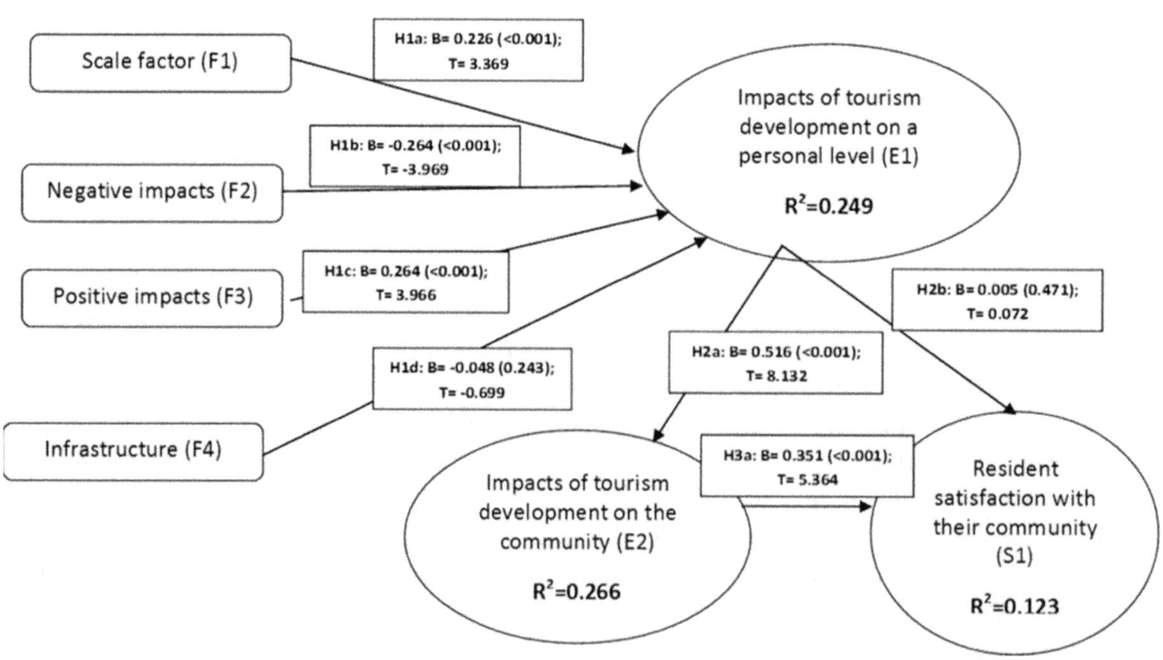

Analysis of the Structural Model

Following the indicated method, we are going to focus both on the *path or* β coefficients (their respective *p-values* in parentheses) and the value of the explained variance (R^2) for the latent dependent variables (Chin, 1998; Falk & Miller, 1992; Leal & Roldán, 2001). Those indicators are reflected in figure 3, with the arrows indicating the causal order of the relationships between the proposed model's latent variables. The Bootstrap resampling technique has been used to calculate the values of the T-Students, that help us undertake the testing of the proposed hypotheses.

As can be observed in figure 3, the variances explained for the endogenous or dependent variables (R^2) are all situated above 0.1 (Falk & Miller, 1992).

Accordingly, it can be affirmed that, except for the hypothesis H1d and H2b (that are rejected), all of the rest are accepted for significance levels $P<0.001$-based on two-tailed t(499). In these (the accepted ones), the beta coefficients in all cases exceed the threshold of 0.2 established by Chin (1998). The model fit and quality indexes offered by WarpPLS 6.0 are provided in the appendix, verifying their suitability.

DISCUSSION

The results obtained by testing the proposed research model, when confronting them with the literature in this study area, allow us to make some observations.

Firstly, it is worth drawing attention to the fact that, in the population considered for our analysis (El Rompido), the traditional classification, defended by authors such as Balbuena (2014) for the factors that influence the perception of the local community on tourism development, is not confirmed, which means evidence of its contingent character. Therefore, there is a new classification for them when it comes to understanding the perception that, in this case, the resident has of tourism development in their town, confirmed by the factor analysis performed. In this particular combination, factors related to the increase in the volume of visitors appear (scale factor) and to the destination's infrastructure. However, as López and Palomino (2008) defend, factors have been found that include positive and negative impacts of tourism. In general, their positive or negative influence is what seems people care about, regardless of their economic, social or environmental character.

Secondly, our model shows significant cause-effect relationships between the scale factors, negative impacts and positive impacts of tourism on the perception the resident has of their effects on a personal level (acceptance of the H1a, H1b and H1c hypotheses), the three with a similar impact (size of effect). Likewise, we can conclude that the construct relating to the infrastructure and its improvement does not influence the perception of the impact of tourism development on a personal level (hypothesis H1d is rejected), a finding that would merit its study to further look into the reasons for this.

Thirdly, the construct that includes the perception of the impacts of tourism development on a personal level shows having a decisive influence (the greatest size of the ones that the model shows), on the one corresponding to the perception of the impacts that this development has on the town they live in, with there being a highly significant and positive relationship between them (acceptance of the H2a hypothesis). This finding seems in line with the approaches of Sirakaya et al. (2002) and Besculides et al. (2002), that establish that a destination's tourism development has a direct impact on certain socio-cultural variables of the resident population that lead to the impacts of that development on the whole local community becoming palpable.

The hypothesis on the (positive) influence of the perception of the impacts on a personal level on resident satisfaction with their local community (H2b) is not accepted, with the conclusion that there is no direct cause-effect relationship between these two constructs, even though there is an indirect one via the perception of the impacts on the community as a whole (H2a) and the influence of this variable on resident satisfaction (H3a). This double finding (i.e. the fact that the perception of the effects of tourism development on an individual level is not reflected in the satisfaction with the community, although nuanced by the influence that the perception of the impacts of that development on the destination as a whole exerts on that level of satisfaction) can entail the feeding back of negative impacts in the terms that have been proposed by Royo and Ruiz (2009) when affirming that a local community which is apathetic about, displeased with, or distrustful of tourism activity can lead to tourists having a negative perception and, therefore, it is a competitive disadvantage for the destination. It is a risk that tourism destination managers must assess, taking the level of tourism development into account as a determinant in residents' perceptions and satisfaction (Vargas et al., 2015).

Therefore, we consider that the results of this research have affirmed the approaches of Díaz and Gutiérrez (2010), as well as the ones by Vargas et al. (2007) and Mendoza and Motenrrubio (2012), regarding the need to undertake, as a next step, new works to find out the attitude of the local community toward tourism development as a way of generating potential competitive advantages for the destination, based on the perceptions analysed here.

CONCLUSION

The impacts of tourism on the local population form a relevant research subject that has been a focus of attention for the scientific community. Although there seems to be a broad consensus on classifying them following their economic, social and environmental nature, when evaluating the perception of the local community there are substantial differences between the leading authors on the subject. The subjective aspect which is implicit in all valuations of this kind means that the traditionally accepted classifications must be reconsidered from a methodological perspective.

In this regard a need arises, which this paper deals with, to design an explanatory model which considers the subjective valuations of the resident population (impacts of tourism, effects on a personal and collective level, satisfaction with the community) and their interrelationships, making us reconsider the typical approach to the local community's perceptions in three directions. Firstly, verifying and analysing the contingent variability existing in the traditional acceptance of the triple impact (economic, social and environmental) in the local community, concluding that the scale factor, understood as the increase in economic activity coming from tourism and its consequences, has enough weight to be treated in a specific way, given its influence on the perception of the impact of tourism development on an individual or personal level. In addition to this factor, there are the traditional positive and negative impacts (although without looking at their nature), both able to influence residents' perception, as well as a new factor related to infrastructure improvement that, despite its potential explanatory power, the hypothesis testing carried out does not confirm in the case studied, perhaps because that improvement has not been noticeable for the informants or even due to dealing with elements linked to public investment that can cause controversy among the population, by contributing to reinforcing the negative aspects of tourism at the destination, such as overcrowding there. In this regard, it would be a good idea to look further into this peculiarity to advance the understanding of the reasons for the lack of consistency in this last relationship. Part of this peculiarity lies in the tourism development model being applied and the phase of the life cycle that the destination is in (Vargas et al., 2015).

Nonetheless, it is noteworthy that resident satisfaction with their community is not directly influenced by the valuation made on how tourism affects them on a personal level, but rather indirectly through the assessment of the consequences on the level of the community as a whole, giving prevalence to the collective dimension of the impacts of tourism development compared to the individual dimension, which, according to Porras, Plaza, and Vargas (2019), influences the interpretation that one can have about their sense of belonging and perception of the quality of life. Although this latter consideration hasn't been explicitly part of the focus of this chapter, it is for this reason that we consider the design of citizen engagement instruments in the destination's tourism policies can contribute to improving the residents' perception of it, by internalising the impacts of tourism development on a personal level.

The findings gathered lead us to underscore the practical implications of the research since the obtained results support the idea of the need to study the factors which have the most influence on the resident populations' perception in each case. Not only due to the differential characteristics that local communities can have but because those factors will depend on the model and intensity of tourism development (growth) that is being applied at the analysed destination, due to which is why its approach, out of necessity, has to be contingent, i.e. beyond just emphasising the positive impacts and trying to mitigate the negative ones. This is especially relevant from a social sustainability perspective through the support of the local community receiving flows of tourists, that can only be won over through their satisfaction and knowledge of the impacts of tourism that influence the perception of those who live there most (permanently), both on an individual (personal) level as well as on a collective one (about the community they are a part of). Establishing mechanisms that allow policymakers to monitor these perceptions over time becomes necessary for improving, via the local community's satisfaction, the competitiveness of the tourism destination.

Finally, as far as its limitations are concerned, we must draw attention to the geographical limits of the sample, as well as the possibility that the interviewees may have been able to introduce bias into their answers due to their relationship (at work level with the tourism industry, for example) with the target territory being studied, all of which constrains the possible generalisations that could be drawn from these conclusions. Besides, the variance extracted in the factor analysis is moderate; and the constructs measured with one single item, related to the perception of impacts (on personal -E1- and community levels -E2-) and community satisfaction (S1), could be addressed multidimensionally to unravel all their complexities.

Consistently, in addition to overcoming these limitations, the application of other techniques of a confirmatory nature, with the capacity to add new pieces of evidence, would open the door to future lines of enquiry. In the same way, future research avenues could consider the mediating role of intrinsic factors of the community that can produce variations in residents' perception, such as age, place of birth, time of residence in the unit of analysis, etc.

ACKNOWLEDGMENT

This research was supported by the Spanish Minister of Economy and Competitiveness (grant number CSO2012-32493).

REFERENCES

Akis, S., Peristianis, N., & Warner, J. (1996). Residents attitudes to tourism development: The case of Cyprus. *Tourism Management, 17*(7), 481–404. doi:10.1016/S0261-5177(96)00066-0

Andereck, K., & Nyaupane, G. (2011). Exploring the Nature of Tourism and Quality of Life Perceptions among Residents. *Journal of Travel Research, 50*(3), 248–260. doi:10.1177/0047287510362918

Andereck, K., Valentine, K., Knopff, R., & Vogot, C. (2005). Resident's perceptions of community tourism impacts. *Annals of Tourism Research, 32*(4), 1056–1076. doi:10.1016/j.annals.2005.03.001

Andriotis, K. (2005). Community Groups' Perceptions of and Preferences for Tourism Development: Evidence from Crete. *Journal of Hospitality & Tourism Research (Washington, D.C.), 29*(1), 67–90. doi:10.1177/1096348004268196

Andriotis, K., & Vaughan, D. (2003). Urban residents' attitudes toward tourism development: The case of Crete. *Journal of Travel Research, 42*(2), 172–185. doi:10.1177/0047287503257488

Ap, J. (1992). Residents' Perceptions on Tourism Impacts. *Annals of Travels Research, 19*(4), 665–690. doi:10.1016/0160-7383(92)90060-3

Balbuena, A. (2014). *La actitud del residente ante los impactos del turismo: El caso de Benalmádena* (Doctoral dissertation). Retrieved from RIUMA, Repositorio Institucional de la Universidad de Málaga: (https://riuma.uma.es/xmlui/handle/10630/7492)

Balbuena, A., Barrera, D., & Hernández, M. (2017). El impacto socio-cultural del turismo en Benalmádena. *International Journal of Scientific Management and Tourism, 3*(1), 31–47.

Belisle, F., & Hoy, D. (1980). The Perceived Impact of Tourism by Residents: A Case Study in Santa Marta, Colombia. *Annals of Tourism Research, 7*(1), 83–101. doi:10.1016/S0160-7383(80)80008-9

Besculides, A., Lee, M., & McCormick, P. (2002). Residents' perceptions of the cultural benefits of tourism. *Annals of Tourism Research, 29*(2), 303–319. doi:10.1016/S0160-7383(01)00066-4

Brida, J., Riaño, E., & Zapata, S. (2012). Percepciones de los residentes acerca de los impactos del turismo de crucero en la comunidad: Un análisis factorial y de clúster. *Cuadernos de Turismo, 29*, 79–107.

Carmines, E., & Zeller, R. (1979). *Reliability and Validity Assessment*. Sage Publications. doi:10.4135/9781412985642

Chen, J. S. (2000). Cross-cultural differences in travel information acquisition among tourist from three Pacific-rim countries. *Journal of Hospitality & Tourism Research (Washington, D.C.), 24*(2), 239–251. doi:10.1177/109634800002400207

Chin, W. (1998). The Partial Least Square Approach to Structural Equation Modeling. In G. A. Marcoulides (Ed.), *Modern Methods for Business Research* (pp. 295–336). Erlbaum Associates.

Covarrubias, R., Vargas, A., & Rodríguez, M. (2010). Satisfacción de residentes con el desarrollo de la actividad turística en los pueblos mágicos de México: Un indicador de competitividad. *Gestión Turística, 14*, 33–54.

Díaz Armas, R., & Gutiérrez Taño, D. (2010). La actitud del residente en el destino turístico de Tenerife: Evaluación y tendencia. *Pasos. Revista de Turismo y Patrimonio Cultural*, *8*(4), 431–444. doi:10.25145/j.pasos.2010.08.039

Dogan, H. Z. (1989). Forms of adjustment: Sociocultural impacts of tourism. *Annals of Tourism Research*, *16*(2), 216–236. doi:10.1016/0160-7383(89)90069-8

Eagly, A. H., & Chaiken, S. (1993). *The Psychology of Attitudes*. Harcourt, Brace, & Jovanovich.

Falk, R., & Miller, N. (1992). *A primer soft modeling*. The University of Akron Press.

Figuerola, M., Fernández, J., Marzal, F., Gómez, D., & Martín, C. (2015). *Medición del impacto del turismo en la Economía, Empleo y el Medio Ambiente en el ámbito local*. Fundación Escuela Organización Industrial.

Fornell, C., & Larcker, D. (1981). Evaluating structural equation models with unobservable variables and measurement error. *JMR, Journal of Marketing Research*, *18*(1), 39–50. doi:10.1177/002224378101800104

García, C. (1999). *Las Peregrinaciones Jacobeas*. San Pablo.

García, F. A., Vázquez, A. B., & Macías, R. C. (2015). Resident's attitudes toward the impacts of tourism. *Tourism Management Perspectives*, *13*(January), 33–40. doi:10.1016/j.tmp.2014.11.002

González-García, R. J., Parra, D., Añó, V., & Calabuig, F. (2016). Percepción de los residentes sobre el impacto del Mundobasket 2014 en Gran Canaria y apoyo a la celebración de eventos deportivos. *Revista Iberoamericana de Psicología del Ejercicio y el Deporte*, *11*(2), 279–288.

Gursoy, D., Jurowski, C., & Uysal, M. (2002). Resident attitudes. A structural modeling approach. *Annals of Tourism Research*, *29*(1), 79–105. doi:10.1016/S0160-7383(01)00028-7

Gutiérrez, D. (2010). *Las actitudes de los residentes ante el turismo* (Doctoral dissertation). Retrieved from RIULL, Repositorio Institucional Universidad de La Laguna: (https://riull.ull.es/xmlui/handle/915/2117)

Hair, J., Hult, G., Ringle, C., & Sarstedt, M. (2017). *A Primer on Partial Least Square Structural Equation Modeling (PLS-SEM)*. Sage.

Henseler, J., Ringle, C. M., & Sarstedt, M. (2015). A new criterion for assessing discriminant validity in variance-based structural equation modeling. *Journal of the Academy of Marketing Science*, *43*(1), 115–135. doi:10.100711747-014-0403-8

Hoijer, B. (2011). Social Representation Theory: A New Theory for Media Research. *Nordicom Review*, *32*(2), 3–16. doi:10.1515/nor-2017-0109

Jurosky, C., & Gursoy, D. (2004). Distance effects on resident's attitudes toward tourism. *Annals of Tourism Research*, *31*(2), 296–304. doi:10.1016/j.annals.2003.12.005

Kline, R. B. (2011). *Principles and practice of structural equation modeling*. Guilford Press.

Ko, D., & Stewart, W. (2002). A structural equation model of resident attitudes for tourism development. *Tourism Management*, *23*(5), 521–530. doi:10.1016/S0261-5177(02)00006-7

Lankford, S., & Howard, D. (1994). Developing a tourism impact attitudes scales. *Annals of Tourism Research, 21*(1), 121–139. doi:10.1016/0160-7383(94)90008-6

Leal, A., & Roldán, J. (2001). Validación de un modelo de implantación de gestión de calidad total y su efecto en los resultados empresariales: un estudio causal predictivo mediante análisis PLS (Partial Least Square). In *Proceedings of the X International Conference of the European Association of Management and Business Economics* (pp. 541-555). Reggio Calabria: AEDEM.

Lindsay, P., & Norman, D. (1977). *Human information processing: an introduction to Psychology.* Academic Press.

Liu, J. C., & Var, T. (1986). Resident attitudes toward tourism impacts in Hawai. *Annals of Tourism Research, 13*(2), 193–214. doi:10.1016/0160-7383(86)90037-X

Long, P. T., Perdue, R. R., & Allen, L. (1990). Rural resident tourism perceptions and attitudes by community level of tourism. *Journal of Travel Research, 28*(3), 3–9. doi:10.1177/004728759002800301

López, G., & Palomino, B. (2008). El turismo como actividad emergente para las comunidades y pueblos indígenas. In A. Castellanos-Guerrero & J. A. Machuca (Eds.), *Turismo, identidades y exclusión* (pp. 31–50). Universidad Autónoma Metropolitana y Casa Juan Pablos.

Madrigal, R. (1995). Residents' perceptions and the role of government. *Annals of Tourism Research, 22*(1), 86–102. doi:10.1016/0160-7383(94)00070-9

Martín de la Rosa, B. (2009). *Turismo en ecosistemas insulares. Antropología en el paraíso. Colección PASOS Edita, 3.* Asociación Canaria de Antropología.

Martín-Duque, C. (2017). Los impactos del turismo en el camino de Santiago Francés: Una aproximación cualitativa. *Methaodos. Revista de Ciencias Sociales, 5*(1), 62–73.

Martínez, M., & Fierro, E. (2018). Aplicación de la técnica PLS-SEM en la gestión del conocimiento: Un enfoque práctico. *Revista Iberoamericana para la Investigación y Desarrollo Educativo, 8*(16), 1–35.

Masa'deh, R. E., Nasseef, M. A., Alshayeb, H., Ojilat, J., & Alshafiee, M. (2017). The Effect of Sport Tourism Management on Support for Tourism Development. *Journal of Management and Strategy, 8*(3), 20–34. doi:10.5430/jms.v8n3p20

Mason, P. (2003). *Tourism Impacts, Planning and Management.* Butterworth-Heinemann.

Mendoza, M., & Motenrrubio, J. (2012). Actitud de la comunidad residente en Acapulco hacia los Spring Breakers y su comportamiento. *Revista Análisis Turístico, 12*(1), 27–38.

Monterrubio, C., Coredero, J., & García, M. (2011). Percepciones de la comunidad local infantil en Huatulco México. Un acercamiento al estudio del cambio sociocultural asociado al turismo. *El Periplo Sustentable, 20*, 149–185.

Monterrubio, J. (2008). Comunidades receptoras y percepciones: Un estudio sobre turismo y sexualidad. *Teoría y Praxis, 5*(5), 145–160. doi:10.22403/UQROOMX/TYP05/11

Monterrubio, J. (2018). *Impactos socioculturales del turismo. Dimensiones culturales, teóricas y empíricas.* Triullas.

Monterrubio, J. C., Mendoza, M. M., Fernández, M. J., & Gullete, G. S. (2011). Turismo y cambios sociales. Estudio cualitativo sobre percepciones comunitarias en Bahías de Huatulco, México. *Cuadernos de Turismo*, *28*, 171–189.

Montoya, O. (2007). Aplicación del análisis factorial a la investigación de mercados: Caso de estudio. *Sciences et Techniques (Paris)*, *XII*(35), 281–286.

Numally, J. (1978). *Psychometric Theory*. MacGraw-Hill.

Nunkoo, R., & Gursoy, D. (2012). Resident's Support for Tourism an Identity Perspective. *Annals of Tourism Research*, *39*(1), 243–268. doi:10.1016/j.annals.2011.05.006

Nunkoo, R., & So, K. K. F. (2016). Residents' support for tourism: Testing alternative structural models. *Journal of Travel Research*, *55*(7), 847–861. doi:10.1177/0047287515592972

Page, S. J., & Hall, C. M. (2014). *The Geography of Tourism and Recreation: Environment, Place and Space*. Routledge.

Ponce, M. D. (2004). Percepción del modelo turístico de sol y playa: El caso del mar menor. *Papeles de Geografía*, *39*, 173–186.

Porras, N., Plaza, M. A., & Vargas, A. (2019). Quality of Life and Perception of the Effects of Tourism: A Contingent Approach. In A. M. Campón, J. M. Hernández, & J. A. Folgado (Eds.), *Best Practices in Hospitality and Tourism Marketing and Management* (pp. 109–132). Springer. doi:10.1007/978-3-319-91692-7_6

Presenza, A., Del Chiappa, G., & Sheehan, L. (2013). Residents' engagement and local tourism governance in maturing beach destinations. Evidence from an Italian case study. *Journal of Destination Marketing & Management*, *2*(1), 22–30. doi:10.1016/j.jdmm.2013.01.001

Ramírez, P., Mariano, A., & Salazar, E. (2014). Propuesta metodológica para aplicar modelo de ecuaciones estructurales con PLS: El caso del uso de bases de datos científicas en estudiantes universitarios. *Revista ADMpg Gestão Estratégica*, *7*(2), 133–139.

Ramón-Cardona, J. (2014). Tipos de oferta turística y actitudes de los residentes: El caso de Ibiza. *Revista Brasileira de Pesquisa em Turismo*, *8*(1), 3–22. doi:10.7784/rbtur.v8i1.704

Ramón-Cardona, J. (2015). Comparativa entre islas de la valoración de distintos tipos de oferta turística por parte de los residentes: El caso de las Islas Baleares (España). *Turismo e Sociedade*, *8*(2), 255–275.

Rasoolimanesh, S. M., Ringle, C. M., Jaafar, M., & Ramayah, T. (2017). Urban vs. rural destinations: Residents' perceptions, community participation and support for tourism development. *Tourism Management*, *60*(June), 147–158. doi:10.1016/j.tourman.2016.11.019

Roldán, J., & Sánchez, M. (2012). Variance-Based Structural Equation Modeling: Guidelines for Using Partial Least Square in Information System Research. In M. Mora, O. Gelman, A. L. Steenkamp, & M. Raisinghani (Eds.), *Research Methodologies, Innovations and Philosophies in Software Systems Engineering and Information Systems* (pp. 193–221). IGI Global. doi:10.4018/978-1-4666-0179-6.ch010

Royo, M., & Ruiz, M. (2009). Actitud del residente hacia el turismo y el visitante: Factores determinantes en el turismo y excursionismo rural-cultural. *Cuadernos de Turismo*, *23*, 214–236.

Santana, A. (1997). *Antropología y turismo. ¿Nuevas hordas, viejas culturas?* Ariel.

Santana, R. (2011). *Estudio sociodemográfico de los impactos del turismo para el desarrollo local del municipio Morón* (Unpublished dissertation). University of Ciego de Ávila, Ciego de Ávila, Cuba.

Sharpley, R. (2014). Host perceptions of tourism: A review of the research. *Tourism Management*, *42*(June), 37–49. doi:10.1016/j.tourman.2013.10.007

Sinclair-Maragh, G. (2017). Demographic analysis of residents' support for tourism development in Jamaica. *Journal of Destination Marketing & Management*, *6*(1), 5–12. doi:10.1016/j.jdmm.2016.03.005

Sirakaya, E., Teye, V., & Sönmez, S. (2002). Understanding residents' support for tourism development in the Central Region of Ghana. *Journal of Travel Research*, *41*(8), 57–67. doi:10.1177/004728750204100109

Suess, C. (2014). *Factors Influencing the Perceived Impacts of Medical Tourism Development on Quality of Life* (Doctoral dissertation). Retrieve from UNLV Theses, Dissertations, Professional Papers, and Capstones: https://digitalscholarship.unlv.edu/thesesdissertations/2302/

Teye, V., Sirakaya, E., & Sönmez, S. (2002). Resident's attitudes toward tourism development. *Annals of Tourism Research*, *29*(3), 668–688. doi:10.1016/S0160-7383(01)00074-3

Vargas, A., Plaza, M. A., & Porras, N. (2007). Desarrollo del turismo y percepción de la comunidad local: Factores determinantes de su actitud hacia un mayor desarrollo turístico. In *Proceedings of the XXI National Conference of AEDEM* (vol. 1, pp. 24-34). Madrid: AEDEM.

Vargas, A., Plaza, M. A., & Porras, N. (2009). Understanding Residents' Attitudes toward the Development of Industrial Tourism in a Former Mining Community. *Journal of Travel Research*, *47*(3), 373–387. doi:10.1177/0047287508322783

Vargas, A., Plaza, M. A., & Porras, N. (2011). Explaining resident's attitudes to tourism. Is it a universal model possible? *Annals of Tourism Research*, *38*(2), 460–480. doi:10.1016/j.annals.2010.10.004

Vargas, A., Porras, N., Plaza, M. A., & Riquel, F. (2008). Tourism and wine-marking industry. Perspective and proposals for development in the Spanish province of Huelva. In *Proceedings of the lASK International Conference, Advance in Tourism Research* (pp. 33-43). Aveiro: IASK.

Vargas, A., & Riquel, F. (2012). Influence of the institutional context on the performance of golf courses, considering the natural environment. *Environmental Engineering and Management Journal*, *11*(11), 2001–2012. doi:10.30638/eemj.2012.249

Vargas, A., Valle, P., Mendes, J. C., & Silva, J. A. (2015). Residents' attitude and level of destination development: An international comparison. *Tourism Management*, *48*(June), 199–210. doi:10.1016/j.tourman.2014.11.005

Vela, M., & Molina, M. (2009). Actitud del residente hacia el turismo y el visitante: Factores determinantes en el turismo y excursionismo Rural-Cultural. *Cuadernos de Turismo*, *23*, 217–236.

Yürük, P., Akyol, A., & Şimşek, G. G. (2017). Analyzing the effects of social impacts of events on satisfaction and loyalty. *Tourism Management, 60*(June), 367–378. doi:10.1016/j.tourman.2016.12.016

Zhang, J., Inbakaran, R., & Jackson, M. S. (2006). Understanding community attitudes towards tourism and host-guest interaction in the urban-rural border region. *Tourism Geographies, 8*(2), 182–204. doi:10.1080/14616680600585455

ADDITIONAL READING

Ap, J., & Crompton, J. (1993). Residents' Strategies for Responding to Tourism Impacts. *Journal of Travel Research, 32*(1), 47–50. doi:10.1177/004728759303200108

Brida, J. G., Disegna, M., & Osti, L. (2014). Residents' perceptions of tourism impacts and attitudes towards tourism policies. *Tourismos, 9*(1), 37–71.

Choi, H.-S. C., & Sirakaya, E. (2005). Measuring Residents' Attitude toward Sustainable Tourism: Development of Sustainable Tourism Attitude Scale. *Journal of Travel Research, 43*(4), 380–394. doi:10.1177/0047287505274651

Jaafar, M., Rasoolimanesh, S. M., & Ismail, S. (2017). Perceived sociocultural impacts of tourism and community participation: A case study of Langkawi Island. *Tourism and Hospitality Research, 17*(2), 123–134. doi:10.1177/1467358415610373

Lee, T. H., & Jan, F.-H. (2019). Can community-based tourism contribute to sustainable development? Evidence from residents' perceptions of the sustainability. *Tourism Management, 70*(February), 368–380. doi:10.1016/j.tourman.2018.09.003

Nunkoo, R., Smith, S., & Ramkissoon, M. (2013). Resident attitudes to tourism: A longitudinal study of 140 articles from 1984 to 2010. *Journal of Sustainable Tourism, 21*(1), 5–25. doi:10.1080/09669582.2012.673621

Özel, Ç. H., & Kozak, N. (2017). An exploratory study of resident perceptions toward the tourism industry in Cappadocia: A social exchange theory approach. *Asia Pacific Journal of Tourism Research, 22*(3), 284–300. doi:10.1080/10941665.2016.1236826

Rasoolimanesh, S. M., Jaafar, M., Kock, N., & Ramayah, T. (2015). A revised framework of social exchange theory to investigate the factors influencing residents' perceptions. *Tourism Management Perspectives, 16*(October), 335–345. doi:10.1016/j.tmp.2015.10.001

Woo, E., Kim, H., & Uysal, M. (2015). Life satisfaction and support for tourism development. *Annals of Tourism Research, 50*(January), 84–97. doi:10.1016/j.annals.2014.11.001

Woo, E., Uysal, M., & Sirgy, M. (2018). Tourism impact and stakeholders' quality of life. *Journal of Hospitality & Tourism Research (Washington, D.C.), 42*(2), 260–286. doi:10.1177/1096348016654971

KEY TERMS AND DEFINITIONS

Environmental Perception: It refers to the process by which humans organize and interpret elements of their environment (for instance, tourism development and its impacts). This perception is affected by the individual's scheme of values, past experiences, expectations, motivations and needs. The understanding of the reality perceived by people (subjective, therefore) is paramount in explaining their behaviour.

Impacts of Tourism: Positive or negative changes in destinations which occur as a consequence of the industry development. They are usually divided into three categories economic, social and environmental effects.

Partial Least Square Method: This is a multivariate statistical regression technique that allows estimating complex cause-effect relationship models with latent variables. It is one of several covariance-based statistical methods which are often referred to as structural equation modelling or SEM.

Resident: A person who lives or has their home in a place.

Residents' Attitudes: The set of residents' beliefs, predispositions, and behaviours (or intended behaviours) toward specific aspects of tourism.

Residents' Satisfaction: The positive feeling about living conditions in the community where they live. These living conditions are affected by tourism development. A variety of dimensions could be considered to assess the overall level of satisfaction, such as satisfaction with public services, with the health system, with the environment, with entertainment opportunities, with the local economy, etc.

Satisfaction: The positive feeling experienced when expectations are met or surpassed.

Social Exchange Theory: In general, it proposes that social behaviour is the result of an exchange process guided to maximize benefits and minimize costs. According to this theory, developed by sociologist George Homans, people weigh the potential benefits and risks of social relationships. In the tourism field, this is the theory most frequently applied by researchers to understand residents' attitudes towards tourism development. Several authors state that the perception of impacts (positive and negative ones) derived from tourism is the principal variable for explaining residents' attitude towards tourism.

ENDNOTES

1 The so-called "Costa de la Luz is formed by the following municipalities: Ayamonte, Isla Cristina, Lepe, Punta Umbría, Palos de la Frontera, Moguer, Almonte and Cartaya.

2 Starting with the active lines in homes (no companies) detected in the CATI, which are estimated to be 647 lines, from the 2011 Census, which estimates in 745 homes in El Rompido.

APPENDIX

Table 5. Confirmatory Factor Analysis (results)

Item	F1	F2	F3	F4
P07	**0,917**	-0,011	-0,343	-0,087
P10	**0,854**	-0,035	0,075	-0,206
P12	**0,864**	-0,006	-0,246	0,174
P13	**0,790**	0,180	-0,110	-0,559
P15	**0,707**	-0,399	-0.227	0,411
P16	**0,834**	0,086	0,217	-0,157
P18	**0,794**	0,110	0,298	0,216
P25	**0,767**	0,182	0,563	0,208
P11	0,432	**0,896**	-0,601	-0,051
P14	-0,075	**0,892**	-0,043	-0,230
P20	-0,003	**0,927**	-0,212	0,101
P21	-0,037	**0,938**	-0,090	0,056
P22	0,012	**0,939**	0,327	0,151
P23	-0,232	**0,976**	0,162	0,167
P24	0,053	**0,924**	0,409	-0,338
P17	0,133	0,105	**0,825**	0,004
P19	-0,119	-0,094	**0,871**	-0,004
P08	-0,000	0,000	-0,000	**1,000**

Note: normalized loadings.
Source: Authors.

Table 6. Individual reliability of each indicator

Constructs	Items	Loadings (λ)
F1: Scale factor	P07	0,710
	P10	0,770
	P12	0,699
	P16	0,703
	P25	0,712
F2: Negative impacts	P14	0,666
	P20	0,805
	P21	0,739
	P22	0,652
	P23	0,739
	P24	0,653
F3: Positive impacts	P17	0,846
	P19	0,846

Source: Authors.

Table 7. Discriminant validity of model constructs

Constructs	F1	F2	F3	F4	E1	E2	S1
F1	**0,719**	0,201	0,526	0,252	0,304	0,104	0,057
F2	0,201	**0,711**	-0,053	-0,137	-0,204	-0,456	-0,276
F3	0,526	-0,053	**0,846**	0,267	0,379	0,309	0,166
F4	0,252	-0,137	0,267	**1,000**	0,138	0,375	0,260
E1	0,304	-0,204	0,379	0,138	**1,000**	0,494	0,127
E2	0,104	-0,456	0,309	0,375	0,494	**1,000**	0,312
S1	0,057	-0,276	0,166	0,260	0,127	0,312	**1,000**

Source: Authors.

Table 8. Model fit and quality indexes

Indexes	Value	Note
Average Path Coefficient (APC)	0,239; P<0,001	Significant
Average R-Squared (ARS)	0,213; P=0,001	Significant
Average Adjusted R-Squared (AARS)	0,204; P=0,001	Significant
Average block VIF (AVIF)	1,339	Ideal (acceptable if <= 5; ideally <= 3,3)
Average Full collinearity VIF (AFVIF)	1,496	Ideal (acceptable if <= 5; ideally <= 3,3)
Tenenhaus GoF (GoF)	0,418	Large (large >= 0,36; medium >= 0,25; small >= 0,1)
Sympson's Paradox Ratio (SPR)	1,000	Ideal (acceptable if >= 0,7; ideally = 1)
R-Squared Contribution Ratio (RSCR)	1,000	Ideal (acceptable if >= 0,9; ideally = 1)
Statistical Suppression Ratio (SSR)	1,000	Acceptable (acceptable if >= 0,7)
NonLinear Bivariate Causality Direction Ratio (NLBCDR)	1,000	Acceptable (acceptable if >= 0,7)

Source: Authors.

Chapter 4
Residents' Subjective Well-Being and Support for Tourism Development:
The Case of Alxa, China

Manuel Alector Ribeiro
University of Surrey, UK

Chen Cui
University of Surrey, UK

Kyle Maurice Woosnam
University of Georgia, USA & University of Johannesburg, South Africa

ABSTRACT

This study aims to identify the factors that influence subjective well-being and their support for tourism development. The questionnaire was first designed and distributed online to residents in Alxa. Data were analysed, and the results indicated that positive economic impacts, positive socio-cultural impacts, and positive environmental impacts influence subjective well-being. Residents' perceived negative economic impacts and negative environmental impacts influence negatively their subjective well-being. Also, positive environmental impacts of the tourism development are the most significant factor that affects residents' subjective well-being. Furthermore, subjective well-being was also found to positively influence residents' support for the tourism development in Alxa. Ultimately, based on the findings, this study proposed and discuss both theoretical and managerial implications, as well as the limitations and directions of further studies.

DOI: 10.4018/978-1-7998-3156-3.ch004

INTRODUCTION

Subjective well-being (SWB) is generally used to evaluate quality of life, and at the same time, within the tourism literature, has been shown to be significantly associated with perceptions of tourism development (Chi, Cai, & Li, 2017; McCabe & Johnson, 2013; O'Dell & Billing, 2005; Woo, Kim, & Uysal, 2015). SWB is regarded as a broad construct, containing both cognitive and affective components (Chi et al., 2017; McCabe & Johnson, 2013). Based on previous research (see Diener, 2009), Chi et al. (2017: p. 210) has defined cognitive and affective well-being as:

The affective well-being reflects individuals' ongoing evaluations of the conditions in their lives. Whereas, the cognitive well-being requires cognitive processing: examining the conditions in their lives, weighing the importance of these conditions, and then evaluating their life on a scale ranging from dissatisfied to satisfied.

Based on extant research, one can argue that a bilateral relationship exists between local residents' SWB and their perceptions of tourism development (Ridderstaat et al., 2016; Woo et al., 2015). More specifically, residents' perspectives pertaining to the development of the tourism industry in a region affects SWB, which then has a significant bearing on level of support locals ascribe to tourism development (Chi et al., 2017; Diener, 2009; Ridderstaat et al., 2016). Prior studies have focused on identifying the factors that affect perceptions of tourism development, but minimal research has been undertaken that examines the relationship between such factors and residents' SWB as well as the relationship between this SWB and reported support for the development of the tourism industry. Therefore, this study aims to bridge this gap (Chi et al., 2017; Diener, 2009; Ridderstaat et al., 2016; Woo et al., 2015), with the intent of exploring the causal relationship between perceived impacts of tourismand residents' subjective well-being, which is proposed to inform individuals' support for the development of the regional tourism industry.

Previous studies have highlighted the importance of various impacts related to tourism development, including perceived economic impacts (Andereck and Nyaupane, 2011; Dyer et al., 2007; Proença & Soukiazis, 2008), sociocultural impacts (Almeida-Garcia, Balbuena-Vásquez, & Cortés-Macias, 2016; Andereck et al., 2005; Besculides et al., 2002; García et al., 2015), as well as environmental impacts (Andereck et al., 2005; Costanza et al., 2007; McCombes et al., 2015; Yu et al., 2011). Most of such work has relied on quantitative data to examine the relationships involving these perceived impacts of tourism. Therefore, in terms of data analysis, regression analysis will be used to test the relationships between residents' SWB and its antecedents and outcomes. Both negative and positive impacts, as well as cognitive and affective components of residents' SWB, along with perceived support for tourism development within Alxa, China will be considered in the proposed model.

LITERATURE REVIEW

Subjective Well-Being and Perceived Quality of Life

The meaning of happiness is comprehensive, involving research from many disciplines such as psychology, sociology, philosophy and ethics (McCabe & Johnson, 2013). Psychology focuses on the description of this state of mind and its causes (McCabe & Johnson, 2013). For example, in psychology, there are multiple terms to describe and define happiness, such as satisfaction and pleasure. Brülde (2007) states that due to the high degree of the daily life of these terms, peoples' understanding is somewhat confused, and happiness can be classified as follows: (1) Happiness defined by an external standard (i.e., happiness is based on a system of values and standards of the observer, not on the subjective judgment of the actor); (2) Happiness defined by non-verbal facial expressions (i.e., happiness as conveyed through positive facial expressions can be judged by comparing positive emotions and negative emotions); (3) Happiness based on individual self-evaluation. It is considered that happiness is the overall evaluation of the quality of life of the evaluator, which is referred to as subjective well-being or SWB (McCabe & Johnson, 2013). According to Diener et al. (1985), SWB is defined as all kinds of evaluations people make about their lives, including positive and negative evaluations, including cognitive assessments of one's life, positive emotions such as happiness, satisfaction, joy and pride, and negative emotions such as pain and worry (Stiglitz et al., 2009). In other words, SWB is an emotional state that human beings realize when they need to be satisfied, and their ideas need to be realized. It is formed by the interaction of psychological factors such as needs (including motivation, desire, interest), cognition, and emotion with external incentives — a complex, multi-layered state of mind. Chi et al. (2017) has pointed out that SWB is a broad concept, consisting of cognitive (evaluative) and affective (experiential) states of mind. Emotional components can be further divided into positive emotions and negative emotions, reflecting the individual's continuous assessment of their living conditions. Cognition refers to life satisfaction and requires cognitive processing (i.e.,monitoringconditions in one's life, weighing the importance of these conditions, and then assessing ones' life, from dissatisfaction to satisfaction. As a cognitive factor, the quality of life is an essential indicator of SWB, a more valid measure of affirmation, and another factor independent of positive and negative emotions.

In order to assess quality of life more preciously by different measurable indicators and reflect perceived differences in quality of life between mulitiple individuals, most scholars advocate defining quality of life from a multidimensional, subjective perspective (Carneiro & Eusébio, 2012; Costanza et al., 2007), for example, considering social and environmental factors (McGillivray, 2007). Khizindar (2012) and Pavli et al. (2015) pointed out that the definition of the quality of life defined as 'quality of life' (QOL) from the subjective perspective, can be easily confused with quality of life defined from an objective perspective. In order to distinguish quality of life from an objective perspective and highlight the individual's subjective cognition and emotions in all aspects of life, this study adopts a subjective form of the construct—perceived quality of life (Yu et al., 2011).

Uysal and Sirgy (2019) point out that quality of life (QOL) indicators can also be used as performance indicators, either independently or in a nested manner with traditional performance indicators for tourism. The authors reviewed and discussed QOL indicators for three selected stakeholders, namely visitors, residents of the communities in which they live, and employees of travel and hotel companies. Chi et al. (2017) mentioned that two theories of "top-down" and "bottom-up" are related to SWB's causal hypothesis. The theory from top-down belongs to personality theory that people have a personality tendency to experience life in a positive way, that is, to have a happy quality (Guo et al., 2017). The bottom-up theory holds that happiness is equal to the simple sum of happy events, assuming that a person's subjective well-being depends on external events, as well as a wide range of contextual and demographic factors (Sirgy, 2010).

Factors Influencing Residents' SWB

Neal et al. (1999) states that Bottom-up Spillover Theory provides an essential basis for explaining why perceptions of tourism impact affects quality of life. According to the theory, life satisfaction from the bottom-up perspective includes three levels: life care satisfaction, life dimension satisfaction, and overall life satisfaction. Overall life satisfaction is a function of the two former levels of satisfaction. According to the theory, perceptions of tourism impacts belongs to the bottom layer, and quality of life perceptions belong to the middle layer. The former will have an impact on the latter (Woo et al., 2015). Such a perspective was highlighted in the work of Neal et al. (2007).

Tourism is considered to be a useful tool in promoting economic development, and it has a positive effect on increasing the economic income of community residents along with driving direct and indirect employment. Andereck and Nyaupane (2011) pointed out that tourism improves the quality of life by improving the living standards of community residents. In SWB related literature, the factors affecting the construct are related to economic, sociocultural and environmental dimensions of life (Kim et al., 2013). Perceptions of the economic impacts of tourism have a significant positive influence on quality of life by potentially increasing the GDP of the destination country. Existing studies show a complex relationship between income and SWB. The "Easterlin paradox" notes that the growth of money does not necessarily lead to the improvement of SWB; as material wealth accumulates, it cannot promote the increase in SWB after a certain stage (Natali et al., 2018). In this regard, social comparison theory divides people into two groups: those who tend to compare themselves with those who do better than themselves, and those who tend to compare themselves with those who do worse than themselves. Moreover, most individuals tend to compare upward (Ferrer-i-Carbonell, 2005). Thus, this psychological trend partly explains the increase in people's income but not necessarily the year-after-year increase in SWB.

In addition to the relationships connecting economic impacts of tourism to quality of life, tourism promotes sociocultural exchanges between community residents and tourists, enhancing the formers' quality of life perceptions. Therefore, sociocultural impacts of tourism have been shown to significantly influence quality of life. Furthermore, attitudes about environmental impact perceptions positively affect perceived quality of life (Gursoy & Rutherford, 2004; Khizindar, 2012; Kim et al., 2013; Yu et al., 2011). Therefore, in the context of this study, the economic, sociocultural and the environmental impacts are factors that influence resident's subjective well-being.

(a) Perceived Economic Impacts

Studies have shown that the level of tourism development in a region is closely related to the quality of life indicators of its residents (Proença & Soukiazis, 2008). Some studies have indicated that the tourism industry does have an economic and sociocultural impact on residents. From the economic perspective, tourism is recognized as a mechanism to foster economic growth and development in developing countries (Choi & Sirakaya, 2005). In addition, the multiplier effect of tourism enhances the economic activities of the business, investment (e.g. Liu, 1987), transportation, infrastructure, accommodation, food and beverage industries and many other service sectors (Proença & Soukiazis, 2008). All of these forms of economic growth and development inevitably change the quality of life of residents.

In general, the economic impact of tourism can be divided into positive and negative impacts. On the one hand, the positive attitude of residents is mainly due to the positive impact of developing tourism (García et al., 2015). Residents can view tourism as a tool to reduce unemployment by creating new jobs, bringing new businesses and creating new investment opportunities (Andrews & Withey, 2012). More importantly, Nzama (2008) pointed out that there is a strong positive relationship between the degree of tourism development of residents and their perception of increased tourism development. Since tourism development usually needs to be measured between economic benefits, cultural or environmental costs, some residents tend to weaken their negative impacts while developing their destinations, emphasizing economic benefits to achieve tourism developing requirements (Cavus & Tanrisevdi, 2003; Dyer et al., 2007). Nzama (2008) pointed out that there is a strong positive relationship between the degree of tourism development of residents and their perception of increased tourism development. Since tourism development usually needs to be measured between economic benefits, cultural or environmental costs, some residents tend to weaken their negative impacts while developing their destinations, emphasizing economic benefits to achieve tourism developing requirements (Cavus & Tanrisevdi, 2003; Dyer et al., 2007).

On the other hand, tourism can also have negative economic impacts. Some studies have pointed out that the development of tourism will also increase the cost of living for residents, including the price of goods and services (Ribeiro, Valle, & Silva, 2013; Ribeiro, Pinto, Silva, & Woosnam, 2018). Secondly, the cost of import fees will also increase. To meet the increasing demand of different tourists from acrosss the globe, destination countries may need to import increased number of different products, which can lead to a rise in import fees (Johnson, Snepenger & Akis, 1994). Also, market inflation (Akis et al., 1996) can lead to an increase in land and housing prices, which drives up the cost of rent. At the same time, Harrill (2004) also mentioned that residents often ignore the seasonal characteristics of tourism, and aside from creation of employment opportunities, it is also necessary to take tourism workers into account, who rely on tourism to survive. Especially in the off-season, those tourism workers may need to find another job or will be unemployed, which can be a burden on the local community and economy.

In general, the overall economic impact on residents or local development is positive (Proença & Soukiazis, 2008). Tourism development can indeed enrich and change the local social structure, and the economic benefits noted above play a large role as to why residents view tourism in a positive light (Akis et al., 1996). In fact, research on economic efficiency and tourism development attitudes indicate a positive relationship between the two (Guo et al., 2017). That said, work connecting perceptions of the economic impacts of tourism to residents' SWB is further needed. This study proposes the first two hypotheses as:

H1: Residents' perceived positive economic impacts of tourism will significantly explain their subjective well-being by *positively influencing* a) *cognitive well-being* and b) *affective well-being*

H2: Residents' perceived negative economic impacts of tourism will significantly explain their subjective well-being by *negatively influencing* a) *cognitive well-being* and b) *affective well-being*

(b) Perceived Sociocultural Impacts

It is quite difficult to separate culture from everyday life. Culture is constantly changing. The continuous development of tourism will inevitably affect the sociocultural dimensions of tourism destinations, such as customs, social life, religious beliefs and values (Megeirhi, Woosnam, Ribeiro, Ramkissoon, & Denley, 2020; Nunkoo & Gursoy, 2012;). These effects have both positive and negative aspects (Besculides et al., 2002). Moreover, the development of tourism is closely related to community residents as their support plays a pivotal role in the development, successful operation and sustainable development. The interaction between residents and tourists will also bring them different feelings (Almeida-García et al., 2016; Moghavvemi et al., 2017; Woosnam et al., 2018).

On the one hand, tourism has played an active and important role in cultural influence due to the collision and exchange between different cultures (Driver, 1991). First of all, residents can communicate with tourists (such as local host families and communities that receive tourists), and they can learn about the external culture and bring their own unique culture to the outside world without going out. This kind of behaviour promotes cultural exchange and can bring the benefits of tolerance and understanding. Secondly, residents showing their cultural behaviour to outsiders can increase pride and cultural identity and bring a stronger sense of national identity (García et al., 2015). Moreover, can also attract interest in historical buildings and archaeological sites, retaining cultural values, which McKean (1978) called for "cultural involution", it can be understood that tourism has saved traditional culture and art from being on the verge of extinction. However, there are also reports that some residents are concerned that tourism and related developments may remove them from the land and take away their culture. Besculides et al. (2002) point out that tourism can lead to cultural adaptation and assimilation, and the influx of tourism language can contribute to the loss of native languages.

On the other hand, the attitude of residents is not as optimistic regarding the social impacts of tourism. According to Almeida-García et al. (2016), based on travel season, public and leisure facilities will become saturated, which can probably lead to traffic congestion and parking problems of the destination country. Andereck et al. (2000) also pointed out that tourism can possibly increase criminal activity, homelessness, and gambling in the destination country as well. García et al. (2015) added that tourism can lead to drug and alcohol problems. Last but not the least, Dyer et al. (2007) also mentioned that tourism increases the possibility of human trafficking problems and the development of the sex tourism industry.

Based on the above-mentioned arguments, it can be concluded that the attitudes of residents are different for the sociocultural impacts of tourism. A significant percentage of cultural impacts of tourism are positive, whereas those surrounding social impacts tend to be negative (Ko & Stewart, 2002). Nawijn and Mitas (2012) pointed out that the social impact of tourism is complex, although residents have a reasonable and consistent influence on the impact, more research is needed to gain a deeper understanding of how perceptions of impacts contribute to individuals' SWB. Therefore, this study proposes the following hypotheses:

H3: Residents' perceived positive sociocultural impacts of tourism will significantly predict their subjective well-being by *positively influencing* a) *cognitive well-being* and b) *affective well-being*

H4: Residents' perceived negative sociocultural impacts of tourism will significantly predict their subjective well-being by *negatively influencing* a) *cognitive well-being* and b) *affective well-being*

(c) Perceived Environmental Impacts

McCombes et al. (2015) pointed out that compared to economic and sociocultural impacts, residents' can be more sensitive to alterations of the environment Tourism contains seeds of its own destruction, tourism can kill tourism, destroying the very environment individuals travelled to experience (Butler, 1980; Ribeiro et al., 2018). The development of most tourism has put additional pressure on the environmental resources on which it depends, affecting the future prospects of the local population and actually affecting visitors' own expectations (Cater, 1991). As such, residents' attitudes concerning environmental impacts can also take the form of positives and negatives. However, residents' insight into the environmental impact of tourism is lagging behind economic and social land cultural aspects, especially in developing countries where there is less information about single environmental impact. In addition, those residing in rural areas are more keen to notice significant environmental impacts (Andereck et al., 2005; Yu et al., 2011).

On the one hand, there are statements that residents believe that tourism is an economically viable alternative to mining and harvesting, more environmentally friendly, and helps to preserve natural resources (Costanza et al., 2007) (such as the establishment of nature reserves), thus protecting the diversity and habitat of flora and fauna. In addition, it can improve the environment around the community and beautify key places throughout destinations (Chi et al., 2017; Woo et al., 2015).

On the other hand, the environmental damage caused by tourism is also of concern to residents (Guo et al., 2017). Inappropriate development of tourism can cause damage to the natural environment, loss of ecological habitat, and reduction of wildlife (Andereck et al., 2005). Also, it can lead to increases garbage production and pollution. Besides, residents may feel the local environment loses its traditional feel and infrastructural layout due to overcrowding and the accumulation of new businesses. Therefore, this study proposes the next two hypotheses concerning perceptions of environmental impacts and SWB:

H5: Residents' perceived positive environment impacts of tourism will significantly explain their subjective well-being by *positively influencing* a) *cognitive well-being* and b) *affective well-being*

H6: Residents' perceived negative environment impacts of tourism will significantly explain their subjective well-being by *negatively influencing* a) *cognitive well-being* and b) *affective well-being*

Subjective Well-Being and Support for Tourism Development

A number of studies have indicated that a bidirectional relationship exists between subjective well-being and support for tourism development (Andereck & Vogt, 2000; Chi et al., 2017; Ridderstaat et al., 2016). Ridderstaat et al. (2016) has explained this relationship as a "*two-way causal chain*", which means tourism development impacts the local economy, community, and environment, meanwhile such impacts will contribute to residents' subjective well-being and further explain individuals' support for tourism development (Ridderstaat et al., 2016). As Woo et al. (2015) mentioned, residents' positive perceptions serve to promote local tourism development, however, negative perceptions will inhibit tourism development as manifested through residents' lack of support for the industry.

Additionally, residents' attitudes and support for tourism development are able to influence tourists' perceptions of the tourism destination (Gallarza et al., 2002). Furthermore, the significance of place image in influencing individual's attitude and behavior has been established in environmental psychology (Lynch, 1960), geography (Bolton, 1992), as well as place and product marketing (Ashworth and Voogd, 1990). Stylidis et al. (2014) have suggested that a more positive local image would enhance residents' support for tourism development. In terms of marketing, Bandyopadhyay and Morais (2005) have pointed out that the inconsistency between the external representation of the destination and the image of the location held by the local community leads to residents' dissatisfaction with the tourism industry. Local individuals may also dissatisfied with the development of tourism image standardization, while they are prefer to support development efforts that promote the uniqueness of the place (Bramwell & Rawding, 1996). However, there is a lack of studies explore the causal relationship between residents' subjective well-being and their support for tourism development (Chi et al., 2017), though a number of researchers have realized the importance of the residents' support for tourism (Andereck & Vogt, 2000; Chi et al., 2017; Ridderstaat et al., 2016).

By considering the bidirectional relationship between subjective well-being and residents' support for tourism development, this study proposes the following hypothesis:

H7: *Residents' subjective well-being (SWB)* (as measured through a) cognitive well-being and b) *affective well-being*) will significantly explain individuals' degree of support for tourism development.

Figure 1. Conceptual model with corresponding proposed hypotheses

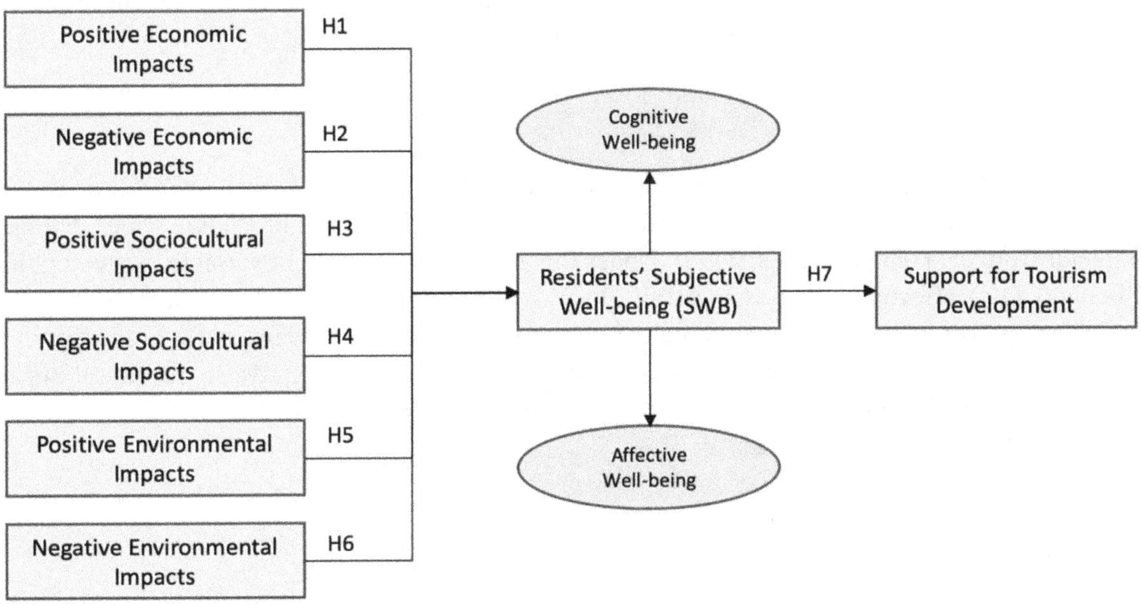

METHODS

Data Collection Process and Sample

Data were collected from residents who live in region od Alxa, north of China in the border with Mongolia. The questionnaire was designed on Qualtrics and conducated with radom sample of residents in Alxa and distributed via the social media platform – WeChat. Wechat is a very popular social media in China. A total of 562 residents in Alxa responded the questionnaire and 411 questionnaires were valid and reliable for data analysis. As stated below in this section, the validity and reliability of online questionnaire are two significant factors to measure the accuracy of data collection. After looking through the questionnaire responses from participants, 151 questionnaires have several missing values and they were discarded. This is due to the fact that participants did not answered all the questions in the questionnaire. A total of 411 valuable questionnaire were used for data analysis.

Measures

The questionnaire included five sections. The first three sections included measures assessing economic, sociocultural and environmental impacts of tourism in Alxa. The 28-item scale used to measure the impacts were based on works developed by Almeida-Garcia, Balbuena-Vásquez, and Cortés-Macias (2015), Ribeiro, Woosnam, Pinto and Silva (2017) and Ribeiro et al. (2013). The fourth section consisted of two scales: one measuring SWB and the other assessing support for tourism development in Axla. Ten items were used to measure SWB; five items for cognitive well-being (Chi et al., 2017) and five items for affective well-being (Anderson et al., 2012). Support for tourism development was measured utilizing five items from the work of Chi et al. (2017) and Liang and Hui (2016). The four scales within these three sections were presented on a 5-pt Likert Scale (1 = strongly disagree; 5 = strongly agree). The final section included demographic (i.e., age, gender, education etc.) and residential questions (i.e., length of residence, experience with tourism in Axla) etc.).

Data Analysis

To assess each of the seven hypotheses within the proposed model, ordinary least squares (OLS) regression was utilized. The OLS model can be defined as a standard linear regression procedure which estimates a parameter from data (Cai & Hayes, 2008). Under this model, the dependent variables were cognitive well-being (COG), affective well-being (AFFECT) and support for tourism development (SUPPORT), measured by the composite mean for each scale. Six independent variables were used as predictors in the model: positive and negative economic impacts of tourism; positive and negative sociocultural impacts of tourism; and positive and negative environmental impacts of tourism. IBM SPSS v.25 was used to carry out the OLS regression analyses.

RESULTS

Demographic Profile

Based on the 411 usable responses, a demographic profile of Alxa residents was created. In the way of age, 66.5% were between the ages of 20 and 40 (e.g., 36.5% between 20 and 29 years of age; 30% between 30 and 39 years of age). Nearly two-thirds of respondents were women (65%) with the remaining portion comprised of men. More than 90% of the participants had completed a college education. For the question related to work experience in the tourism industry, nearly 60% of the participants have job related to Alxa tourism industry, and most of them are living in the Alxa currently. Among their responses, 40% of participants have lived in Alxa for less than 3 years, 50% of them lived in Alxa between 4 to 8 years, and the rest of them lived more than 8 years in the Alxa. However, nearly 40% of participants were born in Alxa and they have some personal experience on the tourism industry development in these years.

Hypothesis Testing

To restate hypotheses from the conceptual model, the first six hypotheses examined the relationships between positive and negative economic, sociocultural, and environmental impacts of tourism and SWB (i.e., cognitive well-being and affective well-being). The final hypothesis considered the role that these two forms of well-being played in explaining residents' support for tourism development. The results are explained in the four subsequent sections pertaining to: 1) the role of perceived economic impacts in explaining SWB; 2) the role of perceived sociocultural impacts in explaining SWB; 3) the role of perceived environmental impacts in explaining SWB; and 4) the relationship between SWB and support for tourism development. Results are presented in Table 1.

Table 1. Results of model analysis

Dependent Variables	Independent Variables	β	*t*-value	*p*	95.0% Confidence Interval		Hypothesis
					LLCI	ULCI	
COG	Positive Economic Impacts	0.17	5.39	.000	0.11	0.23	H1a
	Negative Economic Impacts	-0.04	-1.50	0.13	-0.09	0.01	H2a
	Positive Sociocultural Impacts	0.14	4.00	.000	0.07	0.21	H3a
	Negative Sociocultural Impacts	-0.03	-0.99	.321	-0.08	0.03	H4a
	Positive Environmental Impacts	0.17	6.21	0.000	0.12	0.22	H5a
	Negative Environmental Impacts	-0.01	-0.14	0.887	-0.05	0.05	H6a
	R^2	0.44	$F(6,410) = 53.669, p < 0.0001$				
AFFECT	Positive Economic Impacts	0.18	2.47	0.013	0.04	0.32	H1b
	Negative Economic Impacts	-0.12	-2.15	0.032	-0.24	-0.01	H2b
	Positive Sociocultural Impacts	0.12	1.47	0.140	-0.04	0.27	H3b
	Negative Sociocultural Impacts	-0.01	-0.20	0.843	-0.13	0.10	H4b
	Positive Environmental Impacts	0.28	4.48	0.000	0.16	0.40	H5b
	Negative Environmental Impacts	0.14	2.33	0.020	0.02	0.25	H6b
	R^2	0.179	$F(6,410) = 14.644, p < 0.0001$				
SUPPORT	Positive Economic Impacts	-0.03	-0.54	0.591	-0.13	0.07	
	Negative Economic Impacts	0.03	0.66	0.509	-0.05	0.11	
	Positive Sociocultural Impacts	0.01	0.12	0.904	-0.10	0.12	
	Negative Sociocultural Impacts	0.03	0.66	0.508	-0.05	0.11	
	Positive Environmental Impacts	0.01	0.07	0.943	-0.09	0.09	
	Negative Environmental Impacts	0.02	0.31	0.756	-0.07	0.09	
	COG	-0.01	-0.15	0.880	-0.17	0.15	H7a
	AFFECT	0.49	13.83	0.000	0.42	0.56	H7b
	R^2	0.37	$F(8,410) = 29.864, p < 0.0001$				

(a) Role of Perceived Economic Impacts in Explaining SWB

As presented in Table 2, a significant positive relationship was found between perceptions of positive economic impacts and cognitive well-being ($\beta = 0.17$; $p < 0.001$). Such findings provide support for

Table 2. Summary of results for relationships between perceived economic impacts and SWB

Paths	β	*p*-value	Hypotheses testing
H1a: Positive economic impacts ® cognitive well-being	0.17	0.000	Supported
H1b: Positive economic impacts ® affective well-being	0.18	0.013	Supported
H2a: Negative economic impacts ® cognitive well-being	-0.38	0.134	Not Supported
H2b: Negative economic impacts ® affective well-being	-0.12	0.032	Supported

H1a. However, results only provide marginal support for the **H1b** ($\beta = 0.18$; $p = 0.013$) for the relationship between perceptions of positive economic impacts and affective well-being.

In considering perceptions of negative economic impacts of tourism, findings did not provide support for **H2a** ($\beta = -0.38$; $p > 0.05$) regarding the explanation of cognitive well-being. However, results did confirm **H2b** ($\beta = -0.12$; $p = 0.032$) in terms of affective well-being. Such findings provide mixed support for **H2** overall.

(b) Role of Perceived Sociocultural impacts in Explaining SWB

Table 3 portrays the relationships between perceived sociocultural impacts of tourism and the two forms of SWB. As shown, only one hypothesis was supported. Results demonstrate that perceptions of positive sociocultural impacts significantly explained cognitive well-being ($\beta = 0.14$; $p < 0.001$), supporting **H3a**. Though results indicate a positive relationship between perceived positive sociocultural impacts and affective well-being, it was not significant ($p = 0.14$). Therefore, **H3b** was not supported.

Table 3. Summary of results for relationships between perceived sociocultural impacts and SWB

Paths	β	*p*-value	Hypotheses testing
H3a: Positive sociocultural impacts ® cognitive well-being	0.139	0.000	Supported
H3b: Positive sociocultural impacts ® affective well-being	0.116	0.142	Not Supported
H4a: Negative sociocultural impacts ® cognitive well-being	-0.025	0.321	Not Supported
H4b: Negative sociocultural impacts ® affective well-being	-0.12	0.843	Not Supported

In the way of perceived negative sociocultural impacts of tourism, results revealed a negative relationship with both forms of SWB; however, neither was significant ($p > 0.05$). As a result, **H4a** and **H4b** were not supported.

(c) Role of Perceived Environmental Impacts in Explaining SWB

The proposed relationships between perceived environmental impacts of tourism and SWB demonstrated more significant findings. In fact, three of the four relationships were significant. Results illustrates that perceived positive environmental impacts significantly explained both cognitive well-being ($\beta = 0.17$; $p < 0.001$) and affective well-being ($\beta = 0.28$; $p < 0.001$) As such, both **H5a** and **H5b** were supported.

Table 4. Summary of results for relationships between perceived environmental impacts and SWB

Paths	β	*p*-value	Hypotheses Testing
H5a: Positive environmental impacts ® cognitive well-being	0.17	0.000	Supported
H5b: Positive environmental impacts ® affective well-being	0.28	0.000	Supported
H6a: Negative environmental impacts ® cognitive well-being	0.01	0.887	Not Supported
H6b: Negative environmental impacts ® affective well-being	0.14	0.021	Supported

The influence that perceived negative environmental impacts of tourism has on SWB was more mixed. **H6a** was not supported with an extremely high *p*-value (i.e., 0.89). However, perceived negative environmental impacts of tourism did significantly predict affective well-being (β = 0.14; $p < 0.05$). Such results demonstrate support for **H6b**.

(d) Relationship Between SWB and Support for Tourism Development

The final hypothesis pertained to the relationship between the two forms of SWB and support for tourism development. As demonstrated in Table 5, cognitive well-being did not significantly predict residents' degree of support for tourism development (β = -0.12; $p < 0.88$). Thus, **H7a** was not supported. However, affective well-being did serve as significant predictor of residents' level of support for tourism development (β = 0.49; $p < 0.001$), revealing evidence in support of **H7b**.

DISCUSSION AND IMPLICATIONS

(a) General Discussion

This study aimed to explore and identify the main factors that affect residents' SWB, as well as to examine the causal relationships between residents' SWB and their support for the local tourism industry development. The factors that influence the residents' SWB has been identified based on previous studies (McCabe & Johnson, 2013; Proença & Soukiazis, 2008; Yu et al., 2011). Such factors consist of perceptions of *economic impacts* (Andereck & Nyaupane, 2011; Cavus & Tanrisevdi, 2003; Dyer et al., 2007; Proença & Soukiazis, 2008), *sociocultural impacts* (Almeida-García et al., 2016; Andereck et al., 2005; Besculides et al., 2002), and *environmental impacts* (Andereck et al., 2005; Costanza et al., 2007; McCombes et al., 2015; Yu et al., 2011). Both negative and positive aspects of each factor has been shown to affect residents' SWB (Andereck et al., 2005; Chi et al., 2017; Meng et al., 2010; Yu et

Table 5. Summary of results for relationships between SWB and support for tourism development

Paths	β	*p*-value	Hypotheses testing
H7a: Cognitive well-being ® Support for tourism development	-0.12	0.880	Not Supported
H7b: Affective well-being ® Support for tourism development	0.493	0.000	Supported

al., 2011). Therefore, this study proposed six relevant hypotheses to explore the influence of various perceived impacts on residents' SWB.

According to the tourism literature (Chi et al., 2017; Ridderstaat et al., 2016; Stylidis et al., 2014; Woo et al., 2015),residents' SWB in turn affects the degree to which they support local tourism development. As such, a seventh hypothesis was also examined looking at the relationship between these two constructs Overall, findings from this study describe how various factors affect local residents' SWB, indicate the corresponding importance of each factor for residents' SWB, and illustrate the major role of residents' SWB in predicting their support for tourism development.

(b) Perceived Economic Impacts of Tourism on SWB

Though a number of empirical studies have illustrated that local residents' life satisfaction is related to the perceived economic consequences brought on by tourism development (Archer & Fletcher, 1996; Mathieson & Wall, 1982; Proença & Soukiazis, 2008; Ribeiro et al., 2018; Seetanah, 2011), little research has been undertaken to explore the causal relationship between the constructs. This study bridged such gap.

Previous studies have not specified the negative and positive economic impact on residents' SWB in a region (Chi et al., 2017; García et al., 2015). Our study has done this, and considered two aspects of the residents' SWB (i.e., cognitive well-being and affective well-being) in the process. The only relationship involving perceived economic impacts of tourism and SWB that was not significant included cognitive well-being as the outcome variable (with perceived negative economic impacts as the predictor).

Indeed, the positive economic impact related to regional tourism development would provide positive effects on the cognitive well-being of the local residents. According to priori tourism literature (Andereck & Nyaupane, 2011; Liu et al., 1987), tourism is generally considered an economic development strategy that invites new businesses into the destination, creates new investment opportunities, and provides necessary jobs to local residents. Furthermore, tourism development allows local governments to increase taxes as well as improve the social welfare of residents (Andereck & Nyaupane, 2011). In this vein, residents are likely to realize the benefits of tourism development (Chi et al., 2017; Graham & Pettinato, 2001). The results of this study indicated that those benefits improve residents' cognitive well-being.

On the contrary, the negative economic impact more significantly influenced residents' affective well-being rather than cognitive well-being. Negative economic impacts caused by tourism development include higher price of goods and services (García et al., 2015; Meng et al., 2010; Natali et al., 2018), the cost of import fees (David, 1999), and market inflations (Akis et al., 1996). Our findings revealed that lower levels of residents'affective well-being is a function of larger negative economic impacts related to the tourism development.

(c) Perceived Sociocultural Impacts of Tourism on SWB

Previous literature has illustrated that sociocultural impacts of tourism development would affect residents' SWB (Almeida-García et al., 2016; Andereck et al., 2005; Besculides et al., 2002; García et al., 2015), however little research has explored the relationship between the constructs. By considering the role of both positive and negative aspects of the sociocultural impacts of tourism (Besculides et al., 2002) on the cognitive well-being and affective well-being of residents (Chi et al., 2017; Guo et al., 2017), this study found that only perceived positive sociocultural impacts significantly explained cognitive well-being. Those positive impacts include communication with tourists (Chi et al., 2017; Driver, 1991), cultural exchange (García et al., 2015; O'Dell and Billing, 2005), improving heritage management (Chi et al., 2017), traditional culture protection (Megeirhi et al., 2020; O'Dell & Billing, 2005) and so on (Almeida-García et al., 2016; Chi et al., 2017; Driver, 1991; García et al., 2015; O'Dell & Billing, 2005).

Cognitive well-being requires a process that evaluates residents' life satisfaction with varying weight placed on living conditions (Chi et al., 2017; Diener, 2009), while the affective well-being indicates residents' are constantly evaluating their living conditions (Chi et al., 2017; Diener, 2009). Therefore, cognitive well-being illustrates more of a dynamic process, but affective well-being reflects the status of an individual's perception. Nawijn and Mitas (2012) has described that perceived tourism impacts had a greater influence on cognitive well-being rather than affective well-being. Chi et al. (2017) also demonstrated that the sociocultural impacts of tourism development failed to explain affective component of residents' SWB. As such, the results of this study have provided support for previous literature in demonstrating how these sociocultural impacts have a greater influence on cognitive well-being.

(d) Perceived Environmental Impacts of Tourism on SWB

Following previous research, we also considered the role of perceived environmental impacts on SWB (Andereck et al., 2005; Costanza et al., 2007; McCombes et al., 2015; Yu et al., 2011). Similar to economic impacts, three of the four sub-hypotheses were found to be significant.

Perceptions of positive environmental impacts of tourism has a strong influence on both cognitive and affective well-being of residents. This is due to the fact that many residents believe the tourism industry is generally environmentally friendly and is able to preserve natural resources (e.g., establishing nature reserves, protecting the habitat of fauna) (Costanza et al., 2007). Such perceptions have prompted governments and private companies to develop the tourism industry over manufacturing or construction industries (Costanza et al., 2007). Another explanation as to why perceived positive environmental impacts of tourism explained both cognitive and affective well-being is that the environmental impact is easy for individuals to observe (Chi et al., 2017). These perceptions promote a more positive influence on residents' living conditions. Compared withthe positive environmental impacts of tourism, negative environmental impacts are not as easy to recognized by individuals (Chi et al., 2017). The results of this study also demonstrate this, highlighting that perceived negative environmental impacts only explain cognitive well-being (and not affective well-being). Relevant environmental damages include loss of ecological habitat (Andereck et al., 2005), reduction of wildlife (Andereck et al., 2005), and air or noise pollutions (McCombes et al., 2015).

(e) SWB Impacts on the Supports of Tourism Development

Previous studies have reflected that the relationship between residents' SWB and tourism development is bilateral (Ridderstaat et al., 2016). On the one hand, the development of the tourism industry in a region is able to promote or inhibit the development of the regional economy (Andereck & Nyaupane, 2011; Cavus & Tanrisevdi, 2003; Dyer et al., 2007; Proença & Soukiazis, 2008), improve regional community and traditional culture (Almeida-García et al., 2016; Andereck et al., 2005; Besculides et al., 2002; García et al., 2015), as well as protect the regional environment (Andereck et al., 2005; Costanza et al., 2007; McCombes et al., 2015; Yu et al., 2011). On the other hand, those impacts on residents' SWB would affect their support for the development of the tourism industry (Chi et al., 2017; Ridderstaat et al., 2016; Woo et al., 2015). This study has also bridged the research gap in connecting residents' SWB to their support for tourism development.

Findings revealed that only affective well-being significantly explained residents' support for tourism. This result illustrates that residents are significantly willing to support tourism development when they feel positive perceptions by evaluating the current living condition. The relationship between residents' affective well-being and individuals' support for the tourism development is extremely close. Chi et al. (2017) also summarized a similar conclusion that residents with a higher level of affective well-being are more supportive of local tourism development. However, residents' cognitive well-being is not an important factor in explainin the degree of support one has for tourism development. Such a finding is in line with Chi et al. (2017), that claimed the cognitive well-being level of residents was not significantly related to their degree of support for the development of the regional tourism industry.

Implications

This work has implications for both theory and practice. First and foremost, this research has bridged the gap between perceived tourism impacts and support for tourism developmemt by considering the role SWB plays within the relationship. The exploration and identification of factors that influence the support for tourism development are crucial in contributing to the tourism literature. By understanding how various perceptions of tourism impacts (e.g., economic, sociocultural, and environmental) affect residents' SWB as well as support for tourism development, we are able to move closer to embracing sustainable tourism development by acknowledging the perspective of residents in helping to plan for tourism (Lawton & Weaver, 2015).

Furthermore, previous studies surrounding perceived impacts, SWB, and support for tourism development have primarily focused on western countries (e.g., Nawijn and Mitas (2012) or regions in southern China (e.g., Chi et al. (2017)). This study has broadened the reach of such work by extending into northen China (i.e., Alxa, China) so that researchers are able to understand how the residents' SWB is influenced by tourism impacts in this region. This is a remarkable extension, as the concept of SWB would be different given distinct cultural contexts (Chi et al., 2017; Ng & Diener, 2014).

This work has implications for practitioners as well. Our findings are valuable to local governments, especially tourism authorities, ensuring that they are able to identify and understand how different factors affect not only residents' SWB but also their support for tourism development. The influence between those impacts and the development of the regional tourism industry is bidirectional. The tourism development would brought impacts on regional economy, community and culture, as well as environment. Meanwhile, those impacts would change the residents' perception for the tourism development then further change their supports for the tourism development. Therefore, understanding the relationships between various factors and support for tourism is significant for tourism authorities to design and deliver a tourism strategy that incorporates locals' perspectives when it comes to planning for tourism in Alxa. As such, this study has provided a strong foundation for developing and managing tourism for this town in northern China.

Understanding residents' SWB based on tourism in Alxa is important to inform tourism authorities' decisions to plan for sustainable tourism in the destination. In fact, unhappy regional individuals may reflect their negative feelings toward tourists, which could inhibit tourism development in this region (Nunkoo & Gursoy, 2012; Ribeiro et al., 2018; Su et al., 2016). Therefore, encouraging residents to be involved throughout the tourism decision making process, as well as exploring their perceptions are crucial for effectively developing the tourism industry in this region and helping to ensure such tourism is sustainable.

Limitations and Future Research

This research is not without its limitations; three of which are described below with corresponding opportunities for future research. First, this study focused on only one location. Therefore, findings are limited to this region. Furthermore, sampling was undertaken using a convenience strategy. As a result, findings may be further limited in that they may not be representative of all residents living in Alxa.. Future research may consider undertaking a comparable study in various destinations (across distinct countries) whereby resident participants are secured through random sampling strategies. Such an approach would not only serve to further validate our findings but also provide more representatives samples that would allow for greater generalizability of findings.

Another limitation is that we did not consider the differences of various groups, for instance, in exploring the impacts of tourism or SWB. By further segmenting residents across gender or age, results may be more telling as to reflect more specific distinctions among the resident population. Future work should seek to incorporate demographic variables within the model. More specifically, variables such as gender (i.e., female versus male), length of residence (i.e., newcomer versus long-term), economic dependence (i.e., low versus high), or even personality may be ideal to consider as moderators of the relationship between SWB and support for tourism development. A similar approach has been undertaken by Moghavvemi et al. (2017), Patwardhan et al. (2020), and Ribeiro et al. (2018) most recently.

Furthermore, this research did not consider the indirect effects of economic, sociocultural, and environmental impacts on residents' SWB. As Chi et al. (2017) claims, these indirect effects are valuable to include within models. Primarily, such effects can demonstrate the mediating roles of variables appearing between independent and dependent variables. Most recently, this was undertaken in considering the mediating effects of emotional solidarity within the context of resident-tourist interactions (Aleshinloye, Fu, Ribeiro, Woosnam, & Tasci, 2020). Despite these limitations, this work continues to advance the literature on residents' attitudes in considering the relationships between perceived tourism impacts, SWB, and support for tourism development.

REFERENCES

Akis, S., Peristianis, N., & Warner, J. (1996). Residents' attitudes to tourism development: The case of Cyprus. *Tourism Management, 17*(7), 481–494. doi:10.1016/S0261-5177(96)00066-0

Aleshinloye, K. D., Fu, X., Ribeiro, M. A., Woosnam, K. M., & Tasci, A. D. (2020). The influence of place attachment on social distance: Examining mediating effects of emotional solidarity and the moderating role of interaction. *Journal of Travel Research, ournal of Travel Research, 59*(5), 828-849.

Almeida-García, F., Peláez-Fernández, M. Á., Balbuena-Vazquez, A., & Cortes-Macias, R. (2016). Residents' perceptions of tourism development in Benalmádena (Spain). *Tourism Management, 54*, 259–274. doi:10.1016/j.tourman.2015.11.007

Andereck, K. L., & Nyaupane, G. P. (2011). Exploring the nature of tourism and quality of life perceptions among residents. *Journal of Travel Research, 50*(3), 248–260. doi:10.1177/0047287510362918

Andereck, K. L., Valentine, K. M., Knopf, R. C., & Vogt, C. A. (2005). Residents' perceptions of community tourism impacts. *Annals of Tourism Research, 32*(4), 1056–1076. doi:10.1016/j.annals.2005.03.001

Andereck, K. L., & Vogt, C. A. (2000). The relationship between residents' attitudes toward tourism and tourism development options. *Journal of Travel Research, 39*(1), 27–36. doi:10.1177/004728750003900104

Andrews, F. M., & Withey, S. B. (2012). *Social indicators of well-being: Americans' perceptions of life quality*. Plenum Press.

Archer, B., & Fletcher, J. (1996). The economic impact of tourism in the Seychelles. *Annals of Tourism Research, 23*(1), 32–47. doi:10.1016/0160-7383(95)00041-0

Ashworth, G. J., & Voogd, H. (1990). *Selling the city: Marketing approaches in public sector urban planning*. Belhaven Press.

Bandyopadhyay, R., & Morais, D. (2005). Representative dissonance: India's self and western image. *Annals of Tourism Research, 32*(4), 1006–1021. doi:10.1016/j.annals.2005.02.002

Besculides, A., Lee, M. E., & Mccormick, P. J. (2002). Residents' perceptions of the cultural benefits of tourism. *Annals of Tourism Research, 29*(2), 303–319. doi:10.1016/S0160-7383(01)00066-4

Bolton, R. (1992). 'Place prosperity vs people prosperity'revisited: An old issue with a new angle. *Urban Studies (Edinburgh, Scotland), 29*(2), 185–203. doi:10.1080/00420989220080261

Bramwell, B., & Rawding, L. (1996). Tourism marketing images of industrial cities. *Annals of Tourism Research, 23*(1), 201–221. doi:10.1016/0160-7383(95)00061-5

Brülde, B. (2007). Happiness and the good life. Introduction and conceptual framework. *Journal of Happiness Studies, 8*(1), 1–14. doi:10.100710902-006-9002-9

Butler, R. W. (1980). The concept of a tourist area cycle of evolution: Implications for management of resources. *The Canadian Geographer. Geographe Canadien, 24*(1), 5–12. doi:10.1111/j.1541-0064.1980.tb00970.x

Byrd, E. T., Bosley, H. E., & Dronberger, M. G. (2009). Comparisons of stakeholder perceptions of tourism impacts in rural eastern North Carolina. *Tourism Management, 30*(5), 693–703. doi:10.1016/j.tourman.2008.10.021

Cai, L., & Hayes, A. F. (2008). A new test of linear hypotheses in OLS regression under heteroscedasticity of unknown form. *Journal of Educational and Behavioral Statistics, 33*(1), 21–40. doi:10.3102/1076998607302628

Carneiro, M. J., & Eusébio, C. (2012). Segmentation of the tourism market using the impact of tourism on quality of life. *Tourism & Management Studies, 7*, 91–100.

Cater, E. (1991). Sustainable tourism in the Third World: problems and prospects. *Sustainable tourism in the Third World: problems and prospects.*

Cavus, S., & Tanrisevdi, A. (2003). Resident's Attitudes Towards Tourism Development: A Case Study in Kusadasi, Turkey. *Tourism Analysis*, *7*(3), 259–269. doi:10.3727/108354203108750102

Chi, C. G.-Q., Cai, R., & Li, Y. (2017). Factors influencing residents' subjective well-being at World Heritage Sites. *Tourism Management*, *63*, 209–222. doi:10.1016/j.tourman.2017.06.019

Choi, H.-S. C., & Sirakaya, E. (2005). Measuring residents' attitude toward sustainable tourism: Development of sustainable tourism attitude scale. *Journal of Travel Research*, *43*(4), 380–394. doi:10.1177/0047287505274651

Costanza, R., Fisher, B., Ali, S., Beer, C., Bond, L., Boumans, R., Danigelis, N. L., Dickinson, J., Elliott, C., Farley, J., Gayer, D. E., Glenn, L. M. D., Hudspeth, T., Mahoney, D., McCahill, L., McIntosh, B., Reed, B., Rizvi, S. A. T., Rizzo, D. M., ... Snapp, R. (2007). Quality of life: An approach integrating opportunities, human needs, and subjective well-being. *Ecological Economics*, *61*(2-3), 267–276. doi:10.1016/j.ecolecon.2006.02.023

Diener, E. (2009). The science of well-being: The collected works of Ed Diener. New York: Springer.

Diener, E., Emmons, R. A., Larsen, R. J., & Griffin, S. (1985). The satisfaction with life scale. *Journal of Personality Assessment*, *49*(1), 71–75. doi:10.120715327752jpa4901_13 PMID:16367493

Driver, T. F. (1991). *The magic of ritual: Our need for liberating rites that transform our lives and our communities*. Harper Collins.

Dyer, P., Gursoy, D., Sharma, B., & Carter, J. (2007). Structural modeling of resident perceptions of tourism and associated development on the Sunshine Coast, Australia. *Tourism Management*, *28*(2), 409–422. doi:10.1016/j.tourman.2006.04.002

Ferrer-I-Carbonell, A. (2005). Income and well-being: An empirical analysis of the comparison income effect. *Journal of Public Economics*, *89*(5-6), 997–1019. doi:10.1016/j.jpubeco.2004.06.003

Gallarza, M. G., Saura, I. G., & Garcí, A. H. C. (2002). Destination image: Towards a conceptual framework. *Annals of Tourism Research*, *29*(1), 56–78. doi:10.1016/S0160-7383(01)00031-7

García, F. A., Vázquez, A. B., & Macías, R. C. (2015). Resident's attitudes towards the impacts of tourism. *Tourism Management Perspectives*, *13*, 33–40. doi:10.1016/j.tmp.2014.11.002

Golafshani, N. (2003). Understanding reliability and validity in qualitative research. *Qualitative Report*, *8*, 597–606.

Graham, C., & Pettinato, S. (2001). Happiness, markets, and democracy: Latin America in comparative perspective. *Journal of Happiness Studies*, *2*(3), 237–268. doi:10.1023/A:1011860027447

Guo, A., Guo, Y., Li, H., & Sun, X. (2017a). Relationship between perceived tourism impacts and perceived quality of life of community residents in tourist destinations. *World Regional Studies*, *26*, 115–127.

Gursoy, D., & Rutherford, D. G. (2004). Host attitudes toward tourism: An improved structural model. *Annals of Tourism Research*, *31*(3), 495–516. doi:10.1016/j.annals.2003.08.008

Harrill, R. (2004). Residents' attitudes toward tourism development: A literature review with implications for tourism planning. *Journal of Planning Literature*, *18*(3), 251–266. doi:10.1177/0885412203260306

Johnson, J., Snepenger, D., & Akis, S. (1994). Residents' perceptions of tourism development. *Annals of Tourism Research, 21*(3), 629–642. doi:10.1016/0160-7383(94)90124-4

Khizindar, T. M. (2012). Effects of tourism on residents' quality of life in Saudi Arabia: An empirical study. *Journal of Hospitality Marketing & Management, 21*(6), 617–637. doi:10.1080/19368623.2012.627226

Kim, K., Uysal, M., & Sirgy, M. J. (2013, June). 92013). How does tourism in a community impact the quality of life of community residents? *Tourism Management, 36*, 527–540. doi:10.1016/j.tourman.2012.09.005

Ko, D. W., & Stewart, W. P. (2002). A structural equation model of residents' attitudes for tourism development. *Tourism Management, 23*(5), 521–530. doi:10.1016/S0261-5177(02)00006-7

Landsheer, J., & Boeije, H. (2010). In search of content validity: Facet analysis as a qualitative method to improve questionnaire design. *Quality & Quantity, 44*(1), 59–69. doi:10.100711135-008-9179-6

Lawton, L. J., & Weaver, D. B. (2015). Using residents' perceptions research to inform planning and management for sustainable tourism: A study of the Gold Coast Schoolies Week, a contentious tourism event. *Journal of Sustainable Tourism, 23*(5), 660–682. doi:10.1080/09669582.2014.991398

Liang, Z. X., & Hui, T.-K. (2016). Residents' quality of life and attitudes toward tourism development in China. *Tourism Management, 57*, 56–67. doi:10.1016/j.tourman.2016.05.001

Liu, J., Sheldon, P., & Var, T. (1987). Resident Perceptions of The Environment Impacts of Tourism of Tourism. *Annals of Tourism Research, 14*(1), 17–37. doi:10.1016/0160-7383(87)90045-4

Lynch, K. (1960). *The image of the city*. MIT Press.

Mathieson, A., & Wall, G. (1982). *Tourism, economic, physical and social impacts*. Longman.

Mccabe, S., & Johnson, S. (2013). The happiness factor in tourism: Subjective well-being and social tourism. *Annals of Tourism Research, 41*, 42–65. doi:10.1016/j.annals.2012.12.001

Mccombes, L., Vanclay, F., & Evers, Y. (2015). Putting social impact assessment to the test as a method for implementing responsible tourism practice. *Environmental Impact Assessment Review, 55*, 156–168. doi:10.1016/j.eiar.2015.07.002

Mcgillivray, M. (2007). Human well-being: Issues, concepts and measures. In M. Mcgillivray (Ed.), *Human Well-Being* (pp. 1–22). Palgrave Macmillan. doi:10.1057/9780230625600_1

McKean, P. (1978). Towards a theoretical analysis of tourism: economic dualism and cultural involution in Bali. In V. Smith (Ed.), *Hosts and guests. The anthropology of tourism* (pp. 119–138). University of Pennsylvania Press.

Megeirhi, H. A., Woosnam, K. M., Ribeiro, M. A., Ramkissoon, H., & Denley, T. J. (2020). Employing a value-belief-norm framework to gauge Carthage residents' intentions to support sustainable cultural heritage tourism. *Journal of Sustainable Tourism, 28*(9), 1351–1370. doi:10.1080/09669582.2020.1738444

Meng, F., Li, X., & Uysal, M. (2010). Tourism Development and Regional Quality of Life: The Case of China: 旅游发展和区域生活质量: 中国案例研究. *Journal of China Tourism Research*, *6*(2), 164–182. doi:10.1080/19388160.2010.481602

Moghavvemi, S., Woosnam, K. M., Paramanathan, T., Musa, G., & Hamzah, A. (2017). The effect of residents' personality, emotional solidarity, and community commitment on support for tourism development. *Tourism Management*, *63*, 242–254. doi:10.1016/j.tourman.2017.06.021

Natali, L., Handa, S., Peterman, A., Seidenfeld, D., & Tembo, G. (2018). Does money buy happiness? Evidence from an unconditional cash transfer in Zambia. *SSM - Population Health*, *4*, 225–235. doi:10.1016/j.ssmph.2018.02.002 PMID:29854906

Nawijn, J., & Mitas, O. (2012). Resident attitudes to tourism and their effect on subjective well-being: The case of Palma de Mallorca. *Journal of Travel Research*, *51*(5), 531–541. doi:10.1177/0047287511426482

Neal, J. D., Sirgy, M. J., & Uysal, M. (1999). The role of satisfaction with leisure travel/tourism services and experience in satisfaction with leisure life and overall life. *Journal of Business Research*, *44*(3), 153–163. doi:10.1016/S0148-2963(97)00197-5

Neal, J. D., Uysal, M., & Sirgy, M. J. (2007). The effect of tourism services on travelers' quality of life. *Journal of Travel Research*, *46*(2), 154–163. doi:10.1177/0047287507303977

Ng, W., & Diener, E. (2014). What matters to the rich and the poor? Subjective well-being, financial satisfaction, and postmaterialist needs across the world. *Journal of Personality and Social Psychology*, *107*(2), 326–338. doi:10.1037/a0036856 PMID:25090131

Nunkoo, R., & Gursoy, D. (2012). Residents' support for tourism: An identity perspective. *Annals of Tourism Research*, *39*(1), 243–268. doi:10.1016/j.annals.2011.05.006

Nzama, A. (2008). Socio-cultural impacts of tourism on the rural areas within the World Heritage Sites–The case of KwaZulu-Natal, South Africa. *South Asian Journal of Tourism and Heritage*, *1*, 1–8.

O'dell, T., & Billing, P. (2005). *Experiencescapes: Tourism, culture and economy*. Business School Press DK.

Patwardhan, V., Ribeiro, M. A., Payini, V., Woosnam, K. M., Mallya, J., & Gopalakrishnan, P. (2020). Visitors' Place Attachment and Destination Loyalty: Examining the Roles of Emotional Solidarity and Perceived Safety. *Journal of Travel Research*, *59*(1), 3–21. doi:10.1177/0047287518824157

Pavli, A., Silvestros, C., Patrinos, S., & Maltezou, H. C. (2015). Vaccination and malaria prophylaxis among Greek international travelers to Asian destinations. *Journal of Infection and Public Health*, *8*(1), 47–54. doi:10.1016/j.jiph.2014.07.002 PMID:25129447

Proença, S., & Soukiazis, E. (2008). Tourism as an economic growth factor: A case study for Southern E,uropean countries. *Tourism Economics*, *14*(4), 791–806. doi:10.5367/000000008786440175

Ribeiro, M. A., Pinto, P., Silva, J. A., & Woosnam, K. M. (2017). Residents' attitudes and the adoption of pro-tourism behaviours: The case of developing island countries. *Tourism Management*, *61*, 523–537. doi:10.1016/j.tourman.2017.03.004

Ribeiro, M. A., Pinto, P., Silva, J. A., & Woosnam, K. M. (2018). Examining the predictive validity of SUS-TAS with maximum parsimony in developing island countries. *Journal of Sustainable Tourism*, *26*(3), 379–398. doi:10.1080/09669582.2017.1355918

Ribeiro, M. A., Valle, P. O., & Silva, J. A. (2013). Residents' Attitudes towards Tourism Development in Cape Verde Islands. *Tourism Geographies*, *15*(4), 654–679. doi:10.1080/14616688.2013.769022

Ribeiro, M. A., Woosnam, K. M., Pinto, P., & Silva, J. A. (2018). Tourists' destination loyalty through emotional solidarity with residents: An integrative moderated mediation model. *Journal of Travel Research*, *57*(3), 279–295. doi:10.1177/0047287517699089

Ridderstaat, J., Croes, R., & Nijkamp, P. (2016). A two-way causal chain between tourism development and quality of life in a small island destination: An empirical analysis. *Journal of Sustainable Tourism*, *24*(10), 1461–1479. doi:10.1080/09669582.2015.1122016

Seetanah, B. (2011). Assessing the dynamic economic impact of tourism for island economies. *Annals of Tourism Research*, *38*(1), 291–308. doi:10.1016/j.annals.2010.08.009

Sirgy, M. J. (2010). Toward a quality-of-life theory of leisure travel satisfaction. *Journal of Travel Research*, *49*(2), 246–260. doi:10.1177/0047287509337416

Stylidis, D., Biran, A., Sit, J., & Szivas, E. M. (2014). Residents' support for tourism development: The role of residents' place image and perceived tourism impacts. *Tourism Management*, *45*, 260–274. doi:10.1016/j.tourman.2014.05.006

Su, M. M., Wall, G., & Xu, K. (2016). Heritage tourism and livelihood sustainability of a resettled rural community: Mount Sanqingshan World Heritage Site, China. *Journal of Sustainable Tourism*, *24*(5), 735–757. doi:10.1080/09669582.2015.1085868

Su, X., & Teo, P. (2008). Tourism politics in Lijiang, China: An analysis of state and local interactions in tourism development. *Tourism Geographies*, *10*(2), 150–168. doi:10.1080/14616680801999992

Suh, E., Diener, E., Oishi, S., & Triandis, H. C. (1998). The shifting basis of life satisfaction judgments across cultures: Emotions versus norms. *Journal of Personality and Social Psychology*, *74*(2), 482–493. doi:10.1037/0022-3514.74.2.482

Uysal, M., & Sirgy, M. J. (2019). Quality-of-life indicators as performance measures. *Annals of Tourism Research*, *76*, 291–300. doi:10.1016/j.annals.2018.12.016

Woo, E., Kim, H., & Uysal, M. (2015). Life satisfaction and support for tourism development. *Annals of Tourism Research*, *50*, 84–97. doi:10.1016/j.annals.2014.11.001

Woosnam, K. M., Maruyama, N., Boley, B. B., & Erul, E. (2018). Stereotypes and perceived solidarity in ethnic enclave tourism. *Journal of Tourism and Cultural Change*, *16*(2), 138–154. doi:10.1080/14766825.2016.1242596

Yu, C.-P., Charles Chancellor, H., & Tian Cole, S. (2011). Examining the effects of tourism impacts on resident quality of life: Evidence from rural midwestern communities in USA. *International Journal of Tourism Sciences*, *11*(2), 161–186. doi:10.1080/15980634.2011.11434643

ADDITIONAL READING

Anglim, J., Horwood, S., Smillie, L. D., Marrero, R. J., & Wood, J. K. (2020). Predicting psychological and subjective well-being from personality: A meta-analysis. *Psychological Bulletin, 146*(4), 279–323. doi:10.1037/bul0000226 PMID:31944795

Pyke, J., Pyke, S., & Watuwa, R. (2019). Social tourism and well-being in a first nation community. *Annals of Tourism Research, 77*, 38–48. doi:10.1016/j.annals.2019.04.013

Saayman, M., Li, G., Uysal, M., & Song, H. (2018). Tourist satisfaction and subjective well-being: An index approach. *International Journal of Tourism Research, 20*(3), 388–399. doi:10.1002/jtr.2190

Sirgy, J. M. (2019). Promoting quality-of-life and well-being research in hospitality and tourism AU - Joseph Sirgy, M. *Journal of Travel & Tourism Marketing, 36*(1), 1–13. doi:10.1080/10548408.2018.1526757

Wang, Y.-C., Liu, C.-R., Huang, W.-S., & Chen, S.-P. (2019). Destination Fascination and Destination Loyalty: Subjective Well-Being and Destination Attachment as Mediators. *Journal of Travel Research.* Advance online publication. doi:10.1177/0047287519839777

Woosnam, K. M., Erul, E., & Ribeiro, M. A. (2017). Heterogeneous community perspectives of emotional solidarity with tourists: Considering Antalya, Turkey. *International Journal of Tourism Research, 19*(6), 639–647. doi:10.1002/jtr.2136

Woosnam, K. M., Maruyama, N. U., Ribeiro, M. A., & Joo, D. (2019). Explaining minority residents' attitudes of ethnic enclave tourism from general perceptions of tourism impacts. *Journal of Tourism and Cultural Change, 17*(4), 467–484. doi:10.1080/14766825.2019.1601407

Yolal, M., Gursoy, D., Uysal, M., Kim, H., & Karacaoğlu, S. (2016). Impacts of festivals and events on residents' well-being. *Annals of Tourism Research, 61*, 1–18. doi:10.1016/j.annals.2016.07.008

KEY TERMS AND DEFINITIONS

Affective Wellbeing: Refers to the frequency and intensity of positive and negative emotions and mood.

Cognitive Wellbeing: Refers to domain-specific and global evaluations of life.

Quality of Life: As an individual's perception of their position in life.

Social Exchange Theory: Is a general sociological theory concerned with understanding the exchange of resources between individuals and groups in an interaction situation.

Subjective Wellbeing: Describes how people experience the quality of their lives and includes both affective reactions and cognitive judgments.

Support for Tourism: Is the degree to which residents avaluate the effects of tourism in their community.

Tourism Impacts: Include the effects of tourism on the environment and on destination communities, and its economic contributions.

Chapter 5
Residents' Support for Sustainable Tourism Development in Rural Areas:
The Case of Karuizawa, Japan

Manuel Alector Ribeiro
University of Surrey, UK

Mayu Moritomo
KCS Co. LTD Consulting, Japan

Kyle Maurice Woosnam
University of Georgia, USA & University of Johannesburg, South Africa

ABSTRACT

The present study aims to examine residents' support for sustainable tourism. Karuizawa town was selected due to the increasing number in both domestic and foreign visitors, as well as the existence of seasonal residents. Residents' participation in tourism and their support are vital for sustainable tourism. In doing so, this study primarily used the theory of reasoned action, social exchange theory, and emotional solidarity to examine residents' support for sustainable tourism. Data were collected using qualitative approach, and semi-structured interviews were adopted to explore the relationship between theoretical frameworks. Twenty participants were selected by non-probability sampling to examine differences among residents in tourism-related businesses and those who are not, and among permanent and seasonal residents. Deductive thematic analysis was chosen because of the ability to produce themes from theoretical frameworks. Lastly, an in-depth discussion, as well as theoretical and managerial implications are offered along with the limitations and avenues for future studies.

DOI: 10.4018/978-1-7998-3156-3.ch005

INTRODUCTION

International tourist numbers have ballooned from 435 million in 1990 to 1,322 million in 2017 (UNWTO, 2018). This trend is especially significant in Asia. Although the number of international tourists to Asia and the Pacific (323.7 million in 2017) is roughly half of the number to Europe (671.1 million in 2017), the growth rate since 2000 has been highest in Asia and the Pacific (2.93 times) compared to the Middle East (2.60 times), Africa (2.37 times), Europe (1.71 times), and the Americas (1.61 times) (UNWTO, 2018). Japan is no exception to this dramatic tourism growth. Inbound tourism is developing dramatically, especially from 2014 to 2017, in contrast to the slow growth of outbound tourism (Japan National Tourism Organization or JNTO, 2017). Given the population in Japan is declining (and this trend is especially true in rural areas compared to metropolitan regions), the potential for inbound tourism is high for various rural areas. This is evidenced by the fact that the promotion of inbound tourism and rural rejuvenation through tourism are currently two core policies for the Japanese government (MLIT, 2018a).

To promote inbound tourism, the Japanese government is keen on developing inbound tourism sustainably. For instance, the Ministry of Land, Infrastructure and Transport (MLIT) in Japan has conducted several studies to understand sustainable tourism development from the residents' perspectives by reviewing the international transition of sustainable tourism and international case studies (MLIT, 2018). However, in practice, it seems that the government has not yet entirely implemented sustainable inbound tourism promotion. For example, the main inbound tourism policies established by MLIT appear to be for the preparation of necessary systems that serve to attract international visitors (e.g., investment, laws, DMO establishment, Wi-Fi maintenance, and multilingual services) (MLIT, 2018). Notably, it appears that the benefits for the communities and their involvement in tourism development, which can be essential elements for realising sustainable tourism (Berlin Declaration, 1997; Choi & Shirakaya, 2005; Kitnuntaviwat & Tang, 2008; The United Nations Environment Programme (UNEP) and the World Tourism Organization (WTO), 2005), are not emphasised by the government (MLIT, 2018a). With this in mind, the present study provides more 'sustainable' tourism policies than current ones for Japanese rural areas, including inbound tourism, which are based on local support for and participation in the tourism development. Based on the background and rationale, the aim of the present study is to examine residents' attitudes and their support for sustainable tourism development in a rural area in Japan. Specifically, this study examines four aspects from residents' perspective--perceptions of tourism impacts, feelings towards visitors, support for, and participation in sustainable tourism development.

Study Location and Context

Karuizawa town in the Nagano prefecture was selected as the study location for this research. Three reasons exist for this selection. First, since Karuizawa town has been a popular domestic tourism destination for nearly a century (Uchida, 1989), its residents likely have specific feelings towards visitors and have established perceptions of tourism impacts. Second, the number of international visitors is growing substantially in the town, as demonstrated by the fact that it has hosted the second-largest number of visitors in Nagano prefecture in 2016 (Nagano Prefectural Government, 2017), which leads to a better understanding of residents' feelings towards international visitors and perceptions of tourism impacts. Furthermore, Karuizawa town has two types of residents; permanent and seasonal residents (Takahashi, 2013). Thus, the comparison between permanent and seasonal residents can provide more rigorous suggestions for sustainable rural tourism policies, because the national government encourages people living in metropolitan cities to move to rural areas (Tsutsui, Sakuma & Kasami, 2015).

LITERATURE REVIEW

Residents' Support for Sustainable Tourism Development

It can be argued that the notion of sustainable tourism development is increasingly recognised as one of the important measures for promoting tourism. This concept is derived from sustainable development. The emergence of the sustainability concept is said to be from 1850s in Western countries due to the issues from modernisation (Choi & Sirakaya, 2005), such as the environmental destruction and social inequalities (Sachs, 1999, cited in Choi & Sirakaya, 2005). Subsequently, the World Commission on Environment and Development (WCED) defined sustainable development as 'development that meets the needs of the present without compromising the ability of future generations to meet their own needs' (WCED, 1987, p.41). The WCED (1987) contends that the main aspects of sustainable development are the three pillars of sustainability (i.e., economic, social, and environmental) (Blanchard & Buchs, 2015), equality, long-term growth and a unified approach (Hall & Lew, 1998). The concept of sustainable development is increasingly used by domestic, regional and international policy-makers (Auzins & Vanags, 2012).

Following the late 1970s, the notion of sustainable tourism gained attention given the widespread awareness of negative consequences of tourism (Choi & Shirakaya, 2005; Ribeiro, Pinto, Silva, & Woosnam, 2018). Sustainable tourism also has analogous elements with sustainable development. According to the UNEP and the WTO (2005, p.12), sustainable tourism refers to 'Tourism that takes full account of its current and future economic, social and environmental impacts, addressing the needs of visitors, the industry, the environment and host communities'. Similarly, the Berlin Declaration (1997) states that sustainable tourism should be achieved through benefitting local economies, ensuring environmental sustainability, and respecting local cultures. Also, the unified approach should be achieved among visitors, the tourism industry, and host societies (Sharpley, 2000). In doing so, these three stakeholders should have equal power and responsibility in fostering sustainable tourism (Lane, 1994). These elements are harmonious with the aspects of sustainable development. Thus, it can be argued that the stakeholders with equal power should cooperate to maximise positive economic, socio-cultural and environmental impacts on host societies and visitor satisfaction in the long-term. However, host community members tend to be in a vulnerable position in the tourism decision-making process (Joppe, 1996). It is therefore important to promote community members' involvement in sustainable tourism (Choi & Shirakaya, 2005; Kitnuntaviwat & Tang, 2008).

In terms of research on community support for tourism development, a large number of studies are conducted through quantitative approach, testing extant theories and theoretical frameworks (e.g., Chen & Chen, 2010; Chen & Raab, 2012; Ribeiro, Pinto, Silva, & Woosnam, 2017). The primary theories regarding this topic are the theory of reasoned action (TRA) (e.g., Chen & Raab, 2012; Lepp, 2007; Ribeiro et al., 2017), social exchange theory (SET) (e.g., Chen & Chen, 2010; Chen & Raab, 2012; Gursoy & Rutherford, 2004; Jurowski, Uysal, & Williams, 1997; Kitnuntaviwat & Tang, 2008; Lee, 2013; Nunkoo & Gursoy, 2012; Nunkoo & Ramkissoon, 2011; Ribeiro et al., 2017), and emotional solidarity (e.g., Woosnam & Norman, 2010). SET and emotional solidarity scale (ESS) are used to examine residents' perceptions of tourism impacts and host-guest relationship, respectively (Nunkoo et al., 2013; Woosnam & Norman, 2010).

TRA can be used to clarify the differences between attitudes (i.e., perceptions) and behaviour (i.e., support) by residents. It is important to differentiate these two elements. Judging from Ribeiro et al. (2017), who argued that some studies used 'attitudes' and others used 'behavioural intentions' to calculate community support. The determinants of TRA mainly consist of subjective norms, attitudes, intentions, and behaviours (Fishbein & Ajzen, 1975). In this theory, individual's 'behaviour' is determined by 'intention' (Fishbein and Ajzen, 1975, cited in Chen & Raab, 2012), which is the possibility of whether an individual performs an action or not (Zhang, Inbakaran, & Jackson, 2006). Also, 'intention' is affected by positive or negative 'attitudes' and 'subjective norms', which provides the external pressure to determine whether to perform or not (Zhang et al. 2006). Thus, the combination of SET and TRA is useful for predicting community perceptions of tourism impacts and support for tourism development, as Chen and Raab (2012) and Ribeiro et al. (2017) have advanced. TRA can be used to divide residents' perceptions and support. In contrast, SET can be used to analyse perceptions in greater depth. In the following section, the topic of residents' perceptions towards tourism impacts is explored.

Residents' Perceptions Towards Tourism Impacts

Though the ratio of research surrounding residents' perceptions of tourism impacts conducted without theories is greater (54.3%) than with theories (45.7%), more work as of late is utilizing theoretical frameworks (Nunkoo, Smith, & Ramkissoon, 2013). For this topic, SET is the most widely tested theory (56.3%) (Nunkoo et al., 2013). Following SET, Butler's (1980) Tourist Area Life Cycle (TALC) model (18.8%) and Doxey's (1975) irritation index (11.0%) are used (Nunkoo et al., 2013). However, some researchers question the usefulness of these theories. Lepp (2007) claims that although the Doxey's (1975) theory argues that residents tend to welcome tourists at the initial phase (euphoria) and then move on to apathy (tourists are recognised to be usual), irritation (residents doubt tourism), and finally, antagonism, recognising tourists as problematic (Hunt & Stronza, 2014), the residents showed opposite reactions, such as concern and scepticism in the early stages. Also, Ritchie and Inkari (2006) argue that this theory is oversimplified. In the TALC model, a tourist destination is divided into 'exploration', 'involvement', 'development', 'consolidation', 'stagnation', 'decline' and 'rejuvenation' stages (Butler, 1980) and the negative perceptions by residents is evident in the latter stages (Butler, 1980, cited in Sharpley, 2014). Nevertheless, this theory is also questioned due to the lack of premise that community members are heterogeneous (Sharpley, 2014). Despite some scepticism of SET (García, Vázquez & Macías, 2015; Nunkoo & Gursoy, 2011), this theory is said to be the common theoretical principle (Gursoy & Rutherford, 2004; Lee, Kang, Long, & Reisinger, 2010).

Considering the frequency of testing and its validity, SET is generally the most accepted for this topic. Ap (1992, p.668) defines SET as 'a general sociological theory concerned with understanding the exchange of resources between individuals and groups in an interaction situation.' In a tourism context, the residents who perceive benefits of tourism tend to have positive perceptions of tourism (Lee et al., 2010) and support for tourism development (Látková & Vogt, 2012; Nunkoo & Ramkissoon, 2011). Thus, in many studies, SET, which studies residents' positive (benefits) and negative (costs) perceptions, is used as antecedents for examining support level (Chen & Raab, 2012; Kitnuntaviwat & Tang, 2008; Lee, 2013; Nunkoo & Gursoy, 2011; Ribeiro et al., 2017; Stylidis, Biran, Sit, & Szivas, 2014). Specifically, Gursoy and Rutherford (2004) and Jurowski et al. (1997) divided the benefits and costs into economic, social and environmental, and economic, social and cultural perspectives, respectively, which reflect the three pillars of sustainability (Blanchard & Buchs, 2015).

Residents' Perceptions Towards Positive Tourism Impacts

As for the findings for the residents' perceived tourism impacts, it can be effective to classify the impacts into economic, socio-cultural and environmental dimensions, which are three pillars for sustainable tourism (Berlin Declaration, 1997; UNEP and WTO, 2005). In terms of positive economic effects, Kayat, Sharif and Karnchanan (2013) divide the perceived economic impacts by residents into individual and communal aspects. In accordance with this classification, improvement of residents' income (Gursoy & Rutherford, 2004; Pratt, 2015) and job creation (Andereck & Nyaupane, 2011; McGehee & Andereck, 2004) are more closely related to individually-perceived economic impacts. As for the communal one, the increase in the share of merchandizes and services by local corporations (Pratt, 2015) as a result of generation of additional businesses and investment (Dyer, Gursoy, Sharma, & Carter, 2007; Lindberg & Johnson, 1997), reinforced connection between the tourism industry and supply chain (Pratt, 2015), earnings by tax (Chen & Chen, 2010; Pratt, 2015) and by exchange rates (Pratt, 2015), and the increase in government's receipts (Gursoy & Rutherford, 2004; Kayat et al., 2013) are argued. As for the research about positive economic impacts in Japan, Ishikawa and Fukushige (2007) assess the multiplier effect by tourism in Amami islands, suggesting that approximately US$480 is spent per tourist. However, they do not state the effects on residents. However, Maruyama, Woosnam and Boley (2017) examine the differences between Japanese and Brazilian residents' perceptions towards 'job opportunities by tourism' in Oizumi Town, suggesting that Brazilians tend to perceive job opportunities created by tourism industry more than Japanese. In the case of Shirakawa village, approximately half (47.3%) of the residents perceived tourism as a contributor to 'businesses diversification' (Jimura, 2011).

As for sociocultural benefits, creating additional recreational opportunities (Andereck & Nyaupane, 2011), such as foods (Fredline, 2002), enhancing local pride (Besculides, Lee, & McCormick, 2002) and identity (Esman, 1984), advancing living standards (Perdue, Long and Allen, 1987), stimulating understanding (Dyer et al., 2007), specifically among hosts and guests (Korça, 1996) can be categorised as individual impacts. The maintenance of public infrastructures (Korça, 1996), advancing public security services (Vargas-Sánchez, Plaza-Mejia, & Porras-Bueno, 2009), reinvigoration of local arts (McKean, 1989), and appearance of the local areas (Perdue et al., 1987), preserving cultures (Andereck & Nyaupane, 2011; Brunt & Courtney, 1999) can be categorised as communal ones (Kayat et al., 2013). In the case of Japan, 'recreational opportunities', 'infrastructure improvement', and 'advancement of living standards' are tested by Maruyama et al. (2017), identifying that the same results as economic impacts (i.e., Brazilians have the tendency to perceive more positive socio-cultural impacts). In addition, Jimura (2011) states that the majority (64.0%) of residents considered the 'degree of pride' about their residential area has increased. Finally, Radzuan, Fukami, & Ahmad (2015) found that the vast majority of locals living in Ainokura (75%, 67%) and Kawagoe (98%, 98%) think that it is important to protect both material and immaterial heritage. As for environmental ones, Kayat et al. (2013) argue that the environmental impacts are perceived collectively rather than individually. This is evident in the findings from the existing research, such as the preservation of nature (Andereck & Nyaupane, 2011; Andereck et al., 2005).

Residents' Perceptions of Negative Tourism Impacts

In terms of perceived negative economic impacts, the rise in cost of living (Johnson, Snepenger and Akis, 1994; Long, Perdue and Allen, 1990; McGehee and Andereck, 2004) and taxes (Chen & Chen, 2010), overreliance on tourism industry (Mathieson & Wall, 1982), economic benefits not being evenly distributed to the community (Johnson et al., 1994), and exploitation of economic benefits by external agents (Pratt, 2015) are claimed. Here, these findings can be also classified into individual (rise of the cost of living and taxes) and communal (overreliance on tourism industry, exploitation of economic benefits by external agents, and seasonality) perceptions, as Kayat et al. (2013) state.

Furthermore, in addition to the rise of immoral activities (Andereck et al., 2005; Dyer et al., 2007; Tosun, 2002), such as theft (Belisle & Hoy, 1980), drug usage (Diedrich & García-Buades, 2009; Fredline, 2002; Tosun, 2002), and prostitution (Yan, Xu, & Zhou, 2018), overpopulation (Andereck et al., 2005; Brunt & Courtney, 1999), traffic issues (Andereck et al., 2005; Brunt and Courtney, 1999; Dyer et al., 2007; Fredline, 2002; Mason & Cheyne, 2000; McGehee & Andereck, 2004), noise (Dyer et al., 2007; Mason & Cheyne, 2000), unwelcoming feelings by residents (Liu & Var, 1986), and displacement from initial residential areas (Young, 1983) are claimed as social costs. In relation to cultural ones, transformation of indigenous cultures (Akis, Peristianis & Warner, 1996) were argued. In the case of Japan, Jimura (2011) identified that some (31.3%) residents reported 'disruption of their privacy' by tourists and about half (47.3%) of the locals said, 'community unity had deteriorated'.

For environmental impacts, increases in pollution (Dyer et al., 2007; Eshliki, & Kaboudi, 2017) and garbage (Andereck et al., 2005; Brunt and Courtney, 1999), and the destruction of flora and fauna (Andereck, 1995, cited in Andereck et al., 2005) are confirmed, which can be classified as communal impacts (Kayat et al., 2013). From the above discussion, it can be argued that the perceived impacts of tourism tend to vary according to research location, as well as with the variables used to measure the impacts.

Residents' Feelings Towards Visitors

In addition to residents' support and perceptions, research examining residents' feelings towards visitors has grown in popularity. Indeed, the in-depth comprehension of the host-guest relationship is essential for sustainable tourism (Zhang et al., 2006).

In terms of the theories regarding this topic, Doxey's (1975) irritation index, and Ap and Crompton's (1993) model for resident reactions to tourism are used. As suggested above, the irritation index represents the change of residents' feelings to tourists from 'euphoria' to 'antagonism', as the tourism develops (Hunt & Stronza, 2014). Likewise, Ap and Crompton (1993) have divided residents' reaction to tourism into four phases (i.e., embracement, tolerance, adjustment and withdrawal) (Hunt & Stronza, 2014). The common point of these theories is that as tourism develops, the contact between tourists and residents tends to become unfavourable (Hunt & Stronza, 2014). In addition, Woosnam (2011) argues that social distance, intimacy theory, and contact-hypothesis theory are used to test the level of relationship between hosts and guests. However, these theories can lack the ability to calculate residents' individual and emotional dimensions. Indeed, Woosnam (2012) claims that the perspectives of how residents feel about tourists individually are absent in the existing studies. Furthermore, existing research lacks a deeper and more complex understanding of the relationship (Woosnam, 2011).

To fill in this research gap, the theory of emotional solidarity could be effective. This theory is derived from Durkheim (1915), which focuses on figuring out the largest basic elements of religion (Woosnam,

2011; Woosnam, Norman, & Ying, 2009). The research suggests that sharing 'beliefs' and 'behaviour' among followers of the religion have functions to enhance solidarity (Woosnam, 2011; Woosnam et al., 2009). Subsequently, Collins (1975) added 'interaction' as a component of solidarity (Woosnam et al., 2009). Thus, 'shared belief', 'shared behaviour', and 'interaction' are considered main components of emotional solidarity (Woosnam et al., 2009). This theory is used in other disciplines, such as anthropology, sociology and family research (Woosnam et al., 2009). As of late, more attention has been paid to the construct within the field of tourism (Joo & Woosnam, 2019; Woosnam et al., 2009), despite the recognition of the importance of emotional solidarity to form residents' perceptions (Boley, McGehee, Perdue, & Long, 2014). Based on this, Woosnam et al. (2009) and Woosnam (2011) have investigated the validity of the emotional solidarity in the context of tourism, focusing on correlations between hosts and guests. As a result, both studies have revealed that 'shared beliefs', 'shared behaviours', and 'interaction' can be effective antecedents for residents' emotional solidarity with tourists.

For more specific tool based on this theory, ESS, which is based on three dimensions of welcoming nature (e.g., feeling proud to welcome visitor, visitors' contributions to economy), emotional closeness (e.g., friendship with visitors), and sympathetic understanding (e.g., affinities to visitors) (Woosnam & Norman, 2010), and reliability and validity of this scale was confirmed by Woosnam and Norman (2010). According to Li and Wan (2017), Woosnam (2012), and Woosnam et al. (2016), the more residents feel ESS, the more they support tourism development. In addition, the scale has gained recognition for promoting affective correlation between hosts and guests (Woosnam, Aleshinloye, & Maruyama, 2016), which was lacking in previous studies, as Woosnam and Norman (2010) suggest. As for findings using this scale, welcoming nature (Hasani, Moghavvemi, & Hamzah, 2016; Moghavvemi, Woosnam, Paramanathan, Musa, & Hamzah, 2017) and sympathetic understanding (Moghavvemi et al., 2017) can predict residents' perceptions towards tourism, in the case of Malaysia. Also, Moghavvemi et al., (2017) identified that these degrees are different depending on genders and individual characters. Furthermore, Li and Wan (2017) concluded that emotional solidarity can be a predictor of residents' perceived benefits and support, but not perceived costs, in the case of Macao. Woosnam et al. (2016) examined the construct from the tourists' vantage point, suggesting that the scale can dictate their perceptions of safety and repeat visitation in the WHS in Nigeria. From these outcomes, overall, ESS can be used to predict the level of residents' perceived tourism impacts and support for sustainable tourism development.

METHODOLOGY

Research Questions

Both ESS and SET are effective theories to measure host-guest relationship and residents' perceptions of tourism impacts, respectively. Also, these theories can influence the level of support (i.e., behaviour), which should be clearly distinguished from perceptions (i.e., attitudes), according to TRA. However, the existing research tends not to study how support affects participation (e.g., Andereck et al., 2005; Besculides et al., 2002; Chen & Chen, 2010; Chen & Raab, 2012; Eshliki & Kaboudi, 2017; Lee, 2013; Lee et al., 2010; Lepp, 2007), which is considered the final form residents. Although Eshliki and Kaboudi (2017) and Lepp (2007) did study the effect of residents' perceptions on their participation, their research lacks the relationship between support for tourism development and participation in tourism development. Thus, for the present research, the theoretical model which includes participation is proposed on Figure 1. According to this model, the research questions are provided. It should be noted that this research explores the effects of ESS and SET on support, and the effects of support on participation, rather than examining the relationship among each element. This is because the present research does not employ a statistical approach.

Figure 1. Theoretical model for the present research

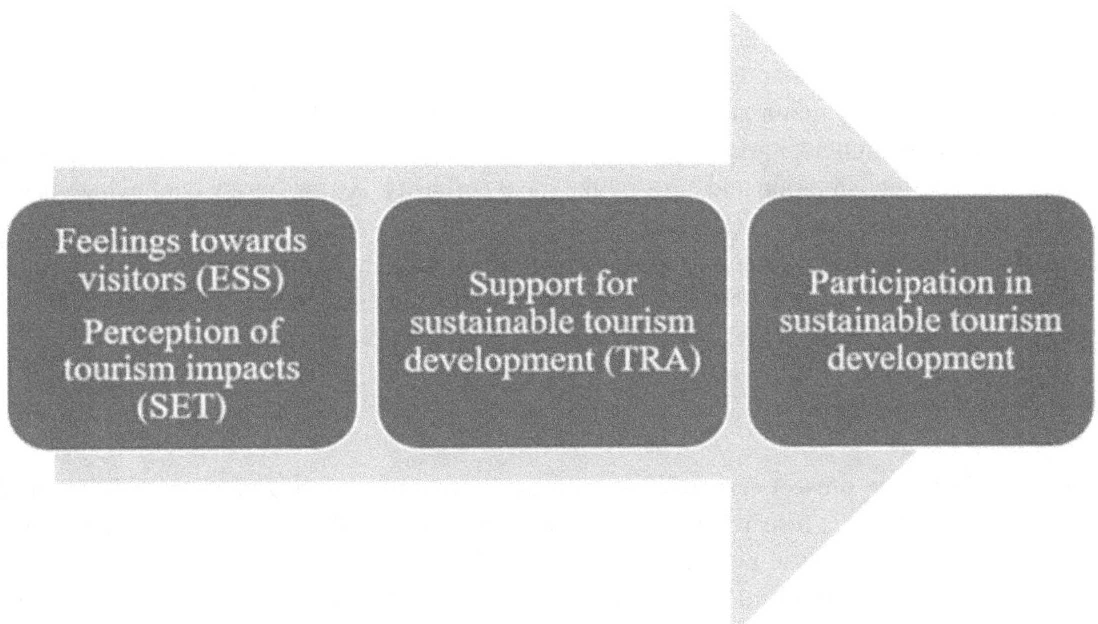

In addition, this research seeks to consider the differences between tourism-related businesses and non-tourism-related businesses (i.e., 'degree of benefits from tourism' (paralleling the work of Andereck et al., 2005) and between permanent and seasonal residents (i.e., 'length of stay' (parallels with Davis et al., 1988; Haley et al., 2005)). Furthermore, since the increase of international visitors is a highly recent phenomenon, it is interesting to explore the differences between tourism overall and inbound tourism. From these contexts, the four research questions posed in this research are:

1. How do residents' feelings to visitors and perceived tourism impacts different depending on industries, and between permanent and seasonal residents in Karuizawa? (theories: ESS, SET) (Objective 1);
2. How do residents' feelings toward visitors and perceived tourism impacts different between tourism overall and inbound tourism in Karuizawa? (theories: ESS, SET) (Objective 1);
3. How do residents' feelings toward international visitors and perceived inbound tourism impacts affect the level of their support for inbound tourism development? (theories: ESS, SET, TRA) (Objective 2);

4. How do residents' feelings toward visitors and perceived tourism impacts affect the level of their support for and participation in sustainable tourism development in Karuizawa? (theories: ESS, SET, TRA) (Objectives 2).

Data Collection Procedure

Secondary Data Collection

According to Saunders, Lewis and Thronhill (2011), cited in Saunders, Lewis, and Thornhill (2012), there are three types of secondary data: documentary data, survey data, and multiple sources. Documentary data refers to both text and non-text data (Saunders, Lewis and Thronhill, 2011, cited in Saunders et al., 2012), and this research uses text documentary data, such as tourism strategies in Karuizawa town, to understand the current situation and aims by the local authority. As for survey data, which is mainly based on the form of questionnaires (Saunders et al., 2012), the national government's census is used for gaining statistical information, such as population and industries in Karuizawa town. Furthermore, in terms of multiple sources, which various data from different sources are mixed (Saunders et al., 2012), academic journals are utilised to review extant theories and research outcomes.

Primary Data Collection

As for primary data, the present research adopts a qualitative approach. This is due to the need for a qualitative method for this research field. Horn and Simmons (2002) argue that it is difficult for a quantitative approach to analyse peoples' perceptions of tourism from wider societal backgrounds. However, the qualitative method enables it (Nunkoo et al., 2013) through understanding the topic more extensively (Sharpley, 2014). Therefore, a qualitative approach can provide deeper understanding of this research field than quantitative one, which can be more suitable for suggesting some new theoretical insights and policy implications. Specifically, semi-structured interviews were utilized. Saunders et al. (2012) argue that there are mainly three approaches for the interview; structured, semi-structured, and unstructured or in-depth interview. The characteristics of each interview, which is classified according to the types of research, are provided in Table 1. Considering that the present research is explanatory in nature, which examines the correlations among variables (i.e. ESS, SET and TRA) (Saunders et al., 2012), semi-structured interviews are the most appropriate.

Table 1. Characteristics of structured, semi-structured and unstructured interviews

	Exploratory	Descriptive	Explanatory
Structured		√√	√
Semi-structured	√		√√
Unstructured	√√		

√√ = more frequent; √ = less frequent
Source: Adapted from *Saunders et al. (2012)*

Figure 2. Classification of seasonal residents

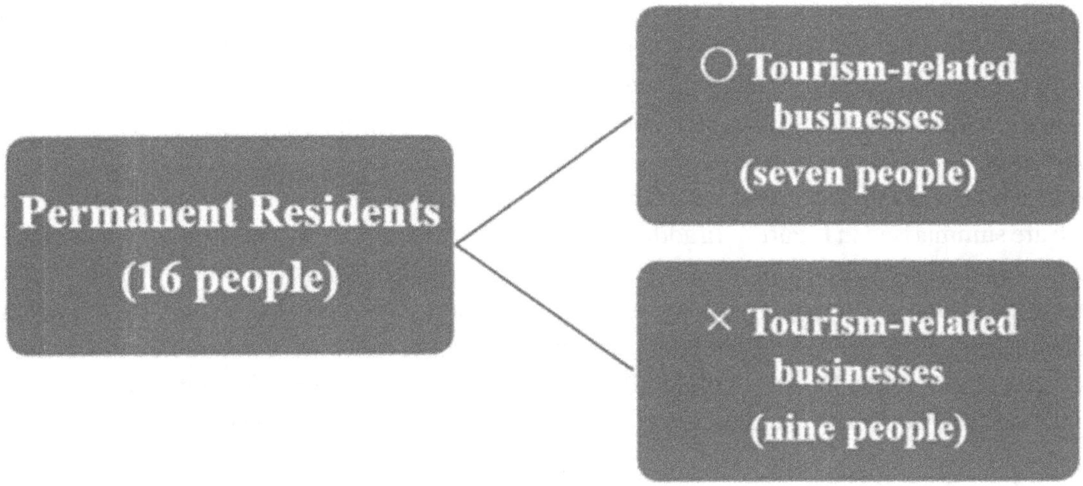

In terms of sampling, the present research adopted non-probability sampling, which allows intended sample selection (Saunders et al., 2012) to understand differences in opinions between residents in tourism-related businesses (i.e., businesses which provide opportunities to interact with visitors on a daily basis, thus hospitality industries and museum for the present research) and those who are not, and among permanent and seasonal residents. In the present study, 'permanent residents' refer to those who are permanently living in or around (in Nagano prefecture) Karuizawa town, and those who are currently have businesses in the town. The permanent residents are divided into those who are engaging in tourism-related businesses and those who are not. 'Seasonal residents' are those who are occasionally living in or around Karuizawa town and those who are paying taxes to the town. Likewise, the seasonal residents are classified into those in tourism-related industries and not. The classification of the samples is shown in Figures 2 and 3. Also, 'visitors' were defined as people who visit Karuizawa for a leisure

Figure 3. Classification of seasonal residents

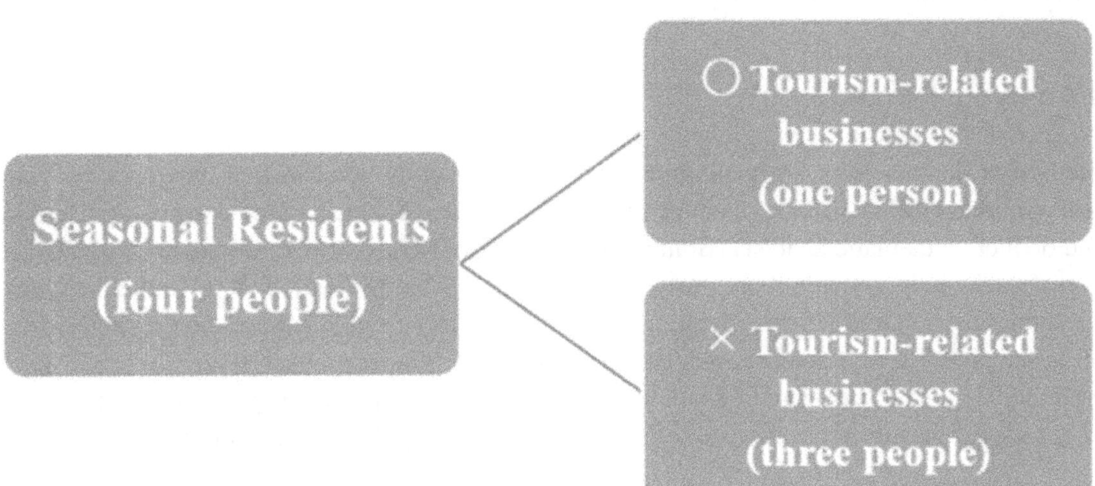

purpose and do not have villas. As for the number of participants, whilst it could be important to continue interviewing until researchers go to 'saturation point', which is the point to gain no newer information (Long, 2007), Saunders (2012), cited in Saunders et al., (2012) suggests that at least 5-25 interviewees should be necessary for semi-structured interviews. Overall, 20 residents were interviewed.

In terms of interview questions, each question was considered based on SET, ESS and TRA and Saunders et al.'s (2012) classification of question styles for a semi-structured interviewing (i.e., open-ended questions, probing questions, and specific and closed-ended questions). The characteristics of each question are summarised in Figure 4. In addition, referring to Mason (2014), the structure of the interview questions was made to begin with asking more straightforward questions (i.e., demographic questions) and end with asking participants' closing remarks. A pilot test was carried out with two residents to understand the content validity of the interview questions. Interviews were conducted in Japanese and then translated into English. Due to the limited time availability of the participants, interviews for the participants C and D, E and F, and M, N and O were conducted together, respectively.

Figure 4. Classification of question styles for a semi-structured interview
Source: *Adapted from Saunders et al. (2012)*

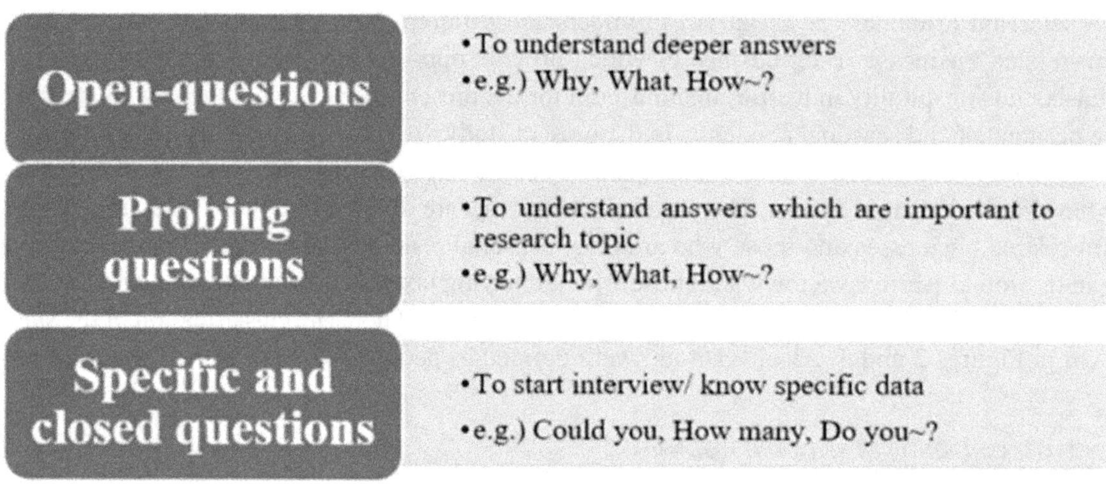

Data Analysis Procedure

The method of data analysis for qualitative approach can mainly be classified into six types; content analysis, grounded theory method, analytic induction, narrative analysis, discourse analysis, and inductive and deductive thematic analysis (Braun & Clarke, 2006; Mason, 2014; Saunders et al., 2012). The features of each analysis method are provided in Figure 5. Among them, deductive thematic analysis was deemed most suitable for this research, because this analytic style enables coding and generating themes based on theoretical frameworks (Braun & Clarke, 2006). Additionally, this analytical approach is most conducive in providing an in-depth understanding of residents' feelings and perceptions through asking quasi-predetermined questions which are based on existing theories. As for more specific analytic

Figure 5. Features of each analysis
Source: *adapted from Braun and Clarke (2006); Mason (2014); Saunders et al. (2012)*

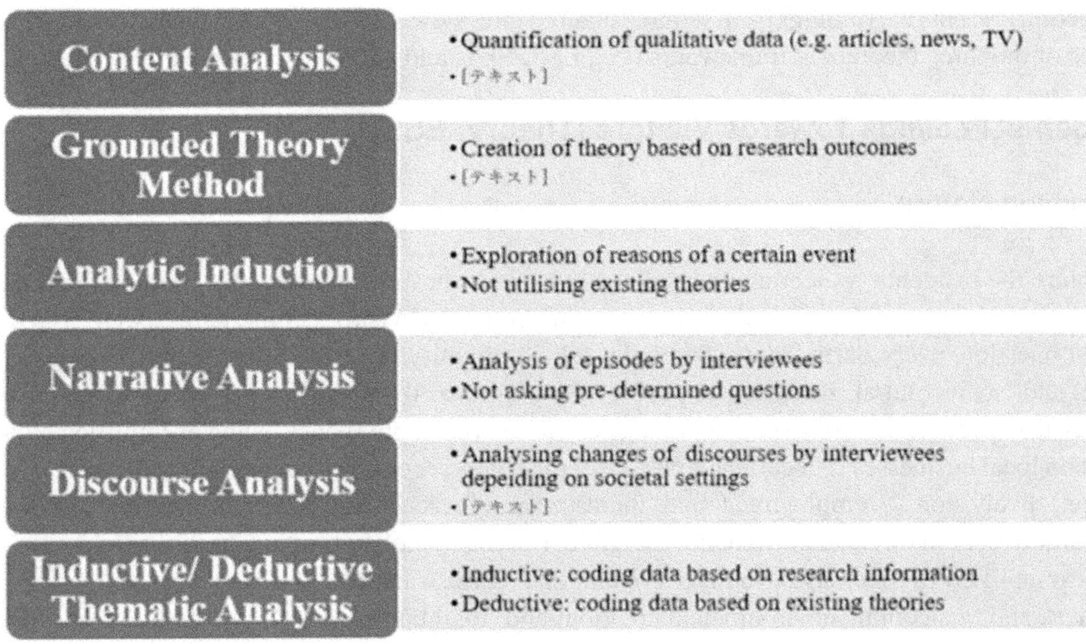

method of deductive thematic analysis, Braun and Clarke (2006) classified the analysis into five stages (see Figure 6). The present research follows this process for data analysis.

Figure 6. Five stages of deductive thematic analysis
Source: *Adapted from Braun and Clarke (2006)*

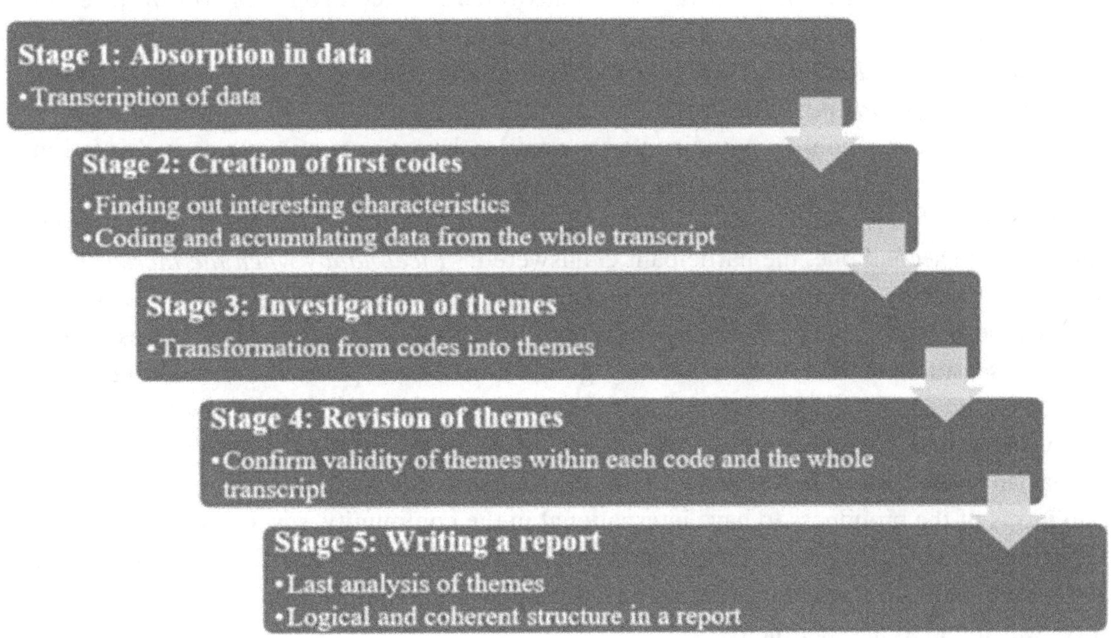

RESULTS

This section presents the findings from semi-structured interviews and thematic analysis, looking through the lens of the three theoretical frameworks (i.e., ESS, SET, and TRA) discussed earlier.

Residents' Feelings Towards Visitors (Theory: ESS)

Welcoming Nature

To explore the residents' welcoming nature, the interview questions 'Do you feel community benefits from having visitors in Karuizawa?' and 'Are you proud that visitors come here?' were asked. For the first question, every participant (20 people) felt community benefits. Specifically, 'economic' (14 people) and 'sociocultural' benefits (six people) were found as themes. Under them, economic benefits were divided into 'general' (five people) and 'specific' (seven people) ones, being exemplified by 'great impact on local businesses', 'benefits for public sector', 'good economic circulation', and 'increase town revenue', 'lively stores', 'employment' and 'increase sales for local businesses', respectively. Moreover, sociocultural benefits were classified into 'tangible' (three people) and 'intangible' (three people) ones, being exemplified by 'increased younger population', 'infrastructure improvement' and 'preservation activities', and 'welcoming environmental creation' and 'maintaining Karuizawa brand', respectively.

As for community benefits from international visitors, 13 participants felt community benefits, which can also be divided into 'economic' (seven people) and 'sociocultural' (three people) benefits. For the economic benefits, 'specific benefits' (four people), which are illustrated by 'economic benefits especially for tourism-related sectors', and 'increase employment', were found out. For sociocultural benefits, 'brand enhancement' (three people) was presented, which can be classified into 'review Karuizawa' and 'promotion worldwide by foreigners'. In terms of the first one, the participant H mentioned: "*I think it might be meaningful to receive foreign visitors because we can review Karuizawa's characteristics to convey its good points to them.*" Also, participant R answered the second one: *'Maybe the foreigners can promote Karuizawa all over the world. They sometimes ask me whether it is ok to take pictures and post them on social media, so I think it is a big advantage that they post the pictures to share them because it allows various people to know Karuizawa.'*

As for the second question, 13 participants answered to have proudness, which is divided into two themes (visitors' and residents' affection to Karuizawa). From the visitors' viewpoint, visitors' 'selection to Karuizawa', 'understanding of Karuizawa', and 'enjoyment in Karuizawa' were conveyed. As an example of the second one, the participant G answered: "*I feel happy when the increasing number of visitors come here and become to like this place through exploring good points of Karuizawa.*" For the residents' perspective, their 'good image towards Karuizawa' was presented. This was illustrated by 'premier four-season mountain resort', 'nature', 'good climate' and 'politician's visitation'.

Conversely, three participants answered no or unsure to feel proudness because of 'limited 'interaction with visitors' and 'limited interest in visitors', which were exemplified by 'usually living outside of Karuizawa', and 'just interest in Karuizawa' and 'limited participation in tourism development', respectively. In terms of the proudness to have international in the community, seven residents answered that they felt proud. Among them, the same theme (i.e., visitors' affection to Karuizawa) and examples (i.e., visitors' enjoyment in, and selecting Karuizawa) as visitors in general were mentioned. Also, 'increased visibility' was found as another theme.

Emotional Closeness

In exploring residents' emotional closeness with visitors, the question 'Do you feel close to and make friends with visitors who come here?' was asked. As for close feelings, six participants answered to have close feelings with visitors, with 'interaction with visitors through work' and 'sharing affection to Karuizawa' as reasons. For instance, participant R mentioned both reasons: "*Yes, I feel close to visitors. Because I can talk to them as customers. We can really enjoy talking about nature and climate here.*" In contrast, six participants answered not to feel close, whose main reasons were 'limited interactions' and 'recognising as customers'. As for the intimate feelings with international visitors, four participants felt close with such individuals. The main reason was 'selecting Karuizawa', which was also found as a reason for proudness. Moreover, the reason for one of the two participants who did not feel close to international visitors was due to 'language barriers'.

In regard to the establishment of friendships, four participants answered that they had made friends due to 'sharing common points', which was conveyed by comments concerning 'same birthplaces', 'visitors' repeat visitation', 'interaction with visitors', specifically 'through work' were main reasons. On the other hand, eight participants did not make friends, which was due to 'limited interaction', 'personality' and 'recognising as customers.' Additionally, one individual (e.g., participant S) claimed to have made friends with international visitors.

Sympathetic Understanding

Residents were also asked whether they had something in common with visitors as well as if they felt affection such individuals. For the first question, 10 participants answered to have common points, which were divided into three themes. One reason was 'sharing affection to Karuizawa', is exemplified by 'nature' and 'healing', which was also found as a reason for close feelings. The second and third reasons were 'both are outsiders' and 'feelings like visitors', respectively.

In contrast, five participants said they did not feel as though they had things in common with visitors. As a reason, participant E mentioned that he 'recognises visitors as customers', which is also found as the reason for not feeling an emotiona closeness with visitors. Moreover, two participants expressed ambivalence regarding commonality. According to these residents, although they can feel 'common points with visitors who like history or nature', there are 'no common points with visitors who go shopping'. Furthermore, five participants felt common they had things in common with international visitors, referring to the same reasons for commonality with tourists in general (i.e., 'sharing affection for Karuizawa' and 'both are outsiders').

As for the second question, 10 participants expressed feeling affection for visitors. One reason was 'sharing affection for Karuizawa', which was also found in emotional closeness and common points in sympathetic understanding. The other reasons were 'good impression of visitors', which is exemplified by visitors' 'politeness', 'enjoyment in Karuizawa' and 'contributions to the town'. In addition, 'doing business with visitors' was found as a third reason.

Conversely, four participants felt no affection or where unclear as to what "affection" meant. Moreover, two participants expressed ambivalent feelings for affection. Although both participants expressed 'affection for many visitors', they felt 'no affection for bad-mannered visitors.' Finally, five participants had affection for international visitors. As reasons, 'doing business with visitors' and 'sharing affection for Karuizawa', which were in accordance with the purposes for affection to visitors in general, as well as 'town rejuvenation', were found.

Residents' Perceptions Towards Tourism Impacts (Theory: SET)

Perception of Positive Tourism Impacts

To explore the residents' perceived benefits, the question 'Could you tell me your opinion about the major benefits you perceive from tourism?' was asked. Overall, the benefit most residents conveyed were related to economic aspects (16 people), compared with sociocultural (nine people) and environmental (one person). In terms of economic benefits, they were divided into general (eight people) (i.e., economic rejuvenation) and specific (eight people) ones. For specific ones, 'revenue for town' (i.e., communal benefits, parallels with findings from Gursoy and Rutherford (2004) and Kayat et al. (2013), 'employment' especially 'for the younger residents (i.e., individual benefits were in line with what Andereck and Nyaupane (2011), Maruyama et al. (2017), and McGehee and Andereck (2004) found. 'High-level income' (i.e., individual benefits, was similar to what Gursoy and Rutherford (2004) and Pratt (2015) reported. Concerning economic benefits from inbound tourism, three participants expressed specific economic benefits regarding 'decreased seasonality' and 'increased sales for local businesses' (i.e., community benefits as Pratt (2015) demonstrated).

As for sociocultural benefits, they were classified into tangible (six participants) and intangible (three participants). For tangible ones, 'increase of population' and 'town planning', were represented by 'good restrictions by town' (Vargas-Sánchez et al., 2009), 'preservation activities' (Andereck & Nyaupane, 2011; Brunt & Courtney, 1999; Radzuan et al., 2015), and 'infrastructure improvement' (Korça, 1996; Maruyama et al., 2017). For instance, the participant Q said the first example: *'I think Karuizawa is a town which develops a common-sense while attracting tourists, because of some restrictions by the town. For example, the stores cannot do businesses after 11 pm and it is prohibited to do sex businesses here.'*

Concerning intangible ones, 'environment to welcome visitors' and 'fostering host-guest interactions' (Dyer et al., 2007; Korça, 1996) were mentioned. In addition, nine participants communicated sociocultural benefits from inbound tourism, which were also divided into tangible (two people) and intangible (eight people) ones. In addition to 'increase of population', 'fostering host-guest interaction' and 'environment to welcome visitors' (which were the same as sociocultural benefits from tourism overall), 'internationalisation' and 'increased visibility worldwide' were mentioned. Regarding environmental benefits, one participant provided 'nature and animal preservation,' which is in line with what Andereck and Nyaupane (2011) and Andereck et al. (2005) highlighted in their works.

Perception of Negative Tourism Impacts

Next, residents were asked about major costs they perceived from tourism. Overall, sociocultural costs (eight people) were the most reported costs, over economic (two people), environmental (four people), and no costs (three people). Economic costs can be considered *current* (one person) and *future* (one person) costs. Participant C mentioned the current one: "*In Karuizawa, there was no bullet train station. So many people used to come here by car and stay at accommodations. [...] However, as the bullet train has been opened, the visitors who use accommodations have decreased. As a result, small accommodations have become bankruptcy.*" In contrast, participant S mentioned his concern for the future economy: "*There has been recently a lot of development of hotels and resort mansions, [...] and I do not think that the number of hotel rooms is sustainable over the long-term, and there is a plenty of hotel spaces available in Karuizawa. [...] So, by building more and more hotels, [...] probably they are going to destroy the economy in the town.*'

Participant F told about the 'disparity of economic benefits' (i.e., current cost): "*Many foreign visitors use bicycles for sightseeing here, so the rental bicycle shops are gaining a lot of benefits from them. [...] But they do not come to our stores so often.*" Such a perspective was also put forth in the work of Johnson et al. (1994). On the contrary, future concern for the 'overdependence on inbound tourism' (Mathieson and Wall, 1982) was mentioned by participant I: "*As many foreign visitors come here, many businesses can gain economic benefits now. But I am worried about these stores' businesses which currently depend too much on foreigners when foreign visitor number will go down in the future.*"

As for sociocultural costs, tangible (seven people) and intangible (two people) costs were confirmed. Explicitly, the tangible ones were represented by 'congestion' (Andereck et al., 2005; Brunt & Courtney, 1999) and 'visitor manners', which was exemplified by 'noise problem' (Dyer et al., 2007; Mason & Cheyne, 2000). For intangible ones, 'popularisation of Karizawa tourism' and 'deterioration of public security' were mentioned. In terms of the socio-cultural costs from inbound tourism, tangible costs were expressed by 11 participants, which were classified into 'visitor manners' and 'congestion'.

In terms of environmental costs, 'nature destruction' (Andereck et al., 2005) was mentioned for both from tourism overall (four people) and from inbound tourism (one people). Specifically, 'accommodation development', 'decreased trees', and 'litter problem', (Andereck et al., 2005; Brunt & Courtney, 1999) were provided.

Finally, three and seven participants said that there were no costs from tourism overall and from inbound tourism, respectively. For both types of tourism, 'no currently-perceived environmental costs', being illustrated by 'good public security', 'no environmental costs', and 'belief that tourism does not bring any costs' were communicated. For inbound tourism, 'good impression to foreign visitors' such as 'understanding of Japanese cultures', and 'recognition of international visitor as an important market' were mentioned.

Residents' Support for Sustainable Tourism Development (Theory: TRA)

To explore residents' support for sustainable tourism development, the following questions were asked: 'Do you support the Karuizawa town's policy to further increase the number of international visitors?;' 'Do you support the town's policy to preserve the local environment further?;' 'Do you support the town's policy to preserve the local cultures further?;' and 'Do you support further tourism industry development in Karuizawa?'

Thirteen participants answered in favour of supporting further inbound tourism development. 'Economic' (five people) and 'sociocultural' (five people) benefits were mentioned as central to their argument. This was conveyed by rationales such as, 'increase more customers', 'limited demand for domestic visitors', and 'good source of revenue', 'good image formation', 'preservation and succession of town', and 'importance of internationalisation.'

For those who do not support further inbound tourism (two people), 'language barriers' and 'concern over lack of inbound tourism policies' were the reasons. The participant S spoke to the second reason: "*The current policy is not clear. [...] I think I will support if it is linked to a vision, strategic goals, and strategic strategies and tactics which are all linked together, [...] but currently there is not.*" In addition, four participants expressed ambivalent feelings about this topic. One reason is the importance of 'development while solving visitor manners issues'. Another reason is that 'although inbound tourism is good for addressing seasonality, there is a risk for standardisation through receiving many foreign visitors'. The third reason provided was that participants will support further development if 'the town can balance revenue and expenditures'.

Moreover, 16 participants supported further environmental conservation. Specifically, the reasons were classified into 'affection for Karuizawa' (e.g., living in Karuizawa for a long time, affection for Karuizawa nature), 'for tourism' (e.g., recognition of nature as main tourist attraction), 'fear of over-construction' (e.g., decreased trees), and 'for good image formation' (e.g., for distinction from other places). In contrast, two participants did not support for it, because of 'the necessity to focus on current conservation' and 'doubt in the government's policies', which 'disconnects between its declaration to conserve nature and action to support constructions'. One participant mentioned ambivalent feelings, due to the necessity for 'balance between development and natural conservation'. As for further cultural conservation, 19 participants supported such endeavours. Reasons provided were: 'because of affection for Karuizawa', 'for tourism', 'for good image formation', and 'for deeper cultural understanding' was answered as reasons.

Finally, 11 participants supported further tourism industry development, with reasons divided into economic (i.e., economic rejuvenation; seven people) and sociocultural (i.e., good image formation; three people) aspects. Specifically, 'reducing seasonality', 'for distinction from other places' and 'increase recognition' were mentioned. However, eight participants were ambivalent for tourism industry development, because of the importance of 'development within the capacity of town', which is exemplified by 'balance between development, and natural and cultural conservation', 'balance between attracting visitors and seasonal residents', 'balance between revenue and expenditures', and 'mainly attracting visitors with a deeper understanding of Karuizawa'.

Residents' Participation in Sustainable Tourism Development

For identifying residents' level of participation in sustainable tourism development, three questions were asked: 'Are you currently participating in any activity related to tourism development in Karuizawa town?'; 'Are you currently participating in any environmental conservation activity?'; and 'Are you currently participating in any local cultural preservation activity?'. The summary of the participants' number regarding these questions was provided in Table 2 Reasons for not participating were, 'busy with other activities', 'retired', 'limited time to stay at Karuizawa', and 'limited promotion by town'.

DISCUSSION AND IMPLICATIONS

Theoretical Implications

Five notable theoretical implications exist for this work. First, although existing studies have attempted to find some attitudinal differences by considering numerous variables, for the present study, the differences of ESS and SET level were derived from different types of tourism (i.e., tourism overall and inbound tourism), which tends to be a lacking from extant research on the topic. Specifically, this research has demonstrated that residents' feelings of the existence of 'bad-mannered visitors' might contribute

Table 2. Local current participation for tourism development, environmental conservation and cultural conservation activities

	Tourism Development (TD)	Environmental Conservation (EC)	Cultural Conservation (CC)
YES	A, J, L, M, P, S	A, C, D, E, L, Q, S, T	P, Q, S
NO	B, C, D, E, G, H, I, K, N, O, Q, R	B, G, H, I, J, K, M, N, O, P, R	A, B, C, D, E, G, H, I, J, K, L, M, N, O, R, T

Note: participant F did not answer.

to individuals feeling less sympathetic understanding of international visitors, as well as, perceiving greater economic and sociocultural costs from inbound tourism. Second, using variables of 'length of stay' (Davis et al., 1988; Haley et al., 2005) and 'benefits from tourism' (Andereck et al., 2005) have provided new insights that 'self-consciousness of outsiders' for seasonal residents, whose length of stay tends to be shorter, can influence more common points with visitors (e.g., sympathetic understanding), 'recognising visitors as customers' can affect less emotional closeness, and the level of 'interaction' can influence all three components of ESS.

Third, although Nunkoo and Ramkissoon (2011) suggest that more perceived benefits can contribute to a greater degree of support for tourism development, the present research has found out that residents who perceived no costs were ambivalent or did not support inbound tourism development, mainly due to the town's policy. Therefore, there is a possibility that appropriate policy-making can enhance residents' support for tourism development. In addition, although the existing research discussed the effect of ESS on perceived benefits and costs (Li and Wan, 2017) and support for tourism development (Woosnam, 2012), the present research has explored of the impact of ESS on support for sustainable tourism development. As a result, this study has revealed that residents who have higher ESS level tend to support sustainable tourism development, which is in keeping with what Woosnam (2012) demonstrated. Finally, this study explored the effect of residents' support level on their participation, which tends to be the lacking area for the existing research (e.g., Chen and Chen, 2010; Chen & Raab, 2012; Lee, 2013; Ribeiro et al., 2017). Specifically, for the present research, despite the relatively high level of support, the participation level is low. Therefore, it can be said that the level of support does not always translate to high levels of participation.

Managerial Implications

As demonstrated from the findings, the level of the residents' support for cultural conservation was the highest (represented by all participants' support) among various options surrounding sustainable tourism development. Also, environmental conservation level tends to be high, due to the support by almost all participants and high awareness for environmental conservation for the participants who did not support. Therefore, the level of support for two, out of three, pillars of sustainable tourism appears to be high. As for the support for tourism industry development (i.e. economic pillar), the higher ESS levels the residents have, the more likely they will support sustainable tourism development. For inbound tourism, the participants whose ESS levels are lower tend not to support or are ambivalent about inbound tourism development. As for SET, it could be important to maximise benefits and minimise costs from tourism. This is because of the need for returning the profits to local society, which is an essential aspect of sustainable tourism (Berlin Declaration, 1997; UNEP & WTO, 2005). Thus, it is required to create policies which enhance residents' ESS level, maximise positive effects and minimise negative ones for promoting sustainable tourism, with utilising inbound tourism. Notably, creating specific and rigorous policies for inbound tourism is required because of the current weak strategies regarding this market (Karuizawa town, 2017) and the existence of residents who do not support inbound tourism development due to reservations about local government policies.

First, regulations should be created to encourage better visitor behaviours, through tourism taxes, for instance. The case in Bhutan shows that the country successfully preserved its environment with a developing economy by imposing tourism taxes on visitors (Schroeder, 2015). Therefore, tourism taxes can be a good option to preserve nature and cultures in Karuizawa, by enticing higher-end visitors. Subsequently, funds collected through such a tax can be used to create new events to increase interaction with domestic and international visitors, by utilising characteristics of Karuizawa. For example, hosting Karuizawa gourmet events and establishing restaurants adjacent to protected natural areas or historically and culturally significant homes (as aspects of sustainable tourism), may be effective. Since Karuizawa is a destination where Western cultures have mixed with Japanese ones due to the existence of Western villa owners (Uchida, 1989), Karuizawa could have a uniquely burgeoning culture, especially one centred on food. However, Karuizawa town currently does not have such types of events for promoting interaction with international visitors (Karuizawa Tourist Association, 2018). Therefore, it is advisable for the town to create such events, through using the tax for operation, marketing and translation services for international visitors as well. Also, it is important to encourage participation by residents at this stage. This is because the residents do not necessarily participate in sustainable tourism-related activities even if they support them, as this research shows. Since main obstacles for participation are time and geographic issues, it could be necessary to create incentives for participation. In this sense, encouraging residents' participation as employees, rather than volunteers, could be necessary because they might be motivated by salaries.

By encouraging participation by both residents and visitors, it is expected to extend the length of stay for seasonal residents, resulting in residents' higher emotional solidarity with visitors, better perception of tourism, and higher support and participation in sustainable tourism development. Specifically, seasonal residents can perceive more specific economic benefits through being employed in the events. Also, it can be expected to reduce perceived current and future economic costs from inbound tourism by enticing international visitors and employing residents who are not in tourism-related businesses at the events. Moreover, perceived economic, sociocultural and environmental costs from tourism can be reduced through a tourism tax by enticing higher-end visitors, who might stay longer, spend more money and have higher social and environmental awareness. Such an approach may serve as a model for other rural tourism destinations in Japan, because it can provide measures for better interactions between residents and domestic and international visitors, support and participation by permanent and seasonal residents. This can be useful for rural Japanese destinations due to the importance of inbound tourism (MLIT, 2018a) and increasing populations from metropolitan to rural areas (Tsutsui, Sakuma, & Kasami, 2015).

CONCLUSION

Limitations and Direction for Further Studies

Despite the theoretical and practical contributions of the present study, this research is not without its limitations First of all, this study has explored residents' feelings, perceptions and support using a small sample of participants, as compared to a quantitative approach. This could lead to the difficulty of understanding various types of residents' opinions. Therefore, as Nunkoo et al. (2013) suggest, a hybridized mixed methods approach may bring about more robust findings in an effort to triangulate and further corroborate results here.

For the specific case of Karuizawa, the present research has failed to explore the effect of residents' attachment to their residential area on the formation of their feelings, perceptions, and support. It is truly difficult to conceive of aspects of pride, emotional closeness, commonality with others as well as support for the environmental and cultural conservation, without considering the place and the extent to which residents are attached to it. Therefore, further research using Place Attachment Theory, which explores residents' feelings to their residential areas (Stylidis, 2018; Woosnam, Aleshinloye, Ribeiro Stylidis, Jiang, & Erul, 2018), should be undertaken. Secondly, the ratio of permanent and seasonal residents is not well-balanced due to the difficulty of finding seasonal residents. Future work may consider undertaking a quantitative approach and ensuring an equal balance of permanent and seasonal residents are sampled. Lastly, because of the relatively cold weather in the winter in Karuizawa (Japan Meteorological Agency, no date) and the mention of seasonality by interview participants, Karuizawa may indeed suffer from seasonality issues. Therefore, there is a need to find differences in residents' opinions between on-season and off-season, as Vargas-Sánchez et al. (2014) has demonstrated.

REFERENCES

Akis, S., Peristianis, N., & Warner, J. (1996). Residents' attitudes to tourism development: The case of Cyprus. *Tourism Management*, *17*(7), 481–494. doi:10.1016/S0261-5177(96)00066-0

Andereck, K. L., & Nyaupane, G. P. (2011). Exploring the nature of tourism and quality of life perceptions among residents. *Journal of Travel Research*, *50*(3), 248–260. doi:10.1177/0047287510362918

Andereck, K. L., Valentine, K. M., Knopf, R. C., & Vogt, C. A. (2005). Residents' perceptions of community tourism impacts. *Annals of Tourism Research*, *32*(4), 1056–1076. doi:10.1016/j.annals.2005.03.001

Ap, J. (1992). Residents' perceptions on tourism impacts. *Annals of Tourism Research*, *19*(4), 665–690. doi:10.1016/0160-7383(92)90060-3

Auzins, A., & Vanags, J. (2012). Analytical assessment of sustainable development concept. *Geomatics*, *8*, 49–53.

Belisle, F. J., & Hoy, D. R. (1980). The perceived impact of tourism by residents a case study in Santa Marta, Colombia. *Annals of Tourism Research*, *7*(1), 83–101. doi:10.1016/S0160-7383(80)80008-9

Berlin Declaration. (1997) The Berlin Declaration on Biological Diversity and Sustainable Tourism. *International Conference of Environment Ministers on Biodiversity and Tourism*. Retrieved from: http://www.univeur.org/cuebc/downloads/PDF%20carte/06.%20Berlin%20declaration.PDF

Besculides, A., Lee, M. E., & McCormick, P. J. (2002). Residents' perceptions of the cultural benefits of tourism. *Annals of Tourism Research*, *29*(2), 303–319. doi:10.1016/S0160-7383(01)00066-4

Blanchard, O., & Buchs, A. (2015). Clarifying Sustainable Development Concepts Through Role-Play. *Simulation & Gaming*, *46*(6), 697–712. doi:10.1177/1046878114564508

Boley, B. B., McGehee, N. G., Perdue, R. R., & Long, P. (2014). Empowerment and resident attitudes toward tourism: Strengthening the theoretical foundation through a Weberian lens. *Annals of Tourism Research*, *49*, 33–50. doi:10.1016/j.annals.2014.08.005

Braun, V., & Clarke, V. (2006). Using thematic analysis in psychology. *Qualitative Research in Psychology*, *3*(2), 77–101. doi:10.1191/1478088706qp063oa

Brunt, P., & Courtney, P. (1999). Host perceptions of sociocultural impacts. *Annals of Tourism Research, 26*(3), 493-515.

Butler, R. W. (1980). The concept of a tourist area cycle of evolution: Implications for management of resources. *The Canadian Geographer. Geographe Canadien*, *24*(1), 5–12. doi:10.1111/j.1541-0064.1980.tb00970.x

Chen, C. F., & Chen, P. C. (2010). Resident attitudes toward heritage tourism development. *Tourism Geographies*, *12*(4), 525–545. doi:10.1080/14616688.2010.516398

Chen, S., & Raab, C. (2012). Predicting resident intentions to support community tourism: Toward an integration of two theories. *Journal of Hospitality Marketing & Management*, *21*(3), 270–294. doi:10.1080/19368623.2011.584268

Choi, H. S. C., & Sirakaya, E. (2005). 'Measuring residents' attitude toward sustainable tourism: Development of sustainable tourism attitude scale'. *Journal of Travel Research*, *43*(4), 380–394. doi:10.1177/0047287505274651

Davis, D., Allen, J., & Cosenza, R. M. (1988). Segmenting local residents by their attitudes, interests, and opinions toward tourism. *Journal of Travel Research*, *27*(2), 2–8. doi:10.1177/004728758802700201

Diedrich, A., & García-Buades, E. (2009). Local perceptions of tourism as indicators of destination decline. *Tourism Management*, *30*(4), 512–521. doi:10.1016/j.tourman.2008.10.009

Dyer, P., Gursoy, D., Sharma, B., & Carter, J. (2007). Structural modeling of resident perceptions of tourism and associated development on the Sunshine Coast, Australia. *Tourism Management*, *28*(2), 409–422. doi:10.1016/j.tourman.2006.04.002

Eshliki, S. A., & Kaboudi, M. (2017). Perception of community in tourism impacts and their participation in tourism planning: Ramsar, Iran. *Journal of Asian Behavioural Studies*, *2*(4), 59–69. doi:10.21834/jabs.v2i4.207

Esman, M. R. (1984). Tourism as ethnic preservation: The Cajuns of Louisiana. *Annals of Tourism Research, 11*(3), 451–467. doi:10.1016/0160-7383(84)90031-8

Fredline, E. (2002). *Social impacts of tourism on the Gold Coast*. Altona: Common Ground Publishing.

García, F. A., Vázquez, A. B., & Macías, R. C. (2015). Resident's attitudes towards the impacts of tourism. *Tourism Management Perspectives, 13*, 33–40. doi:10.1016/j.tmp.2014.11.002

Gursoy, D., & Rutherford, D. G. (2004). Host attitudes toward tourism: An improved structural model. *Annals of Tourism Research, 31*(3), 495–516. doi:10.1016/j.annals.2003.08.008

Haley, A. J., Snaith, T., & Miller, G. (2005). The social impacts of tourism a case study of Bath, UK. *Annals of Tourism Research, 32*(3), 647–668. doi:10.1016/j.annals.2004.10.009

Hall, C. M., & Lew, A. A. (1998). *Sustainable Tourism: A Geographical Perspective*. Addison Wesley Longman Limited.

Hasani, A., Moghavvemi, S., & Hamzah, A. (2016). The impact of emotional solidarity on residents' attitude and tourism development. *PLoS One, 11*(6), 157–624. doi:10.1371/journal.pone.0157624 PMID:27341569

Horn, C., & Simmons, D. (2002). Community adaptation to tourism: Comparisons between Rotorua and Kaikoura, New Zealand. *Tourism Management, 23*(2), 133–143. doi:10.1016/S0261-5177(01)00049-8

Hunt, C., & Stronza, A. (2014). Stage-based tourism models and resident attitudes towards tourism in an emerging destination in the developing world. *Journal of Sustainable Tourism, 22*(2), 279–298. doi :10.1080/09669582.2013.815761

Ishikawa, N., & Fukushige, M. (2007). Impacts of tourism and fiscal expenditure to remote islands: The case of the Amami islands in Japan. *Applied Economics Letters, 14*(9), 661–666. doi:10.1080/13504850500461423

Jimura, T. (2011). The impact of world heritage site designation on local communities - A case study of Ogimachi, Shirakawa–mura, Japan. *Tourism Management, 32*(2), 288–296. doi:10.1016/j.tourman.2010.02.005

Johnson, J. D., Snepenger, D. J., & Akis, S. (1994). Residents' perceptions of tourism development. *Annals of Tourism Research, 21*(3), 629–642. doi:10.1016/0160-7383(94)90124-4

Joo, D., & Woosnam, K. M. (2019). Measuring tourists' emotional solidarity with one another—A modification of the Emotional Solidarity Scale. *Journal of Travel Research*. doi:10.1177/0047287519878503

Joppe, M. (1996). Sustainable community tourism development revisited. *Tourism Management, 17*(7), 475–479. doi:10.1016/S0261-5177(96)00065-9

Jurowski, C., Uysal, M., & Williams, D. R. (1997). A theoretical analysis of host community resident reactions to tourism. *Journal of Travel Research, 36*(2), 3–11. doi:10.1177/004728759703600202

Karuizawa Tourist Association. (2018) *Events*. Retrieved from: https://karuizawa-kankokyokai.jp/event/ (in Japanese)

Karuizawa Town. (2017). *Genetic Strategy for Regional Revitalisation in Karuizawa Town*. Retrieved from: https://www.town.karuizawa.lg.jp/www/contents/1454307204126/index.html

Kayat, K., Sharif, N. M., & Karnchanan, P. (2013). Individual and collective impacts and residents' perceptions of tourism. *Tourism Geographies*, *15*(4), 640–653. doi:10.1080/14616688.2012.762690

Kitnuntaviwat, V., & Tang, J. C. (2008). Residents' attitudes, perception and support for sustainable tourism development. *Tourism and Hospitality Planning & Development, 5*(1), 45-60.

Korça, P. (1996). Resident attitudes toward tourism impacts. *Annals of Tourism Research*, *23*(3), 695–697. doi:10.1016/0160-7383(95)00092-5

Lane, B. (1994). Sustainable rural tourism strategies: A tool for development and conservation. *Journal of Sustainable Tourism*, *2*(1-2), 102–111. doi:10.1080/09669589409510687

Látková, P., & Vogt, C. A. (2012). Residents' attitudes toward existing and future tourism development in rural communities. *Journal of Travel Research*, *51*(1), 50–67. doi:10.1177/0047287510394193

Lee, C. K., Kang, S. K., Long, P., & Reisinger, Y. (2010). Residents' perceptions of casino impacts: A comparative study. *Tourism Management*, *31*(2), 189–201. doi:10.1016/j.tourman.2009.02.011

Lee, T. H. (2013). Influence analysis of community resident support for sustainable tourism development'. *Tourism Management*, *34*, 37–46. doi:10.1016/j.tourman.2012.03.007

Lepp, A. (2007). Residents' attitudes towards tourism in Bigodi village, Uganda. *Tourism Management*, *28*(3), 876–885. doi:10.1016/j.tourman.2006.03.004

Lindberg, K., & Johnson, R. L. (1997). Modeling resident attitudes toward tourism. *Annals of Tourism Research*, *24*(2), 402–424. doi:10.1016/S0160-7383(97)80009-6

Liu, J. C., & Var, T. (1986). Resident attitudes toward tourism impacts in Hawaii. *Annals of Tourism Research*, *13*(2), 193–214. doi:10.1016/0160-7383(86)90037-X

Long, J. A. (2007). Subjects, respondents, participants and collaborators. (2007). In J. A. Long (Ed.), *Researching leisure, sport, and tourism* (pp. 34–46). SAGE Publications Ltd. doi:10.4135/9780857024619

Long, P. T., Perdue, R. R., & Allen, L. (1990). Rural resident tourism perceptions and attitudes by community level of tourism. *Journal of Travel Research*, *28*(3), 3–9. doi:10.1177/004728759002800301

Maruyama, N. U., Woosnam, K. M., & Boley, B. B. (2017). 'Residents' attitudes toward ethnic neighborhood tourism (ENT): Perspectives of ethnicity and empowerment'. *Tourism Geographies*, *19*(2), 265–286. doi:10.1080/14616688.2016.1258432

Mason, P. (2014). *Researching tourism, leisure and hospitality for your dissertation*. Oxford: Goodfellow Publishers.

Mason, P., & Cheyne, J. (2000). Residents' attitudes to proposed tourism development. *Annals of Tourism Research*, *27*(2), 391–411. doi:10.1016/S0160-7383(99)00084-5

Mathieson, A., & Wall, G. (1982). *Tourism, economic, physical and social impacts*. Longman.

McGehee, N. G., & Andereck, K. L. (2004). Factors predicting rural residents' support of tourism. *Journal of Travel Research*, *43*(2), 131–140. doi:10.1177/0047287504268234

McKean, P. (1978). Towards a theoretical analysis of tourism: economic dualism and cultural involution in Bali. In V. Smith (Ed.), *Hosts and guests. The anthropology of tourism* (pp. 119–138). University of Pennsylvania Press.

Ministry of Land, Infrastructure and Transport. (2018). *The White Paper of Tourism*. Retrieved from: http://www.mlit.go.jp/statistics/file000008.html (in Japanese)

Moghavvemi, S., Woosnam, K. M., Paramanathan, T., Musa, G., & Hamzah, A. (2017). The effect of residents' personality, emotional solidarity, and community commitment on support for tourism development. *Tourism Management*, *63*, 242–254. doi:10.1016/j.tourman.2017.06.021

Nagano Prefectural Government. (2017). *The Results of Survey on the Total Number of Foreign Lodgers in 2016* (in Japanese). Retrieved from: https://www.pref.nagano.lg.jp/kankoki/sangyo/kanko/toukei/gaikokujin.html

Nunkoo, R., & Gursoy, D. (2012). Residents' support for tourism: An identity perspective. *Annals of Tourism Research*, *39*(1), 243–268. doi:10.1016/j.annals.2011.05.006

Nunkoo, R., & Ramkissoon, H. (2011). Developing a community support model for tourism. *Annals of Tourism Research*, *38*(3), 964–988. doi:10.1016/j.annals.2011.01.017

Nunkoo, R., Smith, S. L., & Ramkissoon, H. (2013). Residents' attitudes to tourism: A longitudinal study of 140 articles from 1984 to 2010. *Journal of Sustainable Tourism*, *21*(1), 5–25. doi:10.1080/09669582.2012.673621

Perdue, R. R., Long, P. T., & Allen, L. (1987). Rural resident tourism perceptions and attitudes. *Annals of Tourism Research*, *14*(3), 420–429. doi:10.1016/0160-7383(87)90112-5

Pratt, S. (2015). The economic impact of tourism in SIDS. *Annals of Tourism Research*, *52*, 148–160. doi:10.1016/j.annals.2015.03.005

Radzuan, I. S. M., Fukami, N., & Ahmad, Y. (2015). Incentives for the conservation of traditional settlements: Residents' perception in Ainokura and Kawagoe, Japan'. *Journal of Tourism and Cultural Change*, *13*(4), 301–329. doi:10.1080/14766825.2014.952302

Ribeiro, M. A., Pinto, P., Silva, J. A., & Woosnam, K. M. (2017). Residents' attitudes and the adoption of pro-tourism behaviours: The case of developing island countries'. *Tourism Management*, *61*, 523–537. doi:10.1016/j.tourman.2017.03.004

Ribeiro, M. A., Pinto, P., Silva, J. A., & Woosnam, K. M. (2018). Examining the predictive validity of SUS-TAS with maximum parsimony in developing island countries. *Journal of Sustainable Tourism*, *26*(3), 379–398. doi:10.1080/09669582.2017.1355918

Ritchie, B. W., & Inkari, M. (2006). Host community attitudes toward tourism and cultural tourism development: The case of the Lewes District, Southern England. *International Journal of Tourism Research*, *8*(1), 27–44. doi:10.1002/jtr.545

Saunders, M., Lewis, P., & Thornhill, A. (2012). *Research methods for business students* (6th ed.). Pearson.

Schroeder, K. (2015). Cultural values and sustainable tourism governance in Bhutan. *Sustainability, 7*(12), 16616–16630. doi:10.3390u71215837

Sharpley, R. (2000). Tourism and sustainable development: Exploring the theoretical divide'. *Journal of Sustainable Tourism, 8*(1), 1–19. doi:10.1080/09669580008667346

Sharpley, R. (2014). Host perceptions of tourism: A review of the research. *Tourism Management, 42*, 37–49. doi:10.1016/j.tourman.2013.10.007

Stylidis, D. (2018). Residents' place image: A cluster analysis and its links to place attachment and support for tourism. *Journal of Sustainable Tourism, 26*(6), 1007–1026. doi:10.1080/09669582.2018.1435668

Stylidis, D., Biran, A., Sit, J., & Szivas, E.M. (2014). Residents' support for tourism development: The role of residents' place image and perceived tourism impacts. *Tourism Management, 45*, 260-274.

Takahashi, N. (2013). *A Study on the Environmental Change and Conservation of the Summer Retreat Areas in Karuizawa* (in Japanese). Retrieved from: http://nc.heritage.tsukuba.ac.jp/?action=common_download_main&upload_id=578

The United Nations Environment Programme and World Tourism Organization. (2005). *Making Tourism More Sustainable*. Retrieved from: http://www.unep.fr/shared/publications/pdf/dtix0592xpa-tourismpolicyen.pdf

The United Nations World Tourism Organisation. (2017). *UNWTO Tourism Highlights 2017 Edition*. Retrieved from: https://www.e-unwto.org/doi/pdf/10.18111/9789284419029

The United Nations World Tourism Organisation. (2018). *UNWTO World Tourism Barometer*. Retrieved from: https://cf.cdn.unwto.org/sites/all/files/pdf/unwto_barom18_01_january_excerpt_hr.pdf

The World Commission on Environment and Development. (1987). *Our Common Future*. Retrieved from: http://www.un-documents.net/our-common-future.pdf

Tosun, C. (2002). Host perceptions of impacts: A comparative tourism study. *Annals of Tourism Research, 29*(1), 231–253. doi:10.1016/S0160-7383(01)00039-1

Tsutsui, K., Sakuma, Y., & Kasami, K. (2015). Regional Regeneration by Business in Rural Community of In-migrants from Urban Areas' [in Japanese]. *Journal of Association of Rural Planning, 34*(1), 45–50. doi:10.2750/arp.34.45

Uchida, Y. (1989). 'Regarding the establishment of images of 'luxurious summer resort/ cottage destination' in Karuizawa' (in Japanese), *A review of Geography. Ser. A, 62*(7), 495–512.

Vargas-Sánchez, A., Plaza-Mejia, M. D. L. Á., & Porras-Bueno, N. (2009). Understanding residents' attitudes toward the development of industrial tourism in a former mining community. *Journal of Travel Research, 47*(3), 373–387. doi:10.1177/0047287508322783

Vargas-Sánchez, A., Porras-Bueno, N., & de los Ángeles Plaza-Mejía, M. (2011). Explaining residents' attitudes to tourism: Is a universal model possible? *Annals of Tourism Research, 38*(2), 460–480. doi:10.1016/j.annals.2010.10.004

Vargas-Sánchez, A., Porras-Bueno, N., & de los Ángeles Plaza-Mejía, M. (2014). Residents' attitude to tourism and seasonality. *Journal of Travel Research, 53*(5), 581–596. doi:10.1177/0047287513506295

Woosnam, K. M. (2011). Testing a model of Durkheim's theory of emotional solidarity among residents of a tourism community. *Journal of Travel Research, 50*(5), 546–558. doi:10.1177/0047287510379163

Woosnam, K. M. (2012). Using emotional solidarity to explain residents' attitudes about tourism and tourism development. *Journal of Travel Research, 51*(3), 315–327. doi:10.1177/0047287511410351

Woosnam, K. M., Aleshinloye, K. D., & Maruyama, N. (2016). Solidarity at the Osun Osogbo Sacred Grove—A UNESCO world heritage site. *Tourism Planning & Development, 13*(3), 274–291. doi:10.1080/21568316.2015.1104380

Woosnam, K. M., Aleshinloye, K. D., Ribeiro, M. A., Stylidis, D., Jiang, J., & Erul, E. (2018). Social determinants of place attachment at a World Heritage Site. *Tourism Management, 67*, 139–146. doi:10.1016/j.tourman.2018.01.012

Woosnam, K. M., Aleshinloye, K. D., Van Winkle, C. M., & Qian, W. (2014). Applying and expanding the theoretical framework of emotional solidarity in a festival context. *Event Management, 18*(2), 141–151. doi:10.3727/152599514X13947236947428

Woosnam, K. M., & Norman, W. C. (2010). Measuring residents' emotional solidarity with tourists: Scale development of Durkheim's theoretical constructs. *Journal of Travel Research, 49*(3), 365–380. doi:10.1177/0047287509346858

Woosnam, K. M., Norman, W. C., & Ying, T. (2009). Exploring the theoretical framework of emotional solidarity between residents and tourists. *Journal of Travel Research, 48*(2), 245–258. doi:10.1177/0047287509332334

Yan, L., Xu, J., & Zhou, Y. (2018). 'Residents' attitudes toward prostitution in Macau'. *Journal of Sustainable Tourism, 26*(2), 205–220. doi:10.1080/09669582.2017.1338293

Young, B. (1983). Touristization of traditional Maltese fishing-farming villages: A general model. *Tourism Management, 4*(1), 35–41. doi:10.1016/0261-5177(83)90048-1

Zhang, J., Inbakaran, R. J., & Jackson, M. S. (2006). Understanding community attitudes towards tourism and host—Guest interaction in the urban—rural border region. *Tourism Geographies, 8*(2), 182–204. doi:10.1080/14616680600585455

ADDITIONAL READING

Aleshinloye, K. D., Fu, X., Ribeiro, M. A., Woosnam, K. M., & Tasci, A. D. A. (2020). The Influence of Place Attachment on Social Distance: Examining Mediating Effects of Emotional Solidarity and the Moderating Role of Interaction. *Journal of Travel Research, 59*(5), 828–849. doi:10.1177/0047287519863883

Erul, E., Woosnam, K. M., & McIntosh, W. A. (2020). Considering emotional solidarity and the theory of planned behavior in explaining behavioral intentions to support tourism development. *Journal of Sustainable Tourism*, *28*(8), 1158–1173. doi:10.1080/09669582.2020.1726935

Gursoy, D., Ouyang, Z., Nunkoo, R., & Wei, W. (2019). Residents' impact perceptions of and attitudes towards tourism development: A meta-analysis. *Journal of Hospitality Marketing & Management*, *28*(3), 306–333. doi:10.1080/19368623.2018.1516589

Joo, D., & Woosnam, K. M. (2019). Measuring Tourists' Emotional Solidarity with One Another—A Modification of the Emotional Solidarity Scale. *Journal of Travel Research*. Advance online publication. doi:10.1177/0047287519878503

Megeirhi, H. A., Woosnam, K. M., Ribeiro, M. A., Ramkissoon, H., & Denley, T. J. (2020). Employing a value-belief-norm framework to gauge Carthage residents' intentions to support sustainable cultural heritage tourism. *Journal of Sustainable Tourism*, *28*(9), 1351–1370. doi:10.1080/09669582.2020.1738444

Patwardhan, V., Ribeiro, M. A., Payini, V., Woosnam, K. M., Mallya, J., & Gopalakrishnan, P. (2020). Visitors' Place Attachment and Destination Loyalty: Examining the Roles of Emotional Solidarity and Perceived Safety. *Journal of Travel Research*, *59*(1), 3–21. doi:10.1177/0047287518824157

Tara, D., Woosnam, K. M., Ribeiro, M. A., Hehir, C., Boley, B., & Abrams, J. (2020). Individuals' intentions to engage in last chance tourism: Applying the value-belief-norm model. *Journal of Sustainable Tourism*. Advance online publication. doi:10.1080/09669582.2020.1762623

KEY TERMS AND DEFINITIONS

Emotional Solidarity: Is considered the degree of closeness between residents and tourists in a destination.

Social Exchange Theory: Is a general sociological theory concerned with understanding the exchange of resources between individuals and groups in an interaction situation.

Support for Tourism: Is the degree to which residents evaluate the effects of tourism in their community.

Sustainable Tourism: Increases the benefits and to reduce the negative impacts caused by tourism at destinations.

Theory of Reasoned Action: Explains the relationship between attitudes and behaviours within human action.

Tourism Impacts: Include the effects of tourism on the environment and on destination communities, and its economic contributions.

Chapter 6
Perspectives on Community Involvement in the Revitalisation of Cultural and Natural Heritage for Sustainable Tourism:
A Case of an Island in Hong Kong

Dora Agapito
https://orcid.org/0000-0002-2568-3720
Faculty of Economics and Research Centre for Tourism, Sustainability, and Well-Being (CinTurs), University of Algarve, Portugal

Chung-Shing Chan
Department of Geography and Resource Management, The Chinese University of Hong Kong, Hong Kong

ABSTRACT

This chapter explores how local community involvement contributed to reviving cultural and natural heritage as assets for sustaining tourism on a small offshore island in Hong Kong. Yim Tin Tsai is a village established by the Hakka Chan clan and the main economic activity was salt production in the past. After the village became deserted due to villagers' migration, the existing salt fields were revived by the community. This process has increased the interest of visitors to travel to the village to experience the cultural, religious, and natural heritage of the destination. This case study combines in loco observations, interviews with key stakeholders on the island, and consultation of secondary sources. Opportunities and challenges regarding community involvement in tourism and its contribution to the conservation of tangible and intangible heritage assets of this offshore island are discussed.

DOI: 10.4018/978-1-7998-3156-3.ch006

INTRODUCTION

This chapter explores how the local community in a small offshore island in Sai Kung, Hong Kong, revived cultural and natural heritage assets for sustaining tourism on the island. Whilst conserving unique assets in the village, the process of community involvement has increased the interest of visitors to travel to the village to experience the cultural, religious and nature-based heritage of the destination (Agapito & Chan, 2019). This research follows the rationale that community-based tourism has been advocated as one approach to cater for a more sustainable tourism industry (Blackstock, 2005; Okazaki, 2008), and that community participation is pointed out as key for the process of place making (Lew, 2017). Yet, research has been centred in developed parts of the world and case studies conducted at local and more vulnerable contexts are still scant (Friedman, 2010; Samadhi, 2001).

Rural communities around the world face numerous challenges, which include declining and ageing populations, limited economic and social opportunities for residents, vulnerable natural resources and potential loss of community identity (Duxbury & Campbell, 2009; Roberts & Hall, 2001). Yim Tin Tsai is a rural small village which became deserted after all its residents left for finding better conditions of life elsewhere in the 1990s. Recently, members of the community, who share the same cultural identity, grouped and became involved in the process of rejuvenating their ancestral village to both attract visitors and preserve the community identity. Worth noting is the fact that this traditional Hakka village is rich in multiple cultural and natural resources. Recently, some of these resources gained recognition from UNESCO and are deemed unique tourist and educational resources (Agapito & Chan 2019).

Considering extant literature, this chapter will start by reflecting on the dynamics involved in community participation in tourism, followed by theoretical depictions on opportunities and challenges of community-based tourism. The focus will be on the role of community-based tourism in the revitalisation of cultural and natural heritage in small destinations for sustainable tourism. Subsequently, a subsection of the theoretical background will be dedicated to the relationship between community involvement and the process of place making within travel destinations. The research methods approach section will be followed by the presentation of the case under study, as well as the results of the interviews conducted with representatives of the community. The findings of this chapter will allow to discuss the importance of community involvement in the making of small and rural-based travel destinations, such as the case of the off-shore island under study. The discussion will also offer reflections on opportunities and challenges on the relationship between tourism and the conservation of tangible and intangible assets (Ahmad, 2006). The use of multisensory stimuli as a contribution to sustainable tourism (Agapito & Chan, 2019) and the place making of destinations will also be taken into consideration.

This chapter represents the second part of a study developed by the authors in Yim Tin Tsai focused on the contribution of a multisensory approach to responsible management in the context of community-based tourism from the perspective of visitors (Agapito & Chan, 2019). In the first stage of this study, a visitor survey was conducted to identify multi-sensory impressions reported by tourists in relation to their experience with local resources on the island. The referred research suggests that a holistic approach to sensory stimuli based on endogenous resources can contribute to responsible management in community-based tourism. The present study complements the first approach and broadens the research scope by adding a supply-side viewpoint on the topic under study.

COMMUNITY INVOLVEMENT AND SUSTAINABLE TOURISM

Community Participation in Tourism

While community is a complex concept to define, it has been regarded in literature as 'a geographical area, or a group of people with shared origins or interests. If the geographical definition is used, then the community can be defined as citizens within a given locality' (Aas et al., 2005, p.30). Community development focused on involving local communities in decision-making processes was introduced as an approach to rural development during the 1950s and 1960s, which was followed by the promotion of community participation as a good practice worldwide in the 1960s and 1970s (Sebele, 2010). Since the 1980s, tourism literature has stressed the role of the inclusion and involvement of local communities in tourism development based on the rationale that residents are key for tourism sustainable development (Hardy et al., 2002; Sebele, 2010).

Research posits that the participation of the local community is central in ensuring that tourists have a positive experience, while at the same time enabling the community to gain benefits from tourism activities in the destination (Sebele, 2010). The assumption is that community participation lessens negative impacts on the destination and revitalises economies (Hardy et al., 2002). In the same vein, Prentice (1993, p. 218) stresses that community participation in tourism development has become an ideology of tourism planning. Table 1 depicts some of the most used definitions of community participation in literature.

Table 1. Definitions of community participation

Reference	Definition
Arnstein (1969, p.216)	'[…] citizen participation is a categorical term for citizen power. It is the redistribution of power that enables the have-not citizens, presently excluded from the political and economic processes, to be deliberately included in the future.'
Connell (1997, p. 250)	Participation is 'is not only about the more efficient and more equitable distribution of material resources - it is also about the sharing of knowledge and the transformation of the process of learning itself in the service of people's self-development.'
Haywood (1988, p.106)	Community participation is related to 'a process of involving all stakeholders (local government officials, local citizens, architects, developers, business people, and planners) in such a way that decision-making is shared.'
Stone (1989, p.207)	Community participation is related to the design of 'development in such a way that intended beneficiaries are encouraged to take matters into their own hands, to participate in their development through mobilising their resources, defining their own needs, and making their own decisions about how to meet them.'
Tosun, C. (2000, p. 615)	Community participation is a 'tool to readjust the balance of power and reassert local community views against those of the developers or the local authority, or to redefine professionalism, which may determine the conditions of successful participation and prevent manipulation of a community in the participation process.'

Community-Based Tourism: Opportunities and Challenges

Community-based tourism is focused on the involvement of the host community in the process of designing and coordinating tourism development for a more sustainable industry (Blackstock, 2005). Hence, this approach has deemed good practice to achieve sustainable development in rural areas and small cities as travel destinations (Duxbury & Richards, 2019). In practice, it is worth notice that community participation can involve different degrees of involvement (Tosun, 1999), and that this approach to tourism can be ineffective in case residents have no access both the right (decision power) and the means (high transaction costs to implement and maintain programs) (Getz & Jamal, 1994). In addition to time-consuming, other difficulties, such as the level of education, inexperience with regards to business-related issues, insufficient access to funding, as well as conflicting interests, are limitations to community participation in tourism (Addison, 1996). Gray (1985) emphasises that the community needs adequate resources and skills to take part in the decision-making process. According to the author, the power to obtain these can be detained by governments or other stakeholders who often do not regard locals as partners at the same level.

For Okazaki (2008, p.512), despite the existing challenges and barriers to taking into account, the community-based approach is still the best solution within tourism development considering: a) local aspects have a direct influence in the tourist experience; b) local assets, which include the community, natural and cultural-based environments, help to shape destination image; c) public involvement is a driving force to protect the community's natural environment and culture as tourism products, and d) the feasibility and long-term life of tourism plans are linked with the socio-economic development of the host community.

Community-Based Tourism and the Revitalisation of Cultural and Natural Heritage in Small Destinations

Small destinations, especially the ones situated in rural areas, are typically more vulnerable to undesirable impacts of tourism. This aspect is related to environmental, social and cultural values, as well as economic fragilities, which call for a more responsible managerial approach (Croes & Semrad, 2015; Lane, 1994). Subsequently, these areas should focus on facilitating diversified and positive destination experiences by attaining balance in preserving endogenous resources, ensuring residents' quality of life and quality of the tourist experience, while addressing other stakeholders needs (e.g. tourism providers) and, subsequently, aiming at local sustainable development (Kastenholz et al., 2012a; Manente & Minghetti, 2006). The potential of the countryside is well known due to the multisensory nature of diverse assets, such as vegetation, wildlife, water, and natural landscapes, and for being associated with unique cultural heritage (e.g. handicrafts, gastronomy and local customs) (Kastenholz et al., 2012b; Roberts & Hall, 2001; Sebele, 2010). Typically, rural capital assets represent a collection of natural, built, and socio and cultural resources. In a tourism context, visitors both receive satisfaction from rural-based experiences and have an impact on this countryside-based capital (Garrod et al., 2006).

Contemporary definitions of heritage acknowledge the relationships between cultural and natural environments, and the tangible and intangible nature of heritage (Taylor & Lennon, 2011; UNESCO, 2003). As such, current definitions link cultural heritage with material and non-material resources related to the imprint of 'past activities on the environment, which comprise archaeological remains; monuments and sites; cultural landscapes superimposed on the natural environment; local indigenous knowledge systems; folk-life and folklore; and traditional practices and rituals attached to the biophysical environment' (Keitumetse, 2011, p.50). Rossler (2006) highlights the relationship between culture and natural heritage by pointing out that cultural landscapes can be approached as 'the interface between nature and culture, tangible and intangible heritage, biological and cultural diversity.' Hence, these aspects represent people's identity and are 'a symbol of the growing recognition of the fundamental links between local communities and their heritage, humankind and its natural environment' (Rossler, 2006, p. 334).

Community involvement in the process of place making and in attracting visitors can contribute to cultural and heritage responsible management, thus to its conservation. In specific cases, communities can develop ecomuseums, which are community-based projects focus on heritage that eco local identity and sense of place (Davis, 1999). This bottom-up approach results from the community reflection and the residents' empowerment to preserve, interpret and manage their heritage for sustainable development (Murtas & Davis, 2009). Moreover, Moscardo (1996) stresses that adequate 'interpretation can play a critical role in sustainable tourism by educating tourists about the nature of the host region and culture, informing them of the consequences of their actions, enhancing their experience and encouraging them to engage in sustainable behaviors' (p. 378). Community involvement in heritage management is therefore key to attain sustainable tourism, especially in more vulnerable areas such as small rural destinations and off-shore islands (Agapito & Chan, 2019; Keitumetse, 2011).

Moreover, a contemporary strand of research advocates that multisensory stimuli related to endogenous resources can be used to conceptualise and design sensory-informed tourist experiences by involving different stakeholders in a sustainable manner (Agapito et al., 2013, 2014; Kastenholz et al., 2012b). This line of research argues that a holistic approach to multisensory stimuli can contribute to the effective responsible management of community-based tourism projects as a tool to achieve sustainable tourism (Agapito & Chan, 2019). Additionally, niche markets within cultural tourism, such as creative tourism, are suitable for community-based approaches to tourism. Indeed, according to UNESCO, creative forms of tourism are related to 'a travel directed toward an engaged and authentic experience, with participative learning in the arts, heritage, or special character of a place, and it provides a connection with those who reside in this place and create this living culture' (UNESCO, 2006, p.3).

Community-Based Tourism and Place Making

Community involvement in the process of planning and managing public resources or spaces has led naturally to a more people-centred approach to place making that is characterised by being 'community-driven, visionary, function before form, adaptable, inclusive, focused on creating destinations, context-specific, dynamic, trans-disciplinary, transformative, flexible, collaborative, and sociable' (Project for Public Spaces – PPS, 2007). These attributes are rooted in the origin of cultural geography, which is associated with giving sense and meaning to spaces (Massey, 2005; Othman et al., 2013; Rose-Redwood & Alderman, 2011; Wortham-Galvin, 2008).

Notwithstanding, approaches to the making of places can vary. On one hand, the process of place making and value creation may occur in a relatively short period through the planning and development of physical spaces and facilities (Castillo, 2014; Main & Sandoval, 2015; Mazumdar & Mazumdar, 2008; Trudeau, 2006). This relates to a planned and professional design perspective aiming at shaping people's behaviour and perceptions of a particular place (Lew, 2012). Hence, placemaking is often guided by experts such as urban planners, architects, landscape designers and managers in a view to achieving multiple objectives, with a particular focus on the economic dimension (Wyckoff et al., 2015). The outcome of this top-down placemaking process is sometimes unpredictable considering the impacts on the community.

Criticism around the top-down process of placemaking has increasingly been fostered to support an organic and bottom-up approach through the creation and shaping of places with the involvement of local people and communities (Hou & Rios, 2003; Lems, 2016). Lew (2017) distinguished the two approaches by using different written expressions: 'placemaking' (planned, top-down) and 'place-making' (organic, bottom-up). In between these opposed approaches, a mixed approach aims to combine the advantages of professional planning and organic ways of place development. Understanding residents' sense of place, as well as visitors' experiences, are thus key to the process of integrating the top-down and bottom-up approaches of place making (Lew, 2017). As a result, an opportunity contributing to responsible place making and management is to analyse the multisensory nature of local cultural and natural resources available at the destination both from the perspective of residents and visitors (Chan & Agapito, 2019).

Although it has been suggested that smaller destinations and less-tourism oriented places tend to have a better foundation and the characteristics for exhibiting a community-involved and bottom-up place-making process (Lew, 2017), there are still uncertainties especially when places are receiving more attention in the policy agenda for larger-scale development. This can be related to flagship attractions and infrastructure for tourism purposes. In the process of place development and management, it is often unavoidable to rely on top-down planned marketing strategies when the decision or policy-makers turn to adopt such strategies from a touristic viewpoint or perspective (Boisen et al., 2011). However, the cultural "soul" of a place is pivotal for a more organic and people-centred placemaking process. If the decision-makers succeed in capturing local identity and in representing it to foster a livelier and quality community to both local people and visitors, places can become both a sustainable place of residence and travel destination. At the same time, it can become attractive to creative classes and innovative economic activities (Florida, 2002; Kavaratzis & Hatch, 2013; Landry, 2000; Richards, 2014; Thurlow & Jaworski, 2014).

In this vein, it is critical to engage both policy-makers and community members in a more collaborative form of decision-making for place development (Brand & Gaffikin, 2007). Subsequently, stakeholders should balance and trade-off among, sometimes, opposed interests (Boisen et al., 2011; Hultman & Hall, 2012; Kendig & Keast, 2010; Peirce et al., 2011). In this regard, the existing mechanisms of developing places should play a role in empowering locals to climb up the ladder of participation in decision-making processes (Arnstein, 1969). This should be part of place governance that encourages greater equity, participation, inclusiveness and, eventually, boost up the social capital of the place (PPS, 2013). According to some scholarly observations, the theoretical development of place making as reflected in the literature tends to be concentrated in developed parts of the world, such as North America, Europe and Australia (Lew, 2017). This phenomenon calls for more academic contributions to the developing world, especially at their regional and local contexts of places (Friedman, 2010; Samadhi, 2001). Specifically in Hong Kong, although most of the urban or place (re)development has been government-led or developer-initiated (Chung & Ngai, 2007; Development Bureau, 2011), the focus on local resources as elements for placemaking has been also supported (e.g., Blake, 2017; Cartier, 1999; Chan, 2016). Notwithstanding, more attention has been paid to urban areas (Kee, 2017), given the institutional circumstances of Hong Kong (Albrechts & Denayer, 2001; Cheung, 2011) and historical reasons (Knowles & Harper, 2009; Ng, 2009). Yet, there has been a growing local attempt to realise and achieve more locally engaged and community-based place making projects (e.g., Community Project Workshop – CPW, 2013).

RESEARCH METHODS

This study followed a case study approach based on a single case (Yin, 2004). A triangulation of methods was utilised (Brunt et al., 2017) by encompassing: a) in loco overt observations, b) three in-depth interviews conducted with key stakeholders representing the host community in Yim Tin Tsai who were available to participate in the study, and c) secondary sources of information. Since the focus of this study was on community involvement, the scope of interviews included representatives of the host community who have been engaged in the development of tourism in the village. Some stakeholders, such as local authorities (public administration), were not actively engaged in the operational implementation of tourism on the island at the time of data collection; hence, they were not interviewed during this research. At a later stage of data collection, there was an indication that the government could become involved in this process, with an early prediction of an Arts Festival for December 2019 in the village, followed by another two related festivals in 2020 and 2021, respectively. This aspect was left to the future development of this research. Data were collected between May 2018 and March 2019. Secondary sources included previous research developed by the authors on the village from the perspective of visitors (Agapito & Chan, 2019), published news, websites managed by different stakeholders, and published online videos on the destination. Information on the profile of the three interviewees can be found in table 2.

Table 2. Interviewees' profile

Interviewee	Profile
A	Male; Research assistant in the Chinese University of Hong Kong and project director; Role in Yim Tin Tsai: Crowd control, design of tours' contents, and management of the tour guides (Salt and Light Preservation Centre)
B	Male; Retired; Role within visitor management in Yim Tin Tsai: Ecotourism Tour Guide (Indigenous resident and member of the Yim Tin Tsai Village Committee)
C	Female; Retired teacher; Role in Yim Tin Tsai: Tour guide in Yim Tin Tsai and catholic member of the local church (Catholic Diocese of Hong Kong, Sacred Heart Church)

The interviews were conducted by trained interviewers in Cantonese and translated into English. Topics addressed during the interviews, which had a duration of thirty minutes to one hour, can be found in Table 3. The content of the interviews was digitally recorded with the consent of the respondents. A qualitative approach to data analysis was followed by depicting emergent themes from the interviews, a process which had into account the research objective. Given the exploratory nature of the study as well as the limited number of interviews to analyse, the following manual process for data analysis was found to be adequate. First, the researchers became familiar with the data by reading the transcripts, taking notes and generating a preliminary thematic framework. This process was developed independently by two researchers in the first stage. In the second stage, the same researchers discussed the results of the preliminary phase, and adjustments were made to guarantee the suitability of themes against data and the research objective. The topics utilised to assist the interviewers to conduct the interviews were useful in this process but did not dictate the final themes. After reaching consensus on the themes (five themes in our case), these were applied to the three interviews in a systematic way; key phrases illustrating each theme were selected and interpretation was performed (Brunt et al., 2017; Ritchie & Spencer, 1994).

Table 3. Topics addressed during the interviews

Topics
Characterisation of the current tourism in Yim Tin Tsai
Community involved in the tourism development in Yim Tin Tsai
Visitor activities currently performed in Yim Tin Tsai
Resources/assets unique to the village
Visitor management strategies on the island
Use of responsible and sustainable approach to tourism in Yim Tin Tsai
Global visitor experience
Opportunities and challenges for sustainable tourism in the village
Future projects planned for the village

This analysis was complemented by information gathered during the observations on the island and the consultation of secondary sources (Veal, 2011).

FINDINGS

The Case Yim Tin Tsai

Observations on the island and consultation of secondary sources were used to describe the location under analysis. Yim Tin Tsai is an off-shore island in Sai Kung, comprising an area of less than one square mile in the southeast part of Hong Kong. The island has increasingly attracted the interest of visitors, with a focus on local visitors, considering the existing religious-cultural-ecological resources (Su, 2018; Yau, 2016). Historically, Yim Tin Tsai is a traditional Hakka village of the Chan's clan, which has existed for more than 300 years. Differently from other traditional Chinese-style villages, Yim Tin Tsai villagers formed a Catholic community. Catholicism-Hakka culture is, therefore, a distinctive characteristic of this small-scale community, which is currently involved in the tourism development of the island.

In the early years, villagers of the island lived on farming and salt-making ("Yim Tin" means saltpan). In the 1920s, there were nearly 300 residents on the island. Because of the adverse situation of rural life in recent decades, all villagers have migrated to urban areas and overseas for living. While the village was abandoned in the 1990s, some members of Chan's clan, who share the same cultural identity, started to group together and contributed to the rejuvenation of their ancestral village (https://www.youtube.com/watch?v=CXN7yyNNFic). Recently, with its increasing popularity, Yim Tin Tsai has become an emerging travel destination. Since 2013, the number of visitor arrivals on the island has reached between 25,000 to 30,000 per year. The figure peaked at 37,000 visitors in 2016 when the historic saltpans were listed in the 2015 Asia-Pacific Heritage Awards by UNESCO (Agapito & Chan, 2019). These facts indicate an increased effort in community participation in the context of tourism development (Sai Kung District Council, 2011; Su, 2018; Wan, 2004).

In terms of the resources available on Yim Tin Tsai island, the religious features (e.g., St. Joseph's the Great Chapel, which earned an award of merit for cultural heritage conservation by UNESCO in

2005, and the Catholic pilgrimage route), Hakka culture and history (e.g., Heritage Centre and the salt-pans, which earned an award of distinction for cultural heritage conservation in 2015 by UNESCO), as well as ecological resources (e.g., mangrove, and other local floral and faunal species), all provide a unique and multisensory landscape for visitation (Figure 1). The current heritage exhibition centre was built at the site of the former primary school on the island named Ching Po School. The relics and other items at the centre were collected from different parts of the island by a group of local university students, in collaboration with the villagers, and the management committee of the village. The items were displayed systematically in four themes (household, containers, school and agriculture), which were relevant to the activities of the village livelihoods in the past. In 2016, a creative project named YIM-AGE was developed. The village community collaborated with a local non-governmental organisation (Lulu Culture) to display artworks related to community identity throughout the island (e.g., the glass windows displayed at the Hakka houses shown in Figures 2 and 3). The stakeholders in Yim Tin Tsai include the resource and visitor management party (The Salt and Light Preservation Centre) actively involved in the operational implementation of tour guiding in the village, the local government/tourism authority in Hong Kong, visitors and villagers. A Joint Committee including representatives from the Catholic Diocese of Hong Kong, Sacred Heart Church (Sai Kung), The Salt and Light Preservation Centre (in the capacity of a non-governmental organisation), as well as the Yim Tin Tsai Village Committee, was established to plan, manage and maintain the necessary operational procedures on the island with respect to tourism development.

Figure 1. Resources available in Yim Tin Tsai (Adapted from https://yimtintsai.com/)

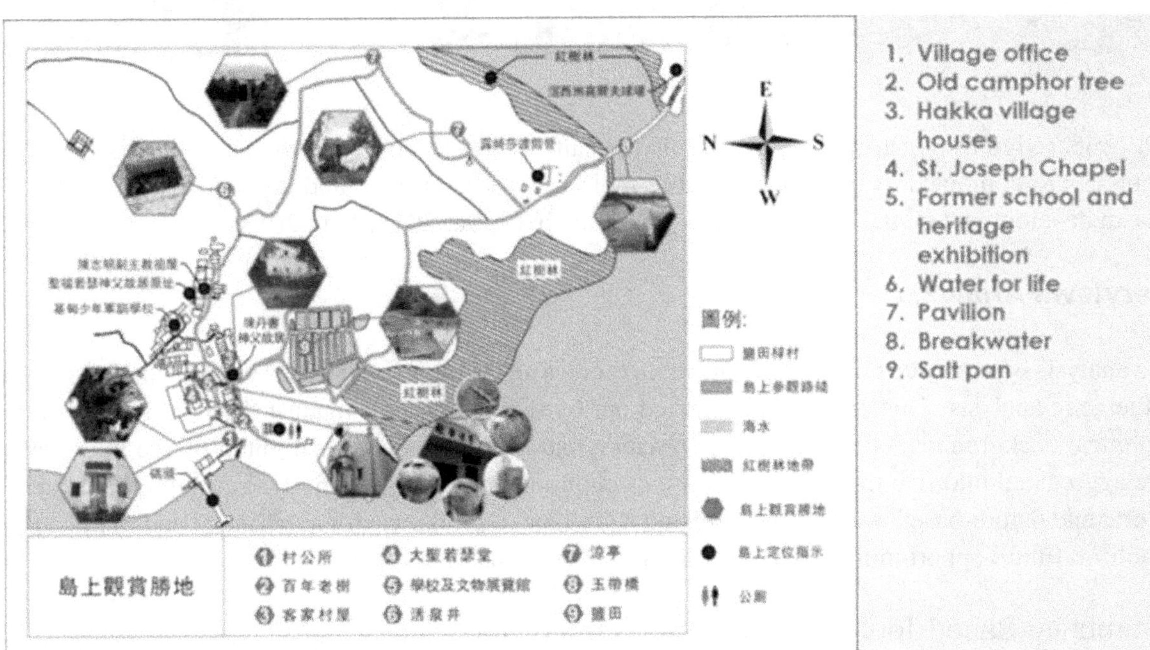

The availability of natural and cultural heritage in Yim Tin Tsai as attractions is similar to other outlying islands and rural environments in Hong Kong. However, this island also offers unique features,

Figure 2. Glass windows under the YIMAGE project (Source: Authors' own)

such as its religious constructions, as well as the salt-pans (Williams, 2018). These diverse facets of resources are combined to become a suitable multisensory landscape and distinctive community for tourism development on the island (Agapito & Chan, 2019; Su, 2018; Yau, 2016).

Interviews Analysis

Data analysis was focused on depicting the main themes emerging from the interviews, which was based on thematic analysis. This process was carried out by two researchers against the research object and theoretical background. As a result of this process, patterns of aspects highlighted by the interviewers were aggregated into five themes, as follows: a) community-based tourism dynamics on the island; b) cultural and nature-based resources and related activities; c) global visitor experience and multisensory stimuli; d) future opportunities; and e) challenges to sustainable tourism.

Community-Based Tourism Dynamics on the Island

The interviewees agree that an approach to community-based tourism is being used on the island, and that the members of the community are highly involved in the development of the village voluntarily. The role of the non-governmental organisation *The Salt and Light Preservation Centre* is set as central in

Figure 3. Glass windows under the YIMAGE project (Source: Authors' own)

the development of tourism and to enhancing the visitor experience, while developing efforts to achieve sustainable local development.

Interviewee a

[…] community includes villagers, church members and volunteers. They are the key part of maintaining tourism. …In the whole developing process, the whole community has also participated in it. […] We have to help in the process of the reception of tourists and control of the activities on the island. […] The Salt and Light Preservation Centre was founded in 2013. It is a non-governmental organisation formed by the villagers. This organisation has to deal with the restoration work on the island and the tourists' activities.

Interviewee b

[…] some villagers came back from overseas and they also influenced a lot of villagers to come back to YTT for conservation.

Interviewee c

Many volunteers are helping there […] There are member meetings […] people are not doing this for money or their own interest.

Cultural and Nature-Based Resources and Related Activities

The locals involved in this research highlight the Hakka cultural-based assets existing in the village as key endogenous resources. In addition to the Hakka houses and art portrayed in the village (e.g. glass windows), some relics are displayed at the old primary school in the village, which was transformed into an exhibition cultural centre. The religious component of the cultural heritage of Yim Tin Tsai is tangibilised by the St Joseph's Chapel, which has unique features in Hong Kong. The nature-based resources related to the salt fields and ecological assets are also pointed out as unique to the island and part of the local identity and intangible heritage. The main activities developed on the island are related to cultural tours, which are led by local guides. These also include specific workshops associated with traditional practices, such as food preparation and salt-making. Moreover, eco-guided tours and school tours, with a strong educational purpose, are also available to visitors.

Interviewee a

Hakka cultural relics left there […]. For example, tools for agricultural activities like plough and stone mills, daily utilities such as decorations, table, ceramics and Kylin [….] Small chapel and a road of reconciliation […] Salt fields […] There are also unique natural resources.

There are new types of tours. Some are about history, religion and nature. […] also activities with a specific topic, such as the mental cultivation through retreats. Another activity is the workshops. Visitors can have the experience of making steamed bun and salt by themselves.

Interviewee b

There are 12 tourist sites in YTT, such as the church, salt field, village houses, Fung Shui wood, wells and graveyard […] In the graveyard, there is an opened altar which is the only one existing in Hong Kong, and the cross inside the altar is special because it involves the art design of burgeon. In the cross, it is also possible to observe a mixture of Hakka and Christian culture, which is difficult to find in the whole South China region. […] There are eco-tours and cultural guided tours.

Interviewee c

Local customs […]. There is a special culture there, which is not just the material aspect. […] There are school tours.

Global Experience and Multisensory Stimuli

The interviewees mention that the visitor experiences on the island involve actively a variety of senses, in addition to visual stimuli. For example, visitors can observe the process of salt production, but they can also touch and taste this local and educational resource. In addition, endogenous plants and trees can trigger the olfactory sense, and local gastronomy is regarded as of multisensory nature. These sensory stimuli are enveloped by the intangible heritage related to Hakka culture present in the village.

Interviewee a

When the visitors first arrive on the island, there are already visual stimuli. When they first see the big bucket of salt and know that they are produced in Hong Kong, they are greatly amazed and this makes them know that Hong Kong can produce salt. Most of the time we only involve verbal interpretation. There is less use of smell in the activities. However, during the guided tour in nature, the guide may sometimes ask the participants to smell the special scent from the plants, such as the plants that can remove the mosquitos. The workshop may also include the making of steamed bun.

Interviewee b

[…] Hakka culture, actually, can be "eaten". We can taste Hakka Chaguo (茶果), Hakka dishes, Hakka rice. There is local food, at the same time, which is also the "Hakka culture". There is a museum and there are a few classrooms inside. We have exhibited some simple agricultural tools. For the 4000 - 5000 history, the Hakka culture and story have kept the tradition. For the heritage or relics, we can touch,

listen to the story, eat and also feel. Moreover, the camphor trees locate at the village, we can smell the camphor, we can touch it, we can see the tree crown and listen to the Hakka cultural story. It is about multi-sensory development.

Future Opportunities

As for future opportunities, participants in the research point out the potential of further developing workshops in the village. Activities related to mindfulness and therapy are seen as adequate to the development of sustainable tourism on the island; also, they can contribution to the revitalisation and preservation of the endogenous resources. As mentioned by one of the interviewees, recently Hong Kong government predicts to organise a new arts festival to release the unique potential of experiencing the culture and natural scenery of the island. This decision implies rising attention by the public sector to releasing the potential of Yim Tin Tsai as a travel destination even though the outcome of such development still has some uncertainties due to the presence of more potential stakeholders, such as investors.

Interviewee a

Due to responsible management, the visitors are now focusing more on the environment [....]. They are no longer making themselves as the centre. This kind of sustainable tourism makes it change into a more in-depth kind of education. This can enhance new mental development. [...] financial return is mainly related to the workshops. The government is also preparing an art festival and it is a 3 years plan.

Interviewee c

The atmosphere is very comfortable and enjoyable.... In recent years more people are talking about multisensory experience and I think it is possible to apply this to the community-based activities... The visitors nowadays are willing to pay more as they think it is worth it. In Yim Tin Tsai there is a unique environment that can be used to develop mindfulness activities, develop our selves, and perform forest therapy. I believe the tours should not be only about science, but more importantly, the story.

Challenges to Sustainable Tourism

Some of the main challenges stressed by the interviewees are regarded to difficulties in accessing funding, time constraints of the locals, who develop work on a voluntary basis, and the lack of managerial skills from the part of volunteers. Subsequently, there is a need for additional training. In addition, there are current constraints in controlling visitors' numbers.

Interviewee a

The people with heart may not have sufficient time as a long period of time is needed for designing the activities. [...] some of the people do not have enough knowledge about that. The [...] financial situation is tight and we need funding. […] tourist control (numbers) is also another challenge […] When there are too many people (visitors), this affects the multisensory experience of the visitors. We are having a passive management method.

Interviewee b

Environmental challenges: The climate is different from the past. When we recover and restore the salt field resources, we face the obstacles. We are lacking skills […] As cultural challenges: It is difficult to understand or feel the culture […] usually, too many tourists come during the hot season (i.e. weekends and holidays). The huge number of visitors can heavily burden the infrastructure (like toilets and the classroom) in YTT.

Interviewee c

The development of tourism is still at a preliminary stage as we do not have enough experience about it. The main concern is still the financial problem of running this and we have to think about how to further develop. It is money, time and manpower. There is the possibility of development when people get through the training process.

DISCUSSION AND RECOMMENDATIONS

The findings of research conducted in Yim Tin Tsai evinced the key role of voluntary community involvement in the revitalisation and conservation of local cultural and natural resources. This process is being undertaken by focusing on a community-based approach to tourism, and the findings revealed the potential of this approach for sustainable tourism and development of this island.

The multisensory nature of the endogenous resources existing in the village (which involve both tangible and intangible heritage), combined with the potential to enhance and diversify the visitor experience, is one of the emergent themes derived from the interviews; these were conducted with members of the community. While this aspect was stressed by villagers in this study, a prior study conducted on the island using a questionnaire to examine sensory experiences as reported by visitors, suggested that a holistic approach to multisensory stimuli could contribute to responsible tourism management within community-based tourism in the village (Agapito & Chan, 2019). In particular, the referred study pointed out 'the potential of the proposed approach to enhance the tourist experience, to make the tourist experience more inclusive, to contribute to heritage conservation, and to boost local identity, sense of place, and place making in the context of community-based tourism' (p. 8). Similar to the host/supply perspective explored in the present study, Agapito and Chan (2019) revealed that gastronomic impressions related to local cuisine, as well as the saltpans, were mentioned by visitors as multisensory resources. These results show similarities between the perspectives of visitors and the host population in relation to this aspect.

In terms of the implications for management on the island, the host perspective suggested that opportunities regarding the process of stimulating different senses can attract visitors with a profile that is aligned with the sustainable objectives of local development. This can be implemented by focusing on unique cultural and natural heritage by developing diversified and creative community-led workshops. This finding is in line with an emergent field of research on creative tourism which is informed: a) by the demands of a tourist segment that is motivated to develop their own creativity/knowledge through active and immersive cultural experiences, and b) by the concerns of contemporary destinations on attaining sustainable competitiveness by focusing on people, local identity and culture (Duxbury & Richards, 2019). This idea is also aligned with a contemporary strand of research which combines design science and tourism. This strand of research focuses on planning stimuli within destination experiences in an innovative, creative and systematic fashion, which can contribute to 'improve people's lives as well as their travel experiences', and subsequently boost people's well-being (Fesenmaier and Xiang, 2017, p. 4). Accordingly, alternative activities based on the tangible and intangible heritage of the village can be developed using multisensory stimuli and technology, which allows visitors to engage with the identity of the island in different environments. These environments (e.g. interpretative centres and virtual environments) can be dislocated from the village, thus relieving pressure on the island in relation to visitor numbers and social and ecological negative impacts.

Triangulation of sources also suggested that cultural and eco-tours conducted by locals act as important activities on the island to ensure adequate heritage interpretation, preservation of local identity and endogenous resources, quality of tourist experience, and perceived authenticity by visitors in the village. Accordingly, research on the tourist experience points out that an adequate interpretation of cultural heritage is pivotal to the quality of the tourist experience, effective management and conservation of heritage (Moscardo, 1996). Inasmuch as representatives of the community believe the interplay between the cultural and natural environments offered in the village is ideal for the development of mindfulness activities, mindful-based activities could be enhanced by adequate interpretation performed by locals. Indeed, quality interpretation can contribute to 'produce mindful visitors; visitors who are active, interested, questioning and capable of reassessing the way they view the world' (Moscardo, 1996, p. 382).

In the event of the tourism interest in the village continues to grow and larger events are organized *in situ*, a potential problem of overtourism related to possible negative impacts derived from exceeding physical, environmental and social-related carrying capacity on the island may rise and influence negatively sustainable development (Newsome et al., 2013). Indeed, whereas overtourism impacts both tourists and residents, this phenomenon can be foreseen and tackled in the management of travel destinations (Milano et al., 2019). As Hong Kong government is expected to organise a new arts festival on the island with the potential to attract a high number of tourists to the village during a short period of time (Legislative Council Panel on Economic Development – LCPED, 2018), solutions helping to prevent and mitigate overtourism issues should be a priority for management.

While, currently, the management on the island follows predominantly an organic approach, a mixed approach by involving local government in a more integrated fashion could be relevant to tackle some of the constraints identified by the villagers, which are aligned with common limitations of community-based projects identified in literature (Addison, 1996). These limitations are related to aspects related to lack of expertise of volunteers, human resources availability, visitors' management and control and access to funding. Notwithstanding, local government involvement in the development of the island and the planning of events should maintain the focus on preserving the local identity, cultural and natural resources of the village. The reason why is because this process would involve additional stakeholders (economically related) and, consequently, more uncertainties. In this light, a mixed approach to place making can be undertaken by travel destinations by balancing the organic bottom-up approach (local initiatives) with the top-down involvement in the marketing and management decision-making process (Lew, 2017).

LIMITATIONS AND FUTURE RESEARCH DIRECTIONS

Although this study had an exploratory nature, it is important to acknowledge that data collection was limited. A triangulation of methods included: a) three interviews developed with representatives of the host community in Yim Tin Tsai who were available at the time of this study, b) observations performed by the researchers in one point in time, and c) available secondary sources. Only representatives that were actively engaged in the development of tourism in the village were considered at this stage of the study. Hence, there is scope to furthering the present research by expanding the number and diversity of stakeholders involved in data collection. This could help to achieve a more comprehensive understanding of the process of revitalisation of cultural and natural heritage for sustainable tourism through community involvement. This research will be expanded by analysing opportunities and challenges to sustainable tourism (emerged themes from the interviews) against the government and visitors' perspectives (both domestic and international). This future research will allow addressing, in particular, the arts festival which was reported as expected in the village and aiming at combining the unique potential of experiencing the culture and natural scenery of the island.

In addition, future research on the topic covered in this study could explore how the focus on a holistic approach to the multisensory stimuli, enveloped by the intangible heritage on off-shore islands and small villages, such as Yim Tin Tsai, could encourage a more inclusive experience for visitors. Indeed, the use of multiple sensory stimulations (both direct and virtual) has the potential not only to tangibilise and boost the experience for tourists with sensory disabilities but also to contribute to continuously co-creating experiences with all visitors and locals (Agapito et al., 2013).

Moreover, quasi-experimental research could be useful to assess the potential of incorporating multisensory stimuli on virtual and interactive environments for educational and visitation purposes. The focus would be on disseminating cultural and natural heritage, as well as community identities, such as the ones found in Yim Tin Tsai. Also, the effectiveness of these alternative environments (e.g virtual environments and interactive visitation centres located on the mainland) could be explored as managerial tools to address challenges to sustainability, such as issues related to the control of tourist numbers, as well as in situations of destination crisis management. Accordingly, this type of research could examine practical solutions to tackle issues such as overtourism or undertourism (Milano et al., 2019) in the context of community-based projects. Also, creative and innovative solutions incorporating technology to attract funding and partnerships with universities and companies (which could generate additional forms of economic benefits to the community) could be explored in future research. Comparative studies with small destinations with similar profiles to Yim Tin Tsai would be relevant to help to generalise results and develop a managerial framework on community-based projects based on the revitalisation of cultural and natural resources for sustainable tourism.

CONCLUSION

This research was based on a case study and concluded that the community involvement in Yim Tin Tsai was key to the revitalisation of local cultural and natural resources for the development of sustainable tourism in this small off-shore island in Hong Kong. The challenges of the current bottom-up approach of management in the village for place making are mostly related to time constraints, access to funding, need for further training on managerial skills and constraints in controlling visitor numbers. Nevertheless, the local community (mostly composed of Hakka clan descendants) managed to raise the attractiveness of the island for visitation by revitalising cultural and natural heritage (such as the salt pans), which is part of the community identity. Hence, facing a desertification phenomenon and a potential loss of community identity for future generations, an organic community-based approach encouraged heritage conservation and sustainable development. The voluntary nature and the high level of community participation were pivotal in this process, considering the vulnerable sociocultural and ecological environment of this island, as well as the unique characteristics of the community. Hence, the research findings are aligned with the rationale that a responsible management approach to tourism in more vulnerable travel destinations should focus on targeting tourism markets and managerial strategies matching specific goals on cultural and nature conservation (Collison & Spears, 2010).

ACKNOWLEDGMENT

This paper is financed by National Funds provided by FCT - Foundation for Science and Technology through project UIDB/04020/2020. We would like to thank Mr. Kwo Fung Shek for his assistance in the process of data collection.

REFERENCES

Aas, C., Ladkin, A., & Fletcher, J. (2005). Stakeholder collaboration and heritage management. *Annals of Tourism Research*, *32*(1), 28–48. doi:10.1016/j.annals.2004.04.005

Addison, L. (1996). An approach to community-based tourism planning in the Baffin region, Canada's Far North – A retrospective. In L. C. Harrion & W. Husbands (Eds.), *Practicing Responsible Tourism* (pp. 296–329). John Wiley and Sons.

Agapito, D., Almeida, H., & Campos, A. C. (2016). Sensations, perception and co-creation in rural tourism experiences. In E. Kastenholz, M. J. Carneiro, C. Eusébio, & E. Figueiredo (Eds.), *Meeting Challenges for Rural Tourism through Co-Creation of Sustainable Tourist Experiences* (pp. 39–60). Cambridge Scholars Publishing.

Agapito, D., & Chan, C. S. (2019). A multisensory approach to responsible management in community-based tourism: A case study in Hong Kong. *Journal of Tourism Quarterly*, *1*(1), 1–13.

Agapito, D., Mendes, J., & Valle, P. (2013). Conceptualizing the sensory dimension of tourist experiences. *Journal of Destination Marketing & Management*, *2*(2), 62–73. doi:10.1016/j.jdmm.2013.03.001

Agapito, D., Valle, P., & Mendes, J. (2014). The sensory dimension of tourist experience: Capturing meaningful sensory-informed themes in South Portugal. *Tourism Management, 42*, 224–237. doi:10.1016/j.tourman.2013.11.011

Ahmad, Y. (2006). The scope and definitions of heritage: From tangible to intangible. *International Journal of Heritage Studies, 12*(3), 292–300. doi:10.1080/13527250600604639

Albrechts, L., & Denayer, W. (2001). Communicative planning, emancipatory politics and postmodernism. In R. Paddison (Ed.), *Handbook of urban studies* (pp. 369–384). SAGE Publications Ltd. doi:10.4135/9781848608375.n24

Arnstein, S. R. (1969). A ladder of citizen participation. *Journal of the American Institute of Planners, 35*(4), 216–224. doi:10.1080/01944366908977225

Blackstock, K. (2005). A critical look at community-based tourism. *Community Development Journal: An International Forum, 40*(1), 39–49. doi:10.1093/cdj/bsi005

Blake, M. K. (2017). Building an unjust foodscape: Shifting governance regimes, urban place making and the making of Chinese food as ordinary in Hong Kong. *The International Journal of Justice and Sustainability, 23*(11), 1047–1062.

Boisen, M., Terlouw, K., & van Gorp, B. (2011). The selective nature of place branding and the layering of spatial identities. *Journal of Place Management and Development, 4*(2), 135–147. doi:10.1108/17538331111153151

Brand, R., & Gaffikin, F. (2007). Collaborative planning in an uncollaborative world. *Planning Theory, 6*(3), 282–313. doi:10.1177/1473095207082036

Brunt, P., Horner, S., & Semley, N. (2017). *Research Methods in Tourism, Hospitality and Events Management*. Sage.

Cartier, C. (1999). The state, property development and symbolic landscape in high-rise Hong Kong. *Landscape Research, 24*(2), 185–208. doi:10.1080/01426399908706558

Castillo, R. (2014). Feeling at home in the "Chocolate City": An exploration of place-making practices and structures of belonging amongst Africans in Guangzhou. *Inter-Asia Cultural Studies, 15*(2), 235–257. doi:10.1080/14649373.2014.911513

Chan, P. (2016). *Place-making – harbourfront for Hong Kong people*. Retrieved September 1, 2019, from https://www.devb.gov.hk/en/home/Blog_Archives1/index_id_180.html

Cheung, P. T. Y. (2011). Civic engagement in the policy process in Hong Kong: Change and continuity. *Public Administration and Development, 31*(2), 113–121. doi:10.1002/pad.597

Chung, C. Y., & Ngai, P. (2007). Neoliberalization and privatisation in Hong Kong after the 1997 financial crisis. *China Review, 7*(2), 65–92.

Collison, F. M., & Spears, D. L. (2010). Marketing cultural and heritage tourism: The Marshall Islands. *International Journal of Culture, Tourism and Hospitality Research, 4*(2), 130–142. doi:10.1108/17506181011045208

Community Project Workshop (CPW). (2013). *The renewal and development of Kwun Tong industrial district study*. Hong Kong: Kwun Tong District Council; Kwun Tong Development and Renewal Task Force.

Connell, D. (1997). Participatory development: An approach sensitive to class and gender. *Development in Practice*, *7*(3), 248–259. doi:10.1080/09614529754486 PMID:12348314

CPW - Community Project Workshop. (2013). *The renewal and development of Kwun Tong industrial district study*. Hong Kong: Kwun Tong District Council; Kwun Tong Development and Renewal Task Force.

Croes, R., & Semrad, K. J. (2015). The relevance of cultural tourism as the next frontier for small island destinations. *Journal of Hospitality & Tourism Research (Washington, D.C.)*, *39*(4), 469–491. doi:10.1177/1096348013491599

Davis, P. (1999). *Ecomuseums: a sense of place*. Leicester University Press.

Development Bureau. (2011). *Urban renewal strategy*. Hong Kong Special Administrative Region Government.

Duxbury, N., & Campbell, H. (2009). Developing and Revitalizing Rural Communities through Arts and Creativity: A Literature Review. *Creative City Network of Canada*. Retrieved September 1, 2019, from https://www.creativecity.ca/database/files/library/rural_communities_arts_2009.pdf

Duxbury, N., & Richards, G. (2019). *A Research Agenda for Creative Tourism*. Edward Elgar. doi:10.4337/9781788110723

Fesenmaier, D. F., & Xiang, Z. (2017). *Design Science in Tourism: Foundations of Destinations Management*. Springer. doi:10.1007/978-3-319-42773-7

Florida, R. (2002). *The rise of the creative class, and how it's transforming work, leisure community and everyday life*. Basic Books.

Friedman, J. (2010). Place and place-making in cities: A global perspective. *Planning Theory & Practice*, *11*(2), 149–165. doi:10.1080/14649351003759573

Garrod, B., Wornell, R., & Youell, R. (2006). Re-conceptualising rural resources as countryside capital: The case of rural tourism. *Journal of Rural Studies*, *22*(1), 117–128. doi:10.1016/j.jrurstud.2005.08.001

Getz, D., & Jamal, T. B. (1994). The environment–community symbiosis: A case of collaborative tourism planning. *Journal of Sustainable Tourism*, *2*(3), 152–173. doi:10.1080/09669589409510692

Gray, B. (1985). Conditions facilitating interorganisational collaboration. *Human Relations*, *38*(10), 911–936. doi:10.1177/001872678503801001

Hardy, A., Beeton, R., & Pearson, L. (2002). Sustainable tourism: An overview of the concept and its position in relation to conceptualisations of tourism. *Journal of Sustainable Tourism*, *10*(6), 475–496. doi:10.1080/09669580208667183

Haywood, K. M. (1988). Responsible and responsive tourism planning in the community. *Tourism Management*, *9*(2), 105–108. doi:10.1016/0261-5177(88)90020-9

Hou, J., & Rios, M. (2003). Community-driven place making. *Journal of Architectural Education, 57*(1), 19–27. doi:10.1162/104648803322336557

Hultman, J., & Hall, C. M. (2012). Tourism place-making: Governance of locality in Sweden. *Annals of Tourism Research, 39*(2), 547–570. doi:10.1016/j.annals.2011.07.001

Kastenholz, E., Carneiro, M. J. A., & Marques, C. P. (2012b). Marketing the rural tourism experience. In R. H. Tsiotsou & R. E. Goldsmith (Eds.), *Strategic marketing in tourism services* (pp. 247–264). Emerald.

Kastenholz, E., Carneiro, M. J. A., Marques, C. P., & Lima, J. (2012a). Understanding and managing the rural tourism experience – The case of a historical village in Portugal. *Tourism Management Perspectives, 4*, 207–214. doi:10.1016/j.tmp.2012.08.009

Kavaratzis, M., & Hatch, M. J. (2013). The dynamics of place brands: An identity-based approach to place branding theory. *Marketing Theory, 13*(1), 69–86. doi:10.1177/1470593112467268

Kee, T. (2017). Innovation in sustainable development: Sustainable place-making strategies in Hong Kong. *The 19th International Conference on Innovation and Sustainable Development Conference Proceedings.*

Keitumetse, S. O. (2011). Sustainable development and cultural heritage management in Botswana: Towards sustainable communities. *Sustainable Development, 19*(1), 49–59. doi:10.1002d.419

Kendig, L. H., & Keast, B. C. (2010). *Community character: Principles for design and planning*. Island Press.

Knowles, C., & Harper, D. (2009). *Hong Kong: Migrant lives, landscapes, and journeys*. University of Chicago Press. doi:10.7208/chicago/9780226448589.001.0001

Landry, C. (2000). *The creative city: A toolkit for urban innovation.* Comedia.

LCPED - Legislative Council Panel on Economic Development. (2018). *2018 Policy Address Policy Initiatives of Commerce, Industry and Tourism Branch, Commerce and Economic Development Bureau.* Retrieved September 1, 2019, from https://www.legco.gov.hk/yr18-19/english/panels/edev/papers/edevcb4-10-2-e.pdf

Lems, A. (2016). Placing displacement: Place-making in a world of movement. *Ethnos, 81*(2), 315–337. doi:10.1080/00141844.2014.931328

Lew, A. A. (2012). Geography and the marketing of tourism destinations. In J. Wilson (Ed.), *The Routledge handbook of tourism geographies* (pp. 181–186). Routledge.

Lew, A. A. (2017). Tourism planning and place making: Place-making or placemaking? *Tourism Geographies, 19*(3), 448–466. doi:10.1080/14616688.2017.1282007

Main, K., & Sandoval, G. F. (2015). Placemaking in a translocal receiving community: The relevance of place to identity and agency. *Urban Studies (Edinburgh, Scotland), 52*(1), 71–86. doi:10.1177/0042098014522720

Manente, M., & Minghetti, V. (2006). Destination management organizations and actors. In D. Buhalis & C. Costa (Eds.), *Tourism business frontiers: Consumers, products and industry* (pp. 228–237). Elsevier Butterworth Heinemann. doi:10.1016/B978-0-7506-6377-9.50032-8

Massey, D. (2005). *For space*. Sage.

Mazumdar, S., & Mazumdar, S. (2008). Religious placemaking and community building in Diaspora. *Environment and Behavior*, *41*(3), 1–31.

Milano, C., Cheer, J., & Novelli, M. (2019). *Overtourism: Excesses, discontents and measures in travel and tourism*. Cabi Publishing. doi:10.1079/9781786399823.0000

Moscardo, G. (1996). Mindful Visitors. *Annals of Tourism Research*, *23*(2), 376–397. doi:10.1016/0160-7383(95)00068-2

Murtas, D., & Davis, P. (2009). The role of the Ecomuseo dei terrazzamenti e della vite, (Cortemilia, Italy) in community development. *Museum & Society*, *7*(3), 150–186.

Newsome, D., Moore, S. A., & Dowling, R. K. (2013). *Natural Area Tourism: Ecology, Impacts and Management* (2nd ed.). Channel View Publications.

Ng, M. N. (2009). *Pilgrimages: Memories of Colonial Macau and Hong Kong*. Hong Kong University Press.

Okazaki, E. (2008). A Community-Based Tourism Model: Its Conception and Use. *Journal of Sustainable Tourism*, *16*(5), 511–529. doi:10.1080/09669580802159594

Othman, S., Nishimura, Y., & Kubota, A. (2013). Memory association in place making: A review. *Procedia: Social and Behavioral Sciences*, *85*, 554–563. doi:10.1016/j.sbspro.2013.08.384

Peirce, J., Martin, D. G., & Murphy, J. T. (2011). Relational place-making: The networked politics of place. *Transactions of the Institute of British Geographers*, *36*(1), 54–70. doi:10.1111/j.1475-5661.2010.00411.x

PPS (Project for Public Spaces). (2007). *What Is Placemaking?* Retrieved September 1, 2019, from https://www.pps.org/article/what-is-placemaking

PPS (Project for Public Spaces). (2013). Stronger citizens, stronger cities: Changing governance through a focus on place. *Project for Public Spaces website*. Retrieved September 1, 2019, from http://www.pps.org/blog/stronger-citizens-stronger-cities-changing-governance-through-a-focus-on-place/

Prentice, R. (1993). Community-driven tourism planning and residents' preferences. *Tourism Management*, *14*(3), 218–227. doi:10.1016/0261-5177(93)90023-E

Ren, H. (2010). *Neoliberalism and culture in China and Hong Kong: The countdown of time*. Routledge. doi:10.4324/9780203844939

Richards, G. (2014). Creativity and tourism in the city. *Current Issues in Tourism*, *17*(2), 119–144. doi:10.1080/13683500.2013.783794

Ritchie, J., & Spencer, L. (1994). Qualitative Analysis for Applied Policy Research. In A. Bryman & R. G. Burgess (Eds.), *Analyzing Qualitative Data* (pp. 173–194). Routledge. doi:10.4324/9780203413081_chapter_9

Roberts, L., & Hall, D. (2001). *Rural tourism and recreation: principles to practice*. CABI Publishing. doi:10.1079/9780851995403.0000

Rose-Redwood, R., & Alderman, D. (2011). Critical interventions in political toponymy. *ACME: An International E-Journal for Critical Geographies*, *10*(1), 1–6.

Rossler, M. (2006). World Heritage cultural landscapes. *Landscape Research*, *31*(4), 333–353. doi:10.1080/01426390601004210

Sai Kung District Council. (2011). *Yim Tin Tsai Village and St. Joseph's Chapel.* Retrieved April 20, 2018, from http://travelinsaikung.org.hk/en/Routes/view.aspx?menu=20&Nid=59

Samadhi, T. N. (2001). The urban design of a Balinese town: Placemaking issues in the Balinese urban setting. *Habitat International*, *25*(4), 559–575. doi:10.1016/S0197-3975(01)00024-8

Sebele, L. S. (2010). Community-based tourism ventures, benefits and challenges: Khama Rhino Sanctuary Trust, Central District, Botswana. *Tourism Management*, *31*(2), 136–146. doi:10.1016/j.tourman.2009.01.005

Sharpley, R., & Roberts, L. (2004). Rural Tourism – 10 Years On. *International Journal of Tourism Research*, *6*(3), 119–124. doi:10.1002/jtr.478

Stone, L. (1989). Cultural Crossroads of Community Participation in Development: A Case from Nepal. *Human Organization*, *48*(3), 206–213. doi:10.17730/humo.48.3.d436524749371702

Su, X. (2018). Yim Tin Tsai village in Hong Kong: a model for effective conservation. *South China Morning Post.* Retrieved April 20, 2018, from https://www.scmp.com/photos/hong-kong/2161879/yim-tin-tsai-village-hong-kong-model-effective-conservation

Taylor, K., & Lennon, J. L. (2011). Cultural landscapes: A bridge between culture and nature? *International Journal of Heritage Studies*, *17*(6), 537–554. doi:10.1080/13527258.2011.618246

Thurlow, C., & Jaworski, A. (2014). 'Two hundred ninety-four': Remediation and multimodal performance in tourist placemaking. *Journal of Sociolinguistics*, *18*(4), 459–494. doi:10.1111/josl.12090

Tosun, C. (1999). Towards a Typology of Community Participation in the Tourism Development Process. *Anatolia*, *10*(2), 113–134. doi:10.1080/13032917.1999.9686975

Tosun, C. (2000). Limits to community participation in the tourism development process in developing countries. *Tourism Management*, *21*(6), 613–633. doi:10.1016/S0261-5177(00)00009-1

Trudeau, D. (2006). Politics of belonging in the construction of landscapes: Place-making, boundary-drawing and exclusion. *Cultural Geographies*, *13*(3), 421–443. doi:10.1191/1474474006eu366oa

UNESCO Creative Network. (2006). *Towards Sustainable Strategies for Creative Tourism*. Discussion Report of the Planning Meeting for 2008 International Conference on Creative Tourism, Santa Fe, NM. Retrieved April 20, 2019, from https://unesdoc.unesco.org/ark:/48223/pf0000159811

Veal, A. J. (2011). *Research Methods for Leisure and Tourism* (4th ed.). London: Financial Times/Prentice Hall.

Wan, C. T. (2004). *From salt pan to resort plan: heritage conservation for the Island of Yim Tin Tsai, Sai Kung* (Unpublished master dissertation). The University of Hong Kong, Hong Kong.

Williams, M. (2018). *Hong Kong's only working salt pans and a Robinson Crusoe moment just a boat ride away from Sai Kung*. Retrieved September 1, 2019, from https://www.scmp.com/lifestyle/travel-leisure/article/2130134/hong-kongs-only-working-salt-pans-and-robinson-crusoe

Wortham-Galvin, B. D. (2008). Mythologies of placemaking. *Places*, *20*(1), 32–39.

Wyckoff, M. A., Neumann, B., Pape, G., & Schindler, K. (2015). *Placemaking as and economic development tool: A placemaking guide*. Land Policy Institute, Michigan State University.

Yau, E. (2016). Abandoned Hong Kong island gets new life as heritage site and ecotourism destination. *South China Morning Post*. Retrieved September 1, 2019, from https://www.scmp.com/lifestyle/travel-leisure/article/1908076/abandoned-island-sai-kung-comes-back-life-heritage-site-and

Yin, R. (2014). *Case study research*. Sage Publications.

ADDITIONAL READING

du Cros, H. (2001). A new model to assist in planning for sustainable cultural heritage tourism. *International Journal of Tourism Research*, *3*(2), 165–170. doi:10.1002/jtr.297

Lussetyowati, T. (2015). Preservation and conservation through cultural heritage tourism: Case study: Musi Riverside Palembang. *Procedia: Social and Behavioral Sciences*, *184*, 401–406. doi:10.1016/j.sbspro.2015.05.109

McKercher, B., Hoa, P. S. Y., & du Cros, H. (2005). Relationship between tourism and cultural heritage management: Evidence from Hong Kong. *Tourism Management*, *26*(4), 539–548. doi:10.1016/j.tourman.2004.02.018

Murphy, P. E. (1985). *Tourism a community approach*. Methuen.

Pederson, A. (2014). *A framework for defining a world heritage path to tourism: Small island developing states (SIDS)*. Ministry of Education, Culture and Science of Netherlands.

Smith, S. (2015). A sense of place: Place, culture and tourism. *Tourism Recreation Research*, *40*(2), 220–233. doi:10.1080/02508281.2015.1049814

Til, V. J. (1984). Citizen participation in the future. *Policy Studies Review*, *3*(2), 311–322. doi:10.1111/j.1541-1338.1984.tb00122.x

Tosun, C. (2006). Expected nature of community participation in tourism development. *Tourism Management*, *27*(3), 493–504. doi:10.1016/j.tourman.2004.12.004

Tussyadiah, P. (2014). Toward theoretical foundation for Experience Design in Tourism. *Journal of Travel Research*, *53*(5), 543–564. doi:10.1177/0047287513513172

KEY TERMS AND DEFINITIONS

Community: A group of citizens within a given geographical area and/or a group of individuals who share origins or interests.

Cultural Heritage: Tangible and intangible resources related to signs of societies' past activities on the environment.

Heritage Conservation: The process of managing heritage assets in a way that sustains, preserves, and, in some cases, enhances its significance.

Natural Heritage: Heritage related to specific natural features, areas involving endemic species (flora and fauna), geological and physiographical formations, and sites of unique natural beauty or conservation.

Place Making: Planning and marketing management decision-making processes involved in shaping destinations that can follow an organic bottom-up approach (local initiatives), top-down involvement (e.g., local government), or a combination of both approaches.

Responsible Management: Approach to management that promotes the sustainable use of local resources by ensuring its long-term use and enjoyment, encouraging responsible interactions between stakeholders, and considering ethical issues.

Sustainable Tourism: Approach to tourism that addresses the needs of the different stakeholders at the destination and considers the current and future economic, sociocultural, and environmental impacts generated by tourism-related activities. This approach aims to mitigate the negative impacts and maximise the positive ones.

Chapter 7
Residents' Perception and Their Support for Tourism Development:
The Case of South Korea

Manuel Alector Ribeiro
University of Surrey, UK

Yo Han Kim
University of Surrey, UK

Kyle Maurice Woosnam
University of Georgia, USA & University of Johannesburg, South Africa

ABSTRACT

The purpose of this study is to examine residents' support for tourism development in South Korea by the lens of social exchange theory and theory of reasoned. Random sampling strategy was used to collect South Koreans residents' opinions about the perceived impacts of tourism and their level of support for tourism development. Data were collected online using a questionnaire a link was sent to residents in South Korean through several social media platforms, and 693 respondents answered the questionnaire. By using regression analysis, the findings showed that residents who perceive positive impacts of tourism more than negative impacts are more likely to support tourism, while those who perceive more negative impacts of tourism are less likely to support tourism. Both personal economic benefits and community attachment were found to be important determinants of residents' perception and support toward tourism. Both theoretical and managerial implications are discussed, as well as the limitations and recommendations for future studies.

DOI: 10.4018/978-1-7998-3156-3.ch007

INTRODUCTION

Tourism brings impacts destinations in both positive and negative ways, thus affecting residents' quality of life (Andereck & Nyaupane, 2011; Woo, Kim, & Uysal, 2015). Residents can perceive tourism as the development tool that provides benefits such as generating more business, providing more employment opportunities and improving the quality of public infrastructures (Andereck & Nyaupane, 2011; Dyer et al., 2007; Gursoy et al., 2002). However, tourism also brings negative impacts such as polluting the local environment and increasing the cost of living in local areas (Andereck et al., 2005; MacGehee & Andereck, 2004).

According to Gursoy and Rutherford (2004), the success of tourism requires active support from residents. According to Murphy (1985, p. 153), "if residents resent or fear tourism, their resistance and hostility can destroy the local industry's potential". In other words, to develop sustainable tourism, we must seek to understand and incorporate residents' perceptions concerning tourism moving forward. Depending on how residents perceive impacts of tourism, they will fall somewhere along the spectrum from opposition to support for tourism development (Nunkoo & Ramkisson, 2011). In other words, residents' support for tourism development can depend on how residents perceive impacts of tourism (Andereck et al., 2005). Many previous studies reported that if residents positively perceive impacts of tourism, they are more likely to support tourism development (Allen et al., 1988; Gursoy et al., 2002; Sheldon & Abenoja, 2001).

On the other hand, negative impacts of tourism can lead residents to have negative perceptions toward tourism, which ultimately translate to less support for tourism development (Choi & Sirakaya, 2005; Nunkoo & Ramkisson, 2011). Furthermore, other studies have reported that residents' active support for tourism development is strongly related to the success of tourism development and also sustainable tourism development (Gursoy & Rutherford, 2004; Gursoy, Jurowski, & Uysal, 2002). According to Andereck et al. (2005), residents' support for tourism development can be based on how residents evaluate and perceive benefits and costs from tourism. Overall, to get residents' support for tourism development, how to bring perceived positive impacts of tourism to residents needs to be considered, while minimizing perceived negative impacts of tourism (Jurowski et al., 1997; Gursoy et al., 2010).

According to Korea Tourism Organization's reports in 2017 and 2018, there are several factors that positively affect the number of international tourists to South Korea. It includes increased international tourists' interest in Korean foods, cultures and music, and the 2018 Olympic Winter Games held in PyeongChang. K-pop (focusing on pop music, dramas, and movies), as of late, has also attracted many international tourists to South Korea. Given these factors, South Korea has experienced an average increase of 6.6% in the number of tourists to South Korea between 1999 and 2018 (e.g., 7 million in 2009; 15 million in 2018). Such growth in visitor numbers are most welcomed by the South Korean government, however, gauging residents' perceptions of tourism impacts and how that translates to overall support for the industry is crucial in ensuring tourism development is as sustainable as possible for the country.

This study will examine positive and negative impacts of tourism that can influence South Korean residents' perception toward tourism and examine how they perceive it. Then it will identify relationships between perceived impacts of tourism and residents' support for tourism development. The constructs of this study are divided into environmental sustainability, personal economic benefits and community attachment that have been reported as important sustainability-related factors for residents' perception and support toward tourism development (Choi & Murray, 2010; Gursoy et al., 2019; Nicholas et al., 2009; Ribeiro et al., 2017).

Social exchange theory (SET) is used in this study to investigate residents' reaction to positive and negative impacts of tourism and how such perceptions affect support for tourism development (Gursoy & Rutherford, 2004). However, SET may not explain the psychological factors such as emotional behaviours that influence residents' perception and support toward tourism (Ribeiro et al., 2017). Therefore, in addition to SET, the Theory of Reasoned Action (TRA) was chosen to explain how residents' attitude affect their support for tourism development (Prayag et al., 2013). So, this study adopted both SET and TRA to examine the relationship between residents' perceived impacts of tourism and their support for tourism development.

That said, the aim of this study is to explore the effects of South Korean residents' perceived impacts of tourism on their support toward tourism development. Impacts of tourism are divided into positive and negative environmental, socio-cultural and economic aspects Also, environmental sustainability, personal economic benefits and community attachment are considered important factors that affect residents' perceptions on their support toward tourism development. So, this study intends to identify how those three factors affect South Korean residents' support toward tourism development. As such, a survey of South Korean residents will be undertaken to meet study objectives.

LITERATURE REVIEW

Residents' Support for Tourism Development

Residents' support for tourism is an important factor in determining the success of tourism development and sustainable tourism development (Gursoy, Jurowski, & Uysal, 2002; Lee, 2013). Previous studies reported that residents' active support is related to the success of tourism development, while tourism developers, the government and policymakers need to be aware of the importance of residents' support for tourism development (Ap, 1992; Choi & Sirakaya, 2005; Gursoy et al., 2002; Gursoy & Rutherford, 2004; Yoon et al., 2001). In other words, it is important to identify and understand factors that influence residents' support for tourism development. Many previous studies showed that residents' support for tourism development is related to residents' perceived impacts of tourism (Deccio & Baloglu, 2002; Nunkoo & Ramkisson, 2012; Ward & Berno, 2011).

In previous studies of residents' perceptions of and support for tourism development, different theories have been applied, such as growth machine theory (Harrill, 2004), community attachment theory (Jurowski, Uysal, & Willams, 1997; McCool & Martin, 1994; Um & Crompton, 1987), social exchange theory (SET) (Ap, 1992; Chen & Hsu, 2001; Gursoy, Chi, & Dyer, 2010; Jurowski & Gursoy, 2004), emotional solidaity theiry (Moghavvemi, Woosnam, Paramanathan, Musa, & Hamzah, 2017; Ribeiro, Pinto, Silva, & Woosnam, 2017; Woosnam, 2012), social representation theory (Fredline & Faulkner, 2000), Weber's theory of rationality and empowerment (Boley et al., 2014), stakeholder theory (Nicholas et al., 2009). However, SET has been frequently applied to examine the relationship between residents' perception toward tourism and their support for tourism development, identifying how residents react to perceived impacts of tourism (Fredline & Faulkner, 2000; Gursoy, Chi, & Dyer, 2009; Andereck et al., 2005; Wang & Pfister, 2008).

Social exchange theory (SET) was originally introduced by Emerson (1976) as a sociological theory, explaining that people evaluate an exchange of resources with the industry, depending on benefits (positive impacts) and costs (negative impacts) that can be gained from the exchange in the industry (Boley et al., 2014). According to Ap (1992, p.668), SET is "a general sociological theory concerned with understanding the exchange of resources between individuals and groups in an interaction situation". Utilising SET's theoretical base, many previous studies successfully investigated the relationship between residents' support and perceived impacts of tourism. Based on SET, previous studies explained that residents are more likely to participate in the exchange with the industry and support tourism development, when they perceive more positive impacts than negative impacts of tourism (Ap, 1992; Gursoy et al., 2002; Haobin et al., 2014; Jurowski et al., 1997; Jurowski and Gursoy, 2004; Ko & Stewart 2002; Lee, 2013; Nunkoo & Ramkissoon, 2011, 2012; Nunkoo, Smith, & Ramkissoon, 2013; Perdue et al., 1990; Wang & Pfister, 2008). On the other hand, based on SET, previous studies reported that residents who perceive more negative impacts than positive impacts of tourism are more likely to be opposed to tourism development (Ap, 1992; Gursoy et al., 2002; Gursoy et al., 2010; Nicholas et al., 2009; Nunkoo & Gurosy 2012; Nunkoo & Ramkissoon, 2010). This means that perceived negative impacts of tourism are negatively related to residents' support for tourism development (Nunkoo & Ramkissoon, 2011).

Some studies reported that in developing countries,many residents do notperceive impacts of tourism negatively because they perceive such impacts as an exchange to gain benefits in the future, while accepting some inconveniences from negative impacts of tourism (Ribeiro et al., 2017; Teye, Sirakaya & Sönmeza, 2002; Var, Kendall & Tarakcioglu, 1985). In addition, if a community relies on tourism heavily, residents can still accept negative impacts of tourism as they understand that their community relies on benefits from tourism (Nunkoo & So, 2016). Previous studies of residents' support for tourism development researched in South Korea proposed similar results by reporting that residents are more likely to support tourism, when they perceive benefits from tourism, while some residents can still bear negative impacts of tourism for future benefits (Kim, Gursoy, & Lee, 2006; Park, Lee & Lee, 2017; Park, Nunkoo, & Yoon, 2015; Yoon, Kim, & Kim, 2015).

Furthermore, the theory of reasoned action (TRA) explains that individuals process information then evaluate and make decisions. It means that individuals are more likely to participate in an exchange, if they have positive perceptions of the exchange, while they are less likely to participate in the exchange, if they have negative perceptions (Ajzen, 1980). So, through TRA the relationship between residents' perception toward impacts of tourism and their support for tourism development can be further explained, while many studies used TRA to examine residents' perception toward tourism (Kwon & Vogt, 2010; Lepp, 2007; Prayag et al., 2013; Ribeiro et al., 2017). According to MacKay and Campbell (2004, p. 445), "The theory of reasoned action not only allows investigators to assess attitudinal and normative influences on intentions, but more importantly it enables examination of why people hold particular attitudes and are influenced by others".

Overall, through SET and TRA, many previous studies have proposed a direct relationship between residents' support for tourism development and perceived positive and negative impacts of tourism. In addition, perceived impacts of tourism are generally divided into economic, socio-cultural and environmental impacts (Ap, 1992; Gursoy et al., 2002; Haobin et al., 2014; Jurowski et al., 1997; Jurowski & Gursoy, 2004; Ko & Stewart 2002; Lee, 2013; Nunkoo & Ramkissoon, 2011, 2012; Nunkoo, Smith, & Ramkissoon, 2013; Perdue et al., 1990; Ribeiro et al., 2017; Wang & Pfister, 2008).

Residents Perceived Positive Impacts of Tourism

Residents' perceptions of tourism is positively affected by the degree to which individuals perceive impacts of tourism as positive. When residents perceive positive impacts of tourism, they tend to have positive perceptions toward tourism and this positive perception leads to residents' support for tourism development (Ap, 1992; Ribeiro et al., 2017). In other words, what positive impacts of tourism make residents have positive perception toward tourism should be identified to gain residents' support for tourism development (Gursoy et al., 2002).

Many studies reported that a direct positive relationship exists between positive economic impacts of tourism and residents' perceptions of tourism. For instance, positive economic impacts of tourism include generating more employment opportunities for residents (Andereck & Nyaupane, 201; King, Pizam, & Milman, 1993; Lee, 2013), generating more business and attracting more investors to local areas (Dyer et al., 2007). Also, tourism improves the local economy and the level of income for residents (Dye et al., 2007; Gursoy et al., 2002). Other studies have also reported that economic benefits from tourism is the main reason residents have positive perceptions about tourism, which then translates to their support for tourism development (Dyer et al., 2007; Gursoy et al., 2002; Gursoy & Rutherford, 2004; Lepp, 2007; Ribeiro et al., 2017).

Tourism brings positive socio-cultural impacts as well that affect residents' perceptions of tourism. Many studies have reported that perceived positive socio-cultural impacts of tourism are positively related to residents' support for tourism (Andereck & Vogt, 2000; Andereck et al., 2005; Besculides et al., 2002). For instance, tourism generates more leisure and cultural activities, entertainment opportunities, while improving public facilities that increase residents' quality of life (Andereck & Vogt, 2000; Brunt & Courtney, 1999; Gursoy et al., 2002; Haobin et al., 2014; Korca, 1996; Tovar & Lockwood, 2008; Yoon et al., 2001). Furthermore, tourism provides an incentive for the preservation of local culture and increases pride of local culture and revival of local cultures (Akis et al., 1996; Andereck et al.,2005; Besculides et al.,2002; Yoon et al., 2001). Furthermore, other studies reported that tourism generates social exchange and interaction between residents and tourists, allowing residents to have opportunities to communicate with tourists (Besculides et al., 2002; Dyer et al., 2007; Kim et al., 2013; Korça, 1996). Based on previous studies, it is evident that positive socio-cultural impacts of tourism bring several positive benefits to residents, contributing to positive perceptions of tourism overall.Based on above discussion, the following hypothesis is proposed:

Hypotheses 1: There is a direct positive relationship between perceived positive impacts of tourism and residents' support for tourism development.

Residents Perceived Negative Impacts of Tourism

On the other hand, there are negative impacts of tourism that affect residents' support for tourism, such as noise pollution,natural environmental pollution, and environmental degradation brough on by tourism development (Choi & Murray, 2010; Ko & Stewart, 2002). When the local environment is compromised by tourism development and tourists, residents perceive impacts of tourism negatively and they are less likely to support tourism, perceiving that tourism is responsible for the degradation of the local natural environment (Andereck et al., 2005; McGehee & Andereck, 2004). Other previous studies have demonstrated similar results, reporting direct negative relationships between perceived negative environmental impacts and residents' support for tourism (Bujosa & Rosselló, 2007; Gursoy, Chi & Dyer, 2010; Lee, 2013; Lindberg & Johnson, 1997; Nunkoo & Ramkissoon, 2011; McGehee & Andereck, 2004). Furthermore, tourism increases the number of tourists and it leads to the overuse of natural local resources such as water (Andereck et al., 2005; Brunt & Courtney,1999). Overall, local areas can be polluted by tourism and residents can suffer from negative environmental impacts of tourism. This contributes to residents having negative perceptions oftourism, which ultimately leads to less residents' support for tourism development (Nunkoo & Gursoy, 2011).

Economic impacts of tourism influence residents' perceptions of and support for tourism positively but there are negative economic impacts to residents as well. Economic impacts of tourism can bring direct negative impacts to residents' quality of living, and when residents perceive negative economic impacts of tourism, they are more likely to be opposed to tourism development (Gursoy et al., 2002). For instance, previous studies reported that tourism raises the cost of living, the price of land and housing, and the price of local goods and services (Brunt & Courtney, 1999; MacGehee and Andereck, 2004; Madrigal, 1993; Saveriades, 2000; Korça,1996; Var et al., 1985). Economic impacts of tourism were found to be one of the strongest factors that affect residents' perceptions of tourism, meaning that negative economic impacts of tourism can bring strong negative impact to residents' perceptions of tourism development (NunKoo & Gursoy, 2011).

Previous studies have reported that several negative socio-cultural impacts contribute to perceptions of tourism overall. For instance, tourism causes traffic congestion (Andereck et al., 2005; Brunt & Courtney, 1999; King et al., 1993), overcrowding in public areas (Ko & Stewart, 2002), and raises crime rates such as drug, gambling and prostitution (Lindberg & Johnson, 1997; Saveriades, 2000; Sheldon & Abenoja, 2001; Tosun, 2002). In addition, according to Teye, Sirakaya and Sönmez (2002), there are residents who perceive social interaction with tourists less positive and it can be assumed that this is because the residents already have negative perception toward tourism. Compared to economic impacts, socio-cultural impacts may not affect residents' perceptions of and support for tourism strongly. However, many previous studies reported that negative socio-cultural impacts of tourism influence residents' support for tourism development negatively (Andereck et al., 2005; Gursoy et al., 2019; McGehee & Andereck, 2004). Based on above discussions, the following hypothesises is proposed:

Hypothesis 2: There is a direct negative relationship between perceived negative impacts of tourism and residents' support for tourism development.

Residents' Perceptions of Environmental Sustainability

Tourism development can improve residents' quality of life but an emphasis needs to be placed on protecting the natural environment destinations for both residents and tourists (Vincent & Thompson, 2002). Several tourism studies have reported that residents' perceptions toward environmental sustainability positively affect individuals' perceptions about tourism (Gursoy et al., 2002; Nicholas et al., 2009). Other studies have reported that residents who consider tourism can give an incentive to preserve natural environment in touristic destinations have positive perceptions of tourism impacts (Perdue et al., 1990). In addition, other studies have claimed that some residents who perceive impacts of tourism positively believe that the industry can increase the awareness of the importance of preservation of environment for both tourists and residents (Akis et al., 1996; Andereck & Nyaupane, 2011; Andereck et al., 2005; Choi & Sirakaya, 2005). According to Olya and Gavilyan (2016), residents who think that tourism brings positive environmental impacts to local areas are more likely to be supportive of tourism. In other words, residents' perception toward environmental sustainability can be positively related to residents' support for tourism development, if residents perceive environmental impacts of tourism positively. This argument is supported by several previous studies, proposing a positive relationship between perceived positive environmental impacts of tourism and residents' perception and support toward tourism development (Bujosa & Rosselló, 2007; Choi & Murray, 2010; Choi & Sirakaya, 2005; Olya & Gavilyan, 2016; Yoon et al., 2001).

On the other hand, several studies reported that residents' perception toward environmental sustainability is negatively related to residents' perception toward impacts of tourism. Residents who value environmental sustainability in local areas perceive impacts of tourism more negatively (Choi & Sirakaya, 2005). Perceived negative environmental impacts include environmental pollution from tourists, destruction of local lands, historical heritages and natural environment from tourism development (Yoon et al., 2001). Furthermore, many previous studies reported that residents who consider environmental sustainability are more likely to perceive impacts of tourism negatively, believing that tourism pollutes the local natural environment (Bujosa & Rosselló, 2007; Choi & Murray, 2010; Olya & Gavilyan, 2016). Based on above discussions, the following hypotheses are proposed:

Hypotheses 3: Residents' perceptions of environmental sustainability is positively related to perceived positive impacts of tourism.

Hypotheses 4: Residents' perceptions of environmental sustainability is negatively related to perceived negative impacts of tourism.

Hypotheses 5: Residents' perceptions of environmental sustainability is positively related to residents' support for tourism development.

Residents' Personal Economic Benefits of Tourism

Previous studies have demonstrated a positive relationship between residents' perceived personal economic benefits from tourism and their perceptions of tourism impacts (Andereck et al., 2005; Ko & Stewart, 2002). For instance, tourism generates more employment opportunities and improves residents' income levels and these kinds of economic benefits are perceived positively by residents, leadingto positive perceptions of tourism overall (Andereck & Nyaupane, 2011; Andereck et al., 2005; Besculides, Lee, & McCormick, 2002; Choi & Sirakaya, 2006; Dyer, Gursoy, Sharma, & Carter, 2007; Gursoy et al., 2002; King et al., 1993; Ko & Stewart, 2002; Saveriades, 2000). Other numerous studies reported that residents who reap economic benefits from tourism are more likely to possess positive perceptions about the impacts of tourism (Boley et al., 2014; Brunt & Courtney, 1999; Nunkoo & So, 2016). This argument is supported by other tourism studies which proposed that residents who get personal economic benefits from tourism perceive tourism positively and it leads to supporting tourism development, while residents who get less or no economic benefits are less likely to be supportive of tourism development (Boley et al., 2014; McGehee & Andereck, 2004; Nunkoo & So, 2016; Vargas-Sánchez et al., 2015). This reveals that a positive relationship exists between personal economic benefits from tourism and residents' support for tourism development (Andereck, 2004; Ap, 1992; Boley, Stzelecka & Woosnam, 2018; Perdue et al., 1990; Ribeiro et al., 2017).

Many previous studies reported that there is a strong positive relationship between personal economic benefits and perceived positive impacts of tourism, meaning that positive economic impacts affect positively residents' perception and support toward tourism. So, it is not difficult to assume that negative economic impacts of tourism can bring negative impact to residents' perception and support toward tourism development. However, several studies could not find significant negative relationship between personal economic benefits and perceived impacts of tourism (Kwon & Vogt, 2010; Ribeiro et al., 2017; Vargas-Sánchez et al., 2015)

Overall, perceived personal economic benefits have a strong impact on perceived impacts of tourism and it can be perceived both positively or negatively, while strongly influencing residents' support for tourism development (Andereck et al., 2005; Ko & Stewart, 2002). Based on above discussions, the following hypotheses are proposed:

Hypotheses 6: Perceived personal economic benefits are positively related to perceived positive impacts of tourism.

Hypotheses 7: Perceived personal economic benefits are negatively related to perceived negative impacts of tourism

Hypotheses 8: Perceived personal economic benefits are positively related to residents' support for tourism development.

Community Attachment

Community attachment is defined as an emotional connection between residents and their communities, whether they be towns, states or countries (Nicholas et al., 2009). Several tourism studies have found that residents who were born in and lived for a long time in the same community have a greater attachment to their community and have positive perceptions about tourism development (Woosnam, Aleshinloye, Ribeiro, Stylidis, Jiang, & Erul, 2018). Also, Jurowski et al. (1997) reported that residents who are more attached to their community perceive economic and social impacts of tourism in a positive way. Other studies proposed that community attachment is positively related to perceived impacts of tourism and it can lead to residents' support for tourism development (Gursoy et al., 2002; Jurowski et al., 1997). In addition, other studies reported that there is a positive relationship between community attachment and individuals' support for tourism development (Gursoy et al., 2019; Gursoy & Kendall, 2006; Lee, 2012; Nicholas et al., 2009). So, it can be said that residents who have a strong attachment to their community have positive perceptions of tourism and support the industry.

On the other hand, according to Jurowski, Uysal, and Williams (1997), community attachment is negatively related to residents' perception toward impacts of tourism. Also, other studies showed that residents who are more attached to their community perceive impacts of tourism negatively, perceiving that negative impacts of tourism affect their community negatively (Deccio & Baloglu, 2002; Um & Crompton, 1987; Nepal, 2008). The negative relationship between community attachment and residents' perceived impacts of tourism can be both significant and insignificant, depending on how much residents are attached to their community (Choi & Murray, 2010; Gursoy & Rutherford, 2004; Gursoy et al., 2019; McCool & Martin 1994). Also, residents who have a strong attachment to their community are more likely to be interested in protecting their community from negative impacts of tourism (Choi & Murray, 2010; Nepal, 2008). So, based on previous studies, it can be assumed that if residents are attached to their community, they are more likely to perceive impacts of tourism negatively than those who are not attached. This assumption is supported by several previous studies, reporting that residents who are attached to their community can less likely to be supportive of tourism development, perceiving tourism negatively (Allen et al., 1988; McCool & Martin, 1994). Based on above discussions, the following hypotheses are proposed:

Hypotheses 9: Community attachment is positively related to perceived positive impacts of tourism.
Hypotheses 10: Community attachment is negatively related to perceived negative impacts of tourism.
Hypotheses 11: Community attachment is positively related to residents' support for tourism development.

The Proposed Model

Based on above discussions, this study proposed 11 hypotheses, each displayed in Figure 1 based on SET and TRA framework. The model sets environmental sustainability, personal economic benefits and community attachment as independent variables. Perceived positive and negative impacts of tourism are consideredmediator variables, while support for tourism development is the dependent variable.

Figure 1. The proposed model of residents' support for tourism development

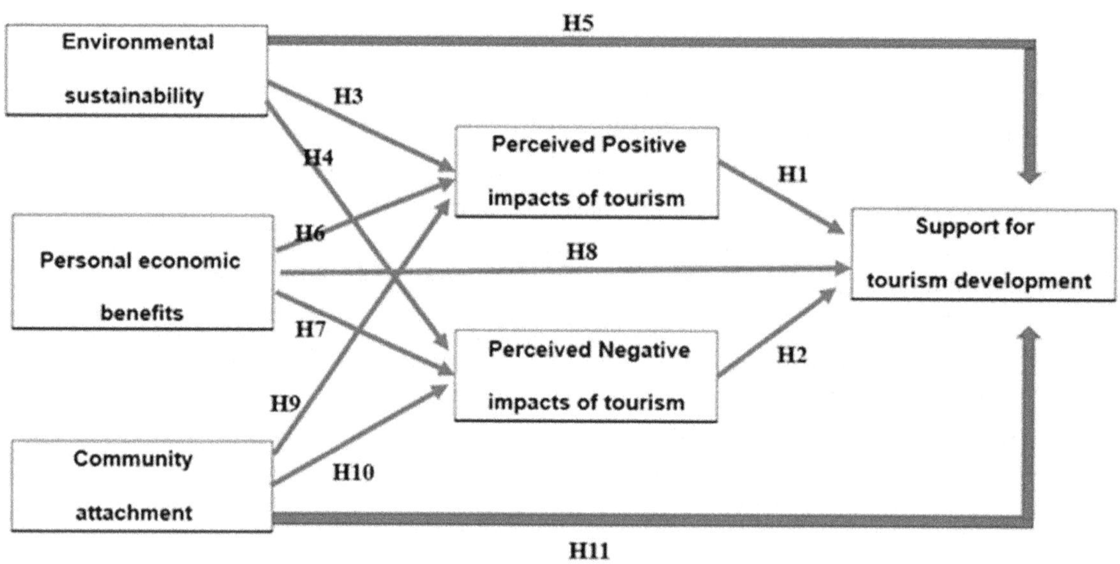

METHODS

Study Setting

The study location selected for this study was South Korea, with a population of 50.4 million inhabitants. The country has $40.096 US per capita, showing a 2.5% increase of projected growth rate and 3.1% GNP growth rate in 2019 (E-Naragipue, 2019; OECD, 2019). South Korea is surrounded by sea and mountains and has four seasons that allow tourists and residents to experience different natural beauty. In general, the rural areas of South Korea are less developed, while the capital city (Seoul) is developed as a large metropolitan city. South Korea is full of numerous cultural heritages sites such as Changdeokgung Palace (UNESCO World Heritage), Deoksugung Palace, Changgyeonggung Palace, Gwangmyeong Cave, etc. (Visit Korea, 2019). South Korea was ranked the top in the UIA Global Meetings Statistics in 2017 as preferred destination for international events, largely due to the country's good record in public safety and security. In addition, South Korea is ranked as the world's most innovative country, obtaining the highest score (89.9%) in the latest Bloomberg innovative index, while having over 5000 years of history, preserved through customs andculture (Korea Convention, 2019). According to the Korea Tourism Organization's report in 2018, between 1999 and 2018, the number of international tourists to South Korea had increased by an annual rate of 6.6% (i.e., 4,659,785 international visitors were received in 1999; 15,346,879 international visitors were received in2018). According to 2017 and 2018 reports from the Korea Tourism Organization, international tourists to South Korea are from various countries, with a majority arriving from other Asian countries such as Japan, China, Taiwan and Vietnam, while the remainder are from the USA, Canada, Russia and various European countries.

Data Collection and Sampling

The data were collected from South Korean residents. This study used a convenience sampling strategy to get as much data from various types of residents (e.g., across age, employment, gender, etc.) Such an approach was taken to secure as much variation in responses concerning perceptions of tourism impacts, environmental sustainability, personal economic benefits, community attachment, and support for tourism development.

The data were collected through online survey using Qualtrics. The questionnaire was spread to acquaintances of the researcher in South Korea by sending them a Qualtrics's anonymous link and it was further spread to their acquaintances. The data collection period had been planned to be over several weeks, but as enough data were collected faster than the researcher had expected, the actual data collection occurred over a one-week period. A total of 693 South Korean respondents completed the questionnaire.

Survey Items

This study employed a quantitative survey design (following the work of Andereck et al. 2005; Boley et al. 2014; Choi & Sirakaya, 2005; Gursoy et al. 2002; Nunkoo & Gursoy 2011; Ribeiro et al., 2017), utilizing questionnaire items and scales from previous studies. The questionnaire was developed in English and reviewed by a tourism professor to check its clarity and validity. Then it was back-translated into Korean and reviewed by bilingual speakers who speak both English and Korean to check its clarity. This was done so thatparticipants were able to read and answer this questionnaire clearly, while guaranteeing that the Korean version provides the same meaning as the original English version (Brislin, 1970).

Six constructs were included in the questionnaire by borrowing items from previous studies. The constructs of the questionnaire included perceived positive and negative impacts of tourism, divided into environmental, socio-cultural and economic impacts. Also, the other constructs included environmental sustainability, personal economic benefits of tourism, community attachment and support for tourism development.

Perceived positive and negative impacts of tourism were measured with 17 items. Two items were used to measure environmental impacts adapted from Choi and Sirakaya, (2005) and Ribeiro et al. (2017). Six items were used that measureeconomic impacts adapted from McGehee and Andereck (2004), Ribeiro et al. (2017), and Perdue et al. (1990). Eight items were utilized to measure socio-cultural impactsborrowed from Choi and Murray (2010), Nunkoo and Gursoy (2011), and Nunkoo and Ramissoon (2011). In addition, to measure South Korean residents' perceptions about K-pop, this study originally designed one new item related to socio-cultural impacts of tourism (e.g., "tourism related to K-pop damages the traditional culture of South Korea").

As to the items of environmental sustainability, four items were used based on the work ofChoi and Murray (2010), Choi and Sirakaya (2005), and Yu, Chancellor and Cole (2011). To measure perceptions ofpersonal economic benefits, four items were adapted from the research by McGehee and Andereck (2004), Ribeiro et al. (2017), and Wang and Pfister (2008). Perceptions of community attachment were measured by four items based on the work of Choi and Murray (2010), Gursoy and Rutherford (2004), and Nicholas, Thapa and Ko (2009). Finally, residents' support for tourism was assessed by six items adapted from Deccio and Baloglu (2002) Gursoy et al.(2002), Nunkoo and Ramkissoon (2011), Ribeiro et al. (2017), Teye et al. (2002), and Wang and pfister (2008). And this study originally designed one more item considering the role of K-pop in tourism development (e.g., "I am against tourism development related to K-pop that attracts international tourists"). All measures were asked on a 5-point Likert Scale of agreement (where 1= strongly disagree; 5 = strongly agree).

Data Analysis

This study employed IBM SPSS v25 for data analysis. Exploratory factor analysis (EFA) was performed with principal component analysis and varimax rotation. All items with factor loadings below0.40 were eliminated in the EFA. Cronbach's alpha was examined to assess the reliability of items. Cronbach's alphas for factors in excess of 0.70 were deemed to demonstrate good reliability (Nunnally, 1978). Correlation analysis was performed to check the correlation between variables, as linear regression analysis was used to test the 11 hypotheses from the proposed conceptual model.

RESULTS

Sample Characteristics

The sample profile is summarized in Table 1. In terms of gender, 65.8% were male. Age was somewhat evenly distributed, with the highest percentage of respondents (34.6%) falling between 30 and 39 years of age. The sample was fairly well educated as 83.6% had attained at least an undergraduate degree. Most individuals were company employees (63.2%).

Table 2 shows the respondents' tourism-related characteristics. As to the question, "Do you earn your living through tourism?", only 54 respondents (7.8%) responded in the affirmative. In addition, as to the question, "Do you have direct contact with tourist as a part of your work?", it was found that the respondents answered 'few times' (n=634, 91.5%) or 'often' (n=59, 8.5%). As to the question, "Do you own any business related to tourism?", the respondents answered 'yes' (n=61, 8.8%) Finally, individuals were asked,"Are you involved in tourism as a source of income or employment?," to which 100 respondents said 'yes' (14.4%) Such findings demonstrate that most respondents do not strong ties to the tourism industry.

Table 1. Sociodemographic characteristics

Measure		Frequency (N=693)	Percentage (100%)
Gender	Male	456	65.8
	Female	237	34.2
Age	Below 20	6	0.9
	20-29	101	14.6
	30-39	240	34.6
	40-49	184	26.6
	50-59	128	18.5
	Order than 60	11	1.6
Highest level of education	Middle school	4	0.6
	High school	42	6.1
	College	24	3.5
	Undergraduate	527	76.0
	Master	94	13.6
	Other	2	0.3
Occupation	Company employee	438	63.2
	Profession	50	7.2
	Self-employed	46	6.6
	Student	38	5.5
	Homemaker	24	3.5
	IT related job	75	10.8
	Official	14	2.0
	Other	8	1.2

Table 2. Respondent's tourism-related characteristics

Measure		Frequency (N=693)	Percentage (100%)
Do you earn your living through tourism?	Yes	54	7.8
	No	639	92.2
Do you have direct contact with tourist as a part of your work?	Often	59	8.5
	Few times	634	91.5
Do you own any business related to tourism?	Yes	61	8.8
	No	632	91.2
Are you involved in tourism as a source of income or employment?	Yes	100	14.4
	No	593	85.6

Exploratory Factor Analysis of The Independent Variables

Table 3 shows the results of the EFA for theindependent variables. Four items pertaining to environmental sustainability, four items concerning personal economic benefits, and four items corresponding tocommunity attachment were included. The item, "it is important to keep traditions and culture of this community" was removed from the community attachment factor due to cross-loading concerns.The factors, environmental sustainability ($\alpha = 0.772$) and personal economic benefits ($\alpha = 0.908$) demonstrated Cronbach's alpha in excess of 0.70. However, community attachment came in at 0.619.

Table 3. Exploratory factor analysis of independent variables

Items	Environmental Sustainability	Personal Economic Benefits	Community Attachment
	α= 0.772	**α= 0.908**	**α= 0.619**
Tourism raises awareness of the importance of environment	0.806	0.250	0.078
I believe that tourism strengthen efforts to protect and improve environment	0.801	0.254	0.099
Community environment needs to be protected now and for the future	0.733	-0.122	0.173
Tourism needs to be developed in harmony with natural environment	0.684	-0.119	0.165
I gain economic benefits from tourism	0.050	0.907	0.049
I find tourism positive because it improves my income	0.075	0.888	0.053
I can get more job opportunities because of tourism	0.036	0.875	0.132
My economic future depends on tourism in South Korea	0.027	0.834	-0.001
I have no attachment to this community	0.052	-0.062	0.776
I would like to live in this community more than in other communities	0.144	0.070	0.709
This community has meaning to me	0.303	0.241	0.705
Eigenvalue Predictable V (%) Cumulative V (%)	2.421 22.013 51.956	3.294 29.944 29.944	1.697 15.430 67.386

Exploratory Factor Analysis of the Dependent Variables

Table 4 summarizes the results of EFA for the dependent variables. Six items were deleted from the positive impacts, five were deleted from negative impacts and one item from support for tourism development—all with factor loadings less than 0.40.. All constructs had Cronbach's alphas in excess of 0.70.

Table 4. Exploratory factor analysis of the dependent variables

Items	Positive Impacts	Negative Impacts	Support for Tourism Development
	α = 0.894	α = 0.822	α = 0.851
Tourism generates more leisure and entertainments	0.834	0.095	0.227
Tourism improves the public infrastructure and facilities	0.828	0.073	0.232
Tourism attracts more investment for South Korea	0.812	-0.012	0.177
Tourism generates more job opportunities	0.803	-0.027	0.222
Tourism help to develop local economy	0.775	0.135	0.221
Tourism provides an incentive for the preservation of local culture	0.640	-0.184	0.247
Tourism causes more traffic congestion	-0.007	0.808	-0.091
Tourism pollutes natural environment	-0.050	0.783	-0.042
Tourism causes overcrowding issue in public areas	0.112	0.748	-0.071
Tourism generates noise pollution	0.161	0.731	-0.056
Tourism overuses natural local resources such as water	-0.145	0.720	-0.074
I am willing to help tourists to enhance their experience	0.263	-0.067	0.821
I am willing to support tourism and would like to see that tourism is developed more	0.275	-0.147	0.815
I would like South Korea to attract more tourists	0.337	-0.152	0.766
I am willing to accept inconvenience in exchange for benefits from tourism development	0.126	-0.192	0.698
I am willing to protect natural environment and resources on which tourism depends	0.209	0.120	0.661
Eigenvalue Predictable V (%) Cumulative V (%)	4.075 25.472 25.472	3.046 19.037 64.312	3.169 19.804 45.276

Correlation Analysis

Table 5 shows the correlations between each construct included within the conceptual model. It was found that there is no significant correlation between negative impacts, environmental sustainability and personal economic benefits. However, the rest are correlated with each other. In addition, there is a negative correlation between negative impacts and community attachment ($r = -0.117, p < 0.01$), while support for tourism development is negatively correlated with negative impacts (r = -0.192**). On the other hand, the rest of the variables are positively correlated with each other. Also, environmental sustainability was positively related to perceived positive impacts of tourism, with a higher correlation than with personal economic benefits and community attachment.

Hypothesis Testing

Linear regression analysis was performed to test the proposed hypotheses (Table 6). It was found that environmental sustainability, personal economic benefits and community attachment were positively

Table 5. Descriptive statistics and correlation analysis

	Mean	SD	ES	PEB	CA	PI	NI	STD
ES	4.100	0.712	1					
PEB	2.802	1.068	**0.167****	1				
CA	3.665	0.722	**0.374****	**0.166****	1			
PI	4.053	0.691	**0.447****	**0.208****	**0.275****	1		
NI	3.306	0.832	-0.046	-0.052	**-0.117****	0.004	1	
STD	3.848	0.669	**0.522****	**0.317****	**0.403****	**0.544****	**-0.192****	1

1. **: $p < 0.01$. SD = Standard deviation
2. ES: Environmental Sustainability, PEB: Personal Economic Benefits, CA: Community attachment, PI: Positive Impacts, NI: Negative Impacts, STD: Support for Tourism development.

related with perceived positive impacts of tourism (Hypotheses 3,6,9). In addition, compared to personal economic benefits (β=0.125, t=3.670, p<0.001) and community attachment (β=0.110, t=3.040, p<0.01), environmental sustainability (β=0.384, t=10.579, p<0.001) had a stronger influence on perceived positive impacts of tourism.

As to the relationship between negative impacts and environmental sustainability, personal economic benefits and community attachment (Hypotheses 4,7,10), the results showed that environmental sustainability (β=0.002, t=0.045, p>0.05) and personal economic benefits (β=-0.034, t=-0.0878, p>0.05) had no significant effect on perceived negative impacts of tourism. So, Hypotheses 4 and 7 were not supported. However, it was found that community attachment (β=-0.112, t=-2.739, p<0.01) had a significant negative influence on negative impacts of tourism. So, Hypotheses 10 was supported. It means that South Korean residents who are attached to their community are more likely to have positive perception toward tourism and less likely to have negative perception toward tourism. This may be because residents believe that although there are some negative impacts of tourism, in the end, tourism can bring more positive impacts to their community, exceeding negative impacts.

Finally, the regression analysis was performed about the relationship between support for tourism development and environmental sustainability, personal economic benefits, community attachment, positive impacts and negative impacts of tourism (hypotheses 1,2,5,8,11). The results showed that environmental sustainability (β=0.274, t=8.485, p<0.001), personal economic benefits (β=0.165, t=5.811, p<0.001), and community attachment (β=0.161, t=5.308, p<0.001) all had a significant positive influence on support for tourism development. In addition, the result showed that positive impacts of tourism had a significant positive impact on support for tourism development (β=0.343, t=10.919, p<0.001), while on the other hand, negative impacts of tourism had a significant negative effect on support for tourism development (β=-0.153, t=5.500, p<0.001). Therefore, Hypotheses 1,2,5,8,11 were all supported.

This result can be explained that if residents perceive negative impacts of tourism, they are less likely to support tourism development. However, perceived positive impacts of tourism increase residents' willingness to support tourism development. Also, the results showed that the residents who are attached to their community, gain personal economic benefits and consider environmental sustainability are more likely to support tourism development. Also, the results showed that perceptions about environmental sustainability significantly explains residents' support for tourism development (more so than ersonal

Table 6. The regression analysis

DV	IVs	B	Standard β	T-value	VIF	Hypotheses
Positive impacts	Environmental sustainability	0.373	0.384	10.579***	1.178	H3 (Supported)
	Personal economic benefits	0.081	0.125	3.670***	1.042	H6 (Supported)
	Community attachment	0.106	0.110	3.040**	1.117	H9(Supported)
	$R^2 = 0.228$. Adj.$R^2 = 0.225$. F (3, 689) = 67.847, $p<0.001$					
Negative impacts	Environmental sustainability	0.002	0.002	0.045 n/s	1.178	H4 (Not supported)
	Personal economic benefits	-0.026	-0.034	-0.878 n/s	1.042	H7 (Not supported)
	Community attachment	-0.130	-0.112	-2.739**	1.177	H10 (Supported)
	$R^2 = 0.015$. Adj.$R^2 = 0.011$. F (3, 689) = 3.465. $p<0.05$					
Support for tourism development	Environmental sustainability	0.257	0.274	8.485**	1.370	H5 (Supported)
	Personal economic benefits	0.104	0.165	5.811**	1.064	H8 (Supported)
	Community attachment	0.149	0.161	5.308**	1.207	H11 (Supported)
	Positive impacts	0.332	0.343	10.919**	1.298	H1 (Supported)
	Negative impacts	-0.123	-0.153	-5.500**	1.017	H2 (Supported)
	$R^2 = 0.477$. Adj. $R^2 = 0.473$. F (5, 687) = 125.139. $p<0.001$.					

Note: *$p<0.05$, **$p<0.01$, *** $p<0.001$, n/s not significant

economic benefits and community attachment). Table 7 shows the overall results of the proposed hypotheses tests.

DISCUSSION, IMPLICATIONS AND LIMITATIONS

Discussion

The current study investigated residents' perceptions of and support for tourism development in South Korea through the lens of SET and TRA. The results of this study found that Hypothesis 1 was supported that proposed a significant positive relationship between perceived positive impacts of tourism and residents' support for tourism development. In other words, South Korean residents are more likely to support tourism, when they erceive positive impacts from tourism. This result is in line with many previous studies concerningresidents' perceptions and support for tourism (Boley, Strzelecka, & Woosnam., 2018; Choi & Murray, 2010; Deccio & Baloglu, 2002; Gursoy et al., 2002; Jurowski et al., 1997; NunKoo & Gursoy, 2011; Ribeiro et al., 2017).

Findings demonstrate support for Hypothesis 2 that proposed a negative significant relationship between perceived negative impacts of tourism and residents' support for tourism development. This conveys that residents in South Korea do not support tourism development when they perceive negative impacts of tourism. It can also mean that the residents do not bear inconveniences from tourism in exchange for the future benefits from tourism. This finding is in line with the results of previous studies that proposed a significant negative relationship between perceived negative impacts of tourism and residents' support for tourism development (Andereck et al., 2005; Choi & Murray, 2010; Dyer et al., 2007; Gursoy & Rutherford, 2004; NunKoo & Ramsissoon, 2011; Teye et al., 2002).

Hypothesis 3, proposing that residents' perceptions of environmental sustainability is positively related to perceived positive impacts of tourism was supported. Additionally, perceived positive impacts of tourism are positively affected by perceptions of environmental sustainability stronger than personal economic benefits and community attachment. This shows that in South Korea, residents consider environmental sustainability more than short-term benefits such as personal economic benefits from tourism. In addition, Hypothesis 4 that proposed a direct negative relationship between residents' perceptions toward environmental sustainability and perceived negative impacts of tourism was not supported . This means that perceptions toward environmental sustainability does not significantly increase negative perceptions of tourism. So, it can be said that South Korean residents who consider environmental sustainability do not tend to have negative perception toward impacts of tourism. In other words, residents who consider environmental sustainability focus on perceived positive impacts more than negative impacts of tourism. In addition to this result, not surprisingly, Hypothesis 5 was supported, proposing that there is a direct positive relationship between residents' perceptions toward environmental sustainability and their support for tourism development. Furthermore, the correlation analysis of this study showed that environmental sustainability is correlated with perceived positive impacts of tourism stronger than personal economic benefits and community attachment. In other words, there are South Korean residents who understand the importance of environmental sustainability and they perceive that tourism needs to be developed, considering environmental sustainability. So, to get more active support from residents, environmental sustainability should be considered in tourism development, minimizing negative environmental impacts from tourism.

Hypothesis 6 that proposed a significant positive relationship between residents' perceived personal economic benefits and perceived positive impacts of tourism was supported. This result is in line with many previous studies (McGehee & Andereck, 2004; Sirakaya et al., 2002; Ribeiro et al., 2017; Vargas-Sánchez et al., 2015). In addition, Hypothesis 7 that proposed a negative relationship between personal economic benefits and negative impacts of tourism was not supported. Because residents are aware of positive economic impacts of tourism bringing positive economic benefits to their quality of living, they are interested in and expect future economic benefits from tourism and have a positive perception toward positive economic impacts of tourism instead of having negative perception toward tourism. In addition, the positive relationship between perceived personal economic benefits and residents' support for tourism (Hypothesis 8) was found to be supported. Overall, the results of Hypothesis 6,7,8 showed that the residents have a high expectation about personal economic benefits from positive economic impacts of tourism by perceiving tourism positively rather than negatively and this positive perception leads to supporting tourism development, expecting that tourism can benefit them economically. This result is in line with many previous studies, proposing that economic gain brings strong positive impacts on perceived impacts of tourism (Andereck et al., 2005; Andereck & Nyaupane, 2011; Ko & Stewart, 2002; Nunkoo & So, 2016; McGehee & Andereck, 2004; Ribeiro et al., 2017).

This study proposed that the positive relationship between community attachment and perceived positive impacts of tourism is would be significant (Hypothesis 9), showing that residents who are attached to their community are more likely to have positive perceptions of positive impacts of tourism. In addition, Hypothesis 10 that proposed the negative relationship between community attachment and perceived negative impacts of tourism was found to be significant, showing that strong community attachment decreases negative perception toward tourism. The results of Hypothesis 9 and 10 showed that if residents have strong attachment to their community, they are more likely to focus on perceived positive impacts rather than focus on negative impacts. This may be because the residents believe that even if negative impacts of tourism affect their community negatively, positive impacts of tourism are perceived more strongly by believing that tourism brings benefits to their community eventually. Also, Hypothesis 11 that proposed the positive relationship between community attachment and residents' support for tourism development was supported. This means that the more residents are attached to their community, the more they support tourism development and this result is in line with (Choi & Murray, 2010; Deccio & Baloglu, 2002; McCool & Martin, 1994).

Implications

This study examined residents' perceptions and support for tourism development in South Korea. Eleven hypotheses were tested, and nine hypotheses were supported. This study applied Social Exchange Theory (SET) and Theory of reasoned action (TRA), allowing the study to examine how South Korean residents perceive and react to impacts of tourism, depending on a costs and benefits perspective. Also, the study examined the relationship between residents' support for tourism and different variables that include positive and negative impacts of tourism, environmental sustainability, personal economic benefits and community attachment. The researcher had assumed that this study researched in South Korea would show different result, compared to other previous studies researched in other countries. However, the overall result of this study ended up demonstrating similar results to previous studies, that South Korean residents are more likely to support tourism, when perceiving positive impacts of tourism. This finding suggests that cultural differences may not affect residents' perceptions and support for tourism.

According to Gursoy and Rutherford (2004), community leaders, tourism developers, local planners and policymakers should consider residents' perceptions for the success of tourism development. In other words, findings of this study have managerial implications for South Korea Tourism organization to understand factors that influence residents' perceptions and support for tourism development. In addition, as this study researched South Korea, the findings of this study can be compared to results from previous studies undertaken in other countries, contributing to further understanding of residents' perceptions of tourism development.

This study found that perceived positive impacts of tourism encourage residents to support tourism, while perceived negative impacts make residents less supportive of tourism development. Although there are negative impacts from tourism, if residents perceive positive impacts of tourism more than negative impacts of tourism, there will be more residents, supporting tourism development. This finding suggests that to get more residents' support for tourism development, South Korea Tourism Organization should consider how to translate impacts of tourism to actual perceived positive impacts for residents. It means that even if there are positive impacts of tourism, residents should perceive positive impacts because actual perceived positive impacts lead to residents' support for tourism development. So, to get active support from residents, Korea Tourism Organization needs to raise residents' awareness of benefits of tourism enough to be perceived. At the same time, this study suggests that Korea Tourism organization needs to consider minimizing perceived negative impacts. Impacts of tourism are divided into environmental, socio- cultural and economic impacts.

The findings of this study demonstrated that environmental sustainability affects residents' perceptions oftourism stronger than personal economic benefits and community attachment. It means that in South Korea there are many residents who are aware of the importance of environmental sustainability in tourism development. This finding suggests that Korea tourism organization needs to minimize destruction of natural environment, when there is a development for tourism such as building more hotels by destroying forests. In addition, as residents perceive that tourists pollute the local environment, tourists need to be educated that they should not pollute the environment. For instance, through official websites of South Korea and Korea tourism organizations, educational information on environmental sustainability can be provided to international tourists (translated into English and other necessary languages).

Many previous studies have found that economic benefits are one of the strongest impacts that influence residents' perceptions and support for tourism (Choi & Muarry, 2010; Gursoy et al., 2012; Nunkoo & Gursoy, 2011; Ribeiro et al., 2017). However, this study found that in South Korea, environmental sustainability affects residents' perceptions and support of tourism stronger than economic benefits. This finding may be the result of cultural differences or the assumption that nowadays more residents are aware of the importance of environmental sustainability. So, this finding suggests that environmental impacts of tourism can be considered more importantly than economic impacts of tourism. So, there is a need of examining what factors affect residents' perception toward environmental sustainability in tourism development because it can be strongly related to residents' support for tourism development.

This study also found that residents who gain economic benefits from tourism generally tend to perceive tourism positively and support tourism development. Economic benefits can be perceived as personal economic benefits and can be perceived as economic benefits for their community and country. For instance, it was found that residents in South Korea perceive tourism brings more investment to South Korea and improves local economy, generating more business. Furthermore, residents also perceive that they can gain personal economic benefits from tourism such as greater job opportunities and improved income. So, this finding suggests that Korea tourism organization needs to consider how to bring economic benefits from tourism to the individual and community level, while guaranteeing that residents perceive economic benefits from tourism.

Also, findings demonstrated that strong community attachment positively affects perceived impacts of tourism and support for tourism development. South Korean residents who are attached to their community tend to perceive tourism positively, perceiving positive impacts more than negative impacts and this perception leads to supporting tourism development. Depending on how much residents are attached to their community, they perceive tourism positively but also negatively. This finding suggests that even if there are negative impacts of tourism, residents who are attached to their community do not perceive negative impacts seriously, while focusing on positive impacts of tourism. So, it can be suggested that Korea Tourism Organization may need to communicate with local residents more to explainwhat positive impacts tourism can bring to their community. This way, residents can understand that tourism can bring more positive impacts to their community, meaning that it can lead to decreasing a conflict between tourism development and residents who perceive that tourism development brings negative impacts to their community.

Overall, findings of this study demonstrated that environmental sustainability, personal economic benefits, community attachment, positive and negative environmental, socio-cultural and economic impacts each influence residents' perceptions of and support for tourism in South Korea. Therefore, Korea Tourism Organization, tourism planners and policy makers should identify residents who are attached to their community, who are concern about environmental sustainability, who have a high expectation of economic benefits from tourism, and who have negative perceptions of tourism impacts. Based on this kind of identification, different tourism development strategies can be applied, depending on types of residents. For instance, when a specific touristic location in South Korea is planned to be developed, information of residents in that area should be gathered. And if the majority of residents identified are concerned with environmental sustainability, tourism development in that area should be planned in eco-friendly ways, focusing on minimizing negative environmental impacts of tourism.

Limitations and Directions for Future Study

This study utilised data collected during the summer months. However, depending on seasonality, there are different tourism-related products and services, while residents' perceptions of tourism can be affected by seasonality (Vargas-Sánchez, Plaza-Mejía and Porras-Bueno, 2008; Ribeiro et al., 2017). This means that if this study collected the data through different seasons, there could be different results. So, there is a need for further research in different seasons. Another limitation of this study is that the data were not collected from various cities in South Korea. Although the participants of the questionnaire are originally from different cities, most of them are currently living in the capital city in South Korea, which is one of the most crowded places in all of South Korea. So, residents who are currently living in rural areas of South Korea may have different perceptions of tourism. This means that findings of this study may not reflect overall perceptions of South Korean residents.

This study investigated residents' perceptions of and support for tourism development in South Korea. So, comparative studies are needed in other countries to gain a better sense of how cultural difference may affect residents' perceptions of and support for tourism. In addition, the findings of this study indicated that there is a negative relationship between negative impacts of tourism and residents' perceptions and support for tourism, while their quality of life is affected by negative impacts of tourism. As such, how tourism development affects the quality of life of residents needs to be investigated for further understanding of the relationship between impacts of tourism and residents' perceptions of and support for tourism. Finally, this study did not investigate psychological factors in detail.Future research should consider how psychological variables (e.g., personality, psychographics, etc.) affect residents' perceptions of tourism in an effort to further our understanding of individuals' support for tourism development.

REFERENCES

Akis, S., Peristianis, N., & Warner, J. (1996). Residents' attitudes to tourism development: The case of Cyprus. *Tourism Management*, *17*(7), 481–494. doi:10.1016/S0261-5177(96)00066-0

Allen, L. R., Long, P. T., Perdue, R. R., & Kieselbach, S. (1988). The impact of tourism development on resident's perception of community life. *Journal of Travel Research*, *27*(1), 16–21. doi:10.1177/004728758802700104

Andereck, K. L., & Nyaupane, G. P. (2011). Exploring the nature of tourism and quality of life perceptions among residents. *Journal of Travel Research*, *50*(3), 248–260. doi:10.1177/0047287510362918

Andereck, K. L., Valentine, K. M., Knopf, R. C., & Vogt, C. A. (2005). Residents' perceptions of community tourism impacts. *Annals of Tourism Research*, *32*(4), 1056–1076. doi:10.1016/j.annals.2005.03.001

Andereck, K. L., & Vogt, C. A. (2000). The relationship between residents' attitudes toward tourism and tourism development options. *Journal of Travel Research, 39*(1), 27–36. doi:10.1177/004728750003900104

Ap, J. (1992). Residents' perceptions on tourism impacts. *Annals of Tourism Research, 19*(4), 665–690. doi:10.1016/0160-7383(92)90060-3

Besculides, A., Lee, M. E., & McCormick, P. J. (2002). Residents' perceptions of the cultural benefits of tourism. *Annals of Tourism Research, 29*(2), 303–319. doi:10.1016/S0160-7383(01)00066-4

Boley, B. B., McGehee, N. G., Perdue, R. R., & Long, P. (2014). Empowerment and resident attitudes toward tourism: Strengthening the theoretical foundation through a Weberian lens. *Annals of Tourism Research, 49*, 33–50. doi:10.1016/j.annals.2014.08.005

Boley, B. B., Stzelecka, M., & Woosnam, K. M. (2018). Resident perceptions of the Economic Benefits of Tourism: Toward a Common Measure. *Journal of Hospitality & Tourism Research (Washington, D.C.), 42*(8), 1295–1314. doi:10.1177/1096348018759056

Brislin, R. W. (1970). Back-Translation for Cross-Cultural Research. *Journal of Cross-Cultural Psychology, 1*(3), 185–216. doi:10.1177/135910457000100301

Brunt, P., & Courtney, P. (1999). Host perceptions of sociocultural impacts. *Annals of Tourism Research, 26*(3), 493–515. doi:10.1016/S0160-7383(99)00003-1

Bujosa, A., & Rosselló, J. (2007). Modelling environmental attitudes toward tourism. *Tourism Management, 26*(3), 688–695. doi:10.1016/j.tourman.2006.04.004

Chen, J. S., & Hsu, C. H. C. (2001). Developing and validating a riverboat gaming impact scale. *Annals of Tourism Research, 28*(2), 459–476. doi:10.1016/S0160-7383(00)00059-1

Choi, H., & Murray, I. (2010). Resident attitudes toward sustainable community tourism. *Journal of Sustainable Tourism, 18*(4), 575–594. doi:10.1080/09669580903524852

Choi, H., & Sirakaya, E. (2005). Measuring Residents' Attitude toward Sustainable Tourism: Development of Sustainable Tourism Attitude Scale. *Journal of Travel Research, 43*(4), 380–394. doi:10.1177/0047287505274651

Choi, H., & Sirakaya, E. (2006). Sustainability indicators for managing community tourism. *Tourism Management, 27*(6), 1274–1289. doi:10.1016/j.tourman.2005.05.018

Deccio, C., & Baloglu, S. (2002). Garfield county resident perceptions of the 2002 Winter Olympics: The spillover effects. *Journal of Travel Research, 41*(1), 45–56. doi:10.1177/0047287502041001006

Dyer, P., Gursoy, D., Sharma, B., & Carter, J. (2007). Structural modelling of resident perceptions of tourism and associated development on the Sunshine Coast, Australia. *Tourism Management, 28*(2), 409–422. doi:10.1016/j.tourman.2006.04.002

Fredline, E., & Faulkner, B. (2000). Host community reactions: A cluster analysis. *Annals of Tourism Research, 27*(3), 763–784. doi:10.1016/S0160-7383(99)00103-6

Gursoy, D., Chi, C. G., & Dyer, P. (2009). An examination of locals' attitudes. *Annals of Tourism Research, 36*(4), 723–726. doi:10.1016/j.annals.2009.06.003

Gursoy, D., Chi, C. G., & Dyer, P. (2010). Locals' attitudes toward mass and alternative tourism: The case of sunshine coast, Australia. *Journal of Travel Research, 49*(3), 381–394. doi:10.1177/0047287509346853

Gursoy, D., Jurowski, C., & Uysal, M. (2002). Resident attitudes: A structural modeling approach. *Annals of Tourism Research, 29*(1), 79–105. doi:10.1016/S0160-7383(01)00028-7

Gursoy, D., & Kendall, K. W. (2006). Hosting mega events: Modeling locals' support. *Annals of Tourism Research, 33*(3), 603–623. doi:10.1016/j.annals.2006.01.005

Gursoy, D., Ouyang, Z., Nunkoo, B., & Wei, W. (2019). Residents' impact perceptions of and attitudes towards tourism development: A meta-analysis. *Journal of Hospitality Marketing & Management, 28*(3), 306–333. doi:10.1080/19368623.2018.1516589

Gursoy, D., & Rutherford, D. G. (2004). Host attitudes toward tourism: An improved structural model. *Annals of Tourism Research, 31*(3), 495–516. doi:10.1016/j.annals.2003.08.008

Haobin Ye, B., Qiu Zhang, H., Huawen Shen, J., & Goh, C. (2014). Does social identity affect residents' attitude toward tourism development? An evidence from the relaxation of the Individual Visit Scheme. *International Journal of Contemporary Hospitality Management, 26*(6), 907–929. doi:10.1108/IJCHM-01-2013-0041

Harrill, R. (2004). Residents' attitudes toward tourism development: A literature review with implications for tourism planning. *Journal of Planning Literature, 18*(3), 251–266. doi:10.1177/0885412203260306

Jurowski, C., Uysal, M., & Williams, D. R. (1997). A theoretical analysis of host community resident reactions to tourism. *Journal of Travel Research, 36*(2), 3–11. doi:10.1177/004728759703600202

Kim, H., Gursoy, D., & Lee, S. (2006). The impact of the 2002 World Cup on South Korea: Comparisons of pre- and post-games. *Tourism Management, 27*(1), 86–96. doi:10.1016/j.tourman.2004.07.010

King, B., Pizam, A., & Milman, A. (1993). Social impacts of tourism: Host perceptions. *Annals of Tourism Research, 20*(4), 650–665. doi:10.1016/0160-7383(93)90089-L

Korça, P. (1996). Residents attitudes toward tourism impacts. *Annals of Tourism Research, 23*(3), 695–726. doi:10.1016/0160-7383(95)00092-5

Korea Convention. (2019). *Top reasons for choosing*. Retrieved from http://www.koreaconvention.org/eng/info/reason.kc

Korea Tourism Organization. (2017). *2017 hankook ganging trend bogosu*. Retrieved from http://kto.visitkorea.or.kr/kor/notice/data/report/org/board/view.kto?id=430336&isNotice=false&instanceId=127&rnum=12

Korea Tourism Organization. (2018). *Sukjaro bonen Hankook ganging* Retrieved from http://kto.visitkorea.or.kr/kor/notice/data/report/org/board/view.kto?id=431195&isNotice=false&instanceId=127&rnum=5

Kwon, J., & Vogt, C. (2010). Identifying the role of cognitive, affective and behavioural components in understanding residents' attitudes toward place marketing. *Journal of Travel Research, 49*(4), 423–435. doi:10.1177/0047287509346857

Lee, T. H. (2013). Influence analysis of community resident support for sustainable tourism development. *Tourism Management, 34*, 37–46. doi:10.1016/j.tourman.2012.03.007

Lepp, A. (2007). Residents' attitudes towards tourism in Bigodi village, Uganda. *Tourism Management, 28*(3), 876–885. doi:10.1016/j.tourman.2006.03.004

Lindberg, K., & Johnson, R. L. (1997). Modeling resident attitudes toward tourism. *Annals of Tourism Research, 24*(2), 402–424. doi:10.1016/S0160-7383(97)80009-6

MacGehee, N. G., & Andereck, K. L. (2004). Factors Predicting Rural Residents' Support of Tourism. *Journal of Travel Research, 43*(2), 131–140. doi:10.1177/0047287504268234

MacKay, K., & Campbell, J. (2004). An examination of residents' support for hunting as a tourism product. *Tourism Management, 25*(4), 443–452. doi:10.1016/S0261-5177(03)00127-4

Madrigal, R. (1993). A tale of tourism in tow cities. *Annals of Tourism Research, 22*(2), 336–353. doi:10.1016/0160-7383(93)90059-C

McCool, S. F., & Martin, S. R. (1994). Community attachment and attitudes toward tourism development. *Journal of Travel Research, 32*(3), 29–34. doi:10.1177/004728759403200305

McGehee, N. G., & Andereck, K. L. (2004). Factors predicting Rural Residents' Support of Tourism. *Journal of Travel Research, 42*(2), 131–140. doi:10.1177/0047287504268234

Moghavvemi, S., Woosnam, K. M., Paramanathan, T., Musa, G., & Hamzah, A. (2017). The effect of residents' personality, emotional solidarity, and community commitment on support for tourism development. *Tourism Management, 63*, 242–254. doi:10.1016/j.tourman.2017.06.021

Murphy, P. E. (1985). *Tourism: A community approach*. Methuen.

Nepal, S. (2008). Residents' Attitudes to Tourism in Central British Columbia, Canada. *Tourism Geographies, 10*(1), 42–65. doi:10.1080/14616680701825123

Nicholas, L. N., Thapa, B., & Ko, Y. J. (2009). Residents' perspectives of a World Heritage Site: The Pitons Management Area, St. Lucia. *Annals of Tourism Research, 36*(3), 390–412. doi:10.1016/j.annals.2009.03.005

Nunkoo, R., & Gursoy, D. (2012). Residents' support for tourism: An identity perspective. *Annals of Tourism Research, 39*(1), 243–268. doi:10.1016/j.annals.2011.05.006

Nunkoo, R., & Ramkissoon, H. (2011). Residents' satisfaction with community attributes and support for tourism. *Journal of Hospitality & Tourism Research (Washington, D.C.), 35*(2), 171–190. doi:10.1177/1096348010384600

Nunkoo, R., & Ramkissoon, H. (2012). Power, trust, social exchange and community support. *Annals of Tourism Research, 39*(2), 997–1023. doi:10.1016/j.annals.2011.11.017

Nunkoo, R., Smith, S. L., & Ramkissoon, H. (2013). Residents' attitudes to tourism: A longitudinal study of 140 articles from 1984 to 2010. *Journal of Sustainable Tourism, 21*(1), 5–25. doi:10.1080/09669582.2012.673621

Nunkoo, R., & So, K. (2016). Residents' Support for Tourism: Testing Alternative Structural Models. *Journal of Travel Research, 57*(7), 847–861. doi:10.1177/0047287515592972

Nunnally, J. C. (1978). *Psychometric theory* (2nd ed.). McGraw-Hill.

OECD. (2019). *South Korea.* Reatrived from https://data.oecd.org/korea.htm#profile-economy

Olya, H., & Gavilyan, Y. (2016). Configurational Models to Predict Residents' Support for Tourism Development. *Journal of Travel Research, 56*(7), 893–912. doi:10.1177/0047287516667850

Park, D. B., Nunkoo, R., & Yoon, Y. S. (2015). Rural residents' attitudes to tourism and the moderating effects of social capital. *Tourism Geographies, 17*(1), 112–133. doi:10.1080/14616688.2014.959993

Park, K., Lee, J., & Lee, T. (2017). Residents' attitudes toward future tourism development in terms of community well-being and attachment. *Asia Pacific Journal of Tourism Research, 22*(2), 160–172. doi:10.1080/10941665.2016.1208669

Perdue, R. R., Long, P. T., & Allen, L. (1990). Resident support for tourism development. *Annals of Tourism Research, 17*(4), 586–599. doi:10.1016/0160-7383(90)90029-Q

Prayag, G., Hosany, S., Nunkoo, R., & Alders, T. (2013). London residents' support for the 2012 Olympic Games: The mediating effect of overall attitude. *Tourism Management, 36*, 629–640. doi:10.1016/j.tourman.2012.08.003

Ribeiro, M. A., Pinto, P., Silva, J. A., & Woosnam, K. M. (2017). Residents' attitudes and the adoption of pro-tourism behaviours: The case of developing island countries. *Tourism Management, 61*, 523–537. doi:10.1016/j.tourman.2017.03.004

Saveriades, A. (2000). Establishing the social tourism carrying capacity for the tourist resorts of the east coast of the Republic of Cyprus. *Tourism Management, 21*(2), 147–156. doi:10.1016/S0261-5177(99)00044-8

Sheldon, P. J., & Abenoja, T. (2001). Resident attitudes in a mature destination: The case of Waikiki. *Tourism Management, 22*(5), 435–443. doi:10.1016/S0261-5177(01)00009-7

Teye, V., Sirakaya, E., & Sönmez, S. (2002). Residents' attitudes toward tourism development. *Annals of Tourism Research, 29*(3), 668–688. doi:10.1016/S0160-7383(01)00074-3

Tosun, C. (2002). Host perceptions of impacts: A comparative tourism study. *Annals of Tourism Research, 29*(1), 231–253. doi:10.1016/S0160-7383(01)00039-1

Tovar, C., & Lockwood, M. (2008). Social impacts of tourism: An Australian regional case study. *International Journal of Tourism Research, 10*(4), 365–378. doi:10.1002/jtr.667

Um, S., & Crompton, J. L. (1987). Measuring resident's attachment levels in a host community. *Journal of Travel Research, 26*(1), 27–29. doi:10.1177/004728758702600105

Var, T., Kendall, K. W., & Tarakcioglu, E. (1985). Resident attitudes towards tourists in a Turkish resort town. *Annals of Tourism Research, 12*(4), 652–658. doi:10.1016/0160-7383(85)90086-6

Vargas-Sánchez, A., Plaza-Mejía, M. A., & Porras-Bueno, N. (2008). Understanding residents' attitudes toward the development of industrial tourism in a former mining community. *Journal of Travel Research, 47*(3), 373–387. doi:10.1177/0047287508322783

Vincent, V., & Thompson, W. (2002). Assessing Community Support and Sustainability for Ecotourism Development. *Journal of Travel Research, 41*(2), 153–160. doi:10.1177/004728702237415

Visit Korea. (2019). *100 Must-Visit Tourist spots in Korea.* Retrieved from http://english.visitkorea.or.kr/enu/TRV/TV_ENG_Must100.jsp#

Wang, Y. A., & Pfister, R. E. (2008). Residents' attitudes toward tourism and perceived personal benefits in a rural community. *Journal of Travel Research, 47*(1), 84–93.

Ward, C., & Bernom, T. (2011). Beyond social exchange theory: Attitudes toward tourists. *Annals of Tourism Research, 38*(4), 1556–1569. doi:10.1016/j.annals.2011.02.005

Woo, E., Kim, H., & Uysal, M. (2015). Life satisfaction and support for tourism development. *Annals of Tourism Research, 50*, 84–97. doi:10.1016/j.annals.2014.11.001

Woosnam, K. M. (2012). Using emotional solidarity to explain residents' attitudes about tourism and tourism development. *Journal of Travel Research, 51*(3), 315–327. doi:10.1177/0047287511410351

Woosnam, K. M., Aleshinloye, K. D., Ribeiro, M. A., Stylidis, D., Jiang, J., & Erul, E. (2018). Social determinants of place attachment at a World Heritage Site. *Tourism Management, 67*, 139–146. doi:10.1016/j.tourman.2018.01.012

Yoon, Y., Gursoy, D., & Chen, J. S. (2001). Validating a tourism development theory with structural equation modeling. *Tourism Management, 22*(4), 363–372. doi:10.1016/S0261-5177(00)00062-5

Yoon, Y., Kim, S., & Kim, S. (2015). Successful and unsuccessful film tourism destinations: From the perspective of Korean local residents, Perceptions of film tourism impacts. *Tourism Analysis, 20*(3), 297–311. doi:10.3727/108354215X14356694891852

Yu, C., Chancellor, H., & Cole, S. (2011). Measuring Residents' Attitudes toward Sustainable Tourism: A Reexamination of the Sustainable Tourism Attitude Scale. *Journal of Travel Research, 50*(1), 57–63. doi:10.1177/0047287509353189

ADDITIONAL READING

Almeida García, F., Balbuena Vázquez, A., & Cortés Macías, R. (2015). Resident's attitudes towards the impacts of tourism. *Tourism Management Perspectives, 13*(0), 33–40. doi:10.1016/j.tmp.2014.11.002

Amuquandoh, F. (2010). Residents' perceptions of the environmental impacts of tourism in the Lake Bosomtwe Basin, Ghana. *Journal of Sustainable Tourism, 18*(2), 223–238. doi:10.1080/09669580903298531

Erul, E., Woosnam, K. M., & McIntosh, W. A. (2020). Considering emotional solidarity and the theory of planned behavior in explaining behavioral intentions to support tourism development. *Journal of Sustainable Tourism, 28*(8), 1158–1173. doi:10.1080/09669582.2020.1726935

Gursoy, D., Ouyang, Z., Nunkoo, R., & Wei, W. (2019). Residents' impact perceptions of and attitudes towards tourism development: A meta-analysis. *Journal of Hospitality Marketing & Management*, *28*(3), 306–333. doi:10.1080/19368623.2018.1516589

Megeirhi, H. A., Woosnam, K. M., Ribeiro, M. A., Ramkissoon, H., & Denley, T. J. (2020). Employing a value-belief-norm framework to gauge Carthage residents' intentions to support sustainable cultural heritage tourism. *Journal of Sustainable Tourism*, *28*(9), 1351–1370. doi:10.1080/09669582.2020.1738444

Ribeiro, M. A., Pinto, P., Silva, J. A., & Woosnam, K. M. (2018). Examining the predictive validity of SUS-TAS with maximum parsimony in developing island countries. *Journal of Sustainable Tourism*, *26*(3), 379–398. doi:10.1080/09669582.2017.1355918

Sinclair-Maragh, G. (2017). Demographic analysis of residents' support for tourism development in Jamaica. *Journal of Destination Marketing & Management*, *6*(1), 5–12. doi:10.1016/j.jdmm.2016.03.005

Vargas-Sánchez, A., Valle, P. O., Mendes, J. C., & Silva, J. A. (2015). Residents' attitude and level of destination development: An international comparison. *Tourism Management*, *48*(0), 199–210. doi:10.1016/j.tourman.2014.11.005

KEY TERMS AND DEFINITIONS

Community Attachment: The extent to which residents of a destination possess cognitive or affective bonds to a place.

Social Exchange Theory: Is a general sociological theory concerned with understanding the exchange of resources between individuals and groups in an interaction situation.

Support for Tourism: Is the degree to which residents evaluate the effects of tourism in their community.

Sustainable Tourism: Increases the benefits and to reduce the negative impacts caused by tourism at destinations.

Theory of Reasoned Action: Explains the relationship between attitudes and behaviours within human action.

Tourism Impacts: Include the effects of *tourism* on the environment and on destination communities, and its economic contributions.

Chapter 8
Different Type of Residents, Different Type of Attitudes?
The Case of Tourism Development in East Timor

Manuel Vong
Dili Institute of Tecnology, East Timor

Patrícia Pinto
https://orcid.org/0000-0002-3153-2830
Faculty of Economics and Research Centre for Tourism, Sustainability, and Well-Being (CinTurs), University of Algarve, Portugal

João Albino Silva
Faculty of Economics and Research Centre for Tourism, Sustainability, and Well-Being (CinTurs), University of Algarve, Portugal

ABSTRACT

This study examines residents' attitudes towards sustainable tourism development in Timor-Leste, a new country that is also in the initial stages of tourism development. In particular, this study intends to understand whether residents with different roles and responsibilities (ordinary residents and residents as leaders in the public, private sectors, and NGOs) have different attitudes about the development of sustainable tourism. In order to accomplish this goal, the sustainable tourism attitude scale (SUS-TAS) is applied. Results have shown that 'ordinary residents' as well as residents with leadership responsibilities in the community are very positive and supportive of tourism development. These results are encouraging for the development of tourism in this country and should be considered by the government, private sectors, and NGOs to promote tourism industries as a contribution to the sustainable development of Timor-Leste.

DOI: 10.4018/978-1-7998-3156-3.ch008

INTRODUCTION

Tourism in small islands should be planned to be compatible with and scaled to the available human capital and physical resources if sustainability in terms of economic, socio-cultural, ecological and political dimensions is to be achieved (Tinsley & Lynch, 2001; Twining-Ward & Butler, 2002). East-Timor is a small island state located in the South Asian and Pacific region. In this country, tourism is considered one of the future priorities for national economic development (Vong, Valle, & Silva, 2014). However, sustainable tourism development cannot be successfully implemented without the involvement of those who are most affected by tourism, namely, the residents (Sheldon & Abenoja, 2001; Woosnam, Maruyama, Ribeiro, & Joo, 2019).

Sustainable tourism development requires a participatory approach involving all residents. However, a local community is far from homogeneous, with different residents having different roles and responsibilities (Aref, Redzuan, & Emby, 2009). This implies that there may be different attitudes regarding the sustainable tourism development process within the community, and this diversity of attitudes may result in different perceptions regarding tourism. Within a local community, two main groups of residents can be identified in terms of their roles and responsibilities: the 'non-leading residents', i.e. those who do not have specific roles and responsibilities as leaders in the community; and the residents with roles and responsibilities as leaders (in the public sector, the private sector and non-governmental organizations (NGOs), named in this study as 'leading residents'. The latter are those involved in decision-making processes for policy, planning, implementation and control of tourism development (Choi & Sirakaya, 2006; Ribeiro, Valle, & Silva, 2013). As Butler (1980, p. 11) suggests, "a change of attitude is required on the part of those who are responsible for planning, developing, and managing tourist area".

The purpose of this study is to examine the different residents' attitudes towards sustainable tourism development in East-Timor, a destination that is not only new but also in the first stage of tourism development. Henceforth, the contribution of this study is twofold. First, although there are many studies about residents' attitudes towards tourism, most of them focus on mature destinations, and few focus on developing destinations, such as Bigodi in Uganda (Lepp, 2007, 2008) and Cape Verde (Ribeiro et al., 2013; Ribeiro, Pinto, Silva, & Woosnam, 2018). East-Timor is in the first stage of tourism development and has been little studied as a destination (Vong et al., 2014; Vong, Pinto, & Silva, 2017). Second, this study intends to understand whether residents with different roles and responsibilities (non-leading residents and leading residents in the public and private sectors and NGOs) have different attitudes about sustainable tourism development. This new approach will allow for a more complete understanding of residents' attitudes and consideration of the different roles residents can play in sustainable tourism development (Richards & Hall, 2002). To achieve this aim, the sustainable tourism attitude scale (SUS-TAS) will be used. The SUS-TAS scale has been proposed by Choi and Sirakaya (2005 & 2006) in order to measure residents' attitudes towards sustainable tourism development.

BACKGROUND

East-Timor as the Research Setting

East-Timor is a small island state (2002) located in the South Asian and Pacific region (Taylor et al., 2011). It emerged from 450 years of Portuguese colonization, three years of Australian and Japanese occupation and twenty-four years of Indonesian occupation, followed by two-and-a-half years of interim United Nations (UN) administration (Myrttinen, 2009; Sousa, 2003). It is located 500km from the north-west of Darwin in Australia and occupies the eastern half of Timor Island, with a total area of approximately 15,000km^2. Indonesia is located in the western part of the aforementioned island.

The territorial administration is divided into 13 districts, 65 sub-districts, 442 towns and 2,220 villages. The National Statistics Directorate (NSD) of East-Timor recorded on Census 2015 that the population of East-Timor totalled 1,179,654 people: 598,921 men and 580,733 women (NSD, 2015). The country has 38 native languages, 34 local languages and 4 international languages. The local languages most widely spoken in the territory are: Tetum, Mambai, Makasai, Kemak, Bunak, Tokodede, Fataluku, Baikenu, Waima'a, Naueti and Galoli. The country's official languages are Tetum and Portuguese. English and Bahasa Indonesia are used as working languages. Timorese people have specific words to express their welcome to guests, such as "*Simu Bainaka*" (in Tetum).

This study was conducted in local communities in three districts – Díli, Baucau and Maliana – which have not only potential tourism resources, but also vocational tourism activities. Díli is the capital of East-Timor and was founded on 10 October 1769; it is the centre of the public administration, education and business activities. Díli is located in the north coast of the territory. Its natural tourist attractiveness stems from the beaches, the island of Atauro, the marine biodiversity such as coral reefs, dolphins and whales and good diving locations. In this area, there are several historical sites. The cultural and religious heritage reveals the struggle for independence in the country (UNWTO, 2007). Baucau is the second city of East-Timor, situated in the eastern part of the territory, located 122 km from Díli and having a population surpassing 122,709 people (NSD, 2015). Maliana is the capital of the Bobonaro District, located 149 km to the southwest of Díli in the western part of the country, near the border with Indonesia. It has 97,153 inhabitants (NSD, 2015).

Tourism Policy and Development in East-Timor

East-Timor has had a separate Ministry of Tourism since 2012. The government is working towards the achievement of the tourism-related goals set in the East-Timor Strategic Development Plan 2011–2030. East-Timor is largely unknown as a tourist destination. It is a challenge for a young, small nation to compete with other destinations in the region (Vong, 2014).

During the last decade of independence (2002–2019), East-Timor, with the support of the United Nations (UN) and in cooperation with international development agencies, has undertaken humanitarian and short-term development projects in all sectors – economic, political, social, cultural and tourism (Vong, 2014). The goals have been to stabilize and mature the process of restructuring the political system, national stabilization, reconciliation, peace and justice, aiming to ensure sustainable development, fight poverty and improve the quality of people's lives.

Available East-Timor tourism statistics are limited. The National Tourism Directorate (NTD) (2018) and National Immigration Directorate (NID) (2018) estimate that 122,857 tourists have arrived in East-Timor in 2017. This number increased by 17% in the following five years with a total of 102,267 tourists in 2013. Ministry of Tourism East-Timor statistics report 122,857 international tourist arrivals in 2017, and tourism receipts were estimated as totalling USD $3,693,980.00 in 2017, referring to expenditures on travel items only, excluding passenger transport items.

The Strategic Development Plan (SDP) for East-Timor (2011-2030) regards tourism as one of the strategic sectors and priorities for the national economic development. SDP 2011-2030 states that the development of tourism should rely on the sustainability of the natural environment, its promotion, protection and on the preservation of cultural values (RDTL, 2019). An increase in the number of hotels, restaurants, apartments and houses around the tourism areas Metiaut, Areia Branca, Praia dos Coqueiros and Bebonuk Dili has been observed in comparison to 2007 (Vong et al., 2014, 2017). Residents have had the opportunity to rent their land, houses and cars, and they have had increased job opportunities. They have also seen an increase in their living standards. Various short-term programmes have been initiated, such as community-based tourism (CBT) and the development of eco-tourism projects in Atauro, Tutuala, Maubisse, Maubara and Com Lautem (Vong et al., 2017).

However, as a post-conflict, low-income country, East-Timor still faces many challenges, as more than 649,065 persons are in vulnerable employment or are unemployed (SEFOPE & DNE, 2018). Most people have limited competence in terms of knowledge, skills and abilities that would be useful to tourism. There is no legal framework for tourism businesses, land ownership or land use. Strategic development planning has not yet been applied to sustain tourism at local, regional and national levels (Taylor et al., 2011; Vong et al., 2014; Vong & Valle, 2013).

THEORETICAL FRAMEWORK

The Importance of Studying Residents' Attitudes

In the last four decades, researchers have been very interested in residents' attitudes towards the impact of tourism development (Sirakaya, Ingram, & Harrill, 2009; Wang, 2019). In these studies, 'attitude' is defined as a psychological tendency that is expressed by the evaluation of a particular entity with which a person can associate a favourable or unfavourable opinion (Sirakaya et al., 2008).

Theoretically, attitude can be defined as a tendency to respond favourably or unfavourably towards a particular object (Fishbein & Ajzen, 1974). Attitudes have several components: cognitive, affective and conative aspects. Many factors can explain attitudes, such as social norms, habits, personality characteristics and situational factors, among others (Fishbein & Ajzen, 1974).

Social exchange theory (SET) is one of the most widely used theories to analyse the residents' attitudes (Vargas-Sánchez, Porras-Bueno, & Plaza-Mejía, 2014). Social exchange theorists suggest that the evolution of the value of an exchange is complex and dynamic (Jurowski, Uysal, & Williams, 1997). This theory refers to the consented actions of individuals based upon the expectation of rewards (Blau, 2009). On the other hand, Homans (1961) defined social exchange as a trading activity, tangible or otherwise, between at least two persons. The viability of the exchange is assured by two main factors: the resultant value of the rewards and costs that do not exceed rewards (Skidmore, 1975). Consequently, residents will support the development of tourism if they perceive the costs as being lower than the benefits (Jurowski et al., 1997).

Residents can be involved in the tourism industry, depending on policy, planning and management of tourism, leading to the improvement of the residents' quality of life in economic, sociocultural, environmental and political aspects of sustainable tourism development (Miskinyte & Smirnova, 2011). Therefore, evaluating residents' attitudes towards the impact of tourism development in local communities is key to understanding the sustainability of the tourism industry (Yoon, Gursoy, & Chen, 2001).

In the tourism lifecycle model, tourism development comprises several stages through which a destination evolves (Butler, 1980): exploration/discovery, investment, development, consolidation, stagnation and, finally, decline or rejuvenation. This study is only focused on the first stage - exploration or discovery because Timor-Leste is still giving its first steps regarding tourism development. The exploration or discovery stage is characterized by a small number of visiting tourists or explorers, making individual travel arrangements and following irregular visitation patterns. Tourists can also be expected to be non-local visitors who have been attracted to the area by its unique or considerable natural and cultural features (Butler, 1980).

In the initial stages of tourism development, the residents' attitudes are positive towards tourism development, in agreement with the euphoria stage characterized by the Doxey Irridex model (1975). In this stage visitors are welcomed, and there is usually little planning or control mechanism (Doxey, 1975). The residents are enthusiastic about tourism and see it as a source of unlimited opportunities; they welcome the stranger and there is a strong feeling of satisfaction associated with the process (Doxey, 1976; Vargas-Sánchez, Porras-Bueno, & Plaza-Mejía, 2009). Opportunities are presented to locals and money flows in along with tourists (Doxey, 1976). Residents demonstrate positive attitudes associated with the belief that tourism creates community development, opportunities for earning income and improved agriculture markets (Lepp, 2007).

Typology of Residents as Community Leaders

This study defines leadership as a process in which a person influences others to accomplish an objective and directs an organization in a way that makes it more cohesive and coherent (O'Halloran, 1992; Wart, 2003). Leadership is a function of the leader, whereby an individual influences and directs a group of individuals to achieve a common goal (Darr, 2011; Northouse, 2013). In other words, the leader is the person who occupies the position of highest hierarchy in a group, having roles and responsibilities concerning decisions, plans, organization and implementation, among others, relevant to a tourist destination (Scott, Laws, Richins, & Agrusa, 2011; Wart, 2003). In general, the leader's attitudes are heterogeneous depending on the roles, control mechanisms, task focus and work coordination, both inside and outside the unit of the organization (Wart, 2003).

In the tourism context, residents with leadership roles are those with important responsibilities in the planning and implementation of tourism development (Aref & Redzuan, 2010). The local leaders play a vital role in addressing sustainable tourism development and represent a mechanism for the capacity building of community for sustainable tourism development (Aref et al., 2009). Sustainable tourism development requires the knowledgeable participation of all relevant organizational leaders (public, private and NGOs), as well as strong community leadership to ensure wide participation and consensus building (WTO, 2004). These leaders are those working in the public sector (the government institutions), the private sector (in tourism and travel companies) and in the non-government organizations (NGOs) (including religious institutions and social institutions) (WTO, 2009).

To achieve sustainable development and optimal tourism industry outcomes, leaders in the public sector and the government have responsibilities of both a legal nature and regulatory nature (Lacy et al., 2002). Product development, marketing and investment in operations related to the tourism industry are some of the responsibilities attributed to leaders in the private sector (tour operators and hotel and restaurant owners, among others) (Lacy et al., 2002). Leaders in NGOs have responsibilities tied to the development of local skills, especially in poor and/or socially deprived areas, hopefully channelling resources into countries with low levels of tourism industry (Scott et al., 2011). Finally, residents acting in leadership roles within tourism organizations and travel industries play important roles in coordinating bottom-up and top-down approaches to tourism planning, development and management (Scott et al., 2011).

Measurement Tools for Residents' Attitudes Towards Tourism: SUS-TAS

Many scholars suggest that residents are major actors in the sustainable tourism development process since they are directly affected by tourism (Butler, 1980; Choi & Sirakaya, 2005; Gunn & Var, 2002). Based on tourism literature, it is important to understand the residents' attitudes toward tourism if successful tourism development is desired. Scales and methods have been developed and applied by tourism scholars to collect data regarding the attitudes towards tourism (Sirakaya et al., 2008).

A multiple measure tourism impact attitude scale (TIAS) was developed in response to the need for a standardized measurement of resident's attitudes towards sustainable tourism development (Bachleitner & Zins, 1999; Lankford & Howard, 1994; Wang & Pfister, 2008). However, the TIAS has some limitations, since it has a specific use and only partially evaluates residents' attitudes towards tourism marketing and sociocultural, geographical, social and economic impacts, among others. Additionally, most of the studies associated with this scale have been conducted in mature tourism destinations. Finally, many TIAS measures do not refer to sustainable tourism development principles when evaluating residents' attitudes towards tourism development impacts, especially in developing countries (Choi & Sirakaya, 2005).

Recognizing the limitations and the paucity of research on residents' attitudes towards sustainable tourism development, Choi and Sirakaya (2005 & 2006) developed the Sustainable Tourism Attitude Scale (SUS-TAS), assisted by a panel of prominent international tourism scholars. SUS-TAS offers a paradigm shift, not only offering a promising instrument for gauging community feelings towards sustainable tourism development (Ribeiro et al., 2018), but also bridging existing paradigms, such as the social exchange theory and the theory of sustainability (Sirakaya et al., 2008).

The present study used the SUS-TAS framework and made use of some additional tools customised to the case of East-Timor, which are appropriate to a new country in the initial stage of tourism development. Although the SUS-TAS model has been developed based on the paradigm shift in Western societies, according to the social exchange theory and the new environmental paradigm (Choi & Sirakaya, 2005; Sirakaya et al., 2009), it incorporates the principles of sustainability, thus becoming useful as an alternative measurement tool for researchers in communities, especially in the developing world, where sustainability is often considered as a panacea for development problems (Cottrell, Vaske, Shen, & Ritter, 2007; Ribeiro et al., 2013, 2018; Sirakaya et al., 2008, 2009; Yu, Chancellor, & Cole, 2011). Previous studies using SUS-TAS tested its applicability in developed destinations in Europe and America (Hung, Sirakaya, & Ingram, 2011; Kvasova, 2011; Prayag, Dookhony-Ramphul, & Maryeven, 2010) or in Africa (Cape-Verde) (Ribeiro, Pinto, Silva, & Woosnam, 2018). All these contexts are consolidated destinations. So, the scale was not previously tested in a country that is giving its first steps as a destination, as East-Timor.

The SUS-TAS integrates seven dimensions: economic, social-cultural, environmental, political or community participation, long-term planning, visitor satisfaction and community-based tourism (Choi & Sirakaya, 2006; Sirakaya et al., 2008).

Research Hypotheses

This study uses the SUS-TAS scale proposed by Choi and Sirakaya (2005) to measure the residents' attitudes towards sustainable tourism development in East-Timor. Since attitudes towards tourism are multi-faceted phenomena involving different impact factors and resulting from complex processes of social exchange between tourist and host communities, residents' perceptions of tourism may have both negative and positive aspects, which include economic, sociocultural, environmental, political and long-term planning, ensuring visitor satisfaction and community-based tourism impacts (Choi & Sirakaya, 2005; Richards & Hall, 2002; Yoon et al., 2001). Therefore, based on this theoretical framework, four research hypotheses are proposed:

Hypothesis 1: Non-leading residents and leading residents in East-Timor report a euphoric attitude towards tourism and positively perceive the benefits of sustainable tourism development.

Hypothesis 2: Non-leading residents and leading residents in East-Timor report a euphoric attitude towards tourism and underestimate the negative impacts of tourism development.

Hypothesis 3: There is a significant difference between non-leading residents and leading residents in terms of their attitudes towards sustainable tourism development.

Hypothesis 4: There is a significant difference between the three groups of residents who are leaders (from the public sector, the private sector and from the NGOs) in terms of their attitudes towards sustainable tourism development.

METHODOLOGY

Population and Sample for Non-Leading Residents

The study was conducted in three districts in East-Timor during 2013: Baucau, Dili and Maliana. The three sample districts were selected because of their geographical location, comprising the three main cities located in the eastern, central and western areas of the territory. These districts have the potential and characteristics best suited to the tourism industry in East-Timor.

The three districts are composed by 3,058 households (NSD & UNFPA, 2010). Baucau has 240 households; Dili has 2.638 households; and Maliana has 180 households. Based on Yamane's (1967) formula, n = N / (N*(e^2)) + 1, with precision (e) = 0.035, and the sample size was calculated as 625 people distributed in different households. A total of 621 valid questionnaires were obtained, distributed among the different districts as follows: Baucau - 150 respondents; Dili - 347 respondents; and Maliana - 124 respondents. The study was conducted in the local communities targeted for interview. Respondents were people of 18 years of age or older.

Population and Sample for Leading Residents

The target population in the three evaluated districts included 156 residents with leading roles in various organizations such as the government, the private sector (hotels, apartments, travel agencies and restaurants) and NGOs (Vong et al., 2014). They were distributed as follows: Baucau - 18 people; Dili - 126 people; and Maliana - 12 people. Based on Yamane's (1967) formula, the sample size was calculated as

131 leaders. In this case, a total of 125 valid questionnaires were obtained, divided as follows: government - 28 respondents; private sector - 69 respondents; and NGOs - 28 respondents. These respondents were distributed among the three districts as follows: Baucau - 17 respondents; Dili - 96 respondents; and Maliana - 12 respondents.

Questionnaire Design and Data Collection

The questionnaire was designed for both non-leading and leading residents. It was structured in two parts. Part one comprised socio-demographic information, including questions about gender, age, education levels, occupation and monthly expenditure (please note that monthly expenditure was used to calculate income due to pre-test results which showed that respondents were not revealing how much income they received every month). Part two was based on the seven dimensions of SUS-TAS (economic, socio-cultural, environmental tourism development impacts; the importance of the political aspect or community participation, long-term planning, ensuring visitor satisfaction and community-based tourism). Items in these dimensions were measured using a five-point Likert scale (ranging from 1 – strongly disagree to 5 – strongly agree). The questionnaire was initially developed in Portuguese. It was subsequently back translated into Tetum (the national language) by the Dili Institute of Technology Language Centre of East-Timor to ensure the consistency and correctness of content.

To collect data, fifteen senior students from the Tourism Department at the Dili Institute of Technology (DIT) were selected from a group of twenty-five to attend data collection training, which was carried out from 12–14 January 2012. The first author of this study conducted this training The questionnaire was pre-tested with 50 people residing in Dili. The contents of the final questionnaire were revised and adjusted based on the results of this pre-test. Data collection took place over two weeks from 20 January to 4 of February 2012. Participants were contacted door-to-door using a random route systematic sampling procedure (Rada & Martins, 2014). In most cases, the questionnaires were self-responded. In some occasions, students collecting data helped to fulfil the questionnaire, especially for older and less qualified residents. The responses from the questionnaires were analysed using the software 'Statistical Package for the Social Sciences' (SPSS). Data analysis relied on descriptive statistics and inferential tests. For each set of items within each dimension, a reliability analysis was carried out.

Sample Profile: Non-Leading Residents

The data below reflect the characteristics of residents in the three research areas: Baucau, Dili and Maliana. A descriptive summary of the sample is presented in Table 1. Regarding the sample population, 50.4% of the non-leading residents in the sample were female, and 53.0% were aged between 20 and 29 years. The education of 52.4% of non-leading residents reached the high school level, and the degree of diversity in the employment variable was high, reflecting the heterogeneity of the livelihood of non-leading residents. Regarding employment status, 20.0% of respondents worked as farmers or fishermen, representing rural communities even though they are living in the capital of the district. Moreover, 17.9% of respondents were full-time employed, working as public servants (among others), whereas 17.6% were self-employed workers, comprising small traders with 'three-wheeled bikes' and 'kiosks' in a private home to attend to the visitors. Most of the respondents (50.2%) reported monthly expenditures of between \$100 and \$300. This information shows that most of the non-leading residents' monthly income is probably between \$100 and \$300 or more.

Table 1. Sample Profile - Non-Leading Residents

Variables	n	%	Variables	n	%
Gender	**621**	**100**	**Employment**	**621**	**100**
Male	308	49.6	Self-employed	109	17.6
Female	313	50.4	Paid employer	67	10.8
Age	**621**	**100**	Student	101	16.3
20-29	327	53	Farmer and Fishermen	124	20.0
30-39	153	25	Full time work	111	17.9
40-49	78	13	Full time home-work	106	17.1
50-59	31	5	Others (retired & Unemployed)	3	0.3
> 60	32	4	**Expenditure**	**621**	**100**
Education	**621**	**100**	<100 USD	247	39.8
Basic school	140	22.5	100 - 300 USD	312	50.2
High school	325	52.4	301 - 600 USD	50	8.1
Bachelor's degree	141	22.7	> 600 USD	12	1.9
Master's and/or doctoral degree	15	2.4			

Sample Profile: Leading residents

Table 2 shows the main socio-demographic characteristics of residents with leadership roles in Baucau, Dili and Maliana. Most of these residents were male: 92.9% were in the public sector, 62.3% were in the private sector and 57.1% were in NGOs. Evaluating all the districts as a whole, a total of 68.0% male respondents was observed. Regarding age, 36% of the residents were aged between 30 and 39 years old, and 41% aged between 40 and 49 years old. Education levels of the leaders mostly reached the level of a Bachelor's degree (75.0%). The majority of residents working as leaders occupied positions such as directors/presidents and managers/assistant managers (78.0%). Based on the list of basic salaries in East-Timor, the monthly salary of 38.0% of residents working as leaders can be calculated as being between 600 and 799 USD and between 600 and 799 USD for 24.0% of the residents. Thus, leaders in the private sector and NGOs receive better salaries than leaders in the public sector.

RESULTS

Based on the SUS-TAS, residents' attitudes towards tourism may be either negative and/or positive under the headings of economic, socio-cultural and environmental dimensions. This scale proposes that the degree of community participation, political involvement, long-term planning, community base-tourism and ensuring visitors' satisfaction are all aspects underpinning the sustainability of the tourism industry. Reliability tests for all items included in each dimension were carried using Cronbach's alpha coefficients. This coefficient is based on the correlations among the items on a single scale, and it is a measure of its internal consistency. Cronbach's alpha coefficients higher than 0.7 are desirable (Nunnally, 1978). The

Table 2. Sample Profile - Leading residents

Variables	Leading residents in Public Sec.		Leading residents in Private Sec.		Leading residents in NGOs	
	n	%	n	%	n	%
Gender	**28**	**100**	**69**	**100**	**28**	**100**
Male	26	92.9	43	62.3	16	57.1
Female	2	7.1	26	37.7	12	42.9
Age	**28**	**100**	**69**	**100**	**28**	**100**
20-29	2	7.1	14	20.3	3	10.72
30-39	6	21.4	34	49.3	11	39.28
40-49	12	42.9	20	29.0	13	46.4
50-59	7	25.0	1	1.4	1	3.6
> 60	1	3.6	0	0.0	0	0.0
Education	**28**	**100**	**69**	**100**	**28**	**100**
Basic school	1	3.6	3	4.3	0	0.0
High school	3	10.7	6	8.7	3	10.7
Bachelor's degree	21	75	55	79.7	18	64.3
Master's and doctoral degree	3	10.7	5	7.3	7	25.0
Job Occupation	**28**	**99.9**	**69**	**100**	**28**	**100**
Administrators and heads of villages	9	32.1	0	0.0	0	0.0
Directors and presidents	16	57.1	10	14.5	12	42.9
Managers and assistant managers	2	7.1	54	78.3	4	14.2
Academics and professionals	1	3.6	5	7.2	12	42.9
Salary	**28**	**100**	**69**	**100**	**28**	**100**
200 - 399 USD	3	11.0	0	0.0	0	0.0
400 - 599 USD	21	75.0	1	1.0	0	0.0
600 - 799 USD	4	14.0	20	30.0	23	82.0
800 - 999 USD	0	0	27	39.0	3	11.0
> 1000 USD	0	0	21	30.0	2	7.0

internal consistency of the SUS-TAS dimensions with Cronbach's alpha coefficients (α) ranged from 0.625 to 0.976. These coefficients can be found in tables 3 to 9.

Economic Impacts

Perceptions about the economic impacts of tourism development in East-Timor are presented in Table 3. This table shows the level of agreement in each item (e.g., the sum of responses in the categories "agree" and "totally agree") and the *p-value* from the Pearson Chi-Square tests of the independence between residents' agreement with tourism impacts and residents' typology. The first test distinguishes between the non-leading residents and leading residents. The second test considers the differences among the three types of leading residents (public sector, private sector and NGOs). Results show that, overall, the four

Table 3. Agreement level with tourism economic impacts

Economic Impacts Items	Non-leading Residents (%)	Leading residents in Public Sec. (%)	Leading residents in Private Sec. (%)	Leading residents in NGOs (%)	Chi-Square Test - Asymp. Sig. (2-sided) for 2 Groups	Chi-Square Test - Asymp. Sig. (2-sided) for 3 Groups
Positive Economic Impacts Items ($\alpha = 0.860$)						
Tourism promotes investment and economic growth	96.5	96.4	89.9	100	0.138	0.142
Tourism promotes job opportunities	92.8	96.4	95.7	96.4	0.185	0.976
An investment in local tourist attractions is positive	91.9	96.4	97.1	100	0.025	0.629
I like tourism because it yields revenues for our community	91.5	89.3	98.6	92.9	0.158	0.124
I believe tourism to be a strong economic contribution to our community	93.4	92.9	98.6	92.9	0.270	0.271
Tourism generates substantial taxes for our local government	87.8	85.7	97.1	71.4	0.745	0.001
I believe tourism to be good for our economy	95.5	92.9	97.1	92.9	0.887	0.544
Tourism promotes the marketing of our local products	95.0	92.9	97.1	96.4	0.637	0.621
Tourism diversifies the local economy	92.6	92.9	94.2	96.4	0.473	0.840
Tourism benefits other sectors in our community	94.7	92.9	95.7	96.4	0.814	0.795
Tourism improves the quality of life	89.2	96.4	95.7	75.0	0.507	0.003
Tourism increases the quality of services in local restaurants, stores and hotels	92.4	96.4	95.7	78.6	0.868	0.012
There are more local shopping opportunities due to tourism activity	85.0	96.4	95.7	64.3	0.272	0.000
Tourism promotes the appearance of my region	77.9	92.9	84.1	78.6	0.085	0.319
There are more parks and recreational areas due to tourism	83.3	96.4	92.8	85.7	0.013	0.316
There is better infrastructure (roads, water supply, electricity, telecommunications, among others) due to tourism	80.8	100	89.9	67.9	0.092	0.001
Negative Economic Impacts Items ($\alpha = 0.675$)						
Tourism increases housing costs	6.4	3.6	2.9	14.3	0.724	0.076
Tourism increases the cost of living (goods and services costs)	4.7	3.6	4.3	14.3	0.416	0.152

groups of residents demonstrate a strong agreement level with the positive economic impacts of tourism and a very low agreement level in what concerns the negative impacts. In most cases, the differences between the leaders and non-leaders and within the three types of leaders are not significant ($p > 0.05$).

Socio-Cultural Impacts

The levels of agreement on the socio-cultural impacts of tourism development in East-Timor are presented in Table 4. Concerning the economic impacts, the four groups of residents show a strong agreement level with the positive socio-cultural impacts and a low agreement level in respect of the negative impacts. In most measures, the differences between leaders and non-leaders are statistically significant ($p < 0.05$). Overall, the results show that most of the "non-leading" residents have less favourable assessments of the positive impacts of tourism than leaders. The former report a stronger concern about the negative impacts. Table 4 shows the opinions within the three types of leaders. Even though, in general, perceptions and leaders' typology are independent ($p > 0.05$), leading residents in the private sector demonstrate a stronger agreement with the negative socio-cultural views of tourism.

Table 4. Agreement level with tourism's socio-cultural impacts

Socio-cultural Impacts Items	Non-leading Residents (%)	Leading residents in Public Sec. (%)	Leading residents in Private Sec. (%)	Leading residents in NGOs (%)	Chi-Square Test - Asymp. Sig. (2-sided) for 2 Groups	Chi-Square Test - Asymp. Sig. (2-sided) for 3 Groups
Positive Socio-cultural Impacts Items ($\alpha = 0.649$)						
There are more services and cultural and recreational activities due to tourism	83.1	92.9	94.2	67.9	0.173	0.001
Tourism promotes understanding of other cultures in the community	81.0	82.1	91.3	82.1	0.099	0.313
Tourism promotes the preservation of local culture	74.2	78.6	84.1	85.7	0.033	0.744
Tourism promotes the maintenance and refurbishment of historical buildings	78.7	85.7	82.6	85.7	0.183	0.895
Negative Socio-cultural Impacts Items ($\alpha = 0.976$)						
Tourists visiting my community affect my quality of life	19.3	3.6	13.0	7.1	0.009	0.315
My quality of life has deteriorated because of tourism activities	23.2	3.6	13.0	3.6	0.000	0.178
I often feel angry because of tourism activities in my community	23.3	3.6	14.5	7.1	0.001	0.228
The community's recreational resources are excessively used by tourists	22.5	3.6	13.0	7.1	0.001	0.315
My community is overburdened by tourism	22.2	3.6	14.5	7.1	0.003	0.228
I'm not at ease with the presence of tourism enterprises in my region	24.0	3.6	13.0	3.6	0.000	0.178
Tourism is growing too fast	18.2	3.6	13.0	7.1	0.019	0.315
I believe that the environment in my community has deteriorated because of tourism	24.8	3.6	13.0	7.1	0.000	0.315
Tourism only benefits a small portion of the residents	23.0	3.6	13.0	7.1	0.001	0.315
Tourism generates traffic and parking issues	21.7	3.6	13.0	7.1	0.002	0.315
Tourism promotes thievery and vandalism	23.0	3.6	13.0	7.1	0.001	0.315
Tourism promotes alcoholism	22.4	3.6	13.0	7.1	0.001	0.315

Environmental Impacts

The results of responses to questions about the environmental impacts of tourism development in East-Timor are presented in Table 5. These results are not very different from those observed for economic and socio-cultural impacts. In fact, the four groups of residents show a strong agreement level with the positive environmental impacts and a low agreement level regarding the negative impacts (overall, quite similar to those observed regarding the socio-cultural impacts). Regarding the positive impacts, and in general, the differences between leaders and non-leaders' opinions and opinions within the three groups of residents were not statistically significant ($p > 0.05$). In contrast, and in terms of negative impacts, variability between the non-leading residents and leading residents was observed ($p < 0.05$). The former is more aware of the negative impacts. Among the three types of residents, the differences in the agreement levels are not significant; leading residents in the public sector and in NGOs display a less critical point of view than leading residents in the private sectors.

The Importance of Political Aspect and Community Participation

Table 6 shows a strong agreement with the importance of political aspects and community participation. The levels of agreement do not vary significantly between the two groups (non-leading residents and

Table 5. Agreement level with tourism environmental impacts

Environmental Impacts Items	Non-leading Residents (%)	Leading residents in Public Sec. (%)	Leading residents in Private Sec. (%)	Leading residents in NGOs (%)	Chi-Square Test - Asymp. Sig. (2-sided) for 2 Groups	Chi-Square Test - Asymp. Sig. (2-sided) for 3 Groups
Positive Environmental Impacts Items ($\alpha = 0.808$)	**98.6**	**100**	**100**	**100**	**0.176**	**1.00**
The community must protect the environment in order to guarantee the future	98.4	100	100	100	0.153	1.00
Environmental diversity must be valued and preserved	98.7	96.4	100	96.4	0.782	0.286
Tourism must protect the environment and the communities	97.7	100	98.6	100	0.291	0.664
Tourism should be developed in tune with the environment and culture	92.8	100	97.1	96.4	0.044	0.629
Tourism development should always protect wildlife and natural habitats	94.7	100	97.1	100	0.073	0.438
Tourism development should promote environmental ethics among each interested part	91.5	92.9	91.3	78.6	0.342	0.146
Environmental standards are needed in order to reduce the impacts of tourism	92.9	100	92.8	92.9	0.548	0.343
I believe that tourism should improve the environment so that future generations could benefit from it	90.0	92.9	94.2	92.9	0.210	0.954
Negative Environmental Impacts Items ($\alpha = 0.939$)	**20.3**	**3.6**	**13.0**	**7.1**	**0.005**	**0.315**
Tourism leads to environmental and landscaping deterioration	18.8	3.6	13.0	7.1	0.013	0.315
Tourism promotes disorganized urbanization growth	21.9	3.6	13.0	7.1	0.002	0.315
Tourism destroys the traditional architecture	21.1	3.6	13.0	3.6	0.001	0.178
Tourism promotes environmental deterioration (air and water pollution, increase in noise)	98.6	100	100	100	0.176	1.00

leading residents), nor do they vary significantly among the three groups (public, private and NGOs) (overall, in the two situations there is independence between the levels of agreement and residents' typology; $p > 0.05$). The exception is the measure relating to the item "tourism-related decisions should only be made by competent authorities", which shows low levels of agreement, especially within leading residents in the public sector and NGOs. In this case, leaders and non-leaders show a significant difference in agreement levels ($p = 0.000$), with the former expressing a higher agreement level (44.4%). Although not statistically significant, the results for other measures show some differences between the three types of leaders. For example, in the item "there is more law enforcement and fire prevention plans due to tourism", the NGO leaders express a more critical point of view with a lower agreement (53.6%) than residents in the public sector (75.0%) or in the private sector (71.0%).

The Importance of Long-Term Planning Aspects

All groups attribute great importance to the topic of long-term planning aspects, especially to the statement that the "tourism industry should ensure the quality of services for future visitors", with agreement levels higher than 89% in the four groups. For these aspects, no significant differences were found between leaders and non-leaders or within the three groups of leaders ($p > 0.05$). Note, however, that, as for most items, leading residents in the NGOs expressed higher levels of agreement.

Table 6. The importance of political terms and community participation

Political and community participation Importance Items ($\alpha = 0.625$)	Non-leading Residents (%)	Leading residents in Public Sec. (%)	Leading residents in Private Sec. (%)	Leading residents in NGOs (%)	Chi-Square Test - Asymp. Sig. (2-sided) for 2 Groups	Chi-Square Test - Asymp. Sig. (2-sided) for 3 Groups
Residents feel proud to belong in the community due to tourism	75.8	71.4	88.4	78.6	0.112	0.115
There is more law enforcement and fire prevention plans due to tourism	73.1	75.0	71.0	53.6	0.245	0.166
Tourism leads to an increase in public services available for residents	72.1	71.4	79.7	78.6	0.209	0.668
Decisions related to tourism should be made by all community members, regardless of social background or education	74.1	67.9	82.6	82.1	0.228	0.244
The full commitment of the community regarding tourism-related decisions promotes successful tourism development	73.8	60.7	73.9	75.0	0.556	0.378
Tourism-related decisions should only be made by competent authorities	44.4	17.9	26.1	17.9	0.000	0.548
Tourism development should benefit all members of the community	86.0	85.7	76.8	85.7	0.137	0.454
Tourism should promote peace and reconciliation between people	95.5	92.9	95.7	89.3	0.368	0.501

The Importance of Ensuring Visitors' Satisfaction Aspects

Table 7. The importance of long-term planning aspects

Long-term Planning Items ($\alpha = 0{,}679$)	Non-leading Residents' (%)	Leading residents in Public Sec. (%)	Leading residents in Private Sec. (%)	Leading residents in NGOs (%)	Chi-Square Test - Asymp. Sig. (2-sided) for 2 Groups	Chi-Square Test - Asymp. Sig. (2-sided) for 3 Groups
The tourism industry should plan ahead	74.9	78.6	78.3	82.1	0.305	0.909
I believe that successful tourism management demands advanced planning	62.8	60.7	69.6	82.1	0.106	0.208
I believe that tourism should be planned in the long term	76.7	64.3	79.7	75.0	0.727	0.281
The tourism industry should ensure the quality of services for future visitors	90.0	92.9	89.9	96.4	0.493	0.547

Regarding aspects relating to ensuring visitors' satisfaction, different types of residents expressed different levels of agreement (Table 8). In the two first statements, "The tourism industry has the responsibility to promote the needs of visitors" and "The community attractiveness is a crucial element in the ecological appeal of visitors", there is not a significant difference between non-leading residents and leading residents ($p > 0.05$). However, for these two statements, opinions vary among the three groups of leaders. Leading residents in the NGOs exhibit a greater agreement level to the statement "The community attractiveness is a crucial element in the ecological appeal of visitors" (92.9%), and most leading residents in the public sector agree with the statement "The tourism industry has the responsibility to promote the needs of visitors" (71.4%). In comparison to the previous statements, "Tourism enterprises should access the satisfaction levels of visitors" shows lower levels of agreement within the four groups of residents. In this case, non-leading residents are more favourable to this statement than leading residents, and the relationship between agreement level and residents' typology is significant ($p < 0.05$).

The Importance of Community-Based Tourism Aspects

Table 8. The importance of ensuring the visitors' satisfaction aspects

Ensuring the Visitors' Satisfaction Items ($\alpha = 0.737$)	Non-leading Residents (%)	Leading residents in Public Sec. (%)	Leading residents in Private Sec. (%)	Leading residents in NGOs (%)	Chi-Square Test - Asymp. Sig. (2-sided) for 2 Groups	Chi-Square Test - Asymp. Sig. (2-sided) for 3 Groups
The tourism industry has the responsibility to promote the needs of visitors	70.4	71.4	69.6	42.9	0.159	0.030
The community attractiveness is a crucial element in the ecological appeal to visitors	65.7	53.6	69.6	92.9	0.234	0.005
Tourism enterprises should access the satisfaction levels of visitors	54.9	46.4	40.6	46.4	0.017	0.806

Community-based tourism aims to maximize the involvement of residents and equalize the distribution among residents of revenues generated from tourism. Table 9 shows that all groups agree or totally agree with community-based tourism aspects (agreement levels in all groups were higher than 85%). However, for the statement "the tourism industry should be obliged to get at least half of its revenues (goods and services) from the local community", the differences between leaders and non-leaders and among the three types of leaders are significant ($p < 0.05$), with the former expressing a slightly lower agreement level (91.5%). Within leading residents, NGO leaders express a somewhat lower level of agreement (85.7%).

Research Hypotheses Testing

Table 9. The importance of community-based tourism aspects

Community-based Tourism Items ($\alpha = 0.759$)	Non-leading Residents (%)	Leading residents in Public Sec. (%)	Leading residents in Private Sec. (%)	Leading residents in NGOs (%)	Chi-Square Test - Asymp. Sig. (2-sided) for 2 Groups	Chi-Square Test - Asymp. Sig. (2-sided) for 3 Groups
The tourism industry should be obliged to get at least half of its revenues (goods and services) from the local community	91.5	100	100	85.7	0.041	0.001
I believe that the tourism industry should have at least half of its employs from the local population	93.1	100	100	85.7	0.118	0.001
The tourism industry should benefit the quality of life of local community members	96.9	100	98.6	92.9	0.691	0.162

Taking all the results into account, Figure 1 summarizes the results (positive or negative impacts addressing the different aspects of the SUS-TAS scale). This figure shows the average agreement level with the set of statements within each dimension considering both the non-leading residents and the leading residents. Figure 2 does a similar analysis, but compares the three groups of leaders.

Figure 1. Non-leading residents' and leading residents' perspectives

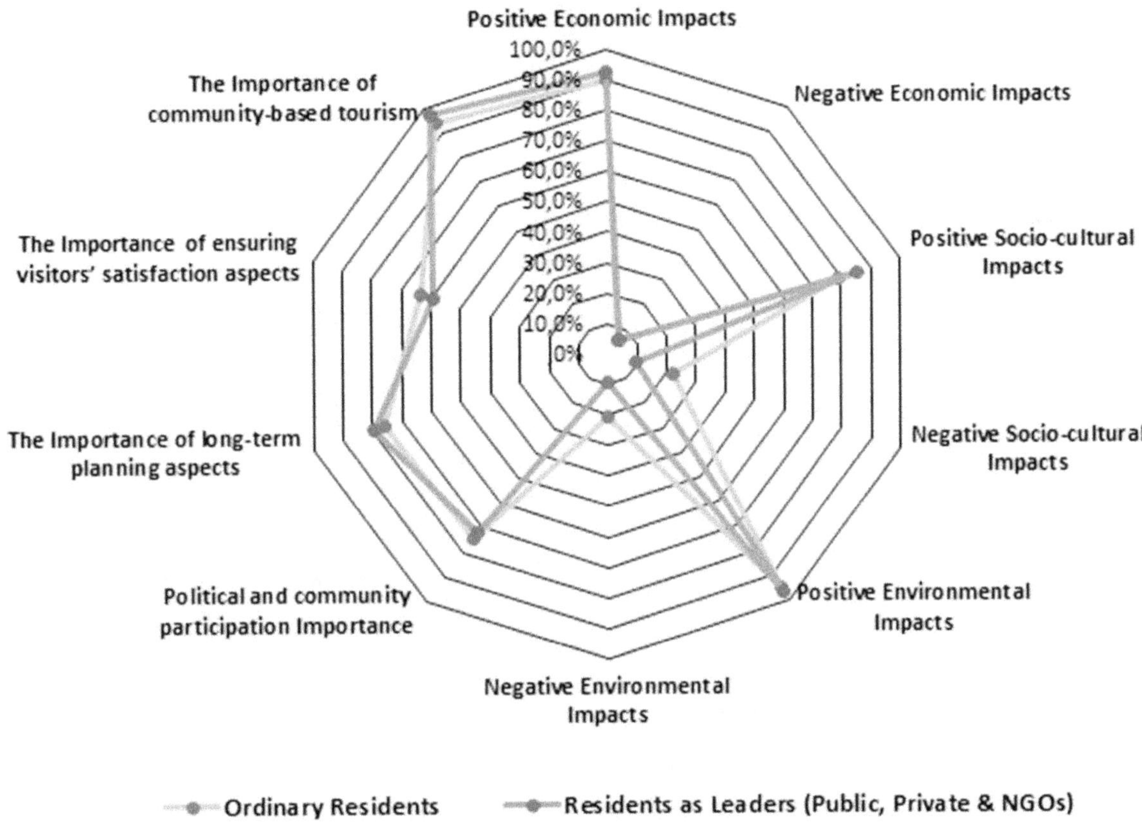

Figures 1 and 2 show that both leading and non-leading residents report high levels of agreement in the SUS-TAS dimensions that reflect the positive impacts of tourism development. These include "positive economic impacts", "positive socio-cultural impacts", "positive environmental impacts", "political and community participation importance", "long term planning aspects", "the importance of ensuring visitors' satisfaction aspects" and "the importance of community-based tourism". Among these dimensions, the former three and "the importance of community based tourism" report particularly high levels of average agreement, in general exceeding 80%. These results validate hypothesis 1, i.e. residents in East-Timor, both leaders and non-leaders, report a euphoric attitude towards tourism, positively perceiving the benefits of sustainable tourism development.

Figures 1 and 2 also highlight that negative impacts deriving from tourism (economic, sociocultural and environmental) are clearly undervalued by residents in this country, even by leading residents. Therefore, hypothesis 2 should be validated, i.e. residents in East-Timor, both leaders and non-leaders, report a euphoric attitude towards tourism, underestimating the negative impacts of tourism development. In this respect, it is interesting to see that non-leading residents are more aware of socio-cultural and environmental aspects, but less concerned with the negative economic impacts than leading residents.

Figure 1 can now be used to evaluate hypothesis 3. Although some differences between non-leading residents and residents with leadership roles were found to be significant within the measures compris-

Figure 2. Three groups of leading residents' perspectives

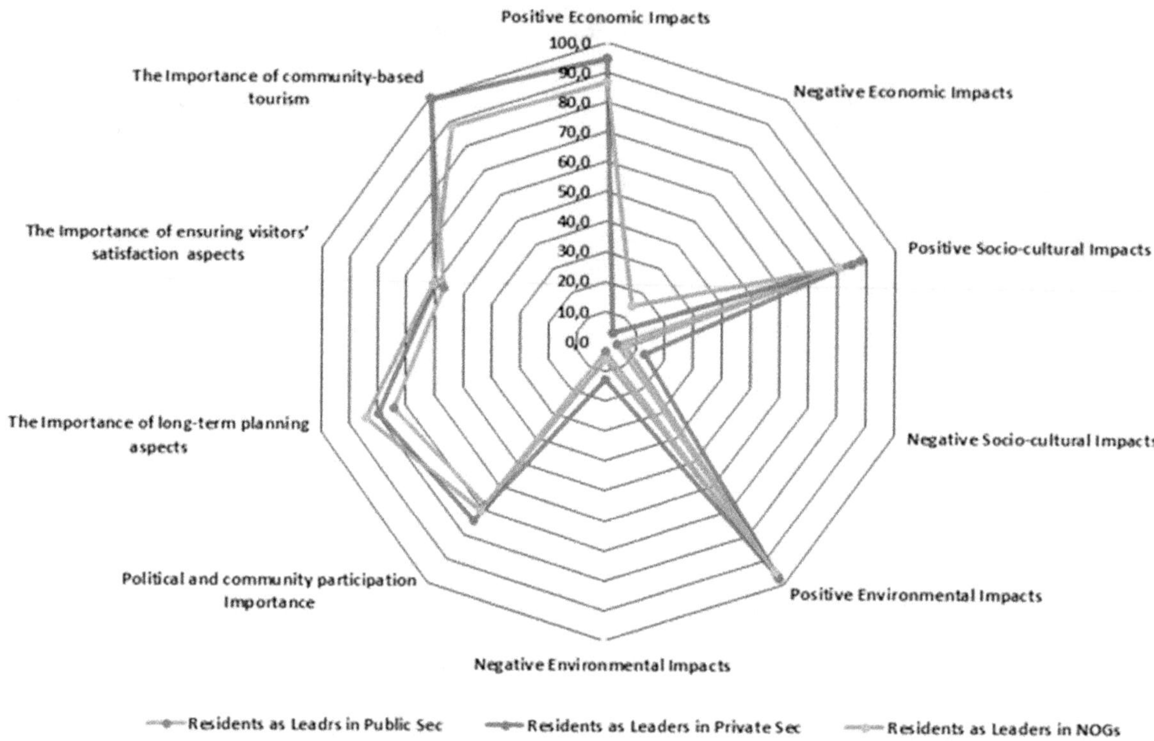

ing each dimension of the SUS-TAS scale (Tables 3–9), Figure 1 shows that, on average, their opinion is quite similar. In actuality, it is only when examining the negative environmental impacts and the negative sociocultural impacts that significant differences could be identified ($p < 0.05$), with non-leading residents expressing higher levels of awareness regarding these problems. In the remaining eight dimensions, opinions are homogeneous. Thus, based on these findings, hypothesis 3, as a whole, should be rejected.

Similar results are found when comparing the three groups of leaders (Figure 2). However, regarding hypothesis 4, some differences among the groups were significant for a few measures forming each dimension of the SUS-TAS scale (Tables 3–9). Figure 2 shows that, on average, their opinions are quite alike. Here and overall, significant differences were also found in two of the 10 dimensions, specifically, the negative economic impacts and the importance of community-based tourism ($p < 0.05$), with leaders from the NGOs being more aware of the former and less supportive of the latter. For this reason, and generally, hypothesis 4 should be rejected.

SOLUTIONS AND RECOMMENDATIONS

Sustainable tourism development requires that the positive attitudes of both non-leading and leading residents be converted in concrete actions and that they work together to address challenges in the development of tourism (Vong, 2014). Overall, these challenges include consensus building by coordination of bottom-up and top-down approaches to planning, developing and managing tourist areas, and strengthening the tourism industry to be more competitive and self-sustainable.

East-Timor is a new country and a small island destination with geographic, natural and socio-cultural weaknesses and vulnerabilities (Vong et al., 2013, 2014). Policy makers should plan the development of tourism to be compatible with the available human capital and physical resources to ensure that tourism remains sustainable in economic, socio-cultural, ecological and political terms. Indeed, the territory has great tourism potential, especially if the opportunity can be taken to design sustainable tourism models and harmonize relationships among the people, culture and nature stand to benefit from tourism (Vong et al., 2017). The role of local leaders is absolutely pivotal in ensuring rational and balanced sustainable tourism development. These policies should be addressed to the development, planning and management of tourism to attain self-sustainability. However, non-leading resident must be receptive and able to collaborate in the implementations of these policies. There is evidence that such conditions are met in East-Timor. Indeed, this study shows that both leading and non-leading residents express a very positive attitude towards sustainable tourism development.

Some policy implications for sustainable tourism development could be recommended which would involve the non-leading residents and leading residents. These policies would support the community participating actively in tourism management and development processes, such as planning, organizing, implementing, monitoring and evaluating tourism initiatives (Vong, 2014).

Firstly, it would be an important priority for the government and policy makers to establish tourism law, rules and regulation enforcements for national and foreign investment, i.e. land use planning, strategic tourism plans and tourism business plan, among others. Secondly, government and policy makers should encourage greater investment in education and training for tourism to improve the quality of knowledge, skills and attitudes of those involved in tourism services and hospitality based on Timorese culture, such as "*Simu Bainaka*". Thirdly, to increase participation of residents or communities in tourism development at the local level, policy makers need to establish tourism policies at the national, district/municipal and local levels. These should the provision of technical and financial support (e.g. small loans for small businesses related to tourism so as to avoid direct subsidies for the community, because tourism is a business activity). Fourthly, it is crucial for policy makers and tourism planners from government departments and ministries, in cooperation with the private sector, NGOs and other relevant sectors, to pay more attention to developing a specific fundamental programme for the tourism sector. This programme could include capacity-building programmes for the community and integrate tourism strategies at the local level (urban and rural) to implement the programme of "one *Aldeia* one product", i.e. eco-tourism, cultural tourism, heritage tourism, religious tourism, maritime tourism and sport tourism.

FUTURE RESEARCH DIRECTIONS

There is a lack of literature addressing sustainable tourism development in developing countries in the first stage of tourism development, particularly in East-Timor. Additionally, there is a lack of statistical data about tourism matters that could inform the policy, action planning and strategy for the sustainable development of tourism in this territory. Thus, more studies should be conducted in other developing countries to determine if the responses are consistent with the research in this work. To be comparable, these studies should also take into account the typology of residents according to their responsibilities and functions. Further research might either extend the analysis conducted in this study, by assessing the antecedents and the consequences the residents' attitudes towards sustainable tourism development in developing counties. The items used in the different dimensions of SUS-TAS should also be refined for destinations in developing countries, since, in some of them, low levels of reliability were found in this study.

CONCLUSION

This study has examined the typology of residents, leading and non-leading residents, and their attitudes, by applying the SUS-TAS scale to the case of East-Timor. The result of the socio-demographic data showed differences between non-leading residents and leading residents in terms of gender, age, education levels, occupation and monthly expenditure. The most important difference is in educational qualifications: 75.0% of the leaders had a bachelor's degree, and 52.4% of non-leading residents had high school qualifications only. Since these residents are effectively different, heterogeneous levels of attitude regarding the tourism development impacts, ate least at some dimensions of the SUS-TAS scale, were expected and tested.

Regarding attitudes, the two groups of residents in East-Timor, leaders and non-leaders, expressed high levels of agreement with the positive SUS-TAS dimensions (economic, socio-cultural, environmental, political, long-term planning, ensuring visitors' satisfaction and community-based tourism) and low levels of agreement with the negative SUS-TAS dimensions (economic, socio-cultural and environmental). In other words, residents report a euphoric attitude towards tourism, positively perceiving its benefits, and they underestimate its negative impacts. The same trends characterize the views of the three groups of leaders. These results are according to what is expected in a destination at the first stage of tourism development and confirm the studies of Doxey (1975, 1976) and Vargas-Sánchez et al. (2009).

While the opinions of non-leading residents and leading residents are quite similar on average, there were found to be some punctual differences that were significant within the measures composing each dimension of the SUS-TAS scale. In respect of the negative environmental impacts and the negative sociocultural impacts, significant differences were identified, with non-leading residents expressing higher levels of awareness regarding these problems. Likewise, the opinions of the three groups of leaders (in public sectors, private sectors and NGOs) are quite alike on average, but some significant differences were found in two measures of the SUS-TAS dimensions among the three groups. In the case of the negative economic impacts and the importance of community-based tourism, leaders from the NGOs were more aware of negative economic impacts and the importance of community-based tourism than leaders from public and private sectors.

In short, results showed that, globally, the attitudes of residents in East-Timor are "euphorically" supportive of tourism development. These results are encouraging for the future development of tourism and should be used by the government, private sectors and NGOs to promote tourism industries that contribute to sustainable development in East-Timor.

ACKNOWLEDGMENT

This paper is financed by National Funds provided by FCT - Foundation for Science and Technology through project UIDB/04020/2020.

REFERENCES

Alkire, S., & Aguilar, G. R. (2011). Least developed countries and the multidimensional poverty index (MPI). Poverty & Human Development Initiative, University of Oxford.

Aref, F., & Redzuan, M. (2010). Community leaders' perceptions towards socio-cultural impacts of tourism on local communities. *Journal of Human Ecology (Delhi, India)*, *29*(2), 87–90. doi:10.1080/09709274.2010.11906251

Aref, F., Redzuan, M., & Emby, Z. (2009). Assessing community leadership factor in community capacity building in tourism development: A Case Study of Shiraz, Iran. *Journal of Human Ecology (Delhi, India)*, *28*(3), 171–176. doi:10.1080/09709274.2009.11906235

Bachleitner, R., & Zins, A. H. (1999). Cultural tourism in rural communities: The residents' perspective. *Journal of Business Research*, *44*(3), 199–209. doi:10.1016/S0148-2963(97)00201-4

Blau, P. M. (2009). *Social Exchange Theory. Exchange and power in social life*. John Wiley and Sons.

Butler, R. W. (1980). The concept of a tourist area cycle of evolution: Implications for management of resources. *The Canadian Geographer. Geographe Canadien*, *24*(1), 5–12. doi:10.1111/j.1541-0064.1980.tb00970.x

Choi, H. C., & Sirakaya, E. (2005). Measuring residents' attitude toward sustainable tourism: Development of sustainable tourism attitude scale. *Journal of Travel Research*, *43*(4), 380–394. doi:10.1177/0047287505274651

Choi, H. C., & Sirakaya, E. (2006). Sustainability indicators for managing community tourism. *Tourism Management*, *27*(6), 1274–1289. doi:10.1016/j.tourman.2005.05.018

Cottrell, S. P., Vaske, J. J., Shen, F., & Ritter, P. (2007). Resident perceptions of sustainable tourism in Chongdugou, China. *Society & Natural Resources*, *20*(6), 511–525. doi:10.1080/08941920701337986

Darr, K. (2011). Introduction to management and leadership concepts, principles, and practices. In R. E. Burke & L. H. Friedman (Eds.), Essentials of management and leadership in public health (p. 7). London: Jones & Bartlett Learning International.

De Lacy, T., Battig, M., Moore, S., & Noakes, S. (2002). *Public / private partnership for sustainable tourism delivering a sustainability strategy for tourism destinations*. Asia Pacific Economic Cooperation.

Doxey, G. V. (1975). A causation theory of visitor-resident irritants: methodology and research inferences. In *The Travel Research Association 6th Annual Conference Proceedings* (pp. 195-198). San Diego, CA: Academic Press.

Doxey, G. V. (1976). When enough is enough: the native are restless in old Niagara. *Heritage Canada, 2*(2), 26-27.

Fishbein, M., & Ajzen, I. (1974). Attitudes towards objects as predictors of single and multiple behavioral Criteria. *Psychological Review, 81*(1), 59–74. doi:10.1037/h0035872

Gunn, C. A., & Var, T. (2002). Tourism planning basics, concepts and cases (4th ed.). New York: Routledge, Taylor & Francis Group.

Homans, G. C. (1961). Social behaviour: Its elementary forms. New York: Whitestable Litho, Straker Brothers, Ltd. Harcourt, Brace & Worl, Inc.

Hung, K., Sirakaya, E., & Ingram, L. J. (2011). Testing the efficacy of an integrative model for community participation. *Journal of Travel Research, 50*(3), 276–288. doi:10.1177/0047287510362781

Jurowski, C., Uysal, M., & Williams, D. R. (1997). A theoretical analysis of host community resident reaction to tourism. *Journal of Travel Research, 36*(3), 3–10. doi:10.1177/004728759703600202

Kvasova, O. (2011). Socio-demographic determinants of eco-friendly tourist attitudes and behaviour. *Tourism Today (Nicosia), 11*, 73–95.

Lankford, S. V., & Howard, D. R. (1994). Developing tourism impact attitude scale. *Annals of Tourism Research, 21*(1), 121-139.

Lepp, A. (2007). Residents' Attitudes towards tourism in Bigodi village, Uganda. *Tourism Management, 28*(3), 876–885. doi:10.1016/j.tourman.2006.03.004

Lepp, A. (2008). Attitudes towards initial tourism development in a community with no prior tourism experience: The Case of Bigodi, Uganda. *Journal of Sustainable Tourism, 16*(1), 5–22. doi:10.2167/jost630.0

Miniterio das Finanças da República Democrática de East-Timor (MF-RDTL). (2019). *State budget*. East-Timor.

Miskinyte, E., & Smirnova, E. (2011). *Effective governance in tourism destination development. Federal Agency for Tourism, Russian Federation & World Tourism Organization*. UNWTO.

Myrttinen, H. (2009). East-Timor: A relapsing "success" story. *Taiwan Journal of Democracy, 5*(1), 219–239.

National Immigration Directorate (NID). (2018). *Estatistica dos visitantes e pedidos de visto em Timor-Leste*. East-Timor.

National Statistics Directorate (NSD). (2015). *East-Timor population and housing census 2015*. NSD.

National Statistics Directorate (NSD) & United Nations Population Fund (UNFPA). (2010). *Highlights of the 2010 census main results in East-Timor*. Dili, East-Timor: Authors.

National Tourism Directorate (NTD). (2018). *Estatistica dos visitantes e turistas em Timor-Leste*. NTD.

Northouse, P. G. (2013). Leadership: theory and practice (6th ed.). Thousand Oaks, CA: Sage Publications, Inc.

Nunnally, J. C. (1978). Assessment of reliability. In *Psychometric theory* (2nd ed.). McGraw-Hill.

O'Halloran, R. M. (1992). Tourism management profiles: Implications for tourism education. *Hospitality Review, 10*(1).

Prayag, G., Dookhony-Ramphul, K., & Maryeven, M. (2010). Hotel development and tourism impacts in Mauritius: Hoteliers' perspectives on sustainable tourism. *Development Southern Africa, 27*(5), 697–712. doi:10.1080/0376835X.2010.522832

Rada, V., & Martín, V. (2014). Random route and quota sampling: Do they offer any advantage over probably sampling methods? *Open Journal of Statistics, 4*(05), 391–401. doi:10.4236/ojs.2014.45038

República Democrática de Timor Leste (RDTL) Ministerio das Finanças (MF). (2019). *Orçamento geral do estado 2019*. Dili, East-Timor: Author.

Ribeiro, M. A., Pinto, P., Silva, J. A., & Woosnam, K. M. (2017). Residents' attitudes and the adoption of pro-tourism behaviours: The case of developing island countries. *Tourism Management, 61*, 523–537. doi:10.1016/j.tourman.2017.03.004

Ribeiro, M. A., Pinto, P., Silva, J. A., & Woosnam, K. M. (2018). Examining the predictive validity of SUS-TAS with maximum parsimony in developing island countries. *Journal of Sustainable Tourism, 26*(3), 379–398. doi:10.1080/09669582.2017.1355918

Ribeiro, M. A., Valle, P. O., & Silva, J. A. (2013). Residents' attitudes toward tourism development in the Cape Verde islands. *Tourism Geographies, 15*(4), 1–21. doi:10.1080/14616688.2013.769022

Richards, G., & Hall, D. (2002). The Community: A sustainable concept in tourism development? In G. Richards & D. Hall (Eds.), Tourism and sustainable community development (2nd ed., pp. 1-13). London: Routledge, Taylor & Francis Group.

Scott, N., Laws, E., Richins, H., & Agrusa, J. (2011). *Tourist destination governance practice, Theory and Issues*. CAB International - MPG Books Group.

Secretario Estado Formação Profissional e Emprego (SEFOPE) e Direção Naciona da Estatistica (DNE). (2018). *East-Timor labour force survey 2018*. Author.

Sheldon, P. J., & Abenoja, T. (2001). Resident attitudes in a mature destination: The case of Waikiki. *Tourism Management, 22*(5), 435–443. doi:10.1016/S0261-5177(01)00009-7

Sirakaya, E., Ekinci, Y., & Kaya, A. G. (2008). An examination of the validity of SUS-TAS in cross-cultures. *Journal of Travel Research, 46*(4), 414–421. doi:10.1177/0047287507308328

Sirakaya, E., Ingram, L., & Harrill, R. (2009). Resident typologies within the integrative paradigm of sustaincentric tourism development. *Tourism Analysis, 13*(5), 531–544. doi:10.3727/108354208788160405

Skidmore, W. (1975). *Theoritical thinking in sociology*. Cambrige University Press.

Taylor, J. G., Gomes, R. A., Rasmussen, T., Ross, M., Rosser, A., Sandbu, M., & van Staveren, T. (2011). East-Timor human development report 2011. United Nations Development Programme.

Tinsley, R., & Lynch, P. (2001). Small tourism business networks and destination development. *International Journal of Hospitality Management, 20*(4), 367–378. doi:10.1016/S0278-4319(01)00024-X

Twining-Ward, L., & Butler, R. (2002). Implementing STD on a small island: Development and use of sustainable tourism development indicators in Samoa. *Journal of Sustainable Tourism, 10*(5), 363–387. doi:10.1080/09669580208667174

UNWTO (United Nations World Tourism Organization), Government of East-Timor, & UNDP (United Nations Development Programme). (2007). Sustainable tourism sector development and institutional strengthening project Volume 1 - Main report. Authors.

Van Wart, M. (2003). Public-sector leadership theory: An assessment. *Public Administration Review, 63*(2), 214–228. doi:10.1111/1540-6210.00281

Vargas-Sánchez, A., Porras-Bueno, N., & Plaza-Mejía, M. de L. Á. (2009). Residents' attitudes towards tourism development in the Spanish province of Huelva. Narodnostopanski Arhiv, International edition, University of Huelva, Spain, 31-63.

Vargas-Sánchez, A., Porras-Bueno, N., & Plaza-Mejía, M. Á. (2011). Explaining residents' attitudes to tourism: Is a universal model possible? *Annals of Tourism Research, 38*(2), 460–480. doi:10.1016/j.annals.2010.10.004

Vargas-Sánchez, A., Porras-Bueno, N., & Plaza-Mejía, M. A. (2014). Residents' attitude to tourism and seasonality. *Journal of Travel Research, 53*(5), 581–596. doi:10.1177/0047287513506295

Vong, M. (2014). *Tourism in Timor-Leste: Contributions to a sustainable development* (Unpublished doctoral dissertation). University of Algarve.

Vong, M., & do Valle, P. O. (2013). A opinião das empresas turísticas sobre o turismo em East-Timor. In J. A. Silva & F. Anjos (Eds.), Revista internacional em língua portuguesa - Turismo (III Série, pp. 215-227). Associação das Universidades de Língua Portuguesa, Lisboa: Clássica, Artes Gráficas S.A.

Vong, M., do Valle, P. O., & Silva, J. A. (2014). Turismo em Timor-Leste: Presente e futuro. In C. Costa, F. Brandão, R. Costa, & Z. Breda (Eds.), *Turismo nos países lusófonos: Conhecimento, estratégia e territótios* (Vol. I, pp. 399–410). Escolar Editora.

Vong, M., Pinto, P., & Silva, J. A. (2017). Diaspora tourism: The case of East-Timor. *Tourism (Zagreb), 65*(2), 218–233.

Wang, W. (2019). The effect of early-life outdoor experiences on residents' attitudes towards sustainable tourism within an urban context. *Journal of Outdoor Recreation and Tourism, 25*, 1–9. doi:10.1016/j.jort.2018.10.002

Wang, Y., & Pfister, R. E. (2008). Residents' attitudes toward tourism and perceived personal benefits in a rural community. *Journal of Travel Research, 47*(1), 84–93. doi:10.1177/0047287507312402

Woosnam, K., Maruyama, N. U., Ribeiro, M. A., & Joo, D. (2019). Explaining minority residents' attitudes of ethnic enclave tourism from general perceptions of tourism impacts. *Journal of Tourism and Cultural Change, 17*(4), 467–484. doi:10.1080/14766825.2019.1601407

WTO (World Tourism Organization). (2004). Indicators of sustainable development for tourism destinations: A guidebook. Madrid, Spain: World Tourism Organization.

WTO (World Tourism Organization). (2009). Handbook on tourism destinations branding with an introduction by Simon Anholt. Madrid, Spain: The World Tourism Organization and the European Travel Commission.

Yamane, T. (1967). *Statistics: An introductory analysis* (2nd ed.). Harper and Row.

Yoon, Y., Gursoy, D., & Chen, J. S. (2001). Validating a tourism development theory with structural equation modeling. *Tourism Management*, *22*(4), 363–372. doi:10.1016/S0261-5177(00)00062-5

Yu, C. P. S., Chancellor, H. C., & Cole, S. T. (2011). Examining the effects of tourism impacts on residents' quality of life: Evidence from rural Midwestern communities in USA. *International Journal of Tourism Sciences*, *11*(2), 161–186. doi:10.1080/15980634.2011.11434643

ADDITIONAL READING

Kwon, J., & Vogt, C. A. (2009). Identifying the Role of Cognitive, Affective, and Behavioral Components in Understanding Residents' Attitudes toward Place Marketing. *Journal of Travel Research*, *49*(4), 423–435. doi:10.1177/0047287509346857

Moniz, A. I. D. de S. A. (2006). A Sustentabilidade do Turismo em Ilhas de Pequena Dimensão: O Caso dos Açores. Departamento das Ciências Económicas e Empresariais na Especialidade de Desenvolvimento Económico e Social e Economia Pública, Universidade dos Açores.

Ribeiro, M. A., Pinto, P., Silva, J. A., & Woosnam, K. M. (2017). Residents' Attitudes and the Adoption of Pro-Tourism Behaviours: The Case of Developing Island Countries. *Tourism Management*, *61*(August), 523–537. doi:10.1016/j.tourman.2017.03.004

Ritchie, J. R. B., & Crouch, G. I. (2000). The Competitive Destination: A Sustainability Perspective. *Tourism Management*, *21*, 1–7.

Tosun, C. (2000). Limits to Community Participation in the Tourism Development Process in Developing Countries. *Tourism Management*, *21*(6), 613–633. doi:10.1016/S0261-5177(00)00009-1

Waqaisavou, T. (2001). Tourism in Fiji: Native Land Owner Attitude and Involvement. *Pacific Rim Real Estate Society Annual Conference - Land Management and Development* (pp. 1-11). Tourism Studies Programme, School of Social and Economic Development, University of the South Pacific, Suva, Republic of Fiji Islands.

Woosnam, K. M., Maruyama, N. U., Ribeiro, M. A., & Joo, D. (2019). Explaining minority residents' attitudes of ethnic enclave tourism from general perceptions of tourism impacts. *Journal of Tourism and Cultural Change*, *17*(4), 467–484. doi:10.1080/14766825.2019.1601407

Zhang, J., Inbakaran, R., & Jackson, M. (2006). Understanding Community Attitudes Towards Tourism and Host—Guest Interaction in the Urban—Rural Border Region. *Tourism Geographies*, *8*(2), 182–204. doi:10.1080/14616680600585455

KEY TERMS AND DEFINITIONS

Attitude: A tendency to respond favourably or unfavourably towards a particular object (Fishbein & Ajzen, 1974).

Destination Lifecycle Model: Tourism development model, proposed by Butler (1980), that recognizes six stages in the tourism product life cycle: exploration, investment, development, consolidation, stagnation and, finally, decline or rejuvenation.

Leadership: A process in which a person influences others to accomplish an objective and directs an organization in a way that makes it more cohesive and coherent (O'Halloran, 1992; Wart, 2003).

Leading Residents: Residents with leadership roles are those with important responsibilities in the planning and implementation of tourism development (in the public sector, the private sector and non-governmental organizations (NGOs).

Non-Leading Residents: Residents who do not have specific roles and responsibilities as leaders in the community.

SUS-TAS: Multiple measure tourism impact attitude scale developed by Choi and Sirakaya (2005, 2006) to measure residents' attitudes towards sustainable tourism development. SUS-TAS measures not only community feelings towards sustainable tourism development, but also existing paradigms, such as the social exchange theory and the theory of sustainability.

Sustainable Tourism: Tourism which presently meets the needs of tourists and residents while protecting and enhancing future opportunities of tourists and residents (WTO, 2004).

Section 2

Chapter 9
Experience Economy Domains in Tourism:
Extending the Knowledge and Understanding of Experiences

Nelson de Matos
School for Management, Hospitality and Tourism (ESGHT) and Research Centre for Tourism, Sustainability, and Well-Being (CinTurs), University of Algarve, Portugal

Julio Mendes
Faculty of Economics and Research Centre for Tourism, Sustainability, and Well-Being (CinTurs), University of Algarve, Portugal

Manuela Guerreiro
https://orcid.org/0000-0002-6398-9712
Faculty of Economics and Research Centre for Tourism, Sustainability, and Well-Being (CinTurs), University of Algarve, Portugal

ABSTRACT

The experience economy captivated the attention of academia and industry managers regarding how to engage and create memorable experiences for customers. The extensive literature developments in recent decades reinforced the need for a timely review of how the main concepts of the experiential marketing construct apply to the tourism industry. This study presents a narrative literature review to provide an overview in an attempt to find common ground, identify potential gaps, and provide suggestions for academia and tourism managers. The chapter provides value by discussing the key topics within marketing and tourism.

DOI: 10.4018/978-1-7998-3156-3.ch009

INTRODUCTION

The global market in general and tourism in particular have changed regarding its dynamics and competition; with many new competitors entering in the market, the pressure increased for brands and tourism organisations (Brakus, Schmitt, & Zarantonello, 2009; Gertner, 2011; Pike, 2009; Qu, Kim, & Im, 2011). One of the answers managers have found in the past to face this pressure has been the use of experience marketing, as an instrument to improve customers' engagement and consumption during experiences. However, the marketing experience theoretical framework reminds marketers and managers of the need to set a stage for consumers to be engaged and involved with the event (i.e., to play a passive and/or active role) while experiencing it, to generate greater satisfaction and long lasting memories (Ali, Hussain, & Ragavan, 2014; Assaker, Vinzi, & O'Connor, 2011; Hudson & Ritchie, 2009; Kim, Ritchie, & McCormick, 2012; Liljander & Strandvik, 1997; Manhas & Ramjit, 2013). In the same context, several authors have viewed emotions as a way to engage and involve consumers (Csikszentmihalyi & Lefevre, 1989; Johanson & Woods, 2008; Kim, 2012; Liljander & Strandvik, 1997; Yoo & MacInnis, 2005).

From this engagement and involvement, consumers will mentally form conceptions, ideas and narratives of the experiences before the actual consumption (e.g., holidays), as if they were at the setting enjoying it (Adams, 2004; Dann, 1996; Lu, Chi, & Liu, 2015). In effect, the seminal paper from Pine and Gilmore (1998) highlighted the need for organisations in general, and tourism in particular, to align their strategy and actions into delivering rich experiences that may contribute to memorable encounters and events (Tao, 2014). Tourism is one of the sectors that has been trying to develop this new economy by promoting images of unique holiday experience (deMatos, 2015).

Despite the previous studies, a clearer definition of the attributes of experience marketing and its domains in the tourism sector is still lacking, due to its characteristics being distinct from other sectors. In effect, despite tourism being a service industry (and services being different from goods, e.g., intangibility, inseparability, heterogeneity and perishability; Wolak, Kalafatis, & Harris, 1998), tourism is all about the consumption of experiences that are "a distinct economic offering, as different from services as services are from goods" (Pine & Gilmore, 1998, p. 97). In this context, when tourists purchase an experience they are acquiring a set of intangible events in which the nature of the benefit is experiential, hedonic or even symbolic, with the intention of having richer experiences that may contribute to long lasting memories (e.g., Gohary, Pourazizi, Madani, & Chan, 2020). Therefore, to contribute to a common ground and understanding of experiences and their associated approaches, this study presents a systematic literature review of the concept. The purpose is to offer a timely and useful review that may help academics and industry practitioners to comprehend this complex and dynamic construct and to contribute to its adoption and implementation in the market.

The review process adopted the Pickering and Byrne (2014) method, and the first step taken was to formulate the research question: "What theoretical frameworks have been applied to the study of tourism experiences?" The study's scope was restricted to the tourism marketing area for two reasons: a) time, to avoid screening manuscripts of other scientific areas that are not related to tourism; and b) the connection between tourism and marketing that extends beyond its promotion. The search was conducted on the 10th of November 2018 on the B-on database (e.g., Martins et al., 2020), and the search strings were: "Experiential Marketing" or "Experience Economy" or "tourism experience", specific terms with the purpose of covering all topics within the area selected (tourism marketing). In addition, the document search was limited to full texts and to those manuscripts that were available in the library collection in the English language.

The list of manuscripts was exported to Mendeley (reference manager). After exporting the documents to Mendeley, an initial screening (covering the title and abstract) of all the articles retrieved was performed by the researcher, in which manuscripts were excluded for certain reasons (e.g., not related to customers' experiences). From the initial screening phase to the eligibility phase, a total of 83 documents (article, conference papers, articles in press, book chapters) were included in the synthesis. The extraction and synthesis was made using an inductive content analysis, a common method in social sciences (Kyngäs, Mikkonen, & Kääriäinen, 2020). In addition, a descriptive analysis of the most relevant characteristics was made to investigate the construct (Reichel, Fuchs, & Uriely, 2009; Volo, 2010).

CONCEPTUALIZATION OF EXPERIENCE

Approaches to the Experience Concept

The early conceptualisation of Pine and Gilmore (1998, p. 98) highlighted the importance of engagement between companies and individuals. However, for some tourism researchers, their focus was more on the spacetime boundaries of experience, with many arguing that it can be considered an escape from the ordinary to the extraordinary, as Volo (2010, p. 301) states, "The tourist experience is defined as an occurrence experienced by a person outside the 'usual environment' and the 'contracted time' boundaries that is comprised of a 'sequence' of the following events: sensation, perception, interpretation, and learning". Since experiences are intrinsically and utterly personal, it is crucial for them to be relevant for the individual (Greene, 2011; Kim & Ritchie, 2014) so that he or she may be able to play an active role as co-creators and not co-producers (Duerden et al., 2015).

A common terminology or definition is lacking, which in part can be explained because of the complexity (Volo, 2009) of the human element or/and due to the composite nature of the industry. For other authors, experiences can be divided into different stages. "An individual's subjective evaluation and undergoing [i.e., affective, cognitive and behavioral] of events related to his/her tourist activities which begins before [i.e., planning and preparation], during [i.e., at the destination], and after the trip [i.e., recollection]" (Tung & Ritchie, 2011, p. 1369). These authors wanted to understand how personal the experiences were, giving emphasis to the personal narratives and the value of mentally engaging consumers (Lee & Gretzel, 2012; Pine & Gilmore, 1998). However, different perspectives do not always complement each other, nor do they encompass all the key elements of an experience. For example, using word cloud generating software (worditout.com), with the purpose of identifying the common terms used by previous authors to define experience, found the most common word was *experience*, which relates to the concept itself, followed by *company*, *contact*, *direct*, *customers*, *between* and *individual* (Figure 1).

In this example, memories, the creation of memorable events, or the other dimensions considered by Pine and Gilmore (1998) in their seminal paper (e.g., entertainment, education, aesthetic or escapism) are not evident. This conceptualisation resulted from different approaches studying the construct, with some addressing it from: a) the perspective of the consumer, "Experiences involve the entire living being. They often result from direct observation and/or participating in the event - whether they are real, dreamlike or virtual" (Schmitt, 1999, p. 60); b) the perspective of business, "An experience occurs when a company intentionally uses services as the stage, and goods as props, to engage individual customers in a way that creates a memorable event" (Pine & Gilmore, 1998, p. 98); c) the perspective of the emo-

Figure 1. Definition of Experiences – main words
Source: Adapted from (Cohen, 1979, 1988; Dann, 1977; Holbrook, 2000; Holbrook & Addis, 2001; Holbrook & Hirschman, 1982; Hosany & Gilbert, 2009; Hosany & Gilbert, 2010; Ismail, 2011; Tsaur, Chiu, & Wang, 2007; Tung & Ritchie, 2011; Tussyadiah & Fesenmaier, 2009; Volo, 2009, 2010; Wang, 1999).

tional and sensorial stimuli, "Emotions provoked, sensations felt, knowledge gained and skills acquired through active involvement with the firm pre, during and post consumption" (Ismail, 2011, p. 208).

As previous studies have shown, these different perspectives extended into other definitional approaches that incorporated elements related with what tourists were really seeking (i.e., authenticity; Cohen, 1979; Ma & Lew, 2012; MacCannell, 1973; Pearce & Moscardo, 1986; Rickly-Boyd, 2012), how tourists wanted to be doing rather than looking (i.e., co-creation; Campos, Mendes, do Valle, & Scott, 2016; Campos, Mendes, Valle, & Scott, 2018) and what tourists wanted to be experiencing (meaningful memories; Ali et al., 2014; Chandralal & Valenzuela, 2013; Tung, Lin, Qiu Zhang, & Zhao, 2017).

Identifying and Extending Experience Dimensions

The multidimensionality of the experience can be traced to point the importance of viewing the experiences from a consumption perspective in which the fantasies, the feelings and the fun were key elements in the process. From Pine and Gilmore's (1998) view, the experience can be conceptualised based on two main dimensions, the customer participation (active, passive) and customer connection (absorption, immersion). Within those two dimensions or axes, they consider the four realms of the experience or four E's: Entertainment (an event or activity with the purpose to entertain), Education (intended to serve or educate), Escapism (to escape from daily routine, daydreaming) and Esthetic (to view or appreciate beauty of a picture, scenario or object). However, the number of dimensions within the experience construct varies (Ismail, Melewar, Lim, & Woodside, 2011). As seen in Table 1, over time, the number of dimensions has continued to increase and diverge, ranging from three (i.e., fantasies, feelings and fun) to nine (i.e., hedonism, refreshment, meaningfulness, involvement, local culture, knowledge, novelty, recollection and vividness) in this case.

Table 1. Dimensions of the Experience

Author/Year	Context	Dimensions
(Holbrook & Hirschman, 1982)	Conceptual	Fantasies, feelings, fun.
(Arnould & Price, 1993)	River rafting	Harmony with nature, communitas, personal growth and renewal.
(Pine & Gilmore, 1998)	Conceptual	Entertainment, education, esthetic, and escapism
(Gentile, Spiller, & Noci, 2007)	Brands	Sensorial, emotional, cognitive, pragmatic, lifestyle and relational.
(Brakus et al., 2009)	Brands	Behavioral, sensorial, intellectual and affective
(Kim, Ritchie, & McCormick, 2010)	Travel experience memories	hedonism, refreshment, meaningfulness, involvement, local culture, knowledge, and novelty.
(DeMatos, 2015)	Destination Experience	Entertainment, education, esthetic, and escapism

Source: Adapted from Ismail (2011)

This once again unclear (dimensional) nature of the experience concept reveals the complexity and diversity of the phenomena within academia and provides clues regarding the plethora of elements that contribute to its characteristics.

Experience Characteristics

In this context, several authors (Cohen, Prayag, & Moital, 2014; Mossberg, 2008; Schmitt, 1999) have mentioned the need to approach the market with a holistic perspective so that memorable experiences may be accomplished (Kim & Ritchie, 2014; Kim, Ritchie, & McCormick, 2012; Kim, 2014). During experiences, individual and dynamic narratives are built (Campos, Mendes, Valle, & Scott, 2015; Grönroos & Voima, 2013; Prebensen & Foss, 2011). In this process, the connections and relationships made between tourists and other people (employees, residents, other customers) impacts on the degree in which they will be able to share their experience and immerse themselves in it (Duerden et al., 2015; Ekman, 1992). The process behind the service is crucial, consequently requiring a high level of choreography so that every encounter (Bertella, 2013; Prebensen & Foss, 2011), every moment of the experience staged, becomes an active element in engaging costumers (Cohen et al., 2014), in awakening sensations (Morgan, Lugosi, & Ritchie, 2010), and releasing emotions (Basoc, 2015; Kwortnik & Ross, 2007) to contribute to making strong memories and learning new things and skills (Greene, 2011; Kessler, 2013). Among all the main characteristics of experiences observed in the literature review (Table 2), the personal nature was highlighted.

In Table 2, it can be seen that the authors stressed that the consumption of goods and services represent more than an economic transaction, it represents the consumers or *individuals' personality*, their lifestyle in which their ideologies are present (Rodrigues, Moreira, & Vitorino, 2013). Individuals get involved in the *co-creation* of staged *events* or *shows* (Campos et al., 2015). These events need to be

Table 2. Experience Characteristics

Item	Authors
Individual personality	(Hart, Kerrigan, & vom Lehn, 2016; Pine & Gilmore, 1998).
Co-creators	(Vargo & Lusch, 2016).
Interesting Staged narrative	(Bertella, 2013; Campos et al., 2015; Greene, 2011; Grönroos & Voima, 2013; Hart, Kerrigan, & vom Lehn, 2016; Kim & Ritchie, 2014; Prebensen & Foss, 2011; Rickly-boyd, 2010).
Engaging	(Cohen et al., 2014; Duerden et al., 2015; Lu et al., 2015).
Sensations	(Morgan et al., 2010).
Emotions	(Basoc, 2015; Kwortnik & Ross, 2007).
Learning	(Greene, 2011; Kessler, 2013).
Social interaction	(Duerden et al., 2015; Ekman, 1992).
Memories	(Ali *et al.*, 2014; Hudson & Ritchie, 2009; Kim, Ritchie, & McCormick, 2012).

Source: Own elaboration

interesting and built upon a narrative (Bertella, 2013; Campos et al., 2015; Greene, 2011; Grönroos & Voima, 2013; Kim & Ritchie, 2014; Prebensen & Foss, 2011).

Individuals, like tourists, need to experience *engaging* events built by tourism organisations to learn and experience enjoyment and *emotions* (Basoc, 2015; Kwortnik & Ross, 2007). For this purpose, the *senses* are key elements for the tourists (Morgan et al., 2010) to make their own interpretations, evaluations and *memories*. In their search of memorable experiences, tourists interact with one another (social interaction) and are exposed to different levels of experiences (i.e., supportive or peak; Quan & Wang, 2004).

DPO (DRIVERS-PROCESSES-OUTCOMES) CONCEPTUAL MODEL

The staging and managing of an experiential event in tourism means that managers at the public and private level need to consider the following (Figure 2):

Figure 2. Experiences Drivers, Process and Outcomes [DPO] Conceptual model
Source: Own elaboration

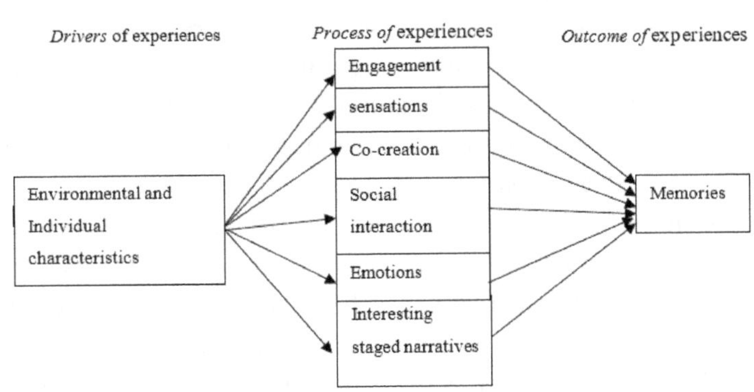

a) The environmental and individual characteristics, since these are the drivers. As mention previously, experiences (in general and particularly in tourism) are about the individuals' or tourists' consumption (Pine & Gilmore, 1998). In effect, the subjective nature of experiences will influence tourists' expectations and perceptions during the experiential process (at the holiday destination; Hart et al., 2016).
b) During the process, managers must develop and provide interesting staged narratives (Bertella, 2013; Campos et al., 2015; Greene, 2011; Grönroos & Voima, 2013; Kim & Ritchie, 2014; Prebensen & Foss, 2011). They also need to provide engagement (i.e., between company and individuals) to encourage individuals into co-creating their own narratives (Rickly-Boyd, 2010). The goal is to lead tourists into experiencing various stimuli during those staged events, to awake their senses and emotions (Basoc, 2015; Morgan et al., 2010). In effect, this process will facilitate tourists to interact with other tourists and will permit them to experience staged events more holistically. This means managers must not only consider the tangible attributes of an experience but also the feelings and auras experienced by the tourists during consumption, since tourists are rational and also emotional (Lee & Chang, 2017; Serazio, 2017).
c) The main outcome of the experiential process should be memories. Managers should strive for tourists to experience memorable events because this is the tourists' aim when going on holidays, to live and experience a memorable event (Ali, Ryu, & Hussain, 2016). Nevertheless, managers should be aware of the degree of subjectivity each tourist will experience during and after the event (i.e., evaluation and recollection of the holidays) due to the composite and complex nature of the experience construct.

DISCUSSION

Tourists seek contrasting experiences from their ordinary life, which makes them eager to embark on diverse holiday experiences. Therefore, the marketing environment is full of messages reinforcing those different experiences, products and brands (Aaker, 1996; Brakus et al., 2009), neglecting to some degree the importance of the consumers' experiences during the experience consumption that occurs during the tourism activity (Addis & Holbrook, 2001; Holbrook & Hirschman, 1982; Morgan, Elbe, & Curiel, 2009; Morgan, Lugosi, & Ritchie, 2010). While not amorphous (Pine & Gilmore, 1998), there is not much agreement in academia regarding its boundaries (Tung & Ritchie, 2011).

A more robust framework of the experience construct is lacking; much of the previous research rests on the extensive theoretical frame in which there is no common ground. As our findings suggest, various conceptualisations (neglecting the purpose of experiences which is to create memories), approaches (i.e., consumer, business, emotional and sensorial stimuli), and dimensions (ranging from three to seven dimensions) are offered by the extant literature, but this literature neglects essential dimensions, better operationalisation of the construct and lastly a common ground.

Carù and Cova (2013) suggest that there is also the lack of operationalisation tools that may capture the essential dimensions of the experience. In this regard, it is important to remember that tourism activities are not solely surrounded by tangible attributes; they also involve in more intangible aspects, such as human senses (Quan & Wang, 2004). According to Huang and Hsu (2010), another relevant aspect that contributed towards the current situation is the researchers' use of different theoretical frameworks to conceptualise experiences, which ultimately created diverging dimensions of the tourist experience. The authors suggest that academics need to pay more attention to the customers' interactions (with other customers) to find the contribution and impact of these relations.

Oh, Fiore and Jeoung (2007, p. 120), on the other hand, claim that researchers have "emphasized understanding what the tourist experience is and how it is formed as a result of visiting a destination". Despite the attention given to experiences, their understanding is not yet complete since the focus on the utilitarian attributes have left limited research regarding the processes, consequences and impacts of the tourists' subjective emotions, mood(s) and relationships (Han & Patterson, 2007). One example can be drawn from the findings of Gohary et al. (2020) that evaluate the impact of memorable experiences on destination satisfaction. In their study, the local culture impact on memorable experiences was not similar to the research by Kim, Ritchie and McCormick (2010). The impact of the various experiences and their level of intensity (supportive or peak; Quan & Wang, 2004) requires not only tourism researchers and tourism marketers to assess experiences holistically (as a whole entity) but also other academics from different areas of knowledge (Vittersø, Vorkinn, Vistad, & Vaagland, 2000). Otto and Ritchie (1996) claim no comprehensive model has been made to explain the fundamental nature of service experience. Most studies only look at experience consumption from a one-end perspective (e.g., pleasant-unpleasant dimensions), limiting the establishment of potential correlations and/or relationships between the experience construct and other constructs (Mano & Oliver, 1993; Russell, 1980).

Huang and Hsu (2010) also consider, in the same way, that although qualitative methods reveal *the vibrant dynamics* of customers' interactions, further investigation is needed. McIntosh and Siggs (2005) also agree with this view and add that researchers have been focused on features and benefits of products and services, neglecting the products of experiential nature. The fact remains that some factors (e.g., experiences personal consumption) are beyond any control, and consequently these factors limit/restrain the role of managers and tourism stakeholders (Axelsen & Swan, 2010).

CONCLUSION

The study sought to find "What theoretical frameworks have been applied to the study of tourism experiences?" Despite the developments made to date that recognise the importance of experiences, more research must be done to find their essence and core elements because memories are not easy to create. Today, as in the past, the challenge for marketers and tourism managers continues to be uncovering the essence of what makes certain experiences memorable, in the eyes of the customers, at a time and space that appears to be outside of the daily *tempo* and ordinary constraints. Despite some agreement regarding the importance of the construct — the literature shows that the evaluation of tourists' experiences is a key factor for any organisation's success — a common language in the academic literature regarding an all-encompassing theoretical framework remains sparse.

The study's literature review verified the complexity of the construct and testified, to a certain degree, the robustness of the body of knowledge from many different perspectives, ideas, concepts and understandings of the experiential world. Different conceptualisations, perspectives, dimensions and characteristics of the construct were discovered. Findings from the literature review contributed to providing academia a better understanding of the experience's versatility and the immaturity of the body of knowledge, enabling the identification and exploration of potential gaps. A DPO conceptual model was presented with the intention of examining the potential relationships between the elements that make an impact on the goal of experiences: memories.

In summary, a more comprehensive understanding of the construct drivers, processes and outcomes about the construct is needed (Khan & Rahman, 2015). The implications to the industry and academia are that managers must contemplate experiences' dynamic and complex nature. Thus, managers and stakeholders can adapt and develop better marketing strategies based on experiential domains. This way they can increase the appeal and added value of products and services and promote a distinct positioning to increase the competitiveness of their organisations.

The main limitation of this study was the focus on all types of manuscripts in the English language available at the online knowledge library (b-on.pt). This means that manuscripts in other languages with other keywords that might have included other dimensions and characteristics were not included in this study. Therefore, future research should consider different search strings, manuscripts subjected to a double-blind review process (to avoid any potential bias) and other scientific areas (e.g., psychology). In addition, academia should place emphasis on more qualitative studies to develop new scales for measuring and conceptualising experiences. In a similar way, the same can be said regarding experiences' emotions, sensations, impacts and effects on customer satisfaction, loyalty and memory.

ACKNOWLEDGMENT

This paper is financed by National Funds provided by FCT- Foundation for Science and Technology through project UIDB/04020/2020.

REFERENCES

Aaker, D. (1996). Measuring Brand Equity Across Products and Markets. *California Management Review*, *38*(3), 102–120. doi:10.2307/41165845

Adams, K. M. (2004). The genesis of touristic imagery : Politics and poetics in the creation of a remote Indonesian island destination. *Tourist Studies*, *4*(2), 115–135. doi:10.1177/1468797604054378

Addis, M., & Holbrook, M. B. (2001). On the conceptual link between mass customisation and experiential consumption: An explosion of subjectivity. *Journal of Consumer Behaviour*, *1*(1), 50–66. doi:10.1002/cb.53

Alagöz, S. B., & Ekici, N. (2014). Experiential Marketing and Vacation Experience: The Sample of Turkish Airlines. *Procedia: Social and Behavioral Sciences*, *150*, 500–510. doi:10.1016/j.sbspro.2014.09.065

Ali, F., Hussain, K., & Ragavan, N. A. (2014). Memorable Customer Experience: Examining the Effects of Customers Experience on Memories and Loyalty in Malaysian Resort Hotels. *Procedia: Social and Behavioral Sciences*, *144*, 273–279. doi:10.1016/j.sbspro.2014.07.296

Ali, F., Ryu, K., & Hussain, K. (2016). Influence of Experiences on Memories, Satisfaction and Behavioral Intentions: A Study of Creative Tourism. *Journal of Travel & Tourism Marketing*, *33*(1), 85–100. doi:10.1080/10548408.2015.1038418

Arnould, E. J., & Price, L. L. (1993). River Magic: Extraordinary Experience and the Extended Service Encounter. *The Journal of Consumer Research*, *20*(1), 24. doi:10.1086/209331

Assaker, G., Vinzi, V. E., & O'Connor, P. (2011). Examining the effect of novelty seeking, satisfaction, and destination image on tourists' return pattern: A two factor, non-linear latent growth model. *Tourism Management*, *32*(4), 890–901. doi:10.1016/j.tourman.2010.08.004

Axelsen, M., & Swan, T. (2010). Designing Festival Experiences to Influence Visitor Perceptions: The Case of a Wine and Food Festival. *Journal of Travel Research*, *49*(4), 436–450. doi:10.1177/0047287509346796

BASOC, Diana-Gabriela (2015). A Theoretical Approach to Experiences in Marketing–Mastered through Experience Marketing. Marketing From Information to Decision, (8), 110-118.

Bertella, G. (2013, Fall). Photography and Northern Lights Tourism in Tromsø, Norway. *The Northern Review*, *37*, 167–186.

Brakus, J. J., Schmitt, B. H., & Zarantonello, L. (2009). Brand Experience: What Is It? How Is It Measured? Does It Affect Loyalty? *Journal of Marketing*, *73*(3), 52–68. doi:10.1509/jmkg.73.3.052

Campos, A. C., Mendes, J., do Valle, P. O., & Scott, N. (2016). Co-Creation Experiences: Attention and Memorability. *Journal of Travel & Tourism Marketing*, *33*(9), 1309–1336. doi:10.1080/10548408.2015.1118424

Campos, A. C., Mendes, J., do Valle, P. O., & Scott, N. (2018). Co-creation of tourist experiences: A literature review. *Current Issues in Tourism*, *21*(4), 369–400. doi:10.1080/13683500.2015.1081158

Carù, A., Caru, A., & Cova, B. (Eds.). (2007). *Consuming experience*. Routledge; doi:10.4324/9780203390498

Chandralal, L., & Valenzuela, F.-R. (2013). Exploring Memorable Tourism Experiences: Antecedents and Behavioural Outcomes. *Journal of Economics. Business and Management*, *1*(2), 177–181. doi:10.7763/JOEBM.2013.V1.38

Cohen, E. (1979). A Phenomenology of Tourist Experiences. *Sociology*, *13*(2), 179–201. doi:10.1177/003803857901300203

Cohen, E. (1988). Authenticity and commoditization in tourism. *Annals of Tourism Research*, *15*(3), 371–386. doi:10.1016/0160-7383(88)90028-X

Cohen, S. A., Prayag, G., & Moital, M. (2014). Consumer behaviour in tourism: Concepts, influences and opportunities. *Current Issues in Tourism*, *17*(10), 872–909. doi:10.1080/13683500.2013.850064

Csikszentmihalyi, M., & Csikszentmihalyi, I. S. (1992). *Optimal experience: Psychological studies of flow in consciousness*. Cambridge University Press.

Csikszentmihalyi, M., & LeFevre, J. (1989). Optimal Experience in Work and Leisure. *Journal of Personality and Social Psychology*, *56*(5), 815–822. doi:10.1037/0022-3514.56.5.815 PMID:2724069

Dann, G. (1977). Anomie, ego-enhancement and tourism. *Annals of Tourism Research*, *4*(4), 184–194. doi:10.1016/0160-7383(77)90037-8

Dann, G. M. S. (1996). Tourists' Images of a Destination - an alternative analysis. *Journal of Travel & Tourism Marketing*, *5*(1/2), 41–55. doi:10.1300/J073v05n01_04

DeMatos, N. (2015). *The impacts of tourism experiences in the destination image. A marketing perspective* (Unpublished doctoral dissertation). University of Algarve.

Duerden, M. D., Ward, P. J., & Freeman, P. A. (2015). Conceptualizing Structured Experiences. *Journal of Leisure Research*, *47*(5), 601–620. doi:10.18666/jlr-2015-v47-i5-6096

Ekman, P. (1992). An argument for basic emotions. *Cognition and Emotion*, *6*(3-4), 169–200. doi:10.1080/02699939208411068

Gentile, C., Spiller, N., & Noci, G. (2007). How to Sustain the Customer Experience. An Overview of Experience Components that Co-create Value With the Customer. *European Management Journal*, *25*(5), 395–410. doi:10.1016/j.emj.2007.08.005

Gertner, D. (2011). Unfolding and configuring two decades of research and publications on place marketing and place branding. *Place Branding and Public Diplomacy*, *7*(2), 91–106. doi:10.1057/pb.2011.7

Gohary, A., Pourazizi, L., Madani, F., & Chan, E. Y. (2020). Examining Iranian tourists' memorable experiences on destination satisfaction and behavioral intentions. *Current Issues in Tourism*, *23*(2), 131–136. doi:10.1080/13683500.2018.1560397

Greene, H. (2011). Freshmen Marketing: A First-Year Experience with Experiential Learning. *Marketing Education Review*, *21*(1), 79–87. doi:10.2753/MER1052-8008210111

Grönroos, C., & Voima, P. (2013). Critical service logic: Making sense of value creation and co-creation. *Journal of the Academy of Marketing Science*, *41*(2), 133–150. doi:10.100711747-012-0308-3

Han, J., & Patterson, I. (2007). An Analysis of the Influence That Leisure Experiences Have on a Person's Mood State, Health and Wellbeing. *Annals of Leisure Research*, *10*(3–4), 328–351. doi:10.1080/11745398.2007.9686770

Hart, A., Kerrigan, F., & vom Lehn, D. (2016). Experiencing film: Subjective personal introspection and popular film consumption. *International Journal of Research in Marketing*, *33*(2), 375–391. doi:10.1016/j.ijresmar.2015.08.005

Holbrook, M. B. (2000). The Millennial Consumer in the Texts of Our Times: Experience and Entertainment. *Journal of Macromarketing*, *20*(2), 178–192. doi:10.1177/0276146700202008

Holbrook, M. B., Chestnut, R. W., Oliva, T. A., & Greenleaf, E. A. (1984). Play as a Consumption Experience: The Roles of Emotions, Performance, and Personality in the Enjoyment of Games. *The Journal of Consumer Research*, *11*(2), 728–739. doi:10.1086/209009

Holbrook, M. B., & Hirschman, E. C. (1982a). The Experiential Aspects of Consumption: Consumer Fantasies, Feelings, and Fun. *The Journal of Consumer Research*, *9*(2), 132. doi:10.1086/208906

Hosany, S., & Gilbert, D. (2009, July 1). Dimensions of Tourists' Emotional Experiences Towards Hedonic Holiday Destinations. Available at SSRN: https://ssrn.com/abstract=1871768

Hosany, S., & Gilbert, D. (2010). Measuring Tourists' Emotional Experiences toward Hedonic Holiday Destinations. *Journal of Travel Research, 49*(4), 513–526. doi:10.1177/0047287509349267

Huang, J., & Hsu, C. H. C. (2010). The Impact of Customer-to-Customer Interaction on Cruise Experience and Vacation Satisfaction. *Journal of Travel Research, 49*(1), 79–92. doi:10.1177/0047287509336466

Hudson, S., & Ritchie, J. R. B. (2009). Branding a memorable destination experience. The case of 'Brand Canada.'. *International Journal of Tourism Research, 11*(2), 217–228. doi:10.1002/jtr.720

Ismail, A. R. (2011). Experience Marketing: An Empirical Investigation. *Journal of Relationship Marketing, 10*(3), 167–201. doi:10.1080/15332667.2011.599703

Ismail, A. R., Melewar, T., Lim, L., & Woodside, A. (2011). Customer experiences with brands: Literature review and research directions. *The Marketing Review, 11*(3), 205–225. doi:10.1362/146934711X589435

Johanson, M. M., & Woods, R. H. (2008). Recognizing the Emotional Element in Service Excellence. *Cornell Hospitality Quarterly, 49*(3), 310–316. doi:10.1177/1938965508316267

Kessler, G. (2013). Collaborative language learning in co-constructed participatory culture. *CALICO Journal, 30*(3), 307–322. doi:10.11139/cj.30.3.307-322

Khan, I., & Rahman, Z. (2015). A review and future directions of brand experience research. *International Strategic Management Review, 3*(1), 1–14. doi:10.1016/j.ism.2015.09.003

Kim, J. H. (2014). The antecedents of memorable tourism experiences: The development of a scale to measure the destination attributes associated with memorable experiences. *Tourism Management, 44*, 34–45. doi:10.1016/j.tourman.2014.02.007

Kim, J. H., & Ritchie, J. R. B. (2014). Cross-Cultural Validation of a Memorable Tourism Experience Scale (MTES). *Journal of Travel Research, 53*(3), 323–335. doi:10.1177/0047287513496468

Kim, J. H., Ritchie, J. R. R., & McCormick, B. (2012). Development of a scale to measure memorable tourism experiences. *Journal of Travel Research, 51*(1), 12–25. doi:10.1177/0047287510385467

Kim, J. H. J. H., Ritchie, J. R. R., & McCormick, B. (2010). Development of a scale to measure memorable tourism experiences. *European Journal of Tourism Research, 3*(2), 123–126. doi:10.1177/0047287510385467

Kim, S. (2012). Audience involvement and film tourism experiences: Emotional places, emotional experiences. *Tourism Management, 33*(2), 387–396. doi:10.1016/j.tourman.2011.04.008

Kwortnik, R. J. Jr, & Ross, W. T. Jr. (2007). The role of positive emotions in experiential decisions. *International Journal of Research in Marketing, 24*(4), 324–335. doi:10.1016/j.ijresmar.2007.09.002

Kyngäs, H., Mikkonen, K., & Kääriäinen, M. (2020). The Application of Content Analysis in Nursing Science Research. In H. Kyngäs, K. Mikkonen, & M. Kääriäinen (Eds.), *The Application of Content Analysis in Nursing Science Research* (pp. 13–21). Springer. doi:10.1007/978-3-030-30199-6_1

Lee, T. H., & Chang, P.-S. (2017). Examining the relationships among festivalscape, experiences, and identity: Evidence from two Taiwanese aboriginal festivals. *Leisure Studies, 36*(4), 453–467. doi:10.1080/02614367.2016.1190857

Lee, W., & Gretzel, U. (2012). Designing persuasive destination websites: A mental imagery processing perspective. *Tourism Management, 33*(5), 1270–1280. doi:10.1016/j.tourman.2011.10.012

Liljander, V., & Strandvik, T. (1997). Emotions in service satisfaction. *International Journal of Service Industry Management, 8*(2), 148–169. doi:10.1108/09564239710166272

Lu, L., Chi, C. G., & Liu, Y. (2015). Authenticity, involvement, and image: Evaluating tourist experiences at historic districts. *Tourism Management, 50*, 85–96. doi:10.1016/j.tourman.2015.01.026

Ma, L., & Lew, A. A. (2012). Historical and geographical context in festival tourism development. *Journal of Heritage Tourism, 7*(1), 13–31. doi:10.1080/1743873X.2011.611595

MacCannell, D. (1973). Staged Authenticity: Arrangements of social space in tourist setting. *American Journal of Sociology, 79*(3), 589–603. doi:10.1086/225585

Manhas, P. S., & Ramjit. (2013). Customer Experience and Its Relative Influence on Satisfaction and Behavioural Intention in Hospitality and Tourism Industry. *South Asian Journal of Tourism and Heritage, 6*(1), 53–68.

Mano, H., & Oliver, R. (1993). Assessing the dimensionality and structure of the consumption experience: Evaluation, feeling, and satisfaction. *The Journal of Consumer Research, 20*(3), 451. doi:10.1086/209361

Martins, L., Varela, M. L. R., Fernandes, N. O., Carmo–Silva, S., & Machado, J. (2020). Literature review on autonomous production control methods. *Enterprise Information Systems, 00*(00), 1–13. doi:10.1080/17517575.2020.1731611

McIntosh, A. J., & Siggs, A. (2005). An exploration of the experiential nature of Boutique accommodation. *Journal of Travel Research, 44*(1), 74–81. doi:10.1177/0047287505276593

Morgan, M., Elbe, J., & Curiel, J. de E. (2009). Has the experience economy arrived? The views of destination managers in three visitor-dependent areas. *International Journal of Tourism Research, 11*(2), 201–216. doi:10.1002/jtr.719

Morgan, M., Lugosi, P., & Ritchie, J. B. (Eds.). (2010). *The tourism and leisure experience: Consumer and managerial perspectives* (Vol. 44). Channel View Publications. doi:10.21832/9781845411503

Mossberg, L. (2008). Extraordinary Experiences through Storytelling. *Scandinavian Journal of Hospitality and Tourism, 8*(3), 195–210. doi:10.1080/15022250802532443

Nasution, R. A., Sembada, A. Y., Miliani, L., Resti, N. D., & Prawono, D. A. (2014). The Customer Experience Framework as Baseline for Strategy and Implementation in Services Marketing. *Procedia: Social and Behavioral Sciences, 148*, 254–261. doi:10.1016/j.sbspro.2014.07.041

Oh, H., Fiore, M., & Jeoung, M. (2007). Measuring Experience Economy Concepts: Tourism Applications. *Journal of Travel Research, 46*(2), 119–132. doi:10.1177/0047287507304039

Otto, J. E., & Ritchie, J. R. B. (1996). The service experience in tourism. *Tourism Management, 17*(3), 165–174. doi:10.1016/0261-5177(96)00003-9

Pearce, P. L., & Moscardo, G. M. (1986). The Concept of Authenticity in Tourist Experiences. *The Australian and New Zealand Journal of Sociology, 22*(1), 121–132. doi:10.1177/144078338602200107

Pickering, C., & Byrne, J. (2014). The benefits of publishing systematic quantitative literature reviews for PhD candidates and other early-career researchers. *Higher Education Research & Development, 33*(3), 534–548. doi:10.1080/07294360.2013.841651

Pike, S. (2009). Destination brand positions of a competitive set of near-home destinations. *Tourism Management, 30*(6), 857–866. doi:10.1016/j.tourman.2008.12.007

Pine, B. J., & Gilmore, J. H. (1998). Welcome to the experience economy. *Harvard Business Review, 76*(4), 97–105. PMID:10181589

Prebensen, N. K., & Foss, L. (2011). Coping and co-creating in tourist experiences. *International Journal of Tourism Research, 13*(1), 54–67. doi:10.1002/jtr.799

Qu, H., Kim, L. H., & Im, H. H. (2011). A model of destination branding: Integrating the concepts of the branding and destination image. *Tourism Management, 32*(3), 465–476. doi:10.1016/j.tourman.2010.03.014

Quan, S., & Wang, N. (2004). Towards a structural model of the tourist experience: An illustration from food experiences in tourism. *Tourism Management, 25*(3), 297–305. doi:10.1016/S0261-5177(03)00130-4

Reichel, A., Fuchs, G., & Uriely, N. (2009). Israeli backpackers. The Role of Destination Choice. *Annals of Tourism Research, 36*(2), 222–246. doi:10.1016/j.annals.2008.11.002

Rickly-boyd, J. M. (2010). The Tourist Narrative. *Tourist Studies, 9*(3), 259–280. doi:10.1177/1468797610382701

Rickly-Boyd, J. M. (2012). Authenticity & aura. A benjaminian approach to tourism. *Annals of Tourism Research, 39*(1), 269–289. doi:10.1016/j.annals.2011.05.003

Rodrigues, F., Moreira, J., & Vitorino, L. (Eds.). (2013). Comportamento do Consumidor: Quando a Neurociência, a Psicologia, a Economia e o Marketing se encontram! Viseu: Psicosoma Editora.

Russell, J. A. (1980). A circumplex model of affect. *Journal of Personality and Social Psychology, 39*(6), 1161–1178. doi:10.1037/h0077714

Schmitt, B. (1999). Experiential Marketing. *Journal of Marketing Management, 15*(1–3), 53–67. doi:10.1362/026725799784870496

Serazio, M. (2017). Branding politics: Emotion, authenticity, and the marketing culture of American political communication. *Journal of Consumer Culture, 17*(2), 225–241. doi:10.1177/1469540515586868

Tao, S.-P. (2014). Experiential marketing and marketing experience: An empirical study of the influence of Summer Boot Camp on military propensity. *Military Psychology, 26*(5–6), 422–433. doi:10.1037/mil0000059

Tsaur, S.-H., Chiu, Y.-T., & Wang, C.-H. (2007). The Visitors Behavioral Consequences of Experiential Marketing. *Journal of Travel & Tourism Marketing, 21*(1), 47–64. doi:10.1300/J073v21n01_04

Tung, V. W. S., Lin, P., Qiu Zhang, H., & Zhao, A. (2017). A framework of memory management and tourism experiences. *Journal of Travel & Tourism Marketing, 34*(7), 853–866. doi:10.1080/10548408.2016.1260521

Tung, V. W. S., & Ritchie, J. R. B. R. B. (2011). Exploring the essence of memorable tourism experiences. *Annals of Tourism Research, 38*(4), 1367–1386. doi:10.1016/j.annals.2011.03.009

Tussyadiah, I. P., & Fesenmaier, D. R. (2009). Mediating Tourist Experiences. *Annals of Tourism Research, 36*(1), 24–40. doi:10.1016/j.annals.2008.10.001

Valle, P. O. D., Silva, J. a., Mendes, J., & Guerreiro, M. (2006). Tourist satisfaction and destination loyalty intention : A structural and categorical analysis. *International Journal of Business Science and Applied Management, 1*(1), 25–44. https://business-and-management.org/library/2006/1_1--25-44--Oom_do_Valle,Silva,Mendes,Guerreiro.pdf

Vargo, S. L., & Lusch, R. F. (2016). Institutions and axioms: An extension and update of service-dominant logic. *Journal of the Academy of Marketing Science, 44*(1), 5–23. doi:10.100711747-015-0456-3

Vittersø, J., Vorkinn, M., Vistad, O. I., & Vaagland, J. (2000). Tourist experiences and attractions. *Annals of Tourism Research, 27*(2), 432–450. doi:10.1016/S0160-7383(99)00087-0

Volo, S. (2009). Experience : A Tourist Based Approach Conceptualizing Experience : A Tourist. *Journal of Hospitality Marketing & Management, 18*(April), 111–126. doi:10.1080/19368620802590134

Volo, S. (2010). Bloggers' reported tourist experiences: Their utility as a tourism data source and their effect on prospective tourists. *Journal of Vacation Marketing, 16*(4), 297–311. doi:10.1177/1356766710380884

Wang, N. (1999). Rethinking authenticity in tourism experience. *Annals of Tourism Research, 26*(2), 349–370. doi:10.1016/S0160-7383(98)00103-0

Wolak, R., Kalafatis, S., & Harris, P. (1998). An investigation into four characteristics of services. *Journal of Empirical Generalisations in Marketing Science, 3*, 22–41.

Yoo, C., & MacInnis, D. (2005). The brand attitude formation process of emotional and informational ads. *Journal of Business Research, 58*(10), 1397–1406. doi:10.1016/j.jbusres.2005.03.011

ADDITIONAL READING

Bogicevic, V., Seo, S., Kandampully, J. A., Liu, S. Q., & Rudd, N. A. (2019). Virtual reality presence as a preamble of tourism experience: The role of mental imagery. *Tourism Management, 74*, 55–64. doi:10.1016/j.tourman.2019.02.009

Bregoli, I. (2020). Experience marketing and tourism: an application of service-dominant logic to AddioPizzo Travel. In Destination Management and Marketing: Breakthroughs in Research and Practice (pp. 1054-1065). IGI global. doi:10.4018/978-1-7998-2469-5.ch058

Bueno, E. V., Weber, T. B. B., Bomfim, E. L., & Kato, H. T. (2019). Measuring customer experience in service: A systematic review. *Service Industries Journal, 39*(11-12), 779–798. doi:10.1080/02642069.2018.1561873

Dhiraj, A., & Ahmad, S. (2019). WOW! Engage, Experience and Enjoy Marketing 4.0. *Communication Today, 10*(1), 163–164.

Jiménez Barreto, J., Rubio, N., & Campo Martínez, S. (2019). The online destination brand experience: Development of a sensorial–cognitive–conative model. *International Journal of Tourism Research, 21*(2), 245–258. doi:10.1002/jtr.2258

Minton, E. A., & Liu, R. L. (2019). It's Only Once, So Let's Indulge: Testing Ordinary vs. Extraordinary Experience within Marketing Messaging, Temporal Distance, and Consumer Indulgence. *The Journal of Consumer Affairs, 53*(3), 1084–1116. doi:10.1111/joca.12219

Mossberg, L. (2007). A marketing approach to the tourist experience. *Scandinavian Journal of Hospitality and Tourism, 7*(1), 59–74. doi:10.1080/15022250701231915

Pappas, N. (2019). The complexity of consumer experience formulation in the sharing economy. *International Journal of Hospitality Management, 77*, 415–424. doi:10.1016/j.ijhm.2018.08.005

KEY TERMS AND DEFINITIONS

Co-Creation: Creation of value by organisations and individual customers in a collaborative way to contribute to more individual experiences.

Engagement: A behavioural response from an individual towards organisations' (products or services) brands, establishing an emotional connection (commitment), during an experience(s).

Experience: Personal and meaningful events or occurrences, outside the usual environment, which can be highly rewarding (physical and psychological) for an individual(s).

Experience Marketing: Individual based events planned by organisations aiming to engage (using the senses and emotions) individuals holistically (i.e., physically, and spiritually) to create memories.

Holistic: The whole or total experience which represents the sum of the various stimulus (parts), interactions and meanings that individuals experience during events (planned or not), interconnected from end-to-end.

Memories: A mental impression retained from an experience.

Tourism Experience: Events, activities and interactions individuals engage outside their residence with organisations, environments, and other people that enable them to create, develop and enjoy personal narrative and memories.

Chapter 10
A Conceptualization of the Senses, Emotions, and Memories in Memorable Tourism Experiences

Rui Mendonça Pedro
Faculty of Economics and Research Centre for Tourism, Sustainability, and Well-Being (CinTurs), University of Algarve, Portugal

Julio Mendes
Faculty of Economics and Research Centre for Tourism, Sustainability, and Well-Being (CinTurs), University of Algarve, Portugal

Nelson de Matos
School for Management, Hospitality and Tourism (ESGHT) and Research Centre for Tourism, Sustainability, and Well-Being (CinTurs), University of Algarve, Portugal

Mário Passos Ascenção
https://orcid.org/0000-0002-0536-0860
Haaga-Helia University of Applied Sciences, Finland

ABSTRACT

This chapter explores a conceptualization of the elements that create a memorable tourism experience – senses, emotions, and memories. A comprehensive and coherent theoretical model is established to explain the complexities involved in the tourism experience creation (tourism experience relational model) and in the memorable tourism experience (conceptual model). The data suggest that, first, the senses developed an important role in the experiential stimuli perception; second, the emotions are fundamental to these experiential stimuli understanding and meaning creation, and third, memory is essential in the codification and storage process and anticipation of meaningful information to future

DOI: 10.4018/978-1-7998-3156-3.ch010

experiences. The conceptual model is based on the five senses (sight, smell, taste, hearing, and touch), three positive emotions (joy, love, and positive surprise), and two memorable elements (recollection and vividness). Future research opportunities in the memorable tourism experience are explored, such as the relation between the sense, emotions, and memories.

INTRODUCTION

In the current global economic context, tourism experiences represent a key value proposal for the tourism sector and specifically for tourists, because they wish to engage in unique experiences (Mendes, Guerreiro, & Matos, 2016). Touristic companies and organizations compete to offer the richest possible experience to appeal to tourists' wishes and desires (Boswijk, Thijssen, & Peelen, 2006; Pine & Gilmore, 1998; Schmitt, 1999). Ayazlar and Arslan (2017) and Stasiak (2013) consider that the tourism industry is one of the areas of excellence in presenting and staging experience according to a commercial structure that is demanding in terms of services and products specificity, as well as intangible and seasonal. The tourism experience is, however, a complex, personal and memorable construct (Pine & Gilmore, 1998; Zhang, Wu, & Buhalis, 2018).

The memorable tourism experience (MTE) allows tourists to revive, remember and revisit the experiential moment through recollection processes that will last forever or continue for a long period. MTE are considered the ultimate goal that tourists want to achieve during their holidays (Tung, Lin, Zhang, & Zhao, 2017), because they are recognized by tourists as satisfactory, enjoyable and pleasurable moments (Chen & Rahman, 2018; Zhang *et al.*, 2017). MTE also contribute to increasing the tourists' intention to revisit, remember and recommend a destination (Kim, 2018).

Several articles have shown the importance of the senses in the tourism experience (Agapito, Mendes, Valle, & Almeida, 2014; Goggin *et al.*, 2017; Lee, Heere, & Chung, 2013; Meacci & Liberatore, 2015), while others have studied the significance of the emotions (Correia, Oliveira, & Pereira, 2017; Hosany, Prayag, Deesilatham, Caušević, & Odeh, 2015; Prayag, Hosany, Muskat, & Del Chiappa, 2017; Shoval, Schvimer, & Tamir, 2018) and some articles have considered the influence of memories (Ali, Hussain, & Ragavan, 2014; Kastenholz, Carneiro, Marques, & Loureiro, 2017; Pezzi & Vianna, 2015; Quadri-Felitti & Fiore, 2013). The relationship between the senses, emotions and memories is, however, a theme on which there has been a lack of attention (Dias, Correia, & Cascais, 2017).

This chapter therefore investigates the role of senses, emotions and memories in the MTE. The objectives of this chapter are fivefold: first, to enhance and identify the state of the art regarding the senses, emotions and memories in the tourism context; second, to clarify the importance of the senses, emotions and memories in the tourism scenario; third, to develop a conceptual framework of the senses, emotions and memories in the memorable tourism experience; fourth, to analyse future research opportunities in this area; and fifth, to clarify the relationship between the senses, emotions and memories in MTE creation.

This chapter is divided into three main sections. The first section introduces the theme and the relevant contributions and articles. In the second section, the background, the authors provide a brief review of the experience paradigm (i.e. dimensions, models and definitions), the tourism experience approach and tourism experience relational model are presented. In the same section, the role of senses, the influence of emotions, the effects of memories in the MTE are considered. the researchers also present the conceptual frameworks to represent the relationship between the senses, emotions and memories in creating an MTE. The paper closes with a conclusion that focuses on the aims of the study.

BACKGROUND: THE EXPERIENCE PARADIGM

The experience paradigm is a crucial issue that is present in several areas, from the economy to tourism, from industry to business and from products to services (Adhikari & Bhattacharya, 2016; Carbone & Haeckel, 1994; Chang & Lin, 2015; Duncan, Fanderl, Maechler, & Neher, 2016; Holbrook & Hirschman, 1982; Rais, Musa, & Muda, 2016; Schmitt, Joško Brakus, & Zarantonello, 2015). Experience as a concept can be outlined as complex process between a customer and the environment in which fantasies, feelings, emotions and fun are involved (Holbrook & Hirschman, 1982). The experience process can also be considered something extraordinary, in which multisensory stimulants, involving all five senses (Carù & Cova, 2003; Shaw, 2005), and emotionally surprising factors are present (Andajani, 2015; Berry, Carbone, & Haeckel, 2002) to fulfil customers' needs and expectations (Gentile, Spiller, & Noci, 2007; Nasution, Sembada, Miliani, Resti, & Prawono, 2014) with the purpose of becoming a memorable event (Chahal, Kaur, & Rani, 2016; Pine & Gilmore, 1998).

For instance, Lemon and Verhoef (2016, p. 74), conceptualize the customer experience as a stage process or "as a cycle journey, (…) an interactive and dynamic process (…) flows from pre-experience to experience and to post-experience (…) between a company and a customer". The experiential process comprehends different stages with distinct customer behaviours. The *pre-experience stage* includes customers' search for information, their personal dreams and wishes reflected in the imagination and mental experiences they anticipate experiencing during their vacations, as well as price analysis between options, interactions with the brand(s) and evaluation of several options in comparison with previous experiences. The *experience stage* covers all of the customer's experience interaction and includes decision-making (choice and ordering), experience consumption, experience creation or co-creation and payment. The *post-experience stage* comprehends the following behaviours: outcome evaluation, satisfaction, loyalty and consideration for future experiences (Arnold & Reynolds, 2003; Lemon & Verhoef, 2016; Peñaloza & Venkatesh, 2006; Tynan & McKechnie, 2009; Verhoef *et al.*, 2009). During the experiential journey, the customer also goes through different experience stages (Chahal *et al.*, 2016; Nasution *et al.*, 2014; Tynan & McKechnie, 2009), with different interactions and behaviours (Baxendale, Macdonald, & Wilson, 2015; Dhebar, 2013; Lemon & Verhoef, 2016) facing different dimensions of the experiential process (Brito, Oliveira, Nobre, Macedo, & Marchetti, 2017; Pine & Gilmore, 1998; Schmitt, 1999).

The experience occurs when the customer is exposed to the consumption (of a service or product) and "this experience journey is composed by dimensions linked to the interaction between a customer to service or a customer to product" (Brito *et al.*, 2017, p. 125). There is no agreement on the dimensional terminology among authors on this subject (Dias *et al.*, 2017; Krishna & Schwarz, 2014; Pine & Gilmore, 1998; Schmitt, 1999). Pine and Gilmore (1998) distinguish two dimensions – customer participation and customer connection – within four realms: entertainment, educational, aesthetic and escapist. Schmitt (1999), in its turn, proposes five dimensions/modules: relate, sense, feel, think and act. Numerous studies have also already contemplated a multidimensional approach. In effect, several dimensions emerge from the tourism experience theme: the sensorial dimension (Dias *et al.*, 2017; Krishna, 2012; Krishna &

Schwarz, 2014; Zatori, Smith, & Puczko, 2018), the emotional dimension (Correia *et al.*, 2017; Dias *et al.*, 2017; Oh, Fiore, & Jeoung, 2007), the memorable dimension (Dias *et al.*, 2017; Hosany & Witham, 2010; Kim, Ritchie, & McCormick, 2012; Oh *et al.*, 2007), the social dimension (Foris, 2014; Ivanovic & Saayman, 2015), the spiritual dimension (Jepson & Sharpley, 2015) and the physical dimension (Cohen & Ben-Nun, 2009; Fernandes & Cruz, 2016; Jepson & Sharpley, 2015).

According to Gentile *et al.* (2007), Kim *et al.* (2012) and Zatori *et al.* (2018), tourism experiences with high sensorial immersion and with stimulation of the five senses is essential to promote the long-term memory process. This is the case because sensorial perception is strongly related to emotional activation, emotional activation and processing involve cognitive functions and, finally, using high emotional and cognitive resources increases memory construction and consolidation (Zatori, 2013). (Dias *et al.*, 2017, p. 190) also state that the tourism experience "must be memorable and evoke emotions: if this does not occur, its future will end soon". Overall, the dimensions are closely linked to sensorial, emotional, rational and social aspects (Dirsehan & Çelik, 2011) through hedonic activities – that is, pleasure, joy, fun and fantasy (Schmitt, 1999) – and during the experiential process, the experience dimensions emerge with an organized structure and are presented according to time, intensity, polarity, action and experience specificity, similar to multidimensional experience models (Carù & Cova, 2003; Chan, To, & Chu, 2016; Fave & Bassi, 2016; Pine & Gilmore, 1998; Schmitt, 1999; Wang, Tang, Liu, & Xing, 2017).

The tourism experience is perceived by tourists according different models (Cohen, 1979; Ek, Larsen, Hornskov, & Mansfeldt, 2008; Gelter, 2010; Jennings, 2005; Killion, 1992; Tussyadiah, 2014). Csikszentmihalyi (1990) was one of the first authors to develop an experiential model: the flow theory or optimal experience. Years later, the same author reformulated flow theory and added task challenge, skill level, subject involvement and commitment to the model (Csikszentmihalyi, 2014). The flow theory or optimal experience "is a psychology state of intense concentration, attention and proper stimulation (…) and occurs when the subject experience the balance between task challenge and skills level (…) which results in a completely involvement, (…) forgetting time, fatigue, and everything else but the activity itself (…) during a transcendental and singular moment" (Csikszentmihalyi, 2014, p. 230). From a less transcendent and more economic point of view, Pine and Gilmore (1998, 1999) developed the four realms model based on four dimensions (entertainment, educational, aesthetic and escapist), the experience involvement level (absorption and immersion) and the experience participation level (active and passive). Schmitt (1999) developed the experiential grid model (ExPros) that relates the five modules (relate, sense, feel, think and act) according to the experience levels of intensity, breadth, depth and linkage. In a distinct, more unidirectional perspective, Carù and Cova (2003) created the *extraordinary* experience model, which underlines that, during the experiential process, "it manages to over-saturate all events to construct a long series of strong emotions and unforgettable and extraordinary experiences" (Carù & Cova, 2003, p. 281).

Presently, one of the most used models of tourism experience is the co-creation model (Hsieh & Chang, 2016; Lin, Chen, & Filieri, 2017; Ramaswamy & Ozcan, 2018). This model allows active participation, tourist interaction during an experience (Cabiddu, Lui, & Piccoli, 2013; Campos, Mendes, Valle, & Scott, 2015; Prebensen, Vitterso, & Dahl, 2013) and improves the experience process outcomes (Buonincontri, Morvillo, Okumus, & van Niekerk, 2017; Campos, Mendes, Valle, & Scott, 2017; Prebensen & Xie, 2017; Ribes-Giner, Perello-Marín, & Díaz, 2016). Meanwhile, the tourism experience must be perceived through a global, multidimensional and holistic approach (Gentile *et al.*, 2007). Thus, despite the various presented conceptual models of the experiential process, it seems that experience remains an elusive and complex construct, linked to the personal domain and multiple dimensions that are yet to be fully understood and explained (Boswijk *et al.*, 2006; Carù & Cova, 2003; Csikszentmihalyi, 2014; Pine & Gilmore, 1998).

Experience is a rich construct in terms of its multiple contents (Adhikari & Bhattacharya, 2016; Lemon & Verhoef, 2016; Stein & Ramaseshan, 2016; Tynan & McKechnie, 2009; Verhoef *et al.*, 2009), dimensions (Chan *et al.*, 2016; Ding & Tseng, 2015; Dirsehan & Çelik, 2011; Pine & Gilmore, 1998; Wang *et al.*, 2017) and explanatory models (Carù & Cova, 2003; Csikszentmihalyi, 1997, 2014; Fave & Bassi, 2016; Gentile *et al.*, 2007; Pine & Gilmore, 1998; Schmitt, 1999). Experience should, however, include a model that addresses the relationship between an organization and a subject throughout the different experiential process stages, touch points and dimensions and finish with experience effects and results to the company and the subject. The experience is an organized, dynamic process that involves interactions between a company and a customer with the purpose of creating extraordinary, personal, emotional and memorable moments to foster customer loyalty (Adhikari & Bhattacharya, 2016; Carù & Cova, 2003; Chan *et al.*, 2016; Csikszentmihalyi, 2014; Ding & Tseng, 2015; Dirsehan & Çelik, 2011; Gentile *et al.*, 2007; Lemon & Verhoef, 2016; Pine & Gilmore, 1998; Stein & Ramaseshan, 2016; Tynan & McKechnie, 2009; Verhoef *et al.*, 2009; Wang *et al.*, 2017).

Some authors consider experiences to be extraordinary moments and transcendent events (Kirillova, Lehto, & Cai, 2017) that consist of "a process of exploring consciousness (...) that is ever-present in human life (...) and how this consciousness interacts with the world" (Chen, Scott, & Benckendorff, 2017, p. 2). Such experiences can be operationalized through the search for new and different encounters to escape the monotony of daily routine, a search for authenticity, local interactions and engagement with natural landscapes (Park & Santos, 2017).

THE TOURISM EXPERIENCE

Tourism experiences provide the source of value in the evolution, dynamics and competition in tourism destinations and for the tourism industry (Ali, Ryu, & Hussain, 2016; Bosangit, Hibbert, & McCabe, 2015; Campos, Mendes, Valle, & Scott, 2016). Given their importance, one of the fundamental concepts in the tourism context is the tourism experience, because it establishes the basis of services and product supply (Fernandes & Cruz, 2016). The perceived quality of tourism experiences has a crucial impact on tourists' satisfaction and loyalty, and it represents an essential factor in tourism management and hospitality industry strategy (Demarco, 2016). Chang (2018, p. 55) notes that "tourism industry revenue increases according to the extent to which consumers enjoy their experiences". The central issue in tourism is, therefore, the tourism experience (Bosangit *et al.*, 2015; Demarco, 2016; Kim, 2010) and its management regarding the capacity to offer, stage, create and co-create such experiences in ever more multisensory, emotional and memorable events (Kim, 2018; Vo Thanh & Kirova, 2018).

According to Kirillova, Lehto and Cai (2017, p. 498) "tourism experiences are distinct from the everydayness because they can become extraordinary, pure, peak or cathartic (…) thus special extraordinary events trigger highly emotional responses, lead to self-exploration, serve for profound intrapersonal changes and are conducive to optimal human functioning". However, the tourism experience concept should include novelty, unusual places (social interaction), loneliness (relaxation) and sensory stimulation (Morgan, 2006; Mossberg, 2007). Larsen (2007) brings the role of memory to the discussion and, more specifically, that the experience event should be iconic enough to enter the long-term memory. Volo (2009, p. 119) describes the experience with a more physiological perspective, noting that a touristic experience happens when "energy reflecting the state of the environment impinges on sensory organs, the energy pattern is transmitted centrally and is interpreted and categorized according to one's knowledge acquired through time and is integrated and may be stored in the form of memory under some conditions".

The main advantages of the touristic experience to tourists are behaviour change and self-discovery (Kirillova *et al.*, 2017), physical wellness and psychological benefits (Chen *et al.*, 2017; Tung *et al.*, 2017), general satisfaction and mental well-being (Fernandes & Cruz, 2016; Martins *et al.*, 2017; Park & Santos, 2017) and the chance to live extraordinary moments or spiritual experiences that evoke memories through cognitive activity (Chen *et al.*, 2017). Tourism quality involves multiple influences such as the tourist, the local population and the touristic product (Nickerson, 2005), so "a quality experience, (…) depends on how each side is treated by the other" (Nickerson, 2005, p. 233).

Tourism Experience Relational Model

The experience relational model always includes a sender, a message sent through a channel, a receiver and then the feedback (see Figure 1). During the creation stage, instruments are delivered and all context staging belongs to the organization (Carù & Cova, 2003; Pine & Gilmore, 1999). The message is established, transmitted and shared throughout the experience between the organization and the tourist based on the tourist's multisensory stimulation and emotional involvement (Andajani, 2015; Carù & Cova, 2003; Shaw, 2005). The feedback comprehends the experiential outcomes between the tourist and the organization, such as satisfaction, attractive moments, extraordinary events and memorable moments (Hosany, Prayag, Van Der Veen, Huang, & Deesilatham, 2017; Kim & Jang, 2016; Kim & Ritchie, 2014; Prayag, Hosany, & Odeh, 2013).

It is clear that, in tourism, the focus is the experience (Duarte, 2012, p. 187), and tourists "want experience more than a product or service because they are more memorable (...) and (...) in the baggage they bring expectations and takes home memories" (Ayazlar & Arslan, 2017, p. 137). Memorable tourism experiences provide competitive advantage to tourists, destinations, hospitality industries and hotel companies among other stakeholders (Kim & Youn, 2017; Tung *et al.*, 2017).

THE MEMORABLE TOURISM EXPERIENCE

"A memorable tourism experience is a past personal travel-related event strong enough to have entered long-term memory" (Larsen, 2007, p. 15). Other authors describe a memorable tourism experience as a sufficiently significant moment that is recalled and reconstructed by the tourist in describing the experience (Tung *et al.*, 2017; Tung & Ritchie, 2011). More recently, Kim *et al.* (2012, p. 13) state that "a tourism experience positively remembered and recalled after the event has occurred". However, other authors note that a tourism experience is a personal journey and only becomes memorable according to the tourist's personal interpretation, meaning assignment and based on culture, beliefs, values and behaviours (Ayazlar & Arslan, 2017; Chen & Rahman, 2018; Tsai, 2016).

According to Park and Santos (2017), this personal journey or process reveals that a memorable tourism experience is a sequential process with multiple stages (pre-experience, experience and post-experience) that is dynamic (interaction with others, with destiny) and personal (built through tourist expectations, needs, motivations, perceptions, emotions, memories). This personal journey is associated

Figure 1. Tourism experience relational model
Source: Own Elaboration

with a subjective sense of well-being, pleasure (hedonism), enjoyment, excitement, self-discovery, meaning, intellectual relevance and connection to something unforgettable and extraordinary (Chandralal & Valenzuela, 2015; Sthapit & Coudounaris, 2018; Tung *et al.*, 2017). However, Kirillova *et al.* (2017) propose that memorable tourism experiences construct and must be considered an unusual event that triggers intense emotional responses and provides a sense of full harmony with the universe. A memorable experience therefore includes a multisensory dimension (Tsai, 2016) and is linked with a sense of well-being, intense positive emotional states, moments of self-discovery, personal meaning and intellectual relevance for tourists (Kirillova *et al.*, 2017; Servidio & Ruffolo, 2016; Sthapit & Coudounaris, 2018). In simpler terms, the memorable experiences aims to develop a strong connection between the experience and the tourist (Tsai, 2016).

The characteristics of the memorable tourism experience are linked to the five senses (Agapito, Valle, & Mendes, 2014; Sthapit, 2017; Sthapit & Coudounaris, 2018), novelty and escape from everyday life (Kastenholz *et al.*, 2017; Kim, 2010; Kim *et al.*, 2012), while at the same time seeking to exceed expectations and build nostalgia (Tung & Ritchie, 2011). The memorable experience includes high positive emotional states, cognitive activity and a sense of freedom, refreshment and spiritual connection to something big or extraordinary (Ali *et al.*, 2014; Kastenholz *et al.*, 2017; Kim, 2010; Kirillova *et al.*, 2017; Tung *et al.*, 2017; Zhang *et al.*, 2017).

The Role of Senses

The senses or the five sensory organs (sight, taste, hearing, smell, and touch) are the channels of communication between the brain and the outside world (Dias *et al.*, 2017). According to Light (2009, p. 12), "the sensory receptors are anatomical structures made up of special cells (nerve cells or neurons) that respond to specific changes in their environment (stimulus) (…) provide to the central nervous system (brain and spinal cord) information about conditions both inside the body (internal homeostasis) and outside the body (environment information) (…) sensory receptors are the physical links between central nervous system and the environment". The sensory receptors can classified in three different ways based on type of sensation or modality, the location of the stimulus to which they respond and their structural complexity (Light, 2009). Sensory stimulus promotes sensations and the sensations create a perception, because the sensory organs activate and transform external stimulus in neural signals (Agapito *et al.*, 2014; Light, 2009). To Agapito *et al.* (2014), the sense most often considered in the tourism field is the analysis of the visual dimension. However, Krishna (2012) notes that tourism product has a multisensory nature, and other sensory stimuli are present, including taste, hearing, smell, and touch, that can be determinants to stimulate other sensory organs and create new and distinct sensations and build new neural signs.

In the same context, but from a marketing perspective, Krishna and Schwarz (2014) state that products or services are designed to appeal to the basic senses, and this seems to play an important role in consumers' emotional engagement with the product or service and the development of purchase desire. A more complex idea of the tourism experience suggests that all five senses or a multisensory approach should be considered in, for example, gastronomic experiences and destination products (Ditoiu & Căruntu, 2014; Jalis, Che, & Markwell, 2014). The concept of the multisensory tourism experience seems to be a relevant and contemporary research subject, because the senses are the main channel for how the tourist collects and uses information from the environment. All tourism destinations have unique and distinct sensory characteristics, which play an important role in creating tourist loyalty (Agapito, Mendes, & Valle, 2013; Dann & Jacobsen, 2003; Jalis *et al.*, 2014; Pawaskar & Goel, 2014).

As Kim and Fesenmaier (2015, p. 2) note, "the sensorial process is the core role of the tourism experience (…) and starts where the environmental stimuli come across the human body's sense organs (…) the emotional and cognitive responses" According to Krishna (2012), the tourism experience can be viewed on four levels: the sensory level and stimulus reception; the perceptive level and stimulus interpretation; the cognitive and emotional level, where the tourist's cognition and emotional response to the world around them takes place; and the action level, where the transformation of attitudes, learning behaviour and memory processes happen.

In the tourism context, Agapito, Mendes, Valle and Almeida (2014) suggest that the sensory dimension of the tourist experience appears to be strongly linked to positive, unique and memorable experiences, and they therefore developed a study to analyse the sensory dimension of the tourist experiences in a rural context, concluding that the most references senses were to sight (26%), hearing (23%), taste (19%), smell (17%) and finally, touch (15%). Each sense is, however, stimulated by different and specific environmental settings and provides different sensory information. The entire sensory component (the multisensory stimulus) allows tourists to achieve an experience more truly meaningful and personalized (Kim & Fesenmaier, 2015) and therefore more memorable. These personalized experiences suggest specific stimuli to provide memorable moments. Tourism experiences that include a multisensory offer therefore appear to contribute to the reinforcement of positive emotions, satisfaction and the creation of long-term memories and destination loyalty (Agapito *et al.*, 2013; Ditoiu & Căruntu, 2014; Pawaskar & Goel, 2014).

The Influence of Emotions

These personalized experiences are present in modern consumer goods and services, which attempt to enable tourists to experience feelings, emotions and stories (Jensen, 2001). Consumers' emotional satisfaction should thus be one of the strengths in an evermore competitive market. It is, however, possible that current tourists not only want a holiday, but also the opportunity to experience new dreams and adventures that embody feelings, spirituality and emotions to satisfy the human nature. Emotions give meaning to the tourist experience, making the experience into something unique, personal, memorable, transcendent, irreplaceable and magnificent, and can manifest themselves before, during and/or after the experience (Prayag *et al.*, 2017). When a tourist, a guest or a visitor looks for new experiences, it is mostly in search of emotions and memories (Kim, 2010).

Damásio (2010) notes that emotions are complex and automatic functions that are characterized by ideas and certain ways of thinking, but emotions are performed on the body, at the external level, through facial expressions and body postures, and at the internal level through endocrinal and visceral changes. Deak (2011) considers emotions to be a complex phenomenon of model behaviour or biological, social and cognitive reaction patterns. This means that emotions can float between the emotional level and the rational level, and this transforms emotions into something more than just a basic and a survival biological behaviour (Dalgleish, 2004; Deak, 2011; LeDoux, 2012). To Deak (2011), emotions are crucial to facilitating understanding of others' cognitive processes.

According to LeDoux (2012), emotions are a biological behaviour necessary for survival and a human self-protection action. However, Perritano (2011) states that emotions can be shaped by culture and society, and characterization of emotions as good or bad can be culturally dependent "in the United States, children are encouraged to take pride in their accomplishments (...) in India, children may heel happier when they have been obedient or helpful", as too are rules about showing emotions "real men don't cry (...) stay calm, don't be angry" (Perritano, 2011, p. 33). Emotions thus have plasticity and adaptability at the cognitive (Deak, 2011), social and cultural levels (Perritano, 2011). Salmela (2014) states that emotions are based on a cognitive characteristic with intentional, rational and voluntary behaviour.

In their study, Pentus, Mehine and Kuusik (2014) changed the product appearance through package design to become more attractive to consumers and found that changes in the visual appearance of the packaging generated positive emotions among consumers, thus unconsciously enhancing likelihood of purchasing. They demonstrated that positive emotions may make consumers more purchasing suggestive, and the emotional response patterns are similar for consumers and tourists (Pentus *et al.*, 2014).

In the tourism context, in a study carried out by Chandralal and Valenzuela (2013), tourists were asked to highlight the emotions that occurred during memorable experiences; the findings indicated that tourists' most common emotions were happiness, enjoyment, surprise, pleasure, satisfaction and excitement, which means that memorable experiences are those that have these as an emotional basis. Güzel (2014) notes that the most satisfying, meaningful and memorable tourist experiences are the ones that cause most emotional excitement. The same author also studied tourist loyalty and the touristic experience and confirmed that tourist's intentional loyalty comes depends on a touristic experience of great emotional excitement (Guzel, 2014).

Consumption of goods or services is currently seen as an emotional decision-making process (Chandralal & Valenzuela, 2013; Guzel, 2014; Pentus *et al.*, 2014). It is important to identify the most pleasant emotions for the consumer and to define the stage of the decision-making process in which the emotions will be most perceptible. The level of emotional arousal will lead the consumer to have the intention to repeat the same emotion. The tourism industry is an economic activity that sells unique, memorable and personal experiences (Pine & Gilmore, 1998). More than experiences, however, the tourist searches for emotions (Bosangit *et al.*, 2015). Emotions in tourism are the key element to the continuous process of renovation, remodelling and recreation that quality tourism requires. Emotions play a key role as mediating and moderating factor of tourism consumption (Pyke, Hartwell, Blake, & Hemingway, 2016). Emotions in the tourism context establish the link between the tourist and the destination, product or service and provide tourism loyalty, help memory building, increase consumption, support the decision-making process and encourage word-of-month advertising (Bosangit *et al.*, 2015; Buda, D'Hauteserre, & Johnston, 2014; Correia *et al.*, 2017).

The Effects of Memories

Humans have two memory structures – the amygdala and the hippocampus (Phelps, 2004). These two distinct brain structures are connected to two independent memory systems with different functions and characteristics. In emotional situations, the amygdala can code and store memory in the hippocampus, and the hippocampus, through the emotional representation and interpretation of events may influence the amygdalar responses when emotional stimuli are found (Phelps, 2004; Rahmann & Rahmann, 1992). Although these are independent memory systems, they operate together when emotion creates the memory links (Phelps, 2004; Rahmann & Rahmann, 1992).

The role of memory in the brain is somewhat mysterious (Marx & Gilon, 2012). Moreover, according to Rahmann and Rahmann (1992) and Marx and Gilon (2012), there are different types of memories: active memory, declarative memory, passive memory, associative memory, short-term memory, long-term memory and super-memory. To Marx and Gilon (2012), memory emerges from the dynamic interaction of three physiological mechanism: neurons, the extracellular neuronal matrix and the dispersion of trace metal cations. This tripartite physiological mechanism involves low energy resources and high-speed information processing, and memories are coded by the neurons, while memory losses occur in the presence of excess dispersion of metal cations (Marx & Gilon, 2012).

Memory formation is a highly dynamic process involving both formation and transformation (Nadel, Hupbach, Gomez, & Newman-Smith, 2012). Memory formation and transformation require an experience process, and memory transformation can involve: a process of selective strengthening of some behaviours through synaptic rescaling or a process of selective survival behaviour; the assimilation of new information into existing knowledge stores; the establishment of new connections within existing knowledge stores; and the updating of an existing episodic memory (Nadel *et al.*, 2012; Rahmann & Rahmann, 1992).

In the tourism scenario, experiences that create positive memories promote the intention to revisit the destination and recommend the destination to family and friends (Tung & Ritchie, 2011). Ali *et al.* (2014) have shown that a positive consumer experience increases positive memories and increases destination loyalty. Halim and Mokhtar (2016) also highlight that positive experiences stimulate positive behaviour, activate positive memory processes, promote loyalty and increase word-of-month intention.

It is possible to conclude that one of the key points of tourism success is the capacity to create positive memories among tourists.

According to Tung and Ritchie (2011), the four dimensions that represent memorable aspects of tourism experiences are affect, expectations, consequentiality and recollection. Affect refers to the positive emotions and feelings associated with the experiences, such as happiness and excitement. Expectations refer to the fulfilment of intentions and/or descriptions of surprises encountered during the trip that were above and beyond tourists' planned agendas. Consequentiality refers to the elements of perceived importance during the trip, such as social relationships, intellectual development, self-discovery, and physical challenges. Recollection refers to telling stories, showing photographs, purchasing souvenirs, re-experience the trip and going back and rebuilding the memory (Tung & Ritchie, 2011).

Conceptual Framework of the Senses, Emotions and Memories in the Memorable Tourism Experience

The conceptual model shows the relationships established between the senses, emotions and memories to create a memorable tourism experience (see Figure 2).

The senses perceive the information (stimuli), the emotions interpret and understand the meaning of that information and, finally, the memory codifies and stores the information, if it is sufficiently important and extraordinary to the person (Agapito *et al.*, 2014; Krishna, 2012; Lee *et al.*, 2013). The sensory configuration of the tourism experience allows multisensory stimulation that increases the construction of positive emotional states and improves the development of memories (Barnes, Mattsson, & Sørensen, 2014; Martins *et al.*, 2017). Multisensory stimulation during the experience increases the creation of memorable experiences with high significance and results in the tourist's physical, emotional and intellectual involvement (Dias *et al.*, 2017; Krishna & Schwarz, 2014).

Once the environment information is perceived through the senses, the emotion stage extracts and explains the meaning of that information (Ekman, 2016). The emotional dimension of the touristic experience should focus on positive emotional states (Barrett, 2017; Correia *et al.*, 2017; Kapoor, 2015; Oh *et al.*, 2007; Tung *et al.*, 2017), because positive emotional states build connections between individual people (Ekman, 2016) or between individuals and places (Correia *et al.*, 2017; Dias *et al.*, 2017), and these are essential in the decision-making processes for choosing a destination or experience (Bimonte & Faralla, 2014; Esteves, Slongo, Esteves, & Barcelos, 2013; Tung & Ritchie, 2011). Emotional states such as joy, love and positive surprise are strongly correlated to emotional activation, emotional satisfaction and, consequently, memory creation, as well as general tourist satisfaction and behaviour intention to recommend or repeat (Chen *et al.*, 2017; Hosany & Prayag, 2013; Kurtz & Welch, 2016; Prayag, Hosany, Muskat, & Del Chiappa, 2017).

The relationship between emotions and memories results from the process of reminiscence – that is, the recollection and vividness of memories (Kim & Youn, 2017). The tourist remembers the experiences through an emotional reconstruction of a particular experience (Lempert & Phelps, 2016; Martins *et al.*, 2017; Servidio & Ruffolo, 2016). Emotions are thus particularly special in terms of quality and quantity during the experience, because they are crucial to the memory process (Correia *et al.*, 2017; Dias *et al.*, 2017). The memory of the experience is the result – that is, the memory encodes and stores only what the tourist perceives with his or her senses as emotionally pleasurable and as a focus of concentration and attention (Colombo & Gold, 2004; Gold, 2004; Mizumori, Yeshenko, Gill, & Davis, 2004; Todd, Talmi, Schmitz, Susskind, & Anderson, 2012). According to Oh *et al.* (2007), the memory encodes and

Figure 2. Conceptual framework of the senses, emotions and memories in the MTE
Source: Own Elaboration

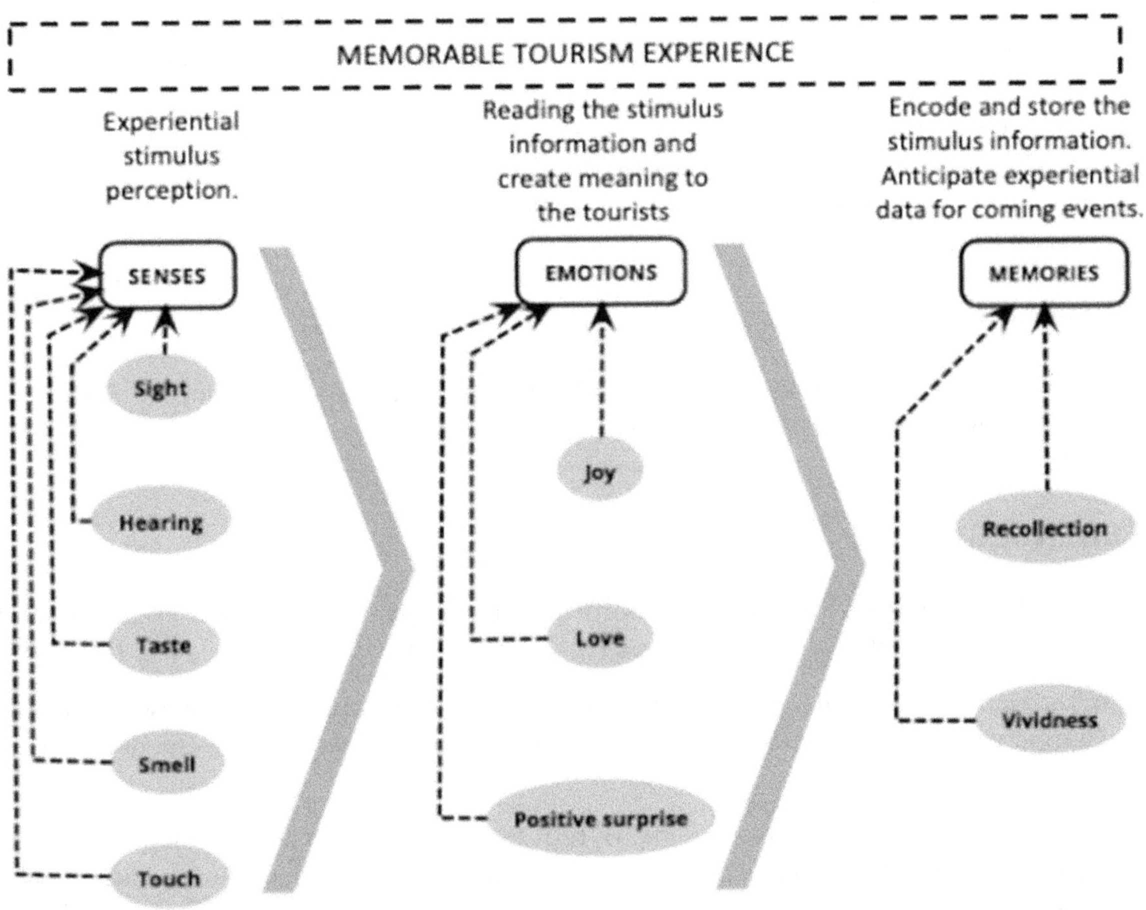

stores memorable moments if the experience involves elements that are wonderful, positive and yield long-term temporal perception.

CONCLUSION

Exploration of the three constructs (senses, emotions and memories) suggests that a memorable tourism experience is the result of a multisensory stimuli and perception, a positive emotional event, high emotional activity, an extraordinary and surprising emotional moment, wonderful and positive memories and a long-term memorable experience (Chen *et al.*, 2017; Correia *et al.*, 2017; Dias *et al.*, 2017; Hosany & Prayag, 2013; Kim, 2014; Kurtz & Welch, 2016; Prayag *et al.*, 2017). In the tourism context, the senses, emotions and memories are an important issue because multisensory stimuli encourage high emotional activity and increase the memory formation (Kim, 2018; Oh *et al.*, 2007; Tung *et al.*, 2017) because all sensory system is strongly related to emotional processes which involve cognitive processing, while the use of high cognitive resources are linked to the formation of memories (Zatori, 2013).

Based on a critical and extensive literature review background, the future research opportunities in this area should explore the relationships between the constructs that lead to the creation of a memorable tourism experience. It is probable that they have a cyclic and sequential relationship: that is, the memory evokes the emotions experienced and these affect the decision-making process and behaviour intention to revisit the destination or repeat the experience (Chen *et al.*, 2017; Hosany & Prayag, 2013; Kurtz & Welch, 2016; Prayag *et al.*, 2017). In this particular investigation, the analyses began with the creation of the memorable tourism experience: sense, emotions and memories. The research gap emerges regarding the relationship between the senses, emotions and memories and the behavioural patterns discovered.

ACKNOWLEDGMENT

This research was supported by the National Funds provided by FCT – Foundation for Science and Technology through project UIDB/04020/2020.

REFERENCES

Adhikari, A., & Bhattacharya, S. (2016). Appraisal of literature on customer experience in tourism sector: Review and framework. *Current Issues in Tourism*, *19*(4), 296–321. doi:10.1080/13683500.2015.1082538

Agapito, D., Mendes, J., & Valle, P. (2013). Exploring the conceptualization of the sensory dimension of tourist experiences. *Journal of Destination Marketing & Management*, *2*(2), 62–73. doi:10.1016/j.jdmm.2013.03.001

Agapito, D., Mendes, J., Valle, P., & Almeida, H. (2014). A contribute to sensory marketing in rural tourist experiences. *PASOS Revista de Turismo y Patrimonio Cultural*, *12*(3), 611–621. doi:10.25145/j.pasos.2014.12.045

Agapito, D., Valle, P., & Mendes, J. (2014). The sensory dimension of tourist experiences: Capturing meaningful sensory-informed themes in Southwest Portugal. *Tourism Management*, *42*, 224–237. doi:10.1016/j.tourman.2013.11.011

Ali, F., Hussain, K., & Ragavan, N. (2014). Memorable Customer Experience: Examining the Effects of Customers Experience on Memories and Loyalty in Malaysian Resort Hotels. *Procedia: Social and Behavioral Sciences*, *144*, 273–279. doi:10.1016/j.sbspro.2014.07.296

Ali, F., Ryu, K., & Hussain, K. (2016). Influence of Experiences on Memories, Satisfaction and Behavioral Intentions: A Study of Creative Tourism. *Journal of Travel & Tourism Marketing*, *33*(1), 85–100. doi:10.1080/10548408.2015.1038418

Andajani, E. (2015). Understanding Customer Experience Management in Retailing. *Procedia: Social and Behavioral Sciences*, *211*(September), 629–633. doi:10.1016/j.sbspro.2015.11.082

Arnold, M., & Reynolds, K. (2003). Hedonic Shopping Motivations. *Journal of Retailing*, *79*(2), 77–95. doi:10.1016/S0022-4359(03)00007-1

Ayazlar, G., & Arslan, R. (2017). A Cross-cultural Investigation of Tourists' Memorable Experiences Between Two Nationalities. *Almatourism - Journal of Tourism. Culture and Territorial Development*, *15*(15), 136–151. doi:10.6092/issn.2036-5195/6558

Barnes, S., Mattsson, J., & Sørensen, F. (2014). Destination brand experience and visitor behavior: Testing a scale in the tourism context. *Annals of Tourism Research*, *48*, 121–139. doi:10.1016/j.annals.2014.06.002

Barrett, L. (2017). *How emotions are made: The secret life the brain*. Houghton Mifflin Harcourt Publishing Company.

Baxendale, S., Macdonald, E., & Wilson, H. (2015). The Impact of Different Touchpoints on Brand Consideration. *Journal of Retailing*, *91*(2), 235–253. doi:10.1016/j.jretai.2014.12.008

Berry, L., Carbone, L., & Haeckel, S. (2002). Managing the total customer experience. *MIT Sloan Management Review*, *43*(3), 85–89. doi:10.1371/journal.pone.0015090

Bimonte, S., & Faralla, V. (2014). Happiness and nature-based vacations. *Annals of Tourism Research*, *46*, 176–178. doi:10.1016/j.annals.2014.02.002

Bosangit, C., Hibbert, S., & McCabe, S. (2015). "If I was going to die I should at least be having fun": Travel blogs, meaning and tourist experience. *Annals of Tourism Research*, *55*, 1–14. doi:10.1016/j.annals.2015.08.001

Boswijk, A., Thijssen, T., & Peelen, E. (2006). A New Perspective on the Experience Economy - Meaningful Experiences. The European Centre for the Experience Economy, The Netherlands, 2, 233–247.

Brito, S., Oliveira, A., Nobre, L., Macedo, Á., & Marchetti, R. (2017). Brand experience dimensions in the Brazilian context. *Race*, *16*(4), 121–146. doi:10.18593/race.v16i1.10527

Buda, D., D'Hauteserre, A., & Johnston, L. (2014). Feeling and tourism studies. *Annals of Tourism Research*, *46*, 102–114. doi:10.1016/j.annals.2014.03.005

Buonincontri, P., Morvillo, A., Okumus, F., & van Niekerk, M. (2017). Managing the experience co-creation process in tourism destinations: Empirical findings from Naples. *Tourism Management*, *62*, 264–277. doi:10.1016/j.tourman.2017.04.014

Cabiddu, F., Lui, T., & Piccoli, G. (2013). Managing Value Co-Creation In The Tourism Industry. *Annals of Tourism Research*, *42*, 86–107. doi:10.1016/j.annals.2013.01.001

Campos, A., Mendes, J., Valle, P., & Scott, N. (2015). Co-creation of tourist experiences: A literature review. *Current Issues in Tourism*, *21*(April), 1–32. doi:10.1080/13683500.2015.1081158

Campos, A., Mendes, J., Valle, P., & Scott, N. (2016). Co-Creation Experiences: Attention and Memorability. *Journal of Travel & Tourism Marketing*, *33*(9), 1309–1336. doi:10.1080/10548408.2015.1118424

Campos, A., Mendes, J., Valle, P., & Scott, N. (2017). Co-creating animal-based tourist experiences: Attention, involvement and memorability. *Tourism Management*, *63*, 100–114. doi:10.1016/j.tourman.2017.06.001

Carbone, L., & Haeckel, S. (1994). Engineering Customer Experiences. *Marketing Management*, *3*(3), 8–19.

Carù, A., & Cova, B. (2003). Revisiting Consumption Experience: A More Humble but Complete View of the Concept. *Marketing Theory*, *3*(2), 267–286. doi:10.1177/14705931030032004

Chahal, H., Kaur, G., & Rani, A. (2016). Exploring the Dimensions of Customer Experience and its Impact on Word-of-Mouths: A Study of Credit Cards. *Journal of Services Research*, *15*(2), 1–33. https://search.proquest.com/abicomplete/docview/1878857032/8DF011A210874C67PQ/4?accountid=132208

Chan, W., To, C., & Chu, W. (2016). Desire for experiential travel, avoidance of rituality and social esteem: An empirical study of consumer response to tourism innovation. *Journal of Innovation & Knowledge*, *1*(1), 24–35. doi:10.1016/j.jik.2016.02.001

Chandralal, L., & Valenzuela, F.-R. (2013). Exploring Memorable Tourism Experiences: Antecedents and Behavioural Outcomes. *Journal of Economics. Business and Management*, *1*(2), 177–181.

Chandralal, L., & Valenzuela, F.-R. (2015). Memorable Tourism Experiences: Scale Development. *Contemporary Management Research*, *11*(3), 291–310. doi:10.7903/cmr.13822

Chang, S. (2018). Experience economy in hospitality and tourism: Gain and loss values for service and experience. *Tourism Management*, *64*, 55–63. doi:10.1016/j.tourman.2017.08.004

Chang, S.-H., & Lin, R. (2015). Building a Total Customer Experience Model: Applications for the Travel Experiences in Taiwan's Creative Life Industry. *Journal of Travel & Tourism Marketing*, *32*(4), 438–453. doi:10.1080/10548408.2014.908158

Chen, H., & Rahman, I. (2018). Cultural tourism: An analysis of engagement, cultural contact, memorable tourism experience and destination loyalty. *Tourism Management Perspectives*, *26*, 153–163. doi:10.1016/j.tmp.2017.10.006

Chen, L., Scott, N., & Benckendorff, P. (2017). Mindful tourist experiences: A Buddhist perspective. *Annals of Tourism Research*, *64*, 1–12. doi:10.1016/j.annals.2017.01.013

Cohen, E. (1979). A Phenomenology of Tourist Experiences. *Sociology*, *13*(2), 179–201. doi:10.1177/003803857901300203

Cohen, E., & Ben-Nun, L. (2009). The Important Dimensions of Wine Tourism Experience from Potential Visitors' Perception. *Tourism and Hospitality Research*, *9*(1), 20–31. doi:10.1057/thr.2008.42

Colombo, P., & Gold, P. (2004). Multiple memory systems. *Neurobiology of Learning and Memory*, *82*(3), 169–170. doi:10.1016/j.nlm.2004.07.008 PMID:15464401

Correia, A., Oliveira, C., & Pereira, R. (2017). From Emotions to Place Attachment. In A. Correia, M. Kozak, J. Gnoth, & A. Fyall (Eds.), *Co-Creation and Well-Being in Tourism* (pp. 163–177). Springer International Publishing. doi:10.1007/978-3-319-44108-5_13

Csikszentmihalyi, M. (1990). Flow - The Psychology of optimal experience. Academy of Management Review. New York: HarperCollins Publishers Inc. doi:10.5465/AMR.1991.4279513

Csikszentmihalyi, M. (1997). Finding Flow - The Psychology of Engagement with Everyday Life. New York: Basic Books.

Csikszentmihalyi, M. (2014). Flow and the Foundations of Positive Psychology. Dordrecht: Springer. doi:10.1007/978-94-017-9088-8

Dalgleish, T. (2004). The emotional brain. *Nature Reviews. Neuroscience*, *5*(7), 583–589. doi:10.1038/nrn1432 PMID:15208700

Damásio, A. (2010). *Self Comes to Mind: Constructing the Conscious Brain*. Random House.

Dann, G., & Jacobsen, J. (2003). Tourism smellscapes. *Tourism Geographies*, *5*(1), 3–25. doi:10.1080/1461668032000034033

Deak, A. (2011). Brain and emotion: Cognitive neuroscience of emotions. *Review of Psychology*, *18*(2), 71–80.

Demarco, D. (2016). Sustainable Urban Development Perspectives in the Era of Tourism Experience. *Procedia: Social and Behavioral Sciences*, *223*, 335–341. doi:10.1016/j.sbspro.2016.05.236

Dhebar, A. (2013). Toward a compelling customer touchpoint architecture. *Business Horizons*, *56*(2), 199–205. doi:10.1016/j.bushor.2012.11.004

Dias, J., Correia, A., & Cascais, T. (2017). Traits in Tourists' Experiences: Senses, Emotions and Memories. In A. Correia, M. Kozak, J. Gnoth, & A. Fyall (Eds.), *Co-Creation and Well-Being in Tourism* (pp. 179–194). Springer International Publishing. doi:10.1007/978-3-319-44108-5_14

Ding, C., & Tseng, T. (2015). On the relationships among brand experience, hedonic emotions, and brand equity. *European Journal of Marketing*, *49*(7/8), 994–1015. doi:10.1108/EJM-04-2013-0200

Dirsehan, T., & Çelik, M. (2011). Profiling online consumers according to their experiences with a special focus on social dimension. *Procedia: Social and Behavioral Sciences*, *24*, 401–412. doi:10.1016/j.sbspro.2011.09.040

Ditoiu, M.-C., & Căruntu, A.-L. (2014). Sensory Experiences Regarding Five-dimensional Brand Destination. *Procedia: Social and Behavioral Sciences*, *109*, 301–306. doi:10.1016/j.sbspro.2013.12.461

Duarte, P. (2012). Tourist experience: Contemporary perspectives. *Journal of Heritage Tourism*, *7*(2), 187–188. doi:10.1080/1743873X.2011.638459

Duncan, E., Fanderl, H., Maechler, N., & Neher, K. (2016). *Customer experience: Creating value through transforming customer journeys* (Vol. 1). McKinsey & Company., doi:10.1057/9780230291775

Ek, R., Larsen, J., Hornskov, S. B., & Mansfeldt, O. K. (2008). A dynamic framework of tourist experiences: Space-time and performances in the experience economy. *Scandinavian Journal of Hospitality and Tourism*, *8*(2), 122–140. doi:10.1080/15022250802110091

Ekman, P. (2016). What Scientists Who Study Emotion Agree About. *Perspectives on Psychological Science*, *11*(1), 31–34. doi:10.1177/1745691615596992 PMID:26817724

Esteves, P., Slongo, L., Esteves, C., & Barcelos, R. (2013). As emoções dos consumidores da terceira idade no processo de escolha de destinos de viagens. *Revista de Administração Da UFSM*, *6*(3), 561–580. doi:10.5902/198346595581

Fave, A., & Bassi, M. (2016). *Flow Experience: Empirical Research and Applications.* doi:10.1007/978-3-319-28634-1_1

Fernandes, T., & Cruz, M. (2016). Dimensions and outcomes of experience quality in tourism: The case of Port wine cellars. *Journal of Retailing and Consumer Services, 31*, 371–379. doi:10.1016/j.jretconser.2016.05.002

Foris, D. (2014). Study Regarding a New Dimension of Tourism Phenomenon – The Political-Administrative Dimension. *Procedia: Social and Behavioral Sciences, 143*, 1078–1082. doi:10.1016/j.sbspro.2014.07.559

Gelter, H. (2010). Total Experience Management – a conceptual model for transformational experiences within tourism. In *The Nordic Conference on Experience 2008. Research, Education and Practice in Media* (pp. 46–78). Vaasa: Medusa Group of Vaasa Consortium of Higher Education.

Gentile, C., Spiller, N., & Noci, G. (2007). How to Sustain the Customer Experience. An Overview of Experience Components that Co-create Value With the Customer. *European Management Journal, 25*(5), 395–410. doi:10.1016/j.emj.2007.08.005

Goggin, C., Please, P., Ridges, M., Booth, C., Simpson, G., Green, R., & Leys, J. (2017). Connecting with Country in Mungo National Park, Australia: A case study to measure the emotional dimension of experience and place attachment. *Local Environment, 22*(10), 1217–1236. doi:10.1080/13549839.2017.1334142

Gold, P. (2004). Coordination of multiple memory systems. *Neurobiology of Learning and Memory, 82*(3), 230–242. doi:10.1016/j.nlm.2004.07.003 PMID:15464406

Guzel, O. (2014). The Dimensions of Tour Experience, Emotional Arousal, and Post-experience Behaviors: A Research on Pamukkale in Turkey. *Procedia: Social and Behavioral Sciences, 150*, 521–530. doi:10.1016/j.sbspro.2014.09.069

Halim, A., & Mokhtar, A. (2016). Creating Memorable Visitor Experiences by Assessing the Satisfaction Level and Behavioural Consequences of Attendees. *Procedia Economics and Finance, 37*, 1–6. doi:10.1016/S2212-5671(16)30084-3

Holbrook, M., & Hirschman, E. (1982). The Experiential Aspects of Consumption: Consumer Fantasies, Feelings, and Fun. *The Journal of Consumer Research, 9*(2), 132. doi:10.1086/208906

Hosany, S., & Prayag, G. (2013). Patterns of tourists' emotional responses, satisfaction, and intention to recommend. *Journal of Business Research, 66*(6), 730–737. doi:10.1016/j.jbusres.2011.09.011

Hosany, S., Prayag, G., Deesilatham, S., Caušević, S., & Odeh, K. (2015). Measuring Tourists' Emotional Experiences: Further Validation of the Destination Emotion Scale. *Journal of Travel Research, 54*(4), 482–495. doi:10.1177/0047287514522878

Hosany, S., Prayag, G., Van Der Veen, R., Huang, S., & Deesilatham, S. (2017). Mediating Effects of Place Attachment and Satisfaction on the Relationship between Tourists' Emotions and Intention to Recommend. *Journal of Travel Research, 56*(8), 1079–1093. doi:10.1177/0047287516678088

Hosany, S., & Witham, M. (2010). Dimensions of Cruisers' Experiences, Satisfaction, and Intention to Recommend. *Journal of Travel Research, 49*(3), 351–364. doi:10.1177/0047287509346859

Hsieh, S., & Chang, A. (2016). The Psychological Mechanism of Brand Co-creation Engagement. *Journal of Interactive Marketing, 33*, 13–26. doi:10.1016/j.intmar.2015.10.001

Ivanovic, M., & Saayman, M. (2015). Authentic economy shaping transmodern tourism experience. *African Journal for Physical, Health Education. Recreation and Dance*, (Supplement), 24–36.

Jalis, M. H., Che, D., & Markwell, K. (2014). Utilising Local Cuisine to Market Malaysia as a Tourist Destination. *Procedia: Social and Behavioral Sciences, 144*, 102–110. doi:10.1016/j.sbspro.2014.07.278

Jennings, G. (2005). Perspectives on quality tourism experiences: An introduction. In G. Jennings & N. P. Nickerson (Eds.), *Quality Tourism Experiences* (pp. 1–21). Elsevier Ltd., doi:10.4324/9780080455778

Jensen, R. (2001). *The Dream Society: How the Coming Shift from Information to Imagination will Transform Your Business* (2nd ed.). McGraw-Hill.

Jepson, D., & Sharpley, R. (2015). More than sense of place? Exploring the emotional dimension of rural tourism experiences. *Journal of Sustainable Tourism, 23*(8–9), 1157–1178. doi:10.1080/09669582.2014.953543

Kapoor, A. (2015). *Consumer Experiences and Emotion Management* (N. Malhotra, Ed.). Business Expert Press.

Kastenholz, E., Carneiro, M., Marques, C., & Loureiro, S. (2017). The dimensions of rural tourism experience: Impacts on arousal, memory, and satisfaction. *Journal of Travel & Tourism Marketing, 35*(2), 1–13. doi:10.1080/10548408.2017.1350617

Killion, G. (1992). *Understanding tourism*. Study Guide.

Kim, J., & Fesenmaier, D. (2015). Designing Tourism Places: Understanding the Tourism Experience through our Senses. In *Tourism Travel and Research Association - Turning Insights into Actions, the Crucial Role of Tourism Research* (p. 5). Travel and Tourism Research Association. Retrieved from https://scholarworks.umass.edu/ttra/ttra2015/Academic_Papers_Oral/19

Kim, J.-H. (2010). Determining the factors affecting the memorable nature of travel experiences. *Journal of Travel & Tourism Marketing, 27*(8), 780–796. doi:10.1080/10548408.2010.526897

Kim, J.-H. (2014). The antecedents of memorable tourism experiences: The development of a scale to measure the destination attributes associated with memorable experiences. *Tourism Management, 44*, 34–45. doi:10.1016/j.tourman.2014.02.007

Kim, J.-H. (2018). The Impact of Memorable Tourism Experiences on Loyalty Behaviors: The Mediating Effects of Destination Image and Satisfaction. *Journal of Travel Research, 57*(7), 856–870. doi:10.1177/0047287517721369

Kim, J.-H., & Jang, S. (2016). Memory Retrieval of Cultural Event Experiences: Examining Internal and External Influences. *Journal of Travel Research, 55*(3), 322–339. doi:10.1177/0047287514553058

Kim, J.-H., & Ritchie, J. (2014). Cross-Cultural Validation of a Memorable Tourism Experience Scale (MTES). *Journal of Travel Research, 53*(3), 323–335. doi:10.1177/0047287513496468

Kim, J.-H., Ritchie, J., & McCormick, B. (2012). Development of a scale to measure memorable tourism experiences. *Journal of Travel Research*, *51*(1), 12–25. doi:10.1177/0047287510385467

Kim, J.-H., & Youn, H. (2017). How to Design and Deliver Stories about Tourism Destinations. *Journal of Travel Research*, *56*(6), 808–820. doi:10.1177/0047287516666720

Kirillova, K., Lehto, X., & Cai, L. (2017). What triggers transformative tourism experiences? *Tourism Recreation Research*, *42*(4), 498–511. doi:10.1080/02508281.2017.1342349

Krishna, A. (2012). An integrative review of sensory marketing: Engaging the senses to affect perception, judgment and behavior. *Journal of Consumer Psychology*, *22*(3), 332–351. Advance online publication. doi:10.1016/j.jcps.2011.08.003

Krishna, A., & Schwarz, N. (2014). Sensory marketing, embodiment, and grounded cognition: A review and introduction. *Journal of Consumer Psychology*, *24*(2), 159–168. doi:10.1016/j.jcps.2013.12.006

Kurtz, J., & Welch, C. (2016). The Emotion of Happiness. In H. L. Meiselman (Ed.), *Emotions Measurement*. Elsevier Ltd. doi:10.1016/B978-0-08-100508-8.00020-5

Larsen, S. (2007). Aspects of a Psychology of the Tourist Experience. *Scandinavian Journal of Hospitality and Tourism*, *7*(1), 7–18. doi:10.1080/15022250701226014

LeDoux, J. (2012). Rethinking the Emotional Brain. *Neuron*, *73*(4), 653–676. doi:10.1016/j.neuron.2012.02.004 PMID:22365542

Lee, S., Heere, B., & Chung, K. (2013). Which Senses Matter More? The Impact of Our Senses on Team Identity and Team Loyalty. *Sport Marketing Quarterly*, *22*(4), 203–213. http://search.ebscohost.com/login.aspx?direct=true&db=bth&AN=93709410&lang=pt-br&site=ehost-live

Lemon, K., & Verhoef, P. (2016). Understanding Customer Experience Throughout the Customer Journey. *Journal of Marketing*, *80*(6), 69–96. doi:10.1509/jm.15.0420

Lempert, K., & Phelps, E. (2016). Affect in Economic Decision Making. In L. F. Barrett, M. Lewis, & J. M. Haviland-Jones (Eds.), *Handbook of Emotions* (4th ed., Vol. 1, pp. 98–112). The Guilford Press.

Light, D. (2009). *The Senses: The Human Body, How It Works*. Chelsea House Publishers.

Lin, Z., Chen, Y., & Filieri, R. (2017). Resident-tourist value co-creation: The role of residents' perceived tourism impacts and life satisfaction. *Tourism Management*, *61*, 436–442. doi:10.1016/j.tourman.2017.02.013

Martins, J., Gonçalves, R., Branco, F., Barbosa, L., Melo, M., & Bessa, M. (2017). A multisensory virtual experience model for thematic tourism: A Port wine tourism application proposal. *Journal of Destination Marketing & Management*, *6*(2), 103–109. doi:10.1016/j.jdmm.2017.02.002

Marx, G., & Gilon, C. (2012). The Molecular Basis of Memory. *ACS Chemical Neuroscience*, *3*(8), 633–642. doi:10.1021/cn300097b PMID:23050060

Meacci, L., & Liberatore, G. (2015). Towards a Senses-Based Model for Experimental Tourism: The YOUTOOSCANY.COM Case. *Proceedings 1st International Conference on Experiential Tourism*, (10), 9–11.

Mendes, J., Guerreiro, M., & Matos, N. (2016). Monitoring the Quality of Tourism Experience. *Proceedings of the Asia Tourism Forum 2016 - the 12Th Biennial Conference of Hospitality and Tourism Industry in Asia, 19*(1), 294–304. 10.2991/atf-16.2016.44

Mizumori, S., Yeshenko, O., Gill, K., & Davis, D. (2004). Parallel processing across neural systems: Implications for a multiple memory system hypothesis. *Neurobiology of Learning and Memory, 82*(3), 278–298. doi:10.1016/j.nlm.2004.07.007 PMID:15464410

Morgan, M. (2006). Making space for experiences. *Journal of Retail & Leisure Property, 5*(4), 305–313. doi:10.1057/palgrave.rlp.5100034

Mossberg, L. (2007). A Marketing Approach to the Tourist Experience. *Scandinavian Journal of Hospitality and Tourism, 7*(1), 59–74. doi:10.1080/15022250701231915

Nadel, L., Hupbach, A., Gomez, R., & Newman-Smith, K. (2012). Memory formation, consolidation and transformation. *Neuroscience and Biobehavioral Reviews, 36*(7), 1640–1645. doi:10.1016/j.neubiorev.2012.03.001 PMID:22465050

Nasution, R., Sembada, A., Miliani, L., Resti, N., & Prawono, D. (2014). The Customer Experience Framework as Baseline for Strategy and Implementation in Services Marketing. *Procedia: Social and Behavioral Sciences, 148*, 254–261. doi:10.1016/j.sbspro.2014.07.041

Nickerson, N. (2005). Some Reflections on Quality Tourism Experiences. In G. Jennings & N. P. Nickerson (Eds.), *Quality Tourism Experiences* (pp. 227–236). Elsevier Ltd.

Oh, H., Fiore, A., & Jeoung, M. (2007). Measuring Experience Economy Concepts: Tourism Applications. *Journal of Travel Research, 46*(2), 119–132. doi:10.1177/0047287507304039

Park, S., & Santos, C. (2017). Exploring the Tourist Experience. A Sequential Approach. *Journal of Travel Research, 56*(1), 16–27. doi:10.1177/0047287515624017

Pawaskar, P., & Goel, M. (2014). A Conceptual Model: Multisensory Marketing and Destination Branding. *Procedia Economics and Finance, 11*(11), 255–267. doi:10.1016/S2212-5671(14)00194-4

Peñaloza, L., & Venkatesh, A. (2006). Further evolving the new dominant logic of marketing: From services to the social construction of markets. *Marketing Theory, 6*(3), 299–316. doi:10.1177/1470593106066789

Pentus, K., Mehine, T., & Kuusik, A. (2014). Considering Emotions in Product Package Design through Combining Conjoint Analysis with Psycho Physiological Measurements. *Procedia: Social and Behavioral Sciences, 148*, 280–290. doi:10.1016/j.sbspro.2014.07.044

Perritano, J. (2011). *Science of Emotions*. Marshall Cavendish Benchmark.

Pezzi, E., & Vianna, S. (2015). Tourist Experience and Experiential Tourism: A study about dimensions of memorable experience. *Turismo Em Análise, 26*(1), 165–187. doi:10.11606/issn.1984-4867.v26i1p165-187

Phelps, E. A. (2004). Human emotion and memory: Interactions of the amygdala and hippocampal complex. *Current Opinion in Neurobiology, 14*(2), 198–202. doi:10.1016/j.conb.2004.03.015 PMID:15082325

Pine, J., & Gilmore, J. (1998). Welcome to the Experience Economy. *Harvard Business Review, 76*(4), 97–105. PMID:10181589

Pine, J., & Gilmore, J. (1999). *The Experience Economy: Work is Theater & Every Business a Stage*. Harvard Business School Press.

Prayag, G., Hosany, S., Muskat, B., & Del Chiappa, G. (2017). Understanding the Relationships between Tourists' Emotional Experiences, Perceived Overall Image, Satisfaction, and Intention to Recommend. *Journal of Travel Research, 56*(1), 41–54. doi:10.1177/0047287515620567

Prayag, G., Hosany, S., & Odeh, K. (2013). The role of tourists' emotional experiences and satisfaction in understanding behavioral intentions. *Journal of Destination Marketing & Management, 2*(2), 118–127. doi:10.1016/j.jdmm.2013.05.001

Prebensen, N., Vitterso, J., & Dahl, T. (2013). Value Co-creation Significance of Tourist Resources. *Annals of Tourism Research, 42*, 240–261. doi:10.1016/j.annals.2013.01.012

Prebensen, N., & Xie, J. (2017). Efficacy of co-creation and mastering on perceived value and satisfaction in tourists' consumption. *Tourism Management, 60*, 166–176. doi:10.1016/j.tourman.2016.12.001

Pyke, S., Hartwell, H., Blake, A., & Hemingway, A. (2016). Exploring well-being as a tourism product resource. *Tourism Management, 55*, 94–105. doi:10.1016/j.tourman.2016.02.004

Quadri-Felitti, D., & Fiore, A. (2013). Destination loyalty: Effects of wine tourists' experiences, memories, and satisfaction on intentions. *Tourism and Hospitality Research, 13*(1), 47–62. doi:10.1177/1467358413510017

Rahmann, H., & Rahmann, M. (1992). *The Neurobiological Basis of Memory and Behavior*. Springer. doi:10.1007/978-1-4612-2772-4

Rais, N., Musa, R., & Muda, M. (2016). Reconceptualisation of Customer Experience Quality (CXQ) Measurement Scale. *Procedia Economics and Finance, 37*(16), 299–303. doi:10.1016/S2212-5671(16)30128-9

Ramaswamy, V., & Ozcan, K. (2018). What is co-creation? An interactional creation framework and its implications for value creation. *Journal of Business Research, 84*, 196–205. doi:10.1016/j.jbusres.2017.11.027

Ribes-Giner, G., Perello-Marín, M., & Díaz, O. (2016). Co-creation Impacts on Student Behavior. *Procedia: Social and Behavioral Sciences, 228*, 72–77. doi:10.1016/j.sbspro.2016.07.011

Salmela, M. (2014). *True Emorions* (R. Ellis & P. Zachar, Eds.). John Benjamins Publishing Company.

Schmitt, B. (1999). Experiential marketing. *Journal of Marketing Management, 15*(1–3), 53–67. doi:10.1362/026725799784870496

Schmitt, B., Brakus, J., & Zarantonello, L. (2015). From experiential psychology to consumer experience. *Journal of Consumer Psychology, 25*(1), 166–171. doi:10.1016/j.jcps.2014.09.001

Servidio, R., & Ruffolo, I. (2016). Exploring the relationship between emotions and memorable tourism experiences through narratives. *Tourism Management Perspectives, 20*, 151–160. doi:10.1016/j.tmp.2016.07.010

Shaw, C. (2005). *Revolutionize Your Customer Experience* (1st ed.). Palgrave Macmillan. doi:10.1057/9780230513457

Shoval, N., Schvimer, Y., & Tamir, M. (2018). Real-Time Measurement of Tourists' Objective and Subjective Emotions in Time and Space. *Journal of Travel Research, 57*(1), 3–16. doi:10.1177/0047287517691155

Stasiak, A. (2013). Tourist Product in Experience Economy. *Tourism (Zagreb), 23*(1), 27–35. doi:10.1016/j.tourman.2015.08.005

Stein, A., & Ramaseshan, B. (2016). Towards the identification of customer experience touch point elements. *Journal of Retailing and Consumer Services, 30*, 8–19. doi:10.1016/j.jretconser.2015.12.001

Sthapit, E. (2017). Exploring tourists' memorable food experiences: A study of visitors to Santa's official hometown. *Anatolia, 28*(3), 404–421. doi:10.1080/13032917.2017.1328607

Sthapit, E., & Coudounaris, D. (2018). Memorable tourism experiences: Antecedents and outcomes. *Scandinavian Journal of Hospitality and Tourism, 18*(1), 72–94. doi:10.1080/15022250.2017.1287003

Todd, R., Talmi, D., Schmitz, T., Susskind, J., & Anderson, A. (2012). Psychophysical and neural evidence for emotion enhanced perceptual vividness. *The Journal of Neuroscience: The Official Journal of the Society for Neuroscience, 32*(33), 11201–11212. doi:10.1523/JNEUROSCI.0155-12.2012 PMID:22895705

Tsai, C.-T. (2016). Memorable Tourist Experiences and Place Attachment When Consuming Local Food. *International Journal of Tourism Research, 18*(6), 536–548. doi:10.1002/jtr.2070

Tung, V., Lin, P., Zhang, H., & Zhao, A. (2017). A framework of memory management and tourism experiences. *Journal of Travel & Tourism Marketing, 34*(7), 853–866. doi:10.1080/10548408.2016.1260521

Tung, V., & Ritchie, J. (2011). Exploring the essence of memorable tourism experiences. *Annals of Tourism Research, 38*(4), 1367–1386. doi:10.1016/j.annals.2011.03.009

Tussyadiah, I. (2014). Toward a Theoretical Foundation for Experience Design in Tourism. *Journal of Travel Research, 53*(5), 543–564. doi:10.1177/0047287513513172

Tynan, C., & McKechnie, S. (2009). Experience marketing: A review and reassessment. *Journal of Marketing Management, 25*(5–6), 501–517. doi:10.1362/026725709X461821

Verhoef, P., Lemon, K., Parasuraman, A., Roggeveen, A., Tsiros, M., & Schlesinger, L. (2009). Customer Experience Creation: Determinants, Dynamics and Management Strategies. *Journal of Retailing, 85*(1), 31–41. doi:10.1016/j.jretai.2008.11.001

Vo Thanh, T., & Kirova, V. (2018). Wine tourism experience: A netnography study. *Journal of Business Research, 83*, 30–37. doi:10.1016/j.jbusres.2017.10.008

Volo, S. (2009). Conceptualizing Experience: A Tourist Based Approach. *Journal of Hospitality & Leisure Marketing, 18*(2–3), 111–126. doi:10.1080/19368620802590134

Wang, L., Tang, H., Liu, D., & Xing, C. (2017). Study on Customer Experience Dimension Based on Online Shopping reviews of MI ' s Products. In *MATEC Web of Conferences* (Vol. 100, pp. 1–7). Academic Press.

Zatori, A. (2013). *Tourism Experience Creation From a Business Perspective.* Corvinus University of Budapest. Retrieved from http://phd.lib.uni-corvinus.hu/801/7/Zatori_Anita_den.pdf

Zatori, A., Smith, M. K., & Puczko, L. (2018). Experience-involvement, memorability and authenticity: The service provider's effect on tourist experience. *Tourism Management, 67*, 111–126. doi:10.1016/j.tourman.2017.12.013

Zhang, H., Wu, Y., & Buhalis, D. (2018). A model of perceived image, memorable tourism experiences and revisit intention. *Journal of Destination Marketing and Management, 8*, 326–336. doi:10.1016/j.jdmm.2017.06.004

ADDITIONAL READING

Bakker, I., van der Voordt, T., Vink, P., & Boon, J. (2014). Pleasure, Arousal, Dominance: Mehrabian and Russell revisited. *Current Psychology (New Brunswick, N.J.), 33*(3), 405–421. doi:10.100712144-014-9219-4

Bjorklund, R. (2010). The Amazing Human Body: The Senses. Marshall Cavendish Benchmark, NY: Tarrytown.

Nalbantian, S., Matthews, P. M., & McClelland, J. L. (2011). The Memory Process: Neuroscientific and Humanistic Perspectives. The MIT Press, Massachusetts: Cambridge. Fayos-Solà, E. & Cooper, C. (2019). The Future of Tourism: Innovation and Sustainability. Springer International Publishing, Switzerland: Cham.

Skavronskaya, L., Moylea, B., Scott, N., & Schaffer, V. (2019). Collecting Memorable Tourism Experiences: How Do 'wechat'? *Journal of China Tourism Research*, •••, 1–23. doi:10.1080/19388160.2019.1656131

Tugade, M. M., Shiota, M. N., & Kirby, L. D. (2014). The Handbook of Positive Emotions. Guilford Press, NY: New York.

Yan, N., & Halpenny, E. A. (2019). Tourists' savoring of positive emotions and place attachment formation: A conceptual paper. *Tourism Geographies*, 1–21. Advance online publication. doi:10.1080/14616688.2019.1647454

key TERMS and definitions

Arousal: Comprehends the autonomic nervous system activation, is the level of mental alertness and physical activity and ranges from sleep through intermediate states of drowsiness and then alertness to frenzied excitement at the opposite extreme (e.g. sleep, inactivity, boredom, and relaxation at the lower end versus wakefulness, bodily tension, strenuous exercise, and concentration at the higher end).

Endocrinal Response: Is a chemical response at internal or endocrinal level caused by hormones released by glands directly into the circulatory system that fostering visceral changes.

Homeostasis: This is the condition of optimal functioning for the organism and includes the steady state of many variables maintained, such as: internal, physical and chemical.

Long-Term Memory: Information is held in mind for a long period of time, between days, months or all life. The long-term memory is an outcome of the information resulting from short-term memory, plus a strong emotional connection, and a repetition process (or recall process).

Memorable Tourism Experience (MTE): Is a tourism experience remembered and recalled after the event has occurred.

Plasticity: Is the ability of the humans to change continually throughout is life as a result of experiences and is the level of individual's adaptability of the environment changes. These changes occur at the brain level (neuroplasticity), emotional, cognition, social, cultural, etc.

Sensation: Is the physical process during which sensory systems respond to a stimulus and provide sensorial data for stimulus perception. The sensation consists in five stages, i.e.: signal recognition, stimulus collection, transduction, transmission via afferent, central processing (brain), transmission via efferent and response/action (biochemical response and motor action).

Sensorial Memory: Know as bodily memory too, is a memory information from the sensorial stimulus inputs information and is held in mind during 0.2 to 2 seconds.

Short-Term Memory: Information is typically held in mind for a relatively short period of time, between 2 to 60 seconds, before being forgotten. This specifically memory is the resulting from sensorial memory, more an emotional outcome and attention.

Tourism Experience Relational Model: Is a model that allows to relate all those involved in the tourist experience establishment, such as: the experience dimensions, experience stages and actions, touchpoints, antecedents and outcomes to defining and framing the tourist experience in a holistic and comprehensive state.

Chapter 11
Measuring Attendees' Cultural Event Experiences:
Insights From 365 Algarve

Magda Wikesjö
https://orcid.org/0000-0002-2612-3741
Faculty of Economics, University of Algarve, Portugal

Manuela Guerreiro
https://orcid.org/0000-0002-6398-9712
Faculty of Economics and Research Centre for Tourism, Sustainability, and Well-Being (CinTurs), University of Algarve, Portugal

Patrícia Pinto
https://orcid.org/0000-0002-3153-2830
Faculty of Economics and Research Centre for Tourism, Sustainability, and Well-Being (CinTurs), University of Algarve, Portugal

Dália Paulo
Loulé's Municipality, Portugal

ABSTRACT

This chapter explores the concept of cultural event experience, as the interaction between an individual and the cultural event environment, altered by the level of engagement with multiple experiential elements and outputs, measured in four dimensions: experience novelty, affective engagement, cognitive engagement, and conative engagement. The purpose is to test the event experience scale (EES) as a valid measurement instrument for cultural event experiences, producing valuable insights for stakeholders and marketing management teams. The empirical study lays on a program of cultural and artistic events, the "365 Algarve," which was launched in a bottom-up approach to involve residents and local cultural groups to increase territory cohesion and to reinforce the brand image of Algarve. Using a sample drawn from the attendees present at those events, the results confirm the validity of EES with a new dimension emerged: experiential learning. The study reveals a possible new measurement instrument, the cultural event experience scale (CEES).

DOI: 10.4018/978-1-7998-3156-3.ch011

INTRODUCTION

Recent research is supporting that culture will play a major role in key fields as social cohesion, welfare, entrepreneurship and sustainability (Blessi et al., 2014). As the world faces enormous environmental and economic challenges, issues of culture and sustainability are moving to the forefront of planning, policy, and programs in communities of all sizes (Duxbury & Jeannotte, 2010). Culture is one of the pillars for sustainable development of societies (Duxbury & Jeannotte, 2010).

Events contribute to the 'soft infrastructure' of places that supports a higher subjective well-being (Richards, 2015). Getz (2008) refers, that some event values, promote community-building, cultural development and even national identities.

Cultural events are more than culture guardians of meaning, they create and deliver it. The shared experience imbedded in cultural events makes them singular (Richards and Palmer, 2010). And the non-instrumental character of cultural experiences leads to authentic personal engagement and social relationships (Blessi et al., 2014).

The Algarve, in the recent past, has been the stage of several cultural events. In 2005, Faro was the National Capital of Culture (Ferreira & Martins, 2005) and in 2007, the Portuguese government, introduced a program of events called *Allgarve* (Guerreiro et al.,2011). The Festival MED-World Music created in 2004 to support the European Football Championship (Grenhas, 2013) is still alive in the city of Loulé.

The 365 Algarve cultural program is a set of cultural and artistic events that was launched in 2016, with a budget of 1,5 million euros funded by a partnership between Culture and Tourism Secretaries and implemented by the Algarve Tourism Board. 365 Algarve's main purpose is to foster territory cohesion in the region, involve residents and local cultural groups in different projects, densify the cultural offer in the low touristic season (from October 2016 to May 2017) and reinforce the brand Algarve. Municipalities (public initiatives) and Cultural Agents (public and private initiatives) in the Algarve are the proponents and creators of the planned events, as a bottom-up approach to initiate and sustain the event program (AA.VV., 2016). Due to its different pack of cultural events (Music, Opera, Performing Arts, Theatre, Visual arts, Music/Gastronomy, Cinema, Animation of Heritage, Literature and Contemporary Circus), it was considered the perfect setting to implement the research project.

The objective of this study is to test the Event Experience Scale (EES) (Geus, Richards & Toepel, 2016) in cultural events by collecting insights from the attendees of the 365 Algarve's events. In Geus, Richards & Toepel's (2016) study respondents were asked to remember events they attended until one year ago. In this research a questionnaire was developed and administrated to attendees, handed live at events and collected *in situ*.

An Exploratory Factor Analysis (EFA) followed by a Confirmatory Factor Analysis (CFA) will be used to inspect the measurement model and check the reliability and validity of the constructs of the EES in the context of the 365 Algarve events.

Organizations in the cultural/art sector need to increase their knowledge of the *who* and *why* of consumer behaviour, to engage in stronger branding and positioning strategies and to make better use of data technology for marketing purposes (Colbert, 2003). Hopefully, CEES will contribute with valid insights to support marketing management decisions regarding cultural events.

BACKGROUND

The Experiential View

"The experiential view" introduced by Holbrook and Hirschman (1982) enters the timeline of consumption research, adding focus on the symbolic, hedonic, and esthetic nature of consumption, enlightening the previous studies of consumer behaviour, where consumer was mainly an information processor.

Pine and Gilmore (1998) saw value in the experience of consumption and presented "The experience economy", hence the consumer experience was a fundamental construct. Individuals were to be engaged in personal and memorable experiences/events. The superlative experiences were designed with four realms: entertainment, education, aesthetic and escapist. Two dimensions of consumer experience are referred, the first dimension is *customer participation*, ranging from passive participation, when customers are observers or listeners, to active participation, when customers have active important roles in the event that produces the experience. The second dimension is *connection*, describing the environmental relationship between customers and the event, ranging from absorption to immersion, depending on the degree of involvement offered by the environment.

In the marketing field, Schmitt (1999) proposes five *Strategic Experiential Modules* approaching the concept of Customer Experience: *sense* (sensory experiences), *feel* (affective experiences), *think* (cognitive experiences), *act* (physical experiences), and *relate* (social/community experiences).

Gentile et al. (2007), drawing from previous literature and technology advances, proposes a multidimensional structure of the Customer Experience, with the following dimensions: *sensory*, *emotional*, *cognitive*, *pragmatical*, *lifestyle* and *relational*. Although similar to the Customer Experience dimensions proposed by Schmitt (1999), there are some differences, the integration of *act* in the *sensory* dimension, the introduction of the *pragmatic* dimension to represent the interaction of humans with objects, and the *lifestyle* dimension, representing experiences related to personal belief and values.

Schmitt (2010) examines the meaning of experience in different contexts: philosophy, psychology, consumer behaviour and marketing. Experience definitions fall in two categories: those referring to the past knowledge and accumulating experiences over time and others referring to present "here and now", like the perceptions, feelings and thoughts of individuals engaged in consumption activities (Schmitt, 2010:70). Boud, Cohen & walker (1993) posit that experience contains judgement, thought and connectedness with other experiences, it is not isolated sensing. Even in its elementary form, it involves perception and it implies consciousness, it always comes with meaning. Schmitt, Brakus & Zarantonello (2014) refer that meaning originate mostly from intellectual and social experiences and Li (2000) concluded that meaning is the only element in common in all definitions of tourism experience.

Csikszentmihalyi (1990) posits that meanings are related to the type of setting and activity, which influences how perceptions of challenge and skill are formed, "flow", or the optimal experience, occurs when challenge and skill meet both above the normal, learning is more likely to occur in the state of arousal, when challenge is higher than skill. Arousal is the intensity of physiological response to a stimulus on the range from calmness to excitement (Oh et al., 2007)

Schmitt (2010) recognises that experience is a multidimensional concept containing *sensory-affective*, *cognitive-intellectual* and *behavior and action-oriented* components, and that empirical studies show "overlapping" of components, causing unidimensional problems for measurement scales of experience.

In the same conceptual line Ellis, Jamal & Jiang (2015) in a "Situated Tourist Experience", define experience as "the integration of environmental stimuli with motivation, emotion, cognition, attention and behaviour during a defined period of time."

In the marketing field, experience economy concepts inspired the development of approaches and scales to measure the experience, several authors designed and tested models in different contexts, Oh, Fiore & Jeoung (2007) measured *experience economy* concepts in tourism. Brakus, Schmitt & Zarantonello (2009) have constructed a *brand experience* scale with four experience dimensions: sensory, affective, intellectual and behavioural. Vila-López & Rodríguez-Molina (2013) studied the *event experience* related to brand experience in the arts context. Khan & Rahman (2017) developed a scale to measure the *hotel brand experience* that covered all the guest touch points.

Considering that primarily the operationalization of the Experience Economy implies creating events, Geus, Richards & Toepel (2016: 277) developed an event experience scale conceiving that event experience is " *an interaction between an individual and the event environment (both physical and social), modified by the level of engagement or involvement, involving multiple experiential elements and outputs (such as satisfaction, emotions, behaviors, cognition, memories and learning) that can happen at any point of the event journey*." This concept supports the construct of the Event Experience Scale (EES) with four dimensions: affective engagement, cognitive engagement, conative engagement and experiencing novelty. These dimensions are consistent to those experiential dimensions suggested by Getz (2008): three inter-related dimensions referring to what individual do (conative), individual emotions (affective) and individual perceptions (cognitive) during an event.

The Event Experiences

Getz (2008) defines an event as an occurrence, a special and meaningful situation, happening at a specific place and time. And Ayob, Wahid, & Omar, (2011) suggest that events must be seen in a contemporary approach, as an experiential concept, there for is important to understand or assess attendees' experiences in the event environment. Lively events become 'time markers' and offer opportunities for social interactions, which is shown to support the subjective well-being (Richards, 2015).

Events can provide a space and time away from ordinary life in which extraordinary experiences can happen (Morgan, 2008). Deepening this approach, Schouten et al. (2007) studied extraordinary experiences which include extreme emotions that are often communal in nature, and transformational for consumers.

Blessi et al. (2014) emphasizes the benefits of communities participating in cultural events, by recognizing that they feel and get subjective empowerment, reinforcing the decision to participate to the gatherings and increasing their subjective enjoyment. They note that in a cultural event context, like a musical performance, the design of the set and the events' program aims to produce emotions on the audience and the most effective performances are those which generate a sense of intimacy. In this ambience, the audience feel emotionally connected with performers, producing 'resilience', a 'sense of wellness' and reaching 'those emotional qualities which have the capacity to enhance people's quality of life.'

On the other hand Colbert (2003) emphasizes that is not necessary to alter the core product but to observe the auxiliary or secondary product as a means of attracting customers and improve their event experience, regarding the importance of the adequacy of the physical and professional environment. "The provision of places to meet, socialize, and wind-down before and after the main performances is also an important but overlooked element. This means that the surrounding area, its residents, and traders play a part in creating the event experience" (Morgan, 2008:92).

Schmitt (2010), in his deep research of experience, concluded that understanding the relation between experience and happiness needs to be addressed so that marketing and public policy makers, can overcome utilitarian value, improve well-being and life quality of consumers and citizens.

One of experience's most researched outcome is satisfaction, Rojas & Camarero (2008) confirmed in their study that cognitive (quality perceived) and emotional (pleasure) are complementary and determinant for attendee satisfaction in a cultural context, emotional aspects can gain or lose importance for cognitive aspects, depending on the accumulated experience of attendee and after more judgement is required. To Giese & Cote (2000) satisfaction is also an emotional or cognitive response, focused for instance in experience or expectations, which occurs at a particular time, and could be after accumulated experiences.

Another important outcome of experience involves learning and education, Cutler & Carmichael (2008) in their revision of literature found four categories of learning and skill development in tourist experiences, all related to personal discoveries, cognitive, affective, psychomotor and personal development (discovery of the self). At the personal level, learning shapes on learners' previous life experience, engages and stimulates reflection on experience, opening horizons to new experience and, thereafter, continuous learning (Andresen, Boud & Cohen, 1999).

Emotions design our vision of the world, our sense of time and space. Imagination and future experience change in relation to current emotional state. Pre-experienced emotions will condition later-experienced emotions, and continuously (re)design tourism experience (Davidson and Milligan 2004). "Measuring traveller's emotions in real time and in natural settings represents an important advancement for the design and management of tourism experiences" (Kim & Fesenmaier 2015:419).

To have a rich social life increases the level of happiness, and the average of experiential purchases are rewarding (Gilovich, 2015),

MAIN FOCUS OF THE CHAPTER

Methodology

Sample and Data Collection

For a two month period (March and April 2017), during the 1st Edition of "365 Algarve" cultural program, attendees of selected events were live invited to respond *in situ* to a questionnaire as part of a marketing research project to characterise the cultural event experience and test the EES.

The sample size calculation used the most conservative estimate for a sample proportion (p=0.5), a 95% confidence level and a maximum sample error of 5%. It was concluded, therefore, that the sample size should be at least 384 respondent attendees.

The selection of events where the questionnaire would be applied was made to obtain a representative sample of the cultural events offered by "365 Algarve". The events, and their typologies (music, theatre, cinema, visual arts, music-gastronomy, heritage animation and performative arts), were randomly selected during the period of the research, considering all events announced in the various typologies. The selected events were spread throughout the Algarve and occurred in different equipment and schedules.

Attendees who were assisting events of 365 Algarve, during the period under research, were invited at random to participate just before the beginning of each event and informed about the survey objectives

Figure 1. Sample distribution by event typology

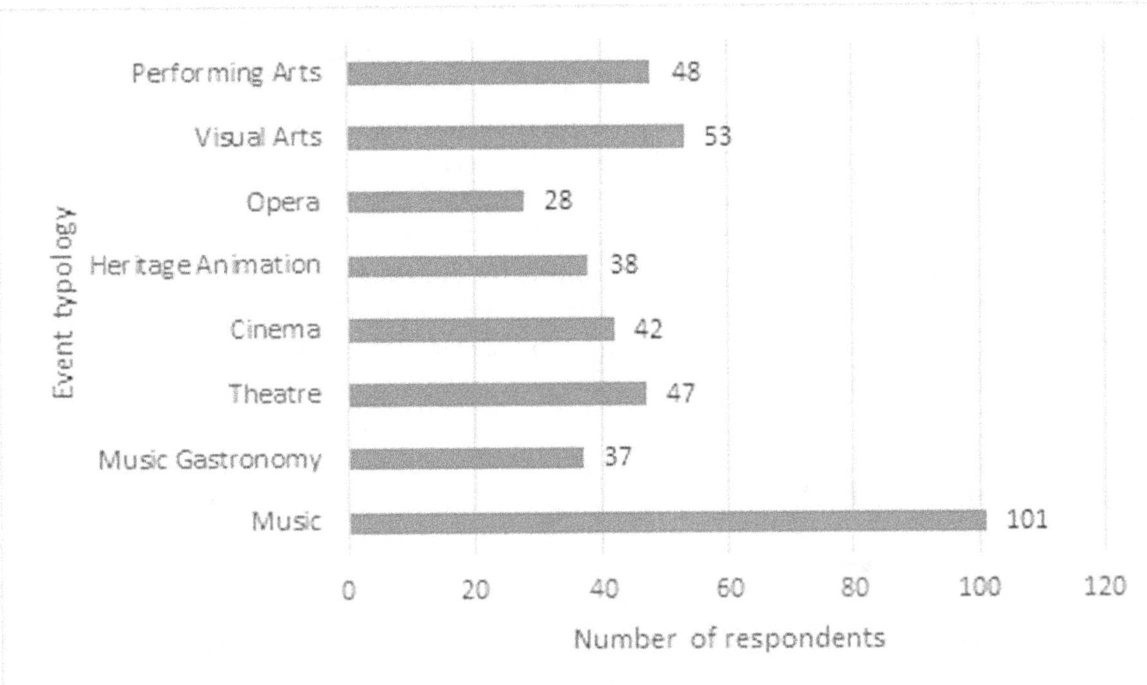

and scope, two interviewers, including the author, handed the questionnaire in the language preferred by the respondent.

Respondents had to fill in the questionnaire *in situ* and hand it back to the interviewer. Younger respondents, under 14 - years old, were not included in the study. Ultimately, a sample of 394 validated questionnaires was obtained.

The questionnaire comprised four sections: section I, with 19 items of the original event experience scale (EES) (Geus, Richards & Toepel, 2016); section II, including 6 items to measure the arousal and memory scales adapted from Oh, Fiore & Jeoung (2007); section III, with 6 items to measure behavioural intentions (adapted from Bonn et al., 2007); and section IV, asking about the sociodemographic profile of the respondent. The questionnaire was translated from English to four languages (Portuguese, Spanish, French and German) and it was revised by native speakers/specialists for each language, an academic in Marketing field and a professional in Cultural Management. The current study concerns sections I and IV.

All items were measured using a 7-point semantic differential or Likert-type scale with anchors of 1 = strongly disagree to 7 = strongly agree.

Data Analysis

Data analysis starts with an exploratory factor analysis (EFA) on the 19 items of EES. This analysis uses the Statistical Package for Social Sciences (SPSS) software, version 22, and intends to find the major of dimension (factors) that characterize EES in the specific setting of cultural events such as the 365 Algarve. Subsequently, confirmatory factor analysis structural, using the software AMOS 20.0 software, was performed to validate the factorial solution provided by the EFA.

Based on literature, a scale with construct validity have to encounter several criteria: (a) content validity- the measurement items are relevant to and representative of the construct's domain; (b) unidimensional - a set of items measure a single construct; (c) reliability – a set of items is internally consistent/homogenous; (d) validity – a set of items measuring a variable correlate or statistically converge together (convergent validity), a set of items measuring a latent variable discriminate from others measuring other variables (discriminant validity): (e) nomological validity – constructs relate to other constructs in a predictable manner. (Churchill, 1979; Gerbin and Anderson, 1988; O'Leary-Kelly and Vokurka, 1998; Ping, 2004).

To perform data analysis, it was used the Statistical Package for Social Sciences (SPSS) software, version 22. An Exploratory Factor Analysis will inspect the structure and internal consistency of the theoretical measurement model, followed by a Confirmatory Factor Analysis to perform the validity test and examine the quality of adjustment of the measurement model, using AMOS 20.0 software with maximum likelihood technique.

ANALYSIS AND FINDINGS

Demographic Information

The final sample included 394 valid questionnaires. As table 1 shows, 62,5% of the total sample were female, and the average age is 48,32 years old (with a standard deviation of 16,6 years old). The youngest respondent was 14 years old and the oldest was 90 years old. Most attendees is higher educated (73,3%) and professionally active (75,9%). Regarding the origins of the attendees they are most Portuguese (59,8%) and permanent residents (69,2%). These last results are in accordance with the positioning of 365 Algarve and the brand positioning of Algarve as tourism destination (Andraz, Silva and Viegas, 2012).

Exploratory Factor Analysis

The sample adequacy for the analysis was confirmed with the Kaiser-Meyer-Olkin measure (KMO = .925) and the Bartlett's test of sphericity [$\chi 2$ (171) = 3136.378; p = .000]. Results indicated that the correlations between items were satisfactorily large to justify the use of EFA. Factors were extracted using the Principal Component Factor Analysis (PCA).

Based on the observation of the eigenvalues and following Kaiser´s criterion, five factors were retained (they had eigenvalues over than 1), explaining 71,16% of the data variability. Table 2 shows the component loadings after rotation (Oblimin with Kaiser Normalization).

The items grouping on the first factor suggested that they represent *experiencing novelty* (distinctiveness of event, unfamiliarity, uniqueness and adventure). Factor 2 represents *cognitive engagement* (cognition/thinking, intellect, interpretation), meaning the understanding of the event (Getz,2008). Factor 3, named as *affective engagement* (excitement, emotional energy, values and intimacy), refers to emotions and moods dimension (Getz,2008);

Factor 4 was designated as *Experiential Learning* (learning and knowledge acquisition), following Pine & Gilmore (1999), and considering that education is one of the principal realms of the experience. Finally, factor 5 was named *conative engagement* (behaviour, active participation, creativity, multisensory, recollection and reflexivity), meaning what persons did during the event and following Getz's (2008)

Table 1. Respondent characteristics (n=394)

Respondent characteristic		Percentage
Gender	Male	37.5
	Female	62.5
Age	Mean	48.32
	Mode	67
	St. Deviation	16.558
	Minimum	14
	Maximum	90
Education level	Primary education	2.1
	Secondary education	24.5
	Higher education	73.3
Profession	Student	10.4
	Active	75.9
	Retired	13.7
Nationality	Portuguese	59.8
	Other	40.2
Residency	Temporary	30.8
	Permanent	69.2

Source: Author

study. These five factors report an adequate internal consistency, with Cronbach's alphas ranging from .777 to .889.

In Figure 2 it can be can observed, in a comparative way, the dimensions and contents obtained with the study (assessing cultural events *in situ*) *versus* those obtained by Geus, Richards & Toepel (2016) (EES). In the current study, a new dimension - *Experiential learning* - emerged.

Confirmatory Factor Analysis

CFA was then applied to validate the theoretical measurement model CEES, resulting from the empirical study, and including the component *Experiential learning* (revealed by the EFA). As shown in Table 3, the results indicate that the data successfully fit the model (Hair et al, 2010). Indeed, χ^2 /df = 2,641 (2 < χ^2 / df < 5 means acceptable fit), RMR = 0.83 (< 0.08 means good fit), RMSEA = 0.065 (< 0.1 means good fit), GFI = 0.917, AGFI = 0,877; NFI = 0,920; CFI = 0.948, IFI = 0.949 (all > 0.90, meaning good fit

Once evaluated the overall fit of the CFA model, the reliability and validity of the constructs were analysed (Hair et al, 2010). As shown in Table 4, the standardized factor loadings of all measurement indicators range from .621 to .918. These values, higher than 0,5, reveal the individual internal consistency of the indicators. Moreover, the composite reliability indexes range from .766 to .900, assuring internal consistency of each construct.

To assess the convergent validity of the constructs, the significance of factor loadings was assessed. As presented in table 4, all factor loadings are statistically significant (p = 0.00). To complement the

Table 2. Summary of EFA (oblique rotation) results for the EES (n=330)

[Concept] Item (during the event...)	*Rotated component loadings*				
	Experiencing novelty	*Cognitive engagement*	*Affective engagement*	*Experiential learning*	*Conative engagement*
[Uniqueness] *I thought this was unique*	.827				
[unfamiliarity] *I experienced things that were unknown to me/I was outside my 'normal life'*	.814				
[Distinctiveness of event] *I had impressions that the events were different from others*	.751				
[Adventure] *I felt a sense of adventure*	.726				
[Cognition/thinking] *I was thinking*		.898			
[intellect] *I used my intellect*		.875			
[Interpretation] *I was explaining/interpreting things for myself*		.824			
[Emotional energy] *I got emotional energy/got emotionally recharged*			.854		
[Excitement] *I was excited*			.848		
[Value(s) creation] *I was aware of my own values*			.645		
[Intimacy] *I experienced intimacy*			.642		
[Learning] *I learned something*				.734	
[Knowledge acquisition] *I acquired new Knowledge*				.720	
[Behaviour] *I was active*					.861
[(Active) participation] *I actively participated*					.851
[Reflexivity] *I reflected on ideas that I got and discussed this with others*					.805
[Creativity] *I was being creative*					.773
[Multisensory] *All my senses were stimulated (taste, sight, hear, smell, touch)*					.699
[Recollection] *I wanted to relive this later (talk about it with others and show pictures)*					.667
Eigenvalues	8.983	1.355	1.174	1.057	.951
Percentage of variance	47.280	7.131	6.181	5.561	5.005
Cronbach's alpha	.801	.846	.777	.904	.889

Source: Author

Figure 2. Theoretical model CEES

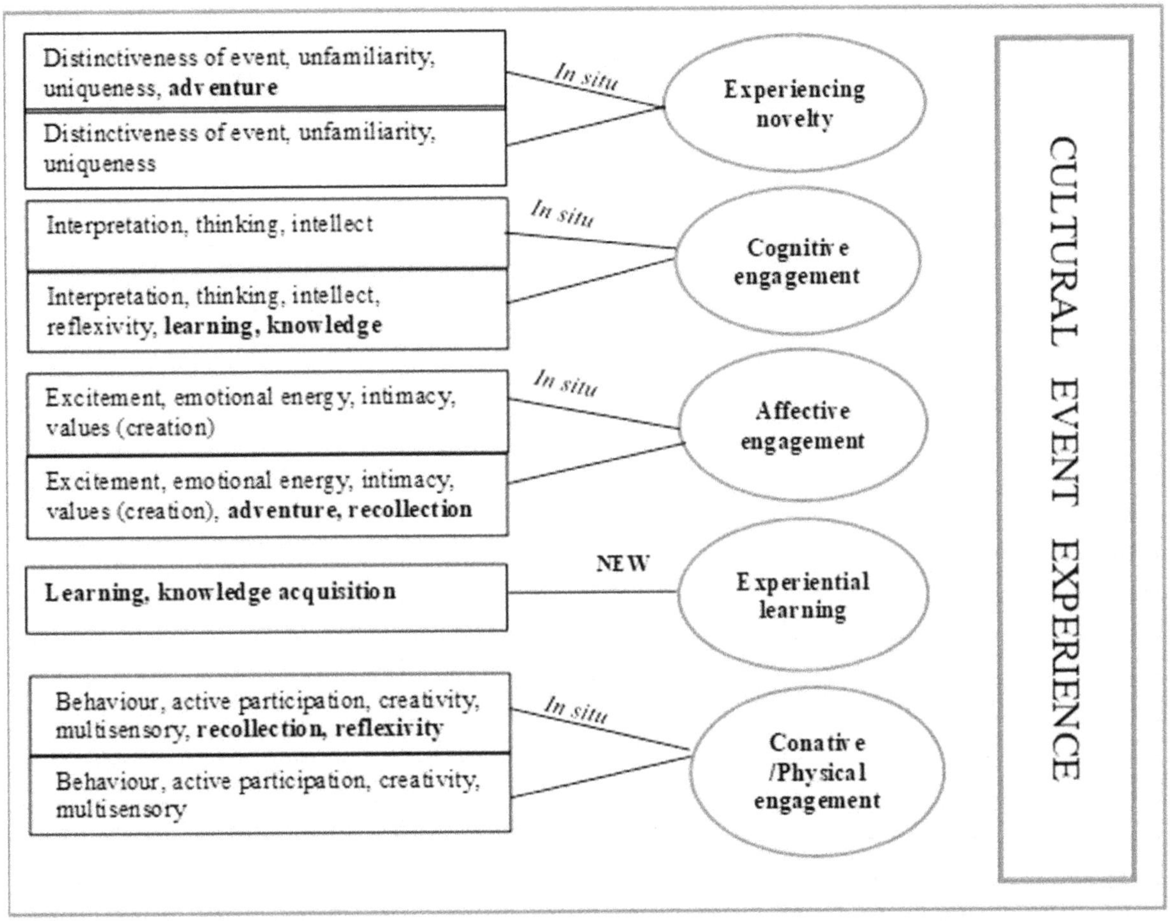

analysis of convergent validity, the average variance extracted (AVE), that should exceed 0.5, was observed. In the study, the AVEs range from .450 to .818, which means that one construct (*Affective engagement* factor) presents an AVE slightly lower than the reference value. Discriminant validity is demonstrated if the squared root of each AVE exceeds the correlations between one construct and the other constructs (Anderson & Gerbin, 1988), which was found in the model (table 5)

Table 3. Goodness-of-fit for the Original and Modified Measurement Model

			Absolute Fit Measures			Incremental Fit Measures		Parsimonious Fit Measures	
	χ2	df	GFI	RMR	RMSEA	AGFI	NFI	CFI	IFI
Model CEES	338,079 P=, 000	128	,917	,083	,065	,877	,920	,948	,949

Note: GFI=goodness of fit; RMSR=root mean square residual; RMSEA=root mean square error of approximation; AGFI=adjusted goodness-of-fit index; NFI=normed fit index; CFI=comparative fit index; IFI=incremental fit index; RFI=relative fit index.

Source: Author

Table 4. Confirmatory Factor Analysis for the measurement Model CEES

Construct/ Indicators	Completely Standardized Loading	t value	Composite Reliability	Average Variance Extracted Estimate
Experiencing novelty			.815*	.526
I thought this was unique	.709			
I experienced things that were unknown to me/ I was outside my 'normal life'	.663	12.648		
I had impressions that the events were different from others	.690	10.920		
I felt a sense of adventure	.829	11.640		
Cognitive Engagement			.869*	.688
I was thinking	.793			
I used my intellect	.856	13.856		
I was explaining/ interpreting things for myself	.838	13.579		
Affective Engagement			.766*	.450
I got emotional energy/got emotionally recharged	.713			
I was excited	.675	14.373		
I was aware of my own values	.621	10.732		
I experienced intimacy	.671	11.570		
Experiential learning			.900*	.818
I learned something	.891			
I acquired new Knowledge	.918	21.256		
Conative engagement			.871*	.530
I was active	.693			
I actively participated	.728	15.286		
I reflected on ideas that I got and discussed this with others	.708	14.299		
I was being creative	.762	13.591		
All my senses were stimulated (taste, sight, hear, smell, touch)	.680	12.500		
I wanted to relive this later (talk about it with others and show pictures)	.790	13.951		

*$p < .001$
Source: Author

SOLUTIONS AND RECOMMENDATIONS

Richards & Lanuza (2017) did a selection of surveys conducted in the framework of the ATLAS Event Experiences Project, showing the use of EES to profile and distinguish event experiences between different locations and types of events. These authors refer that different events produces variations of results across the 18 items of the scale, probably influenced by type of event and its capacity to engage and

Table 5. Correlation matrix for model CEES and square rooted AVE

	Experiencing novelty	*Cognitive engagement*	*Affective engagement*	*Experiential learning*	*Conative engagement*
Experiencing novelty	**.726***				
Cognitive engagement	,485**	**.829***			
Affective engagement	,589**	,527**	**.671***		
Experiential learning	,610**	,504**	,502**	**.905***	
Conative engagement	,716**	,581**	,694**	,666**	**.728***

** Correlation is significant at the 0,01 level (two-tailed).
Source: Author

activate visitors. These surveys showed a high level of emotional and cognitive dimensions stimulation and relatively less for active and novelty dimensions.

The study confirmed the four dimensions of EES (Experiencing novelty, Cognitive engagement, Affective engagement and Conative engagement) and a new dimension has emerged, Experiential learning. Some items migrated in between dimensions: Adventure, Recollection and Reflexivity.

Cronbach Alpha technique was used to measure the reliability (degree of precision or accuracy) of the EES and across the five dimensions, Cronbach alpha ranged from .777 to .889 which validates the internal consistency of dimensions.

The item Adventure, considered a state of mind (Priest & Bunting,1993), migrated from the dimension *Affective engagement* to *Experience novelty*, and the authors find evidence of this association when Priest & Bunting (1993) in their study of perceived risk and competence in adventure activity, imposed that all experiences in the survey "should feel like an adventure (be new and exciting)", the essential constituent of adventure is uncertainty and in the conclusion of the study they left a question: Are novelty and adventure interchangeable concepts? Boyle (2003) refers that participants in "Adventure-Based Training" when placed in unfamiliar environment are stimulated to use mental skills that generally they do not use in a daily basis, also Human (2012) analysing the reports of participants in adventure-based experiences, found that 'discovering uniqueness' was part of the essence of their personal growth. To define adventure, Cambridge Dictionary (2017) gives the following example: "Reading a book should be an adventure, a personal experience full of surprises".

Recollection and Reflexivity both migrated to *Conative engagement,* this dimension refers to actual behaviour or what people do during the event (Getz,2008) and the phrases in the questionnaire evoking these concepts are respectively: I wanted to relive this later (talk about it with others and show pictures), I reflected on ideas I got and discussed this with others. When assessing events onsite like the study did, these phrases suggest actions, things done during the event.

Multisensory (elements), this item maintained its position in the *conative dimension,* was excluded before for presenting a low Cronbach's alpha .61, in current study obtained .699. Sensory stimuli are important for experience recollection (Tung & Ritchie, 2011), consumer motivation is more influenced by subconscious emotional and sensory elements rather than tangible attributes of a product or service (Carbone 1998).

The fourth component that arose from the current PCA was *Experiential Learning,* Learning and Knowledge acquisition, were items grouped before (EES) in *cognitive engagement*, in current study gained dimension significance. The education realm (Pine & Gilmore, 1998) seem to be found. A crucial component of experiential learning is that learners analyse their experience considering prior experience, to extract meaning from it (Andresen, Boud & Cohen, 2000). Learning is influenced by the socio-emotional context in which it occurs (Boud, Cohen & Walker, 1993) Most persons learn when reaching the state of arousal when the challenges are somewhat higher than the skills needed, and they can easily move into "flow" developing their skills (Csikszentmihalyi, 2004).

Gilovich, (2015) argue that finding meaning in experiences contributes to happiness, discussing that social policy makers should target experiential investments providing hedonic value and positive emotions, therefore improving societal well-being.

In a touristic region as the Algarve it was important to have respondents of various nationalities, also an opportunity to test results against the original sample of EES where 94,3% were Dutch.

FUTURE RESEARCH DIRECTIONS

Dew to restrictions of time and budget, this study could not address the full pack of cultural events of 365 Algarve. Art projects like "Lavrar o Mar" involving co-creation processes with residents and attendees, were not included. Hopefully following research will address this gap.

The authors hope to proceed in the research to study the relationships between the five dimensions of CEES with other latent variables like *eventscape* or with memory and behaviour, aiming to enhance the predicting character of the measurement instrument.

Future directions for event experience research suggested by Greg Richards (2019) are the development of improved measurement tools, to explore the relationships between different elements of the event experience, the co-creation of value through event experiences or the application of new technologies to event experience research.

CONCLUSION

Research has repeatedly considered experiences at the core of events, a key concern for event management and design, focusing on how experiences are produced and consumed within events (Richards, 2019). Therefore the assessment approach based on the experiential concept is an efficient way to measure the outcome of an event, help managers to understand the nature of the attendees' personal event experience in event settings, emphasized the importance of the "memorable experience" (Ayob et al., 2011).

To know who the consumers are, what motivates them to select a certain cultural/art event and how the consumers experience the event is critical for Cultural Marketing Management.

As the authors have earlier explored, experience is a precursor of outcomes like memory, behaviour intentions and satisfaction, to measure the experience of consumers is then relevant. No less relevant is to select an adequate experience measurement instrument and the present study has hopefully addressed this issue.

As conclusion the present study points out what Brakus et al. (2009:54) noted in their research for brand experience dimensions: "there is no one-to-one correspondence, such that a certain stimulus type would trigger a certain experience dimension and only that dimension." The positive result obtained for the multisensory item opens the question if other items (among the total of 56 in the exploratory phase of EES) should be recovered or substitute other items and tested in specific event contexts. The emergence of the new component *Experiential Learning* suggests a new instrument of measurement, the Cultural Event Experience Scale (CEES) and the validity test confirms its adequacy to measure event experience in specific cultural settings.

Finally, the authors hope that CEES can deliver useful information, helping marketing management teams to take better decisions concerning the "secondary products" of Arts and target the right publics/ audiences.

ACKNOWLEDGMENT

This chapter is financed by National Funds provided by FCT- Foundation for Science and Technology through project UIDB/04020/2020.

REFERENCES

Adventure. (2017). In *Cambridge Dictionary online*. Retrieved from https://dictionary.cambridge.org/dictionary/english/adventure

Andraz, J. M., Silva, J. A., & Viegas, C. M. (2012). The position of tourism activity in the Algarve's economic structure: An integrated region-specific analysis. *Revista Portuguesa de Estudos Regionais, 29*(1), 15–26.

Andresen, L., Boud, D., & Cohen, R. (1999). Experience – Based Learning. In G. Foley (Ed.), *Understanding Adult Education and Training* (pp. 225–239). Allen & Unwin.

Ayob, N., Wahid, N., & Omar, A. (2011). A measurement model of visitor's event experience within festivals and special events. In *Proceeding of the 5th International Conference of the Asian Academy of Applied Business*, (vol. 9, pp.1-10). Phhom Penh, Cambodia.

Blessi, G. T., Grossi, E., Sacco, P. L., Pieretti, G., & Ferilli, G. (2014). Cultural Participation, Relational Goods and Individual Subjective Well-Being: Some Empirical Evidence. *Review of Economics & Finance*, *4*(8), 33–46.

Bonn, M., Joseph-Mathews, S., Dai, M., Hayes, S., & Cave, J. (2007). Heritage/Cultural Attraction Atmospherics: Creating the Right Environment for the Heritage/Cultural Visitor. *Journal of Travel Research*, *45*(3), 345–354. doi:10.1177/0047287506295947

Boud, D., Cohen, R., & Walker, D. (1993). *Using Experience for Learning*. Open University Press.

Boyle, T. I. (2003). *The Impact of Adventure-Based Training on Team Cohesion and Psychological Development in Elite Sporting Teams*. Parkland, FL: Dissertation.com.

Brakus, J., Schmitt, B., & Zarantonello, L. (2009). Brand experience: What is it? How is it measured? Does it affect loyalty? *Journal of Marketing*, *73*(3), 52–68. doi:10.1509/jmkg.73.3.052

Carbone, L. P. (1998). Total customer experience drives value. *Management Review*, *87*(7), 62.

Churchill, G. A. Jr. (1979). A Paradigm for Developing Better Measures of Marketing Constructs. *JMR, Journal of Marketing Research*, *16*(1), 538–562. doi:10.1177/002224377901600110

Colbert, F. (2003). Entrepreneurship and Leadership in Marketing the Arts. *International Journal of Arts Management*, *6*(1), 30–39.

Cronbach, L. (1951). Coefficient alfa and the internal structure of tests. *Psychometrika*, *16*(3), 297–334. doi:10.1007/BF02310555

Csikszentmihalyi, M. (1990). *Flow: The Psychology of Optimal Experience*. Harper Perennial.

Csikszentmihalyi, M. (2004). *Flow: the secret to happiness*. Retrieved from TED talks Web site: www.ted.com/talks/mihaly_csikszentmihalyi_on_flow/transcript#t-1119507

Cutler, S. Q., & Carmichael, B. A. (2010). The dimensions of the tourist experience. In M. Morgan, P. Lugosi, & B. Ritchie (Eds.), *The Tourism and Leisure Experience: Consumer and Managerial Perspectives* (pp. 3–26). Channel View Publications. doi:10.21832/9781845411503-004

Davidson, J., & Milligan, C. (2004). Embodying Emotion Sensing Space: Introducing Emotional Geographies. *Social & Cultural Geography*, *5*(4), 523–532. doi:10.1080/1464936042000317677

de Rojas, C., & Camarero, C. (2008). Visitors' experience, mood and satisfaction in a heritage context: Evidence from an interpretation centre. *Tourism Management*, *29*(3), 525–537. doi:10.1016/j.tourman.2007.06.004

Duxbury, N., & Jeannotte, S. M. (2010). Culture, Sustainability, and Communities: Exploring the Myths. *Oficina do CES*, *353*, 1–26.

Ellis, G. D., Jamal, T., & Jiang, J. (2015). Fast-Thinking and Slow-Thinking: A Process Approach to Understand Situated Tourist Experiences. *Tourism Travel and Research Association: Advancing Tourism Research Globally*, *36*, 1–14.

Ferreira, A. M., & Martins, A. I. (Eds.). (2005). *O Evento FCNC 2005 e o Turismo*. Faro, Portugal: Centro de Estudos da ESGHT, Universidade do Algarve.

Gerbing, D. W., & Anderson, J. C. (1988). An Updated Paradigm for Scale Development Incorporating Unidimensionality and Its Assessment. *JMR, Journal of Marketing Research*, *25*(2), 186–192. doi:10.1177/002224378802500207

Getz, D. (2008). Event Tourism: Definition, evolution, and research. *Tourism Management*, *29*(3), 403–428. doi:10.1016/j.tourman.2007.07.017

Getz, D. (2014). Foreword. In L. Sharples, P. Crowther, D. May, & C. Orefice (Eds.), *Strategic Event Creation* (p. vi). Goodfellow Publishers Limited.

Geus, S., Richards, G., & Toepel, V. (2016). Conceptualisation and Operationalisation of Event and Festival Experiences: Creation of an Event Experience Scale. *Scandinavian Journal of Hospitality and Tourism*, *16*(3), 274–296. doi:10.1080/15022250.2015.1101933

Giese, J. L., & Cote, J. A. (2000). Defining Consumer satisfaction. *Academy of Marketing Science Review*, *2000*(1), 1–24.

Gilovich, T., Kumar, A., & Jampol, L. (2015). A wonderful life: Experiential consumption and the pursuit of happiness. *Journal of Consumer Psychology*, *25*(1), 152–165. doi:10.1016/j.jcps.2014.08.004

Grenhas, M. C. (2013). *Motivação, Satisfação e Fidelização dos Visitantes do Festival Mediterrâneo de Loulé* (Unpublished master's thesis). University of Algarve, Faro, Portugal.

Guerreiro, M., Valle, P., & Mendes, J. (2011). Allgarve events: Implications for the Algarve Image. *Tourism (Zagreb)*, *59*(2), 183–202.

Hair, J., Anderson, R., Tatham, R., & Black, W. (2010). *Multivariate Data Analysis*. Prentice Hall- Pearson.

Human, L. (2012). Adventure-Based Experiences during Professional Training in Psychology: A Follow-up Study. *South African Journal of Psychology. Suid-Afrikaanse Tydskrif vir Sielkunde*, *42*(4), 586–597. doi:10.1177/008124631204200412

Li, Y. (2000). Geographical consciousness and tourism experience. *Annals of Tourism Research*, *27*(4), 863–883. doi:10.1016/S0160-7383(99)00112-7

Likert, R. (1974). The Method of Constructing an Attitude Scale. In G. M. Maranell (Ed.), *Scalling: a sourcebook for behavioural scientists* (pp. 90–96). Transaction Publishers.

O'Leary-Kelly, S. W., & Vokurka, R. J. (1998). The Empirical Assessment of Construct Validity. *Journal of Operations Management*, *16*(4), 387–405. doi:10.1016/S0272-6963(98)00020-5

Oh, H., Fiore, A. M., & Jeoung, M. (2007). Measuring experience economy concepts: Tourism applications. *Journal of Travel Research*, *46*(2), 119–132. doi:10.1177/0047287507304039

Pine, J., & Gilmore, J. H. (1998). Welcome to the experience economy. *Harvard Business Review*, *76*(4), 97–105. PMID:10181589

Pine, J., & Gilmore, J. H. (1999). *The experience economy: Work is theatre & every business a stage*. Harvard Business School Press.

Ping, R. A. Jr. (2004). On Assuring Valid Measures for Theoretical Models Using Survey Data. *Journal of Business Research*, *57*(2), 125–141. doi:10.1016/S0148-2963(01)00297-1

Priest, S., & Bunting, C. (1993). Changes in Perceived Risk and Competence During Whitewater Canoeing. *Journal of Applied Recreation Research*, *18*, 265–280.

Richards, G. (2015). Eventfulness and the quality of life. *Tourism Today (Nicosia)*, *14*, 23–36.

Richards, G. (2017). Measuring event experiences: An international view. In G. Richards & A. R. Lanuza (Eds.), *Experiencias turisticas de festivals y eventos* (pp. 11–27). Colección Pasos.

Richards, G. (2019). Event experience research directions. In J. Ambrecht, E. Lumdberg, & T. Andersson (Eds.), *A research agenda for event management* (pp. 79–93). Edward Elgar Publishing Limited. doi:10.4337/9781788114363.00015

Richards, G., & Palmer, R. (2010). *Eventful Cities: Cultural Management and Urban Revitalisation.* Elsevier.

Richards, G., & Wilson, J. (2006). Developing creativity in tourist experiences: A solution to the serial reproduction of culture? *Tourism Management, 27*(6), 1209–1223. doi:10.1016/j.tourman.2005.06.002

Schmitt, B. (2010). Experience marketing: Concepts, frameworks and Consumer Insights. *Foundations and Trends in Marketing, 5*(2), 55–112. doi:10.1561/1700000027

The cultural program 365 Algarve. (2016). Cultural Program.

Tung, V. W. S., & Ritchie, J. R. B. (2011). Exploring the essence of memorable tourism experiences. *Annals of Tourism Research, 38*(4), 1367–1386. doi:10.1016/j.annals.2011.03.009

ADDITIONAL READING

Getz, D., & Page, S. (2019). *Event Studies: Theory, Research and Policy for Planned Events* (4th ed.). Routledge. doi:10.4324/9780429023002

Richards, G. (2018). Cultural Tourism: A review of recent research and trends. *Journal of Hospitality and Tourism Management, 36*, 12–21. doi:10.1016/j.jhtm.2018.03.005

Richards, G., & Marques, L. (2018). *Creating synergies between cultural policy and tourism for permanent and temporary citizens*. United Cities and Local Governments.

KEY TERMS AND DEFINITIONS

Construct: The subject matter that one wishes to measure using survey questions, the construct is defined by its dimensions.

Cultural Settings: The global scenarios or environments, where cultural events occur including complementary activities, publics and professionals.

Experience: The individual response to the stimulus of a specific environment during a period of time.

In Situ: This Latin expression was used to emphasise that respondents answered the questionnaire directly at the exit of the cultural event they have attended.

Multidimensional: refers to a concept that is explained by several dimensions, interrelated but distinct in between them.

Planned Event: Considering the tourism field a planned event is designed by professionals with an objective or function, differing from individual and community initiatives.

Chapter 12
Tourists' Awareness of World Heritage:
The Case of Tourists Visiting the Algarve (Portugal)

Claudia Helena Henriques
School for Management, Hospitality, and Tourism (ESGHT), & Research Centre for Tourism, Sustainability, and Well-Being (CinTurs), University of Algarve, Portugal

Isabel Teotónio
School of Management, Hospitality, and Tourism (ESGHT), University of Algarve, Portugal

ABSTRACT

This chapter offers an in-depth analysis of cultural tourism and namely of tourists' awareness of Portuguese World Heritage (WH). In accordance, firstly, there is the discussion of a theoretical framework associated to a cultural role and cultural identity in the sustainable development of tourism is presented. Secondly, it follows a case study approach based on the evaluation of Algarvian tourists' awareness of Portuguese WH. A questionnaire was applied in the Algarve region with the aim to determine the importance of cultural motivation and the level of knowledge regarding Portuguese culture by tourists. The results of this exploratory analysis underline the growing importance of cultural motivation and the recognition of culture's importance in the sustainable development of destinations. However, it also shows the low level of cultural motivation by Algarvian tourists as well as the low level of Portuguese WH knowledge, namely by foreign tourists. Nationality, age, academic background are key variables in tourism knowledge about Portuguese culture.

DOI: 10.4018/978-1-7998-3156-3.ch012

INTRODUCTION

This chapter aims at investigating the tourists' awareness of Portuguese World Heritage (WH), namely by the tourists of the most visited region of Portugal – Algarve, since the region plays an important role in Portuguese tourism and is placed in the top rankings among sun and sea tourism destinations. The importance of the cultural motivation and the level of knowledge regarding Portuguese culture by Algarvian tourists and the tourists interest in knowing more about the Portuguese culture are key purposes of this exploratory study, which intends to underline multiple interactions between tourism and the culture relationship associated to Portugal and its history/identity, based on WH. Additionally, World Heritage Sites (WHS) visitation or Intangible Cultural Heritage knowledge are also discussed.

To fulfil these objectives, the structure of the chapter includes the following sections: theoretical framework, methodology, results, discussion and conclusions.

In the first section, the theoretical framework discusses, on the one hand, the role of the culture in the tourist experience valorisation and, on the other hand, the importance of WH as an identity element. Subsequently, it discusses the importance of the culture's identity in the sustainable development of tourism.

The second section presents the methodological elements, mainly based on a quantitative research sustained by a questionnaire applied to Algarvian tourists. Additionally, it intends to investigate multiple interactions between tourism and culture in Portugal.

Regarding the results discussion, in the third section, the concern is to characterize Algarve's tourist profile and determine its level of "cultural motivation", "knowledge of Portuguese Culture", "intention to know more about Portuguese culture". Specifically associated with World Heritage (tangible and intangible), there is the concern to understand if tourists have visited classified WHS, if they intend to visit them in the future and their respective level of knowledge about each Portuguese WHS.

Finally, discussion and conclusions are presented.

THEORETICAL FRAMEWORK

"Tourism is the world's third largest export category after chemicals and fuels, and ahead of automotive products and food" (UNWTO, 2019, p. 8). In 2018, tourism "accounted for 10.4% of global GDP and (…) 10% of total employment [in the world]. The division of money spent overall is firmly weighted towards the leisure market, which represented 78.5% of the total compared with 21.5% for business spend" (WTTC, 2019: ii). In accordance, as UNWTO Secretary-General Zurab Pololikashv (in UNWTO, 2019, p. 2) underlines, "tourism is helping build better lives for millions of individuals and transforming whole communities" and consequently it is desirable that it continues growing "more in value rather than just in volume" (WTTC, 2019).

Documents such as "Tourism and the Sustainable Development Goals – Journey to 2030 – Highlights" (UNWTO/UN, 2017) and "Baseline Report on the Integration of Sustainable Consumption and Production Patterns into Tourism Policies" (UNWTO, 2019), point out the importance of tourism in the enhancement of sustainable development and the need of "interconnectivity" with other sectors, such as the cultural and creative sector. Once culture is presently recognised as the fourth dimension of sustainability, efforts are being enhanced to its integration across Sustainable Development Goals (SDGs)[1].

The 4th UNWTO/UNESCO (2019) "World Conference on Tourism and Culture: Investing in future generations" (in the sequence of Siem Reap, Muscat and Istanbul Declarations) concentrates "on technology, visitor management models and the approaches" that potentiate the widest possible "range of benefits to both visitors and locals, while safeguarding cultural values and heritage in line with responsible tourism principles and the UNWTO Global Code of Ethics for Tourism" (UNWTO, 2019a).

Tourism is an "encounter" with the "other". Consequently, a broader understanding should "interpret the encounter between destination and tourists as the convergence of emotions and activities being an existentially authentic process of interactions that may lead to experiences for both sides (…). It must also be clear that main or side tourism attractions are just expressions of a destination's culture and not its embodiment" (Gisolf, 2017).

This "encounter" is based on culture(s). Fuchs and Pikkemaat (2019, p. 5) highlight at least four different cultures, namely: "The culture of the sending region, the culture of the receiving region, the leisure (tourism) culture and the organisational culture of the tourism enterprises providing services".

Therefore, it is expected that "the cultural encounter does not take place within culturally homogenous groups", as postulated by Hofstede (1984 as cit. in Fuchs & Pikkemaat, 2019, p. 5), but within over the time varying and changing heterogeneous sub-cultural value-groups (Weiermair and Fuchs, 2003 as cit. in Fuchs & Pikkemaat, 2019, p. 5).

The cultural encounter presupposes the acknowledgement of the importance of identity, which is frequently linked to the tourism of a country and to the formation of powerful symbols and images, as "identifiers" of a particular country, in the framework of a "social construction" (Bandhari, 2014). According to Bandhari (2014, p. 23), the country "identifiers" "were always there and predate the arrival of tourism". However, in order to "produce a country" to its visitors, the identifiers are largely exploited. Consequently, "Rather than simply (re)constructing or destroying previously fixed and stable place identities, tourism contributes to on-going processes of change, whilst at the same time being mediated through the elements of continuity that exist within place identities" (Kneafsey, 1998: 114). As proposed by Oakes (1995 as cit. in Kneafsey, 1998, p. 114), "tourism should, therefore, be thought of as just another process through which localised identities are continually reconstructed". It can be seen as "an example of the unique way in which global-local relations are negotiated within the context of particular places, thus allowing for the maintenance of diversity and difference" (Kneafsey, 1998, p. 112).

In this context, heritage tourism has an important role to play. This type of tourism is considered a "carrier of identity" (Timothy, 2011) and sometimes it "acts as a mechanism of dissemination for nationalism and other ideologically loaded discourses" (Graham et al., 2000 as cit. in Bandhari, 2014). Heritage – tangible and intangible – natural and cultural, should be understood in the context of the complexity and dynamic of culture conceptualization.

Heritage fulfils many functions and has different uses and purposes. It can be a tool for governance, in defining and legitimizing identity, and has often been used in this sense in identity politics in nationalism, becoming "a material embodiment of the spirit of the nation" (Hall 2008, p. 219). For instance, Labadi (2007), in the framework of nation and cultural diversity in discourses on World Heritage, finds that "particular values were used to project carefully constructed images of the past, the nation and cultural diversity and that governments construct models of national collective identity based on monumentality, grandeur and heroism". Indeed, heritage and its discourse is linked to the exercise of power – not only meaning individuals who could state and decorate their power, wealth and social status with collections, but also the "symbolic power to order knowledge, to rank, classify and arrange, and thus to give meaning to objects and things through the imposition of interpretative schemas, scholarship and the authority of connoisseurship" (Hall, 2008, p. 220).

Heritage Tourism and World Heritage List of UNESCO

"Since the 1970s, heritage has been linked to the concept of and a discourse on 'value' for the purpose of conservation, protection and preservation. This new meaning and broader sphere is promoted by UNESCO" (Davison, 2008, p. 32).

The 1972 Convention's adoption introduced into international legislation, "the idea that some heritages in the world were of such importance that they were of value to all humanity, and that responsibility for their management was of more than national significance, even if the primary responsibility remained with individual nations" (UNESCO/ICCROM/ICOMOS/IUCN, 2016, p. 29). This is associated with the enlargement of the heritage concept "from a national and local level to an international level, by embracing the anthropological understanding of heritage as both values and the objects and sites they are embodied" (Davison, 2008, p. 32).

During the negotiations leading to the adoption of the 1972 Convention, "a number of state representatives shared the idea that the scope of that Convention was too narrow and that the actions of the international community in the field of cultural heritage should extend to its immaterial manifestations" (Lenzerini, 2011, p. 104). It followed a growing concern on safeguarding Intangible Cultural Heritage (ICH) and the 2003 Convention for the Safeguarding of ICH in the framework of a dynamic process that is been changing and broadening the understanding of culture. Culture becomes associated not only with 'high culture', but also with 'popular culture' and 'everyday culture'. Beyond its more tangible expressions, intangible expressions and lifestyles gain an increasing importance.

The UNESCO World Heritage List (WHL) is designed to protect global heritage (Frey et al., 2013) and has been continually evolving and growing. Following UNESCO (2019), it includes 1121 Properties; 39 Transboundary; 53 in Danger; 869 Cultural; 213 Natural and 39 Mixed, in 167 States Parties. Concerning the Representative List of the Intangible Cultural Heritage of Humanity, UNESCO (2019a) points out the existence of 508 elements corresponding to 122 countries.

The common feature of all properties inscribed on the WHL is that they meet the requirements for Outstanding Universal Value (OUV). This means: "Outstanding: properties (…) should be exceptional, or superlative - they should be the most remarkable places on earth; Universal: Properties need to be outstanding from a global perspective; Value: value, or the natural and/or cultural worth of a property. This value is determined based on standards and processes set out in the World Heritage Operational Guidelines" (Australian Government, 2019).

To be considered of OUV, a property needs to fulfil "one or more of ten criteria", meet the conditions of integrity, authenticity, and should have an adequate system of protection and management to safeguard its heritage. The ten criteria aggregate culture, nature and mixed (UNESCO, 2019). "Integrity relates to the 'wholeness and intactness' of the heritage property and how it conveys the values it holds (…)." The requirement for authenticity relates only to cultural properties. Broadly, it requires an original, authentic or true cultural expression of the values of a property, in material or conceptual form, and cannot be represented by a copy or recreation" (Australian Government, 2019).

Associate to OUV, as Santa Cruz and López-Guzmán (2017, p. 111) underline, is the fact "that inscription allows understanding that all the persons of the world are owners of (and are responsible for) that place or intangible element". In accordance, "The WHL reinforce the identification, preservation and transmission towards future generations of extraordinary places and monuments from the perspective of history and art" (Saipradist & Staiff, 2008 cit. in Santa Cruz & López-Guzmán, 2017, p. 111).

World Heritage (WH) site status "has often been lauded as a symbol of quality" (Adie, 2017), "coveted brand and seal of approval" (Ryan & Silvanto, 2010, p. 533), "branding" (Timothy, 2011), "labelling" (Yang, Ling & Hang, 2010). The WH brand signals property so irreplaceable that its values must be sustained intact in perpetuity. A primary function of the WH symbol, one element of the WH brand, is to prompt positive visitor emotions and behaviours favoured by management agencies. This is in line with the assumption that "Place identity is a kind of group identity" (Liu & Chang, 2016: 77), once "place can be seen as a group criterion, people who come from the same place is easily to form a group (Liu & Chang, 2016, p. 77). Desires to preserve ecological or architectural characteristics of a place have a direct impact on the strength of place attachment felt by individuals, notably through self-pride and self-identity". This is understandable due to the impressive cultural value of WH that could not be separated from other values, such as economic, social, environmental and associated impacts.

As Caust and Vecco (2017, p. 3) point out, "the UNESCO label has been generally used as a marketing tool to attract more tourists". Considering WH status impacts, "tourism" was considered in several studies cited by LDWHP (sd) such as" Pricewaterhouse Coopers LLP/DCMS 2008", "The Scottish Natural Heritage/Hambrey Consulting Study 2007", "The Professor Remy Prud'Homme/Word Heritage Centre Summary". Among these studies, the PwC, for instance, suggested "that the impact on tourism is marginal (…). Visitors awareness of WHS is often relatively low for existing sites" (LDWHP, sd, p. 6). On the other hand, the study entitled "The Scottish Natural Heritage 2007" suggested that "benefits vary widely between sites 'depending on the resource base of the site, the nature of the local economy, governance structures, and individuals involved". Among the areas of common benefits, the study points out: "Enhanced tourism image and profile" and "stimulus to awareness raising and educational initiatives" (Ibidem, p. 9). Prud'Homme study findings that highlight "tourism as the primary economic consideration, but considers tourism through its relationship with development, heritage and attractiveness of the location". The link between WHS inscription and local economic development is probably quite limited" (Ibidem, p. 10).

More recently, the growing acknowledgment of WHS tourism impacts is leading to a spreading discussion both in the supply and demand side. On the one hand, region residents (Jimura, 2011) or on-site employees (Aa, Groote & Hulgen, 2005) gain a sense of pride from a WHS designation and, on the other hand, it is "hypothesized that the WHS designation may have an impact on the visitor' perceptions of the site, its display and the on-site experience" (Poria, Riechel, & Cohen, 2013, p. 272). For instance, regarding demand side studies, there is the spread recognition that there is a "high" association between the classification of WHS and the rise of tourists visit (Breakey, 2012).

"Studies focusing on visits to WHS or UNESCO's designation process seldom relate to visitors' perceptions of the designation" (Poria, Riechel & Cohen, 2013, p. 272). The studies usually point out that WHS designation often results in an increased demand and that appeal for more planning and management, given the importance of the attraction value (Landorf, 2009).

Considering the relevance of WHS impacts, and that heritage "has a dynamic role in society and harnesses, but also delivers to others, the mutual benefits that such a role can create" (UNESCO/ICCROM/ICOMOS/IUCN, 2016), UNESCO/ICCROM/ICOMOS/IUCN (2016) developed manuals or publications providing guidance for States Parties. Among the thematics of guidance, it should point out the aims to: 1) strengthen the knowledge, abilities, skills and behaviour of people with direct responsibilities for heritage conservation and management; 2) improve institutional structures and processes through empowering decision-makers and policy-makers; and 3) introduce a dynamic relationship between heritage and its context that will lead to greater reciprocal benefits through an inclusive approach, such that outputs and outcomes follow on a sustainable basis.

METHODOLOGY

In order to investigate the Algarve tourists' awareness of Portuguese World Heritage, a theoretical framework is discussed, highlighting the importance of culture in sustainable tourism development, not only in traditional cultural tourist destinations but also in 'sun and sea' tourist destinations, as is the case of the Algarve.

The importance of the cultural motivation and the level of knowledge regarding Portuguese culture by Algarvian tourists, the tourists interest in knowing more about the Portuguese culture, WHS visitation or ICH knowledge, the future intention to visit and know more about Portuguese culture, are all key purposes of this exploratory study, not only in traditional cultural tourist destinations but also in 'sun and sea' tourist destinations.

Simultaneously, a quantitative research sustained by a questionnaire applied to the Algarvian tourists was developed. The questionnaire is composed of 13 questions divided into three parts:

1) The first included three questions, on a Likert Scale (very weak to very high / without opinion) that aimed to measure the level of cultural motivation regarding tourist visits to the Algarve, the degree of knowledge of Portuguese culture and to what extent respondents would like to know more about Portuguese culture;
2) The second contained two questions that considered the Portuguese elements included in the World Heritage List (WHL)[2], the representative list of intangible heritage[3] and the tentative list (inventory of properties which each State Party intends to consider for nomination)[4]. The objectives are to determine whether the tourist visited (Yes/No) the elements of the WHL or if he intends to visit them in the future (Yes/No). They also asked to indicate the degree of knowledge of the elements of these assets (on a scale of 1 – Not familiar with to 5 – Very familiar with);
3) The third and final part refers to the sociodemographic characterization of respondents and includes issues related to gender, age, nationality, level of education, marital status, type of accommodation, length of stay in Portugal and money spent on visiting/travelling to the Algarve).

This questionnaire was applied to 820 tourists in the historical centers of six Algarvian cities, namely Faro, Loulé, Tavira, Olhão, Portimão and Lagos, between September and November of 2018.

Since this is a convenience sample, therefore not random, no inferences are possible for the population. However, the results described in the next point are valid for the sample under study.

For data analysis, several specific statistical methodologies were used: 1) Univariate analysis for sample characterization (sociodemographic data and data related to travel); 2) Bivariate analysis in which the degree of knowledge of heritage by nationality was analyzed and t-tests were performed to verify if the differences found were statistically significant, a 5% theoretical significance will be used. A chi-square test was performed to verify whether or not there is an association between the degree of knowledge of culture and heritage and the variables "nationality", "age" and "level of education".

RESULTS AND DISCUSSION

In Portugal, the total contribution of Travel & Tourism to GDP was "17.3% of GDP in 2017, and is forecast to rise by 5.1% in 2018, and 2.4% pa to EUR44.7bn, 20.5% of GDP in 2028" (WTTC, 2019a, p. 1). These data enhance the tourism importance in the Portuguese economy.

Following IMPACTUR (2019), Algarve region attracted 32% [17,6 million] of overnights stays in Portugal, which according to country of residence, tourists are mainly from United Kingdom (30.3%), Portugal (21.9%) and Germany (10.4%).

The Algarve, known as the "last Mediterranean Riviera" ("última Riviera mediterrânea") (Orlando Ribeiro as cit. in Arruda, 1999, p. 21), has been recognised as an important "sun and sea" and "golf" destination and recently has received several awards, such as "Best Beach European Destination", "Best Golf Destination", "Melhor Região de Turismo do País" ("Best Tourism Association"), among others (WTA, 2018).

An Algarve tourist profile study (2016 as cit. in TP/UALG, 2017, p. 12) reveals "beach" (47%) as the main tourist attractive element of the region. Sport (6%) and relaxing (10%) in holiday time is also considered a motivation to travel to the Algarve. Regarding "culture", Valle, Guerreiro, Mendes and Silva (2011) point out that 87% of Algarve tourists search tourist experiences associated not only with beach, sun and sea; attributes such as rural and natural landscapes, gastronomy, touring and monuments are frequently enunciated. Henriques et al. (2014) also highlighted that cultural and creative companies, public entities (such as Direção Regional da Cultura do Algarve and Comissão de Coordenação e Desenvolvimento Regional), social and cultural associations, considered the importance in developing tourist products based on culture. Among the culture thematics suggested to enhance cultural tourism experiences in the Algarve region, the following are enunciated: "Sagres: Mith and History", "Heritage Islamo-Cristian", "Historic Centres", "Traditions and Creative Expressions", "Cultural Sea", "Mediterranean Diet".

In this framework, policy documents such as "Turismo 2020" [Tourism 2020] (TP-Governo de Portugal, 2015) and "Estratégia Turismo 2027" [Tourism Strtegy] (TP, 2017), point out the importance of culture to tourism. For instance, regarding Algarve, the document underlines the need of tourism products diversification based on distinctive elements ("elementos diferenciadores") (such as "History, culture and tradition" - "História, Cultura e Tradição") and qualifiers elements ("elementos qualificadores") ("authenticty, security, competitive quality"). Tourism strategy to the Algarve point out as resources and attributes ("recursos & ativos") elements such as "culture, traditions, gastronomy" and "natural, cultural and heritage", among others (TP-Governo de Portugal, 2015).

SAMPLE AND SOCIO-DEMOGRAPHIC PROFILE

The profile of the sample can be outlined as follows: the majority are adults (66.5%), male (52.0%), single (38.2%), with a higher education (52.0%), ages between 18 and 50, and foreign nationality (72.6%) of which 29.2% are English, 12.6% are German and 11.8% are French. The hotel was the most used type of accommodation (55.0%), expenses were less than 2,500 euros and eight days was the average stay (see Figures 1 – 7).

Culture Motivation Importance

Figure 1. Age

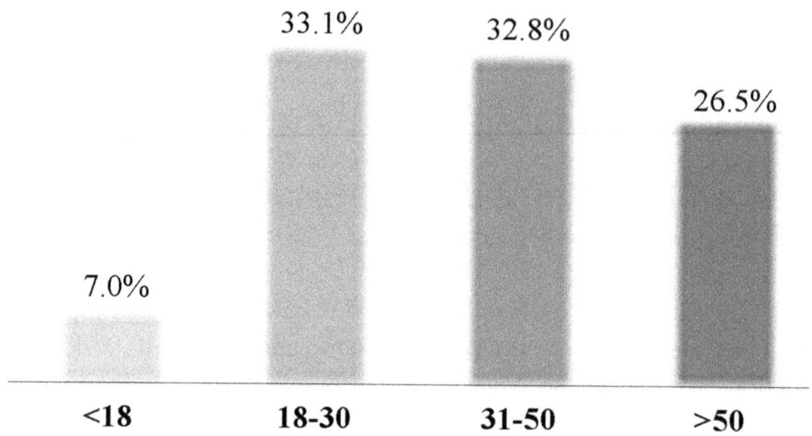

Figure 2. Gender

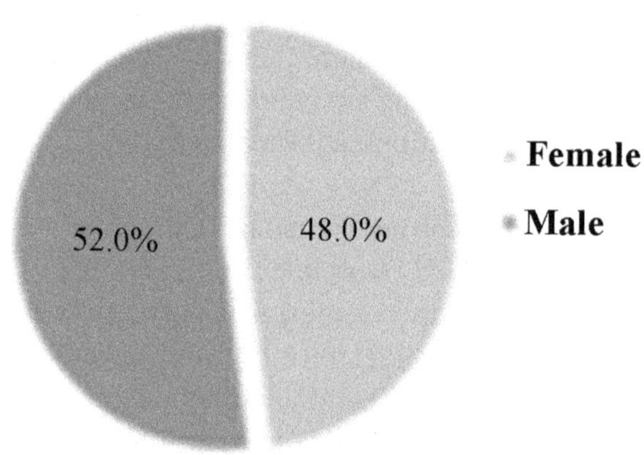

In a descriptive perspective, there seems to be no relevant differences in the motivation to visit the Algarve for both nationals and foreigners. Regarding the question: "To what extent was the visit to the Algarve motivated by culture?" Portuguese and Foreign tourists answered, "very weak" and "weak", respectively 44.2% and 39.8% (see Table 2). As far as young people are concerned, motivation is not clearly cultural, once the answer was "very weak" and "weak" respectively 63.2% (less than 18 years old) and 41.6% (between 18 and 30 years old) (Table 3).

Regarding tourist visitation or contact to Portuguese WH, as an expression and interest to Portuguese culture knowledge, Table 4 shows the five most visited sites (tangible heritage and tentative list) and all elements of intangible heritage. Table 5 shows the future intention of visiting or contacting with world heritage in the future.

Figure 3. Marital status

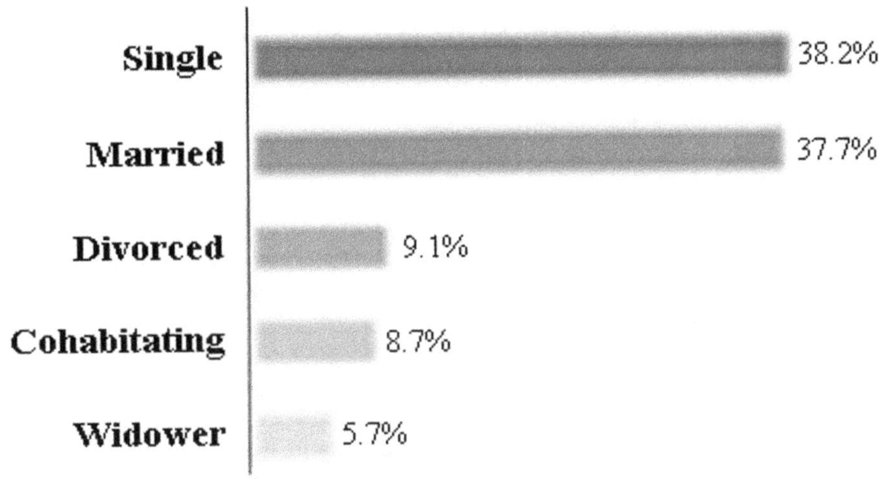

Figure 4. Nationality

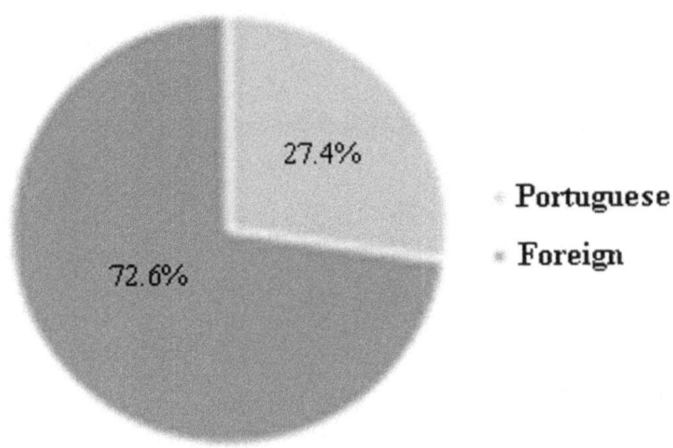

The most visited places are situated in the two major Portuguese cities (Lisbon and Porto) or in their surroundings (for instance: Sintra). Monastery of the Hieronymites and Tower of Belém in Lisbon are the most visited or known WH tangible monument by tourists (62.1%). Regarding the intangible heritage, the Mediterranean Diet reveals a greater degree of contact both globally (68.1%) and by foreigners (62.6%), as far as the Portuguese have had greater contact with Fado (91.8%). Also, as expected, the lowest level of contact occurs with the Manufacture of Cowbells, probably because it develops in a region far from large urban centers. Respondents also put in evidence their intention to visit in the future cultural attributes located mainly in Lisbon, Porto and in its surroundings (namely, Sintra, Mafra and Alto Douro vineyard). Regarding intangible heritage, tourists reveal greater interest in contacting Fado and a minor interest in contacting Manufacture of cowbells.

Figure 5. Level of education

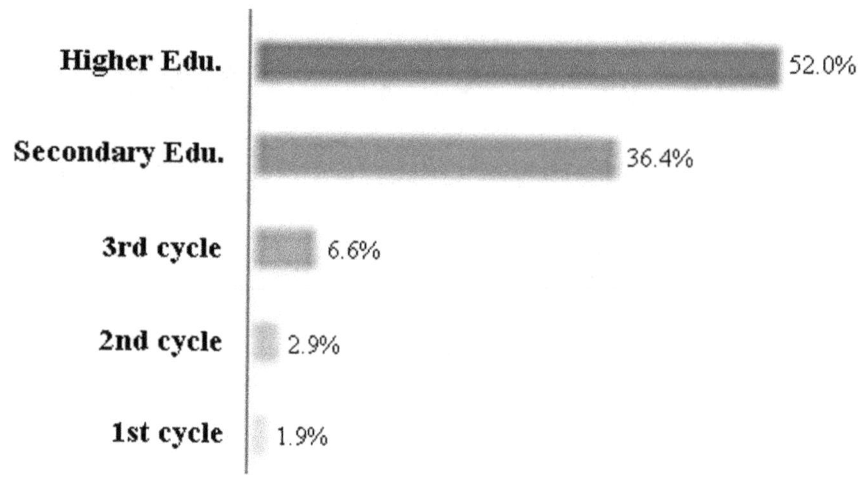

Figure 6. Type of accommodation

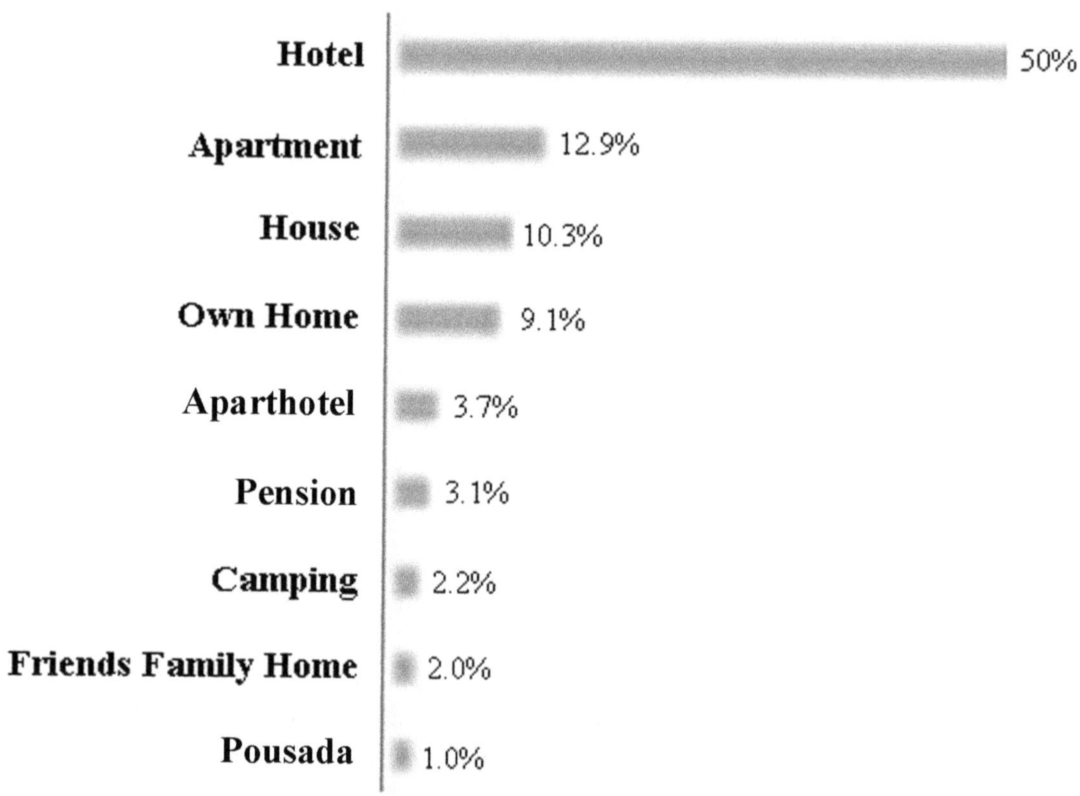

Figure 7. Spent on travel to the Algarve (euros)

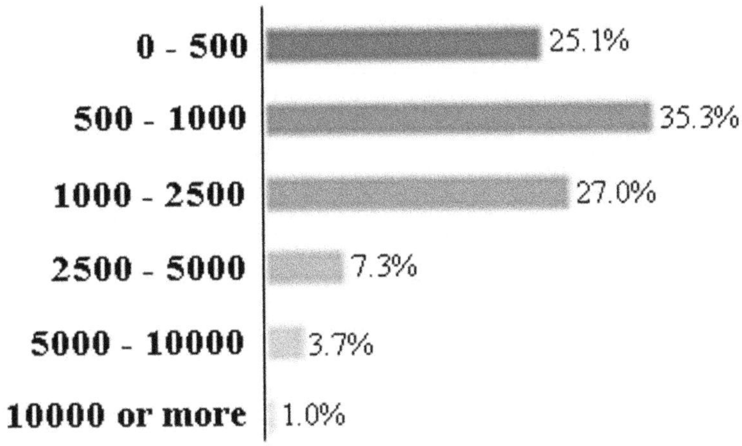

Table 1. Length of stay in Portugal

Mean	Median	Mode
8.3	7	7

Table 2. To what extent the visit to the Algarve was motivated by culture (by nationality)

		Nationality		
		Portuguese	**Foreign**	**Total**
The visit to the Algarve was motivated by culture	Very weak	17.0%	18.0%	17.7%
	Weak	27.2%	21.8%	23.3%
	Middle	**29.9%**	**27.7%**	28.3%
	Pretty much	12.5%	21.3%	18.9%
	Very high	7.1%	9.4%	8.8%
	No opinion	6.3%	1.7%	2.9%
	Total	100.0%	100.0%	100.0%

Regarding the "average level of WH knowledge" of tourists, Table 6 put in evidence its low level, both for nationals and foreigners [the question "average level of knowledge of heritage, by nationality" was classified from 1 (not familiar with) to 5 (very familiar with)].

For the elements of the heritage, the percentage of respondents who revealed a high degree of familiarity (quite or very familiar) with that element is also indicated.

When comparing the top five positions related to tangible heritage, we conclude that the average knowledge is very similar regarding the Portuguese and foreigners. The results reveal a higher knowledge

Table 3. To what extent the visit to the Algarve was motivated by culture - Age

		Age				
		<18	**18-30**	**31-50**	**>50**	**Total**
The visit to the Algarve was motivated by culture	Very weak	**40.4%**	15.3%	14.3%	19.4%	17.8%
	Weak	22.8%	26.3%	22.6%	21.3%	23.5%
	Middle	14.0%	**27.7%**	**30.1%**	**30.6%**	**28.3%**
	pretty much	5.3%	20.4%	22.2%	15.7%	18.7%
	Very high	7.0%	7.3%	10.2%	9.3%	8.7%
	No opinion	10.5%	2.9%	0.8%	3.7%	3.0%
	Total	100.0%	100.0%	100.0%	100.0%	100.0%

Table 4. Visited or contacted with the heritage

Heritage		Nationality		
		Portuguese	**Foreign**	**Total**
Tangible	Monastery of the Hieronymites and Tower of Belém in Lisbon	80.6%	55.2%	62.1%
	Cultural Landscape of Sintra	62.0%	37.0%	43.9%
	Monastery of Batalha	65.0%	34.6%	43.0%
	Historic Centre of Oporto, Luiz I Bridge and Monastery of Serra do Pilar	54.3%	33.7%	39.4%
	Monastery of Alcobaça	58.1%	26.0%	34.8%
Intangible	Mediterranean Diet	82.7%	62.6%	68.1%
	Fado, urban popular song of Portugal	91.8%	53.6%	64.0%
	Cante Alentejo, polyphonic singing from Alentejo	72.9%	29.8%	41.7%
	Manufacture of cowbells	39.0%	12.4%	19.8%
Tentative list	Pombaline Lisbon	67.3%	49.6%	54.4%
	Historical Lisbon, Global City	66.7%	48.0%	53.0%
	Southwest Coast	59.0%	42.1%	46.6%
	Águas Livres Aqueduct	58.5%	34.0%	40.7%
	Head Office and Garden of the Calouste Gulbenkian Foundation	57.0%	29.0%	36.6%

of the heritage located in the main Portuguese cities (Lisbon and Porto) or in their surroundings (Sintra and Alto Douro Wine Region).

4.2.1. Bivariate analysis

In this exploratory study, a bivariate analysis was also considered. For this analysis, *t-test* (to verify if the differences found in the top five positions between the nationalities were statistically significant, a theoretical significance of 5% will be used), *qui-square* independence tests were used (which verify whether or not there is an association for the variables under study, which is considered significant if the significance of the test is lower than the theoretical significance (0.05)) in addition Cramer's V Coefficient

Table 5. Want to visit/contact with the world heritage (by nationality)

Heritage		Nationality		
		Portuguese	Foreign	Total
Tangible	Monastery of the Hieronymites and Tower of Belém in Lisbon	57.6%	62.0%	60.8%
	Cultural Landscape of Sintra	63.4%	57.5%	59.1%
	Historic Centre of Oporto, Luiz I Bridge and Monastery of Serra do Pilar	61.1%	57.8%	58.8%
	Alto Douro Wine Region	63.5%	56.0%	58.2%
	Central Zone of the Town of Angra do Heroismo in the Azores	63.4%	53.5%	56.2%
Intangible	Fado, urban popular song of Portugal	76.0%	68.8%	70.8%
	Mediterranean Diet	69.2%	70.1%	69.8%
	Cante Alentejo, polyphonic singing from Alentejo	66.7%	57.9%	60.3%
	Manufacture of cowbells	43.2%	34.4%	36.9%
Tentative list	Pombaline Lisbon	63.9%	64.3%	64.2%
	Historical Lisbon, Global City	62.9%	62.7%	62.8%
	Southwest Coast	59.7%	58.9%	59.1%
	Royal Building of *Mafra* – Palace, Basilica, Convent, *Cerco* Garden and Hunting Park (*Tapada*)	57.3%	51.8%	53.3%
	Selvagens Islands	60.2%	49.4%	52.5%

as calculated to analyse de degree of association . To avoid violation of the conditions of applicability of the qui-square test, the variable "level of education" was regrouped into three groups (Basic Education, Secondary Education and Higher/University Education) as an alternative to the five previously used (1st cycle, 2nd cycle, 3rd cycle, Secondary Education and Higher Education/University) and the variables "degree of knowledge about Portuguese culture", "level of tangible, intangible knowledge and tentative list" were grouped into three groups (low, medium and high).

Based on the results presented in the table below, only for the Alto Douro Wine Region is the average knowledge of this heritage similar. In the remaining cases it is superior to the Portuguese (Table 7).

Similar tests were carried out for intangible heritage and tentative list elements. The results obtained revealed that in all elements/attributes of intangible heritage there are significant differences (Table 8), again being the superior knowledge for the Portuguese. Similar conclusions can be drawn for the tentative list elements (Table 9).

As expect there is a positive association between "nationality" and "level of education" with all variables representative of "knowledge of culture and heritage". The variable "age" reveals a positive association regarding "knowledge of tangible and intangible heritage", meaning that older people have a greater knowledge of this heritage. It may seem unexpected that "age" does not reveal an association with the "knowledge of Portuguese culture" and "medium knowledge tentative list heritage", which could reveal that even the younger generation is aware of the WH importance and showed some interest in it (Table 10).

Table 6. Average degree of knowledge of tangible, intangible and tentative list of Portuguese heritage by nationality

Heritage		Nationality					
		Portuguese			Foreign		
		Average	Ranking	%	Average	Ranking	%
Tangible (Average = 2.24)	Monastery of the Hieronymites and Tower of Belém in Lisbon	3.34	1	51.50%	2.9	1	35.50%
	Cultural Landscape of Sintra	2.97	2	38.50%	2.37	3	22.50%
	Monastery of Batalha	2.94	3	38.60%	2.24	4	19.50%
	Historic Centre of Évora	2.8	4	34.90%	1.93		11.60%
	Historic Centre of Oporto, Luiz I Bridge and Monastery of Serra do Pilar	2.77	5	35.40%	2.38	2	21.90%
	Alto Douro Wine Region	2.34		21.60%	2.20	5	17.60%
	University of Coimbra – Alta and Sofia	2.68		30.40%	2.05		15.90%
	Monastery of Alcobaça	2.67		26.50%	1.98		13.70%
	Convent of Christ in Tomar	2.58		24.50%	1.86		10.70%
	Historic Centre of Guimarães	2.41		20.10%	1.84		11.20%
	Central Zone of the Town of Angra do Heroismo in the Azores	2.4		21.60%	1.98		13.10%
	Garrison Border Town of Elvas and its Fortifications	2.37		20.50%	1.68		9.20%
	Laurisilva of Madeira	2.12		17.00%	1.74		10.40%
	Prehistoric Rock Art Sites in the Côa Valley and Siega Verde	2.02		14.20%	1.57		7.60%
Intangible (Average = 2.45)	Fado, urban popular song of Portugal	3.74	1	62.3%	2.57	2	23.7%
	Mediterranean Diet	3.39	2	47.8%	2.82	1	32.1%
	Cante Alentejo, polyphonic singing from Alentejo	2.94	3	35.5%	1.87	3	14.2%
	Manufacture of cowbells	2.08	4	17.7%	1.39	4	4.3%
Tentative list (Average = 2.05)	Pombaline Lisbon	3.14	1	48.9%	2.63	1	29.6%
	Historical Lisbon, Global City	3.05	2	44.4%	2.60	2	27.6%
	Águas Livres Aqueduct	2.91	3	35.2%	2.12	4	17.4%
	Southwest Coast	2.80	4	38.2%	2.25	3	17.6%
	Head Office and Garden of the Calouste Gulbenkian Foundation	2.75	5	34.4%	2.09	5	15.6%
	Royal Building of *Mafra* – Palace, Basilica, Convent, *Cerco* Garden and Hunting Park (*Tapada*)	2.52		26.5%	1.88		13.8%
	Mértola	2.43		23.5%	1.60		6.8%
	Levadas of Madeira Island	2.38		22.5%	1.85		12.4%
	Routes of Santiago de Compostela: Routes in Portugal	2.35		23.7%	2.06		15.5%
	Historic Centre of Guimarães and Couros Zone (extension)	2.34		21.7%	1.82		12.1%
	Selvagens Islands	2.34		24.3%	1.70		8.7%
	Sites of Globalization	2.21		19.7%	1.74		12.3%
	Montado, Cultural Landscape	2.20		18.3%	1.60		7.1%
	Sanctuary of Bom Jesus do Monte in Braga	2.17		18.4%	1.76		11.0%
	Dinosaurs INCNITOS	2.11		19.3%	1.53		5.7%
	Ensemble of Álvaro Siza's Architecture Works in Portugal	2.07		19.3%	1.58		7.1%
	Vila Viçosa, Renaissance ducal town	2.06		17.7%	1.60		8.6%
	Route of Magellan. First around the World	2.05		17.7%	1.67		9.5%
	Desert of the Discalced Carmelites and Built Ensemble of the Palace-Hotel in Bussaco	2.01		16.7%	1.68		8.8%
	Roman Production Centre of Fish Salting and Conservation in Tróia	1.97		12.8%	1.61		8.1%
	Mid-Atlantic Ridge	1.95		15.3%	1.73		9.3%
	Bulwarked Fortifications of the "Raia" (Border)	1.89		13.7%	1.60		8.5%

Table 7. T-test for tangible heritage and landscape

Heritage	Value	*p-value*
Monastery of the Hieronymites and Tower of Belém in Lisbon	-4.080	0.000
Cultural Landscape of Sintra	-6.441	0.000
Monastery of Batalha	-8.278	0.000
Historic Centre of Évora	-5.164	0.000
Historic Centre of Oporto, Luiz I Bridge and Monastery of Serra do Pilar	-3.431	0.001
Alto Douro Wine Region	-1.220	0.223

Table 8. T-test for intangible heritage elements

Heritage	Valuer	*p-value*
Fado, urban popular song of Portugal	-11.446	0.000
Mediterranean Diet	-5.101	0.000
Cante Alentejo, polyphonic singing from Alentejo	-9.905	0.000
Manufacture of cowbells	-7.818	0.000

Table 9. T test for heritage of the tentative list

Heritage	Value	*p-value*
Pombaline Lisbon	-4.081	0.000
Historical Lisbon, Global City	-3.834	0.000
Águas Livres Aqueduct	-6.826	0.000
Southwest Coast	-4.912	0.000
Head Office and Garden of the Calouste Gulbenkian Foundation	-6.013	0.000

CONCLUSION

Tourism and culture interconnectivity is growing in the context of a sustainable destination development.

World Heritage classification - tangible and intangible - is an expression of the importance of a cultural recognition. In a tourism perspective, WH classification could enhance a deeper knowledge of different types of culture of visited places and consequently improve the tourists' experience satisfaction. In this framework, the present exploratory study discussed the tourists' awareness of Portuguese World Heritage (WH), namely by the Algarvian tourists, once Algarve is one of the most important tourist regions of Portugal, mainly associated with 'sun and sea motivation'. However, there is a growing institutional/political recognition regarding the need of regional tourism products diversification based on culture, as pointed out in several tourism strategic documents, such as "Turismo 2020" and "Estratégia 2027".

In accordance, it was expected, and confirmed by a quantitative analyse based on a questionnaire, that the majority of Portuguese and foreign tourists reveal a "very weak" and "weak" cultural motivation in visiting the Algarve, namely the younger generation. Simultaneously, this study showed that tourists

Table 10. Average degree of knowledge of Portuguese culture regarding tangible, intangible and tentative list by nationality, age, and level of education

		Value	*p-value*
Degree of knowledge about Portuguese culture	Nationality	79.575	0.000
	Age	10.304	0.112
	Level of education	17.663	0.001
Medium knowledge tangible heritage	Nationality	25.956	0.000
	Age	20.309	0.002
	Level of education	10.153	0.038
Medium knowledge intangible heritage	Nationality	89.709	0.000
	Age	12.924	0.044
	Level of education	14.183	0.007
Medium knowledge tentative list heritage	Nationality	12.73	0.002
	Age	8.726	0.190
	Level of education	14.052	0.007

have more knowledge regarding tangible heritage located in the capital city of the country, than regarding intangible heritage. For instance, monuments, such as the Monastery of the Hieronmytes and Tower of Belém, are highlighted more frequently. Regarding, intangible heritage, Fado and the Mediterranean Diet are the most known (common). These results are expected, once Fado, as an urban traditional song played on the radio and frequently associated to restaurants, animation performances and the Mediterranean Diet, is being disseminated, not only in many restaurant menus in the Algarve, but also through the Mediterranean Diet Route.

Additionally, the research put in evidence that "nationality" is an explanatory variable for the models related to the degree of knowledge of Portuguese culture (WH classification) – once Portuguese resident tourists reveal a higher knowledge of Portuguese culture than foreigners. In accordance, there is also a positive association between "nationality" and "level of education" with all variables representative of "knowledge of culture and heritage". The variable "age" reveals a positive association regarding "knowledge of tangible and intangible heritage" (WHL and ICH), which means that older people have a greater knowledge of this heritage.

ACKNOWLEDGMENT

This paper is financed by National Funds provided by FCT- Foundation for Science and Technology through project UIDB/04020/2020.

REFERENCES

Aa (van der), B., Groote, P. & Hulgen, P. (2005). World heritage as NIMBY? The case of the Dutch part of the Wadden Sea. In D. Harrison, & M. Hichcock (Eds.), *The Politics of World Heritage, Negotiating Tourism and Conservation* (pp. 11-22). Clevedon: Channel View Publications.

Adie, B. (2017). Franchising our heritage: The UNESCO World Heritage brand. *Tourism Management Perspectives*, *24*, 48–53. doi:10.1016/j.tmp.2017.07.002

Arruda, A. M. (1999). Algarve no quadro geocultural do Mediterrâneo Antigo. In *Mª da Graça, Marques (Coord.), O Algarve da Antiguidade aos nossos dias (elementos para a sua história)*. Edições Colibri.

Australian government. (2019). *Understanding World Heritage: what is outstanding universal value?* https://www.environment.gov.au/heritage/about/world-heritage/outstanding-universal-value

Bandhari, K. (2014). *Tourism and National Identity: Heritage and Nationhood in Scotland. Series Tourism and Cultural Change: 39*. Channel View Publications. doi:10.21832/9781845414498

Breakey, N. M. (2012). Study in of World Heritage visitors: The case of the remote Riversleigh Fossil Site. *Visitor Studies*, *15*(1), 82–97. doi:10.1080/10645578.2012.660845

Caust, J., & Vecco, M. (2017). Is UNESCO World Heritage recognition a blessing or burden? Evidence from developing Asian countries. *Journal of Cultural Heritage*, *27*, 1–9. doi:10.1016/j.culher.2017.02.004

Culture2030Goal. (2019). *Culture in the Implementation of the 2030 Agenda*. Author.

Davison, G. (2008). Heritage: from patrimony to pastiche. In G. Fairclough, R., Harrison, J., Jameson, Jr., & J. Schofield (Eds.), The Heritage Reader, (pp.31-41). Routledge.

Frey, B. S., Pamino, P., & Steiner, L. (2013). Explaining the Word Heritage List: An Empirical Study. *International Review of Economics*, *60*(1), 1–19. doi:10.100712232-013-0174-4

Fuchs, M., & Pikkemaat, B. (2019). Destination Image Analysis - A Cross-Cultural Segmentation Approach. Rearchgate.net/publication/330841068_Destination_Image_Analysis-A_Cross-Cultural_Segmentation_Approach

Gisolf, M. (2017). *Tourists' Profiles and Lifestyles. Tourism Theories. Tourist Profiles and Sustainability*. https://www.tourismtheories.org/?p=999

Hall, C. (2008). *Tourism Planning: Policies, Processes and Relationships*. Pearson Education.

Henriques, C., Guerreiro, M., & Mendes, J. (2014). *Turismo Cultural no Algarve*. RTA/UALG.

IMPACTUR. (2019). *Indicadores CIITT/UALG IMPACTUR*. http://www.ciitt.ualg.pt/impactur/dt_anuais.asp?nut=17

Jimura, T. (2011). The impact of World Heritage Site designation on local communities – A case study of Ogimachi, Shirakawa-mura. Japan. *Tourism Management*, *32*(2), 288–296. doi:10.1016/j.tourman.2010.02.005

Kneafsey, M. (1998). Tourism and Place Identity: A case-study in rural Ireland. *Irish Geography*, *31*(2), 111–123. doi:10.1080/00750779809478623

Labadi, S. (2007). Representations of the nation and cultural diversity in discourses on World Heritage. *Journal of Social Archaeology*, *7*(2), 147–170. https://www.academia.edu/516837/Representations_of_the_nation_and_cultural_diversity_in_discourses_on_World_Heritage

Landorf, C. (2009). Managing for sustainable tourism: A review of six cultural world heritage sites. *Journal of Sustainable Tourism*, *17*(1), 53–70. doi:10.1080/09669580802159719

LDWHP. (n.e.). *World Heritage Status. Is there opportunity for economic gain. Research and Analysis of the Socio Economic Impact Potential of UNESCO World Heritage Site Status.* RACES, RC &TBR Rebanks Consulting Ltd and Trends Business Research Ltd. Research and Creative Enterprise Services (RACES). http://icomos.fa.utl.pt/documentos/2009/WHSTheEconomicGainFinalReport.pdf

Lenzerini, F. (2011, February). Intangible Cultural Heritage: The Living Culture of Peoples. *European Journal of International Law*, *22*(1), 101–120. doi:10.1093/ejil/chr006

Liu, Y., & Cheng, J. (2016). Place Identity: How Tourism Changes Our Destination. *International Journal of Psychological Studies*, *8*(2).

Poria, Y., Reichel, A., & Cohen, R. (2013). Tourists perceptions of World Heritage Site and its designation. *Tourism Management*, *35*, 272–274. doi:10.1016/j.tourman.2012.02.011

Ryan, J., & Silvanto, S. (2010). World Heritage Sites: The Purposes and Politics of Destination Branding. *Journal of Travel & Tourism Marketing*, *27*(5), 533–545. doi:10.1080/10548408.2010.499064

Santa Cruz, F., & López-Guzmán, T. (2017). Culture, Tourism and World Heritage Sites. *Tourism Management Perspectives*, *24*, 111–116. doi:10.1016/j.tmp.2017.08.004

Timothy, D. J. (2011). *Cultural Heritage and tourism: An Introduction.* Chanel View Publications.

TP. (2017). *Estratégia Turismo 2027.* https://estrategia.turismodeportugal.pt/

TP/Governo de Portugal. (2015). *Turismo 2020: Plano de Ação para o Desenvolvimento do Turismo em Portugal.* http://www.turismodeportugal.pt/Portugu%C3%AAs/turismodeportugal/apresentacoes/Documents/Turismo-2020-Sintra.pdf

TP/UALG. (2017). *Perfil do turista que visita o Algarve (2016).* https://issuu.com/turismo_algarve/docs/perfil_do_turista_2016_relatorio_fi

UNESCO. (2019). *World Heritage List.* https://whc.unesco.org/en/list/

UNESCO. (2019a). *Browse the Lists of Intangible Cultural Heritage and the Register of good safeguarding practices.* UNESCO.

UNESCO/ICCROM/ICOMOS/IUCN. (2016). *Gestão do patrimônio mundial cultural* [International Council on Monuments and Sites, International Union for Conservation of Nature, International Centre for the Study of the Preservation and Restoration of Cultural Property]. https://unesdoc.unesco.org/ark:/48223/pf0000244283

UNWTO. (2019). *Baseline Report on the Integration of Sustainable Consumption and Production Patterns into Tourism Policies.* https://www.e-unwto.org/doi/book/10.18111/9789284420605

UNWTO. (2019). *International Tourism Highlights 2019 Edition.* https://www.e-unwto.org/doi/pdf/10.18111/9789284421152

UNWTO/UN. (2017). *Tourism and the Sustainable Development Goals – Journey to 2030 – Highlights.* https://www.e-unwto.org/doi/book/10.18111/9789284419340

UNWTO/UNESCO. (2019). *4th World Conference on Tourism and Culture: Investing in future generations.* https://www.unwto.org/4th-unwtounesco-world-conference-tourism-and-culture-investing-future-generations

Valle, P., Guerreiro, M. M., Mendes, J., & Silva, J. (2011). Tourist satisfaction and destination loyalty intention: A structural and categorical analysis. *International Journal of Business Science and Applied Management*, 25–44.

WTA. (2018). *Site World Travel Awards.* https://www.worldtravelawards.com/

WTTC. (2019a). *Travel & Tourism Economic Impact 2019 – Portugal.* https://www.wttc.org/-/media/files/reports/economic-impact-research

WTTC. (2019). *Travel & Tourism Economic Impact 2019 – World.* https://www.wttc.org/-/media/files/reports/economic-impact-research/regions-2019/world2019.pdf

Yang, C. H., Lin, H.-L., & Han, C. C. (2010). Analysis of international tourists arrivals in China: The role of world heritage sites. *Tourism Management, 316*(6), 827–837. doi:10.1016/j.tourman.2009.08.008 PMID:32287732

ADDITIONAL READING

Cohen, S. (2010). Searching for escape, authenticity and identity: Experiences of lifestyle travellers. In M. Morgan, P. Lugosi, & J. R. B. Ritchie (Eds.), *The Tourism and Leisure Experience: Consumer and Managerial Perspectives* (pp. 27–42). Channel View Publications. doi:10.21832/9781845411503-005

Cruz-Milán, O. (2017). *Plog's Model of Typologies of Tourists. The SAGE International Encyclopedia of Travel & TourismPublisher.* SAGE Publications. Available from https://www.researchgate.net/publication/331315044_Plog's_Model_of_Typologies_of_Tourists

King, L. M., & Halpenny, E. (2014). Communicating the World Heritage brand: Visitor awareness of UNESCO's World Heritage symbol and the implications for sites, stakeholders and sustainable management. *Journal of Sustainable Tourism, 22*(5), 768–786. doi:10.1080/09669582.2013.864660

Ng, S. I., Lee, J. A., & Soutar, G. N. (2006). Tourists' intention to visit a country: The impact of cultural distance. *Tourism Management, 28*(6), 1497–1506. doi:10.1016/j.tourman.2006.11.005

Richards, G. (2002). Tourism attraction systems: Exploring cultural behavior. *Annals of Tourism Research, 29*(4), 1048–1064. doi:10.1016/S0160-7383(02)00026-9

KEY TERMS AND DEFINITIONS

Algarve: A geographic and cultural region of the South Europe – Portugal.

Cultural Tourism: Type of tourism associated to the "cultural encounter".

Sustainability: Associated to principles such as conservation, preservation, equity, long term, partnerships enhancement (see Agenda 2030).

World Heritage Value: Value associated to the designation of Outstanding Universal Value. A property that fulfil "one or more of ten criteria" (see UNESCO) and meet the conditions of integrity, authenticity, and should have an adequate system of protection and management in order to promote its safeguard.

ENDNOTES

1. See the "Culture in the Implementation of the 2030 Agenda: A report by the culture 2030 goal campaign" (Culture2030Goal, 2019).
2. Monastery of the Hieronymites and Tower of Belém in Lisbon; Cultural Landscape of Sintra; Monastery of Batalha; Historic Centre of Évora; Historic Centre of Oporto, Luiz I Bridge and Monastery of Serra do Pilar; Alto Douro Wine Region; University of Coimbra – Alta and Sofia; Monastery of Alcobaça; Convent of Christ in Tomar; Historic Centre of Guimarães; Central Zone of the Town of Angra do Heroísmo in the Azores; Garrison Border Town of Elvas and its Fortifications; Laurisilva of Madeira; Prehistoric Rock Art Sites in the Côa Valley and Siega Verde.
3. Fado, urban popular song of Portugal; Mediterranean Diet; Cante Alentejano, Polyphonic singing from Alentejo; Manufacture of cowbells.
4. Pombaline Lisbon; Historical Lisbon, Global City; Águas Livres Aqueduct; Southwest Coast; Head Office and Garden of the Calouste Gulbenkian Foundation; Royal Building of *Mafra* – Palace, Basilica, Convent, *Cerco* Garden and Hunting Park (*Tapada*); Mértola; Levadas of Madeira Island; Routes of Santiago de Compostela: Routes in Portugal; Historic Centre of Guimarães and Couros Zone (extension); Selvagens Islands; Sites of Globalization; Montado, Cultural Landscape; Sanctuary of Bom Jesus do Monte in Braga; Dinosaurs INCNITOS; Ensemble of Álvaro Siza's Architecture Works in Portugal; Vila Viçosa, Renaissance ducal town; Route of Magellan. First around the World; Desert of the Discalced Carmelites and Built Ensemble of the Palace-Hotel in Bussaco; Roman Production Centre of Fish Salting and Conservation in Tróia; Mid-Atlantic Ridge; Bulwarked Fortifications of the "Raia" (Border).

Chapter 13
Creative Tourism as an Inductor of Co-Creation Experiences:
The Creatour Project in the Algarve

Sónia Moreira Cabeça
Research Centre for Tourism, Sustainability, and Well-Being (CinTurs), University of Algarve, Portugal

Alexandra R. Gonçalves
https://orcid.org/0000-0003-3796-1801
School of Management, Hospitality, and Tourism (ESGHT), & Research Centre for Tourism, Sustainability, and Well-Being (CinTurs), University of Algarve, Portugal

João Filipe Marques
Faculty of Economics and Research Centre for Tourism, Sustainability, and Well-Being (CinTurs), University of Algarve, Portugal

Mirian Tavares
https://orcid.org/0000-0002-9622-6527
Faculty of Human and Social Sciences and Research Centre in Arts and Communication (CIAC), University of Algarve, Portugal

ABSTRACT

Creative tourism is characterized by the opportunity that gives visitors to develop their creativity through active participation in learning experiences that are characteristic of their destination. That implies a relationship between visitors and residents in which the exchange of knowledge is central. Being an interactive process, the learning and doing of a creative tourism activity is an exercise of self-affirmation that expresses the creative potential of each participant. That is to say, creative tourism experiences are co-created by tourists and their hosts. Tourists and residents are co-designers in creative tourism, as the CREATOUR experiences prove. In this chapter, the authors intend to present the project activities in the Algarve. After contextualizing co-creation within the creative processes, their research is centred on the creative experiences and discusses the opportunities resulting from co-created activities. They also look at the survey applied to the participants in the experiences to better understand their feelings

DOI: 10.4018/978-1-7998-3156-3.ch013

towards co-creation and creative processes.

Tourism demands new metaphors based on being, doing touching and seeing, rather than just "seeing". (Perkin and Thorns, 2001, p. 189)

INTRODUCTION: CREATIVE TOURISM

The search for the humanization of relationships and the creation of affective bonds is a phenomenon that can be observed in several areas of our daily life. This statement seems rather accurate when addressed to a certain type of tourists, willing "to break themselves entirely from their home lives and engage sincerely with a different way of living" (Amadeus Traveller Trend Observatory [ATTO], 2015, p. 4) when travelling. In fact, the demand for unique experiences, emotions, cultural information and active participation when travelling is a major trend in tourism worldwide (Pine & Gilmore, 1999; ATTO, 2015; Canalis, 2016; European Travel Commission [ECT], 2017; Turismo de Portugal, 2017). Destinations are valued by their story, their symbolism, the emotions they raise, and what they make the tourist feel. Tourists seek to learn more about the culture of their destination and to have a sense of the places visited. They can also cognitive, affective, and socially benefit from their tourism experiences, developing emotional links with the places visited and with the people that inhabit it (Gu & Ryan, 2008; Ilincic, 2014).

The pattern of tourism consumption described above meets the creative tourism definition: "[to] travel directed toward an engaged and authentic experience, with participative learning in the arts, heritage, or special character of a place, and it provides a connection with those who reside in this place and create this living culture' (United Nations Educational, Scientific and Cultural Organisation [UNESCO], 2006, p. 3). Active participation, creative self-expression, learning, and community engagement (Cabeça, Gonçalves, Marques & Tavares, 2019b) are key aspects of a creative tourism experience, which allow destinations to use cultural and social assets and its human capital as driving forces. This sustainable small-scale tourism – that presupposes the use of local resources – provides an immersion into local culture and interaction, and encourages self-expression and an intensive social interaction between visitors and local residents.

The strong interaction between residents and tourists in creative tourism is mainly driven by the learning process that this type of tourism presupposes. Going beyond a mere learning "lesson", in creative tourism the tourists learn and do; are taught and produce. In this process, the creativity and the opportunity for self-expression must be guaranteed. The residents, as the practitioners and rulers of their traditional know-how (Cabeça, 2016) have also an active role, providing and co-creating the experiences with visitors.

BACKGROUND

Co-Creation Processes in Tourism

To experience a creative tourism activity is to go beyond the purchase of commodities, goods and services and enter a path of "immersion and participation" (Pine & Gilmore, 1999) that satisfies the tourist's needs for knowledge, inspiration, creativity and self-expression. "When a person buys a service, *he* purchases a set of intangible activities carried out on *his* behalf. But when *he* buys an experience, *he pays* to spend time enjoying a series of memorable events that a company stages-as in a theatrical play-to engage him in a personal way" (*ibid*, p. 2). Giving residents a central role in the experience, creative tourism shifts the attention from the gaze of the tourism to its body (Cohen & Cohen, 2012): the active participation of both, residents and visitors, is imperative to create the experience. Rather than learning something with the tradition bearers, the tourist creates something. Furthermore: he does not create alone, but along with the residents.

When allowed to use their creativity and express themselves while participating in a learning experience provided by residents, tourists are co-designers of their own experiences (Binkhorst & Dekker, 2009). That is to say that the interaction with local residents generates the tourist's own experience (Prahalad & Ramaswamy, 2004), which is the very concept of co-creation.

Co-creation became an answer to the tourist's demand of unique experiences (Binkhorst, 2008), giving the activities a "meaning and purpose", once they make use of the tourist's skills, talents, and potential (Kashdan, Biswas-Diener & King, 2008), and place participants at the centre of the experience. To co-create is, therefore, to give active roles to participants in the experience. It is not centred in the promoter, neither in the product (Binkhorst, 2008), but in the very participants and in the context in which the interaction between residents and tourists occurs.

When both sides help to create the touristic offer, the result is an experience that allows participants to self-express while doing something. In that sense, the result – tangible and intangible – is unique and distinctive from every other experience. Such outcome is only possible through co-creation: the tourist learns and does something, making use of the local traditions and with the cooperation of local residents, generating a "sense of the place".

It was stated that this type of tourists seeks unique experiences: they want to meet residents, learn about local culture and use creativity while expressing themselves through the development of activities that are characteristic of their destination. To interact, to learn, to do, to self-express, seem to play important roles in the co-creation process. These were key aspects that guided us as we surveyed the tourists who participated in the CREATOUR's creative tourism activities.

The CREATOUR Project

CREATOUR is a multidisciplinary research and incubation project that aims to develop creative tourism destinations in small towns and rural areas of Portugal. Working with several partners that offer creative tourism activities, the project helped these pilots to develop their experiences and now studies and monitors the creative tourism activities that are taken place.

In 2016 and 2017, 40 pilots were chosen to cooperate with the researchers. From North to South Portugal, pilots were selected taking into account the proposal's cultural value, its creative nature and diversity, the capacity to capture tourists' interest, and the benefits that such activities could bring to local communities.

Although creativity can be a mere backdrop for tourism (Richards, 2011), CREATOUR pilots were encouraged to create unique active experiences. More than offer a learning activity that meets the tourist's demand for knowledge, they were encouraged to develop activities that make use of the tourist's creative potential. This shift is from being inspired to being a part of the destination, which is possible through co-creation. Therefore, the CREATOUR experience offers us the possibility to look at co-creation as a fundamental element of creative tourism. Co-creation is, in fact, a prerequisite for the definition of what is a creative experience: experiencing, interacting and sharing, with all parties involved – residents and tourists – being full participants and not passive subjects (Binkhorst, 2008).

In the Algarve region, 9 entities are part of the CREATOUR network, and are being monitored by the Research Centre for Tourism, Sustainability and Well-Being - CinTurs, the research centre responsible for the region's experiences. Cooking classes, design and illustration workshops, walking and tasting routes, mornings at the farm, palm, clay and copper workshops are some of the tourism experiences (that bond visitors and residents in co-creation processes) implemented in the Algarve. Visitors have the opportunity to learn – and do – traditional crafts, cuisine, and other local know-how, and even become a saltpan worker or a shepherd for one day. Discovering the territory by learning more about it and following routes, improving their skills in different areas (art, traditional know-how, rural knowledge), tourist's interact and work along with the local community (artists, craftsmen, entrepreneurs, etc.) making use of their creativity and the opportunity for self-expression along the way. Rather than being a community based tourism, the experience is, in fact, a "community-created tourism" (Idziak, Majewski & Zmyślony, 2015), once residents are a key part of the tourism experience and participate in the process of its development, generating "community-created products" (*ibid*). We dare to go further: the outcome of the creative tourism initiatives is created by both resident and visitors, being an all-participants-created tourism.

CREATIVE EXPERIENCES IN THE ALGARVE

The Algarve is one of the most touristic region of Portugal. The "sun and sea" product is the main attraction, and tourism is the main economic driver, employing half of the population (according to the Tourism Board and the National Statistical Institute). The focus on this type of tourism, however, has conditioned innovation and the diversification of the regional economy (CCDR, 2015, p. 31), with the tourism flow being highly concentrated in coastal areas and the summer season, and tourism generating socioeconomic imbalances within the region (RTA, 2014).

In fact, Algarve has the highest seasonality rate of the country (INE / Turismo de Portugal) despite the growth in overnight stays outside the high season. Most visitors come in July and August to particular coastal areas of the region. The tourism performance is, therefore, highly differentiated between the several municipalities of the region, between the coastal and the interior zones. In 2018, according to INE, Algarve registered 4732165 guests in tourist accommodation establishments in the region. Most stayed in coastal areas: almost 39% of visitors chose the municipality of Albufeira, followed by Loulé (14%) and Portimão (12.5%). In opposite, municipalities from the interior, without beaches, were less visited: Alcoutim had 0.03% of the total guests, São Brás de Alportel 0.07%, Monchique 0.54%, Aljezur 0.78% and Castro Marim 0.,86% (INE, 2018).

The growing demand for new tourism products with higher added value, associated with nature, culture, heritage, local products; useful and unique (Jensen, 1999; Gu & Ryan, 2008), offers an opportunity for tourism hosts in the Algarve to develop more sustainable and less seasonal tourism products that can lead tourists to less visited places; a contemporary exigence towards sustainability. Algarve can offer unique experiences, irreproducible elsewhere. In order to do so, the region must make use of its territorial capital (natural, physical, symbolic, cultural, human and spatial) (Barata, Molinari, Marsh, & Cabeça, 2017) and "to establish a link in the mind of the visitor between particular manifestations of culture and creativity and specific locations" (Richards & Wilson, 2005, p. 7).

To link the creative activities to their specific territories, placing them, was a particular important stage in the development of the CREATOUR's experiences, once the concept of "experience" itself is closely linked to the spirit of the place and to local people (Binkhorst, 2008). As a result, experiences are related to the context in which they occur and are linked to the culture, history, know-how, and tradition of the destination (*ibid*). Culture can be the "key activity" (Martin & Woodside state, 2008) that solidifies the decision to visit the destination. By linking the culture of the Algarve with creative activities, making use of local resources, we can capture the identity of each place, which creates distinctiveness, involve community in the tourism offer – creating a privileged relationship between tourists and local residents in shared experiences – and promote local culture through the tourist immersion into the local culture (Cabeça; Gonçalves; Marques & Tavares: 2019a). It can also bring sustainability for the region (*ibid*), once when the tourism products integrate local communities in their planning; economic, environmental, socio-cultural benefits can be achieved (Simpson, 2008).

The CREATOUR activities in the Algarve state the strong relation between offers and the territory where they take place. Pilot initiatives aim to revitalise traditional arts and crafts; promote and rescue spaces and places; evoke memories; foster creativity; develop new products; introduce the natural, historical and cultural heritage of the Algarve... They presuppose interaction, a "to do" action, and rely on the active participation of the residents. In this sense, they are co-creation oriented, involving place-based resources, and linking tourists and communities in learning processes. Tourists learn and do: they cook, meet, pick, aknowledge, create… Tourists learn with residents, and while doing so, they have the opportunity of expressing themselves, applying what they have learned about their destination. And when places allow visitors and residents to feel being in a special place, both "will enjoy emotional experiences, will perceive a characteristic atmosphere and will intensely appreciate the history and the stratified culture of the place" (Lorentzen, 2009, p. 460).

It is also important to state that most of these activities are taking place in small areas of the region and, when developed in cities, they address local products (culture, nature, food, knowledge) that are not included in the mainstream offer. Also, they take place in less visited areas: for instance, although Faro is the gateway to the region for many, the city itself is not much explored (and only 6% of Algarve's total guests stay in the council).

BOOSTING CREATIVITY THROUGH CO-CREATION

Since the beginning of the CREATOUR Project, researchers have been participating in creative tourism initiates developed by their partners. In order to better understand who are the visitors attending the creative tourism experiences, the project also developed a survey to be applied to participants in the network's activities. The survey aims to assess socioeconomic profiles, consumption habits, satisfaction levels, motivations and key elements for the success of the offers, as well as to assess the impacts of the activities for the territories and communities.

Since August 2017 until today, 124 surveys were applied in the Algarve region in several activities promoted by different entities: in saltpan workshops that invited visitors to be salt workers for one day, learning to catch the salt with the workers; several palm workshops in which the artisans teach and help visitors to make their own craft (rip of, entwining, knitting, making ropes and even a palm coaster); in an experience that invited visitors to be shepherds for one morning (milking goats with the goatherd); in a cooking class were a traditional recipe was made by the participants (and had a previous fieldtrip

Table 1. Creative Tourism Pilots from the Algarve Region

Pilot Project	Where	Approach	Goals	Resources	Promoters / providers	Learning process
Tertúlia Algarvia – Algarve Cooking Vacations	Faro and rural outskirts of the town	Provide participants with skills related to traditional recipes	Experience local culture through gastronomy and local heritage	Traditional recipes with local products	Local entrepeneurs and masters	To cook
Eating Algarve Tours – Food and Cultural Tours	Lagos & Loulé	Bring tourists to the community daily life spaces	Break seasonality and increase the days spent in the destination	Partnership with local restaurants and monuments	Local enterpreneurs and residents	To meet and to taste
In Loco – Tasting Algarve	Throughout all Algarve, specially low density areas	From nature to dish: learning the origin and transformation of products	Raise awareness to the Algarve's culture and history; promote Mediterranean Diet	Variety of each's place Diet	Local community	To know / be aware
Espírito da Terra - Pick, walk and create	Boliqueime	Allow each participant to get to know and enjoy local farm products and create with natural resources	Promote agriculture in a sustainable way; recover mills, and river site	Farm's products,, landscape and surroundings	Local farmers and community	To pick and to create
Odiana – Traditional know-how	Castro Marim	Integrate tourists in local culture through activities characteristics of the region	Revitalize and promote local traditions involving local community	Natural and cultural resources that provide unique experiences	Local community	To know and experience
Barroca – Culture and design experiences	Silves and small areas of the council	Photography workshops and visual experiences linking culture, gastronomy, and visual culture,	Valorize local culture and history; local development	Small places in the interior region	Local enterpreneurs	To express
Loulé Criativo – Traditional workshops	Loulé	Palm, clay and coppersmith workshops with local artisans in the artisan's space	Recover spaces; bring income to artisans; training; revitalize arts and crafts; incubation of entrepreneurs	Traditional know-how placed in their historical place	Local artisans and entrepreneurs	To do / create
Proactivetur – creative experiences	Smallest areas in the Loulé Council	Creative experiences program on arts, crafts and rural knowledge.	Training; network strengthening; territories' attractiveness; preserve knowledge	Local community knowledge	Local community	To do / create
Ciência Viva do Algarve – Walking route	Faro	Routes to explore the city, making use of scientific, artistic and historical knowledge	To value the ludic-scientific offer; create a tool to be used by others	Historical places and scientific knowledge	Local association	To think and to understand

Source: CREATOUR Project, 2019

to better understand the ingredients used); during the mornings at a local farm where tourists pick their own ingredients to make a joint meal; in wool and carpentry workshops provided at the farm with materials that are accessible in the place; in a visual and design experience in which participants made an illustrated interpretative pathway; in a cane workshop and a basketry workshop; and in a walking route where tourists had to overcome several challenges making use of the recently acquired knowledge about the place. We, ourselves, also learned and experienced some of the activities. The survey, along with the researchers' observations while participating in the experiences, provided us some interesting data.

Table 2. General data about the Inquired Participants in CREATOUR Activities

Inquired Participants: General Data	
Residence	77.4% Portugal 19.4% Outside Portugal
Nationality	70.2% Portuguese 24.2% Other European Countries
Gender	76.2% Female 23.8% Male
Age	From 11 until 76 years old 45% aged between 37 to 46
Marital status	Most single (49%) or married / in a marital relationship (41%)
Educational qualification:	64.9% with a Bachelor's Degree or further studies (Master or PhD)
Socio professional situation	Major categories: 19.3% Professional 18.5% Technicians and associate professionals 13.4% Managers 10.1% Retired

Source: CREATOUR Project, 2019

72.4% of the inquired tourists were participating in a creative experience for the first time, and 80% visited the place with the primary reason of attending the activity.

When asked about the motives that lead to the selection of the CREATOUR experiences, researchers learned that their originality, the cultural driven aspect and the possibility of stimulating participant's creativity were determining factors for visitors. The question was pre coded and presented a number of potential reasons. Participants were invited to classify each one according to the following scale: disagree completely, disagree, neither agree or disagree, agree, and agree completely. It was also possible to choose an unlisted motive. Few used that possibility, but the ones who did it, indicated curiosity, learning and well-being as reasons.

It seems that tourists seek to learn more about the culture of their destination, and to have a sense of the places visited, valorising unique experiences which are not reproducible elsewhere (Gu & Ryan, 2008). They also seem to fit the "conscious" and "aware" tourist pattern (Tan, Kung & Luh, 2013), looking for opportunities to boost their creativity.

None of the respondents who answered the question disagreed about the originality of the experience and 93.9% were driven by this aspect. The choice, for 91.2% of respondents, was also determined by the fact that the activity was "culturally motivating" and for being fun, although this second aspect seems rather less important (only 53.1% completely agreed on this, while 64% completely agreed on the cultural aspect). Another of the most mentioned factors, in 91% of the cases (and it was determinant in 63.1% of the cases), was the possiblity of stimulating their creativity. "To accompany someone" and "I know the promoter of the activity" were the less relevant factors. In fact, less than half of the participants have considered them important factors. Another interesting outcome, provided by complementary questions, reveals that most tourists have purposely traveled to participate in the experience, rather than being on the spot and used their own means to travel, indicating that, if interested, they are willing to travel to the venue, even if they did not have the purpose of visiting it. According to the survey results, it was the participation in the activity that brought most visitors to the site, indicating that these creative tourism activities can, in fact, be an attraction factor for their localities.

The full result provided by the answers on motivations (valid percentage) can be consulted below. The lighter the graphic, the largest percentage of positive answers (agree and agree completely). The full items were: "It is culturally motivating", "It permitted interaction with other participants", "It enabled me to meet and interact with the local community", "It is original", "It is suitable for the whole family", "Because of its location", "For being fun", "To stimulate my creativity", "To accompany someone" and "I know the promoter of the activity".

Figure 1. Tourist's reasons for selecting CREATOUR experience
Source: CREATOUR survey, 2017-2019

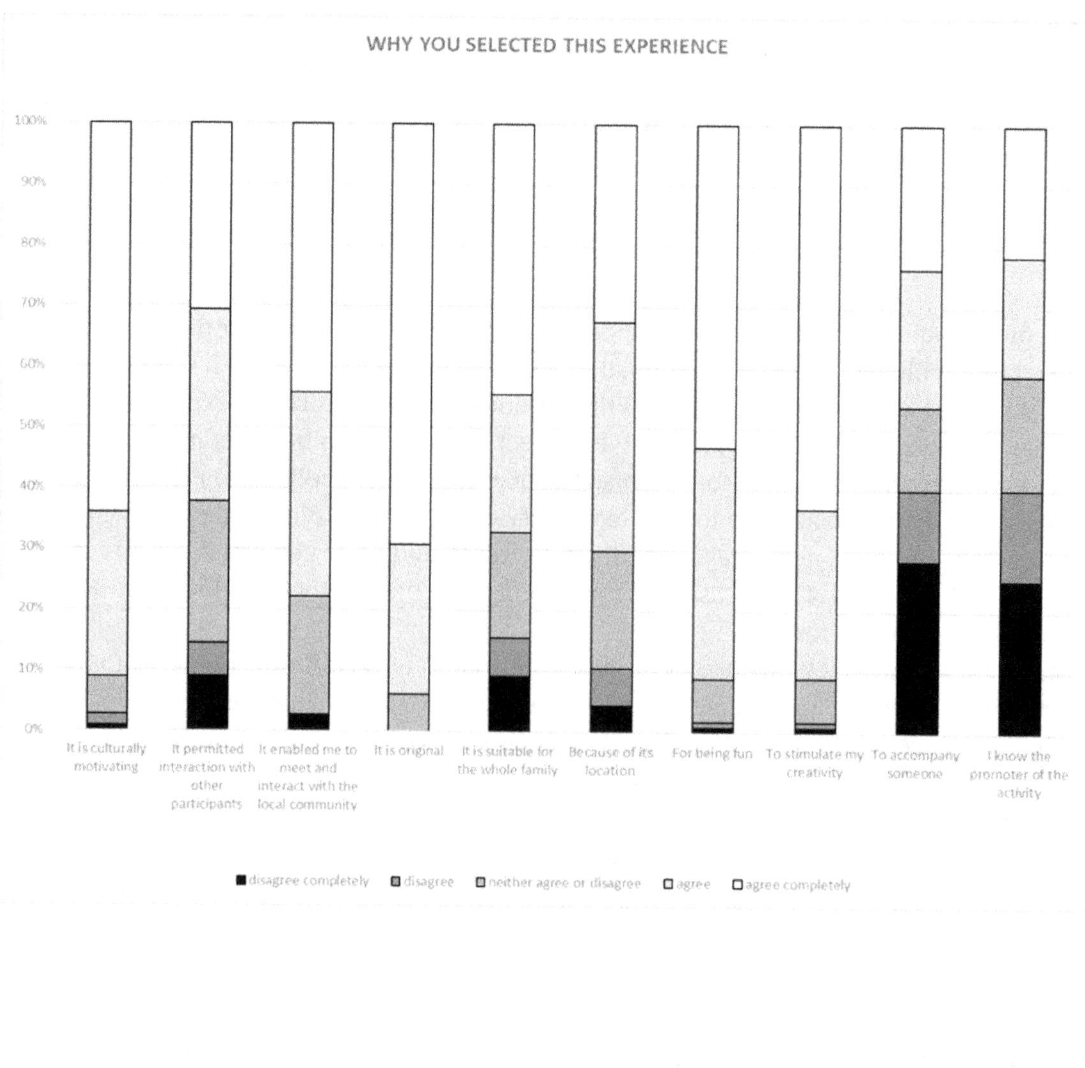

Most respondents have generally rated the experience in which they participated as good or very good, with no experiences rated as bad or less good. When asked about the experiences' outcomes (using the same evaluation scale mentioned above), 95.8% of the participants agreed that they have learnt more about local culture. Although amusement was not the main key factor when choosing the experience, tourists had fun (98.2%): the experience exceeded the expectations on this matter. 94% of participants tried a new activity. In another question, respondents expressed their willingness to attend other activities related to handicraft (cork, pottery, embroidery, weaving…); artistic and multimedia expressions (painting, photography, design, music…); nature (observation, walking…); local know-how and techniques, and traditional activities; gastronomy and local production; local history, archeology, heritage and culture; and other learning and doing activities. Joint activities carried out in the CREATOUR network might, therefore, benefit promoters, participants and communities.

We have stated that visitors move easily to attend activities that they like, and that CREATOUR activities can be attractive assets to territories. But in order to effectively benefit from this occasional increase of visitors, territories need strategies that can extend the visitors' stay. In fact, "I interacted with the local community", and "I contributed to the local community" were the aspects that less visitors agreed with. A lower percentage of agreement on these community impact items might be explained by the low number of overnight stays, meals and on-site purchases done by participants. It is also possible that, in addition to the members of the local community with whom they contacted during the learning processes (artisans and other local experts), participants did not actually meet other local residents. There is a need to expand the contact with local community. A network of partnerships and the diversification of offerings (accommodation, services, activities...) would be important to increase the length of the visitor's stay in the territory. Studies indicate that, when developing a common strategy involving a range of offerings and diverse actors, seasonality can decrease (D'Angella & Go, 2009) and inequalities between places resulting from the different capacity of local communities to embark on a development process can be mitigated (Shucksmith, 2018). Community involvement may also, as Gu and Ryan (2008) point out, strengthen the sense of belonging to a place and lead to greater interest from public authorities in maintaining it.

Although interaction with other participants was not one of the most determining factors when choosing the experience, 86.2% of respondents admitted they have met interesting people. 87.8% of the participants learned to do something and 82.5% acquired new skills. To interact, to learn, and to do, important aspects of the co-creation process, are also key factors in these creative experiences. It is important to emphasize the cultural dimension of the experiences, the main reason to choose the activity and its main outcome as well.

Visitors found the activities original, creative, enriching, and stimulating. Taking into account the percentage of "agree completely" answers – that measure a degree of full agreement – the results show the participants' satisfaction. 74.1% absolutely agreed that their activity had been original and enriching; 71.3% considered it fully creative and 68.4% completely stimulating. Once originality also stood out as an important motivation for the experience choice, we observe the correspondence between expectation and the outcome of the activities in this domain.

The researcher's observation and participation in the CREATOUR experiences meets the results provided by the survey. Researchers attended several activities: palm, tin, and copper workshops, photography classes, interpretive paths, culinary experience… Such active participation and learning experience certified the importance of a culturally stimulating environment and learning as a key factor of satisfaction. The need to give visitors the opportunity to self-express is also emphasized. Learning is

Figure 2. Tourists' perception about the CREATOUR experience outcomes
Source: CREATOUR survey, 2017-2019

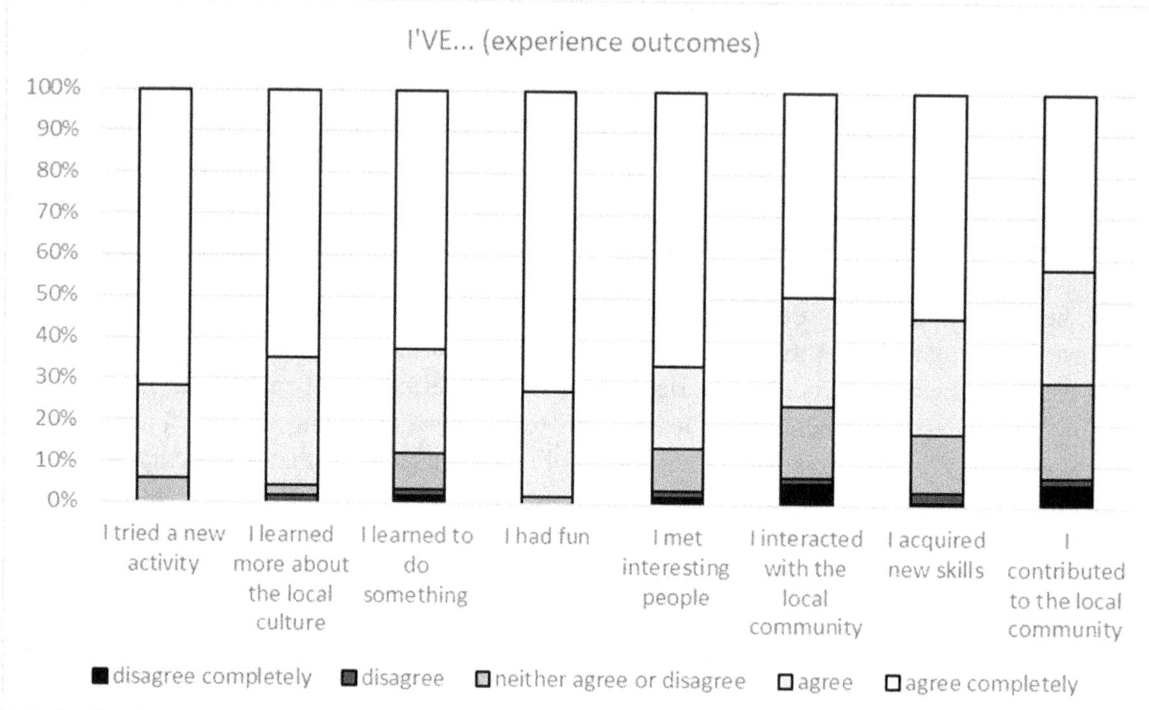

particularly important: more than seeing doing, it is doing that broadens the creative dimension of the experience. Therefore, the fundamental role of both residents and visitors in the design of creative experiences is highlighted: the interaction between them is, as evidenced by several studies (Ilincic, 2014; Miettinen, 2005), a key factor for visitor's satisfaction and indispensable in creative tourism. That is to say that co-creation plays an indispensable role.

Salt workers, tinsmiths, coppersmiths, shepherds, artisans and other members of the local community who have the necessary skills to carry out the activities, play key roles, once the learning process cannot be disconnected from human, cultural and social resources (Galani- Moutaf, 2013): the recipe is not made without the knowledgeable of a traditional cuisine chef, the history of a place is not known without the presence of an informed resident, the participant does not pick the salt flower without the instructions of a saline worker; and in order to produce a copper bracelet, a palm braid, or a brass piece, one must be mentored by a skilled artisan. On the other hand, creative experiences cannot be detached from the context in which they occur (Lazzeroni et al, 2013; Thomas, Harvey & Hawkins, 2013), and from the "sense of place". They promote an immersion in the local culture; are territorialized experiences.

Co-created, context related, and implying active participation from all participants and the acquisition of new skills: these are features that characterize creative tourism. Some key factors in the initiatives' development can also be underlined: the doing process (doing, rather than seeing something being done, enlarges the dimension of creative tourism and the co-creation process); a culturally stimulating environment; the opportunity for self-expression, and self-learning; and the co-creation process (heritage

Figure 3. Emotions raised by the CREATOUR experience
Source: CREATOUR survey, 2017-2019)

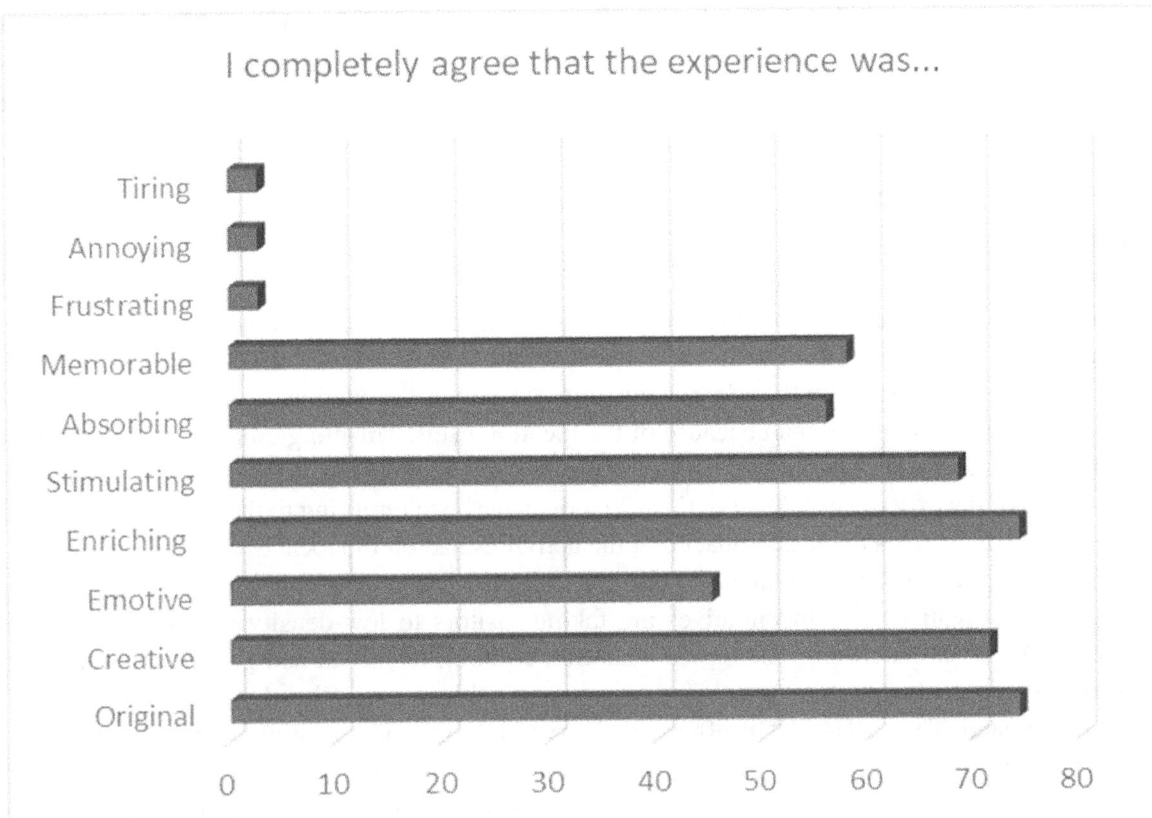

bearers, as the practitioners and rulers of their traditional know-how, must provide and co-create the creative experiences with visitors).

Moving our gaze to the residents, our research also revels the benefit of co-created activities to local communities: "The artisans are a fundamental part of the local culture and authenticity we want to bring in the programs", explains Proactivetur, one of the CREATOUR pilots. The first level of a creative tourism activity development is to learn more about the artisan history and the craft relation with the territory and culture. Then, the activity is designed and discussed with the artisan: what tourists can do, what is necessary to do it, how long can it take, etc… In fact, the "discovery of history" by local groups and their participation in the processes of identity construction (Fabre, 2001) is also an increasingly phenomenon that might help to market a destination and to establish the link between offers, territories, and their historical contexts, involving communities, local networks, and visitors (Lazzeroni et al, 2013). Residents' empowerment can also be achieved (Miettinen, 2005). "Most of them [artisans] come from a rural context, [a] socially and economic disadvantage background", meaning that the activities bring important financial incomes. Moreover, it is often the valorization of artisans and local knowledge that motivate the development of creative and co-created experiences. In the case of Loulé Criativo, another CREATOUR partner, creative tourism was thought to be an instrument of rural development, and a means to valorize heritage and local identity. With organized workshops in 5 areas – crafts, art, heritage, music and gastronomy – it was the municipality intention "not to let the masters die with their wisdom", hence the need for training in traditional arts and crafts and the rehabilitation of the spaces formerly devoted to these arts and crafts. Loulé Criativo trained artisans to revitalize the traditional arts, opened doors (palm, clay and copper houses) and gave conditions to local artisans to develop their activities and enhance their knowledge, perpetuating them. So, creative tourism is not just about providing the unique experience that tourists demand: it is also about valorizing the human capital of the destination.

CONCLUSION

The results of the survey applied so far reinforce the idea that the nature of the activities carried out at the destination may be decisive for the choice of the destination itself (Martin & Woodside, 2008). In the case of CREATOUR activities in the Algarve, visitors chose their destination considering the activities they could develop there, rather than because of the location itself. Traveling on their own willing (not to accompany someone or "persuaded" by the organizers), visitors are motivated by the cultural and original nature of the experiences. Most of these visitors have been returning to their starting point after the activity, which explains the little impact that the activities had on the local economy: few overnight stays, meals served and products bought.

Nevertheless, creative tourism initiatives are taking visitors to low-density and less well-known territories in Algarve, underlining the region's natural, cultural, historical and geographical diversity. It is, therefore, a potential strategy to achieve economic wealth in territories that little benefited from the tourism demand. The CREATOUR offers in Algarve highlight the potential of a tourism that is not solely focused on the product "sun and sea" and contribute to the discovery of an Algarve that is still unknown to most of its visitors, enhancing local development, reducing seasonality and contributing to the affirmation of local identities. In turn, participants in the CREATOUR network activities fit the profile of the cultural driven visitor, seeking cultural immersion and involvement in learning processes.

The identity of each place seems to play an important role as well: there are no unique experiences without unique, distinctive territories. It is the use of local resources and the involvement of communities that makes the experience irreproducible elsewhere. Creative tourism, developing territorialized initiatives, links heritage and people to places. It creates opportunities to participate in co-creation and learning processes that includes active roles for both residents and tourists and makes use of local resources.

It is recognised that "current definitions of creative tourism are all supply-led, such as by tourism boards, service providers, and various industry practitioners, without considering the tourists' viewpoint although tourists are seen as co-creators of the experiences." (Tan et al, 2013, p. 154). The CREATOUR Project brings a new perspective to the creative tourism definition, once it is centred in the co-creation process and in the dynamic relation between all participants. Furthermore, CREATOUR's point of view emphasises affect, emotions and feelings, often a neglected tourism research area (Cohen & Cohen, 2012): the experiences are oriented towards the tourist / resident relationship and the aroused emotions. And in the relationship between participants, which presupposes intense interaction, the co-creation process assumes particular relevance.

CREATOUR initiatives are, therefore, in line with co-creation processes, once they boost the participants' creativity, in an enriching and stimulating experience that goes beyond the mere gaze. They meet participants' expectations and feelings, being engaging experiences, and opportunities to develop new experiences, to learn and to self-express. Tourists attending the experiences were seeking for original activities that stimulate their creativity and that were culturally motivating. Through stimulating opportunities for new learning, participants found in creative tourism the originality and creativity they were searching for.

The importance of co-creation in creative tourism activities is also stated. Creative tourism is induced and generates co-creation, and an experience is not creative unless it bears some co-creation process involving tourists and residents as full participants and not passive subjects (Binkhorst, 2008). Experiencing and interacting are keywords. Bearing in mind the definition of creative tourism, and the fact that a co-created experience contains all the people and things that are needed to provide the experience environment (Binkhorst & Dekker, 2009), it is clear that co-creation is a prerequisite for the definition of what a creative experience means and what it presupposes: local community and stakeholders' involvement, place-based assets, active participation and learning processes (that imply doing rather than just seeing), usage of particular knowledge and traditions, self-expression... Co-creation, and creative tourism, make use of the spirit of the place to turn each experience unique and irreproducible elsewhere.

ACKNOWLEDGMENT

CREATOUR (project nº 16437) is funded under the Joint Activities Programme of PORTUGAL 2020, by COMPETE2020, POR Lisboa, POR Algarve and Fundação para a Ciência e Tecnologia. This paper is financed by National Funds provided by FCT- Foundation for Science and Technology through project UIDB/04020/2020.

REFERENCES

Amadeus Traveller Trend Observatory. (2015). *Future Traveller Tribes 2030, Understanding Tomorrow's Traveller*. Retrieved from: https://amadeus.com/documents/en/retail-travel-agencies/research-report/amadeus-future-traveller-tribes-2030-report.pdf

Barata, F. T., Molinari, F., Marsh, J., & Cabeça, S. M. (Eds.). (2017). *Creative Innovation and Related Living Lab Experiences: A Mediterranean Model*. UNESCO Chair.

Binkhorst, E. (2008). Turismo de cocreación, valor añadido en escenarios turísticos. *Journal of Tourism Research / Revista de Investigación en Turismo, 1*(1), 40-48.

Binkhorst, E., & Dekker, T. D. (2009). Agenda for Co-Creation Tourism Experience Research. *Journal of Hospitality Marketing & Management, 18*(2-3), 2–3, 311–327. doi:10.1080/19368620802594193

Cabeça, S. M. (2016). *Estrutura e processo de formação das Forma Cultural: o Caso do Cante Alentejano*. UNESCO Chair.

Cabeça, S.M., Gonçalves, A.R., Marques, J.F., & Tavares, M. (2019a). Mapping Intangibilities in Creative Tourism Territories through Tangible Objects: a methodological approach for developing creative tourism offers. *Tourism & Management Studies, 15*(SI), 42-49.

Cabeça, S. M., Gonçalves, A. R., Marques, J. F., & Tavares, M. (Eds.). (2019b). *Book of Abstracts of the 3rd CREATOUR International Conference and Creative Tourism Showcase: Creative Tourism Dynamics: Connecting Travellers, Communities, Cultures, and Places*. Faro: Universidade do Algarve.

Canalis, X. (2016, Mar.). Certifican la muerte del Turismo de Experiencias. *Hosteltur, Innovación Turística*.

Clokes, P., & Jenkins, H. C. (1989). Cracking the Canyon with the Awesome Foursome: Representations of Adventure Tourism in New Zealand. *Environment and Planning. D, Society & Space*, *16*(2), 185–218. doi:10.1068/d160185

Cohen, E., & Cohen, S. A. (2012). Current sociological theories and issues in tourism. *Annals of Tourism Research*, *39*(4), 2177–2202. doi:10.1016/j.annals.2012.07.009

Coordination Commission of Development for the Region of the Algarve [CCDR Algarve]. (2015). *RIS3 - Algarve 2014-2020 Estratégia regional de investigação e inovação para a especialização inteligente*. CCDR.

D'Angella, F., & Go, F. M. (2009). Tale of two cities' collaborative tourism marketing: Towards a theory of destination stakeholder assessment. *Tourism Management*, *30*(3), 429–440. doi:10.1016/j.tourman.2008.07.012

European Travel Commission. (2017). *European Tourism Trends & Prospects Q4/2017, ETC*. Retrieved from: https://etc-corporate.org/uploads/reports/ETC-Quarterly-Report-Q4-2017_Final-Public.pdf

Fabre, D. (2001). L'histoire a changé de lieux. In *A. Bensa & D. Fabre (dirs.), Une histoire à soi. Figurations du passé et localités*. Éd Maison des Sciences de l'Homme. doi:10.4000/books.editionsmsh.2952

Galani-Moutaf, V. (2013). Rural space (re) produced – Practices, performances and visions: A case study from an Aegean island. *Journal of Rural Studies*, *32*, 103–113. doi:10.1016/j.jrurstud.2013.04.007

Gonçalves, A. R. (2008). As Comunidades Criativas, o Turismo e a Cultura. *Revista Dos Algarves*, *17*, 10–18.

Gu, H., & Ryan, C. (2008). Place attachment, identity and community impacts of tourism—The case of a Beijing hutong. *Tourism Management*, *29*(4), 637–647. doi:10.1016/j.tourman.2007.06.006

Idziak, W., Majewski, J., & Zmyślony, P. (2015). Community participation in sustainable rural tourism experience creation: A long-term appraisal and lessons from a thematic villages project in Poland. *Journal of Sustainable Tourism*, *23*(8-9), 8–9, 1341–1362. doi:10.1080/09669582.2015.1019513

Ilincic, M. (2014). Benefits of Creative Tourism - The Tourist Perspective. In G. Richards & A.P. Russo (Eds.), Alternative and Creative Tourism. Barcelona: ATLAS - Association for tourism and Leisure Education.

Jensen, R. (1999). *The dream society: how the coming shift from information to imagination will transform your business*. McGraw-Hill.

Kashdan, T. B., Biswas-Diener, R., & King, L. A. (2008). Reconsidering happiness: The costs of distinguishing between hedonics and eudaimonia. *The Journal of Positive Psychology*, *3*(4), 219–233. doi:10.1080/17439760802303044

Lazzeroni, M., Bellini, N., Cortesi, G., & Loffredo, A. (2013). The Territorial Approach to Cultural Economy: New Opportunities for the Development of Small Towns. *European Planning Studies*, *21*(4), 4. doi:10.1080/09654313.2012.722920

Lorentzen, A. (2009). Cities in the experience economy. *European Planning Studies*, *17*(6), 829–845. doi:10.1080/09654310902793986

Martin, D., & Woodside, A. (2008). Grounded Theory of International Tourism Behavior. *Journal of Travel & Tourism Marketing*, *24*(4), 245–258. doi:10.1080/10548400802156695

Miettinen, S. (2005). Creative tourism as tool for local empowerment. In G. Richards & J. Wilson (Eds.), From Cultural Tourism to Creative Tourism. Part 4 (Changing Experiences - the development of Creative Tourism). ATLAS.

Perkins, H., & Thorns, D. (2001). Reflections on Urry's tourist gaze in the context of contemporary experience in the antipodes. *International Sociology*, *16*, 185–2004. doi:10.1177/0268580901016002004

Pine, B. J., & Gilmore, J. H. (1999). The Experience Economy: Work is Theatre and Every Business a Stage. Harvard Business School Press.

Prahalad, C. K., & Ramaswamy, V. (2004). *The future of competition: Co-creating unique value with customers*. Harvard Business School Press.

Região de Turismo do Algarve. (2014). *Plano de Marketing Estratégico para o Turismo do Algarve 2015-2018*. Retrieved from https://www.turismodeportugal.pt/SiteCollectionDocuments/estrategia/Estrategias-Regionais-Algarve/Plano-Marketing-Estrategico-Turismo-Algarve-2015-2018.pdf

Richards, G. (2011). Creativity and tourism. The state of the art. *Annals of Tourism Research*, *38*(4), 1225–1253. doi:10.1016/j.annals.2011.07.008

Richards, G. (2016). The challenge of creative tourism. *Ethnologies (Québec)*, *38*, 1–2, 31–42.

Richards, G., & Raymond, C. (2000). Creative Tourism. *ATLAS News*, *23*, 16–20.

Richards, G., & Wilson, J. (2005). Changing Places - the spatial challenge of creativity. In G. Richards & J. Wilson (Eds.), *From Cultural Tourism to Creative Tourism. Part 3 - Changing Places - the spatial challenge of creativity* (pp. 7–8). ATLAS.

Shucksmith, M. (2018). Re-imagining the rural: From rural idyll to Good Countryside. *Journal of Rural Studies*, *59*, 163–172. doi:10.1016/j.jrurstud.2016.07.019

Simpson, M. C. (2008). Community Benefit Tourism Initiatives - A conceptual oxymoron? *Tourism Management*, *29*(1), 1–18. doi:10.1016/j.tourman.2007.06.005

Tan, S. K., Kung, S. F., & Luh, D. B. (2013). A Model of Creative Experience in Creative Tourism. *Annals of Tourism Research*, *41*, 153–174. doi:10.1016/j.annals.2012.12.002

Thomas, N. J., Harvey, D. C., & Hawkins, H. (2013). Crafting the Region: Creative Industries and Practices of Regional Space. *Regional Studies*, *47*(1), 75–88. doi:10.1080/00343404.2012.709931

Turismo de Portugal. (2017). *Relatório de Sustentabilidade*. Retrieved from https://travelbi.turismodeportugal.pt/pt-pt/Documents/Sustentabilidade/Estrat%C3%A9gia%20de%20Sustentabilidade/relatorio-sustentabilidade-2017.pdf

United Nations Educational, Scientific and Cultural Organisation. (2006). *Towards Sustainable Strategies for Creative Tourism. Discussion Report of the Planning Meeting for 2008 International Conference on Creative Tourism, Santa Fe, New Mexico, U.S.A., October 25–27, 2006*. UNESCO Creative Cities Network. Retrieved from: https://unesdoc.unesco.org/ark:/48223/pf0000159811

ADDITIONAL READING

Bakas, F. E., & Duxbury, N. (2018). Development of rural areas and small cities through creative tourism: The CREATOUR project. *Anais Brasileiros de Estudos Turísticos*, 8(3), 74–84. Available at: http://ojs2.ufjf.emnuvens.com.br/abet/article/view/13869

Barcelona Turisme Creatiu. (2010). *Creative Tourism Study*. FUSIC.

Cabeça, S. M., Gonçalves, A. R., Marques, J. F., & Tavares, M. (2018). Contributions of creative tourism to sustainable development: The CREATOUR research and incubation project. *Proceedings of HERITAGE 2018 – 6th International Conference on Heritage and Sustainable Development* (pp. 1671–1680). Granada, Spain.

Carvalho, R. M. F., Costa, C. M. M., & Ferreira, A. M. A. P. (2019). Review of the theoretical underpinnings in the creative tourism research field. *Tourism & Management Studies*, 15(SI), 11–22.

Duxbury, N., & Richards, G. (Eds.). (2019). *A Research Agenda for Creative Tourism*. Edward Elgar Publishing. doi:10.4337/9781788110723

Gonçalves, A. (2008). As comunidades criativas, a cultura e o turismo. *Dos Algarves: A Multidisciplinary e-Journal*, 17, 10–18.

Marques, L., & Richards, G. (Eds.). (2012). Special journal issue: Exploring creative tourism. *Journal of Tourism Consumption and Practice*, *4*(2).

Organization for Economic Co-operation and Development. (2009). *The Impact of Culture on Tourism*. OECD.

Richards, G., & Wilson, J. (2006). Developing creativity in tourist experiences: A solution to the serial reproduction of culture? *Tourism Management*, *27*(6), 1209–1223. doi:10.1016/j.tourman.2005.06.002

United Nations World Tourism Organisation. (2018). *Report on Tourism and Culture Synergies*. Madrid: UNWTO. Available at: https://unesdoc.unesco.org/images/0015/001598/159811e.pdf

KEY TERMS AND DEFINITIONS

Algarve: Region in South Portugal, well known as a "sun and sea" destination. It is one of the most touristic regions of the country and tourism is its main economic driver.

Co-Creation in Tourism: interaction process between residents and tourists in which both have active roles in the tourism experiences.

Creative Tourism: A type of tourism that develops the tourists' creativity through their active participation in experiences which are characteristic of their destination.

CREATOUR Pilots: Cultural and creative organizations that offer creative tourism activities and are partners of the CREATOUR Project.

CREATOUR Project: Portuguese research project that aims to develop creative tourism destinations in small towns and rural areas of Portugal.

Incubation Project: A project that focus in collaborative processes, in which the implementation of activities by the research partners is a key exercise.

Place-Based Tourism: The use of the destinations' cultural, social and human capital in the tourism offers. Provides an immersion into local culture and intense interaction between tourists and local residents. Strongly linked with the territory.

Chapter 14
Operant Resources of Young Independent Travelers

Ana Cláudia Campos
https://orcid.org/0000-0002-5816-5137
Faculty of Tourism and Hospitality, European University of Lisbon, Portugal & Research Centre for Tourism, Sustainability, and Well-Being (CinTurs), University of Algarve, Portugal

Sofia Almeida
https://orcid.org/0000-0003-4561-6776
Faculty of Tourism and Hospitality, European University of Lisbon, Portugal & Centre for Geographical Studies, IGOT, University of Lisbon, Portugal

ABSTRACT

In this chapter, the construct of operant resource is explored in the context of young independent travel. The main purposes of the study are to analyze young travelers' perceptions about independent travel and operant resources that contribute the most to its success. A qualitative design was adopted, given the explorative nature of the research. Online semi-structured interviews were conducted to a sample of young travelers, using the snowball sampling procedure. Qualitative data analysis was conducted subsequently and main findings indicate that freedom, being in control, and authenticity are predominant feelings within this sample. They also show that most valued operant resources in the context of independent travel are informational and human. Conclusions drawn contribute to a better understanding of independent travel by youths and may help tourism businesses and destinations to strategically approach this market segment.

INTRODUCTION

The S-D Logic proposed by Vargo and Lusch (2004, 2006, 2008, 2017) postulates a new approach to value creation, in which consumer participation is fundamental. In the experience economy, this importance is even more significant, because the consumption experience is *phenomenologically and subjectively lived by the individual* (Campos, Mendes, Valle & Scott, 2015; Mannell & Iso-Ahola, 1987; Ye, Tussyadiah & Fesenmaier, 2009), and depends on the application of the consumer's operant resources. Value

DOI: 10.4018/978-1-7998-3156-3.ch014

creation in this context is the effect of the consumer's contribution to the creation of the experience and value the contextual appraisal of the lived experience (value-in-context, value-in-experience). Operant resources are the set of intangible resources of the various firm's agents and levels that make value creation possible (Madhavaram & Hunt, 2008). From the consumer's perspective, operant resources are considered the knowledge and diverse individual capabilities that influence their greater or lesser capacity to create the consumption experience (Prebensen, Vittersø & Dahl, 2013) and relate to others in interactive environments (Ramaswamy & Ozcan, 2018). The youth market is growing, and there is increasing interest in understanding this segment or sub-groups of it, which adopt an independent style of travel (Farahani & Sukmajati, 2011). Young travelers are known for their preference for independent travel, active participation, and adventure (WTO, 2008). They are also digital natives, hence skilled in the use of digital technologies, interfaces and devices (Starcevic & Konjikusic, 2018). Therefore, the application and development of skills for self-planning the trip and selecting itineraries and engaging activities at the destination are vital. As there is need to understand better the youth's segment (Buffa, 2015; Cavagnaro, Staffieri & Postma, 2018; Hyde & Lawson, 2003), the primary purpose of this study is thus to explore the operant resources involved in the construction of their tourism experience when this experience takes place independently of the mass tourism supply chain and the consumer is this young independent traveler, who prefers to venture in the trip neglecting traditional, pre-organized tourism packages, circuits, and tours. A qualitative approach was adopted to achieve this explorative aim, and accordingly, semi-structured online interviews were conducted to a group of young independent travelers using a snowball sampling procedure. Findings from the study show that sense of freedom and sense of control are predominant in this group and that informational and human resources are the most valued.

The structure of this chapter is as follows: the theoretical sections address main topics related to youth tourism and thus reviews issues such as critical statistics, segment dimension, and interest from tourism associations worldwide. From the initial quantitative approach, the review continues briefly by addressing the characteristics of these consumers in terms of profiles, types, benefits sought, and key motivations. The theoretical part of the chapter ends by highlighting the resources of young travelers in the context of the S-D Logic framework and discourse. This study sheds light on the construct of the *operant resource* not only by introducing and explaining this concept but also it argus *operant resources* as a pertinent basis of the analysis of young travelers. This chapter presents and discusses the accomplished empirical research and clarifies its applied methodology. It continues by representing findings and finally concludes by highlighting key insights gained, its contribution and the limitations of the study.

The Appeal of the Youth Market Segment

Despite the discussion on what counts as *youth tourism* (Farahani & Sukmajati, 2011), in 2008, the World Tourism Organization (WTO) has reached a working definition. According to WTO, youth travel is defined as encompassing all independent trips which last less than one year by motivated young people with the age of 16 to 29 who desire to experience other cultures, to build life experience or to benefit from formal and informal learning opportunities outside their usual environment (WTO, 2008). However, this definition has not yet led to consistent policies and marketing approaches towards this segment (Dionysopoulou & Mylonakis, 2013; Yousaf, Amin, & Santos, 2018). The travel of the youth segment in tourism encompasses different types derived by various reasons, namely, work, education, leisure, or as a volunteer (Richards, 2015).

In 2011, WTO estimated that approximately 20% of the 940 million international tourists traveling around the world in 2010 were young people. Another report indicates that the youth tourism market represents approximately 23% of the 1,235 million tourists traveling each year (WTO, 2008) globally. Additionally, WTO forecasts that in 2020 the number of international youth trips will point to 370 million, which not only presents a significant opportunity for future growth in the travel industry (WTO, 2016) but also claims that youth travel becomes an attractive segment to any tourism destination (WTO, 2008). Changes in values and lifestyle of affluent societies may be identified as a reason for this trend, and the appeal remains as these young travelers represent the markets of the future (Vukic, Kuzmanovic & Stankovic, 2014).

As a response to the constant growth of the youth travel segment and the need to strategically position destinations, data collection about these travelers, their habits, and profiles developed across countries and continents. Concurrently, specific academic research is on the rise, as many studies are emerging, addressing particular topics of interest (Richards, 2015). Current empirical research is exploring attitudes, behaviors, decision-making processes, and preferences of young travelers (Hyde, 2008; Hyde & Lawson, 2003; Buffa, 2015). Some tourism-based agencies and associations are established and collaborated to generate useable knowledge to destinations and tourism businesses (WTO & WYSE, 2011), while others started developing research to uncover missing information about these consumers of tourism (Table 1). These organizations assume that by identifying the motives of young travelers, they can assist them to achieve sought benefits. One of these entities' roles is to measure the weight of the youth segment in the national economy and to extrapolate its potential growth. Although there are many associations around the world focused on young travelers, according to UNWTO's and WYSE Travel Confederation' report (2011), among European countries, only Spain and Austria are developing knowledge on this segment (Table 1).

At the international level, the report highlights the missions of the associations in Russia, Australia, China, and South Africa (WTO & WYSE, 2011). Additionally, it identifies governmental projects or non-profit organizations projects which for instance help youth to experience traveling abroad, or by adopting the responsible tourism perspective, e.g., the *Federation EIL* in Switzerland facilitates standard educational exchange programs; or *STA travel* in Thailand which is a global travel experience specialist advises young people to have holidays and adventures abroad. In Melbourne, Australia, *Conservations Volunteers'* mission is to rehabilitate the degraded landscape and return it to full environmental health; *ESTS* (Egyptian Student travel services) is a non-profit organization who helps in the promotion of educational, cultural and social travel among youth and students in Egypt; and *Mundo Joven* is the largest youth travel organization in Mexico who provides both outbound and inbound products services.

TRAVEL PROFILE OF YOUNG TRAVELERS

Not all young travelers travel independently in the same way. The segment of the young is not homogeneous, as it includes different generations, consumption profile, behavior, motivations, and relationship with technology (Buffa, 2015; Starĉević & Konjikušić, 2018). But as long as they share similar demographics, go through the same societal changes and learn to use identical technology, the segment assumedly also shares some common perspectives and behaviors towards tourism (Moscardo, 2009; Pendergast, 2009). For those who had access to better education, and witnessed the changing roles and structures within the family group (Buffa, 2015), trips are sophisticated and have a lasting impact on

Table 1. Organizations developing research on young travelers

Name of Association	Youth Travel Data	Mission/Goals
Spanish General Secretariat of Tourism (SGST) (Spain)	In 2010, more than 700 000 foreign nationals travelled to Spain to study (mainly coming from France, Germany and USA)	Investing in the Youth Travel segment allows market diversification, and increase profitability, de-seasonality, and dissemination of the Spanish language through foreign students.
Australian Tourism Export Council (ATEC) (Australia)	The Youth market to Australia accounts for 26% of total markets	The objective is to ensure that the youths travelling today and tomorrow have the opportunity to discover and immerse themselves in the Australian experience.
Wyse Travel Confederation (Russia)	-	To develop quality international work programs. They are transformational, life enhancing and are a hugely positive experience for the majority of young people who participate in them, as well as the communities which host them.
Dragon Tail (China)	The Youth Travel market is represented by people born in 70s, 80s and 90s. 95% of Chinese tourists are unsatisfied with current travel products and services available both domestically and internationally	Young travelers' attention is captured through (i) social media, which is the most influential medium in modern China; and (ii) the internet, which became the most important channel to connect with affluent consumers all over China. The strategy is to create multi-directional developments, such as in-depth and theme tours.
South Africa Youth Travel Confederation (SAYTC) (South Africa)	A study shows that although relatively small in terms of total tourism arrivals, the backpacking and youth travel market represents a high yield and is a high growth market with significant development impacts in the country	Through research and strategic partnerships, South Africa has managed its share of the Youth Travel market and positioned itself as a responsible tourism destination.
Cape Town Tourism (South Africa)	Top backpacking source market is the UK. Preferred destination is Cape Town and Kruger Park. At the local level, main characteristics of the Youth Travel market are: no gender differences; more than 68% below the age of 30; more than 50% hold undergraduate degree and 27,3% are post-graduate	To develop the backpacker up-market travel experience. To maintain the positioning of South Africa as the most popular destination for volunteer tourism.
Modul Vienna University (Vienna)		Travelling helps young travelers to pass to the adulthood stage and develops their feelings of independence and the ability to solve problems

Source: (Authors adapted from WTO and WYSE, 2011)

their personal and professional lives (Richards & Wilson, 2003). The WYSE Travel Confederation's research (2008) shows that young travelers often spend more time in the destination than other tourists (higher length of stay); they probably return and give more value to the travel destination and tourism over their lifetime (Richards & King, 2003). Many of them are adventurous and pioneers who desire to discover new destinations, and thus contribute to the visibility of the places they are visiting and draw attention to the destination's cultural benefits. Inclination towards sustainable tourism practices has also been noticed among this cohort (Buffa, 2015).

Young travelers are tech-savvy consumers, as most of them were already born in the new digital technology paradigm. The pervasive availability and use of the Internet enable continuous connectivity through digital devices, thereby granting them access to information regularly. They expect tourism companies and destinations to provide them with good quality and free Internet at all times and places (Yousaf et al., 2018). They quickly understand digital technologies and are quite skilled in using multiple devices (Starčević & Konjikušić, 2018). Therefore, self-empowerment and confidence are consequences of their lifestyle, and values are anchored on easily accessible information platforms and tools found in contemporary affluent societies (Starčević & Konjikušić, 2018).

For young travelers, traveling is a way of living; it is a familiar activity and part of lifestyle (Starčević & Konjikušić, 2018); a way of meeting new people and learning from new cultures (Buffa, 2015). Moreover, traveling is considered as an experience of self-development and building the identity, as travel experiences help youths to develop their personality. The youngs are *experience seekers* who may travel for extended periods searching for culture, adventures, and relaxation and are frequently repeaters of the destination (Richards and Wilson, 2003). They have an interest in local cultures, desire to know about local people and their ways of life. Interaction with community members is also one of the youth's preferences (Yousaf et al., 2018). This preference has been argued as an inclination towards principles of sustainable tourism practice (Buffa, 2015). However, these travelers are not a rigidly homogeneous group, as the adoption of various traveling choices impacts consumption behavior differently; therefore, it influences tourism businesses and destinations as well.

TYPES OF YOUNG TRAVELERS

Literature suggests that young travelers can be split into many sub-groups: travelers, tourists, volunteers, language students, exchange students and interns (Richards, 2015), demonstrating different attitudes towards mobility, tourism, motivations, purposes and travel behavior regarding information search, planning, and decision making (Buffa, 2015; Hyde & Lawson, 2003; Richards, 2015; WTO, 2008). Though not all young travelers adhere to a mode of independent travel, they all are offspring of the contemporary mobility phenomenon, adopting one of the different shapes of the nomad (Richards, 2015). Despite this, a trend towards increase and diversity in youth mobility (Richards & Wilson, 2003) will grant independent travel particularly important. Richards' study (2015) focused on youth mobility and contemporary modes of nomadism. This study's findings represent three prominent types of young travelers: the global nomads, the backpackers, and the flashpackers. These travelers live the mobility phenomenon quite differently, as their relationship with the destination, other travelers, the host community, and digital technology reveals significant differences in their attitude and behavior.

According to Hyde (2008) and Hyde and Lawson (2003), independent travel represents a contemporary trend in tourism concurrent with the decline in popularity of package travel and practice deserves more attention from researchers. Independent travelers are those who take trips independently, i.e., without pre-booking a trip or purchasing a package tour, or they may book only a minimum of their transportation and accommodation arrangements before departure (Hyde & Lawson, 2003; Lawson, 2008). More specifically, "those vacation travelers who have booked only air travel with the travel retailer would be considered independent travelers; those vacation travelers who have booked their travel or accommodation through the Internet would also be regarded as independent travelers (Hyde & Lawson, 2003: 13). Theoretically, Hyde and Lawson (2003) argue that definition of independent travel should be based on the traveler behavior rather than the type of distribution channel used, and thus characterized by a style of travel marked by flexibility in the itinerary and a significant degree of freedom about where to travel within a destination region. According to the authors, the independent profile is looking for the unexpected and willing to take risks, paying little consideration to rigid planning and itinerary of the trip. This type of traveler has the soul of an explorer and prefers to organize their itinerary themselves rather than requiring full service from a professional travel agency. In turn, the backpacker profile shows a preference for budget accommodation and is a frequent user of public transportation (Kelly, 2001). This traveler is a serendipity-friendly tourist, as it values movement, mobility, and also flexible itineraries (Ateljevic & Doorne, 2001). Backpackers travelers count as a specific group within independent travelers (Hyde and Lawson, 2003), and are framed as a traveler in the opposition of the tourist, who is being caught in the constraints of the travel industry and pre-designed travel packages (Richards & Wilson, 2003). Based on prior research, Richards (2015) defined backpackers as individuals that tend to gather in a group that forms enclaves and develops a specific culture revolving around personal stories. His empirical research corroborated the tendency to create a backpacker community "supported by the regular exchange of information and face-to-face contact with other travelers" (p. 349).

BENEFITS SOUGHT BY YOUNG TRAVELERS

Benefits sought by young travelers are of three types: individual, collective, and economic (WTO, 2011). Individual benefits concern individual growth of the young person: what he or she learns from the experience or the level of absorption of explicit knowledge about the host culture or the overall destination. Collective benefits involve the group; it refers to knowing different people around the world as well as the local communities. Cultural knowledge is present when travelers are eager to discover local habits, routines, and traditions. This aspiration of exploration and learning the unknown world has been termed the *Ulysses factor* (Anderson, 1970 as cited by Hyde & Lawson, 2003). Lastly, economic benefits are reachable by discovering new destinations and creating new partnerships that emerge from local interaction facilitated by this discovery. Therefore, economic benefits are also observed through the emergence of new types of tourism (Demeter & Bratucu, 2014).

MOTIVATIONAL FACTORS

Pearce (1990) identified three critical motivations of contemporary backpackers: extending one´s education, escaping from dayly life pressures, and engaging in occasional work to extend travel time. A fourth motivation is related to the emergence of a subculture focused on the pursuit of health and outdoor activities. Richards and Wilson (2003), on the other hand, found that the primary motivations for young tourists to travel are to explore other cultures, to seek excitement, and to increase knowledge by encountering different people and places. Young travelers' trips usually combine at least two different main motives, more frequently relaxation and excitement, or lazing on the beach and cultural encounters. Richards & Wilson (2003) concluded that different traveler types show significant differences in terms of motivation. Backpackers, in particular, tend to emphasize items related to experience seeking and are relatively less concerned about relaxation. Results also indicated the existence of four main motivational factors for young travelers, which are broadly characterized as *experience-seeking, relaxation seeking, sociability, and contributing to the destination*. More recently, Buffa (2015) examined common motivations behind the selection of the travel destination, namely: *discovery of new cultures, discovery of new landscapes, contemplation of natural and artistic heritage, contact with the local community*, and *contact with na-*

Table 2. Different types of independent travelers

Author	Independent Traveler	Backpacker Traveler
Anderson (1970)	The traveler is not looking for anything in particular and is not greatly concerned with what discovers. He's a true explorer	
Ateljevic & Doorne (2001)		The backpacker is characterized by movement and mobility
Elsrud (2001)	Risk and adventure separates long-term independent travelers from their bad cousins: the tourists.	
Hyde & Lawson (2003)	The traveler experiences an evolving itinerary, is willing to take risks in selecting vacation elements, and possesses a desire to experience the unplanned.	Backpackers are one sub-group of the independent travel market
Kelly (2001)		Usually these travelers travel year-round. They are keen on using public transportation. They only carry essentials with them.
Loker-Murphy & Pearce (1995)		Show great interest in meeting other people and like to connect to the local culture
Morrison, Hsieh, & O'Leary (1993)	The ones who make their own transportation and accommodation arrangements, choosing not to buy prearranged packages or tours. Accept prearranged itineraries or flexible itineraries	
Richard & Wilson (2003)		Shows a preference for budget accommodation, and places emphasis on meeting other travelers. The travel schedule is independently organized and flexible. Individuals are aged between 20 to 24 and prefer longer rather than short holidays. Also show preference for informal and participatory holiday activities.
Richards (2015)		They are 'tribal nomads', congregating in specific enclaves to share the stories that form the shared culture of the tribe. They form a self-sufficient and coherent nomadic group
Riley (1988)		Likely to be middle class, at a juncture in life, somewhat older than the earlier travelers on average. Holds college education. However, the backpacker is not an aimless drifter. Usually travels under flexible timetables and itineraries.
Tourism New Zealand (2002)	Those with no travel bookings other than an international air ticket. Additional bookings are not purchased as part of a traveler retailer package	

Source: (Adapted from Hyde and Lawson, 2003)

ture. The shaping of varied types of independent travel is a response to diverse motivations. Demeter and Bratucu (2014) also report educational tourism, volunteering, work travel, cultural exchange, sports and adventure tourism, and leisure tourism as attractive types of tourism to young people.

Table 3. Travel benefits for young independent travelers

Institution	Individual Benefits	Collective Benefits	Economic Benefits
UNWTO and Wyse Travel Confederation	a new form of learning a source of career development a means of self-development a part of personal identity	meeting other people knowing other cultures	
Official Tourism Organizations (OTO)		encourages more tourists to visit a country Meeting local people Supports cultural industries and also social life of destination	increase the economic impact of tourism spreading new areas for tourism fighting seasonal effects stimulate innovation develop new partnerships
UNWTO (2012)	-	develops the interaction with local people and cuisine develops unique food experiences such as: farm-to-table dining experiences; gastronomic routes; cooking classes; wine tasting; food events and fairs	develops the concept of food tourism develops the concept of student tourism
Richard and Wilson (2003)		builds bridges between people and cultures. concentrates spending in local communities	spend longer periods travelling than other tourists, spreading economic contribution further in terms of both time and space develop prototypes for the new tourist by setting trends
Horak & Weber (2000)			create new attractions and help to establish destinations
Krippendorf (1987)	travel offers the individual a sense of freedom and self-determination not available in everyday life		

Source: (Authors adapted from WTO, 2011)

S-D LOGIC AND THE YOUNG TRAVELERS' RESOURCES FOR CREATION OF THE INDEPENDENT TRAVEL EXPERIENCE

In the context of evolved, experienced and empowered consumers, research on tourist resources has been developing to understand and deepen the co-creation perspective of tourism experiences (Campos et al., 2015; Prebensen & Foss, 2011; Prebensen et al., 2013). As long as tourism travel is undertaken voluntarily and desired for attaining hedonic or eudemonic purposes, this makes it a particular type of consumption; therefore, there is a specific need to analyze the behavior and perception of the consumer (Prebensen et al., 2013).

Both S-D Logic and theory on experience co-creation (Vargo & Lusch, 2004; Prahalad & Ramaswamy, 2004) stimulated a line of research focused on the tourism experience, exploring how individuals create experiential value by taking an active part in the design and consumption of their travel experience

(Bertella, 2014; Campos, Mendes, Valle & Scott, 2016; Mathiesen, 2013; Prebensen and Foss, 2011; Prebensen & Xie, 2017). Research has shown that engaging tourists in the design and consumption of their travel experiences have a twofold benefit and outcome: from the tourist perspective, the affective, cognitive, and behavioral dimensions of engagement decisively contribute to the memorability of the overall experience (Schmitt, 1999; Campos et al., 2016), thus bringing personal value and meaning to the individual; from the suppliers point of view, the higher the tourist's experience memorability, the higher the competitiveness of businesses which facilitate memorable experiences (Neuhofer, Buhalis, & Ladkin, 2012).

The S-D Logic framework developed initially by Vargo & Lusch in several publications (2004, 2006, 2008), as well as posterior theoretical frameworks which propose alternative approaches to firms in the current business environment (Grönroos & Voima, 2013), made it possible to discuss value creation in the context of service systems. In particular, Vargo & Lusch (2004) explored value creation and co-creation in service exchange systems, where value is created through the participation of different actors, and each actor contributes to the value co-creation process by integrating its resources with others. S-D Logic elaborates on a broad notion of service by emphasizing the application of competences in deeds, processes and performances, which benefit "another entity or the entity itlself" (Vargo & Lusch, 2004). The concept of *operant resources* was introduced to identify the set of competencies, knowledge, and skills that are required to act upon other types of resources, namely operand resources, such as equipment, machinery, or raw materials (Vargo & Lusch, 2008; Vargo & Lusch, 2017). Operant resources, by their own very nature, are intangible, continuous, and dynamic (Vargo & Lusch, 2004).

Although resource theory is not a new topic, S-D Logic opens a further discussion on what is a resource (Madhavaram & Hunt, 2008). Resources have been theoretically analyzed in several frameworks, namely, the resource-based view of the firm or the resource-advantage theory (Madhavaram & Hunt, 2008). Resources underpin "the tangible and intangible entities available to the firm that enable it to produce efficiently and/or effectively a market offering for some market segment(s)" (Madhavaram & Hunt, 2008: 68). In this respect, resource theory developed by examining and categorizing types of resources and their relative importance to the firm's outputs, competitiveness. Resources are defined by Barney (1991:101) as "all assets, capabilities, organizational processes, firm attributes, information, knowledge, etc., controlled by a firm that enables the firm to conceive of and implement strategies that improve its efficiency and effectiveness." This definition includes resource categories such as financial, physical, legal, human, organizational, informational, and relational resources (Madhavaram & Hunt, 2008). Firm's outperformance, competitiveness, and dynamism depend on the ability of the firm to combine efficiently and effectively those varied resources and competences.

S-D Logic represents an evolution in marketing theory (Madhavaram & Hunt, 2008) by focusing on operant resources as a source of value creation and firm competitive advantage (Lusch & Vargo, 2004; Lusch, Vargo & O'Brien, 2007; Vargo & Lusch, 2017). In S-D Logic, firms are service systems (Vargo, Maglio, & Akaka). Vargo & Lusch (2008) integrated the concept of service systems into S-D Logic so that analysis of value creation could expand to all sorts of value configurations, including individuals, families, firms, departments, cities, societies, nations, i.e. all entities "that exchange to improve their own state of being" (Vargo & Lusch, 2008: 5). Further theoretical developments led to the concept of service ecosystem to describe the networked foundations of interactions among participant actors at various levels of aggregation (Vargo & Lusch, 2017). Components of the system are all the people, technology, value propositions, and shared information interacting at various levels, able to be represented by a specific network arrangement, better understood as a holistic experience (Vargo & Lusch, 2017). Systems are as big as the number of actors involved in value co-creation: they range from one individual as it interacts with others, to the global economy. As all system parties are involved in interactions that put resources into collective use, value, is always co-created. Systems are inherently relational, and their resources are applied and integrated for the benefit of all actors. A primary assumption of S-D Logic is to consider the customer as a system's beneficiary who co-creates value by integrating his/her specific resources into the value co-creation process. Value is described as an effect of customer participation in the process (value-in-use) (Vargo and Lusch, 2004) and determined by the beneficiary (Vargo, Maglia & Akaka, 2008). The value thus acquires subjective and contextualized meaning, which has to be understood in terms of the customer's context or frame of reference (value-in-context) (Vargo et al., 2008). In short, value is not exchanged but is co-created, and the customer determines it through participation "in the process of acquisition, usage, and disposal" (Vargo et al., 2008: 148). Operant resources and their effortful and combined application in systems is what contributes to the customer's determination of value. Recent research emphasized and considered experiences as the actual contexts where interactions occur, and value is co-created (Prahalad & Ramaswamy, 2004). Theoretically, co-creation diverged from S-D Logic by focusing on co-created consumption experiences (Ballantyne, Williams, & Aitken, 2011; Ramaswamy & Gouillart, 2010) rather than concentrating on system processes to co-create value. Co-creation theory evolved to integrate and emphasize new co-creation environments, characterized by heterogeneous relations and driven by digital technologies available in support interfaces, platforms and devices (Ramaswamy & Ozcan, 2018).

Tourism studies began to analyze S-D Logic and co-creation approaches to explore the tourism experience against the circumstances of more engaged, sophisticated, empowered, demanding, and participative tourists (Andrades & Dimanche, 2014; Bertella, 2014; Binkhorst and Den Dekker, 2009; Campos et al., 2015; Kreziak & Frochot, 2011; Minkiewicz, Evans & Bridson, 2013; Prebensen & Foss, 2011; Sfandla & Björk, 2012), adhering to the claim that tourism is a system where all intervening parties are creators of value, including the tourist. More recent work addressed the tourism experience by investigating tourist resources and their applications in experiences. In the tourism literature, these have been identified variedly with *attention* (Campos et al., 2016), *active physical and mental participation* (Campos et al., 2017; Minkiewicz et al., 2013; Prebensen, 2017; Prebensen, Kim, & Uysal, 2015; Prebensen & Xie, 2017), *creativity* (Richards & Wilson, 2006; Tan, Luh, & Kung, 2014), *effort* (Prebensen et al., 2013), *imagination* (Minkiewicz et al., 2013; Prebensen & Foss, 2011), *interpersonal interaction* and *involvement* (Campos; Mendes, Valle & Scott, 2017; Prebensen & Foss, 2011; Prebensen et al., 2013), *interest* and *motivation* (Prebensen and Foss, 2011; Prebensen, Kim, & Uysal, 2015; Prebensen, Woo, Chen, & Uysal, 2013), *learning* and *knowledge* (Prebensen & Foss, 2011), *mastering* and *self-efficacy* (Prebensen & Campos, 2017; Prebensen, Campos & Blomstervik, 2018; Prebensen & Xie, 2017), *money* (Prebensen et al., 2013), *role play* (Bertella, 2014; Mathiesen, 2013), and *time* (Prebensen & Foss, 2011, Prebensen et al., 2013).

Despite this growing attention, the topic of tourist resources has not yet been explored in the context of independent youth travel. This research argues that the theoretical framework of S-D Logic, by characterizing the individual as a service system which co-creates value through interaction with others, as well as postulating the *sine qua non*-condition of application and integration of the customer's operant resources to create experiential and personalized value, is pertinent to study young travelers competences to travel independently. In this respect, the distinction made by resource theory between informational, human, and relational resources is pivotal to this understanding. Young travelers live in digitalized societies, and the use of digital technologies is pervasive in their lives, both as individuals and as travelers in a globally accessible world. Additionally, better-quality education has brought specific competencies (e.g., information and knowledge) to young travelers, which enhances their confidence and sense of competence. Lastly, by experiencing new patterns of mobility, youths are, as never before, able to apply the relational competencies necessary to better connect with the world and the communities they encounter during their travels. The empirical research conducted attempts at uncovering young travelers' perceptions about how competent they are to plan and organize their trips.

METHODOLOGY

Context

This study aims to develop knowledge of the resources that young travelers use when traveling independently, i.e., without using travel packages provided by tourism-related companies, during all stages of the overall tourism experience. The young travelers' operant resources in this research refer to *all the individual's resources applied in the planning and consumption of the independent travel experience*. This working definition is derived from the S-D Logic proposed by Vargo and Lusch (2004; 2006), as well as from research undertaken on co-creation in tourism (Prebensen et al., 2013). In this study, three categories of operant resources namely, *informational, human* and *relational* will be analyzed based on the previous literature about the resource concept (Alves, Ferreira & Fernandes, 2016; Barney, 1991; Constantin & Lusch; 1994; Hunt 2004; Madhavaram & Hunt, 2008) (Table 4).

To deeply understand the youth traveler segment and to explore how do young travelers as active individuals perform independently, this research adopts a qualitative approach and method as a way to value their voice as creators of the travel experience (Finn, Elliot-White, & Walton, 2000; Jennings, 2005; Taylor, Bogdan, & DeVault, 2015).

Additionally, there is a lack of research focusing on perceptions of young travelers about operant resources applied to their independent travels. The data were collected through the semi-structured interview via email. Accordingly, for implementation of the process, Meho's (2006) guidelines were adopted. Moreover, invitations to the interview were sent to participants via email to ensure their interest in the study.

All contacts between researchers and participants were in Portuguese. Also, instructions were communicated in Portuguese. Although the interview was written in English as early invited participants were fluent in English, it was assumed that other participants might not be Portuguese. Therefore, clarifications during the process were allowed and communicated. The structure of the interview is as follows:

Firstly, it starts with a section of seven point-Likert scaled questions, from anchors 1 (=strongly disagree) to 7 (=strongly agree), covering the three types of operant resources selected for analysis. Table 4 shows summarized information relating to the category of resources, including elements, adapted items, and literature sources.

Secondly, the interview includes two questions addressing the participants' perspectives on their independent travels and the relative importance of the different types of operant resources to their positive closure. They were allowed to elaborate on the answers freely. Questions were stated as: (1) How do you feel like traveling independently? And (2) Which type of resources do you believe contribute the most to the success of independent travel?

Table 4. Operant resources, items and literature references

Resource Category	Elements	Items	Literature
Informational	Specific knowledge and information Access to information	• I search for information about the different aspects of the trip, before and during • I have a good level of knowledge on the selected destination	From Alves et al. (2016) on Participation Behaviour From Alves et al. (2016) on Customer Expertise
Human	Efforts (financial, physical, time, overall), Expertise and Self-Efficacy	*Financial resources* • I have my own cash resources to make my holiday trip • I can always rely on financial support from my family • I can always have access to a bank loan to finance my trip • The price I paid for this trip was very good *Physical resources* • In this trip, I've been physically active • I'm usually well equipped for my trip • I usually engage in physical activities during the trip • I feel physically able when participating in touristic activities *Time resources* • The time I spent preparing and undertaking this trip was very reasonable *Expertise* • I understand well all the different aspects involved in the planning of my travel experience • I feel confident about planning my own trip *Self-Efficacy* • I'm always able to resolve difficult situations, before, during or after the trip • It is easy for me to put my intentions into practice and attain my objectives • I am confident that I can deal efficiently with unexpected events • When confronted with a problem, I am generally able to find different solutions • I remain calm when facing difficulties because I trust in my capacities to deal with situations *Overall effort* • All efforts I've made for this trip were very worthwhile	Barney (1991) on **Financial Resources** Barney (1991) and Campos, Mendes, Valle & Scott (2017) on **Physical Resources** Prebensen, Vittersø & Dahl (2013) on **Price, Time,** and **Effort** From Alves et al. (2016) on **Customer Expertise** From Alves et al. (2016) on **Self Efficacy**
Relational	Refers to relationships with others in context	• When contacting with companies during my trip, I provide adequate and necessary information to ensuring good service • When contacting with companies during my trip, I carry out what is requested from me • When contacting with companies during my trip, I have an agreeable attitude towards the staff • During my trip, I give the company my opinion and ideas about the service • During my trip, interacting with others makes me want to try new things • During my trip, interacting with others makes me curious about other parts of the world • During my trip, interacting with others provides me with the opportunity to speak to new persons • During my trip, I'm always meeting new people	From Alves et al. (2016) on Bridging Social Capital

Source: (Authors)

Thirdly, the last section focused on personal information, including age group, gender, occupation, and country of origin. One variable included in the questionnaire was psychographic (motivation to travel). Additional variables were behavioural and related to last trip taken. Specifically they were: travel group, information sources used, selected accommodation type, and world region visited.

Sampling

The sample of young independent travelers was obtained using a purposive method, which fits the study's aim of reaching a similar demographic and attitudinal group (Jennings, 2005; Morse, Olson, & Spiers, 2002), however which is difficult for outsiders to reach (Finn, Elliot-White, & Walton, 2000; Kirchherr & Charles, 2018). Notably, the snowball technique was chosen as youths usually form socially intensive and extensive groups either physically and/or virtually, in that way reaching other members participants to the study more efficiently and practically (Meho, 2006; Naderifar, Goli, & Ghaljaie, 2017). Participants included in the survey integrate the range group of 18-29 years old (WTO, 2008).

Email Interviews Procedures

Procedures involved developing informed consent from prospective participants. Detailed information about the research was provided, and permission was obtained via email. Only after receiving approval, the participant was included in the sample group (Meho, 2006). Early invited participants then facilitated access to email contacts of more prospective participants who might probably agree to participate in the survey. This process eliminated avoiding, deleting, or neglecting of invitation, which might cause insufficient participation barriers. This method also reduced the researchers' costs of time in the recruitment stage.

Email Interviews Schedule

The process firstly initiated by emailing the consent forms together with the study's details and instructions and receiving permission from participants to enroll in the research. Secondly, on the next date, to minimize non-response from participants and to facilitate the process, the questions sent via email while they were embedded in the email message as youths tend to value speed and easiness of process. The length of the schedule was three weeks for the whole process, from collecting consent statements and responses to sending clarification emails to participants. Meanwhile, to develop the process seamlessly, probing emails were also addressed, some of them targeted to respondents who previously required clarification.

Data Analysis

The snowball procedure allowed the researchers to reach a sample of 15 interviewees. As the study is explorative, the sample used to analyze the research questions is considered sufficient for the research.

Interviews collected were first assigned a code number and then inspected one by one. A database was built to insert all data relative to scaled questions and participants' personal information. The two open questions, however, were treated separately and subsequently. The analysis followed a deductive procedure, as the topics were defined before the interview. Hence, analysis concerned understanding of these participants' perceptions about independent travel and resources believed as most important to independent travel. In this qualitative study, data analysis was conducted according to the requirements of research reliability and validity (Golden-Biddle & Locke, 1993). Accordingly, the researchers were knowledgeable of all contacts made with participants, and also jointly analyzed and clarified the data in a shared effort of interpretation. The *verbatim* quotes presented here were selected according to the *criterium* of relevance to the research topic and questions.

RESULTS

Sample Characterization

All the interviewees are above 18 years; they hold higher education degrees and are currently employed. They are Portuguese except the one who is from the UK. Ten respondents are female, and five are males. Most cited travel party members are the partner/spouse (11), and friends (9). Among the many motivations for traveling independently, the most significant ones mentioned by the participants are: to visit friends and relatives (12), to have a good time with friends (11), to increase self-knowledge (11), to experience excitement (10), to explore other cultures (10), to relax physically (9), and to relax mentally (8).

Concerning information sources used to organize independent trips, findings indicate that the Internet (14), relatives and friends (12), and guide books (11) were the most common channels used to access information relevant to planning the travel. In terms of types of accommodation used by these young travelers, friends, and relatives' accommodations (8) and B&B establishments (11) were found the most frequent. The world regions visited by these participants are well distributed around the globe; however, European countries (5) and long haul Asian destinations (6) were the most visited ones.

Feelings and Perceptions towards Independent Travel

The interviews were constructive to explore these participants' ideas about why traveling independently is highly appealing for them. Consensual among them was, perhaps not surprisingly, the sense of *freedom* gained from independent traveling. The sense of *freedom* was specified in terms of its potential to personalize the travel experience, to improvise flexible itineraries and activities, and to extend the length of stay of the trip contextually:

Traveling independently provides for a very freeing experience. [male]

Independent travel allows me to have the freedom to make my own decisions regarding what to do in my chosen destination. [female]

Traveling independently gives you the freedom to do whatever you feel most suitable for you. One day you can walk 20km while sightseeing a whole city and the other you can find a beach or pool and just relax all day. When you have a fixed program for your holidays is always harder to disconnect from a sense of "obligation." [male]

Personalization of the travel was understood as customization, tailoring of the experience, and described through expressions such as "personalize my wishes," "choosing my own route," "tailoring experiences to my liking," "making my own decisions" or "to do whatever feels most suitable for me," which are somewhat substitutions to individualism. Also, the sense of freedom linked to the perception of experiential authenticity and meaningful experience, as travel becomes the reflection of what indeed interests and is meaningful to the traveler. As one says,

I hate group excursions, and mostly I hate the feeling I get when I notice that the experience is not real. [female]

Or the other one, stating that

I prefer going at my own pace and seeing things that are not "touristic." [female]

Prominent in the findings was also a sense of *control*. Control was perceived as very important to the participants. It was directly identified by three participants (*I'm more in control, it is a way to control, when I'm in control of the plans*) and indirectly mentioned by the other four. For those who referred to it indirectly, control was described as *independence from agents and tourism companies, independent decision making, depending only on oneself*, and *adjusting plans as necessary.* Most frequent descriptions convey this positive feeling that control over the trip is reasonable. However, control was perceived as ambiguously when related to risk, safety, and mastering. Being in control of the travel, whether by selecting itineraries, activities or duration of the trip, brings conflicting associated feelings to participants: some consider that as control enables to reduce risk, stronger feelings of safety are expected; other participants, on the contrary, recognize that higher independence increases risk, thus potentially compromising the sense of confidence and mastering. As this participant said:

I feel like I'm more in control of what is going to happen on my trip, but also that if something doesn't go the way it was supposed to, it might be more difficult for me to find a solution because I might not have the resources or the knowledge to solve them. [female]

In general, perceptions about independent travel and its associated feelings are highly positive. Personalization is perceived significantly relevant in this context and described as freedom, namely, the freedom to plan, to improvise, to choose, to meet personal interests and needs. Experience personalization is closely linked to reliable and significant information. Information seems to not only facilitate the tailoring of the experience but also to support informed decision making.

Planning my own trips allows me to access more information about the place I am going to visit. [female]

I prefer to do my own research. [female]

Although independent travel seems, to the participants, highly connected to perceptions of individual freedom, personal identity, and interests, it also appears to enhance the value of others. It is noteworthy to mention that *others* are individuals who are relevant to the young traveler's trip. In this category, participants indicated the valuable contribution of friends and relatives to recommend destinations, activities, or places, as close relationships allow access to essential and reliable information.

Traveling independently allows me to weigh different people's opinions about what is best to visit and see. [male]

Furthermore, independent travel enhances the opportunities of casual encounters with others along the way or at the destination. These are socializing experiences with the potential to serve the required information to travelers. Through locals and other travelers, participants find that independent travel is a way to live more authentic tourism experiences.

Although sometimes it may be challenging to meet new people to socialize or obtain guidance and opinions about potential activities during my travels, it also quite rewarding when you meet interesting new people. [male]

Types of Resources Important to Independent Travel

The essential resources for independent travel identified by these participants are *informational* and *humanistic*. As young tourists are born into the digital paradigm, the Internet is the leading way to access diversified information. The Internet is said to be extensively used during the different stages of the tourism experience to collect data, most notably during the planning stage. In this stage, time becomes an essential factor in thorough research about the destination, alternatives, and required destination activities.

Be open to all opinions and take the time necessary to plan the trip. Online researches usually are very helpful when preparing for a trip. [female]

I think that research is one of the most exciting parts of the experience and enables you to have a 360º view of what you can explore. [female]

Information is found mostly on websites (e.g., *Youtube*), blogs, online tourism intermediaries (e.g., *Booking, Momondo*), online tourist guides, communities of tourists, and social media platforms (e.g., *Instagram*). Thus, the information collected is vital in building prerequisite wholesome knowledge about the destination. Travel guides (e.g., *Lonely Planet*) and recommendations by other people have been noted as relevant sources of information. In general, these participants believe that they have enhanced a beneficial level of knowledge about the destination during the planning stage of the trip.

A good base planning and the rest will come naturally. [male]

Humanistic resources have been considered all those pertaining to the individual, including financial and physical resources, time, expertise, self-efficacy, and overall effort. The general impression about perceptions of valuable humanistic resources to independent travel is that financial resources are determinant to the success of the trip. Regarding this category, one participant said that financial capacity allows people to better deal with inherent flexibility in independent travel. Besides, financial resources ensure that the travel basics are secured. Another participant also states that to engage in independent travel, one may face inherent unforeseen situations which require to feel comfortable with the availability of adequate financial resources, thus saying:

There are always unforeseen situations on trips, so in order to guarantee a positive experience and avoid tension situations, you have to be comfortable in terms of money available. [female]

These young travelers are financially independent and employed in the early stages of the career. They are educated and affluent, and traveling is part of their consumption habits. According to these participants, travel consumption becomes possible as they are currently employed while receiving financial support from the family or requesting money from the bank or financial institutions to arrange a trip is not frequent. Only four participants among all who have been interviewed admitted traveling utilizing a bank loan.

Generally, physical participation is a valued resource in this type of travel. The category involves tangible items required to take the trip, such as equipment, and perception about physical condition and ability to engage in activities. These young tourists believe that they have proper physical resources, which is necessary to the success of the trip, and they are physically capable of facing challenges involved in the touristic activities.

Time is an essential resource at two levels. The first level refers to the time that travelers spend on planning their trips and doing research on the destination. In this level, they focus on the detail of the travel, for instance, what to do, what risks to avoid, what to eat, where to sleep, what to buy. Secondly, time is vital, while an independent traveler adopts a high degree of flexibility and variability at the destination in terms of expansion of the whole duration of the trip or only a part of it.

... giving yourself plenty of time in each city/location allows for this flexibility. [male]

The young travelers interviewed showed confidence in their capacities to accept the challenge of preparing and taking their tourism experience independently. The perception of expertise is very positive. They had a collective feeling of confidence and the understanding which they gained by involving in the different parts of a tourism experience. They are familiar with digital technologies and believe the knowledge they have developed about the destination will lead them to success:

I feel like I'm doing the smart choice: choosing my own route, places to visit that fit with my wishes, needs, and budget. [female]

Similarly, reports regarding self-efficacy also show a very positive tone, and all participants highly valued self-efficacy facets in the context of independent travel. Responses convey an awareness of the potential increase of risk in this type of travel; however, travelers also show confidence in the ability to overcome problems or unexpected situations. Feeling independent (from tourism suppliers or others) is not only about being free but also about being able to protect from potentially hazardous situations and being effective in securing one's interests, goals, and needs. Additionally, the creative potential necessary to address the trip's flexibility and unpredictability was also valued by these participants.

Relational resources refer to travelers' capabilities towards social interaction, which are instrumental in meeting social needs (Madhavaram & Hunt, 2008). This interaction can occur between the traveler and tourism firms and indicate how the individual communicates with the firm's employee. Moreover, it applies to relationships between the traveler and other travelers on vacation or relationships between the traveler and the local community. For these participants, individualism and autonomy in decision making do not discard socializing needs and habits. Indeed, they find the exchange of information crucial to the success of the travel, either by providing the required information so that quality service can be delivered or by taking the appropriate steps as recommended by tourism professionals and qualified staff. Suggestions from locals or other travelers are considered as an essential element of an authentic and meaningful destination experience:

[Traveling independently] allows me to weigh the different people's opinions about what is best to visit and see. [male]

Besides that, relational resources are a way to explore personal interests, feelings of curiosity towards foreign cultures and heritage, and to meet and interact (socializing) with new people. Interpersonal interaction and social relations in independent travel are perceived as challenging, but also rewarding, as asserted by one interviewee as follows:

It is challenging to meet new people to socialize or obtain guidance/opinions on potential activities during my trips; it also quite rewarding when you meet interesting new people. [male]

The ability to share personal stories and memoirs with friends, relatives, and travel communities is significant in the post-consumption stage of the experience, and seen as habitual behavior:

After going on a trip, I always create with must-sees and dos (and don'ts too) and share with my friends. [female]

SOLUTIONS AND RECOMMENDATIONS

This study's results expectedly will provide insights into the young independent travelers' segment, using Vargo and Lusch's (2004, 2008) S-D logic framework, by drawing attention to the concept of *operant resources*. Research on young independent travelers is still scarce (Buffa, 2015), and it is needed in particular to address issues to the segment as it represents a specific group of consumers (Hyde, 2008). This research was conducted to explore young travelers' operant resources applied in the creation of their mobile experiences (Binkhorst & Den Dekker, 2009).

Young tourists are highly supportive of independent forms of tourism despite variations in the number of preparation efforts ahead to the travel (Hyde & Lawson, 2003; Richards, 2015). The participants in this study strongly highlighted the association of independent travel with the feelings of freedom and control. Tourism is a highly personal experience, and this particular aspect of tourism may be emphasized through personalization of the experience. According to this study's participants, for the experience to be perceived as personalized, the individual must feel in control of the process of planning, the combination of experiential elements, the duration of the trip, creative solutions to unexpected problems, *must-dos* and *don'ts*, etc. The results are not surprising; in fact, they are consistent with prior research conducted so far on independent young travelers (Hyde & Lawson, 2003; Richards & Wilson, 2003). Young consumers have grown in affluent societies in which tourism became part of the lifestyle, and individual freedom and consumer empowerment emerged as basic premises of social life. This lifestyle has already been linked to a trend towards individualism in contemporary societies (Buffa, 2015), and young travelers are prominent actors in it. The research findings, in this regard, are supported by prior research linking independent travel to the aspiration to freedom and liberation from daily routines (Hyde & Lawson, 2003).

Additionally, individualism, freedom, and personalization of the travel all connect to a desire to be in control (Buffa, 2015). Information and knowledge play a crucial role in this respect. As long as the information is available, young tourists find themselves capable of harnessing own resources (expertise, mastering, and self-efficacy) to build a sound knowledge base which allows them to gather the necessary touristic elements and design the experience along the way, given the basics of the trip (flights and accommodation). The positive perception about their capabilities and resources provides the necessary psychological strength and confidence to put the travel enterprise on the motion and master it from beginning to end. The experience value is assumedly high for these participants (Prebensen & Xie, 2017). Again, this study's findings are consistent with the profile of the new tourists described in prior research (Quest, 1990; Yousaf et al., 2018). In these participants, personal resources are seen as much important as informational, if not more, to have control over the travel. Most notably, a positive perception of self-efficacy and own competence to plan, design, solve problems, adapt to circumstances, look for, and welcome diversity during the trip.

The search for authenticity was clearly among the fundamental ideas of this group. Independence, control, and personalization of the experience are conditions to find authentic tourist experiences. Interviewees showed concern about living a *real* experience, to be free of *tourist traps*, to avoid *highly touristic* experiences. And authenticity may be found through investing in personal resources as in non-staged touristic interactions with locals or other travelers.

Findings on feelings about traveling independently are consistent with appraisals about own operant resources. Humanistic and informational resources were highly valued by the interviewees. With independence comes the responsibility to create knowledge to inform a decision, and information becomes a fundamental requirement of the process. Despite financial resources being specified as a critical condition to travel, most notable are thus the resources related to personal capacities and skills, particularly those such as time, expertise, and self-efficacy. Time invested in research may be perceived as challenging and rewarding as more personalized and authentic experiences expectedly follow.

Relational resources were also considered contributing to a positive independent travel, and participants are confident, persuaded of their social skills, in particular interpersonal skills, which enable them to relate to tourism firms, other visitors, and local communities. Thus, relational resources hold value to these interviewees in two ways: firstly, a utilitarian value linked to the practical facet of the travel experience, and secondly, an experiential value, derived from the contribution locals and other visitors may give for the individual to reach a deeper level of understanding of the destination.

The significant theoretical contribution of this research to the study of independent travel by young people is the application of Vargo and Lusch's (2004) S-D Logic framework by use of the construct of *operant resources*. Indeed, research on this segment and types of tourism preferred to these travelers is increasing; however, prior research has concentrated on examining profiles, motivations, planning, decision making, or behavior (as in Buffa, 2015; Hyde & Lawson, 2003; Richards, 2015). This study focused on analyzing and discussing perceptions of these travelers relative to their feelings about independent travel, but especially to their capabilities to organize their trip independently. To understand how they perceive their competences and how they explore their abilities to live a personalized and authentic destination experience was the primary intended outcome of the research.

Managerially, this study's results are useful for Destination Management Companies (DMO), which aim at strategically capturing the youth segment. Insights from this research may help inform strategic decisions, such as positioning or product development targeted at these consumers. Being aware of which operant resources are more valued by youths and their relative importance will make communication more accessible and more efficient. In particular, if sustained by digital technologies, as they use them intensively. So, based on this research, managers may better stimulate youths' intangible resources, and by that helping them in enhancing experience value. In this light, partnerships and collaboration among public organizations and businesses may be considered as a way to achieve this enhancement. Emphasis should be placed on the development of smart technologies, platforms, devices and processes which would strategically be disseminated in destination key points, attractions and visitor services, thus matching young travelers proficient use of digital technologies. Design of augmented (AR) and virtual reality (VR) support services are already proving successful strategies to stimulate informed interpretation and knowledge sharing between destination providers and tourists, which are crucial tools to sustain successful independent travel (Femenia-Serra, Neuhofer, & Ivars-Baidal, 2019). Summing up, youth travelers search for personalization; they want to be traveling independently and having control over their options and experiences. Therefore, firms need to identify their expectations and attempt to delight them by facilitating and co-creating amazing experiences.

FUTURE RESEARCH DIRECTIONS

The study's primary limitation, though, is that a non-representative sample of participants was chosen to perform the analysis. This means that results are not generalizable. The second is the exclusive qualitative approach and design of the research. Through a quantitative study using a probabilistic sample method, these limitations would be overcome. Furthermore, assuming young travelers are not an entirely homogeneous group of consumers, a more extensive group for analysis might uncover differences with the potential to influence behavior differently.

In this study, the participants included using the snowball sampling method were all holders of a degree in higher education, Portuguese (but one), and currently employed. The way people value time and money, for instance, maybe influenced by academic background and current professional status (employed/unemployed), and thus also tourism consumption and preferred style of travel. Moreover, under the assumption that variety underlines independent youth travel profiles and experiences, the travel experience might bring light to significant variations in how youths perceive the relative contribution of operant resources to the overall tourism experience. Specifically, additional variables that might play a role in affecting differently use of operant resources in independent travel could be considered, such as cultural backgroung of youths. Further quantitative research would also be welcome if empirically tested the influence of operant resources of these tourists on experience outcomes, such as satisfaction, experience value, repeat visit, or memorability. Comparative studies could be developed under the assumption that different segments of this heterogeneous group of consumers apply differently to their operant resources. Moreover, as independent travel by youths takes into consideration ability to interact with service providers, additional research could explore intermediates behavior best suited to different segments of youths and most relevant operant resources in the context of these interactions. Independent as they are, youths may welcome a proactive co-creative attitude from providers which may deserve further exploration. Thus, the theoretical framework based on the S-D Logic may bring further insights into this appealing, and on the rise, market segment.

CONCLUSION

This research aimed at understanding youths' independent travels by exploring the theoretical framework of S-D Logic which explains value creation through application of actors' operant resources in service exchange systems. The appeal towards independent travel requires specific capabilities that enable planning, organizing and managing the different stages of the trip without relying on packages or pre-arranged products and services. Young travelers are technologically skilled and informed consumers and spend time and effort in activities involved in the trip. They appreciate being in control and the sense of freedom that comes with it. Independent travel, for them, is a way to personalize the trip and to adjust all facets of it to their preferences and interests. Examination of antecedents that may explain this research's findings, exploration of cultural variances that may influence heterogeneity of young travelers in their travel profile or to what extent additional operant resources may contribute to a higher quality independent travel are topics worthy of further examination. Destinations and business managers are urged to pay closer attention to this segment of consumers who look at travel and tourism as a lifestyle consumption experience, in which digital technologies play an increasing important role. Decision makers interested in this market segment are challenged to better understand their skills and competences and match these consumers' technological expertise by evolving towards a smarter tourism framework.

ACKNOWLEDGMENT

This paper is financed by National Funds provided by FCT- Foundation for Science and Technology through project UIDB/04020/2020.

REFERENCES

Alves, H., Ferreira, J. J., & Fernandes, C. I. (2016). Customer's operant resources effects on co-creation activities. *Journal of Innovation and Knowledge*, *1*(2), 69–80. Advance online publication. doi:10.1016/j.jik.2016.03.001

Andrades, L., & Dimanche, F. (2014). Co-creation of experience value: A tourist behaviour approach. In M. Chen & J. Uysal (Eds.), *Creating experience value in tourism* (pp. 95–112). CABI. doi:10.1079/9781780643489.0095

Ballantyne, D., Williams, J., & Aitken, R. (2011). Introduction to Service-Dominant Logic: From propositions to practice. *Industrial Marketing Management*, *40*(2), 179–180. doi:10.1016/j.indmarman.2010.06.025

Barney, J. (1991). Firm resources and sustained competitive advantage. *Journal of Management*, *17*(1), 99–120. doi:10.1177/014920639101700108

Bertella, G. (2014). The co-creation of animal-based tourism experience. *Tourism Recreation Research*, *39*(1), 115–125. doi:10.1080/02508281.2014.11081330

Binkhorst, E., & Den Dekker, T. (2009). Agenda for co-creation tourism experience Research. *Journal of Hospitality Marketing & Management, 18*(2), 311–327. doi:10.1080/19368620802594193

Buffa, F. (2015). Young tourists and sustainability. Profiles, attitudes, and implications for destination strategies. *Sustainability, 7*(10), 14042–14062. doi:10.3390u71014042

Campos, A. C., Mendes, J., Valle, P. O., & Scott, N. (2015). Co-creation of tourist experiences: A literature review. *Current Issues in Tourism*. Advance online publication. doi:10.1080/13683500.2015.1081158

Campos, A. C., Mendes, J., Valle, P. O., & Scott, N. (2016). Co-creation experiences: Attention and memorability. *Journal of Travel & Tourism Marketing, 33*(9), 1309–1336. Advance online publication. doi:10.1080/10548408.2015.1118424

Campos, A. C., Mendes, J., Valle, P. O., & Scott, N. (2017). Co-creating animal-based tourist experiences: Attention, involvement and memorability. *Tourism Management, 63*, 100–114. doi:10.1016/j.tourman.2017.06.001

Cavagnaro, E., Staffieri, S., & Postma, A. (2018). Understanding millennials' tourism experience: Values and meaning to travel as a key for identifying target clusters for youth (sustainable) tourism. *Journal of Tourism Futures, 4*(1), 31–42. doi:10.1108/JTF-12-2017-0058

Constantin, J. A., & Lusch, R. F. (1994). *Understanding resource management*. The Planning Forum.

Demeter, T., & Bratucu, G. (2014). Typologies of youth tourism. *Bulletin of the Transilvania University of Brasov. Series V, Economic Sciences, 7*(56), 115–122.

Dionysopoulou, P., & Mylonakis, J. (2013). Youth tourists' profile and their travel choices as influenced by social media networks. *European Journal of Hospitality and Tourism Research, 1*(3), 22–35.

Farahani, B. M., & Sukmajati, D. (2011). Youth travelers as a growing market in urban tourism; potencies and challenges. In *Proceedings Environmental Talk*. Toward A Better Green Living Conference.

Femenia-Serra, F., Neuhofer, B., & Ivars-Baidal, J. A. (2019). Towards a conceptualization of smart tourists and their role within the smart destination scenario. *Service Industries Journal, 39*(2), 109–133. doi:10.1080/02642069.2018.1508458

Finn, M., Elliott-White, M., & Walton, M. (2000). *Tourism and leisure research methods: Data collection, analysis, and interpretation*. Longman.

Grönroos, C., & Voima, P. (2013). Critical service logic: Making sense of value creation and co-creation. *Journal of the Academy of Marketing Science, 41*(2), 133–150. doi:10.100711747-012-0308-3

Hyde, K. F. (2008). Independent Traveler Decision-Making. In A. Woodside (Ed.), *Advances in Culture, Tourism and Hospitality Research* (Vol. 2, pp. 43–151). Emerald Group Publishing Limited., doi:10.1016/S1871-3173(08)02003-X

Hyde, K. F., & Lawson, R. (2003). The nature of independent travel. *Journal of Travel Research, 42*(1), 13–23. doi:10.1177/0047287503253944

Jennings, G. (2005). Interviewing: a focus on qualitative techniques. In B. W. Ritchie, P. Burns, & C. Palmer (Eds.), *Tourism Research Methods: Integrating Theory with Practice* (pp. 99–117). CABI Publishing. doi:10.1079/9780851999968.0099

Kirchherr, J., & Charles, K. (2018). Enhancing the sample diversity of snowball samples: Recommendations from a research project on anti-dam movements in Southeast Asia. *PLoS One*, *13*(8), e0201710. doi:10.1371/journal.pone.0201710 PMID:30133457

Kreziak, D., & Frochot, I. (2011). Co-construction de l'expérience touristique: Les stratégies des touristes en stations de sport d'hiver [Co-construction of the tourism experience: Tourists strategies in ski resorts]. *Décisions Marketing*, *64*, 23–33. doi:10.7193/DM.064.23.33

Lusch, R. F., Vargo, S. L., & O'Brien, M. (2007). Competing through service: Insights from service-dominant logic. *Journal of Retailing*, *83*(1), 5–18. doi:10.1016/j.jretai.2006.10.002

Madhavaram, S., & Hunt, S. D. (2008). The Service-Dominant Logic and a Hierarchy of Operant Resources: Developing Masterful Operant Resources and Implications for Marketing Strategy. *Journal of the Academy of Marketing Science*, *36*(1), 67–82. doi:10.100711747-007-0063-z

Mannell, R. C., & Iso-Ahola, S. E. (1987). Psychological nature of leisure and tourism experience. *Annals of Tourism Research*, *14*(3), 314–331. doi:10.1016/0160-7383(87)90105-8

Mathisen, L. (2013). Staging natural environments: A performance perspective. *Advances in Hospitality and Leisure*, *9*, 163–183. doi:10.1108/S1745-3542(2013)0000009012

Meho, L. I. (2006). E-Mail Interviewing in Qualitative Research: A Methodological Discussion. *Journal of the American Society for Information Science and Technology*, *57*(10), 1284–1295. doi:10.1002/asi.20416

Minkiewicz, J., Evans, J., & Bridson, K. (2013). How do consumers co-create their experiences? An exploration in the heritage sector. *Journal of Marketing Management*, 1–30. doi:10.1080/0267257X.2013.800899

Moscardo, G., & Beckendorff, P. D. (2009). Mythbusting: Generation Y and Travel. In P. Benckendorff, G. Moscardo, & D. Pendergast (Eds.), *Tourism and Generation Y* (pp. 16–26). CABI Publishing. doi:10.1079/9781845936013.0016

Naderifar, M., Goli, H., & Ghaljaie, F. (2017). Snowball Sampling: A Purposeful Method of Sampling in Qualitative Research. *Strides in Development of Medical Education*, *14*(3). Advance online publication. doi:10.5812dme.67670

Neuhofer, B., Buhalis, D., & Ladkin, A. (2012). Conceptualising technology enhanced destination experiences. *Journal of Destination Marketing and Management*, *1*(1–2), 36–46. doi:.jdmm.2012.08.001 doi:10.1016/j

Pendergast, D. (2009). Getting to know the Y generation. In Tourism and generation Y (pp. 1–15). Wallingford, UK: CABI Publishing. doi:10.1079/9781845936013.0001

Prahalad, C. K., & Ramaswamy, V. (2004). Co-creation experiences: The next practice in value creation. *Journal of Interactive Marketing*, *18*(3), 5–14. doi:10.1002/dir.20015

Prebensen, N. K. (2017). Staging for value co-creation in nature-based experiences: the case of a surfing course at Surfer's Paradise, Australia. In N. K. Prebensen, J. S. Chen, & M. S. Uysal (Eds.), *Co-creation in tourism Experiences* (pp. 36–49). Routledge. doi:10.4324/9781315645919-4

Prebensen, N. K., & Campos, A. C. (2017). Operant resources in experiential consumption: self-efficacy (SE) versus perceived behavioral control (PBC). *7th ATMC Conference Advances in Tourism Marketing Conferences: The Art of Living Together*.

Prebensen, N. K., Campos, A. C., & Blomstervik, I. (2018). Facilitating for Co-creation in experiential consumption through perceived behaviour control and self-efficacy. *Tourism & Management Studies International Conference 2018*.

Prebensen, N. K., & Foss, L. (2011). Coping and co-creating in tourist experiences. *International Journal of Tourism Research, 13*(1), 54–67. doi:10.1002/jtr.799

Prebensen, N. K., Kim, H., & Uysal, M. (2015). Cocreation as moderator between the experience value and satisfaction relationship. *Journal of Travel Research*. Advance online publication. doi:10.1177/0047287515583359

Prebensen, N. K., Vittersø, J., & Dahl, T. I. (2013). Value co-creation significance of tourist resources. *Annals of Tourism Research, 42*, 240–261. doi:.annals.2013.01.012 doi:10.1016/j

Prebensen, N. K., Woo, E., Chen, J. S., & Uysal, M. (2013). Motivation and involvement antecedents of perceived value of the destination experience. *Journal of Travel Research, 52*(2), 253–264. doi:10.1177/0047287512461181

Prebensen, N. K., & Xie, J. (2017). Efficacy of co-creation and mastering on perceived value and satisfaction in tourists' consumption. *Tourism Management, 60*, 166–176. doi:10.1016/j.tourman.2016.12.001

Quest, M. (1990). *Howarth Book of Tourism*. Macmillan. doi:10.1007/978-1-349-11687-4

Ramaswamy, V., & Gouillart, F. (2010). *The power of co-creation*. Free Press.

Ramaswamy, V., & Ozcan, K. (2018). What is co-creation? An interactional creation framework and its implications for value creation. *Journal of Business Research, 84*, 196–205. doi:10.1016/j.jbusres.2017.11.027

Richards, G. (2015). The new global nomads: Youth travel in a globalizing world. *Tourism Recreation Research, 40*(3), 340–352. doi:10.1080/02508281.2015.1075724

Richards, G., & King, B. (2003). International Youth Travel. *Travel & Tourism Analyst, 6*(December).

Richards, G., & Wilson, J. (2003). *New horizons in independent youth and student travel. A report to the international student travel confederation (ISTC) and the association of tourism and leisure education (ATLAS)*. International Student Travel Confederation.

Richards, G., & Wilson, J. (2006). Developing creativity in tourist experiences: A solution to the serial reproduction of culture? *Tourism Management, 27*(6), 1209–1223. doi:10.1016/j.tourman.2005.06.002

Schmitt, B. (1999). Experiential marketing. *Journal of Marketing Management, 15*(1–3), 53–67. doi:10.1362/026725799784870496

Sfandla, C., & Björk, P. (2012). Tourism experience network: Co-creation of experiences in interactive processes. *International Journal of Tourism Research*. Advance online publication. doi:10.1002/jtr1892

Starcevic, S., & Konjikusic, S. (2018), Why Millenials as digital travelers transformed marketing strategy in tourism industry. *International Thematic Monograph Tourism in Function of Development of the Republic of Serbia - Tourism in the Era of Digital Transformation*. University of Kragu. Available at SSRN: https://ssrn.com/abstract=3280320

Tan, S.-K., Kung, S.-F., & Luh, D.-B. (2013). A model of "creative experience" in creative tourism. *Annals of Tourism Research*, *41*, 153–174. doi:10.1016/j.annals.2012.12.002

Tapia, S. A. (2018). Youth tourism? Traveling in the biographies of young people of popular neighborhoods of Buenos Aires. *Revista Brasileira de Pesquisa em Turismo*, *12*(3), 104–123. doi:10.7784/rbtur.v12i3.1508

Taylor, S. J., Bogdan, R., & DeVault, M. (2015). *Introduction to qualitative research methods: A guidebook and resource*. John Wiley & Sons.

Vargo, S. L., & Lusch, R. F. (2004). Evolving to a new dominant logic for marketing. *Journal of Marketing*, *68*(1), 1–17. doi:10.1509/jmkg.68.1.1.24036

Vargo, S. L., & Lusch, R. F. (2006). Service-dominant logic: Reactions, reflections and refinements. *Marketing Theory*, *68*(3), 281–288.

Vargo, S. L., & Lusch, R. F. (2008). Service-dominant logic: Continuing the evolution. *Journal of the Academy of Marketing Science*, *36*(1), 1–10. doi:10.100711747-007-0069-6

Vargo, S. L., & Lusch, R. F. (2017). Service-dominant logic 2025. *International Journal of Research in Marketing*, *34*(1), 46–67. Advance online publication. doi:10.1016/j.ijresmar.2016.11.001

Vargo, S. L., Maglio, P. P., & Akaka, M. A. (2008). On value and value co-creation: A service system and service logic perspective. *European Management Journal*, *26*(3), 145–152.

Vukic, M., Kuzmanovic, M., & Stankovic, M. K. (2014). Understanding the Heterogeneity of Generation Y's Preferences for Travelling: A Conjoint Analysis Approach. *International Journal of Tourism Research*. Advance online publication. doi:10.1002/jtr.2015

World Tourism Organisation and World Youth Student & Educational Travel Confederation (UNWTO and WYSE Travel Confederation). (2008). *Youth travel matters: understanding the global phenomenon of youth travel*. WTO.

World Tourism Organization and World Youth Student & Educational Travel Confederation (UNWTO and WYSE Travel Confederation). (2011). *The power of youth travel*. WTO.

Ye, Y., Tussyadiah, I. P., & Fesenmaier, D. R. (2009). *Phenomenological Capture of Tourist Experience*. Paper presented at Tourism Experiences: Meanings, Motivations, Behaviours Conference, Preston, UK.

Yousaf, A., Amin, I., & Santos, J. A. C. (2018). Tourists' motivations to travel: A theoretical perspective on the existing literature. *Tourism and Hospitality Management*, *24*(1), 197–211. doi:10.20867/thm.24.1.8

ADDITIONAL READING

Akaka, M., & Vargo, S. (2014). Technology as an Operant Resource in Service (Eco)Systems. *Information Systems and e-Business Management, 12*(12), 367–384. doi:10.100710257-013-0220-5

Constantin, J. A., & Lusch, R. L. (1994). *Understanding Resource Management.* The Planning Forum.

Cova, B., & Dalli, D. (2009). Working consumers: The next step in marketing theory? *Marketing Theory,* 9(3), 315e339.

Fesenmaier, D. R., & Gretzel, U. (2004). Searching for experience: Technology- related trends shaping the future of tourism. In K. Weiermaier & C. Mathies (Eds.), *The tourism and leisure industry: Shaping the future* (pp. 285–299). Haworth Hospitality.

Grissemann, U. S., & Stokburger-Sauer, N. E. (2012). Customer co-creation of travel services: The role of company support and customer satisfaction with the co-creation performance. *Tourism Management,* 33(6), 1483e1492. . doi:10.1016/j.tourman.2012.02.002

Grönroos, C., & Gummerus, J. (2014). The service revolution and its marketing implications: Service logic vs service-dominant logic. *Managing Service Quality, 24*(3), 206–229. Advance online publication. doi:10.1108/MSQ-03-2014-0042

Pine, B., & Gilmore, J. H. (1999). *The experience economy.* Harvard Business School Press.

Prahalad, C. K., & Ramaswamy, V. (2004). *The future of competition: co-creating unique value with customers.* Harvard Business School Press.

Strandvik, T., & Heinonen, K. (2015). Essentials of Customer Dominant Logic. In J. Gummerus & C. von Koskull (Eds.), *The Nordic School, Service Marketing and Management for the Future* (pp. 111–127). CERS, Hanken School of Economics.

KEY TERMS AND DEFINITIONS

Co-Creation: Creation of value supported by a collaborative process that connects participant actors of a system in interactive experiences to achieve common goals, which may include product idea, process design, or active consumption.

Independent Travel: Any trip organized by the individual without pre-booking or purchase of a package tour, or booked with a minimum of travel services, e.g. transportation or accommodation, before departure.

Operand Resource: The firm's tanglible resources upon which an action is performed in order to create value or benefit, e.g. equipment, machinery, raw materials.

Operant Resource: The firm's intangible resources which are applied to operand resources, usually understood as human knowledge, skills and competences. They are dynamic and constinuously evolving, thus at the basis of the firm's competitive advantage.

S-D Logic: Service dominant logic is a marketing theory developed by S. Vargo and R. Lusch in several works to explain the process of value creation through exchanges of services among actors of a system.

Youth Market Segment: The segment of travelers aged between 16 to 29, as defined by the World Tourism Organization in 2008.

Chapter 15
The Unfolding Theories on Destination Image, 1990–2020:
A Content Analysis Approach

Fatima Lampreia Carvalho
https://orcid.org/0000-0002-1189-8843
Faculty of Economics and Research Centre for Tourism, Sustainability, and Well-Being (CinTurs), University of Algarve, Portugal

Bernadete Dias Sequeira
Faculty of Economics and Interdisciplinary Centre of Social Sciences (CICS.NOVA), University of Algarve, Portugal

Magda Wikesjö
https://orcid.org/0000-0002-2612-3741
Faculty of Economics, University of Algarve, Portugal

Célia M. Q. Ramos
https://orcid.org/0000-0002-3413-4897
ESGHT and Research Centre for Tourism, Sustainability, and Well-Being (CinTurs), University of Algarve, Portugal

ABSTRACT

This chapter examines a 30-year collection of tourism destination image (TDI) components and scale attributes sourced from tourism and hospitality studies to assess the scientific field at hand. From the 1990s, it was clear that cognitive, affective, and overall (holistic) images were involved in destination image formation. Cognitive psychology is crucial to explain tourists´ behaviour, but tourism studies are not simply a branch of cognitive psychology. This study taps into the multidisciplinary character of destination image. At stake are propositions and theories applicable in tourism such as the theory of self-congruity, self-concept, brand signal theory, destination source credibility, novelty seeking theory, and brand equity theory. The intention of the present chapter is to map its theoretical ground and underlying hypotheses by means of a systematic literature review and point out sub-studied dimensions of TDI.

DOI: 10.4018/978-1-7998-3156-3.ch015

INTRODUCTION

Early published research in tourism studies singled out the complexity of *tourism destination image* (TDI). The notion of TDI has been extensively applied in marketing and behavioural sciences to represent people's perceptions of products, objects, behaviours, and events driven by beliefs, feelings, and impressions (Baloglu & Brinberg, 1997; Crompton, 1979). Several authors strived to conceptualise TDI (Lawson and Baud-Bovy, 1977; Crompton, 1979; Assael, 1984; Phelps, 1986; Gartner & Hunt, 1987; Moutinho, 1987; Coshall, 2000; Kim & Richarson, 2003; Lai & Li, 2015; Kislali, Kavaratzis & Saren, 2019) to respond major questions on how images emerge and what impact they have on tourism.

TDI tend to be defined as "mental/total impressions/perceptions held by tourists about a destination" being used out of a practical concern because it is more natural to ask tourists "What is your impression of Destination A?" rather than "What is your image of Destination A?" (Lai & Li, 2015:10). Existing definitions of TDI tend to be connotative definitions (e.g. genus-and-differentia definition) such as a feeling, thought, or idea, associated in one's mind or imagination with a specific destination. Addressing the conceptual problem of TDI from a definition perspective (Gerring 1997; Kimiecik & Harris 1996), Lai and Li (2015) proposed a new definition by adopting a procedure derived from definition theory in logic studies. Results showed that TDI was "defined mainly as the mental/total impressions/perceptions held by tourists pertaining to a destination" (Lai & Li, 2015: 1). They also found that TDI definitions were quasi-theoretical in type and created by roughly following the connotative definition technique.

The intention of the present chapter is to single out the most relevant theoretical constructs that cutting-edge tourism and management studies have used since 1990. This literature review arises from previous systematic reviews by Sirakaya and Woodside (2005), Zhang, Fu and Cai Lu (2014), Li, Ali and Kim (2015), Shankar (2018), Picazo and Moreno-Gil (2019) and Chan and Peters (2020).

Main Contribution

In coherence with this *Handbook of Research on Resident and Tourist Perspectives on Travel Destinations* this chapter seeks to map the most relevant theoretical constructs that underpin TDI in tourism and management studies, especially from the perceived image perspective. This study highlights the multidisciplinary character of TDI which is situated in the bounds of the Science of Psychology, the Science of Marketing, Philosophy, Sociological theory, Social exchange theory, Public Management, Business Management (Stakeholder Model), Geography and Linguistics, to mention just a few. Another contribution is to demonstrate the need for more studies on the topic, suggesting future research avenues.

BACKGROUND

Since the first studies in the early 1970s, TDI was approached from multiple perspectives (Campubri, Guia & Comas, 2014). Authors like Gallarza, Saura and Garcia (2002), Pike (2002) and Tasci, Gartner, and Cavusgil (2007) later integrated these perspectives and systematized the research field.

Destination image is a key topic of investigation in the marketing scholarship in tourism studies (Cherifi, Smith, Maitland, & Stevenson, 2014; Fu, Ye, & Xiang, 2016; Stepchenkova & Li, 2013; Sun, Ryan, & Pan, 2015). From the 1970s', TDI gained salience in models of travel decision-making because it was viewed as an important aspect of promoting information about a specific destination, "since frequently tourists make travel decision at home, before leaving, rather than on impulse along the way" (Matos, 2014: 81).

One broadly accepted theory of this literature was the relevance of TDI in both inducing tourists' decision-making and behaviour (Baloglu & McCleary, 1999; Bigné, Sánchez, & Sanz, 2009; Crompton, 1979; Echtner & Ritchie, 1991; Jenkins, 1999; Kim & Richardson, 2003; Larsen & George, 2004; Pike & Ryan, 2004; Stabler, 1992; Tasci & Gartner, 2007). The influence of TDI on after-decision-making behaviours was neglected in previous studies (Chen & Tsai, 2007). Following the marketing perspective, Lee, Lee & Lee. (2005) demonstrated that individuals having a favourable TDI would perceive their on-site experiences positively, which in turn would lead to greater satisfaction levels and behavioural intentions. Chen and Tsai (2007:1116) thus proposed an integrated tourist behaviour model by including destination image and perceived value into the ''quality–satisfaction–behavioural intentions'' paradigm.

Destination Image: Concepts, Definitions, and Components

TDI is multisensory and primarily visual (Xiong, Hashim & Murphy, 2015; Lai & Li, 2015). Reynolds (1965:69) proposed that "an image is actually the result of a more complex process. It is the mental construct developed by the consumer based on a few selected impressions among the flood of total impressions; it comes into being through a creative process in which these selected impressions are elaborated, embellished, and ordered". Hunt (1971) clarified destination images as impressions that persons hold about a state in which they do not reside. Later, Hunt (1975) confirmed the theory that those are perceptions held by potential visitors about an area.

Lawson and Bond-Bovy (1977) conceptualised TDI as an expression of knowledge, impressions, prejudice, imaginations, and emotional thoughts an individual has of a specific object or place. Crompton (1979:18) contributed towards the theory's development noting that TDI are the "sum of beliefs, ideas and impressions that a person has of a destination". To Gartner (1989:16) TDI is closely connected to destination attributes and it is considered "a complex combination of various products and associated attributes". Calantone et al. (1989:25) further defined TDI as "perceptions of potential tourist destinations" to what Reilly (1990:21) added "not individual traits…but the total impression an entity makes". Fakeye and Crompton (1991:10) articulated TDI as a theory of representation noting that it is the "representation of an area into the potential tourist's mind". In the view of Echtner and Ritchie (1993:3), TDI is "composed of perceptions of individual attributes (such as climate, accommodation facilities, friendliness of the people) as well as more holistic impressions (mental pictures or imagery) of the place". TDI may be referred to as the visual or mental impression of a place, a product, or an experience held by the general public (Milman & Pizam, 1995:21). To Alhemoud and Armstrong (1996:76) it "comprises the ideas or conceptions held individually or collectively of the destination". TDI refers to a sum of associations and pieces of information connected to a destination, which include multiple components of the destination and personal perception (Murphy, Pritchard & Smith, 2000).

Other researchers (Fakeye & Crompton, 1991; Gartner, 1994; King, Chen & Funk, 2015) elaborated on how TDI take shape and evolve. In this vein, a TDI formation and subsequent change over time is interdependent. Mackay and Fesenmaier (1997: 538) added that a TDI is a composite of various products (attractions) and attributes woven into a total impression". Bigné, Sánchez, and Sánchez (2001) raised a different argument on TDI bringing to the limelight the notion of subjectivity, as the image of a destination consists of the subjective interpretation of reality made by the tourist. Like Bigné et al (2001) various other authors conceptualised TDI in terms of a subjective interpretation of a destination made by individuals which influences tourist behaviour (Beerli & Martín, 2004, Bigné, Sánchez, & Sanz, 2009; Bosque & Martín, 2008; Cai, Wu, & Bai, 2004; Chi & Qu, 2008; Hunt, 1975; Tasci & Gartner, 2007; Agapito, Valle & Mendes, 2013). Alcañiz et al. (2009: 716) agreed with such views when they emphasised that the concept "consists of all that the destination evokes in the individual; any idea, belief, feeling or attitude that tourists associate with the place". Darbellay and Stock (2012) reinforced the idea that TDI is troubled by subjectivity which stands as an obstacle to consensus regarding its theoretical substance.

Lai and Li (2015) noted that the nature and scope of TDI remained vague and for this reason carried a study to address this conceptual problem from a modernist perspective. Thus, TDI can be defined as "a voluntary, multisensory, primarily picture-like, qualia arousing, conscious, and quasi-perceptual mental (i.e., private, nonspatial, and intentional) experience held by tourists about a destination. This experience overlaps and/or parallels the other mental experiences of tourists, including their sensation, perception, mental representation, cognitive map, consciousness, memory, and attitude of the destination." (Lai & Li, 2015: 10).

It was Gartner (1993, 1996) who referred to image development, proposing that destination images are established by three hierarchically interrelated components: cognitive, affective, and conative. Baloglu and McCleary (1999) presented a highly cited framework according to which, the image has two main mechanisms, known as "cognitive" and "affective" components which affect the establishment of what the authors have named the "global" image. Yet, a few authors (Duarte, Folgado-Fernández & Hernández-Mogollón, 2018; Hernández-Mogollón et al., 2018) have acknowledged that cognitive component may play a central function in TDI formation.

The cognitive or perceptual component of a DI is also known as the "designative component" (Alcaniz, Garcia & Blas, 2009:716) which denotes the beliefs and knowledge about the perceived attributes of the destination. It refers to cognitive perception which is the way in which a visitor perceives the attributes of a tourism destination (Baloglu & McCleary, 1999; Beerli & Martín 2004). These attributes comprise the place's landscape, attractions, services, and infrastructure. From this cognitive perspective, a place fascinates people based on the attractiveness of the architectural and natural landscapes, the well-being provided by its facilities, and the specific quality of its attractions (Choi, Chan & Wu, 1999).

The affective component of a DI describes feelings toward a destination (Beerli & Martín, 2004; Konecnik & Gartner, 2007). The affective or evaluative component of a DI consists in the individual's feelings towards the destination (Alcaniz, Garcia & Blas, 2009:716) The Psychological component contains more abstract and intangible characteristics such as atmosphere or friendliness. The affective component of the image thus alludes to a person's feelings toward and emotional responses to a destination (Baloglu & Brinberg, 1997; Shani & Wang, 2011, Stylidis, Shani & Belhasen, 2017). It becomes operational during the evaluation stage of destination selection (Gartner, 1993). The affective component of the image was captured by means of four semantic differential scales (unpleasant-pleasant, sleepy-arousing, gloomy-exciting, and distressing-relaxing), used by Baloglu and McCleary (1999) and Chew and Jahari (2014), amongst other authors.

The overall perception is considered to be a third component of the image, that may be similar to, or different from, the cognitive or affective perceptions of the destination (Alcaniz, Garcia & Blas, 2009:716, Baloglu & Love, 2005; Baloglu & McCleary, 1999). The evaluation of the overall image and that of its components can be different, both should be measured in order to understand the positioning of a destination (Ahmed, 1991). The overall image has been deemed as "the only variable with a signiðcant iñuence on the tourist's willingness to revisit and to recommend the destination" (Alcaniz, Garcia & Blas, 2009: 721).

The conative component of a DI covers action, the individual's concrete behaviour or intention to revisit and endorse the destination to others (Bigné et al., 2001; Gartner, 1993; Konecnik & Gartner, 2007; Pike & Ryan, 2004; Tasci & Gartner, 2007; Tasci et al., 2007; Agapito, Valle & Mendes, 2013). The same dimension encompasses the spread positive word of mouth (Baker & Crompton, 2000). The conative aspect of TDI is also meant to be related to loyalty (Bigné et al., 2001, 2009; Bosque & Martín, 2008; Cai et al., 2004; Chen & Gursoy, 2001; Chi & Qu, 2008; Li et al., 2010; Oppermann, 2000; Petrick, 2004).

With the components of DI defined, it is important to question how and why the above components of a DI influence each other. The hypothesized direction of the relationship between the cognitive and affective components was already scrutinized in previous destination image models (e.g., Baloglu & McCleary, 1999; Beerli & Martin, 2004; Li, Cai, Lehto, & Huang, 2010; Lin et al., 2007; Papadimitriou et al., 2018; Billig, 2006; Rollero & Piccoli, 2010; Baloglu & McCleary, 1999; Boo & Busser, 2005; Russel & Pratt, 1980; Beerli & Martin, 2004; Li, Cai, Lehto, & Huang, 2010; Lin et al., 2007). The cognitive and affective components of a TDI work to influence the overall image of a destination in the mind of tourists (Baloglu & McCleary, 1999; Beerli & Martín, 2004). Affective component can operate as a mediating variable between cognitive image and overall destination image (Papadimitriou et al. (2018: 506). Consumer's cognitive perceptions of a destination have a direct effect on familiarity (Marks and Olson, 2001) which mediates the effect of cognitive perception and affective evaluation on tourism destination reputation (Artiga, Montero & Yrigoyen, 2015:149). Considering the composite nature of the tourism destination image, several studies have argued that it is important to analyse the image components separately in order to understand how people evaluate and behave in relation to places (Baloglu & Brinberg, 1997; Cai et al., 2004; Gartner, 1993; Russell, 1980; Agapito, Valle & Mendes, 2013).

The image that is actively and purposefully promoted by destination marketers is therefore the 'projected' image, formed by a complex and dynamic network of agents acting as information sources (e.g. visitors or residents, and marketing organizations), and this may often be contrasted to the "perceived" image (Kokosalakis, Bagnall, Selby & Burns, 2006; Kim & Lehto, 2013; Kislali, Kavaratzis & Saren, 2019), which is the image formed in peoples' minds (Beerli & Martín, 2004; Kislali, Kavaratzis & Saren, 2019). Perceived image is divided into two major components: the "primary" image, which is the image formed through an individual's first-hand experience of a destination (i.e., after the visit) and the "secondary" image, which is formed through information generated by various other sources beyond the individual's own experience (Jenkins, 1999; Gartner, 1994; Tasci, Gartner, & Cavusgil, 2007). Secondary image is further divided into organic and induced images based on the source of information about the destination, a distinction first used by Gunn (1972). Beerli and Martín (2004a) also developed a model that includes both primary and secondary information sources and stimuli influencing the forming of pre-visit perceptions and post-visit evaluations of destinations.

In fact, there had been few empirical studies aimed at analysing which forces influence the formation of an individual's image of a given destination (Baloglu and McCleary, 1999, Beerli & Martín, 2004a ; Mackay & Fesenmaier, 1997; Mercille, 2005; Tasci & Gartner, 2007; Frias, Rodriguez, Castaneda, Sabiote and Buhalis, 2012). Several authors agree that it is crucial to assess the characteristics of a persons' destination image and the image formation process rather than assessing the content of images.

Measurement scales are a pervasive method of choice in several studies here analysed and mapped out. Albeit the literature highlighted the dominance of structured, word-based approaches (Jenkins, 1999; Dolnicar & Grün, 2013), it is recognized a lack of a universally accepted, valid, and reliable scale for the measurement of TDI (Beerli & Martin, 2004a).

Previous studies of TDI relied on content analysis (CA) – a familiar observational research method in social sciences (Berg, 2009; Krippendorff & Bock, 2009) which can be applied to systematically evaluate the symbolic content of academic articles on tourism destination image (TDI) and to identify different interpretations of theories and conceptual issues of TDI. In the tourism field it is used to codify and analyse specialised literature, and to organise interviews given by tourists and other stakeholders in the tourism industry (Hall & Valentin, 2005; Camprubí & Coromina, 2016), tourism destination marketing, tourists' experience, tourists' motivation, and tourism concept (Baron & Kenny, 1986; Agustí, 2018; Marine-Roig & Ferrer-Rosell, 2018; Stepchenkova & Morrison, 2006; Stepchenkova & Zhan, 2013); Llodrà-Riera et al., 2015; Rodrigues, Correia, Kozak & Tuohino, 2015; Akroush, Jraisat, Kurdieh, AL-Faouri, Qatu, 2016; Li, 2015; Camprubí & Coromina, 2016). Some of those relied on the critical discourse analysis perspective (Stamou & Paraskevopoulos, 2004).

METHODOLOGY

A systematic review is the process of searching evidence on a topic (e.g. tourism destination image) and choosing quality articles to appraise, synthesize and draw conclusions (Cooper, Booth, Varley-Campbell, Britten & Garside, 2018). The present systematic review summarizes different research papers in tourism studies with the aim to guarantee more accurate and reliable evidence-based results, with the support of the CA methodology (São José & Amado, 2016). This study opted for an interpretative CA (Qiu & Zou, 2004) as it seemed suitable help the mapping of studies on destination image to describe and explain the connections between concepts with basis on a quantitative content analysis of word frequency in a collection of literature concerned with TDI and its dimensions.

The current chapter is mindful of previous critique regarding the application of CA in tourism studies (Camprubí & Coromina, 2016). With a view to follow best-practice guidelines for the application of CA in the tourism field, this chapter follows recommendations regarding the analysis of topical issues, sampling method, objectivity, systematisation, reliability.

The literature search evolved in two steps. Firstly, the authors used the consolidated four dimensions of TDI – "overall image", "cognitive image", "affective image" and "conative image" to codify 250 selected documents according to word frequency of such keywords. Secondly, the researchers carried a second assortment of sources based on documents already classified under the four TDI dimensions. This second selection aimed at articles which included the keyword "theory" to a significant extent and retained only 30 documents in which the keyword "theory" was crucial.

The bibliographic data base *Web of Science* was used for indexing and abstracting multidisciplinary academic journals to sample information with basis on clearly defined keywords "tourism destination" and "image". Regarding the reliability measures on the frequency of words like "cognitive", "affective", "conative" and "overall" dimensions of TDI, this study worked adopted the exact same words as Nvivo 12 Plus nodes: (1) Cognitive dimension, (2) Affective dimension, (3) Conative dimension, (4) Overall dimension. All contents were attributed to the four nodes by means of word frequency and automatic coding of the elected keywords. Further, contents already selected were codified as theoretical foundations of TDI. These contents were identified via the NVivo 'word search" function aiming at the words – 'theory' OR 'theoretical' in narrow context. Every theory mentioned in 30 of the 250 sources was selected for analysis and became a sub-node, as presented in Figure 2. This approach should guarantee the objectivity of the study and permit its replicability by other scholars.

This chapter follows the guidelines by the Center for Reviews and Dissemination (CRD) at the University of York (CRD 2009) and the Preferred Reporting Items for Systematic Reviews and Meta-Analyses (PRISMA) (Moher et al., 2009).

Inclusion and Exclusion Criteria

In the initial stages of this systematic review, 664 journal articles (out of 4,362 identified with the help of Web of Science – WoS) with citation average ranging from 50.95 per year (e.g. Baloglu & McLeary, 1999) and records with no citations.. All records with less than 3 average citation per year were excluded as well as were excluded any records in connection to non-relevant Web of Science Categories (e.g. Chemistry and Clinical Neurology). In a second stage of exclusion, all sources with no relevant mention to "theory" were dismissed.

Search Strategy

WoS is a publisher-independent global citation database that performs the role of an article index and indexes contents from thousands of journals and conference proceedings in the Social Sciences, sciences, arts, and humanities. Table 1 summarizes the process and main outcomes.

Table 1. Destination Image study selection process

Identification Publications on **topic** 'destination image' with title 'destination image', timespan (1990-2020) identified through database searching (Web of Science = 4,362) *Total identified= 4,362*
Screening Number of records after duplicates were removed and publications without words 'destination' and 'image' in title eliminated (Web of Science = 664) *Removed*= 5,863 records missing words "destination" and "image" in title
Further Records excluded Excluded articles with less than 3 average citations per year Excluded 8 journals in Web of Science categories Chemistry medicinal or Clinical neurology or Computer science artificial intelligence or Computer science information systems or Engineering electrical electronic or Food science technology or surgery) *Total excluded* =395
Eligibility 1 250 full text articles assessed (WoS) for eligibility *Full text articles included*=250
Eligibility 2 250 full text articles assessed (WoS) for eligibility if included keyword "theory" Full text articles excluded =220
Included 30 full text articles included in qualitative synthesis and analysis

Source: authors' own elaboration

A Brief Overview of the Included Studies

The semi-final list of academic articles included in the systematic review contained 250 journal articles identified via the WoS, after preliminary exclusions. From 2010, there has been a great deal of interest in the conceptualization, case studies and measurement of the various aspects of tourism destination image. Although the full list of the 250 articles is not provided in this chapter's reference list due to word limit restrictions, it is relevant to inform the reader that the Web of Science report shows four types of information about the selected articles for the period 1990-April 2020: the h-index (or Hirsch index) that measures the impact of a scientist rather than a journal (63); Sum of Times Cited, which refers to total number of citations to all items in the results set (17015); Average citations per year concerning academic articles on "Tourism destination" and "image" which raised from 114 in 2006, to 783 in 2010, and 2682 in 2015, to 7,517 in 2019. The numbers demonstrate a flare of interest on those topics dealt with in the present chapter. Yet, interest on theoretical aspects of TDI was not so prevalent.

With basis on Web of Science report, the five most cited works on destination image between January 1990 and February 2020 are presented on Table 2.

Table 2. The five most cited works on destination image

Authors	Total of citations up to 2020.	Average citations per year
Baloglu & McCleary (1999)	1121	50,95
Beerli & Martin (2004a)	800	47,06
Chi, Qu & Hailin (2008)	754	58,00
Chen & Tsai (2007)	742	53,00
Gallarza, Saura & Garcia (2002)	692	36,42

Source: Web of Science report.13/3/2020. Authors' own.

In the next section, this chapter turns to a content analysis performed on a selection of 30 articles with focus on theories used in the study of tourism destination image.

MAIN FOCUS OF THE CHAPTER

As discussed, tourism experts have studied and theorised TDI from multiple perspectives. Those perspectives integrate knowledge, feelings, global impressions, and values about a destination (Baloglu & McCleary 1999; Espelt & Benito, 2005; Huang & Cai, 2010; Lai & Li, 2015). Based on the content analysis of 250 academic articles on TDI, their mentions to diverse underlying theories was salient in 30 articles with more than 1% of their contents dedicated to theoretical issues concerning TDI (Figure 1).

For Lai and Li (2015:12) a definition perspective found in logic studies (e.g., Gerring 1997; Kimiecik & Harris 1996) would be required for the building of a systematic knowledge in tourism studies so that specialists in this scientific field can create and use definitions such as TDI based on an ultimate ratio-

Figure 1. Emphasis on the theoretical bases of TDI in tourism literature, 1990-2020
Source: N Vivo 12 Plus – Authors' own

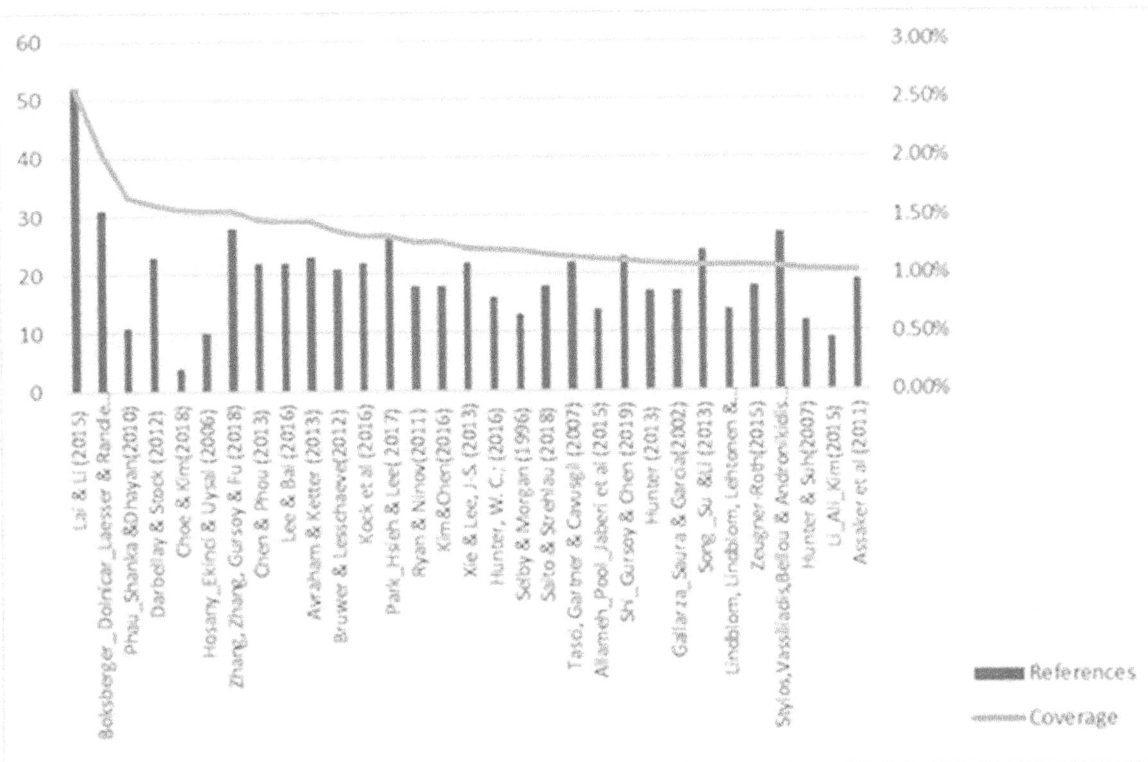

nality. This theoretical strategy would guarantee the development and circulation of better definitions which could influence the validity of research instruments used in TDI studies.

Other authors adopted a different approach to theoretical issues in TDI by empirically testing hypotheses that circulated in the literature. Boksberger, Dolnicar, Laesser and Randle (2011:454) for example focused on the *self-congruity* theory, defined by Sirgy and Su (2000:340) as "the match between the destination visitor image and tourists' self-concept (actual, ideal, social) and ideal social self-image".

Phau, Shanka and Dhayan (2010:760) proposed that the notion of *perceived TDI* could be a sound theoretical foundation for the assessment of potential tourists' image of Mauritius prior to visitation. The authors relied on the theory of reasoned action and the theory of planned behaviour which emphasise behavioural intentions as representative of the degree of conscious effort that a person exerts in order to perform a behaviour.

Daberllay and Stock (2012) posited that tourism embodies a scientiðc object that permits a privileged perspective on human societies and institutes a certain vantage point. The authors demonstrated that an interdisciplinary approach ðts with the cognitive project and the complex object of TDI because it "articulates the double movement of disciplinary specialisation and the awareness of an autonomous logic of the touristic that has been emerging within tourism studies"(Daberllay & Stock, 2012:455).

Zhang, Zhang, Gursoy and Fu (2018) relied on the *framing effects* theory to examine the inñuence of message framing and how a match or mismatch between message framing and individuals' regulatory focus can inñuence their destination image perceptions and visit intentions. Those authors provided empirical evidence that framing of marketing messages could have a signiðcant impact on travellers' decision making and destination selection process.

The theoretical mapping bellow (Figure 2) displays some of the theories underpinning the dimensions of tourism destination image.

Figure 2. TDI- Mapping of Disciplinary Specialisation
Source: NVivo Authors' own.

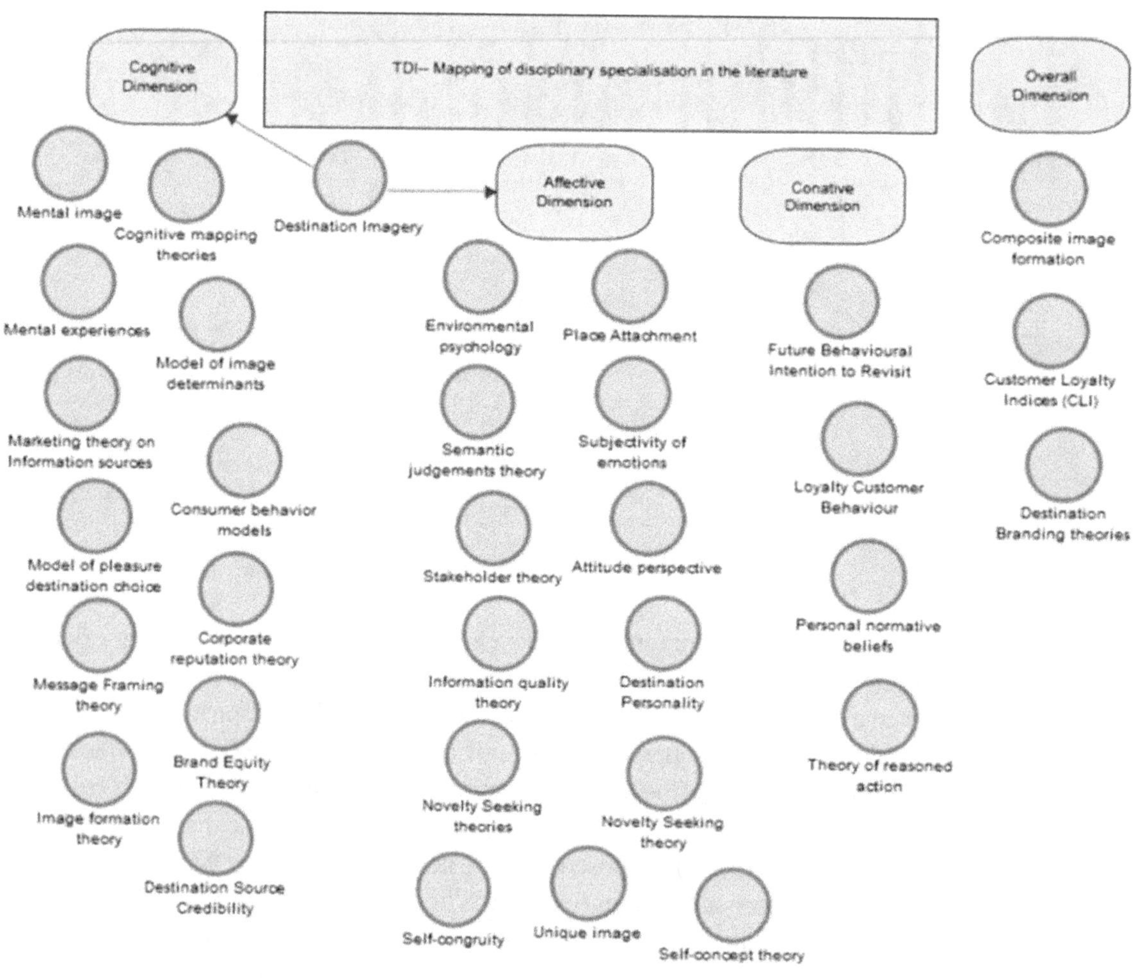

Figure 2 maps out the 33 theoretical references that brings into a new light what Martin, Beerli and Nazzareno (2017) identified as a widespread agreement around notions of TDI in the tourism marketing literature (Baloglu & Brinberg, 1997; Baloglu & Mangaloglu, 2001; Baloglu & Mccleary, 1999a; Baloglu & Mccleary, 1999b; Beerli & Martín, 2004a; Beerli & Martín, 2004b; Chen & Uysal, 2002; Gartner, 1993; Hosany et al.; 2006, Pike, 2009; Pike & Ryan, 2004).

Most of those theories interpreted the image are the result of three closely interrelated components. Those components are (1) perceptual/cognitive, which is related to the beliefs of individuals on the attributes that characterize a destination; (2) emotional/affective, which refers to emotional response or the feelings that individuals express about the place; and (3) global, which corresponds to the overall positive or negative impression of the place. As argued before, psychology is a disciplinary area "with immense importance for the study of tourism" (Skavronskaya, Scott, Moyle, Le, Hadinejad, Zhang, Gardiner, Coghlan & Shakeela, 2017: 222).

Accepting the stark influence of psychology in tourism theories, one can say that the production of hypotheses in tourism studies can be divided into two main approaches the science of behaviour and mind - the *behaviourist approach* and the *cognitive approach* (or appraisal process) – "the behaviourist approach involves study of the relationship between stimuli and behaviour only, whereas the cognitive approach… seeks to understand the mental processes occurring between stimuli and behaviour"(Skavronskaya et al. 2017:222). The question is that many tourism papers refer to emotion or experience but mainly from the behavioural perspective and often with different meanings or expectations when it is not always useful, to correlate tourism stimuli with consequences of hedonic consumption. Alba and Williams, (2013) emphasised the need to consider the subjective mental processes that link stimuli to the consequences of hedonistic consumption. Authors such as Ma, Gao, Scott, and Ding (2013) discussed how a cognitive approach is more suitable to discuss the appraisal process.

Conscious of the above remarks on the status of psychology in tourism hypothesis generation, the next paragraphs revisit some of the key theories applicable in tourism studies, as explored in the present review.

Theory of Self-Concept

Rosenberg (1979) defined *self-concept* as the sum of the individual's thoughts and feelings with reference to oneself as an object. The self-concept is shaped in an interaction process between an individual and others whilst the individual endeavours self-enhancement in the interaction process. As proposed, individuals can have multiple 'situational' selves that become salient in different social situations. (Aaker 1999; Markus & Kunda 1986; Schenk & Holman 1980).

Theory of Self-Congruity, Self-Concept, (Boksberger et al, 2011. Beerli et al. 2007)

Sirgy and Su (2000) defined the concept of *self-congruity* as the degree of congruence between destination image and tourists' *self-image*. A person's self-image comprises their actual, ideal, and social selves and the destination's consistency with a person's actual or ideal self-image tends to inñuence pre-visit preferences and post-visit intentions, which involves satisfaction, revisit intention and intention to recommend. A better match between destination image and self-image of a tourist will strengthen the probability of visitation because destination choice is inñuenced by self-congruity (Litvin & Goh, 2002).

Beerli, Meneses and Gil (2007:575) established that "self-congruity is less important in the choice of a destination when people are familiar with it. This is because the decision is based on more rational criteria". Based on the operationalisation of *self-congruity* theories, tour operators, industry associations, and local authorities gained additional tools to reach appropriate decisions when positioning a destination's communication actions based on attributes that should resemble the potential tourists' image as closely as possible. Segmenting strategies should also be more efficient with this extra knowledge of moderating effects on congruity.

Brand Personality

Boksberger, P., Dolnicar, S., Laesser, C. & Randle, M. (2011) investigated whether the theory of *self-congruity* held for destination choice choosing a mature market (Switzerland) for their investigation. Their study revealed that choosing a very strict operationalization of *self-congruity* resulted in a very small proportion of cases being self-congruent. Zhang, Fu, Cai & Lu (2014: 215) further highlighted that the concept of *self-congruity* is a standpoint that complements the well-known cognitive, affective, and overall perspectives. The authors presented evidence that together with other image variables (cognitive image, affective image, overall image and cognitive-affective joint image), attitudinal loyalty, behavioural loyalty and composite loyalty (*self-congruity variables*) affect tourist loyalty (Zhang et al. 2014:218). . Other researchers contributed to *self-congruity* studies (e.g. Ekinci, Sirakaya-Turk & Preciado, 2013; Nicola, 2011; Usakli & Baloglu, 2011; Hung & Petrick, 2011, 2012; Sameer & Martin, 2012).

Brand Equity Theories

Brand equity refers to a set of brand properties and weaknesses linked to a brand, its name and symbol (Aaker, 1996). The literature on marketing and *brand equity* deploys different perspectives of *brand equity* and the two major standpoints on the matter are the *cognitive psychology view* of marketing and the *information economics perspectives.* One chief argument here is that consumers (e.g. tourists) are likely to be loyal to brands and this is where the notion of *brand equity* becomes relevant.

In cognitive psychology, marketing research studies define *brand equity* as the added value a brand gives to a product (Farquhar, 1989; Erdem & Swait, 1998). This added value can be viewed and analysed from the perspective of either the consumer (e.g. a holidaymaker) or the firm (e.g. flight companies and hotel chains). Keller (1993) gave prominence to brand awareness and associations as key foundations of *consumer-based brand equity.* Erdem and Swait, (1998: 132) reminded that "brand associations, brand awareness, perceived quality, brand loyalty, and other proprietary brand assets (e.g., patents) underlie brand equity". Yet, this perspective of consumer *brand equity* (entrenched in *cognitive psychology)* does not distinguish the informational aspects of the marketplace as the *information quality theory* does when it looks at the dynamic interaction between firms and consumers. Oh & Hsu (2014:156) also noted that the *brand equity* literature, "lacks research efforts to address potential cross-cultural variations".

Brand Signal Theory /Signalling Theory

Brand equity is also explained from the *signalling theory* point of view associated to information economics. Erdem and Swait (1998) advanced this perspective on the value (or equity) ascribed to brands by consumers. One of the main advantages of using the *signalling theory* to frame the notion of brand equity in tourism studies is that it can explain firms' signalling to consumers as well the hinting between firms because this viewpoint stresses the role of *credibility* as the main determinant of consumer-based *brand equity*. The credibility of a brand (a *signal* of the product's position) may increase perceived quality and decrease information costs and the risk perceived by consumers.

Destination Source Credibility

Veasna et al (2013) hypothesized the relationships among four constructs - destination source credibility, destination image, and destination attachment as antecedents of destination satisfaction. The relationships between those elements were examined for a sample of 398 international tourists visiting a world heritage tourism destination (Angkor Wat) and a famous skyscraper (Taipei 101). The Structural Equation Model (SEM) indicated that *destination source credibility* and destination image could "affect tourist perceptions of destination satisfaction with regard to destination attachment" (Veasna et al. 2013:511). Veasna et al (2013:513) proposed that when tourists trust credible destination sources, this factor is likely to exert a persuasive effect on their opinions about destination image.

Veasna et al. (2013:512) singled out a few examples of literature considering attachment and tourist involvement to be predictors of satisfaction and loyalty intentions in tourism destination contexts (e.g., Hwang, Lee, & Chen, 2005; Spry et al., 2011; Yuksel, Yuksel, & Bilim, 2010). Prayag and Ryan (2012), for example, explored the relationships among destination image, place attachment, personal involvement, and visitors' satisfaction as antecedents of loyalty for a sample of international visitors staying in hotels on the island of Mauritius. A structural model indicated that destination image, personal involvement and place attachment are antecedents of visitors' loyalty, but this relationship is mediated by satisfaction levels.

Novelty Seeking Theory

Novelty-seeking theory "provides a strong theoretical foundation in explaining destination choice behaviour [and]… variety and novelty seeking share the same conceptual foundation—that consumers seek optimal levels of stimulation, thus influencing their choice of behaviour" (Assaker & Hallak, 2013:602). Certain customers switch products and make new purchases despite being satisfied with their original purchase (Keaveney, 1995; Reichheld, 1993). The search for novelty is an innate quality in certain travellers (Cohen 1979; Lee & Crompton 1992) and plays an important role in tourists' decision-making process (Petrick 2002). The few studies that have examined this area (Niininen, Szivas, & Riley 2004; Barroso, Armario, & Ruiz 2007; Jang & Feng 2007; Assaker, Vinzi, & O'Connor 2011) proposed that there is a negative relationship between variety/novelty-seeking tendencies and intentions to revisit the destination in the future. Assaker, Vinzi, and O'Connor (2011) found that novelty has a significant negative effect on the intercept of return intention trajectory.

Destination Imagery Theory

Josianssen, Assaf, Woo and Kock (2016:4) defined the *destination imagery* (DY) concept as "an individual's or a group of individuals' diverse *cognitive* and *affective* associations relating to a destination" and not "the understanding of imagery that is taken by a branch of the psychology literature to mean conscious phenomenological experiences of objects or events (e.g., mentally rotating a car to see what is on the other side; mentally re-experiencing the latest holiday trip)". A reasonably full account of DY should then be multidimensional. Josianssen et. al, (2016:10) cautioned that "DI is an outcome of DY, and future research should draw on this notion to investigate the nature of this relationship in more detail".

FUTURE RESEARCH DIRECTIONS

This chapter used a systematic literature review to map and report on the progress made several authors to elaborate a clear definition for TDI. It argues that whilst the approach to TDI from new theoretical perspectives and the dialogue with the various theories underlying this paradigm has expanded and has increased the complexity of this construct, the dimensions of TDI still needs more research projects which include both existing and novel theoretical paradigms and new behavioural change interventions based on perspectives such as (1) Future behavioural intention to revisit, (2) Loyalty customer behaviour, (3) Personal normative beliefs and (4) Theory of reasoned action and others novel approaches may also give rise to interventions that are likely to be effective if they target causal determinants of tourists' behaviour and behaviour change. New research projects will need to explain the theoretical mechanisms of change in tourists' behaviour. New theories may need to be developed and tested by means of interventions that are theoretically informed. It is our view that theory-based interventions in tourism studies will facilitate an understanding of what works across different contexts, populations, and behaviours.

Future research work on image theory still need to produce a systematized structure to properly conceptualise and operationalise the TDI construct, testing the scheme imagery-image duality model proposed by Josiassen et al (2016). As mentioned before, there is a need to develop and consolidate a single definition of image, which can redeem the destination image literature from being "atheoretical and non-scientific" (Tasci, Gartner & Cavusgill, 2007:217). The present chapter exposed the need for more investment on modelling destination imagery and destination image as formative constructs.

It would also be advisable for future research to place emphasis on new experimental research on managerial implications of targeted investment to stimulate *conative* aspects of a destination image. Such studies could be useful to explain how personal normative beliefs connect to future behavioural intentions and customer loyalty. Overall destination image studies should also consider the need for more causal investigations approaching interrelation between branding, composite image formation, and unique image formation in diverse destination categories. Furthermore, tourism studies still require research comparing cognitive, affective, overall and conative images associated to a destination at different points in time being those favourable or unfavourable conjunctures for the tourism industry.

Authors have noticed that some areas of study such as risk and uncertainty in tourism are not contemplated in the 250 sources selected to analyse the unfolding theories on TDI (1990-2020). Given the events affecting the tourism industry in 2020, such the worldwide Covid-19 pandemics, it is notable that none of the sources contemplated the several different aspects of the uncertainty and risk within the tourism field in relation their impact on destination image. Risk management in connection with tourism has been associated to internal negative effects such as lack of security and attacks on tourists whilst they are abroad, and risks associated to external negative threats such as terrorism. However, there is a wide range of uncertainty factors that affect DI.

CONCLUSION

This chapter has mapped the most relevant theoretical constructs that underpin the cognitive, affective, conative, and overall aspects of TDI in tourism and management studies. This study highlighted the multidisciplinary character of TDI and demonstrated that TDI theories are still to approach for example how a proactive crisis management by tourism boards and hotel industry would affect the recovery of trust, which is a key factor of successful TDI.

This systematic review of literature should equip tourism industries and scholars with knowledge of how tourists' TDI is conceptualized and on which theories it relies on. Hopefully, it can contribute to better conceptualize TDI construct in future studies.

ACKNOWLEDGMENT

This paper is financed by National Funds provided by FCT- Foundation for Science and Technology through project UIDB/04020/2020.

This work is financed by national funds through FCT - Foundation for Science and Technology, I.P., within the scope of the project «UIDB / 04647/2020» of CICS.NOVA - Centro Interdisciplinar de Ciências Sociais da Universidade Nova de Lisboa.

REFERENCES

Aaker, D. A. (1991). *Managing brand equity*. Free Press.

Aaker, D. A. (1996). Measuring Brand Equity Across Products and Markets. *California Management Review, 38*(3), 102-120.

Aaker, J. L. (1999). The Malleable Self: The Role of Self-Expression in Persuasion. *JMR, Journal of Marketing Research, 36*(February), 45–57. doi:10.1177/002224379903600104

Agapito, D., Oom do Valle, P., & Costa Mendes, J. (2013). The Cognitive-Affective-Conative Model of Destination Image: A Confirmatory Analysis. *Journal of Travel & Tourism Marketing, 30*(5), 471–481. doi:10.1080/10548408.2013.803393

Agustí, P. (2018). Characterizing the location of tourist images in cities. Differences in user-generated images (Instagram), official tourist brochures and travel guides. *Annals of Tourism Research, 73*, 103–115. doi:10.1016/j.annals.2018.09.001

Ahmed, Z. U. (1991). The inñuence of the components of a state's tourist image on product positioning strategy. *Tourism Management, 12*(4), 331–340. doi:10.1016/0261-5177(91)90045-U

Alba, J. W., & Williams, E. F. (2013). Pleasure principles: A review of research on hedonic consumption. *Journal of Consumer Psychology, 23*(1), 2–18. doi:10.1016/j.jcps.2012.07.003

Alcañiz, E. B., García, I. S., & Blas, S. S. (2009). The functional-psychological continuum in the cognitive image of a destination: A confirmatory analysis. *Tourism Management, 30*(5), 715–723. doi:10.1016/j.tourman.2008.10.020

Alhemoud, A. M., & Armstrong, E. G. (1996). Image of Tourism Attractions in Kuwait. *Journal of Travel Research, 34*(4), 76–80. doi:10.1177/004728759603400413

Assael, H. (1984). *Consumer Behavior and Marketing Action*. Pub. Co.

Assaker, G., & Hallak, R. (2013). Moderating Effects of Tourists' Novelty-Seeking Tendencies on Destination Image, Visitor Satisfaction, and Short- and Long-Term Revisit Intentions. *Journal of Travel Research, 52*(5), 600–613. doi:10.1177/0047287513478497

Assaker, G., Vinzi, V. E., & O'Connor, P. (2011). Examining the effect of novelty seeking, satisfaction, and destination image on tourists' return pattern: A two factor, non-linear latent growth model. *Tourism Management, 32*(4), 890–901. doi:10.1016/j.tourman.2010.08.004

Baker, D., & Crompton, J. (2000). Quality, satisfaction and behavioral intentions. *Annals of Tourism Research, 27*(3), 785–804. doi:10.1016/S0160-7383(99)00108-5

Baloglu, S. (1997). The relationship between destination images and sociodemographic and trip characteristics of international travellers. *Journal of Vacation Marketing, 3*(3), 221–233. doi:10.1177/135676679700300304

Baloglu, S., & Brinberg, D. (1997). Affective images of tourism destinations. *Journal of Travel Research, 35*(4), 11–15. doi:10.1177/004728759703500402

Baloglu, S., & Love, C. (2005). A Cognitive–Affective Positioning Analysis of Convention Cities: An Extension of the Circumplex Model of Affect. *Tourism Analysis, 9*(4), 299–308. doi:10.3727/108354205789807210

Baloglu, S., & Mangaloglu, M. (2001). Tourism destination images of Turkey, Egypt, Greece, and Italy as perceived by US-based tour operators and travel agents. *Tourism Management, 22*(1), 1–9. doi:10.1016/S0261-5177(00)00030-3

Baloglu, S., & McCleary, K. W. (1999). A model of destination image formation. *Annals of Tourism Research, 26*(4), 868–897. doi:10.1016/S0160-7383(99)00030-4

Baron, R., & Kenny, D. (1986). The moderator-mediator variable distinction in social psychological research: Conceptual, strategic, and statistical consideration. *Journal of Personality and Social Psychology, 51*(6), 1173–1182. doi:10.1037/0022-3514.51.6.1173 PMID:3806354

Barroso, C. C., Armario, E. M., & Ruiz, D. M. (2007). The influence of market heterogeneity on the relationship between a destination's image and tourists' future behaviour. *Tourism Management, 28*(1), 175–187. doi:10.1016/j.tourman.2005.11.013

Beerli, A., & Martin, J. D. (2004a). Factors influencing destination image. *Annals of Tourism Research, 31*(3), 657–681. doi:10.1016/j.annals.2004.01.010

Beerli, A., & Martin, J. D. (2004b). Tourists' characteristics and the perceived image of tourist destinations: A quantitative analysis – a case study of Lanzarote, Spain. *Tourism Management, 25*(5), 623–636. doi:10.1016/j.tourman.2003.06.004

Beerli, A., Meneses, G. D., & Gil, S. M. (2007). Self-Congruity and Destination Choice. *Annals of Tourism Research, 34*(3), 571–587. doi:10.1016/j.annals.2007.01.005

Berg, B. (2009). *Qualitative Research Methods for the Social Sciences*. Allyn & Bacon.

Bigné, E., Sánchez, I., & Sanz, S. (2009). The functional psychological continuum in the cognitive image of a destination: A confirmatory analysis. *Tourism Management*, *3*(5), 715–723. doi:10.1016/j.tourman.2008.10.020

Bigné, J. E., Sanchez, M. I., & Sanchez, J. (2001). Tourism image, evaluation variables and after purchase behaviour: Inter-relationship. *Tourism Management*, *22*(6), 607–616. doi:10.1016/S0261-5177(01)00035-8

Billig, M. (2006). Is my home my castle? Place attachment, risk perception, and religious faith. *Environment and Behavior*, *38*(2), 248–265. doi:10.1177/0013916505277608

Boksberger, P., Dolnicar, S., Laesser, C., & Randle, M. (2011). Self-congruity theory: To what extent does it hold in tourism? *Journal of Travel Research*, *50*(4), 454–456. doi:10.1177/0047287510368164

Boo, S., Busser, J., & Baloglu, S. (2009). A model of customer-based brand equity and its application to multiple destinations. *Tourism Management*, *30*(2), 219–231. doi:10.1016/j.tourman.2008.06.003

Bosnjak, M., Sirgy, M. J., Hellriegel, S., & Maurer, O. (2011). Postvisit destination loyalty judgments: Developing and testing a comprehensive congruity model. *Journal of Travel Research*, *50*(5), 496–508. doi:10.1177/0047287510379159

Bosque, I., & Martín, H. (2008). Tourist satisfaction: A cognitive-affective model. *Annals of Tourism Research*, *35*(2), 551–573. doi:10.1016/j.annals.2008.02.006

Cai, L. A., Wu, B., & Bai, B. (2004). Destination image and loyalty. *Tourism Review International*, *7*(3–4), 153–162.

Calantone, R. J., Di Benedetto, C. A., Hakam, A., & Bojanic, D. C. (1989). Multiple Multinational Tourism Positioning Using Correspondence analysis. *Journal of Travel Research*, *28*(2), 25–32. doi:10.1177/004728758902800207

Centre for Reviews and Dissemination, University of York (CDR) Systematic Reviews: CDR's Guidance for undertaking reviews in health care. (n.d.). Downloaded on 20/03/2020 Available at: https://www.york.ac.uk/media/crd/Systematic_Reviews.pdf

Chan, C.-S., & Peters, M. (2020). Investigating brand strategy gap in Hong Kong: First impressions and unique attractions. *Asian Geographer*, 1–22. doi:10.1080/10225706.2020.1729211

Chen, C. F., & Tsai, D. (2007). How destination image and evaluative factors affect behavioral intentions? *Tourism Management*, *28*(4), 1115–1122. doi:10.1016/j.tourman.2006.07.007

Chen, J., & Gursoy, D. (2001). An investigation of tourists' destination loyalty and preferences. *Journal of Contemporary Hospitality Management*, *13*(2), 79–85. doi:10.1108/09596110110381870

Chen, N., Ji, S., & Funk, D. C. (2015). An extended study on destination image decay of sport tourists over time. *Journal of Destination Marketing & Management*, *2*(4), 241–252. doi:10.1016/j.jdmm.2013.11.001

Cherifi, B., Smith, A., Maitland, R., & Stevenson, N. (2014). Destination images of non-visitors. *Annals of Tourism Research*, *49*, 190–202. doi:10.1016/j.annals.2014.09.008

Chew, E. Y. T., & Jahari, S. A. (2014). Destination image as a mediator between perceived risks and revisit intention: A case of post-disaster Japan. *Tourism Management*, *40*, 382–393. doi:10.1016/j.tourman.2013.07.008

Chi, C., & Qu, H. (2008). Examining the structural relationships of destination image, tourist satisfaction and destination loyalty: An integrated approach. *Tourism Management*, *29*(4), 624–636. doi:10.1016/j.tourman.2007.06.007

Choi W.M., Chan, A. & Wu, J. (1999). A qualitative and quantitative assessment of Hong Kong's image as a tourist destination. *Tourism Management, 20*, 361-365.

Chon, K.-S. (1992). Self-Image/Destination Image Congruity. *Annals of Tourism Research*, *19*(2), 360–376. doi:10.1016/0160-7383(92)90090-C

Cohen, E. (1979). A Phenomenology of Tourist Experiences. *Sociology*, *13*(2), 179–201. doi:10.1177/003803857901300203

Cooper, C., Booth, A., Varley-Campbell, J., Britten, N., & Garside, R. (2018). Defining the process to literature searching in systematic reviews: A literature review of guidance and supporting studies. *BMC Medical Research Methodology*, *18*(1), 85. doi:10.118612874-018-0545-3 PMID:30107788

Coshall, J. T. (2000). Measurement of Tourists' Images: The Repertory Grid Approach. *Journal of Travel Research*, *39*(1), 85–89. doi:10.1177/004728750003900111

Darbellay, F., & Stock, M. (2012). Tourism as complex interdisciplinary research object. *Annals of Tourism Research*, *39*(1), 441–458. doi:10.1016/j.annals.2011.07.002

Dolnicar, S., & Grün, B. (2013). Validly Measuring Destination Image in Survey Studies. *Journal of Travel Research*, *52*(1), 3–13. doi:10.1177/0047287512457267

Duarte, P., Folgado-Fernández, J. A., & Hernández-Mogollón, J. M. (2018). Measurement of the impact of music festivals on destination image: The case of a Womad festival. *Event Management*, *22*(4), 517–526. doi:10.3727/152599518X15264726192451

Echtner, C. M., & Ritchie, J. R. B. (1991). The meaning and measurement of destination image. *Journal of Tourism Studies*, *2*(2), 2–12.

Echtner, C. M., & Ritchie, J. R. B. (1993). The measurement of destinations image: An empirical assessment. *Journal of Travel Research*, *31*(4), 3–13. doi:10.1177/004728759303100402

Ekinci, Y., Sirakaya-Turk, E., & Preciado, S. (2013). Symbolic consumption of tourism destination brands. *Journal of Business Research*, *66*(6), 711–718. doi:10.1016/j.jbusres.2011.09.008

Erdem, T., & Swait, J. (1998). Brand equity as a signalling phenomenon. *Journal of Consumer Psychology*, *7*(2), 131–157. doi:10.120715327663jcp0702_02

Erdem, T., & Swait, J. (2004). Brand credibility, brand consideration, and choice. *The Journal of Consumer Research*, *31*(1), 191–198. doi:10.1086/383434

Espelt, N. G., & Benito, J. A. D. (2005). The social construction of the image of Girona: A methodological approach. *Tourism Management*, *26*(5), 777–785. doi:10.1016/j.tourman.2004.04.004

Fakeye, P. C., & Crompton, J. L. (1991). Image Differences between Prospective, First-Time, and Repeat Visitors to the Lower Rio Grande Valley. *Journal of Travel Research, 30*(2), 10–16. doi:10.1177/004728759103000202

Farquhar, P. H. (1989). Managing brand equity. *Marketing Research, 1*, 24–33.

Frias, D. M., Rodriguez, M. A., Castaneda, A., Sabiote, C. M., & Buhalis, D. (2012). The Formation of a Tourist Destination's Image via Information Sources: The Moderating Effect of Culture. *International Journal of Tourism Research, 14*(5), 437–450. doi:10.1002/jtr.870

Fu, H., Ye, B. H., & Xiang, J. (2016). Reality TV, audience travel intentions, and destination image. *Tourism Management, 55*, 37–48. doi:10.1016/j.tourman.2016.01.009

Gallarza, M. G., Saura, I. G., & Garcia, H. C. (2002). Destination image: Towards a conceptual framework. *Annals of Tourism Research, 29*(1), 56–78. doi:10.1016/S0160-7383(01)00031-7

Gartner, W. (1993). Image formation process. In M. Uysal & D. Fesenmaier (Eds.), *Communication and channel systems in tourism marketing* (pp. 191–215). The Haworth Press.

Gartner, W. C. (1989). Tourism Image: Attribute Measurement of State Tourism Products Using Multidimensional Scaling Techniques. *Journal of Travel Research, 28*(2), 16–20. doi:10.1177/004728758902800205

Gartner, W. C. (1994). Image formation process. *Journal of Travel & Tourism Marketing, 2*(2-3), 191–216. doi:10.1300/J073v02n02_12

Gerring, J. (1997). Ideology: A Definitional Analysis. *Political Research Quarterly, 50*(4), 957–994. doi:10.1177/106591299705000412

Gunn, C. A. (1972). *Vacationscape. Designing tourist regions.* Taylor & Francis.

Hall, C. M., & Valentin, A. (2005). Content analysis. In B. W. Ritchie, P. Burns, & C. Palmer (Eds.), *Tourism research methods: Integrating theory with practice.* Cabi Publishing. doi:10.1079/9780851999968.0191

Hernández-Lobato, L., Solis-Radilla, M. G., Moliner-Tena, M. G., & Sanchez-Garcia, J. (2006). Tourism Destination Image, Satisfaction and Loyalty: A Study in Ixtapa-Zihuatanejo, Mexico. *Tourism Geographies, 8*(4), 343–358. doi:10.1080/14616680600922039

Hosany, S., Ekinci, Y., & Uysal, M. (2006). Destination image and destination personality: An application of branding theories to tourism places. *Journal of Business Research, 59*(5), 638–642. doi:10.1016/j.jbusres.2006.01.001

Hung, K., & Petrick, J. F. (2011). The role of self- and functional congruity in cruising intentions. *Journal of Travel Research, 50*(1), 100–112. doi:10.1177/0047287509355321

Hung, K., & Petrick, J. F. (2012). Testing the effects of congruity, travel constraints, and self-efðcacy on travel intentions: An alternative decision-making model. *Tourism Management, 33*(4), 855–867. doi:10.1016/j.tourman.2011.09.007

Hunt, J. D. (1971). *Image: A Factor in Tourism.* Colorado State University Press.

Hunt, J. D. (1975). Image as a factor in tourism development. *Journal of Travel Research, 13*(3), 1–7. doi:10.1177/004728757501300301

Hunter, W. C., & Suh, Y.-K. (2007). Multimethod research on destination image perception: Jeju standing stones. *Tourism Management, 28*(1), 130–139. doi:10.1016/j.tourman.2005.06.013

Hwang, S.-N., Lee, C., & Chen, H.-J. (2005). The relationship among tourists' involvement, place attachment and interpretation satisfaction in Taiwan's national parks. *Tourism Management, 26*(2), 143–156. doi:10.1016/j.tourman.2003.11.006

Jenkins, O. H. (1999). Understanding and measuring tourist destination images. *International Journal of Tourism Research, 1*(1), 1–15. doi:10.1002/(SICI)1522-1970(199901/02)1:1<1::AID-JTR143>3.0.CO;2-L

Josiassen, A., Assaf, A. G., Woo, L., & Kock, F. (2016). The imagery–image duality model—An integrative review and advocating for improved delimitation of concepts. *Journal of Travel Research, 55*(6), 789–803. doi:10.1177/0047287515583358

Keller, K. L. (1993). Conceptualizing, measuring, and managing customer-based brand equity. *Journal of Marketing, 57*(1), 1–22. doi:10.1177/002224299305700101

Kim, H., & Richardson, S. L. (2003). Motion picture impacts on destination images. *Annals of Tourism Research, 30*(1), 216–237. doi:10.1016/S0160-7383(02)00062-2

Kim, S., Lee, K. Y., Shin, S., & Yang, S. (2017). Effects of tourism information quality in social media on destination image formation: The case of Sina Weibo. *Information & Management, 54*(6), 687–702. doi:10.1016/j.im.2017.02.009

Kim, S., & Lehto, X. Y. (2013). Projected and perceived destination brand personalities: The case of South Korea. *Journal of Travel Research, 52*(1), 117–130. doi:10.1177/0047287512457259

Kim, S. S., McKercher, B., & Lee, H. (2009). Tracking tourism destination image perception. *Annals of Tourism Research, 36*(4), 715–718. doi:10.1016/j.annals.2009.04.007

Kimiecik, J. C., & Harris, A. T. (1996). What Is Enjoyment? A Conceptual/Definitional Analysis with Implications for Sport and Exercise Psychology. *Journal of Sport & Exercise Psychology, 18*(3), 247–263. doi:10.1123/jsep.18.3.247

King, C., Chen, N., & Funk, D. C. (2015). Exploring destination image decay: A study of sport tourists' destination image change after event participation. *Journal of Hospitality & Tourism Research (Washington, D.C.), 39*(1), 3–31. doi:10.1177/1096348012461547

Kirstin, H., & Christoph, B. (2010). Image ðt between sport events and their hosting destinations from an active sport tourist perspective and its impact on future behaviour. *Journal of Sport & Tourism, 15*(3), 215–237. doi:10.1080/14775085.2010.513147

Kislali, H., Kavaratzis, M., & Saren, M. (2019). Destination image formation: Towards a holistic approach. *International Journal of Tourism Research*, 1–11.

Kokosalakis, C., Bagnall, G., Selby, M., & Burns, S. (2006). Place image and urban regeneration in Liverpool. *International Journal of Consumer Studies*, *30*(4), 389–397. doi:10.1111/j.1470-6431.2006.00520.x

Konecnik, M., & Gartner, W. (2007). Customer-based brand equity for a destination. *Annals of Tourism Research*, *34*(2), 400–421. doi:10.1016/j.annals.2006.10.005

Krippendorff, K., & Bock, M. (2009). *The content analysis reader*. Sage.

Ladhari, R., Souiden, N., & Ladhari, I. (2011). Determinants of loyalty and recommendation: The role of perceived service quality, emotional satisfaction and image. *Journal of Financial Services Marketing*, *16*(2), 111–124. doi:10.1057/fsm.2011.10

Lai, K., & Li, Y. (2012). Core-periphery structure of destination image concept, evidence and implication. *Annals of Tourism Research*, *39*(3), 1359–1379. doi:10.1016/j.annals.2012.02.008

Lai, K., & Li, Y. (2015). Tourism destination image: Conceptual problems and definitional solutions. *Journal of Travel Research*, *55*(8), 1065–1080. doi:10.1177/0047287515619693

Larsen, G., & George, V. (2004). *The Social Construction of Destination Image – A New Zealand Film Example.* Working Papers, 4/01.February.

Lawson, F., & Baud-Bovy, M. (1977). *Tourism and Recreational Development*. Architectural Press.

Lee, C. K., Lee, Y. K., & Lee, B. K. (2005). Korea's destination image formed by the 2002 World Cup. *Annals of Tourism Research*, *32*(4), 839–858. doi:10.1016/j.annals.2004.11.006

Lee, T. H., & Crompton, J. (1992). Measuring Novelty Seeking in Tourism. *Annals of Tourism Research*, *19*(4), 732–751. doi:10.1016/0160-7383(92)90064-V

Lee, T. H., & Crompton, J. (1992). Measuring Novelty Seeking in Tourism. *Annals of Tourism Research*, *19*(4), 732–751. doi:10.1016/0160-7383(92)90064-V

Li, M., Cai, L. A., Lehto, X. Y., & Huang, J. (2010). A missing link in understanding revisit intention—The role of motivation and image. *Journal of Travel & Tourism Marketing*, *27*(4), 335–348. doi:10.1080/10548408.2010.481559

Litvin, S. W., & Goh, H. K. (2002). Research Note: Self-Image Congruity: A Valid Tourism Theory? *Tourism Management*, *23*(1), 81–83. doi:10.1016/S0261-5177(01)00065-6

Lodrà-Riera, I., Martínez-Ruiz, M. P., Jiménez-Zarco, A. I., & Izquierdo-Yusta, A. (2015). A multidimensional analysis of the information sources construct and its relevance for destination image formation. *Tourism Management*, *48*, 319–328. doi:10.1016/j.tourman.2014.11.012

Ma, J., Gao, J., Scott, N., & Ding, P. (2013). Customer delight derived from theme park experiences: The antecedents of delight based on cognitive appraisal theory. *Annals of Tourism Research*, *42*(1), 359–381. doi:10.1016/j.annals.2013.02.018

MacKay, K. J., & Fesenmaier, D. R. (1997). Pictorial element of destination in image formation. *Annals of Tourism Research*, *21*(3), 537–565. doi:10.1016/S0160-7383(97)00011-X

MacKay, K. J., & Fesenmaier, D. R. (2000). An Exploration of Cross-Cultural Destination Image Assessment. *Journal of Travel Research, 38*(4), 417–423. doi:10.1177/004728750003800411

Marine-Roig, E., & Ferrer-Rosell, B. (2018). Measuring the gap between projected and perceived destination images of Catalonia using compositional analysis. *Tourism Management, 68*, 236–249. doi:10.1016/j.tourman.2018.03.020

Marks, L. J., & Olson, J. C. (2001). Toward a cognitive structure conceptualization of product familiarity. *Advances in Consumer Research. Association for Consumer Research (U. S.), 8*, 145–150.

Markus, H., & Kunda, Z. (1986). Stability and malleability of the self-concept. *Journal of Personality and Social Psychology, 51*(4), 858–866. doi:10.1037/0022-3514.51.4.858 PMID:3783430

Martin, H. S., & del Bosque, I. A. R. (2008). Exploring the cognitive-affective nature of destination image and the role of psychological factors in its formation. *Tourism Management, 29*(2), 263–277. doi:10.1016/j.tourman.2007.03.012

Martín-Santana, J. D., Beerli-Palacio, A., & Nazzareno, P. A. (2017). Antecedents and consequences of destination image gap. *Annals of Tourism Research, 62*, 13–25. doi:10.1016/j.annals.2016.11.001

Matos, N. M. S. (2014). *The Impact of Tourism Experiences in the destination image. A Marketing Perspective* (Unpublished doctoral dissertation). University of Algarve, Faro.

McCartney, G., Butler, R., & Bennett, M. (2008). A Strategic Use of the Communication Mix in the Destination Image-Formation Process. *Journal of Travel Research, 47*(2), 183–196. doi:10.1177/0047287508321201

Mercille, J. (2005). Media effects on image – The case of Tibet. *Annals of Tourism Research, 32*(4), 1039–1055. doi:10.1016/j.annals.2005.02.001

Moher, D., Liberati, A., Tetzlaff, J., Altman, D. G., & the The PRISMA Group. (2009). Preferred reporting items for systematic reviews and meta-analyses: the PRISMA Statement. *PloS Medicine, 6*. Retrieved from: http://www. Plosmedicine.org/article/fetchObject.action?uri=info

Moreno-Gil, S., & Martín-Santana, J. (2015). Understanding the image of self-Contained and serviced apartments: The Case of sun and Beach destinations. *Journal of Hospitality & Tourism Research (Washington, D.C.), 39*(3), 373–400. doi:10.1177/1096348013491594

Moutinho, L. (1987). Consumer Behaviour in Tourism. *European Journal of Marketing, 21*(10), 5–44. doi:10.1108/EUM0000000004718

Murphy, P., Pritchard, M. P., & Smith, B. (2000). The destination product and its impact on traveller perceptions. *Tourism Management, 21*(1), 43–52. doi:10.1016/S0261-5177(99)00080-1

Nicola, E. S. S. (2011). The relevance of visitors' nation brand embeddedness and personality congruence for nation brand identiðcation, visit intentions and advocacy. *Tourism Management, 32*(6), 1282–1289. doi:10.1016/j.tourman.2010.12.004

Papadimitriou, D., Apostolopoulou, A., & Kaplanidou, K. (2015). Destination personality, affective image, and behavioral intentions in domestic urban tourism. *Journal of Travel Research, 54*(3), 302–315. doi:10.1177/0047287513516389

Papadimitriou, D., Kaplanidou, K., & Apostolopoulou, A. (2018). Destination Image Components and Word-of-Mouth Intentions in Urban Tourism: A Multigroup Approach. *Journal of Hospitality & Tourism Research (Washington, D.C.), 42*(4), 503–527. doi:10.1177/1096348015584443

Petrick, J. (2004). Are loyal visitors desired visitors? *Tourism Management, 25*(4), 463–470. doi:10.1016/S0261-5177(03)00116-X

Phau, I., Shanka, T., & Dhayan, N. (2010). Destination image and choice intention of university student travellers to Mauritius. *International Journal of Contemporary Hospitality Management, 22*(5), 758–764. doi:10.1108/09596111011053846

Phelps, A. (1986). Holiday destination image: The problem of assessment. *Tourism Management, 7*(3), 168–180. doi:10.1016/0261-5177(86)90003-8

Picazo, P., & Moreno-Gil, S. (2019). Analysis of the projected image of tourism destinations on photographs: A literature review to prepare for the future. *Journal of Vacation Marketing, 25*(1), 3–24. doi:10.1177/1356766717736350

Pike, S. (2002). Destination image analysis – a review of 142 papers from 1973 to 2000. *Tourism Management, 23*(5), 541–549. doi:10.1016/S0261-5177(02)00005-5

Pike, S., & Ryan, C. (2004). Destination positioning analysis through a comparison of cognitive, affective, and conative perceptions. *Journal of Travel Research, 42*(4), 333–342. doi:10.1177/0047287504263029

Prayag, G. (2009). Tourists' Evaluations of Destination Image, Satisfaction, and Future Behavioral Intentions-The Case of Mauritius. *Journal of Travel & Tourism Marketing, 26*(8), 836–853. doi:10.1080/10548400903358729

Prayag, G., & Ryan, C. (2011). The relationship between the 'push' and 'pull' factors of a tourist destination: The role of nationality – an analytical qualitative research approach. *Current Issues in Tourism, 14*(2), 121–143. doi:10.1080/13683501003623802

Prayag, G., & Ryan, C. (2012). Antecedents of Tourists' Loyalty to Mauritius: The Role and Influence of Destination Image, Place Attachment, Personal Involvement, and Satisfaction. *Journal of Travel Research, 51*(3), 342–356. doi:10.1177/0047287511410321

Qiu, J. P., & Zou, F. (2004). Research about Content Analysis Method. *Journal of Library Science in China, 2*, 14–19.

Reynolds, W. H. (1965). The role of the consumer in image building. *California Management Review, 7*(Spring), 69–76. doi:10.2307/41165634

Rodrigues, A., Correia, A., Kozak, M., & Tuohino, A. M. (2015). Lake-destination image attributes: content analysis of text and pictures. In A. Correia, J. Gnoth, M. Kozak, & A. Fyall (Eds.), *Marketing Places and Spaces (Advances in Culture, Tourism and Hospitality Research)* (Vol. 10, pp. 293–314). Emerald Group Publishing Limited. doi:10.1108/S1871-317320150000010022

Rollero, C., & Piccoli, N. (2010). Place attachment, identification, and environment perception: An empirical study. *Journal of Environmental Psychology*, *30*(2), 198–205. doi:10.1016/j.jenvp.2009.12.003

Rosenberg, M. (1979). *Conceiving the Self*. Basic Books.

Russell, J. (1980). A circumplex model of affect. *Journal of Personality and Social Psychology*, *39*(6), 1161–1178. doi:10.1037/h0077714

Sameer, H., & Martin, D. (2012). Self-image congruence in consumer behavior. *Journal of Business Research*, *65*(5), 685–691. doi:10.1016/j.jbusres.2011.03.015

São José, J. M., & Amado, C. A. F. (2016). On studying ageism in long-term care: A systematic review of the literature. *International Psychogeriatrics*, 1–15. PMID:27852342

Schenk, C. T., & Holman, R. H. (1980). A sociological approach to brand choice: The concept of situational self-image. *Advances in Consumer Research. Association for Consumer Research (U. S.)*, *7*, 610–614.

Schenk, C. T., & Holman, R. H. (1980). A Sociological Approach to Brand Choice: The Concept of Situational Self Image. *Advances in Consumer Research. Association for Consumer Research (U. S.)*, *7*, 610–614.

Shani, A., Chen, P.-J., Wang, Y., & Hua, N. (2010). Testing the Impact of a Promotional Video on Destination Image Change: Application of China as a Tourism Destination. *International Journal of Tourism Research*, *12*, 116–133.

Shankar, R. S. (2018). Destination Personality and Destination Image: A Literature Review. *The IUP Journal of Brand Management*, *15*(4).

Sirakaya, E., & Woodside, A. G. (2005). Building and Testing Theories of Decision Making by Travellers. *Tourism Management*, *26*(6), 815–832. doi:10.1016/j.tourman.2004.05.004

Sirgy, M. J., & Su, C. (2000). Destination Image, Self-Congruity and Travel Behaviour: Toward an Integrative Model. *Journal of Travel Research*, *38*(2), 340–352. doi:10.1177/004728750003800402

Skavronskaya, L., Scott, N., Moyle, B., Le, D., Hadinejad, A., Zhang, R., Gardiner, S., Coghlan, A., & Shakeela, A. (2017). Cognitive psychology and tourism research: State of the art. *Tourism Review*, *72*(2), 221–237. doi:10.1108/TR-03-2017-0041

Souiden, N., Ladhari, R., & Chiadmi, N. E. (2017). Destination personality and destination image. *Journal of Hospitality and Tourism Management*, *32*, 54–70. doi:10.1016/j.jhtm.2017.04.003

Stabler, M. J. (1992). Modelling the Tourism Industry: A New Approach. In M. T. Sinclair & M. I. Stabler (Eds.), *the Tourism Industry: An International Analysis* (pp. 15–43). CAB.

Sun, M., Ryan, C., & Pan, S. (2015). Using Chinese Travel Blogs to Examine Perceived Destination Image: The Case of New Zealand. *Journal of Travel Research*, *54*(4), 543–555. doi:10.1177/0047287514522882

Tasci, A., & Gartner, W. (2007). Destination image and its functional relationships. *Journal of Travel Research*, *45*(4), 413–425. doi:10.1177/0047287507299569

Tasci, A. D. A., Gartner, W. C., & Cavusgil, S. T. (2007). Conceptualization and operationalization of destination image. *Journal of Hospitality & Tourism Research (Washington, D.C.)*, *31*(2), 194–223. doi:10.1177/1096348006297290

Thomson, M., MacInnis, D. J., & Park, C. W. (2005). The ties that bind: Measuring the strength of consumers' emotional attachments to brands. *Journal of Consumer Psychology*, *15*(1), 77–91. doi:10.120715327663jcp1501_10

Xiong, J., Hashim, N. H., & Murphy, J. (2015). Multisensory image as a component of destination image. *Tourism Management Perspectives*, *14*, 34–41. doi:10.1016/j.tmp.2015.03.001

Yuksel, A., Yuksel, F., & Bilim, Y. (2010). Destination attachment: Effects on customer satisfaction and cognitive, affective and conative loyalty. *Tourism Management*, *31*(2), 274–284. doi:10.1016/j.tourman.2009.03.007

Zhang, H., Fu, X., Cai, L. A., & Lu, L. (2014). Destination image and tourist loyalty: A meta-analysis. *Tourism Management*, *40*, 213–223. doi:10.1016/j.tourman.2013.06.006

ADDITIONAL READING

Herrero, A., San Martín, H., Garcia de los Salmones, M. M., & Collado, J. (2017). Examining the hierarchy of destination brands and the chain of effects between brand equity dimensions. *Journal of Destination Marketing & Management*, *6*(4), 353–362. doi:10.1016/j.jdmm.2016.05.001

Horner, S., & Swarbrooke, J. (2016). *Consumer Behaviour in Tourism*. Routledge. doi:10.4324/9781315795232

Kladou, S., Kavaratzis, M., Rigopoulou, I., & Salonika, E. (2017). The role of brand elements in destination branding. *Journal of Destination Marketing & Management*, *6*(4), 426–435. doi:10.1016/j.jdmm.2016.06.011

Kumar, V., & Nayak, J. K. (2014). The measurement & conceptualization of destination personality. *Tourism Management Perspectives*, *12*, 88–93. doi:10.1016/j.tmp.2014.09.002

Pike, S. (2016). *Destination marketing essentials*. Routledge.

San Martin, H., & Del Bosque, I. A. R. (2008). Exploring the cognitive–affective nature of destination image and the role of psychological factors in its formation. *Tourism Management*, *29*(2), 263–277. doi:10.1016/j.tourman.2007.03.012

Tan, W.-K., & Wu, C. E. (2016). An investigation of the relationships among destination familiarity, destination image and future visit intention. *Journal of Destination Marketing & Management*, *5*(3), 214–226. doi:10.1016/j.jdmm.2015.12.008

Veasna, S., Wu, W.-Y., & Huang, C.-H. (2013). The impact of destination source credibility on destination satisfaction: The mediating effects of destination attachment and destination image. *Tourism Management*, *36*, 511–526. doi:10.1016/j.tourman.2012.09.007

KEY TERMS AND DEFINITIONS

Affective Component of a Destination Image: An individual's feelings and emotions raised by tourist destinations affected by the motivations of tourists.

Cognitive Components of Destination Image: An individual's unique knowledge, perceptions, and beliefs about the individual attributes of a destination which are observable, descriptive and measurable.

Conative Component of Tourism Destination Image: An individual's concrete behaviour or intention to revisit and endorse the destination to others, which is related to loyalty.

Destination Image: The expression of all objective knowledge, prejudices, imagination, and emotional thoughts of an individual or group about a location being a predictor of behavioural intentions, such as destination choice and visit intention

Overall Component of Destination Image: An image that people develop about a tourist destination, which is a combination of cognitive and affective components and deemed as the only variable with a significant influence on the tourist's willingness to revisit and to recommend the destination.

Chapter 16
Insular Tourism:
Profile and Consumption Patterns of an International Football Event Held in Madeira Island

Margarida Mascarenhas
Faculty of Human Kinetics, University of Lisbon, Portugal & Research Centre for Tourism, Sustainability, and Well-Being (CinTurs), University of Algarve, Portugal

Bruno Rodrigues
Faculty of Human Kinetics, University of Lisbon, Portugal

Ivo Sousa-Ferreira
https://orcid.org/0000-0001-5526-3594
Statistics and Applications Center, Faculty of Sciences, University of Lisbon, Portugal

Elsa Pereira
School of Education and Communication, Research Centre for Tourism, Sustainability, and Well-Being (CinTurs), University of Algarve, Portugal

ABSTRACT

This chapter aimed to analyse the profile and patterns of consumption of non-resident spectators of a sporting event that took place in an insular tourist city: Funchal (Madeira Island, Portugal). Data collection was performed at the entrance of the football stadium where 108 (n) spectators were randomly surveyed. Data were analysed based on descriptive and inferential statistics. Results showed that on average, each tourist spent 231 euros in the city; the highest cost was due to accommodation and secondly to tickets; 'arrival time at the stadium' was the variable that most positively influenced the total expenditure; individuals with higher education/wage levels had a greater predisposition to spend more money on food/drink and accommodation. As spectators' opinion on the city/stadium showed high levels of satisfaction, the event improved the visitors' perception and image of the city. This study highlights how sporting events can increase the insular tourism. Future research should replicate this study in similar events hosted in comparable destinations.

DOI: 10.4018/978-1-7998-3156-3.ch016

INTRODUCTION

Over the last decades, the research that has been developed around the theme of the sporting events' economic impacts has provided a varied range of studies (Agha & Taks, 2015; Dixon, Backman, Backman, & Norman, 2012; Gómez-Bantel, 2015; Kelley, Harrolle & Casper, 2014; Ratten, 2016b). As stated by Sanz, Moreno, and Camacho (2012), sports events have aroused high interest among scholars for their ability to provide both positive and negative impacts.

The literature has emphasized the importance of this type of studies (Carneiro, Breda, & Cordeiro, 2016; George & Swart, 2012; Ratten, 2016a), because it allows the: a) knowledge of the impact of an event that has already occurred; and b) analysis of potential advantages and disadvantages in a future scenario of hosting an event. Although the studies carried out provide a great deal of information on the economic impacts of mega-events (Getz, 2008), there is a gap regarding the knowledge of these impacts generated in the host region by a special event (George & Swart, 2012; Getz, 2008). In fact, much less attention has been given to the analysis of the economic impact of small-scale sporting events (Gibson, Kaplanidou, & Kang, 2012).

Regional development is an important focuses of sporting events, being the subject of analysis by local councils in view of regional policies' framework (Kwiatkowski, 2016) and tourism development issues (Agha & Taks, 2015). The motivations to host events include improving the perception of the image of the city / destination (Pereira, Mascarenhas, Flores, & Pires, 2015), using the event as a catalyst for economic regeneration with sociocultural benefits, job creation, marketing benefits and the development of local infrastructures (Diedering & Kwiatkowski, 2015; Thomson, Schlenker, & Schulenkorf, 2013).

The entry of 'new money' from the influx of non-resident to the sports event is of paramount importance. It is therefore imperative to understand the attractiveness of this type of events, both to draw conclusions about the consumers' profile and to better visitors' capturing in future events (Crompton, 2006; Gratton, Shibli, & Coleman, 2006; Gursoy, Chen, & Chi, 2014; Halpenny, Kulczycki, & Moghimehfar, 2016). Thus, it is possible to create bases for the loyalty of visitors to the destination, and consequently to promote its economic sustainability. It is necessary to understand what influences the spectator and what desires seeks to satisfy for a correct organization and definition of appropriate marketing strategies. If an event is adapted to what spectators are looking for, it is expectable that in the medium/long term it will generate an effect of economic benefits and catalysts of local tourism (Auger, 2014).

In this sequence, the identification of market niches will allow the adaptation of the characteristics of the event to the desires of the spectators, providing a better experience of the event - and by association, of the destination - thus strengthening visitors' retention (Barajas, Coates, & Sánchez-Fernandez, 2016; George & Swart, 2012). Given the importance of knowing who visits and what benefits these events provide, both for tangible and intangible values (Kaplanidou, Kerwin, & Karadakis, 2013), these studies aim to give decision makers an important tool for the realization of marketing strategies according to the visitors' niches. For example, one of the challenges of this line of research is to realize that if a visitor buys a service/product whose production is not entirely originated in the region of the event, there will be a 'leak' out of the community of that money (Crompton, 2006; Ziakas & Costa, 2011). As such, to ensure the maximization of economic benefits, it will be up to the event's organization - in cooperation with local businesses/suppliers - to devise a strategy to ensure that all products and services provided at the event come from its host region (Chalip, 2004).

This chapter aims to analyse the profile and patterns of consumption of non-residents spectators of a sporting event that took place in an eminently insular tourist city - Funchal (Madeira Island [MI],

Portugal). The event that confronted the national football teams of Portugal and Sweden placed the city in the spotlight of the most diverse media, as it had the presence of the European champion and the best football player in the world - Cristiano Ronaldo, five times gold ball and born in the region that hosted the event. The national football team of Portugal is the current European champion, having won the Euro 2016 held in France. Moreover, in the last two decades, the Portuguese team has been present in the final stages of international competitions, such as the Euro 2004 final and the semi-finals of the 2006 World Cup and Euro 2012.

In short, there is a need to comprehend the profile and consumption patterns of non-resident spectators of the typology and sport of the event under study, who are responsible for injecting 'new money' into the community so that future marketing strategies can be better grounded. The results of the study should help local governments - in cooperation with the decision makers of the events organization - to disseminate the social and psychological characteristics of the event, promoting the return of visitors and thus, the destination's economic sustainability.

BACKGROUND

Historical and cultural links have intrinsically connected the sports and tourism sectors (Devine, Boyd, & Boyle, 2010). The sport phenomenon is seen as attractive and mobilizing people (Jiménez-Naranjo, Coca-Pérez, Gutiérrez-Fernández, & Sánchez-Escobedo, 2016) because of its dynamic character and social importance (Nunes, 2010). Sports tourism is perceived as a way of promoting the image of a tourist destination (Gammon, Ramshaw, & Wright, 2017; Preuss, Seguin, & O'Reilly, 2007). Due to this potential, there is a great interest for the niche of tourists whose motivation to travel focuses on the sporting events (Hassan & Connor, 2009; Preuss et al., 2007). According to Weed and Bull (2009), the rapid growth of this market niche in the tourism sector is due to the democratization of sports given the large investments made in infrastructures and the increasingly easy mobility of travellers. In addition, as defended by Devine, Boyd and Boyle (2011), the 'sport and tourism' binomial allows a wide range of possibilities for the creation of collaboration between the areas, fundamentally fostering partnerships with local companies.

As Schnitzer, Schlemmer and Kristiansen (2017) point out, a sporting event represents an opportunity to boost the use of local infrastructures as well as the development of new destination's trends - i.e., sporting events can leverage the regeneration of the image of a tourist destination (Pereira et al., 2015). Some authors (Briedenhann, 2011; Schnitzer et al., 2017) exalt the ability of sporting events tourism to catalyse economic progress due to infrastructures' investment, growth in the number of tourists and the development of services in the various sectors affected by the event. As M. Saayman and Saayman (2014) defend, it is extremely important to have a precise knowledge about who does sports tourism by attending sporting events, in order to define assertive strategies for this market niche and to devise policies that can effectively stimulate local consumption.

The analysis of the consumption patterns of spectators attracted by the event is the way to assess the measurement of the economic impact of an event, being necessary to discover their desires and behaviours in order to better maximize the economic revenues of future events (Jiménez-Naranjo et al., 2016). As highlighted by Salgado-Barandela, Barajas and Sánchez-Fernández (2017), the organizations have an increased concern about the understanding of the various factors that affect this phenomenon, since only this way it is possible to instigate the intensification of the economic benefits, thus constituting a topic of growing interest.

In order to draw conclusions about the aspects sought by tourists, it is necessary to understand the motivations that are the basis of sports tourism (Lourenço, 2008). The new money comes from the consumption of the affluence resulting from the event, namely from non-residents, as they bring money from outside the community. As argued by Gratton (2005), if local businesses are able to cooperate with one another, and with local authorities and the organization of the event, in order to avoid leakages of this money - particularly, through the acquisition of products/services from local suppliers - the local economy is fostered, providing the state with a source of income through taxation. It is the local administration's responsibility to ensure that a portion of this money is channelled into the promotion of new events and the upgrading of infrastructures in order to attract new tourists (Crompton, 2006). Government tourism policies tend to look to the organization of sports competitions as a source of benefits for the events' hostage city, namely by combating seasonality and promoting the economic sustainability of the destination (Carneiro et al., 2016; Halpenny et al., 2016). In this regard, Briedenhann (2011) highlights the opportunity to regenerate the destination's image, the catalytic role of events in improving the infrastructure and development of the hosting city and still as a chance for personal development and the acquisition of new skills.

Since the estimation of the events' economic impact presupposes an anticipation of the expenditure incurred by the respective spectators, the knowledge of the expenses incurred by the tourists in similar past events constitutes one of the most important factors in this evaluation (Preuss et al., 2007). According to Alexandris and Kaplanidou (2014), once the features and patterns of consumption are delineated, it is possible to define a successful marketing strategy, based on a broad understanding of the spectators' wishes. In this way, it is intended to substantiate the process of strategic decision making on the best way to grow this niche market. According to Ratten (2016a), there is a need for optimization of marketing strategies to better define and plan the organization of sporting events, in order to maximize the visitor's experience and consequently, the positive benefit in terms of the economic impact on the hosting city.

Jiménez-Naranjo et al. (2016) emphasize the attention that should be given to the niche of tourists whose primary motivation is centred on sporting events. Alongside the large events, several authors (Agha & Taks, 2015; Pereira et al., 2015; Pereira, Mascarenhas, Flores, Chalip, & Pires, 2019) also highlight the importance of smaller scale events, given the possibility of promoting the destination as a tourist region and creating local wealth by virtue of the event's organization. Briedenhann (2011) as well as Sánchez-Fernández and Rodríguez-Campo (2012), affirm that the increase of publicity and media coverage are an opportunity for the regeneration of the local image, as well as for the attraction of industry and capital. In turn, Ratten (2016a) argues that identifying the characteristics of market niches behaviour is fundamental to the creation of marketing policies and strategies, both for the event itself and for the hosting city.

The analysis of the economic impact of a sporting event is difficult because of the diversity of variables to be studied, such as the local authority's expenses to host the event, the duration of the visitors' stays, the daily consumption of the tourists and the spectator's profile of each type of event (Mules & Dwyer, 2005). According to Gratton (2005), there is still the possibility of two rounds of benefits, since, as the author points out, in the first instance there will be more money within the community by the expenditure of non-residents, and at a later stage, a use of this money for the acquisition of more products/services within the community.

In this line of investigation, it is usual to analyse the demographic data of the tourist population attracted by the event, in order to draw the profile of the individuals who actually attended the event, thus including characterization variables such as age, gender, educational level, among others. The aggregation of this type of information allows defining a profile, with the intention of effectively targeting the marketing campaigns.

As widely argued in the literature (Gratton, 2005; Jiménez-Naranjo, et al., 2016; Quintal, Paipe, Felipe, & Carvalho, 2016;), determining the economic impact of a sporting event requires accounting for the different categories of expenditure incurred by sports tourists (i.e., consumers residing outside the place where the event is held), in particular regarding to accommodation, food and drink, traveling, tickets, among others. Thus, due to the analysis of the payment of the different expenses made by the event's non-resident consumers, as well as the degree of satisfaction regarding the characteristics of the event - the stadium and the tourist destination itself - it is possible to assess the set of wishes that this important touristic niche market intends to satisfy.

METHOD

Study Context

Tourism is of major importance and growing trend in the region of Madeira Island (MI). Recently, World Travel Wards distinguished Madeira with two outstanding awards – 'Europe's Leading Island Destination 2017' and 'World's Leading Island Destination 2017' - illustrating the prominence given to the tourism sector by the regional government in cooperation with companies to ensure the satisfaction of those who visit the region. The tourism relevance of this insular destination has also been demonstrated by statistics published by local authorities (Regional Board of Statistics of Madeira [RBSM], 2018), pointing to a 7.8% increase in total revenues from overnight stays (in 2017 compared to 2016), corresponding to a total of revenues from tourism in this region of approximately 407.4 million euros in 2017.

The values described above not only emphasize the economic impact of the MI tourism sector but also the entire business environment of the region, as many other sectors indirectly benefit from tourism flows through the provision of services/products to visitors. In this way, it is of great importance to attract individuals to visit the city of Funchal. The actions taken by the local government around this football sports event, as well as the high media attention, can be important means for improving and renewing the city's image, creating new opportunities to attract tourists, increasing visitor's loyalty and promoting economic sustainability of the destination.

The 'new money' that enters the city of Funchal has two origins: a) on the one hand, from inbound tourism that, as previously evidenced, has a great impact for the region; and b) on the other hand, through domestic tourism from the rest of the MI region. As regards the assistance of the sporting event under study, both are sources of tourist flow responsible for bringing 'new money' to the hosting city.

Characterization of the Sporting Event

The game was sold out of the stadium (10415 spectators), according to news from the site of the Portuguese public television channel (RTP) in its edition of March 28, 2017. Several reasons may explain the great turnout of spectators at this event: a) in 2016, the Portuguese Football Team was the European

Champion; b) the presence of Madeiran Cristiano Ronaldo - five times winner of the Golden Ball ('FIFA World Player of the Year' award); and c) for about 15 years this team did not play a game on MI.

The regional government took advantage of the mediation contiguous to the event and inaugurated the Madeira Airport - Cristiano Ronaldo - capitalizing on the great focus placed on MI. As pointed out by several authors (Pereira et al., 2015; Williams & Chinn, 2010), this initiative is a means of promoting the tourist destination by local authorities, taking advantage of the opportunity of the event's media coverage to foster infrastructures which may give rise to the organization of other events in the near future.

In the context of hosting this specific football event, it is also worth mentioning another action taken by the local government in cooperation with the Sport Marítimo Club: the requalification of the Marítimo Stadium, making it accredited as a "Premium Stadium" of level 1 of the Portuguese Professional Football League, which made it possible to host international matches. This fact demonstrates the cities' commitment to support the construction of infrastructures that enable the hosting of events, leveraging the promotion and attractiveness of the destination (Williams & Chinn, 2010).

Data Collection

With a confidence level of 95% and a margin of error of 9.4%, 108 non-resident spectators were randomly surveyed in the city of Funchal. Data collection was performed on March 28th, 2017, at the entrance to the Marítimo Stadium (located in the city of Funchal, MI), before the start of the friendly match between the national teams of football of Portugal and Sweden.

In order to assure the confidence and total understanding about the purpose of the survey, both the anonymity and the confidentiality of the data were guaranteed and the purpose of its realization was described to the respondents. In addition, the questionnaire header contained the logo of the Faculty of Human Kinetics of the University of Lisbon. Four university students previously trained for this purpose supported data collection.

Tool

This study applied the questionnaire adapted and validated to the Portuguese context by Quintal et al. (2016), and originally validated by Preuss et al. (2007). The questionnaire comprises the following analysing dimensions (Table 1): i) Spectator profile; and ii) Spectator consumption patterns.

Table 1. Dimensions and variables

Dimensions	Variables
Spectator Profile	• Gender; Age; Education level; Monthly income • Who did you come with; Plans to visit the city; Overnight stay in the city; Football Practice; Other sports practice; Follow national games; Follow international games • Opinion regarding stadium quality; Opinion regarding city restaurants; Opinion regarding city attractions; Intention of future visit; City recommendation • Time in the city; Arrival at the stadium
Spectator consumption patterns	• Food/drink • Shopping • Accommodation • Travel cost • Ticket to the game

Data Analysis

Processing of data was performed using the Statistical Package for Social Sciences (SPSS) for Windows, version 24.0. Data were analysed based on descriptive and inferential statistics. More precisely, multiple linear regression analysis and the application of non-parametric tests were performed (Marôco, 2014).

RESULTS

Spectator Profile

In order to understand the niche of the tourist population attracted by this football event between the national teams of Portugal and Sweden, which took place in Funchal, the first objective of this study is to assess the profile of non-resident spectators in the host city. For this purpose, 108 spectators were surveyed. Table 2 shows the results obtained in the variables that characterize the sample of the non-residents spectators' profile.

Table 2. Descriptive statistics of demographic variables

Variables		Frequency (n=108)	Percentage (%)
Gender	Male	61	56.5%
	Female	47	43.5%
Origin of the non-residents	Mainland visitors	69	63.9%
	Foreigners visitors	39	36.1%
Age	< 20	9	8.3%
	[20 - 29]	24	22.2%
	[30 - 39]	32	29.6%
	[40 - 49]	34	31.5%
	[50 - 59]	5	4.6%
	≥ 60	4	3.7%
Education level	Primary	21	19.4%
	Secondary	51	47.2%
	Higher	36	33.3%
Monthly income (€: euros)	< 500€	16	14.8%
	[500€ - 999€]	46	42.6%
	[1000€ - 1499€]	25	23.1%
	[1500€ - 2249€]	15	13.9%
	[2250€ - 2999€]	4	3.7%
	[3000€ - 4449€]	2	1.9%

The sample is made up of 61 men and 47 women, aged between 18 and 68 years. The age group most represented (83.3%) in the event refers to individuals between the ages of 20 and 49 years. In addition, there is a lower affluence of individuals over 50 (4.6% for [50-59] years old and 3.7% for ≥ 60 years old), as well as for younger people (8.3% of individuals < 20 years old). With regard to the educational level, more than 80% completed the secondary education level and only 19.4% reported having primary education. Concerning the monthly income, the majority of the surveyed population (65.7%) claimed to earn between 500 and 1499 euros, 14.8% said they live with less than 500 euros and 19.5% declared to earn more than 1500 euros. Regarding to the origin of the visitors, 63.9% of the non-resident spectators who attended the event came from other cities in Portugal, thus integrating the domestic tourism flow. Among the international spectators, the following most represented countries were identified: Sweden (51.3%), Ireland (17.9%), Brazil (17.9%) and Belgium (12.8%). Taking into account the size of each subsample - i.e., 'domestic tourists' and 'international tourists' – a separate characterization of the 108 non-resident spectators, in terms of their origin, wasn't carried out. This is a limitation of the present study.

The results obtained from the variables related to the spectator's behaviour are displayed in Table 3. Regarding the variable 'Who did you come with', all respondents went to the event accompanied by friends (64.8%), family (25.9%) and boyfriend/girlfriend (9.3%). It is also noteworthy that a majority of 50.9% of the spectators visited the city motivated by attending the football game, while the rest stated that they would visit Funchal regardless of the existence of the event. In terms of overnight stays, 60.2% did not spend the night in Funchal, compared to 39.8% who slept in the city for at least one night.

Concerning the sports practice, 25.9% of those surveyed said they had never had any type of physical activity related to football. Conversely, 74.1% of the respondents stated that they practice/practiced football, even if it is/was sporadic. When asked about practicing another sport, non-practitioners' responses drop to 11.1%, revealing that there is a greater percentage of individuals practicing other sport (88.9%) on a regular / occasional basis. Regarding the assistance of national games, a majority of 55.6% of respondents said they always follow the games, which is accentuated in the case of following the international games (66.7%). With reference to the results of the variable 'Time in the city', most of the spectators (58.3%) stated that they had arrived in Funchal more than two hours before the start of the football game. In addition, a large proportion of these spectators went directly to the stadium after arriving in the city, since results for the 'Arrival at the stadium' variable indicate that 47.2% of the sample said they had entered the stadium at least two hours in advance.

Since the majority of respondents showed an interest in following national football matches, and even more so in international football games, the next step was to identify whether sporting practice would lead to greater interest in following sporting events or would otherwise undermine such interest (Aljandali, 2016). For this, Spearman's correlation coefficient was calculated to measure the association between the following variables: 'Football practice', 'Other sports practice', 'Follow national games' and 'Follow international games'.

According to Table 4, the results suggest that there is: a) a poor negative correlation between the variables 'Other sports practice' and 'Follow international games' (-0.236); and b) a high positive correlation between the variables "Follow national games" and "Follow international games" (0.744). Thus, the variables 'Other sports practice' and ' Follow international games' vary in opposite directions, signifying that an individual who practices other sport (i.e., other than football) is less interested in following international football matches. In contrast, the 'Follow national games' and 'Follow international games' variables vary in the same direction, revealing that spectators accompanying national matches are much more likely to follow international football matches as well. For the remaining pairs of variables there

Table 3. Descriptive statistics of the variables related to the spectator's behaviour

Variables		Frequency (n=108)	Percentage (%)
Who did you come with	Friends	70	64.8%
	Boyfriend/Girlfriend	10	9.3%
	Family	28	25.9%
Plans to visit the city	No	55	50.9%
	Yes	53	49.1%
Overnight stay in the city	No	65	60.2%
	Yes	43	39.8%
Football practice	Never	28	25.9%
	Rarely	11	10.2%
	Sometimes	42	38.9%
	Often	15	13.9%
	Always	12	11.1%
Other sports practice	Never	12	11.1%
	Rarely	24	22.2%
	Sometimes	29	26.9%
	Often	30	27.8%
	Always	13	12.0%
Follow national games	Never	35	32.4%
	Rarely	0	0%
	Sometimes	4	3.7%
	Often	9	8.3%
	Always	60	55.6%
Follow international games	Never	16	14.8%
	Rarely	4	3.7%
	Sometimes	6	5.6%
	Often	10	9.3%
	Always	72	66.7%
Time in the city (min: minutes)	< 30 min	5	4.6%
	[31 min - 60 min]	13	12.0%
	[61 min - 90 min]	13	12.0%
	[91 min - 120 min]	14	13.0%
	≥ 120 min	63	58.3%
Arrival at the stadium (min: minutes)	< 30 min	5	4.6%
	[31 min - 60 min]	25	23.1%
	[61 min - 90 min]	13	12.04%
	[91 min - 120 min]	14	12.96%
	≥ 120 min	51	47.2%

Table 4. Correlation analysis of sports behaviour variables (using Spearman coefficient)

		Football practice	Other Sports practice	Follow national games	Follow international games
Football Practice	Correlation Coefficient	1.000	.074	.053	.125
	Sig. (2-tailed)	.	.446	.589	.199
Other sports practice	Correlation Coefficient		1.000	-.069	**-.236***
	Sig. (2-tailed)		.	.475	.014
Follow national games	Correlation Coefficient			1.000	**.744***
	Sig. (2-tailed)			.	.000
Follow international games	Correlation Coefficient				1.000
	Sig. (2-tailed)				.

*The correlation is significant at the 0.05 level.

is an extremely weak and non-significant correlation. These results corroborate the study of Quintal et al. (2016), also carried out in the Portuguese context.

As can be seen from the results in Table 5, the view of the spectators about the characteristics of the sports infrastructure where the event occurred is largely positive. Thus, 'Opinion regarding stadium quality' was rated as 'very good' (63%) and 'good' (37%), and 76.9% of the respondents rated its animation as 'good/very good'.

Regarding local conditions, the answers show high levels of satisfaction, since 100% of the sample classifies the restaurants and attractions of the city as 'good/very good'. The combination of the results obtained for the stadium and the local conditions is clearly positive, illustrating that the spectators experienced a very evocative general experience. The other results shown in Table 5 further reinforce that outcome: when heard about the intention of revisiting the city, as well as recommending the destination to other individuals, a unanimous opinion (100%) is seen around the 'yes'.

Spectator Consumption Patterns

In order to measure visitor consumption patterns, when analysing the economic impact of a sporting event, it is necessary to consider the expenditures made in the city. Table 6 shows the results obtained for each of the five distinct categories considered in this research: 'Food/drink'; 'Shopping'; 'Accommodation'; 'Travel cost'; and 'Ticket to the game'.

The results show that the 'Accommodation' category had the highest expenditure: on average, non-resident spectators spent € 186.02 on sleeping in the city. The second largest source of expenditure was the cost of admission (an average of 29.54 euros) to the stadium to attend the sporting event. Depending on the bench area, the ticket prices for the football match were set at 20, 25 and 150 euros. On the other hand, given the geographical proximity of the cities of MI, on average, each visitor had a traveling cost of 8.41 euros. Lastly, the 'Food/drink' and 'Shopping' categories recorded low expenditure amounts, averaging 5.70 euros and 1.44 euros, respectively.

Analysing the total expenses of this sporting event, it is verified that 108 non-resident spectators contributed with an increase of 24 960 euros for the economy of the city of Funchal, meaning that, on

Table 5. Descriptive statistics of the spectators' opinion variables

Variables		Frequency (n=108)	Percentage (%)
Opinion regarding stadium quality	Very bad	0	0%
	Bad	0	0%
	Don't have	0	0%
	Good	40	37.0%
	Very good	68	63.0%
Opinion regarding stadium animation	Very bad	0	0%
	Bad	0	0%
	Don't have	25	23.1%
	Good	54	50.0%
	Very good	29	26.9%
Opinion regarding city restaurants	Very bad	0	0%
	Bad	0	0%
	Don't have	0	0%
	Good	54	50.0%
	Very good	54	50.0%
Opinion regarding city attractions	Very bad	0	0%
	Bad	0	0%
	Don't have	0	0%
	Good	23	21.3%
	Very good	85	78.7%
Intention of future visit	No	0	0%
	Unlikely	0	0%
	Don't know	0	0%
	Likely	0	0%
	Yes	108	100%
City recommendation	No	0	0%
	Unlikely	0	0%
	Don't know	0	0%
	Likely	0	0%
	Yes	108	100%

average, each visitor funded with 231,11 euros for such economic growth. Thus, accommodation expenses accounted for the largest share of overall spending by non-resident spectators, strengthening the importance of staying overnight in the city. Finally, in addition to examining the average costs associated with each category of spectator expenditures, it is also important to analyse the respective standard deviations, where there is a great dispersion of the costs incurred by each visitor.

Table 6. Descriptive statistics of spectators' expenditures

	Average	Standard deviation	Total
Food/drink	5,70 €	4,61 €	616,00 €
Shopping	1,44 €	6,83 €	156,00 €
Accommodation	186,02 €	287,16 €	20 090,00 €
Travel cost	8,41 €	12,90 €	908,00 €
Ticket to game	29,54 €	29,45 €	3 190,00 €
Total	**231,11 €**	**301,40 €**	**24 960,00 €**

In order to identify the variables with the most significant influence on the spectators' total expenditure (response variable), a multiple linear regression model was adjusted based on the following predictor variables: 'Gender'; 'Age'; 'Education level'; 'Monthly income'; 'Time in the city'; and time of 'Arrival at the stadium'. Table 7 shows the estimates obtained for the various parameters associated with the model. To evaluate the quality of the adjustment of this model, the coefficient of determination (R^2=0.492) was used, meaning that the adjusted model explains nearly half of the data variability.

Table 7. Estimates of the parameters associated with the multiple linear regression model

	Not standardized coefficients		Standardized coefficients		
Variables	**B**	**Standard error**	**Beta**	**t**	**Sig.**
(Constant)	-477.837	139.454		-3.426	**0.001***
Gender	38.595	43.631	0.064	0.885	0.378
Age	4.824	2.303	0.193	2.095	**0.039***
Education level	61.482	42.469	0.146	1.448	0.151
Monthly income	-9.152	24.626	-0.034	-0.372	0.711
Time in the city	0.474	0.688	0.112	0.689	0.493
Arrival at the stadium	2.710	0.788	0.565	3.438	**0.001***

*The regression coefficient (effect variable) is significant at the 0.05 level.

Analysing the regression coefficient estimates, there are two variables presenting a statistically significant effect ($p < 0.05$) on total visitor spending - 'Age' and 'Arrival at the stadium'. As the results illustrate, both variables positively influenced the sample's total expenditure, although the effect of the variable 'Arrival at the stadium' (0.565) is much higher than that of the 'Age' variable (0.193). As showed by data, the sooner an older non-resident spectator reaches the stadium, the higher the consumption.

In order to analyse the distribution of the total expenditures in each of the categories of the demographic variables – 'Gender', 'Education Level' and 'Monthly income' - (Landau & Everitt, 2003), the Wilcoxon-Mann-Whitney nonparametric tests ('Gender') and Kruskal-Wallis ('Education Level' and 'Monthly income') were applied (see Table 8). The application of these tests allows a more specific

Table 8. Nonparametric tests to evaluate the distribution of expenditure among the variables' categories

	Wilcoxon-Mann-Whitney Test		Kruskal-Wallis Test			
	Gender		Education Level		Monthly Income	
Hypothesis	Sig.	Decision	Sig.	Decision	Sig.	Decision
Food/drink	0.587	Not rejected H_0	**0.034***	Reject H_0	**0.008***	Reject H_0
Shopping	0.662	Not rejected H_0	0.866	Not rejected H_0	0.803	Not rejected H_0
Accommodation	0.418	Not rejected H_0	**0.002***	Reject H_0	**0.034***	Reject H_0
Travel Costs	0.091	Not rejected H_0	0.181	Not rejected H_0	0.060	Not rejected H_0
Tickets to game	0.055	Not rejected H_0	0.157	Not rejected H_0	0.446	Not rejected H_0

H_0: The distribution of the expenditure is the same among the categories of the variable.
*Significant at the 0.05 level.

identification of the target niche market because it characterizes the spectators' profile that contribute most to the total expenses associated with the sporting event.

Regarding 'Gender', the results indicate that in no category of expenditure there are significant differences between men and women, assuming that the spending distribution among gender is the same. However, in the case of the variable 'Education level', there are signs that the distribution of spending on 'Food/drink' and 'Accommodation' is different for the diverse levels. For the specific determination of the levels of this variable that present significant differences, post hoc tests were used to perform the pairwise comparisons (more precisely through the nonparametric Dunn-Bonferroni test). Thus, in terms of the distribution of expenditure in 'Food/drink', the results reveal significant differences (at the 0.05 level) only between the 'Secondary education - Higher education' pair, whereas in 'Accommodation' costs there are significant differences between the pairs 'Primary education - Secondary education' and 'Primary education - Higher education'. Therefore, the results illustrate that there is a greater predisposition to spend on 'Food/drink' and 'Accommodation' by individuals belonging to higher levels of education compared to less educated spectators. Finally, since the 'Monthly income' variable also shows that only the distribution of expenditure in 'Food/drink' and 'Accommodation' is statistically significant, the statistical procedure was similar to that of the variable 'Educational level', but now to find the statistically significant differences across the ranges of the 'Monthly income' variable. For the distribution of the expenditure in 'Food/drink', statistically significant differences occur only in the pair '<500€ and [1500€ - 2249€]', while in the cost on 'Accommodation' this difference only appears in the pair '[500€ - 999€] and [1000€ - 1499€]'. Consequently, the data show that those with a higher wage level are more likely to spend more money on 'Food/drink' and 'Accommodation', than those who receive a lower salary.

DISCUSSION

Since more than 83% of respondents claimed to be between 20 and 49 years of age, the studied sports event mostly attracted active-aged individuals, corroborating the findings of previously developed studies (Buning, Cole, & McNamee, 2016; Dixon et al., 2012; Quintal et al., 2016). In addition, as a minority (about 17%) of the sample is made up of individuals aged 50 years and over as well as spectators under 20 years old, there is still a small capacity of this event to entice both the older and the younger. According to Biscaia, Correia, Rosado and Maroco (2013), dedicated young fans tend to recognize the event's sponsors when repeatedly exposed to the brand. Considering the majority of respondents are aged between 20 and 49 years, local companies have an excellent opportunity to make their products/ services known, attracting new customers. In this way, through the promotion of new forms of local attraction, seasonality can be blurred, paving the way for the development of the local economy. Regarding gender, there is only a slight predominance of the male population, since about 43% of respondents are women, highlighting the homogeneity of the spectators of the event. However, compared to previous studies (Buning et al., 2016; Dixon et al., 2012; Quintal et al., 2016), the number of women among the respondents is higher, which underlines a distinctive aspect of this sporting event.

According to the findings of this study, the promotion of the event should focus on individuals who regularly follow football matches, since there is a high and directly proportional correlation between the variables 'Follow national games' and 'Follow international games' (.744). This means that the sports event's organization should intensively promote it among individuals who are truly passionate about football, since national game fans are more motivated to watch international games (and vice versa). It should also be pointed out that only 14.8% of the spectators in the sample declared that they never attended international matches, indicating that their presence in this specific event could have been motivated just by the companionship or by the event's festive nature.

Knowing from previous studies the importance of the relationship between sports events and the destination image (Briedenhann, 2011; Pereira et al., 2015), it was important to identify the perception of the event's consumers about the experience in general. In this regard, the results show a very positive opinion about the quality and animation of the stadium, which could promote loyalty to the destination and, consequently, sustainability of the local economy. It should be noted that this appreciation was further reinforced by the opinion regarding the tourist attractions and restaurants of the city of Funchal, obtaining a 'good/very good' rating from all the respondents. However, the events' organization should give more visibility to the animation of the stadium, since 23.1% of the sample omitted opinion about it.

In this event it was also found that individuals who spend the night in the city have a greater impact on the injection of 'new money' than the spectators who do not spend at least one night at the destination. As mentioned by Dixon et al. (2012), the category of 'Accommodation' is expected to take the lead in total cost accounting, and in this study approximately 40% of non-resident spectators stayed at least one night in Funchal. This finding, also observed by Jiménez-Naranjo et al. (2016), may be relevant to planning and therefore one of the factors to consider in conducting similar future events. It should be noted that all tickets for the event were bought at the Marítimo Stadium, as this was the only means of purchase made available by the organization, which may have hampered the access to tickets by non-residents in the city of Funchal. The frenetic rhythm of the purchase of tickets may be due to what Gómez-Bantel (2016) claims to be local identification with the team, being a source of pride for the people of the destination and high adherence to the event by local residents.

The total of expenditures (€: 24 960,00), calculated for the sample, expresses the event's capacity in the collection of 'new money', that is, wealth for the city of Funchal provided by the sporting event. In a simplified way, it was observed that, on average, each non-resident who attended the sport event of football of Portugal-Sweden added 231.11 euros to the local economy. The distribution of that value by the different types of expenses showed that the phenomenon of a sporting event goes far beyond the revenue obtained from the sale of the tickets and highlighted the importance of staying overnight in the city, since accommodation was the category that most contributed to the total expenditure incurred by the sample. Local businesses - businesses directly connected to the event and service/product providers throughout the visitors' stay - have the opportunity to obtain an additional source of income. Being integrated in the widely positive general experience of the visitors, it is necessary to emphasize the fundamental role that the local companies play in their city's general perception.

As already pointed out by Quintal et al. (2016), the importance of demographic characteristics in total expenses was also reinforced, showing a greater impact on the formation of that expenditure by individuals with higher salary and education. However, regarding the demographic variable 'Gender', the results of this study did not corroborate the findings of Cannon and Ford (2002), because it was not significantly related to the total expenditure of the spectators. Also, the results point out that the 'Time of arrival at the stadium' was the variable with the greatest positive influence on total expenditure. Thus, as highlighted by Chalip and McGuirty (2004) as bundling sport events with the host destination, the event's organization should promote activities before the game's start, encouraging cross-referencing between non-resident spectators and local products/services. Such activities will not only reinforce the primary motivation for immediate cash expenditures, but will also intensify the desire for revisiting and recommending the destination. Since, even at a lower weight, the age variable also had a positive effect on total visitor costs, it is in the interest of the local economy that the event also attracts older non-resident viewers. For example, the existence of a concert with a singer or a musical group from the 70s/80s (or musical tributes to great stars who marked this musical era) could be a strong attraction factor of the older age group.

The sporting event is a factor of tourist attraction and incorporates the experience of the tourists, tempting the visitors to spend more time and therefore more money in the community, as highlighted by Preuss et al. (2007). In this sequence, strategies should be designed to attract individuals to stay overnight in the city, through the dissemination of the event to individuals who fit the spectator's profile traced in this study. As examples, to maximize the economic gains of the city, tickets could be sold in places close to hotels and packages should be created at travel agencies including ticket, lunch/dinner at a restaurant and one or more overnight stays at a hotel. According to Kelley et al. (2014), who argue that consumption varies according to the context, these strategies could not only increase the accessibility of non-residents to the event's tickets, but cumulatively intensify their experience in the destination and consequently, their immediate consumption, plans to revisit and recommendation to family/friends. In this sequence, it could be important to have a broader vision regarding the events' hosting by the city of Funchal. For example, as defended by Ziakas and Costa (2011), creating an event portfolio to leverage the destination brand.

The televised broadcast of the event, and its constant reference to the city where it took place, serves as a great flagship of what the destination has to offer its visitors. The televising of the games to the most diverse zones of the globe, takes sports to millions and millions of fans, expanding the exhibition of the event's associated brands. An event of this nature creates a prime opportunity for local businesses to advertise their brand in new markets. In this sense, it is necessary not to lose the chance to spread the satisfaction of a well-known fan's desire - either as event's visitor or television viewer - to create loyalty and thus promote the economic sustainability of the destination. Studies on the expansion of sporting events and the sports phenomenon to the global scale (Ratten, 2016a) are important in order to spread the economic impact not only to visitors, but to all those interested in the event.

In cooperation with event organizers, local governments must adapt policies in line with what is intended by hosting the event (Agha & Taks, 2015). It is up to the event's organization to maximize the economic impact, and in the case of small-scale events, to plan the event in order to make the most of the existing infrastructures. As an active participant in the sporting phenomenon, local companies must create conditions to satisfy the desires of the fans, guaranteeing the financial profit and reinforcing the image of the destination by improving the general experience of the spectator.

CONCLUSION

The purpose of this study was to outline the profile of non-resident spectators of a football event that confronted the national teams of Portugal and Sweden, as well as to know their consumption patterns throughout their stay in the city of Funchal. Thus, it was intended to accurately assess the niche market on which the marketing strategies should be given greater attention, in order to better adapt the event, promoting local consumption and satisfying visitors (Lita & Ma'ruf, 2015).

Data revealed that the largest source of 'new money' came from accommodation and the time of arrival at the stadium was the major inducer of visitor spending. These findings suggest that organizers should encourage early arrival at the stadium, for example, promoting stadium's fan-friendly entertainment zones to boost the interaction of local visitors and residents in order to publicize local customs, products/services and attractions. Other examples to increase non-resident time spent at the stadium could be the creation of playful activities (such as visits to a museum/art exhibition located in the stadium) and the promotion of activities in partnership with local entities (such as local food/drink tastings). In this way, the improvement of the perceived experience would reinforce the enlargement of the destination's image, especially as the respondents' opinion on the stadium and local conditions was clearly positive.

It should also be noted that pre-event activities should be primarily promoted among individuals with higher levels of education and salary, since the data revealed a greater predisposition to spend money on 'Food / drink' and 'Accommodation' by these spectators. Another possible initiative would be the availability of tickets in places accessible to non-residents in the event's hosting city, such as at online ticket offices that simultaneously advertise the pre-event activities, or travel agencies that provide travel packages that include the ticket, the night at the hotel, dining in typical restaurants or with stunning views, etc.

Although this study helps to explain the profile of the consumers of this type of sporting event, as well as the variables that most influence the spending of money in the destination, it is still necessary that more studies focus on this theme. Future research should study open events held in the city of Funchal, where the number of non-resident spectators is not limited by the installed capacity of the infrastructure where the event takes place or by the constraint of the sale of tickets. Additionally, to support the results obtained in the present study and to 'encourage benchmarking among destinations' (Getz & Page, 2016, p. 612), it is proposed to replicate it in similar events hosted in comparable insular destinations. Finally, although the negative and positive impacts of an event (or an events' portfolio) can coexist in a variety of domains (economic, social, political, environmental, etc.), it is at the social level that these impacts are especially felt (Yürük, Akyol, & Şimşek, 2017), as the flows of visitors to a destination often impose potential sacrifices on the quality of life of local residents. As concluded by Yürük et al. (2017), in order to increase the satisfaction and loyalty of the participants of the events, it is necessary to mitigate concerns related to the residents' quality of life, so efforts must be made to leverage social benefits and minimize social costs. To overcome this limitation in the present study, future research on economic impacts must simultaneously address the social concerns of local residents and the actions taken to improve their quality of life.

In general, this chapter contributed to a deepening of the knowledge to better adapt the sporting events in order to capture new individuals outside the host community, also showing the contributions that the sport events can bring to the increase of the insular tourism.

ACKNOWLEDGMENT

This paper is financed by National Funds provided by FCT- Foundation for Science and Technology through project UIDB/04020/2020 and UID/MAT/00006/2019.

REFERENCES

Agha, N., & Taks, M. (2015). A theoretical comparison of the economic impact of large and small events. *International Journal of Sport Finance*, *10*(3), 199–216.

Alexandris, K., & Kaplanidou, K. (2014). Marketing sport event tourism: Sport tourist behaviors and destination provisions. *Sport Marketing Quarterly*, *23*(3), 125–126.

Aljandali, A. (2016). *Quantitative Analysis and IBM® SPSS® Statistics: A Guide for Business and Finance*. Springer. doi:10.1007/978-3-319-45528-0

Auger, D. (2014). The visitor profile at a major sporting event: the 42nd Final of the Quebec Games (Jeux du Québec) in l'Assomption, Quebec, Canada. *Society & Leisure / Loisir & Société*, *37*(1), 151-162.

Barajas, A., Coates, D., & Sánchez-Fernandez, P. (2016). Beyond retrospective assessment. Sport event economic impact studies as a management tool for informing event organization. *European Research on Management and Business Economics*, *22*(3), 124–130. doi:10.1016/j.iedee.2015.05.001

Biscaia, R., Correia, A., Rosado, A. F., Ross, S. D., & Maroco, J. (2013). Sport sponsorship: The relationship between team loyalty, sponsorship awareness, attitude toward the sponsor, and purchase intentions. *Journal of Sport Management*, *27*(4), 288–302. doi:10.1123/jsm.27.4.288

Briedenhann, J. (2011). Economic and tourism expectations of the 2010 FIFA World Cup. A resident perspective. *Journal of Sport & Tourism*, *16*(1), 5–32. doi:10.1080/14775085.2011.568085

Buning, R. J., Cole, Z. D., & McNamee, J. B. (2016). Visitor expenditure within a mountain bike event portfolio: Determinants, outcomes, and variations. *Journal of Sport & Tourism*, *20*(2), 103–122. doi:10.1080/14775085.2016.1239547

Cannon, T. F., & Ford, J. (2002). Relationship of demographic and trip characteristics to visitor spending: An analysis of sports travel visitors across time. *Tourism Economics*, *8*(3), 263–271. doi:10.5367/000000002101298106

Carneiro, M. J., Breda, Z., & Cordeiro, C. (2016). Sports tourism development and destination sustainability: The case of the coastal area of the Aveiro region, Portugal. *Journal of Sport & Tourism*, *20*(3-4), 305–334. doi:10.1080/14775085.2016.1220863

Chalip, L. (2004). Beyond impact: A general model for sport event leverage. In B. Ritchie & D. Adair (Eds.), *Sport tourism: Interrelationships, impacts and issues* (pp. 226–252). Channel View.

Chalip, L., & McGuirty, J. (2004). Bundling sport events with the host destination. *Journal of Sport & Tourism, 9*(3), 267–282. doi:10.1080/1477508042000320241

Crompton, J. L. (2006). Economic impact studies: Instruments for political shenanigans. *Journal of Travel Research, 45*(1), 67–82. doi:10.1177/0047287506288870

Devine, A., Boyd, S., & Boyle, E. (2010). Unravelling the complexities of inter-organisational relationships within the sports tourism policy arena. *Journal of Policy Research in Tourism, Leisure & Events, 2*(2), 93–112. doi:10.1080/19407963.2010.482269

Devine, A., Boyd, S., & Boyle, E. (2011). Towards a theory of collaborative advantage for the sports tourism policy arena. *International Journal of Public Sector Management, 24*(1), 23–41. doi:10.1108/09513551111099208

Diedering, M., & Kwiatkowski, G. (2015). Economic impact of events and festivals on host regions. Methods in practice. *Polish Journal of Sport and Tourism, 22*(4), 247–252. doi:10.1515/pjst-2015-0033

Dixon, A., Backman, S., Backman, K., & Norman, W. (2012). Expenditure-based segmentation of sport tourists. *Journal of Sport & Tourism, 17*(1), 5–21. doi:10.1080/14775085.2011.635017

Gammon, S., Ramshaw, G., & Wright, R. (2017). Theory in sport tourism: Some critical reflections. *Journal of Sport & Tourism, 21*(2), 69–74. doi:10.1080/14775085.2017.1319515

George, R., & Swart, K. (2012). International tourists' perceptions of crime-risk and their future travel intentions during the 2010 FIFA World Cup in South Africa. *Journal of Sport & Tourism, 17*(3), 201–223. doi:10.1080/14775085.2012.734060

Getz, D. (2008). Event tourism: Definition, evolution, and research. *Tourism Management, 29*(3), 403–428. doi:10.1016/j.tourman.2007.07.017

Getz, D., & Page, S. J. (2016). Progress and prospects for event tourism research. *Tourism Management, 52*, 593–631. doi:10.1016/j.tourman.2015.03.007

Gibson, H. J., Kaplanidou, K., & Kang, S. J. (2012). Small-scale event sport tourism: A case study in sustainable tourism. *Sport Management Review, 15*(2), 160–170. doi:10.1016/j.smr.2011.08.013

Gómez-Bantel, A. (2015). Football clubs as symbols of regional identities. *Soccer and Society, 17*(5), 692–702. doi:10.1080/14660970.2015.1100430

Gratton, C. (2005). *The economic impact of sport tourism at major events*. Elsevier Butterworth-Heinemann.

Gratton, C., Shibli, S., & Coleman, R. (2006). The economic impact of major sports events: A review of ten events in the UK. *Socialist Review, 54*(2), 41–58. doi:10.1111/j.1467-954X.2006.00652.x

Gursoy, D. S., Chen, J., & Chi, C. (2014). Theoretical examination of destination loyalty formation. *International Journal of Contemporary Hospitality Management, 26*(5), 809–827. doi:10.1108/IJCHM-12-2013-0539

Halpenny, E., Kulczycki, C., & Moghimehfar, F. (2016). Factors effecting destination and event loyalty: Examining the sustainability of a recurrent small-scale running event at Banff National Park. *Journal of Sport & Tourism, 20*(3-4), 233–262. doi:10.1080/14775085.2016.1218787

Hassan, D., & Connor, S. O. (2009). The socio-economic impact of the FIA World Rally Championship 2007. *Sport in Society, 12*(6), 709–724. doi:10.1080/17430430902944134

Jiménez-Naranjo, H. V., Coca-Pérez, J. L., Gutiérrez-Fernández, M., & Sánchez-Escobedo, M. C. (2016). Cost-benefit analysis of sport events: The case of World Paddle Tour. *European Research on Management and Business Economics, 22*(3), 131–138. doi:10.1016/j.iedee.2015.04.001

Kaplanidou, K., Kerwin, S., & Karadakis, K. (2013). Understanding sport event success: Exploring perceptions of sport event consumers and event providers. *Journal of Sport & Tourism, 18*(3), 137–159. doi:10.1080/14775085.2013.861358

Kelley, K., Harrolle, M., & Casper, J. (2014). Estimating consumer spending on tickets, merchandise, and food and beverage: A case study of a NHL Team. *Journal of Sport Management, 28*(3), 253–265. doi:10.1123/jsm.2012-0275

Kwiatkowski, G. (2016). Composition of event attendees: A comparison of three small-scale sporting events. *International Journal of Sport Finance, 11*(2), 163–180.

Landau, S., & Everitt, B. (2003). *A Handbook of Statistical Analysis using SPSS*. CRC Press. doi:10.1201/9780203009765

Lita, R. P., & Ma'ruf. (2015). Relationship model among sport event image, destination image, and tourist satisfaction of Tour de Singkarak in West Sumatera. *Journal of Economics, Business, and Accountancy Ventura, 18*(1), 91-102.

Lourenço, R. (2008). *Turismo de prática desportiva. Estudo de caso: Os percursos pedestres e os termalistas clássicos em Monfortinho* [Tourism of sports practice. Case study: The pedestrian and classical thermal trails in Monfortinho] (Unpublished master's thesis). University of Beira Interior, Covilhã, Portugal.

Marôco, J. (2014). *Análise estatística com o SPSS statistics* [Statistical analysis with SPSS statistics]. ReportNumber.

Mules, T., & Dwyer, L. (2005). Public sector support for sport tourism events: The role of cost-benefit analysis. *Sport in Society, 8*(2), 338–355. doi:10.1080/17430430500087864

Nunes, P. (2010). *Desporto, turismo e ambiente: O turismo de Natureza como pólo de atração turística complementar ao produto Sol & mar na sub-região do Litoral Alentejano* [Sports, tourism and environment: Nature tourism as a tourist attraction pole complementary to the Sun & beach product in the Alentejo Litoral sub-region] (Unpublished doctoral dissertation). University of Lisbon, Lisbon.

Pereira, E., Mascarenhas, M., Flores, A., Chalip, L., & Pires, G. (2020). Strategic leveraging: Evidences of small-scale sport events. *International Journal of Event and Festival Management, 11*(1), 69–88. doi:10.1108/IJEFM-07-2018-0046

Pereira, E., Mascarenhas, M., Flores, A., & Pires, G. (2015). Nautical small-scale sports events portfolio: A strategic leveraging approach. *European Sport Management Quarterly*, *1*(15), 27–47. doi:10.1080/16184742.2015.1007883

Preuss, H., Seguin, B., & O'Reilly, N. (2007). Profiling major sport event visitors: The 2002 Commonwealth Games. *Journal of Sport & Tourism*, *12*(1), 5–23. doi:10.1080/14775080701496719

Quintal, G., Paipe, G., Felipe, J. L., & Carvalho, M. J. (2016). Strategic tool to estimate the consumption patterns of non-residents spectators at sporting events (Liga NOS): Adaptation to Portuguese reality. *Brazilian Business Review*, *15*(2), 102–120. doi:10.15728/edicaoesp.2016.6

Ratten, V. (2016a). Sport innovation management: Towards a research agenda. *Innovation (North Sydney, N.S.W.)*, *18*(3), 238–250. doi:10.1080/14479338.2016.1244471

Ratten, V. (2016b). The dynamics of sport marketing: Suggestions for marketing intelligence and planning. *Marketing Intelligence & Planning*, *34*(2), 162–168. doi:10.1108/MIP-07-2015-0131

Regional Board of Statistics of Madeira. (2018). *Estatísticas do Turismo da Região Autónoma da Madeira. Resultados Provisórios* [Tourism Statistics of the Autonomous Region of Madeira. Interim Results]. Author.

Saayman, M., & Saayman, A. (2014). Appraisal of measuring economic impact of sport events. *S.A. Journal for Research in Sport Physical Education and Recreation*, *36*(3), 151–181.

Salgado-Barandela, J., Barajas, A., & Sánchez-Fernández, P. (2017). Economic impact of sport: Topic of growing interest for the scientific literature. *Revista Internacional de Medicina y Ciencias de la Actividad Física y del Deporte*, *17*(68), 729–755.

Sánchez-Fernández, P., & Rodríguez-Campo, M. L. (2012). Sports events as strategic tool of citymarketing: The case of basketball Galicia. *Revista Intercontinental de Gestão Desportiva*, *2*(1), 16–29.

Sanz, V. A., Moreno, F. C., & Camacho, D. P. (2012). Social impact of a major athletic event: The Formula 1 Grand Prix of Europe. *Cultura, Ciencia y Deporte*, *7*(19), 53–65. doi:10.12800/ccd.v7i19.23

Schnitzer, M., Schlemmer, P., & Kristiansen, E. (2017). Youth multi-sport events in Austria: Tourism strategy or just a coincidence? *Journal of Sport & Tourism*, *21*(3), 179–199. doi:10.1080/14775085.2017.1300102

Thompson, A., Martin, A., Gee, S., & Geurin, A. (2016). Fans' perceptions of professional tennis events' social media presence. *Communication & Sport*, *5*(5), 579–603. doi:10.1177/2167479516650442

Thomson, A., Schlenker, K., & Schulenkorf, N. (2013). Conceptualizing sport event legacy. *Event Management*, *17*(2), 111–122. doi:10.3727/152599513X13668224082260

Weed, M., & Bull, C. (2009). *Sports tourism: participants, policy and providers* (2nd ed.). Elsevier.

Williams, J., & Chinn, S. J. (2010). Meeting relationship marketing goals through social media: A conceptual model for sport marketers. *International Journal of Sport Communication*, *3*(4), 422–437. doi:10.1123/ijsc.3.4.422

Yürük, P., Akyol, A., & Şimşek, G. G. (2017). Analyzing the effects of social impacts of events on satisfaction and loyalty. *Tourism Management, 60*, 367–378. doi:10.1016/j.tourman.2016.12.016

Ziakas, V., & Costa, C. A. (2011). Event portfolio and multi-purpose development: Establishing the conceptual grounds. *Sport Management Review, 14*(4), 409–423. doi:10.1016/j.smr.2010.09.003

ADDITIONAL READING

Amador, L., Campoy-Muñoz, P., Cardenete, M. A., & Delgado, M. C. (2017). Economic impact assessment of small-scale sporting events using Social Accounting Matrices: An application to the Spanish Football League. *Journal of Policy Research in Tourism, Leisure & Events, 9*(3), 230–246. doi:10.1080/19407963.2016.1269114

Bazzanella, F. (2019). Perceptions and role of tourist destination residents compared to other event stakeholders in a small-scale sports event. The case of the FIS World Junior Alpine Ski Championships 2019 in Val di Fassa. *Sustainability, 2019*(11), 6909. doi:10.3390u11246909

Connell, J., Page, S. J., & Meyer, D. (2015). Visitor attractions and events: Responding to seasonality. *Tourism Management, 46*, 283–298. doi:10.1016/j.tourman.2014.06.013

Dickson, G., Milne, S., & Werner, K. (2018). Collaborative capacity to develop an events portfolio within a small island development state: The Cook Islands. *Policy Research in Tourism. Leisure and Events, 10*(1), 69–89. doi:10.1080/19407963.2017.1409751

Duglio, S., & Beltramo, R. (2017). Estimating the economic impacts of a small-scale sport tourism event: The case of the Italo-Swiss Mountain Trail CollonTrek. *Sustainability, 2017*(9), 343. doi:10.3390u9030343

Malchrowicz-Mośko, E., & Poczta, J. (2018). A Small-scale event and a big impact - Is this relationship possible in the world of sport? The meaning of heritage sporting events for sustainable development of tourism - Experiences from Poland. *Sustainability, 2018*(10), 4289. doi:10.3390u10114289

Mariani, M., & Giorgio, L. (2017). The 'Pink Night' festival revisited: Meta-events and the role of destination partnerships in staging event tourism. *Annals of Tourism Research, 62*, 89–109. doi:10.1016/j.annals.2016.11.003

Scholtz, M. (2019). Does a small community (town) benefit from an international event? *Tourism Management Perspectives, 31*, 310–322. doi:10.1016/j.tmp.2019.05.006

Toudert, D., & Bringas-Rábago, N. L. (2018). Reciprocity between soccer events and visited destination in a border context. *Journal of Travel & Tourism Marketing, 35*(5), 611–621. doi:10.1080/10548408.2017.1401029

Wood, L., Snelgrove, R., Legg, J., Taks, M., & Potwarka, L. R. (2018). Perspectives of event leveraging by restaurants and city officials. *International Journal of Event and Festival Management, 9*(1), 34–50. doi:10.1108/IJEFM-01-2017-0003

KEY TERMS AND DEFINITIONS

Destination Loyalty: Extremely important for the sustainability of tourist destinations, it is determined by a wide range of factors, which may or may not be mediated by satisfaction, such as destination's unique attributes, destination's image, co-creation of pleasant emotional experiences during the visit, desire for learning opportunities, etc.

Football Fan: An enthusiastic and active follower of football; a person truly passionate about football; a person extremely interested in accompanying a football team and / or football icons.

Leveraging of Sport Events Tourism: Articulated set of strategies aimed at maximizing competitiveness and developing a tourist destination through hosting sporting events.

Sport Event Consumer's Profile: Characterization of variables associated with the individuals (e.g., sport participants, spectators, etc.) and their consumption of products and services during/in the time of the events.

Sport Event's Economic Impact: The net change that occurs in the economy of a city, country or region, resulting from hosting a sporting event; multiple effects - direct, indirect and induced - give rise to this economic impact of expenditure.

Sport Tourism: The consumption (active / passive) of sport (recreational / competitive) integrated in a touristic system; e.g., a spectator who travels outside of his residence to attend a sporting event is a passive consumer of recreational sport tourism.

Tourism Seasonality: The seasonality of tourist destinations results from the concentration of tourist activities in short periods of time (i.e., a few months); this peak in tourist demand causes inefficient use of tourist infrastructures and overload in the ecological, social and cultural systems of destinations.

Chapter 17
Tourism and Sex Outside the "Sex Tourism Paradigm":
Tourists' Perspectives and Practices in a Sun and Beach Destination

Milene Lança
Faculty of Economics and Research Centre for Tourism, Sustainability, and Well-Being (CinTurs), University of Algarve, Portugal

Fernando Bessa Ribeiro
Sociology Department of the Social Sciences Institute and Interdisciplinary Centre of Social Sciences (CICS.NOVA), University of Minho, Portugal

ABSTRACT

The relationship between tourism and sex has been intensively explored over the past decades. The great majority of the studies examining this relationship focus disproportionately on commercial sex tourism as if all the sexual encounters in tourism context are related to prostitution, commercialization, trafficking, inequalities, and disrespect for the human rights. Despite these contributions, the purpose of this chapter is to show that most of the romantic and sexual encounters that occur in a tourism setting are positive, gratifying, and nonpecuniary. Therefore, these relationships should be analyzed outside "sex tourism paradigm." Results from 29 in-depth interviews with tourists in the Algarve (South Portugal) confirm the strong relationship between holidays and romantic and sexual behaviors, due to the liminal nature of tourism. In addition, results show different perspectives and behaviors according to gender, due to the continuous sexual double standard.

DOI: 10.4018/978-1-7998-3156-3.ch017

INTRODUCTION

The literature is relatively unanimous about the sexual nature of tourism. Tourism and sex have commonalities in providing the satisfaction of basic psychological and physiological needs of human beings (Cansel, Ekiz & Bavik, 2009). Moreover, sex is part of life and if people participate in sexual activities at home, then certainly one must expect them to participate in sex when they travel (McKercher & Bauer, 2003). In fact, it is supposed they would engage in sex more frequently or even with more intensity due to the liminal nature of tourism and the anonymity provided by the destination (Zhang & Xu, 2019).

Most of the studies produced about the relationship between tourism and sex has focused, however, on commercial sex tourism, child sex tourism, tourist orientated prostitution, human trafficking, gender inequalities, violation of the human rights and the spread of HIV as a result of sex tourism, especially in the less developed countries (Carter & Clift, 2000; Cohen, 1982; De Albuquerque, 1998; Kempadoo, 1999; Michel, 2009; O'Connell-Davidson, 2001; Oppermann, 1999; 1998; Piscitelli, 2004; Ribeiro & Sacramento, 2006; Sanchez-Taylor, 2000, among others).

Despite the great importance of these studies, helping raise awareness of legitimate issues, the commercialization of sex is only a small portion of the total spectrum of human sexual behavior during tourism (Carr & Poria, 2010; McKercher & Bauer, 2003; Ryan & Hall, 2001). Sex is an inherent element of tourism, not a specific segment (Cantalice, 2009) and most of the romantic and sexual encounters that happen in tourism context are satisfying and nonpecuniary (McKercher & Bauer, 2003; Oppermann, 1999). For many people, especially couples or families on vacation, sex is an «incidental» activity that represents the continuation of their regular sex lives, although possibly at a higher frequency and intensity (McKercher & Bauer, 2003). In fact, the vast majority of people who have sex when traveling, do it with their regular partners or with new found partners in a short but consensual and non-commercial relationship (*Idem*).

The purpose of this chapter is thus to examine tourists' perspectives and practices related to sex and romance in one of the most recognized 'sun and beach' tourism destinations in Europe - the Algarve (south Portugal). The Algarve is not a sex tourism destination, meaning tourists do not come to the region with the purpose of engaging in commercial sex. At least, the results of this study show this is not their motivation for traveling. Nevertheless, the Algarve is a destination for families, couples, and friends on vacation, where sex plays a central role, far away from home and the daily life constraints. As stated by McKercher and Bauer (2003: 13), "(…) since most tourism involves traveling with one's partner and family, it is also apparent that most sexual activities away from home involve lasting relationships. Such activities have generated little academic interest because they are neither controversial nor out of the ordinary". Therefore, there is a need to understand tourists' sexual behaviors outside the "sex tourism paradigm", integrating a new perspective, that of "sex and the sexual during people's leisure and tourism experiences" (Carr & Poria, 2010), (see Figure 1).

The results are part of the first author's PhD thesis and suggest a strong relationship between vacation trip and romantic and sexual encounters, both in terms of new involvements and in strengthening the lasting relationships. The liminal nature of tourism provides new opportunities to engage in sex and romance with the usual partners or with new found partners at the destination. Because vacation time is limited, tourists perform sex and romance in an «extra-ordinary» way, creating good memories about the experiences and the destination.

In addition, the findings reveal different patterns according to gender. Although less visible, sexual double standard still constrains perspectives and practices, in the case of women (Connell, 1987; Jackson

& Scott, 2002). The chance of engaging in a romance or sexual adventure without future commitment is mainly part of the male imaginary. In contrast, women tend to relate vacation trips to the ideal of investing in the lasting relationships.

Some practical and managerial implications are also discussed, regarding the future of the Algarve as a tourism destination. Taking into account that the quality of experiences is primarily determined by the intimacies that exist between people at the place (Trauer & Ryan, 2005), sexual and romantic experiences may play a decisive role in tourists' loyalty to the Algarve.

THEORETICAL BACKGROUND

Relationship Between Tourism and Sex

The studies produced under the social sciences domain are relatively unanimous in assuming the strong romantic, erotic and even sexual nature of tourism, either as a set of initial motivations for travel, either as an inherent part of the travel itself. Tourism and sex are closely related and sex is an undeniable element of tourism (Cohen, 1971). If people participate in sexual activities at home, they should have sex when traveling. In fact, one must expect them to engage in more sexual experiences or, at least, more intense experiences due to the liminal nature of tourism (McKercher & Bauer, 2003; Ryan & Hall, 2001; Zhang & Xu, 2019).

Notwithstanding, many authors declare that the relationship between tourism, sex, love and romance has not only been under-investigated but clearly under-theorized (Agustín, 2010; Berdychevsky, Gibson & Poria, 2013; Cabezas, 2009; Carr & Poria, 2010; Frank, 2007; Frohlick, 2010; Jordan & Aitchison, 2008; Leheny, 1995; McKercher & Bauer, 2003), and, as stated by Ryan and Hall (2001), there is also a lack of methodological coherence between the different streams of research about this articulation.

In the promotional tenets of tourism marketing "sun, sand, sea and sex" (Campbell, Perkins & Mohammed, 1999; Crick, 1989; Hobson & Dietrich, 1994; Thomas, 2005), the first three – sun, sand and sea – have been widely investigated and do not represent major ethical or epistemological problems. But the same does not occur with the fourth – sex – meaning the loving and the sexual behavior in tourism context. Sex does not seem to deserve the same scientific interest as well as other attitudes and behaviors during tourism experiences (Berdychevsky, Poria & Uriely, 2013; Marques, 2016; Poria & Carr, 2010).

Tourism research tends to ignore sex, which is still a kind of taboo in this area. And when some authors decide to study the relationship between tourism and sex, there is a tendency to focus on a certain moralism (Carr & Poria, 2010; Frohlick, 2010). Problems concerning the susceptibility of the topic, which may cause embarrassment to participants, also shape in the career of the researcher studying human sexuality and may compromise its credibility and reputation (Donnan & Magowan, 2010; Hammond & Kingston, 2014). As a result, there is a lack of scientific production, especially the one which integrates theory and empirical evidences from human sexuality, without falling into the «temptation» of denouncement (Weis, 2002).

Due to this moralism, most of the studies - especially those which appeared since the 1990s - aim to deplore the situations of sexual exploitation happening in the poorest countries. The focus of these studies has almost always been on commercial sex relationships in tourism destinations and their undesirable nature, as described by the World Tourism Organization (Ryan, 2002). Many of these studies address sexual interactions between tourists and local people, highlighting power issues and gender inequalities

(Cabezas, 2004; Cohen, 1982; Sanchez-Taylor, 2001; 2006) and ignoring, for instance, the sexual activity that occurs among tourists (Marques & Lança, 2016; McKercher & Bauer, 2003).

The travel motivated by commercial sex or what has been called "sex tourism" has captured the attention of social scientists (Carr & Poria, 2010; Clif & Carter, 2000; Kempadoo, 1999; McKercher & Bauer, 2003). Much of the literature produced so far, comprises the analysis - and the denouncement - of tourists' behaviors traveling for commercial sex (Cohen, 1982, 1988; Frohlick, 2010; Hall, 1992; Oppermann, 1999, Sanchez-Taylor, 2006; Silva & Blanchette, 2005; Truong, 1990). The emphasis placed in this area has been so intense that "sex tourism" is virtually used as a synonym for "commercial sex tourism" (Legardinier & Villette, 2007; McKercher & Bauer, 2003), and although this position is controversial (Brents, Jackson & Hausbeck, 2010; Cabezas, 2004; Frohlick, 2010; Herold, Garcia & DeMoya, 2001; Oppermann, 1999; Ryan, 2000; Ryan & Kinder, 1996), hereafter the authors designate this perspective as the "sex tourism paradigm" (Marques & Lança, 2016).

Regardless the importance of these contributions, it is necessary to assume that the relationship between tourism and human sexual behavior extends far beyond the narrow confines of the sex trade (McKercher & Bauer, 2003; Oppermann, 1999; Ryan, 2000). According to Oppermann (1999), in the relationship between tourism and sex, it is necessary to consider the purpose of travel, the length of time, the kind of relationship established between people, the nature of the sexual encounter and who falls in which category of travel. In addition, McKercher and Bauer (2003) argue that in this relationship, it should be taken into account the role sex and romance play in the motivation for travel, the nature of the encounter and the role played by the tourism industry as a facilitator of romantic and sexual encounters.

The role sex and romance play in the decision to travel is the first element to be considered. They can range from one extreme where they are the central objective of the trip, to the other extreme, where they play no role in the decision to travel. Sex and/or romance can be an expected outcome of the trip, as in the case of couples' romantic escapes, honeymoons or even organized commercial sex tours; they can be anticipated or hoped, as in the case of spring break travels; or they might be an aspirational outcome, especially if tourists seek romance on vacation, as in the case of «cruises for singles». They might also be unplanned, serendipitous cases. For the vast majority of people travelling, sexual activity and romance lie at some point of these extremes but specially for couples or families on vacation, they are incidental activities that represent the continuation of regular sex lives, although possibly at a higher frequency and intensity (McKercher & Bauer, 2003; Zhang & Xu, 2019).

The nature of the encounter is the second element to think over. Considerable part of the literature on tourism and sex tends to focus on the unequal and exploitative nature of the encounter between tourists and their sexual partners (Carter & Clift, 2000; Cohen, 1982; De Albuquerque, 1998; Kempadoo, 1999; Michel, 2009; O'Connell-Davidson, 2001; Piscitelli, 2004; Sanchez-Taylor, 2000, among others). Undeniably, many of the sexual activities associated with commercial sex tourism are exploitative. But, as some authors pointed out, prostitution and other forms of commercial sex represent only a small portion of the sexual activity during tourism (Carr & Poria, 2010; McKercher & Bauer, 2003; Oppermann, 1999; Ryan & Hall, 2001). Therefore, the nature of the encounter can also be described as a continuum from a mutually beneficial and positive experience to a negative and exploitative involvement.

The role played by the tourism industry as a facilitator is the third element to be examined. The various components of the tourism industry play a determinant role in facilitating the sexual and romantic encounters between tourists and their partners. They can play an indirect role, through the setting, the mood, and the context for encounters, or a more direct role, in providing the venue, the infrastructures and possible partners (McKercher & Bauer, 2003). The most important role that tourism, as a phenomenon, plays in sexual or romantic relationships is that it offers a liminal environment away from home and the daily life constraints, reducing inhibitions and providing increased opportunities for sex and romance (*Idem*).

Liminality as a Facilitator of Sex and Romance

The concept of liminality (from Latin *limen*, meaning threshold) is currently one of the most widely used within sociology of tourism. Turner (1974a; 1974b) relied on Van Gennep's thinking (1960 [1909]) about the structure of the "rites of passage", and Durkheim's (1914) distinction between the "sacred" and the "profane", to argue that during ordinary life, people live in a social, economic, and political structure that determines their actions. However, Turner (1974a; 1974b) pointed out that periodically, there is a chance to «escape» these constraints, through the "rites of passage". The "rites of passage" are forms of celebration that enable individuals to behave in a completely different way. This means they move between the normal, structured and «ordinary» states of "profane" everyday life and their temporary dissolution into unstructured, «extra-ordinary» and "sacred" situations, possible through these forms of celebration. These situations subsequently evolve into reintegration in the everyday structure, but in the form of a new social or personal status.

The separation and reintegration process follows three ritual stages: (i) separation (rites of separation), which can be social, symbolic or spatial; (ii) the moment of transition, when the individual, having left the structure of belonging, has not yet entered another one (transition rites); (iii) and reintegration (rites of incorporation), usually in the form of another social category. The great «novelty» in this process concerns the temporal marginality between the exit and the reintegration in the structure. For a moment, the individual is in "betwixt and between" (Turner, 1974b), being this intermediate state what Van Gennep (1960 [1909]) calls liminal state.

Some authors (Graburn, 1977; Jafari, 1987; Ryan & Hall, 2001; Selänniemi, 2003) argue that tourism comprises the same three stages of the rites of passage. The act of traveling represents a symbolic liminal process whereby tourists go away from a familiar place (separation) to a destination (liminal space-time) and then return to the original place (reintegration), (McKercher & Bauer, 2003). This liminal space-time is a kind of «void», where social norms and rules are suspended and occur a symbolic «death» of the person who existed before traveling. This gives tourists the opportunity to do and to express things that they would otherwise suppress.

Moreover, during a vacation trip, individuals adopt new behavioral patterns, gain a degree of freedom and anonymity and share a sense of unity with others in the same situation – what Wagner (1977) has called "communitas". Liminality explains why activities such as spontaneity, social togetherness and participation in role reversal activities are more likely to occur when traveling than at home (McKercher & Bauer, 2003). At the same time, different sexual behaviors, either in choice of partners, frequency, intensity, or perceptions about sexuality can also be explained by the liminal nature of tourism.

In the same direction, Selänniemi (2003) states that tourism can be interpreted as a fourfold transition/transgression. First, it is a *spatial transition*, from home to somewhere else. Second, it is a *temporal transition*, away from everyday time, the time of work, home, routine and clock. Third, it is a *mental transition*, where the latent «other» may come forth and reveal characteristics that self-control and social control keep hidden during the everyday life. Fourth, it is a *sensorial/sensual transition*, where the senses are much more aware of the stimulations. Thus, the spatial-temporal transition/transgression from home and everyday life to a tourism destination changes the psychological state, the social order and the bodily state or the way individuals perceive and experience the surroundings (*Idem*).

Tourism is therefore a liminal space-time where is possible to realize fantasies and desires that are denied during the everyday life (Franklin, 2003). This transition/transgression, associated with the loosening of social control and anonymity, allow different mental and sensorial attitudes, a characteristic that tourism shares with love, romance and sexuality (Marques & Lança, 2016). Thereby, it is possible to argue that tourism allows the same escape opportunities as the "socioerotic life" described by Weitman (1998). Both are opportunities to leave the profane of everyday order and momentarily enter a sacred or festive order – in other words, the manifestations of human corporeality in which sensuality and pleasure play a leading part (Gisolf, 2013; Johnston & Longhurst, 2010; Selänniemi, 2003).

Sexual Double Standard and Surveillance as a Constraint

Despite the fact that tourism is a facilitator of the sexual and romantic encounters, it is also evident the persistence of some puritan forces which continue to shape what are the acceptable behaviors for men and women. Sexual double standard seems to persist even after decades of feminist critiques (Kalish & Kimmel, 2011) and regardless the increasing acceptance of equality concerning sexual life (Bozon, 2004). It positions men who have many sexual partners (positively) as a 'stud' or 'player' and women (negatively) as a 'slut', 'skank' or 'whore' (Farvid, Braun & Rowney, 2017; Milhausen & Herold, 1999).

Sexual double standard does not come from individuals' anatomical or biological characteristics, although they may serve as justification for some discourses about the supposed men's "sexual urgency" (Ellis, 2003; Ribeiro *et al.*, 2007) and the lack of sexual appetite in the case of women (Kinsey, 1998 [1953]; Nicolson & Burr, 2003; Wilson, 1975; 1978). It comes from the normative and political actions over the surveillance of bodies and sexuality, based on ideological perceptions of gender order (Connell, 1987; Crawford & Popp, 2003). The same is to say that sexual double standard is a normative product of the intersection between gender as a structure, and sexuality as a relational practice (Butler, 1999; 2011).

According to sexual double standard, sexual pleasure within a relationship or outside it continues to be associated with male needs. In contrast, sexual pleasure in the case of women, even though is currently desirable, is still full of taboos (Bozon, 2015). It is not because women do not want to achieve sexual pleasure but because their experiences and accounts continue to be profoundly shaped by social moralism.

Women are in a particularly delicate position. Sexual double standard limits their perceptions and practices concerning sexuality, as well as their affirmation as sexual citizens (Jackson & Scott, 2004). Despite the continuous feminist fights for gender equality, women continue to be described as "emotional, whimsical beings, whose thought-processes do not move along rational lines" (Giddens, 1992: 17). This kind of stereotypes do not match modern reality but they still persist, shaping mentalities, discourses and «feeding» what Bourdieu (2001) has called "masculine domination".

In tourism context, there is also scrutiny and surveillance of sexuality, as well as conservative reactions to sexual practices. When people leave their own homes and places of residence to travel somewhere else, they leave behind some of the social baggage, what in turn, enables them to be freer and more relaxed (McKercher & Bauer, 2003). However, the surveillance of individuals in tourism context – as in the societies in general – supported and (re)created by sexualized and embodied discourses (Foucault, 1977), still constrain tourists' perspectives and practices related to sex and romance, especially in the case of women (Jordan & Aitchison, 2008). Above all, in tourism context there is the same principle of gender distinction materialized in the sexual double standard: the same which determines that women seek love and romance while men prefer *tout court* sex (Dahles & Bras, 1999; Jeffreys, 2003; Pruitt & LaFont, 1995).

The previous concepts are relevant for understanding the articulation between gender, sexuality and tourism outside the "sex tourism paradigm". This articulation is explained in the conceptual framework (Figure 1).

Figure 1. Conceptual Framework
Source: Own elaboration

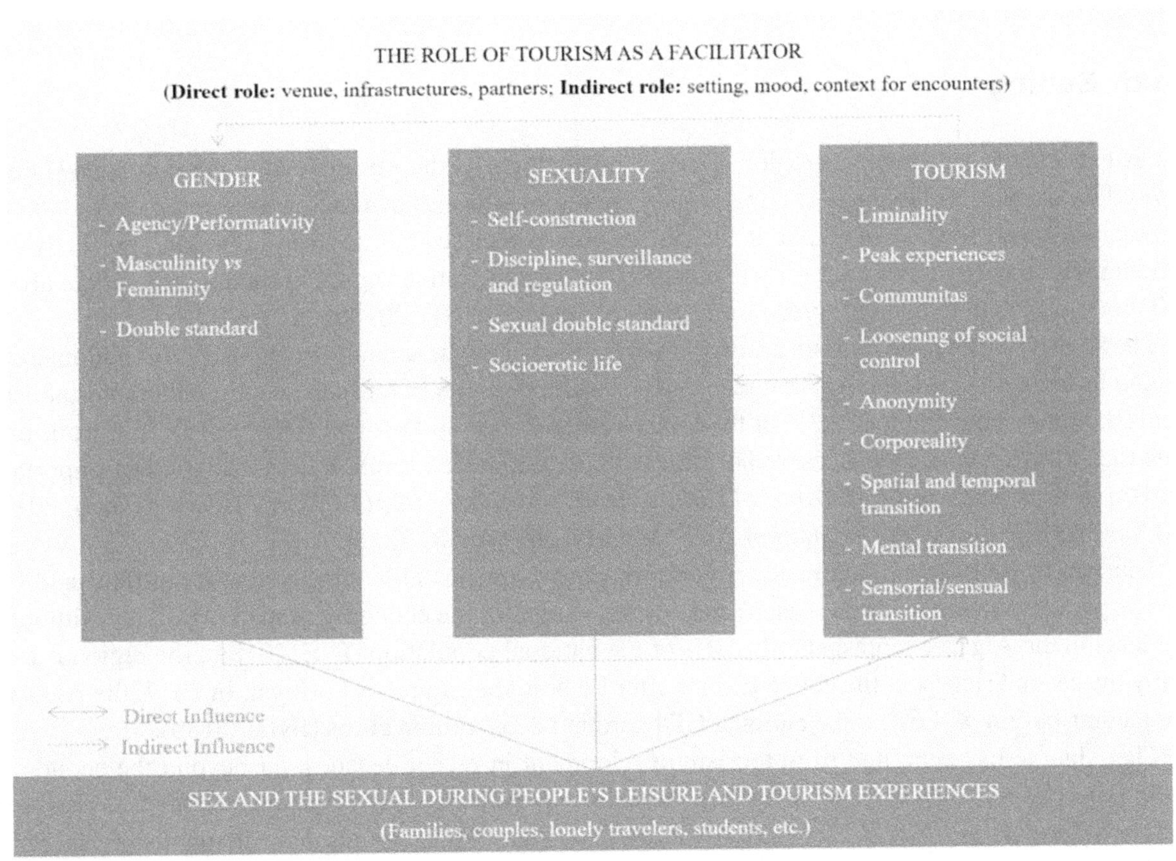

METHODOLOGY

With the aim of understanding tourists' perspectives and practices concerning sex and romance, this research uses a qualitative approach. As stated by Tolman, Hirschman and Impett (2005: 4), "qualitative research combines the power of stories with methodological rigor, providing policy makers with important information about the complexity of problems and suggesting possible solutions", which is particularly relevant in the case of human sexuality.

Qualitative research, from the epistemological point of view, is focused on intentions and perceptions. It is about investigating ideas, discovering meanings in individual actions and social interactions, throughout the individuals' opinions. From the methodological point of view, it is based on the inductive method as the researcher aims to reveal the intention and the purpose of the action, by studying it in its meanings. From the theory-practice point of view, the goal is to contribute for the description and the understanding of concrete situations, thus the theory is interpretive or constructionist (Flick, 2014).

All methodologies have, however, some limitations. In the case of qualitative methodology, the main limitations relate to its eventually subjective character, which depends on the researcher's interpretation, and the difficulty of generalization (Creswell, 2015). As Patton (1990) pointed out, the human factor is its greatest strength but also its major weakness. What is central, according to Creswell (2015), is to prevent moral values and judgements from interfering with the results. At the same time, researcher must be able to ensure transparency, error elimination and appropriate selection of the research methods (Bryman & Bell, 2011).

Study Setting

The Algarve is the southernmost region of mainland Portugal, at the western end of the European Union. Its borders are bounded in the north by the Alentejo region, the east by the Spanish region of Andalusia, and the south and west by the Atlantic Ocean. Its area comprises a little over 5,000 km^2, perfectly coinciding with the district of Faro (city that is also the administrative capital of the region), where about 440 thousand inhabitants live, spread over 16 municipalities (INE, 2018a).

The Algarve is the main tourism destination in Portugal, sought annually by thousands of national and foreign tourists. In 2018, the Algarve received 4.7 million guests in tourism accommodation establishments, being responsible for 33.3% of the total overnight stays in Portugal (INE, 2019). In Europe and even the world scenario, the Algarve has been widely recognized through the attribution of important tourism awards, such as *The Europe's Leading Beach Destination 2017* (World Travel Awards, 2017) and *The World's Best Golf Destination 2018* (World Golf Awards, 2018).

Tourism is the main economic activity in the Algarve, responsible for the creation of thousands of jobs (especially during the high season) and consequently, for the economy of many families. Although residents in the Algarve represent only 4.3% of the national population (INE, 2018a), the region is currently the second richest in the country, right after Lisbon, the capital of Portugal. In 2017, the Algarve had a contribution of 4.6% to the national GDP, around 20.0 million euros (INE, 2018b).

The Algarve has benefited from tourism development in recent decades, far beyond the economic dimension. Nevertheless, the region has been unable to avoid the excessive reliance on the seasonal 'sun and beach' product (PENT, 2012; PROT Algarve, 2004). One of the region's main distinguishing features is its natural resources, particularly the Mediterranean climate: long hot summers, soft short winters, and over 300 sunny days a year, with an average temperature of the highest in Europe.

Despite the importance of tourism, its future is seen with some consternation. The Algarve remains an attractive destination but it is over dependent in 'sun and beach' (Silva, 2019). Excessive specialization around this product as well as the concentration of financial and human resources on a limited number of activities can lead to the sector's decline. In the meantime, tourists' increasing sophistication, new trends in the sector, and the negative impacts of a high concentration in just two months per year, point out the need to complement 'sun and beach' with other offerings, in order to create differentiation, to face trends of seasonality, and to allow greater competitiveness with similar destinations (PENT, 2012; PROT Algarve, 2004).

Concerning nightlife industry, romantic venues or hotspots for romantic or sexual encounters in the Algarve - the tourism industry characteristics that matter for the study - the official statistics only refer to the number of existing beverage establishments. In 2018, there were 2,581 beverage establishments in the region, of which: 646 bars, 263 other beverage establishments without shows and 53 beverage establishments with dance space (INE, 2020). In fact, nightlife may serve as an "euphemism for sex" (Hobbs, Pattalung & Chandler, 2011; Singh & Hart, 2007; Sönmez *et al.*, 2013), showing the need to catalog nightlife establishments, as other forms of entertainment which may be relevant to attract and retain tourists. When one types «Algarve nightlife» on Google, there are around 862,000 references to bars, clubs, gentlemen's clubs, discos and other kind of nightlife entertainment across the region, in an appeal for fun and conviviality. However, there is no official data gathering all this information.

With regard to honeymoon, wedding packages or romantic escapes, the data is also not organized or cataloged, showing that the Algarve tourism industry is not managing properly the great potential of these market segments. Stakeholders are working by themselves in the promotion and sale of this kind of events, and it is possible to observe many company advertisements along the Web. There are hotels selling «luxurious honeymoon suites», «romantic restaurants» and «scenic spots» in all-inclusive packages (Butler, n.d.; *Oyster*, 2020), local wedding planners selling "a superior and completely customised service experience for wedding & event" (*Algarve Events*, 2020; *Algarve Weddings*, 2016) and specialized companies for hen parties and stag parties, offering boat trips, water activities and other sports, open bars, karaoke, dance music and, obviously, strip shows (*Algarve Fun*, 2017a).

The Algarve is being sold as "the ideal place for a fun night out, ranging from sports bars to the debauchery of ridiculously cheap happy hour prices at the rhythm of the beat of some of the best DJ's in the world" (*Algarve Fun*, 2017b), especially in the cities of Albufeira and Portimão, and the neighborhood of Vilamoura (Loulé municipality). It is possible to argue that tourism operators strongly promote «making new friends», «companionship», and the «chance to meet new people» - in other words, love, romance and sex - by creating a sense of heightened sexual anticipation but using, as stated by McKercher and Bauer (2003: 7), "quite different synonyms that are more socially acceptable".

Data Collection

This chapter presents data collected through in-depth interviews using a semi-structured guide, composed of 51 questions resulting from the literature review and prepared in two languages (Portuguese and English).

Tourists in the Algarve, mainly from Portugal and the United Kingdom (the most representative nationalities in the region[1]), were invited to participate in the study during the summers of 2014 e 2015. Tourists were approached at the Algarve beaches, located in 8 of the 16 municipalities of the region: Lagos (Luz Beach), Portimão (Rocha Beach); Silves (Armação de Pêra); Albufeira (Oura and São Rafael Beaches); Loulé (Gigi, Dunas Douradas, Falésia, Júlias, Quarteira and Vale do Lobo Beaches); Faro (Faro Island); Tavira (Tavira Island); and Castro Marim (Praia Verde).

The interviews were conducted on a convenience sample (Creswell, 2015), considering that most of the individuals showed embarrassment in talking about the most intimate aspects of their lives. The contact with tourists and the conquest of their willingness to participate emerged from a certain empathy created after explaining the objectives of the study. In some cases, it was necessary to request the participants to move to quieter areas of the beach, in order to ensure some privacy. All interviews were recorded in digital audio format, after the consent of the participants. In total, 29 interviews were recorded, until the information's saturation point (Flick, 2014; Glaser & Strauss, 1967).

It is worth remembering that despite the increasing number of studies about human sexuality, this is a delicate subject. In a broad sense, the researcher has to deal with people's discourses and not with the absolute truth about their sexual practices or even their thoughts and opinions (Aboim, 2013). Talking about sex and sexuality may be more common nowadays in Western societies but, at the same time, these topics can compromise, as Goffman (1959) would say, the image we want to show to others. For this reason, it is extremely important to read between the lines of what is told, thinking the narratives much beyond a collection of stories about sexual lives.

Data Analysis

The interviews were transcribed and analyzed using NVivo Qualitative Data Analysis, version 12. Language uniformization was necessary to allow a comparison between speeches in the software, since the interviews were made in English (three interviews) and Portuguese (26 interviews). This uniformization was made in two steps: first, the English interviews were translated into Portuguese and all the data were categorized; second, the most relevant speeches in Portuguese were translated into English while the English speeches were maintained in the original, in order to elaborate this chapter. Nicknames were used to protect participants' identity.

The data were analyzed using content analysis method, specifically interpretative analysis according to the recommendations of McCracken (1988), Bardin (1997) and Guerra (2010). Content analysis is a data treatment and technique which allows, on the one hand, the description of participants' opinions regarding a theme and, on the other hand, the construction of valid inferences about the data. In Bardin's (1997) opinion, there are four types of content analysis: categorical, evaluation, enunciation and expression. The analysis can be based also on a closed procedure (text analysis and coding are based on a set of categories that are part of the theoretical framework), open procedure (the analysis of the text gives rise to categories or concepts not originally foreseen) or mixed procedure (the two types of approach are adopted). This research uses the categorical content analysis and the mixed procedure.

RESULTS

Examining sexual perspectives and practices requires knowledge about social and cultural frameworks. Rather than understanding sexuality as something dependent on innate desires, it must be observed within the social order, its rhythms and agendas (Ferreira, 2010; Gagnon & Simon, 1974). Therefore, the results will be observed through the intersection between individual characteristics - with a special focus on gender issues - and the answers regarding sexual perspectives and practices. Tourists' individual characteristics are presented in Table 1.

The proportion of female respondents (55.2%) is higher than male (44.8%) and most of the respondents are aged from 18 to 34 years old (75.8%). Another 17.4% are 35 – 49 years old, and those 50-64 and 65 and over both accounts 3.4%. The great majority are Portuguese (82.9%), two are British (6.9%) one is Russian (3.4%), one is Angolan (3.4%) and one is South-African (3.4%). The major presence of Portuguese tourists in the set of interviews results not only from favorable communication features, but also from some cultural characteristics ascribed to Portuguese people, such as willingness to help, solidarity and welcoming nature (Cabecinhas & Amâncio, 2004).

Table 1. Sociodemographic Characteristics

		Respondents (n = 29)	
		Frequency	**Percentage (%)**
Gender	Male Female	13 16	44.8 55.2
Age	18-24 25-34 35-49 50-64 Over 65	10 12 5 1 1	34.4 41.4 17.4 3.4 3.4
Marital Status	Single Married/Living together Divorced Widowed	19 6 2 2	65.6 20.6 6.9 6.9
Nationality	Portuguese British Russian Angolan South-African	24 2 1 1 1	82.9 6.9 3.4 3.4 3.4
Country of Residence	Portugal United Kingdom Norway Netherlands France Angola South Africa	19 5 1 1 1 1 1	65.6 17.4 3.4 3.4 3.4 3.4 3.4
Academic Degree	Elementary Secondary Graduation PhD	4 9 15 1	13.8 31.0 51.8 3.4

Source: Own elaboration.

Most of the interviewees (n=19) live in Portugal (65.6%) and ten in foreign countries such as the United Kingdom (17.4%), Norway (3.4%), the Netherlands (3.4%), France (3.4%), Angola (3.4%) and South Africa (3.4%). Most of them have higher education levels (55.2%), including graduation and PhD; nine individuals completed the secondary education (31.0%) and four the elementary level (13.8%). Most are single (65.6%), some are married or are living together (20.6%) and few are divorced or widowed (6.9%, respectively).

Concerning the main objective of the study, the results show significant information. First, they allow to look closely at a phenomenon which is original at the regional basis, in Portugal and in other countries. Second, they provide accurate information about the aspects that most influence the relationship between tourism and sex in a 'sun and beach' destination such as the Algarve. Table 2 illustrates which are the main categories and subcategories underlying this relationship[2].

This table ranks the aspects that most contribute to ***liminality*** - which, in turn, explain all the sexual and romantic perspectives and practices during tourism experience (Graburn, 1977; Jafari, 1987; McKercher & Bauer, 2003; Ryan & Hall, 2001; Selänniemi, 2003) -, based on the total number of coded references (n=524).

Table 2. The Coding Process

Categories	Subcategories/Codes	Frequency (n)	Percentage (%)	Male Coded References (n)	Female Coded References (n)
Sensorial/Sensual Transition (n=188)	Corporeality	10	5.3	5	5
	Frequency of sex	29	15.4	13	16
	Intensity of sex	30	16.0	14	16
	Willingness to get involved in a vacation relationship	57	30.3	24	33
	Dissolution of lasting relationships	29	15.4	12	17
	Investment in lasting relationships	33	17.6	15	18
Mental Transition (n=172)	Communitas	29	16.9	14	15
	Peak experiences				
	Alcohol consumption	46	26.7	22	24
	Drugs consumption	11	6.4	8	3
	Paid sex	28	16.3	13	15
	Sex without condom	28	16.3	13	15
	Multiple sex partners	30	17.4	15	15
Spatial and Temporal Transition (n=164)	Loosening of social control/ Anonymity	39	23.8	16	23
	Destination characteristics	26	15.8	10	16
	Sun, sand and sea	28	17.1	11	17
	Time compression	39	23.8	18	21
	Role of tourism industry				
	Nightlife	26	15.8	12	14
	Entertainment	6	3.7	1	5

Source: Own elaboration based upon statistical frequency data from NVivo 12.

The results show a slight advantage of ***sensorial/sensual transition*** (Selänniemi, 2003) over other categories explaining the liminal nature of tourism. In the total interviews, *sensorial/sensual transition* obtained the highest number of coded references (n=188), meaning that tourism provides opportunities to experience the surroundings in a more intense way, since the senses are much more aware of the stimulations (*Idem*).

Ordering the aspects that most contribute to *sensorial/sensual transition* in tourism context, the most relevant are:

The willingness to get involved in a vacation relationship (30.3%);
The investment in lasting relationships (17.6%);
The intensity of sex (16.0%);
The frequency of sex (15.4%);
The dissolution of lasting relationships (15.4%);

Corporeality (5.3%).

The willingness to get involved in a vacation relationship is pointed out mainly by single individuals. This kind of relationship, in the interviewees' opinion, contributes to enhance tourism experience, as well as to enjoy vacation time, to have fun, to bring some adrenaline and to create good memories about holidays and the destination (McKercher & Bauer, 2003). Some of the participants believe that vacation relationships might become lasting relationships:

I like to be romantic but I also like to be crazy, like everyone else. I think it's good and it's part of the youth. I think it is more common in younger ages but it is especially good for learning and appreciate life. [Henrique, male, 18 years old, single].

I've had some vacation relationships, it's a life experience. We cannot expect too much. When I think about getting involved with someone, I don't think it will be just a holiday romance, I think it may extend and become a lasting relationship. We never know… [Inês, female, 28 years old, single].

If I have a vacation relationship, in a special place (…), I will have a vacation to be remembered. [Mário, male, 21 years old, single].

However, when asked about the possibility of having a vacation relationship while in the Algarve, few respondents showed that willingness. In this case, the majority are men and single. For many tourists, it is more of a fantasy than a desire to be materialized and in a broader sense, the results show that perspectives are often different from the undertaken practices. It might be also because some of the interviewees are married or have a lasting relationship, or even because some would like to find someone for a more serious relationship. Vacation relationships are considered to be «funny», but in reality, they are not the kind of relationship the interviewees want for life. Above all, the Algarve is a destination for families, couples or friends, where sex plays a central role, especially within a lasting relationship.

The investment in lasting relationships means the opportunity to spend more time with the usual partner, without the daily life constrains of work, routine and clock (McKercher & Bauer, 2003; Selänniemi, 2003). For the interviewees, vacation trips mean the escape of routine, a moment to strengthen the ties with the partner, to do different things in a relaxed atmosphere, and this opinion is unanimous regardless of marital status:

It's good, we just have to get out of our everyday environment. It's always good! Although we got along very well for 32 years, it always tastes good. I am fortunate of having an excellent husband. [Helena, female, 53 years old, married].

People are more relaxed, more uninhibited, they really want fun. I think it's good because the tension that can exist in the relationship, can dissipate during vacation time. It is an escape from the daily routine. [Alexandre, male, 25 years old, single].

Notwithstanding, it is mainly women who point vacation trips as the ideal moment to invest in their relationships:

If people go together on vacation, they will enjoy leisure time together. For sure, it will contribute to strengthening the relationship. They spend more time together and they escape the routine. [Leonor, female, 30 years old, living together].

Accordingly, the image of the Algarve has been crafted in the tourists' mind to create an aura of sexual and romantic excitement (McKercher & Bauer, 2003), through some scenic spots like the beach and nature, or even the nightlife with the open-air bars. As stated by one of the interviewees,

I am not sure if it is typical of the Portuguese culture, but when we go out at night, we see a very romantic atmosphere. When we go to bars, especially the open-air bars with terraces - there are many in the Algarve - we see love and romance everywhere. It is very attractive and appealing, and it's easy to get involved by this atmosphere. [John, male, 32 years old, single].

The intensity of sex is, therefore, highlighted by some participants. Being away from home, routines and the stress contributes to sexual experiences to be more intense (McKercher & Bauer, 2003; Ryan & Hall, 2001; Zhang & Xu, 2019). In this case, single individuals are the ones for whom vacation time contributes to more intense sex, even if they travel with partners or they find new ones at the destination:

On vacation everything is better. We want to have fun and relax. (…) If we have sex in an exotic place, this makes the holidays so much better. (…) But it depends on the person, obviously. Personally, my favorite sexual intercourses are the ones I have on vacation. (…) We do different things, sometimes even very strange things. (…) I was with a girl who wanted to have sex on the balcony, literally hanging on the balcony, looking at the pool. She told me she wanted to have that kind of adrenaline, which was so much more intense with the fear of the height. It was really funny. [John, male, 32 years old, single].

During vacation trips, we are in such a relaxed atmosphere. (…) It can be at home, before or after we leave for dinner or something else… there is nothing that forces us to have sex at that particular time. (…) We simply enjoy the moment. [Alice, female, 20 years old, single].

The frequency of sex is also mentioned, either within a marital context or a lasting relationship. In fact, tourism provides the opportunity for sexual intercourses to be more frequent (McKercher & Bauer, 2003; Ryan & Hall, 2001; Zhang & Xu, 2019). During tourism, people have more time, are more relaxed and want to have fun:

When I am doing tourism with my boyfriend, we are together every single day. In the place where we live, we don't use to sleep together and I don't see him every day. So, the frequency of sex is much higher on vacation. [Beatriz, female, 23 years old, single].

Sexual intercourse is higher on vacation. We are free, we have no worries, we are living the moment peacefully, taking advantage of every opportunity. Then, maybe with a dinner out, a break to see the sunset, all the things we can't appreciate in the daily life, all lead to love, romance and sex. [Diogo, male, 38 years old, living together].

The dissolution of lasting relationships is pointed out by the participants as not having a direct correlation with the holidays. In the interviewees' opinion, the dissolution happens when the relationship is already deteriorated, which has nothing to do with the holidays. In this case, vacation trips just act as a turning point, emphasizing the differences between people because they spend more time together (McKercher & Bauer, 2003):

If the relationship is not going well or if people are used to routines, then perhaps, vacation trips may contribute to the end of the relationship. It can make you wonder: why am I with this person? I don't really love him. [Susana, female, 25 years old, single].

It depends on the relationship's structure. If it is not a solid relationship, if there is no pure love… then holidays can contribute to separation (…). [Ana, female, 45 years old, widowed].

Corporeality, although is the less mentioned subcategory, contributes to the meaning of *sensorial/ sensual transition*. As described by Selänniemi (2003), tourism gives the opportunity for manifestations of human corporeality, in which sensuality and pleasure play a leading part. Besides, it is important to notice that in a bathing suit everybody looks more or less the same – it is the physical status that starts to rule rather than the social one (Gisolf, 2013). The manifestation of corporeality is, in fact, one of the most interesting areas to analyse liminality during tourism, precisely because there is a greater investment in the body, in the senses, in sensuality and even in sexuality (Johnston & Longhurst, 2010; Gisolf, 2013):

The skin is more exposed, we see many bikinis and, in general, people are happier. They are more open to talk and do things that are typical of hot places, like going to the beach or just relaxing. [John, male, 32 years old, single].

The Algarve is a hot place. People go to the beach, they are almost naked which, in turn, is linked to sensuality. It is easier to look at the person's body and it can facilitate physical attraction. [Beatriz, female, 23 years old, single].

It is important to mention that *corporeality* was the unique subcategory with the same number of coded references for men and women (n=5). This result shows it is valued regardless of gender order or double standards (Connell, 1987; Crawford & Popp, 2003).

Mental transition was the second most coded category (172 coded references). For the participants, tourism means also the possibility of changing the state of mind, to behave in a completely different way, and to reveal hidden characteristics of the self (Selänniemi, 2003). Above all, it means the opportunity to live peak experiences Weitman (1998) and to share the moment with individuals in the same situation (communitas) (Wagner,1977). Ranking the aspects that most contribute to *mental transition* in tourism context, the most relevant are:

Peak experiences (83.1%);
Communitas (16.9%).

During tourism, individuals have the opportunity to do different things, to live experiences that are usually denied during the everyday life, in other words, to live **peak experiences**. Some authors refer to the unruly, transgressive or excessive behaviors in tourism context that are related to the unusual consumption of food, alcohol or drugs (Dahles, 1998; Koc, 2013; Shields, 1990). Accordingly, the interviewees reveal that one of the most common behaviors in tourism context is the often-excessive alcohol consumption, which is closely related to sexual intercourse:

Too much alcohol makes it easier for people to get involved, to spend the night together. People get more uninhibited. [Filipe, male, 24 years old, single].

Alcohol makes people more uninhibited and more promiscuous. It is much easier to seduce a woman if she is under the influence of alcohol. [Inês, female, 28 years old, single].

Occasionally, tourism means also drug abuse in an attempt to enhance the mood, although drug use is less reported by the participants than alcohol consumption. Men admit most to take drugs while on vacation, but drugs and alcohol consumption are both correlated to higher levels of sexual intercourse and, therefore, to higher sexual risk behaviors (Josiam *et at.*, 1998).

Sexual risk behaviors were analyzed in relation to the willingness for having multiple sexual partners, sex without condom or going to a prostitute. Despite the fact that the speeches were coded very similarly, the results show a major tendency for sexual involvement with multiple partners than the willingness to have sex without condom or going to a prostitute. It is important to notice that sexual involvement with multiple partners is mainly part of the male imaginary:

I have two or three friends in the Algarve, it's about to happen. I think it is possible. [Manuel, male, 43 years old, divorced].

No, because of my personal values. I think one partner is enough (…). It means realizing an attraction, raising self-esteem ... but two partners, I think I would start to look at myself in a different way and I wouldn't feel good. [Maria, female, 22 years old, single].

Communitas is also mentioned by the participants as a significant aspect of liminality during tourism (McKercher & Bauer, 2003; Wagner, 1977). It means a sense of communion, equality and closeness among individuals sharing the same space-time. However, for the interviewees, communitas means above all, that people are usually more trusting during tourism experiences. The fact that they are less concerned about life and look similar due to the absence of social status, makes trust easier during vacation time:

During summer, in a tourism setting, we are all on a very similar mood, which means to enjoy the surroundings, weather, sun, friends, food… since the common elements are bigger, there is a greater tendency for trusting other people. In cities, people are more suspicious. We are conditioned by who the person is, what he/she does… it can be limiting. [Bruno, male, 46 years old, divorced/in a new relationship].

Spatial and temporal transition was the third coded category (164 coded references). It means the transition in terms of space, when the person leaves home and arrives at a tourism destination, and the transition in terms of time, away from the everyday time of work, home, routine and clock. At the tourism destination time has no significance or, at least, it is a different time, where everything can be experienced more slowly (Selänniemi, 2003). According to the interviewees, the features that most contribute to *spatial and temporal transition* in tourism context are:

Time compression (23.8%);
The loosening of social control/Anonymity (23.8%);
The role of tourism industry (19.5%);
Sun, sand and sea (17.1%);
The destination characteristics (15.8%).

Time compression means the increasing speed of living. During the ordinary life, time concerns the constraints that shape the routines (Hall, 2004) and although time is not a reference for work and routines during tourism, allowing everything to be experienced more slowly, there are other aspects in which time counts, such as sex and romance. In fact, for this kind of experiences, vacation time is perceived as very limited, so the «normal» steps for sexual and romantic encounters are usually overtaken:

I think it's different on vacation because you really don't have time for all these steps. I think it's faster when you're on vacation. You know someone, you want to flirt that night (...), you're only there for a week... you have to speed up the romance. Sex probably also happens faster. When you're not on vacation or when you live in the same country, I think it's all slower. [Susana, female, 25 years old, single].

The loosening of social control/Anonymity is another important aspect of *spatial and temporal transition* during tourism. According to the participants, the loosening of social control is mainly related to anonymity. In fact, in a tourism context, nobody knows the tourists and therefore nobody can judge them. Anonymity is extremely important for the sexual and romantic experiences, allowing things to be done in a more relaxed and liberal way (Marques & Lança, 2016; Zhang & Xu, 2019):

In the place where I live, everyone knows me and they usually say: "you have been with this or that girl". Here, I'm just with them [friends] but they won't tell anybody because they do the same [laughs]. [Pedro, male, 21 years old, single].

The role of tourism industry has been described by scholars as a strong facilitator for the sexual and romantic encounters between tourists and their partners (McKercher & Bauer, 2003; Oppermann, 1999; Ryan, 2000; Zhang & Xu, 2019). In fact, tourism brings a rupture with the daily life, giving individuals the opportunity to do what they would not usually do. At the same time, tourism industry provides the venue, the infrastructures and even the sexual or romantic partners. In the case of this study, the role of tourism industry as a facilitator is related, above all, with the nightlife. Going to bars and clubs, consuming alcohol or other drugs is linked to a greater disposition for sexual intercourse. In the participants' opinion, nightlife and alcohol consumption go beyond other forms of entertainment in the pursuit of casual sex:

The excess of alcohol, the heat of the moment, the conviviality, and the fact that people don't known each other in holiday destinations. All of this make people more uninhibited and more comfortable. [Rodrigo, male, 30 years old, married].

I definitely think alcohol is a great driver for people to communicate on vacation. I don't necessarily say at the romance level, but certainly for sex. People become more promiscuous when they drink alcohol. [Susana, female, 25 years old, single].

Sun, sand and sea is one of the major attractions of the Algarve, being the reason why so many tourists choose the region for vacations (Hobson & Dietrich, 1994; Marques, 2016; PENT, 2012; Silva, 2019). Simultaneously, as stated by the interviewees, the beach is a privileged place to meet people and make new friends:

The beach is a good place to meet people with whom we can go out at night. [Afonso, male, 18 years old, single].

Beyond sun, sand and sea, other **destination characteristics** are pointed out by the participants regarding natural resources, the weather, and the romantic and relaxed atmosphere. All these features contribute to romantic and sexual encounters:

I live in a country [Scotland] where the temperature is completely different, the weather is different and people are much more «closed». They stay at home and they are uncommunicative too. In the Algarve, there are a lot of parties and what it involves, namely alcohol and other things that set us free, it is easier to get in a new relationship. [Matilde, female, 27 years old, single].

The romantic atmosphere of the Algarve, with the sunset, the heat, everything is wonderful. I would say that when people are happier, they socialize more. It happens more here than in England, for instance. Everything seems more romantic here. [Emma, female, 27 years old, single].

Another characteristic pointed out by the interviewees is the presence of several nationalities in the region, introducing a different dimension to vacation relationships in the Algarve. This aspect was not foreseen by the researchers, since phenomena like "sexotism", "ethnosexual relationships" or the attraction for the "exotic other" (Michel, 2009) have been analyzed regarding the economic/power asymmetries between tourists and their sexual partners in the poorest regions of the world (Altman, 2001; Bandyopadhyay & Nascimento, 2010). However, as stated by Michel (2009) "sexotism" designates a kind of growing trend consisting of seeking sexual gratification with people from different cultures and it is not necessarily associated with sex tourism. Likewise, the results of this study show that nationality or culture are important dimensions of the articulation between sex and tourism, outside the "sex tourism paradigm":

The Algarve is a mix of different cultures. There are people from all over the world coming here for holidays and people end up wanting to meet other people. I think for those who want to meet people, to have some adventure, the Algarve is the ideal place. [Bárbara, female, 34 years old, married].

CONCLUSION

This chapter contributes to the literature about the relationship between tourism and sex by showing this relationship does not exhaust in the commodification features, as it has been argued by the "sex tourism paradigm". The liminal nature of tourism provides the opportunity to engage in sex and romance in a more relaxed and liberal way, far away from home and the daily life constrains, which is not necessarily related to the commodification of sex. Previous studies have already identified the need of understanding human sexual behaviors during tourism experiences, which are considered to be "neither controversial nor out of the ordinary", by integrating a new perspective, that of "sex and the sexual during people's leisure and tourism experiences".

The results of this study show, on the one hand, that the Algarve is not a sex tourism destination, meaning tourists do not come to the region for commercial sex purposes. In fact, the interviewees strongly deny the use of paid sex. They come to the Algarve mainly for the weather and the beaches, although they also enjoy the party atmosphere, the relaxation and the opportunity to have sex or being romantic during vacation time. In other words, the results show a strong relationship between vacation trips and romantic and sexual encounters, far beyond the narrow confines of the sex trade.

On the other hand, the results show that tourists' perspectives and practices regarding sex and romance are, in fact, «extra-ordinary». The frequency and the intensity of sex during vacation time are higher when compared to the daily life, due to the absence of routines, anonymity and the loosening of social control. By entering the "sacred" situation of tourism, individuals have the opportunity to live "peak experiences", which may include the unusual consumption of alcohol or drugs, the involvement with multiple sexual partners, and the sharing moments with individuals in the same situation (communitas).

The results also reveal different behaviors according to gender. Sexual double standard still constrains perspectives and practices. If not at the practice level, for sure, at the discourse level, in the case of women. Thus, the chance of having a romantic or sexual intercourse with an occasional partner or with multiple partners, is mainly part of the male imaginary. In contrast, women relate vacation trips to the ideal of investing in the lasting relationships.

In short, the Algarve is a tourism destination where sex and romance play a central role, both in terms of strengthening the lasting relationships or in finding a new, short but consensual and nonpecuniary relationship. Therefore, it is necessary and even desirable that the tourism industry and its stakeholders to explore the great potential of romance, eroticism and sexual adventure. Examples may include collaborative relations and networking regarding the creation of well-organized honeymoon and wedding packages, couple getaways or romantic programs. So far, there is no official data about such activities or enterprises, showing that the Algarve tourism industry is not managing properly the great potential of these market segments.

The importance of informal social relationships between local entrepreneurs in creating and implementing this kind of tourism activities is crucial to invite more tourists and retain others. Taking into account that the quality of experiences is primarily determined by the intimacies that exist between people at the place, sexual and romantic experiences may play a decisive role in tourists' loyalty to the Algarve.

ACKNOWLEDGMENT

This chapter is financed by National Funds provided by FCT - Foundation for Science and Technology, through the project UIDB/04020/2020.

REFERENCES

Aboim, S. (2013). *A Sexualidade dos Portugueses*. Lisbon: Fundação Francisco Manuel dos Santos.

Agustín, L. (2010). The (crying) need for different kinds of research. In M. Ditmore, A. Levy, & A. Willman (Eds.), *Sex Work Matters: Exploring Money, Power, and Intimacy in the Sex Industry* (pp. 23–27). Zed Books.

Algarve, P. R. O. T. (2004). Regional Spatial Planning Plan for the Algarve.: Vol. II. *Characterization and Diagnosis*. Algarve Regional Commission for Coordination and Development.

Algarve Events. (2020). *Wedding planner*. Retrieved from: https://algarveevents.com/weddingplanner.php

Algarve Fun. (2017a). *Algarve Stag Dos*. Retrieved from: https://www.algarvefun.com/algarve-stag-dos

Algarve Fun. (2017b). *Best resorts & towns for stag dos and parties in the Algarve*. Retrieved from: https://www.algarvefun.com/algarve-stag-dos/best-resorts-towns-stag-dos-parties-algarve/

Algarve Hen Parties. (n.d.). Retrieved from: https://www.algarvefun.com/algarve-hen-parties

Algarve Weddings. (2016). Retrieved from: https://algarve-weddings.co.uk/

Altman, D. (2001). *Global Sex*. University of Chicago Press. doi:10.7208/chicago/9780226016047.001.0001

Bandyopadhyay, R., & Nascimento, K. (2010). "Where fantasy becomes reality": How tourism forces made Brazil a sexual playground. *Journal of Sustainable Tourism*, *18*(8), 933–949. doi:10.1080/09669582.2010.497220

Bardin, L. (1997). *Análise de Conteúdo*. Lisbon: Edições 70.

Berdychevsky, L., Gibson, H., & Poria, Y. (2013). Women's sexual behavior in tourism: Loosening the bridle. *Annals of Tourism Research*, *42*, 65–86. doi:10.1016/j.annals.2013.01.006

Berdychevsky, L., Poria, Y., & Uriely, N. (2013). Sexual behaviour in women's tourist experiences: Motivations, behaviors, and meanings. *Tourism Management*, *35*, 144–155. doi:10.1016/j.tourman.2012.06.011

Bourdieu, P. (2001). *Masculine Domination*. Stanford University Press.

Bozon, M. (2004). *Sociologia da Sexualidade*. Editora FGV.

Bozon, M. (2015). Sexualité et genre. In J. Laufer, C. Marry, & M. Maruany (Eds.), *Masculin Féminin: Questions Pour les Sciences de l'Homme* (pp. 169–186). PUF.

Brents, B., Jackson, C., & Hausbeck, K. (2010). *The Sate of Sex: Tourism, Sex, and Sin in the New American Heartland*. Routledge.

Bryman, A., & Bell, E. (2011). *Business Research Methods* (3rd ed.). Oxford University Press.

Butler, F. (n.d.). *Best honeymoon places in the Algarve*. Retrieved from: https://www.myguidealgarve.com/lists/best-honeymoon-places-in-the-algarve

Butler, J. (1999). *Gender Trouble: Feminism and Subversion of Identity*. Routledge.

Butler, J. (2011). *Bodies That Matter: On the Discursive Limits of Sex*. Routledge. doi:10.4324/9780203828274

Cabecinhas, R., & Amâncio, L. (2004). Estereótipos sociais e assimetria simbólica: três estudos com jovens angolanos e portugueses. *Proceedings of the VIII Luso-Afro Brazilian Congress of Social Sciences*.

Cabezas, A. (2004). Between love and money: Sex, tourism, and citizenship in Cuba and the Dominican Republic. *Signs (Chicago, Ill.)*, *29*(4), 987–1015. doi:10.1086/382627

Cabezas, A. (2009). *Economies of Desire: Sex and Tourism in Cuba and the Dominican Republic*. Temple University Press.

Campbell, S., Perkins, A., & Mohammed, P. (1999). Come to Jamaica and feel all right: tourism and the sex trade. In K. Kempadoo (Ed.), *Sun, Sex and Gold: Tourism and Sex Work in the Caribbean* (pp. 125–156). Rowman & Littlefield Publishers.

Cansel, A., Ekiz, E., & Bavik, A. (2009). Sex tourism in Northern Cyprus: Investigating the current situation. *Tourism Analysis*, *14*(5), 677–689. doi:10.3727/108354209X12597959359374

Cantalice, T. (2009). Feminismo, mercado de sexo e turismo: Reflexões sobre as múltiplas faces e interpretações do sexo mercantil. *Bagoas - Estudos Gays. Género e Sexualidades*, *2*(3), 145–178.

Carr, N., & Poria, Y. (2010). Introduction: provision, marketing and consumption of sex and the sexual in the leisure and tourism environment. In N. Carr & Y. Poria (Eds.), *Sex and the Sexual During People's Leisure and Tourism Experiences* (pp. 1–16). Cambridge Scholars Publishing.

Carter, S., & Clift, S. (2000). Tourism, international travel and sex: themes and research. In S. Clift & S. Carter (Eds.), *Tourism and Sex: Culture, Commerce and Coercion* (pp. 1–19). Pinter.

Clift, S., & Carter, S. (Eds.). (2000). Tourism and Sex: Culture, Commerce and Coercion. London: Pinter.

Cohen, E. (1971). Arab boys and tourist girls in a mixed Jewish-Arab community. *International Journal of Comparative Sociology*, *12*(4), 217–233. doi:10.1177/002071527101200401

Cohen, E. (1982). Thai girls and farang men: The edge of ambiguity. *Annals of Tourism Research*, *9*(3), 403–428. doi:10.1016/0160-7383(82)90021-4

Cohen, E. (1988). Tourism and AIDS in Thailand. *Annals of Tourism Research*, *15*(4), 467–486. doi:10.1016/0160-7383(88)90044-8

Connell, R. W. (1987). *Gender and Power*. Stanford University Press.

Crawford, M., & Popp, D. (2003). Sexual double standards: A review and methodological critique of two decades of research. *Journal of Sex Research*, *40*(1), 13–26. doi:10.1080/00224490309552163 PMID:12806528

Creswell, J. (2015). *Educational Research: Planning, Conducting, and Evaluating Quantitative and Qualitative Research* (5th ed.). Pearson Education, Inc.

Crick, M. (1989). Representations of international tourism in the social sciences: Sun, sex, sights, savings, and servility. *Annual Review of Anthropology, 18*(1), 307–344. doi:10.1146/annurev.an.18.100189.001515

Dahles, H. (1998). Redefining Amsterdam as a tourist destination. *Annals of Tourism Research, 25*(1), 55–69. doi:10.1016/S0160-7383(97)00061-3

Dahles, H., & Bras, K. (1999). Entrepreneurs in romance: Tourism in Indonesia. *Annals of Tourism Research, 26*(2), 267–293. doi:10.1016/S0160-7383(98)00098-X

De Albuquerque, K. (1998). Sex, beach boys and female tourists in the Caribbean. In B. M. Dank & R. Refinetti (Eds.), *Sexuality and Culture* (pp. 87–111). New Brunswinck.

Donnan, H., & Magowan, F. (2010). *The Anthropology of Sex*. Berg.

Durkheim, E. (1914). Le dualisme de la nature humaine et ses conditions sociales. *Scientia, 15*, 206–221.

Ellis, A. (2003). *Sex Without Guilt in the 21st Century*. Barricade Books.

Farvid, P., Braun, V., & Rowney, C. (2017). 'No girl wants to be called a slut!': Women, heterosexual casual sex and the sexual double standard. *Journal of Gender Studies, 26*(5), 544–560. doi:10.1080/09589236.2016.1150818

Ferreira, P. M. (2010). A actividade sexual: frequência, regularidade e inactividade. In *P. M. Ferreira & M. V. Cabral (Orgs.), Sexualidades em Portugal: Comportamentos e Riscos* (pp. 19–54). Bizâncio.

Flick, U. (2014). *An Introduction to Qualitative Research* (5th ed.). Sage.

Foucault, M. (1977). *Discipline and Punish: The Birth of the Prison*. A. Lang.

Frank, K. (2007). Playcouples in paradise: touristic sexuality and lifestyle travel. In M. Padilla, J. Hirsh, M. Laboy, R. Sember, & R. Parker (Eds.), *Love and Globalization: Transformations of Intimacy in the Contemporary World* (pp. 163–185). Vanderbilt University Press.

Franklin, A. (2003). *Tourism: An Introduction*. Sage.

Frohlick, S. (2010). The sex of tourism? Bodies under suspicion in paradise. In J. Scott & T. Selwyn (Eds.), *Thinking Through Tourism* (pp. 51–70). Berg.

Gagnon, J., & Simon, W. (1974). *Sexual Conduct. The Social Sources of Human Sexuality*. Aldine Publishing.

Giddens, A. (1992). *The Transformation of Intimacy: Sexuality, Love & Eroticism in Modern Societies*. Polity Press.

Gisolf, M. (2013). Liminality and tourism. *Tourism Theories*. Retrieved from: http://www. tourismtheories.org/?p=700

Glaser, B., & Strauss, A. (1967). *The Discovery of Grounded Theory: Strategies for Qualitative Research*. Aldine.

Goffman, E. (1959). *The Presentation of Self in Everyday Life*. Anchor Books.

Graburn, N. (1977). Tourism: the sacred journey. In V. Smith (Ed.), *Hosts and Guests: The Anthropology of Tourism* (pp. 17–31). University of Pensilvania Press.

Guerra, I. (2010). *Pesquisa Qualitativa e Análise de Conteúdo: Sentidos e Formas de Uso*. Princípia Editora.

Hall, C. (1992). Sex tourism in Southeast Asia. In D. Harrison (Ed.), *Tourism and the Less Developed Countries* (pp. 65–74). Belhaven Press.

Hall, C. (2004). *Tourism and mobility*. Presented at the CAUTHE 2004: Creating Tourism Knowledge. Retrieved from: https://ourarchive.otago.ac.nz/bitstream/handle/ 10523/738/Mobility_CAUTHE.pdf?sequence=2&isAllowed=y

Hammond, N., & Kingston, S. (2014). Experiencing stigma as sex work researchers in professional and personal lives. *Sexualities*, *17*(3), 329–347. doi:10.1177/1363460713516333

Herold, E., Garcia, R., & DeMoya, T. (2001). Female tourists and beach boys: Romance or sex tourism? *Annals of Tourism Research*, *28*(4), 978–997. doi:10.1016/S0160-7383(01)00003-2

Hobbs, J., Pattalung, P., & Chandler, R. (2011). Advertising Phuket's nightlife on the internet: A case study of double binds and hegemonic masculinity in sex tourism. *Journal of Social Issues in Southeast Asia*, *26*(1), 80–104. doi:10.1355j26-1e PMID:21919275

Hobson, J., & Dietrich, U. (1994). Tourism, health and quality of life: Challenging the responsibility of using the traditional tenets of sun, sea, sand and sex in tourism marketing. *Journal of Travel & Tourism Marketing*, *3*(4), 21–38. doi:10.1300/J073v03n04_02

INE. (2014). *Tourism Statistics 2013*. Statistics Portugal.

INE. (2018a). *Statistical Yearbook of Algarve Region 2017*. Statistics Portugal.

INE. (2018b). *Tourism Statistics 2017*. Statistics Portugal.

INE. (2019). *Statistical Yearbook of Algarve Region 2018*. Statistics Portugal.

INE. (2020). *Number of Beverage Establishments in the Algarve 2018: Database of Statistics Portugal*. Retrieved from: https://www.ine.pt/xportal/xmain?xpid=INE&xpgid=ine_indicadores&userLoadSave=Load&userTableOrder=10765&tipoSeleccao=1&contexto=pq&selTab=tab1&submitLoad=true

Jackson, S., & Scott, S. (2002). *Gender: A Sociological Reader*. Routledge.

Jackson, S., & Scott, S. (2004). Sexual antinomies in late modernity. *Sexualities*, *7*(2), 233–248. doi:10.1177/1363460704042166

Jafari, J. (1987). Tourism models: The sociocultural aspects. *Tourism Management*, *8*(2), 151–159. doi:10.1016/0261-5177(87)90023-9

Jeffreys, S. (2003). Sex tourism: Do women do it too? *Leisure Studies*, *22*(3), 223–238. doi:10.1080/026143603200075452

Johnston, L., & Longhurst, R. (2010). *Space, Place and Sex: Geographies of Sexualities*. Rowman & Littlefield Publishers.

Jordan, F., & Aitchison, C. (2008). Tourism and the sexualisation of the gaze: Solo female tourists' experiences of gendered power, surveillance and embodiment. *Leisure Studies*, *27*(3), 329–349. doi:10.1080/02614360802125080

Josiam, B., Hobson, J., Dietrich, U., & Smeaton, G. (1998). An analysis of the sexual, alcohol and drug related behavioural patterns of students on spring break. *Tourism Management*, *19*(6), 501–513. doi:10.1016/S0261-5177(98)00052-1

Kalish, R., & Kimmel, M. (2011). Hooking up: Hot hetero sex or the new numb normative? *Australian Feminist Studies*, *26*(67), 137–151. doi:10.1080/08164649.2011.546333

Kempadoo, K. (Ed.). (1999). *Sun, Sex and Gold: Tourism and Sex Work in the Caribbean*. Rowman & Littlefield Publishers.

King, N., & Horrocks, C. (2010). *Interviews in Qualitative Research*. Sage Publications.

Kinsey, A. (1998). *Sexual Behavior in the Human Female*. Indiana University Press. (Original work published 1953)

Koc, E. (2013). Inversionary and liminoidal consumption: Gluttony on holidays and obesity. *Journal of Travel & Tourism Marketing*, *30*(8), 825–838. doi:10.1080/10548408.2013.835669

Legardinier, C., & Villette, S. (2007). O turismo sexual. In C. Ockrent & S. Treiner (Eds.), *O Livro Negro da Condição das Mulheres* (pp. 490–496). Difel.

Leheny, D. (1995). A political economy of Asian sex tourism. *Annals of Tourism Research*, *22*(2), 367–384. doi:10.1016/0160-7383(94)00082-4

Marques, J. F. (2016). Sun, sand, sea and sex, tourism. In J. Jafari & X. Honggen (Eds.), *Encyclopedia of Tourism*. Springer. doi:10.1007/978-3-319-01384-8_536

Marques, J. F., & Lança, M. (2016). Para além da vida quotidiana. Amor e sexualidade em contexto turístico: Resultados preliminares de uma pesquisa exploratória no Algarve (Portugal). *Anais Brasileiros de Estudos Turísticos*, *6*(2), 7–22.

McCracken, G. (1988). *The Long Interview*. Sage. doi:10.4135/9781412986229

McKercher, B., & Bauer, T. (2003). Conceptual framework of the nexus between tourism, romance and sex. In T. Bauer & B. McKercher (Eds.), *Sex and Tourism. Journeys of Romance, Love and Lust* (pp. 3–18). The Haworth Hospitality Press.

Michel, F. (2009). *Planeta Sexo. Turismos Sexuais, Mercantilização e Desumanização dos Corpos*. Campo das Letras.

Milhausen, R., & Herold, E. (1999). Does the sexual double standard still exist? Perceptions of university women. *Journal of Sex Research*, *36*(4), 361–368. doi:10.1080/00224499909552008

Nicolson, P., & Burr, J. (2003). What is 'normal' about women's (hetero)sexual desire and orgasm? A report of an in-depth interview study. *Social Science & Medicine*, *57*(9), 1735–1745. doi:10.1016/S0277-9536(03)00012-1 PMID:12948581

O'Connell-Davidson, J. (2001). The sex tourist, the expatriate, his ex-wife and her 'other: The politics of loss, difference and desire. *Sexualities*, *4*(1), 5–24. doi:10.1177/136346001004001001

Oppermann, M. (Ed.). (1998). *Sex Tourism and Prostitution: Aspects of Leisure, Recreation and Work*. Cognizant Communication Corporation.

Oppermann, M. (1999). Sex tourism. *Annals of Tourism Research*, *26*(2), 251–266. doi:10.1016/S0160-7383(98)00081-4

Oyster. (2020). *Most romantic hotels in the Algarve*, Portugal. Retrieved from: https://www.oyster.com/algarve/hotels/roundups/most-romantic-hotels-in-algarve-faro-district-portugal/

Patton, M. (1990). *Qualitative Evaluation and Research Methods* (2nd ed.). Sage.

PENT. (2012). *National Strategic Tourism Plan: Horizon 2013-2015*. Ministry of Economy and Employment.

Piscitelli, A. (2004). On «gringos» and «natives»: Gender and sexuality in the context of international sex tourism in Fortaleza, Brazil. *Vibrant*, *1*(1/2), 87–114.

Poria, Y., & Carr, N. (2010). What have we done and where should we go? In N. Carr & Y. Poria (Eds.), *Sex and the Sexual During People's Leisure and Tourism Experiences* (pp. 181–196). Cambridge Scholars Publishing.

Pruitt, D., & LaFont, S. (1995). For love and money: Romance tourism in Jamaica. *Annals of Tourism Research*, *22*(2), 422–440. doi:10.1016/0160-7383(94)00084-0

Ribeiro, F., & Sacramento, O. (2006). A ilusão da conquista: Sexo, amor e interesse entre *gringos* e *garotas* em Natal (Brasil). *Cronos (Valencia)*, *7*(1), 161–172.

Ribeiro, M., Silva, M. C., Schouten, J., Ribeiro, F., & Sacramento, O. (2007). *Vidas na Raia: Prostituição Feminina em Regiões de Fronteira*. Edições Afrontamento.

Ryan, C. (2000). Sex tourism: paradigms of confusion? In S. Clift & S. Carter (Eds.), *Tourism and Sex: Culture, Commerce and Coercion* (pp. 23–40). Cassel.

Ryan, C. (2002). Equity, management, power sharing and sustainability – issues of the 'new tourism'. *Tourism Management*, *23*(1), 17–26. doi:10.1016/S0261-5177(01)00064-4

Ryan, C., & Hall, M. (2001). *Sex Tourism: Marginal People and Liminalities*. Routledge.

Ryan, C., & Kinder, R. (1996). Sex, tourism and sex tourism: Fulfilling similar needs? *Tourism Management*, *17*(7), 507–518. doi:10.1016/S0261-5177(96)00068-4

Sanchez-Taylor, J. (2000). Tourism and "embodied" commodities: Sex tourism in the Caribbean. In S. Clift & S. Carter (Eds.), *Tourism and Sex: Culture, Commerce and Coercion* (pp. 41–53). Pinter.

Sanchez-Taylor, J. (2001). Dollars are a girl's best friend? Female tourists' sexual behaviour in the Caribbean. *Sociology*, *35*(3), 749–764. doi:10.1177/S0038038501000384

Sanchez-Taylor, J. (2006). Female sex tourism. A contradiction in terms? *Feminist Review*, *83*(1), 42–59. doi:10.1057/palgrave.fr.9400280

Selänniemi, T. (2003). On holiday in the liminoid playground: place, time, and self in tourism. In T. Bauer & B. McKercher (Eds.), *Sex and Tourism. Journeys of Romance, Love and Lust* (pp. 19–31). The Haworth Hospitality Press.

Shields, R. (1990). The 'system of pleasure': Liminality and the carnivalesque at Brighton. *Theory, Culture & Society*, *7*(1), 39–72. doi:10.1177/026327690007001002

Silva, A., & Blanchette, T. (2005). "Nossa Senhora da Help": Sexo, turismo e deslocamento transnacional em Copacabana. *Cadernos Pagu*, *25*(25), 249–280. doi:10.1590/S0104-83332005000200010

Silva, J. A. (2019, Jan. 3). O turismo no Algarve. Desafios em tempos de incerteza – Parte I. *Sulinformação*.

Singh, J., & Hart, S. (2007). Sex workers and cultural policy: Mapping the issues and actors in Thailand. *The Review of Policy Research*, *24*(2), 155–173. doi:10.1111/j.1541-1338.2007.00274.x

Sönmez, S., Apostolopoulos, Y., Theocharous, A., & Massengale, K. (2013). Bar crawls, foam parties, and clubbing networks: Mapping the risk environment of a Mediterranean resort. *Tourism Management Perspectives*, *8*, 49–59. doi:10.1016/j.tmp.2013.05.002

Thomas, M. (2005). 'What happens in Tenerife stays in Tenerife: Understanding women's sexual behaviour on holiday. *Culture, Health & Sexuality*, *7*(6), 571–584. doi:10.1080/13691050500256807 PMID:16864223

Tolman, D., Hirschman, C., & Impett, E. (2005). There is more to the story: The pace of qualitative research on female adolescent sexuality in policy making. *Sexuality Research & Social Policy*, *2*(4), 4–17. doi:10.1525rsp.2005.2.4.4

Trauer, B., & Ryan, C. (2005). Destination image, romance and place – an application of intimacy theory in tourism. *Tourism Management*, *26*(4), 481–491. doi:10.1016/j.tourman.2004.02.014

Truong, T. (1990). *Sex, Money and Morality: Prostitution and Tourism in Southeast Asia*. Zed Books.

Turner, V. (1974a). *Dramas, Fields & Metaphors. Symbolic Action in Human Societies*. Cornell University Press.

Turner, V. (1974b). *O Processo Ritual. Estrutura e Anti-Estrutura*. Vozes.

Van Gennep, A. (1960). *The Rites of Passage. A Classical Study of Cultural Celebrations*. University of Chicago Press. (Original work published 1909)

Wagner, U. (1977). Out of time and place: Mass tourism and charter trips. *Ethnos*, *42*(1-2), 38–52. doi:10.1080/00141844.1977.9981130

Weis, D. (2002). The need to integrate sexual theory and research. In M. Wiederman & B. Whitley (Eds.), *Handbook for Conducting Research on Human Sexuality* (pp. 7–24). Lawrence Erlbaum Associates.

Weitman, S. (1998). On the elementary forms of the socioerotic life. *Theory, Culture & Society*, *15*(3-4), 71–110. doi:10.1177/0263276498015003005

Wilson, E. (1975). *Sociobiology: The New Synthesis*. Harvard University Press.

Wilson, E. (1978). *On Human Nature*. Harvard University Press.

World Golf Awards. (2018). Retrieved from: https://worldgolfawards.com/winners/2018

World Travel Awards. (2017). Retrieved from: https://www.worldtravelawards.com/award-europes-leading-beach-destination

Zhang, H., & Xu, H. (2019). A structural model of liminal experience in tourism. *Tourism Management*, *71*, 84–98. doi:10.1016/j.tourman.2018.09.015

ADDITIONAL READING

Bauer, I. (2014). Romance tourism or female sex tourism? *Travel Medicine and Infectious Disease*, *12*(1), 20–28. doi:10.1016/j.tmaid.2013.09.003 PMID:24332659

Frohlick, S., & Johnston, L. (2011). Naturalizing bodies and places: Tourism media campaigns and heterosexualities in Costa Rica and New Zealand. *Annals of Tourism Research*, *38*(3), 1090–1109. doi:10.1016/j.annals.2011.01.012

Johnston, L. (2001). (Other) bodies and tourism studies. *Annals of Tourism Research*, *28*(1), 180–201. doi:10.1016/S0160-7383(00)00012-8

Lança, M., Marques, J. F., & Pinto, P. (2017). Liminality and the possibilities for sex and romance at an international bike meeting: A structural modeling approach. *Tourism & Management Studies*, *13*(1), 18–26. doi:10.18089/tms.2017.13103

Lett, J. Jr. (1983). Ludic and liminoid aspects of charter yacht tourism in the Caribbean. *Annals of Tourism Research*, *10*(1), 35–56. doi:10.1016/0160-7383(83)90114-7

Simoni, V. (2010). *Seducing bodies: dance-sex (dis)continuities in touristic Cuba*. Conference Proceedings of the 1st Biannual TOCOCU Research Network 'Tourism and Seductions of Difference'. Lisbon, 9-12 September 2010. Retrieved from: https://www.academia.edu/1511660/Seducing_Bodies

Spencer, A., & Bean, D. (2017). Female sex tourism in Jamaica: An assessment of perceptions. *Journal of Destination Marketing & Management*, *6*(1), 13–21. doi:10.1016/j.jdmm.2016.10.002

Wonders, N., & Michalowski, R. (2001). Bodies, borders, and sex tourism in a globalized world: A tale of two cities – Amsterdam and Havana. *Social Problems*, *48*(4), 545–571. doi:10.1525p.2001.48.4.545

KEY TERMS AND DEFINITIONS

Liminal Nature of Tourism: A spatial, temporal, mental, and sensorial border that is crossed when someone is touring, allowing to adopt completely different behaviors of that of the daily life.

Romance Tourism: The belief that in tourism context, men seek *tout court* sex, while women are more interested in love and romance.

Sex Tourism Paradigm: The tendency to mistake sexual relationships during tourism as "sex tourism" or "commercial sex tourism."

Sexual Double Standard: The belief the men and women should have different sexual behaviors.

ENDNOTES

1. In 2018, the Algarve received 4.7 million guests in tourism accommodation establishments of which 28.8% from Portugal and 23.5% from the United Kingdom. In 2013, when the sample for this study was calculated, the guests in the Algarve were likewise from Portugal (33.4%) and the United Kingdom (25.4%) (INE, 2014).
2. The categories derive from the literature review about the articulation between tourism and sex (Franklin, 2003; Marques & Lança, 2016; Selänniemi, 2003). The subcategories derive also from the literature review, but consist in the interviewees' report classifications. These specific themes often cover many different cases and reveal the core concepts and elements of the research (King & Horrocks, 2010).

Chapter 18
Electronic Games as a Touristic Promotional Tool

Fábia Esteves
Faculdade de Economia, Universidade do Porto, Portugal

Pedro Quelhas Brito
INESC-Tec, Faculdade de Economia, Universidade do Porto, Portugal

ABSTRACT

According to the World Tourism Organization (UNWTO), in 2017 tourism had the greatest international growth in seven years, and in 2018, international tourism grew 5% reaching the mark of 1.4 billion, a figure reached two years earlier than predicted. At the same time, in the last 40 years, the video game industry has grown steadily, with games beginning to be seen as one of the primary sources of entertainment. However, there are still few studies analyzing the impact of advertising tourist destinations on digital platforms such as video games. The use of video games in the tourist context may be an inspirational tool, supporting the development of new advertising strategies for tourism marketing. Although the connection between tourism and cinema is widely documented, little research has demonstrated a credible correlation between video games and tourists' attitude towards destinations.

INTRODUCTION

According to the World Tourism Organization (UNWTO), in 2017 tourism had the most exceptional international growth in seven years (UNWTO, 2018), and, in 2018, International tourism grew 5%, reaching the mark of 1.4 billion, two years earlier than predicted (UNWTO, 2019). At the same time, in the last 40 years, the video game industry has grown steadily in value and in public, with games beginning to be seen as one of the primary sources of entertainment for families (Simon, 2018).

However, there are still few studies analyzing the impact of advertising tourist destinations on digital platforms such as video games. Hence, the placement of destination images in electronic games is a topic that deserves attention and research since the subject is still in an initial state of research. Also, the use

DOI: 10.4018/978-1-7998-3156-3.ch018

of video games in the tourist context may be an inspirational tool, supporting the development of new advertising strategies for tourism marketing.

Although the connection between tourism and cinema is widely documented in the literature, little research has demonstrated a credible correlation between video games and the attitude of individuals towards tourist destinations. Traditionally, these two fields are scarcely connected in the literature, and only take into account the digital experience of tourist experiences (Huang, Backman, Backman, & Moore, 2013; F. Xu, Tian, Buhalis, & Weber, 2013).

This chapter intends to present a global vision on the theme of video games and tourist destinations, emphasizing the characteristics of video games, such as image and sound. At the same time, concepts present in the literature are presented, such as imagination and emotion, and their possible relationship with the attitudes of individuals towards tourist destinations.

VIDEO GAME PRODUCT PLACEMENT

The expansion and development of technology through content diversification and large-scale game development has led to the identification of the industry as a strategic area for significant investment. However, video games imply a considerable investment in research and development, regarding mobility (Storz, Riboldazzi, & John, 2015), design (Brandão et al., 2014; Norman & Verganti, 2014) and technology innovation (Beinisch & Paunov, 2005; Koch & Artmayr, 2019).

Video games are significant drivers of the economy (Crandall & Sidak, 2006; Simon, 2018), motivating marketers and developers in a constant search for new advertising and solutions on these platforms. This search happens due inclusion of advertising in video games being a way to reduce the overall costs associated with production, dropping the individual cost of each game (Nelson, Keum, & Yaros, 2004; Siemens, Smith, & Fisher, 2015; Simon, 2018; Vashisht & Royne, 2016).

The history of video games is closely connected to the placement of in-game advertising (IGA) a concept similar to the traditional idea of product placement, but using video games (Mau, Silberer, & Constien, 2008; Yang, Roskos-Ewoldsen, Dinu, & Arpan, 2006). Video games have evolved significantly over time, not only technologically, but also culturally and aesthetically, and this evolution is accompanied by IGA (Behm-Morawitz, 2017; Nelson, 2005; Verberckmoes, Poels, Dens, Herrewijn, & De Pelsmacker, 2016).

Product placement and advertising in video games are usual, and their study began in the early 21st century by Nelson et al. (2004). In general, besides making the game realistic, product placement allows its recognition through variables such as memory, attitudes, and feelings (Cauberghe & De Pelsmacker, 2010; Dardis, Schmierbach, & Limperos, 2012a; Glass, 2007; Lull, Gibson, Cruz, & Bushman, 2016; Martí-parreño, Bermejo-berros, & Aldás-manzano, 2017; Nelson, 2002; Nelson et al., 2004; Taghipour, 2017).

However, the literature concerning touristic advertisements focuses mostly on physical product placement in tourism-oriented environments (Weaver, 2007), on the impact of destination image placement on the media (Fu, Ye, & Xiang, 2016; Su, Huang, Brodowsky, & Kim, 2011; Tessitore, Pandelaere, & Van Kerckhove, 2014); virtual and augmented reality (Yung & Khoo-Lattimore, 2019) and in video games using a destination (Ballagas, Kuntze, & Walz, 2008; Dubois, Griffin, Gibbs, & Guttentag, 2020; F. Xu et al., 2013). It is still necessary to create knowledge regarding the impact of destination image and sound placement in video games.

Product placement is a complex construct. It involves variables such as size, placement, and the number of appearances, duration of the presentation, position on the screen, the relation between the placement and the game, and the placement mode. All of these variables are closely related to the memory of the user (Chaney, Hosany, Wu, Chen, & Nguyen, 2018; Dardis, Schmierbach, & Limperos, 2012b; Gross, 2010; Gupta & Lord, 1998; Law & Braun-LaTour, 2000; Lull, Gibson, Cruz, & Bushman, 2018; Schneider, Systems, & Cornwell, 2005; Terlutter & Capella, 2013). It is already known, for instance, that larger placement is more easily perceived than smaller ones (Chaney et al., 2018; Grigorovici & Constantin, 2004; Schneider & Cornwell, 2005). Besides, the location of the placement on the game screen affects information processing (Green & Bavelier, 2006; Jeong & Biocca, 2012; Peters & Leshner, 2013).

Regarding interactivity, a central position leads to greater recognition and memory, making these ads more effective than those more peripheral (Jeong & Biocca, 2012; M. Lee & Faber, 2007; Schneider & Cornwell, 2005). Also, the duration of exposure is directly related to the strength of the psychological reaction to IGA (Cauberghe & De Pelsmacker, 2010; Li, Edwards, & Lee, 2002; Molesworth, 2006; Redondo, 2012).

It is essential, however, to point out that the consistency of IGA about game content has positive effects on the sense of realism perceived by the players. And also on the logic of intrusion of IGA and subsequent attitudes and behaviors regarding the advertised products (Chang, Yan, Zhang, & Luo, 2010; M. Lee & Faber, 2007; Lewis & Porter, 2010; Nelson et al., 2004; Vashisht, 2019; Verberckmoes et al., 2016).

Hence, both the prominence and the consistency of IGA affect the memory and attitudes of individuals towards product placement.

INDIVIDUALS AND VIDEO GAMES

Escapism was introduced as an explanation for the use of entertainment media in the 50s. The concept was described as a temporary escape from every day-life onto the entertainment media world. Escapism suggests that media may meet the needs of individuals, distracting them from every day-life, through substitute gratifications (Katz & Foulkes, 1962). Escapism may be explained as a tendency to run away from a current situation or environment, and it is an essential motivation for entertainment consumption.

The concept focuses on the assumption that diversion and distraction from real-life circumstances are desires felt by many people in many situations, and that the entertainment media may serve the purpose of recreation very effectively (Klimmt, 2008). Although escapism is closely related to the uniquely human faculty of imagination, not requiring a specific negative situation to fix, it plays an essential role in breaking stagnation, often being a cure for boredom (Calleja, 2010; Singer & Pope, 2012; Yustisiana, 2018).

Video games are considered as escapist for enhancing the loss of the sense of time, but they are also sought by individuals because they are appealing entertainment. While other entertainment media are also attractive, the feedback cycle that games establish with the player increases the ability of players to engage because of their need for action (Calleja, 2010). Research suggests that individuals play video games as a form of escapism (Calleja, 2010, 2013; C. Y. Chen & Chang, 2019; Deleuze et al., 2019; Sherry, 2004).

This escapism associated with video games, according to Calleja (2010), happens for two reasons: First, video games propose a seductive unreality; second, they are perceived as games and distractions opposite to seriousness and work. Nonetheless, video games, while meeting most forms of escapist desire, are forms of escapism like any other engaging activity. That is if you consider video games a kind of

Figure 1. Reasons that lead people to play video games

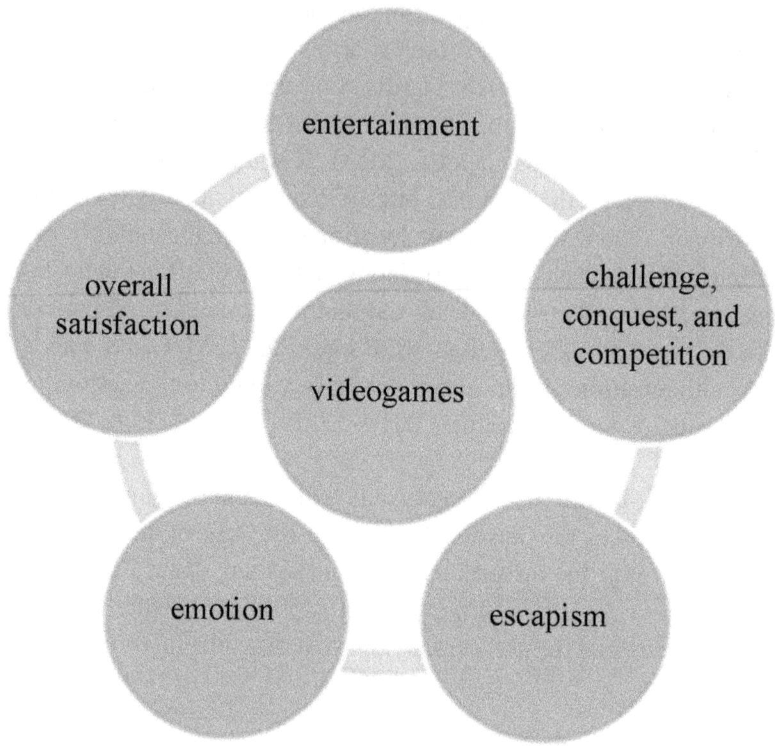

simulated experience, their most distinctive escapist trait is the ability to provide a variety of designed experiments (Calleja, 2010).

The reasons why individuals play video games go beyond escapism. The literature indicates that, generally, people use video games to satisfy the particular needs of pleasure and fun. This is the case of challenge, conquest and competition (Engl & Nacke, 2013; Sepehr & Head, 2018), the fun associated with the gaming experience (Kiili, 2005; Sweetser & Wyeth, 2005), the experience of emotion (Frome, 2019; Lucas & Sherry, 2004) and overall satisfaction (Allen & Anderson, 2018; Huang et al., 2013; Rieber, 1996).

On the other hand, positive experiences retrieved from games, such as interaction or fun (Choi & Kim, 2004; Hsu & Lu, 2004; Maitland et al., 2018), are a vital motivation for involvement in the activity. These positive experiences are associated with the loyalty to the game (Choi & Kim, 2004; M. Lee & Faber, 2007), to the game intentions (Hsu & Lu, 2004), and also have positive consequences on the mood of the individuals (Klimmt, Hartmann, & Frey, 2007; Villani et al., 2018).

Concerning the escapism itself, when combined with tourism, it is in no way a modern phenomenon, nor should it be a personal or social alienation (Smith, 2003).

Mental images or fantasies are used with the purpose of adaptive escapism. Yet this recognition often leads to over-concern about the more negative aspects of daydreaming or imaginative abilities. There are many situations where healthy escapism may be the most useful method to deal with reality. Thus, it is possible to adopt attitudes towards possible or impossible with impunity, through imagery (Singer & Pope, 2012).

The concept of Imagery, derived from the word 'image' (Yustisiana, 2018) is as a form of information processing, including conditioned sensory representations (images) in the working memory, which are used in the same way that the perceptions of external stimuli (Goossens, 2000). Imagery can influence individuals, increasing their satisfaction (MacInnis & Price, 1990), positively impacting their attitudes (Nadkarni, Hasbach, Thys, Crockett, & Schnacker, 2017; Rossiter & Percy, 1980), and developing positive attitudes towards consumer products (W. Lee, Gretzel, & Law, 2010; Miller & Stoica, 2003a; Rossiter & Percy, 1980).

Consumers tend to develop and use detailed imagery during the period immediately before vacations, making mental imagery an essential part of vacation anticipation (Garrod, 2009), especially for consumers looking to invest more time and money in the activity (MacInnis & Price, 1990), and more probably the consumers will choose destinations that implicated or stimulated imagery (S. B. Kim, Kim, & Bolls, 2014).

VIDEO GAMES AND TOURISM MARKETING

Matos, Mendes, and Pinto (2015) describe five stages of the tourism experience: pre-visit, travel to the destination, destination, homecoming trip, home. In all these stages, the imagery of individuals can be worked on, which is particularly important in the early stages.

To use video games platforms as a way of advertising destinations, several entities responsible for Tourism Marketing in their cities and countries have already developed video games about their respective places.

A cultural heritage discovery video game – 'REXplorer' - the first mobile game of a city in the world, available in Regensburg, Germany, aims to extend the experience of the Regensburg Museum throughout the city, making it a museum as well. Players enjoy the game via a GPS signal, making their tourist experience more fun, engaging, and educational ("Regensburg Experience," 2019). This game is based on a location, offering a way to experience points of interest.

The Tourism Authority of Thailand issued a series of interactive games with the brand 'Smile Land', incorporating images based on the national icons of the country, allowing the players to experience Thailand in a fun way. Over 200 points of interest can be found in the games, through the process of transformation onto a virtual traveler in Thailand (Tourism Authority of Thailand, 2011).

In 2011, Tourism Ireland issued a game on Facebook, intended to capitalize on the popularity of games such as FarmVille and CityVille. The tool, considered promotional, challenged players to create their idyllic town in the country. Upon completion of the tasks (such as recommending the game to friends), players could win a holiday in Ireland. Mark Henry, marketing director at Tourism Ireland, stated: "*Given that Tourism Ireland currently has a fanbase of a quarter of a million people around the globe, and each of the fans has, in turn, an average of 130 friends, Tourism Ireland can potentially engage with 32.5m people worldwide through our new Ireland Town game*" (Silicon Republic, 2011).

Cape Town, in South Africa, used the Facebook platform to publicize the city as a tourist destination through a video game released in September 2012. Thousands of people played the interactive game in which the first prize was a trip with all expenses paid to Cape Town. The second prize was a week vacation to a domestic destination and the third prize long-haul flights. With this strategy, reservations in Cape Town increased by 118% (Netimperative - latest digital marketing news, 2013).

There are several video games that, although not based on the promotion of a specific tourist destination, end up presenting the locations. This is the case of the game Grand Theft Auto and Assassin's Creed where several real locations are used as the game scenario. In these games, it is possible for players to digitally experience places like the Bixby Canyon Bridge, the Watts Towers, or even the city of Paris.

The use of video games can benefit tourism marketing. Developing a game design based on real environments can potentially increase destination recognition and attract tourists. Video games share similar traveling motivation elements to those identified in the movies and should be considered a tourism engine. Tourism institutions responsible for destinations should ponder video games as a platform to motivate tourists (Dubois & Gibbs, 2018). Before the visit, individuals form images of the destination by the information collected in an organic or induced way (Govers, Go, & Kumar, 2007; Gunn, 1988; H. Kim & Stepchenkova, 2015; Nadkarni et al., 2017).

Since imagery contributes to the decision-making process, it is also at this stage that mental images play a central role (MacInnis & Price, 1987, 1990). In pre-visit, the experience begins through the dreams and expectations generated as to what will be experienced and lived at the destination. Also, in the first stage of pre-visit, factors as the media and other stakeholders shape the information sources searched by the consumers and available to be processed by them (Matos et al., 2015).

When a product is advertised in a video game, it is more than just an advertisement. Product placement increases the credibility and reality of the scene or environment in which it is inserted (H. Chen & Haley, 2014), blending in with the background (Lehu, 2007). The rapid increase in the quantity and variety of product placement approaches allows further study of this technique in different situational contexts (Tiwsakul, Hackley, & Szmigin, 2005). In this context, product placement in tourism marketing is a singular topic because, unlike other markets, it does not need to be noticeable, allowing resistance decrease from individuals.

The lack of resistance of individuals to product placement may be due to the ignorance of such placement, which allows an unconscious influence. However, when they become aware of that placement, they are also able to decide whether to resist the purchase intent or not (Tessitore & Geuens, 2013). In the case of placing tourist destinations in video games, the player's lack of knowledge of the reason for the game scenario allows for an apparent lack of resistance.

STARTING THE RESEARCH

Among the different topics that could be addressed, within the scope of destination image placement, in video games, it was decided to start with three variables: (1) Typology of video game image | Imagery and Imaginary, (2) Presence of people in the game environment | The individual and emotion, (3) Background music in video games | Music and emotion.

The ongoing research aims to understand the topic in question. To analyze the chosen variables, a video game was developed. It takes place in a tourist destination (Southeast Asia), with three manipulations: Image typology (photographic vs. illustration), Melody typology (typical of video games vs. typical of calm destination vs. typical of festive destination), and Presence of people in the game scenario (with the presence of people vs. without the presence of people).

Imagery and Imaginary

Imagery was defined by Lutz and Lutz (1978) as "*a mental event involving visualization of a concept or relationship*" (p. 611). Mental images are a way of processing information, defined as "*(1) a process (not a structure) by which (2) sensory information is represented in working memory.*" (MacInnis & Price, 1987, p. 473). Imagery allows individuals to retain in their consciousness past experiences, possible future actions, as well as the present moment. The ability to imagine inspires the emotions and passions of individuals, motivating their lives (Calleja, 2010).

Destination marketing needs to develop the imagery of individuals at all stages of consumption due to the intangibility that characterizes tourism and services in general. Individuals cannot experience the so desired vacation before visitation. Thus, imagery becomes an essential element during the decision-making process, during consumption, and after the experience (Matos et al., 2015).

Tourist destinations should have their image built as such. Therefore, the essential qualities of the site must be molded into images that are attractive to individuals (Jenkins, 1999; Tasci & Gartner, 2007; Tasci, Gartner, & Cavusgil, 2007). Imagery is one of the significant human mental events as it allows remembering, planning, and making decisions for the future (Pearson, Naselaris, Holmes, & Kosslyn, 2015). Imagination offers individuals the possibility to distance themselves temporarily from the present moment, a defining element of escapism.

Many cases of escapism are merely ways to exercise imagination abilities (Calleja, 2010). The use of imagery in the processing of advertising messages increases the emotional response of the consumer to the ad, as well as the ad and brand evaluations (Bone & Ellen, 1992). When individuals engage in mental image processing, they experience an approach to destiny through imagination (W. Lee & Gretzel, 2012). Previous and newer research on imagery has provided evidence that their processing can positively influence consumers decision outcomes (Babin & Burns, 1997; Ha, Huang, & Park, 2019; Park & Yoo, 2020).

The concepts of imagery and destination image coexist and are positively related (Josiassen, Assaf, Woo, & Kock, 2016). In the long run, the individual must have a destination image that matches the real characteristics. Otherwise, he will not be satisfied after the visit (Fakeye & Crompton, 1991; Tasci & Gartner, 2007), will not recommend the destination in his word-to-mouth, and will not return to the site (MacKay & Fesenmaier, 1997; Milman & Pizam, 1995). However, it is widely recognized that the correspondence between the disclosed and actual image need not be entirely accurate. Tourism marketing tends to be most effective when tangible essential qualities of the destination are merged with images of exoticism, paradise, and dream fulfillment (Tuohino & Pitkänen, 2004).

In the literature, there are three main ways of obtaining mental images: pictorial stimuli, particular verbal stimuli, and imagery instructions (Burns, Biswas, & Babin, 1993; Miller & Stoica, 2003b; Paivio, 1971). In the manipulation of pictorial material (an image or an illustration), there is a two-dimensional illustration, including at least one element non-alphabetic, non-numeric, and nonarithmetic (Lutz & Lutz, 1978).

In 1969, Paivio emphasized the research between concrete images and abstract images, concluding that concrete images are more effective in generating memories than abstract ones. Babin and Burns (1997) compared the effects of concrete and abstract images. The authors defined concrete images like those in which the object is easily identifiable, and abstract images like those in which the object is not so easy to identify.

Figure 2. Imagery Creation

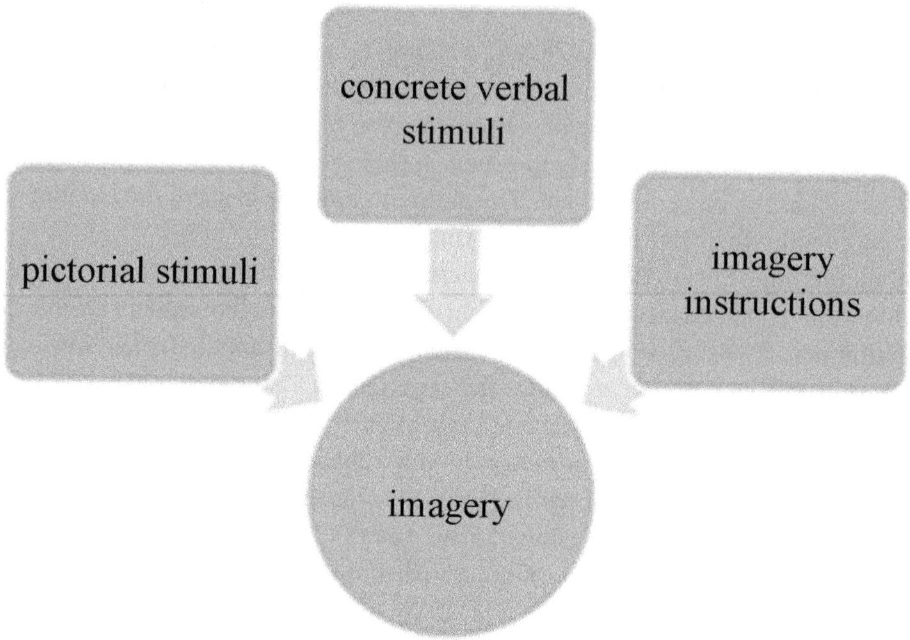

They concluded that an ad containing a concrete image of a product is more effective in stimulating vivid processing of visual images, favorably influencing attitude towards the ad and the brand when compared to an advertisement with a considerably less concrete image or one without Image.

The use of photographic and artistic images continues to attract the attention of the authors. Miller and Stoica (2003b) addressed the impact of photographic and artistic images in the advertising context, concluding that if advertisers wish to draw attention to a print ad, a vivid artistic interpretation with product-related information might be quite effective.

However, if the purpose of the ad is to draw the public to a website through a mental imagery strategy, photographs should be preferred to abstract artistic representations. Photographs produce more favorable attitudes and behavioral intentions. Similarly, Walters, Sparks and Herington (2007) concluded that the presence of concrete images in the dissemination of tourist destinations contributes to the elaboration and quality of the perspective of consumption of individuals.

For this study, the researchers created a fictional image of an illustrative Southeast Asian beach as a publicity stimulus for a video game. In the study, the photographic image representing the region was, in the first stage, developed by designers. The construction characteristics of the image should be appropriate for use in a video game. Subsequently, the image was processed through illustration techniques by an illustrator. Ultimately, two images were obtained, exactly alike, one being a digitally manipulated photography, and the other one an illustration.

Two groups of college students, aged between 17 and 25 years, participated in the study. The first group with 51 elements viewed the photograph while the second group with 50 elements viewed the illustration. The image was not at any time identified as illustrative of Southeast Asia. Of the sample, 29% of the individuals identified the image as representative of Southeast Asia, but there was no statistically significant difference in destination identification between the photography and illustration groups.

Figure 3. Destination Presentation - Photographic vs. Illustrated

Other countries also commonly identified as beach destinations were identified as possible destinations: Hawaii (14%), Caribbean (14%), Brazil (9%) and Maldives (8%).

As expected, there were significant statistical differences between photography and illustration, demonstrating that photography is considered more realistic. Also, it was considered more attractive as a tourist destination, in terms of design and as a whole.

Concerning imagination and the mental image of the individuals, the sample was questioned through an open-ended question of what comes to mind while watching the image. Through the analysis of the answers, seven new binary variables were created: Positive Feelings; Sensory stimuli; Human Characterization of the Site; Material Characterization of the Site; Activities; Family Company; Company of Friends. Examples of features identified as imagination:

- Positive Feelings: "*The idea of relaxation, peace, happiness, and tranquility*","(...) *Tranquility and immediate desire to go there to visit, to relax and rest.*"
- Sensory Stimuli: "*A calm environment where we can only hear the sea and animals.*", "*I imagine the sound of the waves, of the seagulls, the smell of saltwater.*"
- Human Characterization and/or physical location: "*A hotel room with a heavenly view, a gourmet breakfast and around me mostly tourists*"; "*Nature in its pure state with no human interference.*"
- Vacation Activities: "*In my view, the water must be warm, suitable for a few dives, and also by boat, it would be possible to reach deeper waters.*"; "*Comes to mind the feeling of calm, freedom, curiosity to know and explore the small islands.*"
- Company of Family and/or friends: "*relaxation; to be relaxed on the beach; socializing with friends.*"

However, no statistically significant difference was detected in the mental characterization of the images, photographic or illustrated.

Individuals and Emotion

Research on emotions started many years ago, and the Basic Emotion Theory of Ekman (Ekman et al., 1987) was a breakthrough. One of the main ways Ekman (1992) used to distinguish an emotion from another was through the facial expression of the individual. Evidence showed that there are distinct universal facial expressions for the five basic emotions: anger, fear, pleasure, sadness and disgust.

The evidence is based on the high agreement between literate cultures on the labeling of the meanings of these expressions and studies on the actual expression of deliberate and spontaneous emotions, as well as on the association of expressions with the interactive social context. Addressing the physiological level, there is evidence of distinctive patterns of activity of the nervous system for anger, fear and disgust. Besides, according to the author, there may be a distinct pattern for sadness. However, new developments and evidence have challenged this theory, especially the statement that emotions are distinct phenomena, easy to produce, and easily recognizable (Barrett, 2006; Barrett, Mesquita, & Gendron, 2011; Russell, 2003). However, recent approaches propose that different emotions do not have singular and unique manifestations in the facial expressions of individuals (Mesquita, 2010; Russell, 2003).

Facial expressions are often classified as observable indicators of the underlying emotional state of individuals (Ekman, 1993; Tarnowski, Kołodziej, Majkowski, & Rak, 2017), which may subsequently influence the observer (Marsh, Ambady, & Kleck, 2005; Stins et al., 2011). For example, there is a high consistency that a smiling person is in a good mood (Reisenzein, Studtmann, & Horstmann, 2013). However, two different perspectives explain why facial expressions evolved as a form of nonverbal communication. The traditional and predominant hypothesis sees facial expressions as statements of internal emotional states of individuals. Also, some cross-cultural studies support the theory of universality, suggesting that some emotions are universally perceived through facial expressions (Ekman, 1993; Izard, 1994).

Smiling facial expressions are used as signs of happiness since the association of smiles with positive affective states encourage favorable perceptions from the observers. The demonstration of certain emotional expressions leads other individuals to attribute specific characteristics to those who express those emotions (Hareli, Shomrat, & Hess, 2009; Tarnowski et al., 2017). However, facial expressions have a cultural and contextual component that mediates their influence. Although emotions portrayed through facial expressions may not be universal, there is a similar impact of these expressions, especially in developed societies. Thus, facial expressions of individuals have a high social and emotional value and is an significant component of social interaction (Crivelli & Fridlund, 2018; Crivelli, Jarillo, Russell, & Fernández-Dols, 2016; Crivelli, Russell, Jarillo, & Fernández-Dols, 2017, 2016).

Social presence, analyzed through the degree to which an individual is aware of another person, has been studied in different contexts, particularly in websites use, explaining the positive effects of the presence of images of people (Hassanein & Head, 2007). People smiling are widely used by marketers to produce positive impressions on individuals. The smile is an already established strategy, in the relation with individuals (Y. H. Lee & Lim, 2010), and smiling faces have long been omnipresent in ads (Petroshius & Crocker, 1989; Wang, Mao, Li, & Liu, 2017).

Marketers have long been supporting images of "friendly faces" to create a positive attitude toward products since the images of people create a cozy and friendly atmosphere and induce emotional responses (Riegelsberger, 2003; Riegelsberger, Angela, & McCarthy, 2003; Salo & Karjaluoto, 2007). Also, the perception of the social presence at a digital level can positively influence the trust and the intentions of the user in an online context (Gefen & Straub, 2003). Besides, the perception of high levels of social presence positively affects the perceived usefulness, reliability, and enjoyment of shopping websites, leading to more favorable attitudes by consumers (Hassanein & Head, 2007). Dormann (2001) also suggests that image effectiveness, through an emotional or social display, maybe a key factor for the success of the digital activity.

Emotional contagion may be defined as: "*the tendency to mimic the verbal, physiological, and/or behavioral aspects of another person's emotional experience/expression, and thus to experience/express the same emotions oneself*" (Hsee, Hatfield, Carlson, & Chemtob, 1990, p. 328). This primitive emotional contagion process referred to as the mood of an individual infects other individuals via an automatic process (Neumann & Strack, 2000) and develops through a continuous process of (1) Mimicry, (2) Feedback and (3) Contagion, which drives the emotion in the affected individual. The three identified processes are visceral, automatic, and unconscious psychological mechanisms, defined as follows (Hatfield, Cacioppo, & Rapson, 1993):

1. Mimicry: It is an automated process, almost instantaneous and unconscious, to imitate the emotions expressed by another individual. Although mainly induced by facial expression, tone, and voice posture also influence the process.
2. Feedback: Addresses the fact that individuals tend to have sincere feedback on expressed emotions. In other words, forcing a facial expression of a particular emotion may convey that emotion.
3. Contagion: Individuals draw inferences about their emotional states based on emotional expressions and behaviors evoked in them by the emotional state of others.

However, emotional contagion is not limited to direct interaction with humans. It extends to the expression of emotions conveyed through stills images of human facial expressions (Wild, Erb, & Bartels, 2001), video clips (Hess & Blairy, 2001), robots (J. Xu, Broekens, Hindriks, & Neerincx, 2015) and even social networks (Kramer, Guillory, & Hancock, 2014).

Emotional contagion thus allows the visualization of facial expressions to affect a person's mood. If they are expressionless, facial expressions will not induce any mood swings (Friedman & Riggio, 1981). Advertising uses images of people to create positive attitudes toward products or brand trust (Riegelsberger et al., 2003). Although tested on many websites, neither the presence of photos nor the reliability of a portrayed person, had a significant effect. The presence of photos reduced the ability of participants to identify suppliers with good and bad reputations.

The perceived reliability of underperforming vendors increased while that of reputable vendors decreased (Riegelsberger et al., 2003). The positive effect of emotional contagion can have a favorable bias effect on the attitude of individuals towards products. That is, individuals who experience a positive (happy) effect tend to have more favorable attitudes toward products than those with neutral humor (Batra & Stayman, 1990).

The use of emotional agents in video games intends to make the experience of the users more engaging and funny. The virtual characters that during the video game express positive emotions when the individual wins a game, and the negative emotions when they lose are perceived as more affectionate, pleasant, reliable, and supportive (Brave, Nass, & Hutchinson, 2005). It was also identified that during a video game, when a "virtual human" is empathetic regarding the individual, it is generally perceived more positively than the agent that is not empathetic regarding the individual (Brave et al., 2005; Hone, 2006; Klein, Moon, & Picard, 2002).

Returning to the images developed, images of people were included in the fictional scenarios: photographic images and illustrated images; the images of people were produced by designers, in the first stage, to develop the photographic image, and subsequently, an illustrator developed the illustration. A sample group composed of 101 college students aged 17 to 25 participated in the study. The first group with 50 elements watched the image with the presence of people, while the second group with 51 ele-

Figure 4. People in the game environment - Absence vs. Presence

ments watched an image without the presence of people. The image was not identified as illustrative of Southeast Asia.

The sample group was asked to identify the depicted destination. There were statistically significant differences (significance value if 0.018) with the manipulated images. The presence of people in the image decreased the identification of the scenario as belonging to the Southeast Asian region. That is, the presence of people in the scene diminished the perception of individuals regarding the portrayed destination.

The sample was also requested to describe what came to mind by watching the image, freely through an open question. Based on the analyzed responses, it was possible to create seven new variables: Positive Feelings, Sensory Stimuli, Human Characterization of the Site, Material Characterization of the Site, Activities, Family Company, and the Company of Friends.

The Chi-Square Test analyzed the association between the two conditions under study and the answers through the new variables. For the condition 'people', there was a significant association for the variables Sensory Stimuli (significance value 0.006); Material Characterization of the Site (significance value 0.007); Family presence (significance value 0.000); Friends presence (significance value 0.007).

1. Sensory Stimulation: In the condition of the presence of people in the image, sensory stimuli were reported 6 times. However, for the image with no presence of people, this number increased to 18. This validates the assumption that the presence of people in the presented scenario decreases the perception regarding sensory stimuli. Some examples of sensory stimuli identified were *"The sound of the sea and the wind. Laughter and people talking." "(...) Fine sand and very hot."*
2. Material Characterization of the Site: The material characteristics of the represented site were referred to by 60 elements from the sample, whereas in the group with the condition "presence of people", material characteristics were reported 23 times. In the image with no presence of people, this number increased to 37, validating that the presence of people in the image decreases the perception of the material characteristics of the presented scenario. Some examples of material characterizations of the place include: *"As I watch the images I imagine a heavenly destination with a long sandy beach that ends with the sea connection that remains calm.", "Incredible landscapes and a beautiful beach."*

3. Family Company: The reference to the company of family happened 15 times, 93% of which in the group with the presence of people in the image. It thus validates that in the presented scenario, the presence of people increased the family reference. Some examples of family company identification include: "*A good place to relax, but also to bring the family.*", "*I watch several people taking a vacation on a paradise island, resting with their family, and taking pictures.*"
4. Company of Friends: The reference to the presence of friends was reported 13 times, 85% of which in the group whose condition was the presence of people. It thus validates that the presence of people in the image increases the reference to the presence of friends. Some examples of references to friends include: "*comes to mind moments with friends (…) that during class or work are not so frequent*", "*I immediately imagine a relaxed moment with family or with friends.*"

In the advertising media, happy people generate positive attitudes from individuals, creating a warm and friendly environment, thus inducing positive emotional responses (Riegelsberger, 2003; Riegelsberger et al., 2003; Salo & Karjaluoto, 2007). Hence, the aspect of the people portrayed in the image intended to suggest happiness and well-being. To validate this feature, the perception of the sample group regarding the people from the image was analyzed. The analysis was performed by sampling 50 individuals and presenting them 4 questions on a Likert scale between 1 and 7:

1. Happiness – I find the people portrayed in the image are: Not Happy (1) - Very Happy (7)
2. Attraction – I find the people portrayed in the image are: Not Attractive (1) - Very Attractive (7)
3. Friendliness – I find the people portrayed in the image: Not Friendly (1) - Very Friendly (7)
4. Beauty – I find the people portrayed in the Image: Not Beautiful (1) – Very Beautiful (7)

An Exploratory Factor Analysis was performed to analyze the items of perception of people's appearance in the scenario, with the extraction of factors using the Principal Component Analysis method. With a value of Kaiser-Meyer-Olkin (KMO) equal to 0,760, regarded as average (Marôco, 2011) and Bartlett's test for sphericity with a p-value <0.001, indicating that data is suitable for Factor Analysis. One component was identified, pulchritude, with eigenvalues greater than 1, comprised of the four variables, with a Cronbach's Alpha of 0.822 and total variance explained of 65,92%.

The new variable, with an average of 5.5, when compared between the two types of image (photographic and illustrated), demonstrates there are no statistically significant differences between the averages of the two groups. The characterization of the people in the images is similar, regardless of whether they are photographic or illustrated. This is very important because the intention to use the images in a video game requires that the perception of the people included in the scenario is similar in both types of images.

Music and Emotion

Human beings are in constant communication through sound, allowing sound to be always active and easily reached. Besides, sound and music can stimulate emotions. The challenge is now to select the emotions and feelings one wishes to convey (Daj, 2013). Although individuals attribute meanings to sound, especially to music, as sensory expressions (Hou, Zhao, & Zheng, 2019; Hultén, 2011), music is often seen only as an effective stimulus, neglecting its potential as an information vehicle. However, music conveys meaning to individuals, which is possible to analyze beyond the affective qualities and may even contribute to the purchase (Hou et al., 2019; Kotler, 1973).

One of the main issues in research, concerning music and emotion, is the genuine emotional responses from listeners - emotivist position – or if listeners only perceive the emotions expressed by music - the cognitivist position. In their study, Lundqvist, Carlsson, Hilmersson and Juslin (2009) measured the self-reported emotion, the facial muscle activity and the autonomic activity of the participants while listening to happy or sad music, showing that happy music generates more happiness and less sadness than sad music.

The finding that the induced emotion in the listener was the same as the emotion expressed in music, is consistent with the notion that music can influence emotions through a process of emotional contagion. One of the most cited reasons for listening to music is that it induces strong feelings, being used by listeners in many different emotional ways in various life contexts (Juslin & Laukka, 2004). However, the same person, in different market segments, may have different emotional experiences for the same song (Hou et al., 2019)

Musical coherence is also an essential aspect of the behavior of individuals. It affects consumer behavior both in a non-conceptual way, as it affects the cognitive preparation of semantic networks in memory, and operationally, when, for instance, consistent with the country of origin of a product or with the connotations of the utilitarian or social identity of a product. Specific genres of music (e.g., classical) activate related concepts in memory (e.g., expensive, sophisticated, formal, polite), which influence memory, perception, and the choice of products. The presence of a specific genre of music during the initial product exposure improves the subsequent recall of related (congruent) products when compared to unrelated products. It also affects how much participants are willing to pay for congruent products (North, Sheridan, & Areni, 2016). A common conclusion is that the congruence between the type of music and certain products can influence the perceptions of individuals and the respective sales of the product (North et al., 2016).

Music can be an influential tool for humor and perception of individuals, making products and brands more attractive. In the study presented by North, Hargreaves, & McKendrick (1999), French and German songs were used to understand their effect on the decisions of individuals concerning wine purchase. The study demonstrated that when French music was playing, the French wine had an increase in sales and similarly for German music and wine.

Music is essential regarding the influence it has on individuals. These are not only capable of understanding the feelings conveyed by music, but also experience authentic emotions associated with physiological, behavioral, and subjective changes (Krumhansl, 1997; Lundqvist et al., 2009). At a cognitive level, empathy plays a vital role in the emotions induced by music, as is the case of emotional contagion (Overy & Molnar-Szakacs, 2009) or cognitive empathy (Scherer & Zentner, 2001).

Cognitive empathy may be one of the main ways by which music induces emotions (through the central nervous system). Based on the emotion conveyed by the artist (tone of voice, facial expressions, gestures), individuals perceive the emotions expressed by the artists, and feel safe to deliberately respond with similar emotions (Scherer & Zentner, 2001).

In the cinema, music can influence individuals' sympathy for characters. Besides, it is considered an antecedent of concern and empathic accuracy towards the character (Hoeckner, Wyatt, Decety, & Nusbaum, 2011). Regarding sound in video games, individuals have now abandoned passive participation in the video game and want to be active participants.

Thus, popular video games should lead to an immersive gaming experience to achieve a particularly powerful influence on the actions and attention of the player (Zhang & Fu, 2015). Although image and video are of great importance to video games, it is essential to acknowledge that audio also plays a significant role (Collins, 2009; Williams, Cowling, & Murphy, 2019; Zhang & Fu, 2015).

Sometimes the music implemented in a video game is not directly related or highly relevant to the game, intending only to support the narrative through a dimension of affection. In this case, a less responsive approach to the melody in the video game may be preferable (Nispen, Huiberts, Donders, & Koch, 2011). It is important to note, however, that the inclusion of audio in the video game enables the players to improve the emotional response through interactivity, such as accentuating danger, success, failure, and other elements of the game (Berndt, 2010).

The tested condition was the "Music" associated with the scenes, on a sample group of 47 elements composed of college students between 18 and 34 years old. Two typical Southeast Asian melodies were selected one quiet and another festive. The group watched the video game with no associated control as if it were a movie. Each element watched the film in a double sequence, randomly, with different melodies in the first and second presentation, a quiet melody and a festive one. 57% of the sample considered the quiet melody more suitable for the image, and only 43% chose the "Festive" melody.

To analyze the reasons that led to the identification of a particular melody as more appropriate to the image, the elements of the sample group answered an open answer question concerning the reasons for their response. The group indicated the characteristics of the melodies as the reason leading to the choice. Further analysis of the answers allowed identifying the interpretation of the decisions associated with concepts such as:

1. A feeling of tranquility, calm and/or relaxation
2. A feeling of joy, fun and/or energy

The characterization of the melodies as "quiet" and "festive" was corroborated by the analysis of the answers to the open question. Through the Pearson's correlation coefficient (0.002), it is possible to conclude there is a significant association between the type of music considered as "quiet" and the identification of feelings of tranquility, quiet and/or relaxation. Of the 22 elements of the sample that identified feelings of tranquility, calm and/or relaxation, 18 did so for the "quiet" melody.

Figure 5. Melody Placement in the video game

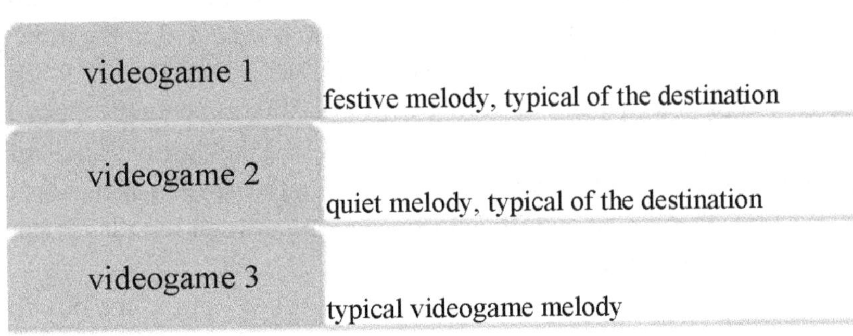

The same happens with the melody considered "festive". Pearson's correlation coefficient of 0.000 allows concluding there is a significant association between the type of music and the identification of feelings of joy, fun and/or energy. Of the 15 elements that identified feelings of pleasure, fun and/or energy, all of them did so for the "festive" melody.

In a subsequent test, and with a sample of 146 elements, college students from 17 to 28 years old tested the condition "Music" associated with the video game. Since the game was short, with a short learning period, it was conceived and designed by the researchers to be easy to understand and play. The sample group was questioned on a Likert scale from 1 to 7, regarding the ease of the game, being 1 "easy" and 7 "difficult". Of the 146 elements in the sample, the response mode was 1, with an average of 2.2.

The game takes place in the developed scenario, where a player-controlled plane will have to pick up a series of balloons that will award points. No obstacle alters the operation of the game, and there is no way to lose. That is, the game will progress equally regardless of the performance of the player. In this way, it was possible to ensure that the sample group elements were exposed to the same conditions during the same period. For this purpose, the game was presented in three distinct strands in the music condition:

– Typical video game melody: 49 sample elements
– Typical destiny melody - Quiet: 49 sample elements
– Typical destiny melody - Festive: 48 sample elements

To analyze the internal consistency of the "player type" in the sample group, the Cronbach's alpha of the four questions was measured through a Likert scale of 0 to 7 points, characterizing the player:

1. I am extremely competent/ able to play video games.
2. I have an in-depth knowledge of video games.
3. Compared to other players, I consider myself superior in video game playing
4. When my friends need to know more about video games, they ask me

With a Cronbach's alpha value of 0.874, there was a good internal consistency of the factors (<0.7) and the variable "Expertise" was created. Applying the Levene's test allowed revealing the homogeneity of variances between the three groups: (1) Video game Music, (2) Festive Destination Music and (3) Quiet Destination Music. It was concluded that the variances are equal in the three groups since the significance associated with the test is higher than 0.05 [F (2.143 = 0.053), p=0.949].

After the game, the sample was questioned about which destination/country they identified with the game scenario. Unlike earlier, where 29% of the sample identified the destination displayed in the static image as Southeast Asia, only 17% of the sample identified Southeast Asia as the destination portrayed in the video game. Besides, there is no significant association between the identification of Southeast Asia and the melody of the game, since $p > 0.05$.

The sample group answered a question concerning the adequacy of the melody to the video game through a Likert scale: I consider the melody 1 - Poorly suited to the video game - 7 - Very suitable to the video game. It intended to assess the homogeneity of the variance between groups - Video Game Music, Festive Destination Music, and Quiet Destination Music. The Levene's test allowed concluding that the variances are similar in the three groups since the significance associated with the test is higher than 0.05 [F(2.143=2.386), p=0.096]. These results allowed us to analyze the melodies through

ANOVA, recognizing there are statistically significant differences regarding the melody identified as more appropriate for the video game.

The subsequent application of the Tukey's test allowed identifying a statistically significant difference between the averages of the melodies: typical of video games (5.02), and quiet (4.04). The melody typical of video games was considered more suitable to the video game in question. The adequacy of the festive melody did not present statistically significant differences when compared with the other two melodies under analysis, with an average of 4.56.

CONCLUSION, LIMITATIONS, AND IMPLICATIONS FOR THE FUTURE OF DESTINATION MARKETING

Three conditions were analyzed to assess their impact on tourist destination images. The image typology used did not obtain significant differences in the analyzed points. Regarding the presence of people on destination scenarios, this resulted important concerning the imagination associated with the presented scenario.

It was possible to demonstrate that the presence of people triggered the imagination associated with friends and family. This may be of great interest for destinations targeted to families or groups of friends looking for a place of gathering and leisure to enjoy their holidays ("*I immediately imagine a relaxed moment with family or friends.*").

On the other hand, it was also possible to conclude that the presence of people in the destination scenario decreases the identification of sensorial stimuli and the material characterization of the site, reducing the emphasis they receive. Regarding the insertion of melodies in touristic video games it was possible to conclude that the preference between melodies changes when moving from a mere destination video to an interactive video game, even though the image and duration are the same.

In the analyzed case, at the time of video presentation, 57% of the sample considered the Quiet Melody more appropriate to the picture. However, in the analysis with the same melody in the video game, the Quiet Destination Melody obtained a lower classification when compared to the Festive Melody. Background music in video games has proven to be an indispensable part of their success. In addition, video game designers and gamers expect video games to be as engaging as possible (Zhang & Fu, 2015).

According to Nelson (2002), product placement in a video game differs from placement in movies or television, regarding their ability to increase interactivity and engagement. Hence, the existence of differences between the preferences evidenced between the video and the video game may be analyzed in the light of interactivity.

Concerning the motivation to travel, video games share similar elements with the movies and should be considered a factor to encourage tourism. It is essential for destination marketing to realize that video games are a platform of dissemination for tourists (Dubois & Gibbs, 2018). Since the literature is yet very scarce, it is necessary to develop further research crossing two realities: destination marketing and video games.

The video game industry has grown in value from $ 70.6 billion in 2012 to $ 137.9 billion in 2018. This growth should reach $ 180.1 billion by 2021 (Statista.com, 2019). Therefore, it may be assumed that video games will remain in society and that destination-marketing strategies should be developed to take advantage of this market.

Concerning the limitations of this study, it is an early study still under development, and focusing only on one tourist destination, through a convenience sample. The video game use was only assessed for the manipulation of "music", reducing the possibility to analyze the two other manipulations - the type of image and presence of people.

There should be future developments and updates of this work, ensuring broader conclusions and thus making a more extensive contribution to the literature connecting video games and destination marketing.

REFERENCES

Allen, J. J., & Anderson, C. A. (2018). Satisfaction and frustration of basic psychological needs in the real world and in video games predict internet gaming disorder scores and well-being. *Computers in Human Behavior*, *84*, 220–229. doi:10.1016/j.chb.2018.02.034

Babin, L. A., & Burns, A. C. (1997). Effects of Print Ad Pictures and Copy Containing Instructions to Imagine on Mental Imagery That Mediates Attitudes. *Journal of Advertising*, *26*(3), 33–44. doi:10.1080/00913367.1997.10673527

Ballagas, R., Kuntze, A., & Walz, S. P. (2008). Gaming tourism: Lessons from evaluating REXplorer, a pervasive game for tourists. Lecture Notes in Computer Science, 5013, 244–261.

Barrett, L. F. (2006). Are Emotions Natural Kinds? *Perspectives on Psychological Science*, *1*(1), 28–58. doi:10.1111/j.1745-6916.2006.00003.x PMID:26151184

Barrett, L. F., Mesquita, B., & Gendron, M. (2011). Context in emotion perception. *Current Directions in Psychological Science*, *20*(5), 286–290. doi:10.1177/0963721411422522

Batra, R., & Stayman, D. M. (1990). The Role of Mood in Advertising Effectiveness. *The Journal of Consumer Research*, *17*(2), 203. doi:10.1086/208550

Behm-Morawitz, E. (2017). Examining the intersection of race and gender in video game advertising. *Journal of Marketing Communications*, *23*(3), 220–239. doi:10.1080/13527266.2014.914562

Beinisch, Y., & Paunov, C. (2005). *Broadband Content: The online computer and video game industry.* OECD Report.

Berndt, A. (2010, Apr.). Diegetic Music: New Interactive Experiences. *Game Sound Technology and Player Interaction*, 60–77.

Bone, P. F., & Ellen, P. S. (1992). The Generation and Consequences of Communication-Evoked Imagery. *The Journal of Consumer Research*, *19*(1), 93–104. doi:10.1086/209289

Brandão, R., Lins, P., Lima, F., Arruda, A., Neves, A., & Correia, W. (2014). Design e inovação tecnológica na indústria de video games: Nintendo, um estudo de caso. *SBC - Proceedings of the SBGames*, 1101–1110.

Brave, S., Nass, C., & Hutchinson, K. (2005). Computers that care: Investigating the effects of orientation of emotion exhibited by an embodied computer agent. *International Journal of Human-Computer Studies*, *62*(2), 161–178. doi:10.1016/j.ijhcs.2004.11.002

Burns, A. C., Biswas, A., & Babin, L. A. (1993). The Operation of Visual Imagery as a Mediator of Advertising Effects. *Journal of Advertising*, *22*(2), 71–85. doi:10.1080/00913367.1993.10673405

Calleja, G. (2010). Digital games and escapism. *Games and Culture*, *5*(4), 335–353. doi:10.1177/1555412009360412

Calleja, G. (2013). Narrative Involvement in Digital Games. *Foundations of Digital Games.*

Cauberghe, V., & De Pelsmacker, P. (2010). Advergames: The Impact of Brand Prominence and Game Repetition on Brand Responses. *Journal of Advertising*, *39*(1), 5–18. doi:10.2753/JOA0091-3367390101

Chaney, I., Hosany, S., Wu, M. S. S., Chen, C. H. S., & Nguyen, B. (2018). Size does matter: Effects of in-game advertising stimuli on brand recall and brand recognition. *Computers in Human Behavior*, *86*, 311–318. doi:10.1016/j.chb.2018.05.007

Chang, Y., Yan, J., Zhang, J., & Luo, J. (2010). Online In-Game Advertising Effect: Examining the Influence of a Match Between Games and Advertising. *Journal of Interactive Advertising*, *11*(1), 63–73. doi:10.1080/15252019.2010.10722178

Chen, C. Y., & Chang, S. L. (2019). Moderating effects of information-oriented versus escapism-oriented motivations on the relationship between psychological well-being and problematic use of video game live-streaming services. *Journal of Behavioral Addictions*, *8*(3), 564–573. doi:10.1556/2006.8.2019.34 PMID:31328955

Chen, H., & Haley, E. (2014). Product placement in social games: Consumer experiences in China. *Journal of Advertising*, *43*(3), 286–295. doi:10.1080/00913367.2013.858086

Choi, D., & Kim, J. (2004). Why People Continue to Play Online Games : In Search of Online Contents. *Cyberpsychology & Behavior*, *7*(1), 11–25. doi:10.1089/109493104322820066 PMID:15006164

Collins, K. (2009). Generating Meaningful Sound: Quantifying the Affective Attributes of Sound Effects for Real-Time Sound Synthesis in Audio-Visual Media. *AES 35th International Conference*, 11–16.

Crandall, R. W., & Sidak, J. G. (2006). *Video Games: Serious Business for America's Economy*. Academic Press.

Crivelli, C., & Fridlund, A. J. (2018). Facial Displays Are Tools for Social Influence. *Trends in Cognitive Sciences*, *22*(5), 388–399. doi:10.1016/j.tics.2018.02.006 PMID:29544997

Crivelli, C., Jarillo, S., Russell, J. A., & Fernández-Dols, J. M. (2016). Reading emotions from faces in two indigenous societies. *Journal of Experimental Psychology. General*, *145*(7), 830–843. doi:10.1037/xge0000172 PMID:27100308

Crivelli, C., Russell, J. A., Jarillo, S., & Fernández-Dols, J. M. (2016). The fear gasping face as a threat display in a Melanesian society. *Proceedings of the National Academy of Sciences of the United States of America*, *113*(44), 12403–12407. doi:10.1073/pnas.1611622113 PMID:27791137

Crivelli, C., Russell, J. A., Jarillo, S., & Fernández-Dols, J. M. (2017). Recognizing spontaneous facial expressions of emotion in a small-scale society of papua New Guinea. *Emotion (Washington, D.C.)*, *17*(2), 337–347. doi:10.1037/emo0000236 PMID:27736108

Daj, A. (2013). Economic and Legal Aspects of Introducing Novel Ict Instruments: Integrating Sound Into Social Media Marketing - From Audio Branding To Soundscaping. *Bulletin of the Transilvania University of Brasov. Series V, Economic Sciences*, *6*(2), 15–24.

Dardis, F. E., Schmierbach, M., & Limperos, A. M. (2012a). The Impact of Game Customization and Control Mechanisms on Recall of Integral and Peripheral Brand Placements in Video games. *Journal of Interactive Advertising*, *12*(2), 1–12. doi:10.1080/15252019.2012.10722192

Dardis, F. E., Schmierbach, M., & Limperos, A. M. (2012b). The Impact of Game Customization and Control Mechanisms on Recall of Integral and Peripheral Brand Placements in Video games. *Journal of Interactive Advertising*, *12*(2), 1–12. doi:10.1080/15252019.2012.10722192

Deleuze, J., Maurage, P., Schimmenti, A., Nuyens, F., Melzer, A., & Billieux, J. (2019). Escaping reality through video games is linked to an implicit preference for virtual over real-life stimuli. *Journal of Affective Disorders, 245*, 1024–1031.

Dormann, C. (2001). Seducing consumers, evaluating emotions. *Joint Proceedings of IHM-HCI 2001, 2*, 10–14. Retrieved from papers2://publication/uuid/C5C62B4E-D574-4063-9AEB-76E7C1C01AA5

Dubois, L. E., & Gibbs, C. (2018). Video game–induced tourism: A new frontier for destination marketers. *Tourism Review*, *73*(2), 186–198. doi:10.1108/TR-07-2017-0115

Dubois, L. E., Griffin, T., Gibbs, C., & Guttentag, D. (2020). The impact of video games on destination image. *Current Issues in Tourism*, 1–13. doi:10.1080/13683500.2020.1724082

Ekman, P. (1992). An Argument for Basic Emotions. *Cognition and Emotion*, *6*(3–4), 169–200. doi:10.1080/02699939208411068

Ekman, P. (1993). Facial expression and emotion. *The American Psychologist*, *48*(4), 384–392. doi:10.1037/0003-066X.48.4.384 PMID:8512154

Ekman, P., Friesen, W. V., O'Sullivan, M., Chan, A., Diacoyanni-Tarlatzis, I., Heider, K., Krause, R., LeCompte, W. A., Pitcairn, T., Ricci-Bitti, P. E., Scherer, K., Tomita, M., & Tzavaras, A. (1987). Universals and Cultural Differences in the Judgments of Facial Expressions of Emotion. *Journal of Personality and Social Psychology*, *53*(4), 712–717. doi:10.1037/0022-3514.53.4.712 PMID:3681648

Engl, S., & Nacke, L. E. (2013). Contextual influences on mobile player experience - A game user experience model. *Entertainment Computing*, *4*(1), 83–91. doi:10.1016/j.entcom.2012.06.001

Fakeye, P. C., & Crompton, J. L. (1991). Image Differences between Prospective, First-Time, and Repeat Visitors to the Lower Rio Grande Valley. *Journal of Travel Research*, *30*(2), 10–16. doi:10.1177/004728759103000202

Friedman, H. S., & Riggio, R. E. (1981). Effect of individual differences in nonverbal expressiveness on transmission of emotion. *Journal of Nonverbal Behavior*, *6*(2), 96–104. doi:10.1007/BF00987285

Frome, J. (2019). Interactive Works and Gameplay Emotions. *Games and Culture*, *14*(7–8), 856–874. doi:10.1177/1555412019847907

Fu, H., Ye, B. H., & Xiang, J. (2016). Reality TV, audience travel intentions, and destination image. *Tourism Management*, *55*(February), 37–48. doi:10.1016/j.tourman.2016.01.009

Garrod, B. (2009). Understanding the relationship between tourism destination imagery and tourist photography. *Journal of Travel Research*, *47*(3), 346–358. doi:10.1177/0047287508322785

Gefen & Straub. (2003). Managing User Trust in B2C e-Services. *E-Service Journal*, *2*(2), 7.

Glass, Z. (2007). The Effectiveness of Product Placement in Video Games. *Journal of Interactive Advertising*, *8*(1), 1–27. doi:10.1080/15252019.2007.10722134

Goossens, C. (2000). Tourism information and pleasure motivation. *Annals of Tourism Research*, *27*(2), 301–321. doi:10.1016/S0160-7383(99)00067-5

Govers, R., Go, F. M., & Kumar, K. (2007). Virtual destination image a new measurement approach. *Annals of Tourism Research*, *34*(4), 977–997. doi:10.1016/j.annals.2007.06.001

Green, C. S., & Bavelier, D. (2006). Effect of action video games on the spatial distribution of visuospatial attention. *Journal of Experimental Psychology. Human Perception and Performance*, *32*(6), 1465–1478. doi:10.1037/0096-1523.32.6.1465 PMID:17154785

Grigorovici, D. M., & Constantin, C. D. (2004). Experiencing Interactive Advertising beyond Rich Media: Impacts of Ad Type and Presence on Brand Effectiveness in 3D Gaming Immersive Virtual Environments. *Journal of Interactive Advertising*, *4*(3).

Gross, M. L. (2010). Advergames and the effects of game-product congruity. *Computers in Human Behavior*, *26*(6), 1259–1265. doi:10.1016/j.chb.2010.03.034

Gunn, C. A. (1988). *Vacationscape: Designing tourist regions*. Van Nostrand Reinhold.

Gupta, P. B., & Lord, K. R. (1998). Product Placement in Movies: The Effect of Prominence and Mode on Audience Recall. *Journal of Current Issues and Research in Advertising*, *20*(1), 47–59. doi:10.1080/10641734.1998.10505076

Ha, S., Huang, R., & Park, J. S. (2019). Persuasive brand messages in social media: A mental imagery processing perspective. *Journal of Retailing and Consumer Services*, *48*(February), 41–49. doi:10.1016/j.jretconser.2019.01.006

Hareli, S., Shomrat, N., & Hess, U. (2009). Emotional Versus Neutral Expressions and Perceptions of Social Dominance and Submissiveness. *Emotion (Washington, D.C.)*, *9*(3), 378–384. doi:10.1037/a0015958 PMID:19485615

Hassanein, K., & Head, M. (2007). Manipulating perceived social presence through the web interface and its impact on attitude towards online shopping. *International Journal of Human-Computer Studies*, *65*(8), 689–708. doi:10.1016/j.ijhcs.2006.11.018

Hatfield, E., Cacioppo, J. T., & Rapson, R. L. (1993). Emotional contagion. *Current Directions in Psychological Science*, *2*(3), 96–100. doi:10.1111/1467-8721.ep10770953

Hess, U., & Blairy, S. (2001). Facial mimicry and emotional contagion to dynamic emotional facial expressions and their influence on decoding accuracy. *International Journal of Psychophysiology*, *40*(2), 129–141. doi:10.1016/S0167-8760(00)00161-6 PMID:11165351

Hoeckner, B., Wyatt, E. W., Decety, J., & Nusbaum, H. (2011). Film Music Influences How Viewers Relate to Movie Characters. *Psychology of Aesthetics, Creativity, and the Arts*, *5*(2), 146–153. doi:10.1037/a0021544

Hone, K. (2006). Empathic agents to reduce user frustration: The effects of varying agent characteristics. *Interacting with Computers*, *18*(2), 227–245. doi:10.1016/j.intcom.2005.05.003

Hou, J., Zhao, X., & Zheng, J. (2019). The impact of consistency between the emotional feature of advertising music and brand personality on brand experience. *Journal of Management Analytics*, *6*(3), 250–268. doi:10.1080/23270012.2019.1613684

Hsee, C. K., Hatfield, E., Carlson, J. G., & Chemtob, C. (1990). The effect of power on susceptibility to emotion contagion. *Cognition and Emotion*, *4*(4), 327–340. doi:10.1080/02699939008408081

Hsu, C. L., & Lu, H. P. (2004). Why do people play on-line games? An extended TAM with social influences and flow experience. *Information & Management*, *41*(7), 853–868. doi:10.1016/j.im.2003.08.014

Huang, Y. C., Backman, S. J., Backman, K. F., & Moore, D. W. (2013). Exploring user acceptance of 3D virtual worlds in travel and tourism marketing. *Tourism Management*, *36*, 490–501. doi:10.1016/j.tourman.2012.09.009

Hultén, B. (2011). Sensory marketing: The multi-sensory brand-experience concept. *European Business Review*, *23*(3), 256–273. doi:10.1108/09555341111130245

Izard, C. E. (1994). *Innate and universal facial expressions: Evidence from developmental and cross-cultural research. Psychological Bulletin*. American Psychological Association.

Jenkins, O. (1999). Understanding and measuring tourist destination Images. *International Journal of Tourism Research*, *1*(1), 1–15. doi:10.1002/(SICI)1522-1970(199901/02)1:1<1::AID-JTR143>3.0.CO;2-L

Jeong, E. J., & Biocca, F. A. (2012). Are there optimal levels of arousal to memory? Effects of arousal, centrality, and familiarity on brand memory in video games. *Computers in Human Behavior*, *28*(2), 285–291. doi:10.1016/j.chb.2011.09.011

Josiassen, A., Assaf, A. G., Woo, L., & Kock, F. (2016). The Imagery–Image Duality Model: An Integrative Review and Advocating for Improved Delimitation of Concepts. *Journal of Travel Research*, *55*(6), 789–803. doi:10.1177/0047287515583358

Juslin, P. N., & Laukka, P. (2004). Expression, Perception, and Induction of Musical Emotions: A Review and a Questionnaire Study of Everyday Listening. *Journal of New Music Research*, *33*(3), 217–238. doi:10.1080/0929821042000317813

Katz, E., & Foulkes, D. (1962). On the Use of the Mass Media as "Escape": Clarification of a Concept. *Public Opinion Quarterly*, *26*(3), 377–388. doi:10.1086/267111

Kiili, K. (2005). Digital game-based learning: Towards an experiential gaming model. *The Internet and Higher Education*, *8*(1), 13–2. doi:10.1016/j.iheduc.2004.12.001

Kim, H., & Stepchenkova, S. (2015). Effect of tourist photographs on attitudes towards destination: Manifest and latent content. *Tourism Management*, *49*, 29–41. doi:10.1016/j.tourman.2015.02.004

Kim, S. B., Kim, D. Y., & Bolls, P. (2014). Tourist mental-imagery processing: Attention and arousal. *Annals of Tourism Research*, *45*, 63–76. doi:10.1016/j.annals.2013.12.005

Klein, J., Moon, Y., & Picard, R. W. (2002). This computer responds to user frustration: Theory, design, and results. *Interacting with Computers*, *14*(2), 119–140. doi:10.1016/S0953-5438(01)00053-4

Klimmt, C. (2008). Escapism. In The International Encyclopedia of Communication. Academic Press.

Klimmt, C., Hartmann, T., & Frey, A. (2007). Effectance and control as determinants of video game enjoyment. *Cyberpsychology & Behavior*, *10*(6), 845–847. doi:10.1089/cpb.2007.9942 PMID:18085976

Koch, S., & Artmayr, P. (2019). Stability and development of user innovation strategies for video game producers. *European Journal of Innovation Management*, *ahead-of-print*(ahead-of-print). Advance online publication. doi:10.1108/EJIM-05-2019-0117

Kotler, P. (1973). Atmospherics as a Marketing Tool. *Journal of Retailing*, *49*(4), 48–64.

Kramer, A. D. I., Guillory, J. E., & Hancock, J. T. (2014). Experimental evidence of massivescale emotional contagion through social networks. *Proceedings of the National Academy of Sciences of the United States of America*, *111*(29), 8788–8790. doi:10.1073/pnas.1320040111 PMID:24889601

Krumhansl, C. L. (1997). An exploratory study of musical emotions and psychophysiology. *Canadian Journal of Experimental Psychology/Revue Canadienne de Psychologie Expérimentale*, *51*(4), 336–353.

Law, S., & Braun-LaTour, K. A. (2000). I'll have what she's having: Gauging the impact of product placements on viewers. *Psychology and Marketing*, *17*(12), 1059–1075. doi:10.1002/1520-6793(200012)17:12<1059::AID-MAR3>3.0.CO;2-V

Lee, M., & Faber, R. J. (2007). Effects of Product Placement in On-Line Games on Brand Memory: A Perspective of the Limited-Capacity Model of Attention. *Journal of Advertising*, *36*(4), 75–90. doi:10.2753/JOA0091-3367360406

Lee, W., & Gretzel, U. (2012). Designing persuasive destination websites: A mental imagery processing perspective. *Tourism Management*, *33*(5), 1270–1280. doi:10.1016/j.tourman.2011.10.012

Lee, W., Gretzel, U., & Law, R. (2010). Quasi-Trial Experiences through Sensory Information on Destination Web Sites. *Journal of Travel Research*, *49*(3), 310–322. doi:10.1177/0047287509346991

Lee, Y. H., & Lim, E. A. I. C. (2010). When good cheer goes unrequited: How emotional receptivity affects evaluation of expressed emotion. *JMR, Journal of Marketing Research*, *47*(6), 1151–1161. doi:10.1509/jmkr.47.6.1151

Lehu, J.-M. (2007). *Branded entertainment: Product placement & brand strategy in the entertainment business*. Kogan Page Publishers.

Lewis, B., & Porter, L. (2010). In-game advertising effects: Examining player perceptions of advertising schema congruity in a massively multiplayer online role-playing game. *Journal of Interactive Advertising, 10*(2), 46–60. doi:10.1080/15252019.2010.10722169

Li, H., Edwards, S. M., & Lee, J.-H. (2002). Measuring the Intrusiveness fo Advertisements: Scale Development and Validation. *Journal of Advertising, 31*(2), 37–48. doi:10.1080/00913367.2002.10673665

Lucas, K., & Sherry, J. L. (2004). Sex differences in video game play: A communication-based explanation. *Communication Research, 31*(5), 499–523. doi:10.1177/0093650204267930

Lull, R. B., Gibson, B., Cruz, C., & Bushman, B. J. (2016, May). Killing Characters in Video Games Kills Memory for In-Game Ads. *Psychology of Popular Media Culture.*

Lull, R. B., Gibson, B., Cruz, C., & Bushman, B. J. (2018). Killing characters in video games kills memory for in-game ads. *Psychology of Popular Media Culture, 7*(1), 87–97. doi:10.1037/ppm0000108

Lundqvist, L.-O., Carlsson, F., Hilmersson, P., & Juslin, P. N. (2009). Emotional responses to music: Experience, expression, and physiology. *Psychology of Music, 37*(1), 61–90. doi:10.1177/0305735607086048

Lutz, K. A., & Lutz, R. J. (1978). Imagery-Eliciting Strategies: Review and Implications of Research. *Advances in Consumer Research. Association for Consumer Research (U. S.), 5*, 611–620.

MacInnis, D. J., & Price, L. L. (1987). The Role of Imagery in Information Processing: Review and Extensions. *The Journal of Consumer Research, 13*(4), 473–491. doi:10.1086/209082

MacInnis, D. J., & Price, L. L. (1990). An Exploratory Study of the Effects of Imagery Processing and Consumer Experience on Expectations and Satisfaction. *Advances in Consumer Research. Association for Consumer Research (U. S.), 17*, 41–47.

MacKay, K. J., & Fesenmaier, D. R. (1997). Pictorial element of destination in image formation. *Annals of Tourism Research, 24*(3), 537–565. doi:10.1016/S0160-7383(97)00011-X

Maitland, C., Granich, J., Braham, R., Thornton, A., Teal, R., Stratton, G., & Rosenberg, M. (2018). Measuring the capacity of active video games for social interaction: The Social Interaction Potential Assessment tool. *Computers in Human Behavior, 87*, 308–316.

Marsh, A. A., Ambady, N., & Kleck, R. E. (2005). The effects of fear and anger facial expressions on approach- and avoidance-related behaviors. *Emotion (Washington, D.C.), 5*(1), 119–124. doi:10.1037/1528-3542.5.1.119 PMID:15755225

Martí-parreño, J., Bermejo-berros, J., & Aldás-manzano, J. (2017). Product Placement in Video Games: The Effect of Brand Familiarity and Repetition on Consumers ' Memory. *Journal of Interactive Marketing, 38*, 55–63. doi:10.1016/j.intmar.2016.12.001

Matos, N., Mendes, J., & Pinto, P. (2015). The Role of Imagery and Experiences in the Construction of a Tourism Destination Image. *Journal of Spatial and Organizational Dynamics, 3*(2), 65–84.

Mau, G., Silberer, G., & Constien, C. (2008). Communicating brands playfully: Effects of in-game advertising for familiar and unfamiliar brands. *International Journal of Advertising, 27*(5), 827–851. doi:10.2501/S0265048708080293

Mesquita, B. (2010, Nov.). Emoting: A contextualized process. *The Mind in Context*, 83–104.

Miller, D. W., & Stoica, M. (2003a). Comparing the effects of a photograph versus artistic renditions of a beach scene in a direct-response print ad for a Caribbean resort island: A mental imagery perspective. *Journal of Vacation Marketing*, *10*(1), 11–21. doi:10.1177/135676670301000102

Miller, D. W., & Stoica, M. (2003b). Comparing the effects of a photograph versus artistic renditions of a beach scene in a direct-response print ad for a Caribbean resort island: A mental imagery perspective. *Journal of Vacation Marketing*, *10*(1), 11–21. doi:10.1177/135676670301000102

Milman, A., & Pizam, A. (1995). The Role of Awareness and Familiarity with a Destination: The Central Florida Case. *Journal of Travel Research*, *33*(3), 21–27. doi:10.1177/004728759503300304

Molesworth, M. (2006). Real brands in imaginary worlds: Investigating players' experiences of brand placement in digital games. *Journal of Consumer Behaviour*, *5*(4), 355–366. doi:10.1002/cb.186

Nadkarni, N. M., Hasbach, P. H., Thys, T., Crockett, E. G., & Schnacker, L. (2017). Impacts of nature imagery on people in severely nature-deprived environments. *Frontiers in Ecology and the Environment*, *15*(7), 395–403. doi:10.1002/fee.1518

Nelson, M. R. (2002). Recall of Brand Placements in Computer / Video Games. *Journal of Advertising Research*, *42*(2), 80–92. doi:10.2501/JAR-42-2-80-92

Nelson, M. R. (2005). Exploring Consumer Response to "Advergaming.". In C. Haugtvedt, K. Machleit, & R. Yalch (Eds.), *Online Consumer Psychology. Understanding and Influencing Consumer Behavior in the Virtual World* (pp. 167–223). Lawrence Erlbaum Associates, Inc.

Nelson, M. R., Keum, H., & Yaros, R. (2004). Advertainment or Adcreep: Game players' attitudes toward advertising and product placements in computer games. *Journal of Interactive Advertising*, *5*(1), 3–21. doi:10.1080/15252019.2004.10722090

Netimperative - latest digital marketing news. (2013). *The MyCapeTownHoliday.com*. Retrieved October 27, 2019, from http://www.netimperative.com/2013/02/video-viral-of-the-week-facebook-users-get-virtual-holiday-in-cape-town/

Neumann, R., & Strack, F. (2000). "Mood contagion": The automatic transfer of mood between persons. *Journal of Personality and Social Psychology*, *79*(2), 211–223. doi:10.1037/0022-3514.79.2.211 PMID:10948975

Nispen, T. Van, Huiberts, S., Donders, S., & Koch, S. (2011). *The Nln-Player : a System for Nonlinear Music in Games, 2009–2012*. Academic Press.

Norman, D. A., & Verganti, R. (2014). Incremental and radical innovation: Design research vs. technology and meaning change. *Design Issues*, *30*(1), 78–96. doi:10.1162/DESI_a_00250

North, A. C., Hargreaves, D. J., & McKendrick, J. (1999). The influence of in-store music on wine selections. *The Journal of Applied Psychology*, *84*(2), 271–276. doi:10.1037/0021-9010.84.2.271

North, A. C., Sheridan, L. P., & Areni, C. S. (2016). Music Congruity Effects on Product Memory, Perception, and Choice. *Journal of Retailing*, *92*(1), 83–95. doi:10.1016/j.jretai.2015.06.001

Overy, K., & Molnar-Szakacs, I. (2009). Being together in time: Musical experience and the mirror neuron system. *Music Perception, 26*(5), 489–504. doi:10.1525/mp.2009.26.5.489

Paivio, A. (1969). Mental imagery in associative learning and memory. *Psychological Review, 76*(3), 241–263. doi:10.1037/h0027272

Paivio, A. (1971). Imagery and Verbal Processes. Holt, Rinehart & Winston.

Park, M., & Yoo, J. (2020). Effects of perceived interactivity of augmented reality on consumer responses: A mental imagery perspective. *Journal of Retailing and Consumer Services, 52.*

Pearson, J., Naselaris, T., Holmes, E. A., & Kosslyn, S. M. (2015). Mental Imagery: Functional Mechanisms and Clinical Applications. *Trends in Cognitive Sciences, 19*(10), 590–602. doi:10.1016/j.tics.2015.08.003 PMID:26412097

Peters, S., & Leshner, G. (2013). Get in the Game: The Effects of Game-Product Congruity and Product Placement Proximity on Game Players' Processing of Brands Embedded in Advergames. *Journal of Advertising, 42*(2–3), 113–130. doi:10.1080/00913367.2013.774584

Petroshius, S. M., & Crocker, K. E. (1989). An empirical analysis of spokesperson characteristics on advertisement and product evaluations. *Journal of the Academy of Marketing Science, 17*(3), 217–225. doi:10.1007/BF02729813

Redondo, I. (2012). The effectiveness of casual advergames on adolescents' brand attitudes. *European Journal of Marketing, 46*(11), 1671–1688. doi:10.1108/03090561211260031

Regensburg Experience. (2019). Retrieved October 27, 2019, from https://rex.julienbiere.de/

Reisenzein, R., Studtmann, M., & Horstmann, G. (2013). Coherence between emotion and facial expression: Evidence from laboratory experiments. *Emotion Review, 5*(1), 16–23. doi:10.1177/1754073912457228

Rieber, L. P. (1996). Seriously Considering Play: Designing Interactive Learning Environments Based on the Blending of Microworlds, Simulations, and Games. *Educational Technology Research and Development, 44*(2), 43–58. doi:10.1007/BF02300540

Riegelsberger, J. (2003). Interpersonal cues and consumer trust in e-commerce. *Conference on Human Factors in Computing Systems - Proceedings*, 674–675. 10.1145/765891.765923

Riegelsberger, J., Angela, S. M., & McCarthy, J. D. (2003). Shiny happy people building trust?: photos on e-commerce websites and consumer trust. *Proceedings of the Conference on Human Factors in Computing Systems*, 121–128. 10.1145/642611.642634

Rossiter, J. R., & Percy, L. (1980). Attitude change through visual imagery in advertising. *Journal of Advertising, 9*(2), 10–16. doi:10.1080/00913367.1980.10673313

Russell, J. A. (2003). Core Affect and the Psychological Construction of Emotion. *Psychological Review, 110*(1), 145–172. doi:10.1037/0033-295X.110.1.145 PMID:12529060

Salo, J., & Karjaluoto, H. (2007). A conceptual model of trust in the online environment. *Online Information Review, 31*(5), 604–621. doi:10.1108/14684520710832324

Scherer, K., & Zentner, M. (2001). Emotional effects of music: Production rules. *Music and Emotion: Theory and Research*, 361–392.

Schneider, L.-P., & Cornwell, B. T. (2005). Cashing in on crashes via brand placement in computer games. *International Journal of Advertising*, *24*(3), 321–343. doi:10.1080/02650487.2005.11072928

Schneider, L.-P., Systems, B., & Cornwell, T. B. (2005). Cashing in on crashes via brand placement in computer games. *International Journal of Advertising*, *24*(3), 321–343. doi:10.1080/02650487.2005.11072928

Sepehr, S., & Head, M. (2018). Understanding the role of competition in video gameplay satisfaction. *Information & Management*, *55*(4), 407–421. doi:10.1016/j.im.2017.09.007

Sherry, J. L. (2004). Flow and media enjoyment. *Communication Theory*, *14*(4), 328–347. doi:10.1111/j.1468-2885.2004.tb00318.x

Siemens, J. C., Smith, S., & Fisher, D. (2015). Investigating the Effects of Active Control on Brand Recall Within In-Game Advertising. *Journal of Interactive Advertising*, *15*(1), 43–53. doi:10.1080/15252019.2015.1021432

Silicon Republic. (2011). *IrelandTown*. Retrieved October 27, 2019, from https://www.siliconrepublic.com/play/tourism-ireland-launches-facebook-game-ireland-town

Simon, J. P. (2018). Triggering the emergence of digital ecosystems: The role of mobile and video games in emerging economies. *Digital Policy. Regulation & Governance*, *20*(5), 449–478. doi:10.1108/DPRG-03-2018-0008

Singer, J., & Pope, K. (2012). *The Power of Human Imagination: New Methods in Psychotherapy*. Springer Science & Business Media.

Smith, M. (2003). Holistic holidays: Tourism and the reconciliation of body, mind and spirit. *Tourism Recreation Research*, *28*(1), 103–108. doi:10.1080/02508281.2003.11081392

Statista.com. (2019). *Video game market value worldwide from 2012 to 2021*. Retrieved November 19, 2019, from https://www.statista.com/statistics/292056/video-game-market-value-worldwide/

Stins, J. F., Roelofs, K., Villan, J., Kooijman, K., Hagenaars, M. A., & Beek, P. J. (2011). Walk to me when I smile, step back when I'm angry: Emotional faces modulate whole-body approach-avoidance behaviors. *Experimental Brain Research*, *212*(4), 603–611. doi:10.100700221-011-2767-z PMID:21698468

Storz, C., Riboldazzi, F., & John, M. (2015). Mobility and innovation: A cross-country comparison in the video games industry. *Research Policy*, *44*(1), 121–137. doi:10.1016/j.respol.2014.07.015

Su, H. J., Huang, Y. A., Brodowsky, G., & Kim, H. J. (2011). The impact of product placement on TV-induced tourism: Korean TV dramas and Taiwanese viewers. *Tourism Management*, *32*(4), 805–814. doi:10.1016/j.tourman.2010.06.020

Sweetser, P., & Wyeth, P. (2005). GameFlow: A model for evaluating player enjoyment in games. *Computers in Entertainment*, *3*(3), 1–23. doi:10.1145/1077246.1077253

Taghipour, A. (2017). Product Placement in Mobile Games and Thai Students Attitude Toward the Brand Recalling. *PEOPLE: International Journal of Social Sciences, 3*(2), 2279–2294. doi:10.20319/pijss.2017.32.22792294

Tarnowski, P., Kołodziej, M., Majkowski, A., & Rak, R. J. (2017). Emotion recognition using facial expressions. *Procedia Computer Science, 108*, 1175–1184. doi:10.1016/j.procs.2017.05.025

Tasci, A. D. A., & Gartner, W. C. (2007). Destination Image and Its Functional Relationships. *Journal of Travel Research, 45*(4), 413–425.

Tasci, A. D. A., Gartner, W. C., & Cavusgil, S. T. (2007). Measurement of destination brand bias using a quasi-experimental design. *Tourism Management, 28*(6), 1529–1540. doi:10.1016/j.tourman.2007.02.009

Terlutter, R., & Capella, M. L. (2013). The Gamification of Advertising: Analysis and Research Directions of In-Game Advertising, Advergames, and Advertising in Social Network Games. *Journal of Advertising, 42*(2–3), 95–112. doi:10.1080/00913367.2013.774610

Tessitore, T., & Geuens, M. (2013). PP for "product placement" or "puzzled public"? The effectiveness of symbols as warnings of product placement and the moderating role of brand recall. *International Journal of Advertising, 32*(3), 419–442. doi:10.2501/IJA-32-3-419-442

Tessitore, T., Pandelaere, M., & Van Kerckhove, A. (2014). The Amazing Race to India: Prominence in reality television affects destination image and travel intentions. *Tourism Management, 42*, 3–12. doi:10.1016/j.tourman.2013.10.001

Tiwsakul, R., Hackley, C., & Szmigin, I. (2005). Explicit, non-integrated product placement in British television programmes. *International Journal of Advertising, 24*(1), 95–111. doi:10.1080/02650487.2005.11072906

Tourism Authority of Thailand. (2011). *Smile Land*. Retrieved October 27, 2019, from http://www.smilelandgame.com

Tuohino, A., & Pitkänen, K. (2004). The transformation of a neutral lake landscape into a meaningful experience-interpreting tourist photos. *Journal of Tourism and Cultural Change, 2*(2), 77–93. doi:10.1080/14766820408668170

UNWTO. (2018). *Tourism Highlights 2018 Edition International Tourism Trends 2017*. UNWTO.

UNWTO. (2019). *International Tourism Highlights International tourism continues to outpace the global economy*. UNWTO.

Vashisht, D. (2019). Effect of interactivity and congruence on brand advocacy and brand acceptance. *Arts and the Market,* (ahead-of-print).

Vashisht, D., & Royne, M. B. (2016). Advergame speed influence and brand recall: The moderating effects of brand placement strength and gamers' persuasion knowledge. *Computers in Human Behavior, 63*, 162–169. doi:10.1016/j.chb.2016.05.022

Verberckmoes, S., Poels, K., Dens, N., Herrewijn, L., & De Pelsmacker, P. (2016). When and why is perceived congruity important for in-game advertising in fantasy games? *Computers in Human Behavior*, *64*, S29–S38. doi:10.1016/j.chb.2016.07.062

Villani, D., Carissoli, C., Triberti, S., Marchetti, A., Gilli, G., & Riva, G. (2018). Video games for Emotion Regulation: A Systematic Review. *Games for Health Journal*, *7*(2), 85–99. doi:10.1089/g4h.2017.0108 PMID:29424555

Walters, G., Sparks, B., & Herington, C. (2007). The effectiveness of print advertising stimuli in evoking elaborate consumption visions for potential travelers. *Journal of Travel Research*, *46*(1), 24–34. doi:10.1177/0047287507302376

Wang, Z., Mao, H., Li, Y. J., & Liu, F. (2017). Smile big or not? Effects of smile intensity on perceptions of warmth and competence. *The Journal of Consumer Research*, *43*(5), 787–805.

Weaver, A. (2007). Product Placement and Tourism-Oriented Environments: An Exploratory Introduction. *International Journal of Tourism Research*, *284*(4), 275–284. doi:10.1002/jtr.612

Wild, B., Erb, M., & Bartels, M. (2001). Are emotions contagious? Evoked emotions while viewing emotionally expressive faces: Quality, quantity, time course and gender differences. *Psychiatry Research*, *102*(2), 109–124. doi:10.1016/S0165-1781(01)00225-6 PMID:11408051

Williams, D. A. H., Cowling, P. I., & Murphy, D. T. (2019). Emotional Congruence in Video Game Audio. Encyclopedia of Computer Graphics and Games, 1–3.

Xu, F., Tian, F., Buhalis, D., & Weber, J. (2013). Marketing tourism via electronic games: Understanding the motivation of tourist players. *5th International Conference on Games and Virtual Worlds for Serious Applications, VS-GAMES 2013*. 10.1109/VS-GAMES.2013.6624235

Xu, J., Broekens, J., Hindriks, K., & Neerincx, M. A. (2015). Mood contagion of robot body language in human robot interaction. *Autonomous Agents and Multi-Agent Systems*, *29*(6), 1216–1248. doi:10.100710458-015-9307-3

Yang, M., Roskos-Ewoldsen, D., Dinu, L., & Arpan, L. (2006). The Effectiveness of "in-Game" Advertising: Comparing College Students' Explicit and Implicit Memory for Brand Names. *Journal of Advertising*, *35*(4), 143–152. doi:10.2753/JOA0091-3367350410

Yung, R., & Khoo-Lattimore, C. (2019). New realities: A systematic literature review on virtual reality and augmented reality in tourism research. *Current Issues in Tourism*, *22*(17), 2056–2081. doi:10.1080/13683500.2017.1417359

Yustisiana, R. A. (2018). Elements of Nature As the Object of Escapism in Maya Angelou'S "Woman Work.". *Journal of Language and Literature*, *18*(02), 158–167. doi:10.24071/joll.2018.180207

Zhang, J., & Fu, X. (2015). The Influence of Background Music of Video Games on Immersion. *Journal of Psychology & Psychotherapy*, *05*(04).

ADDITIONAL READING

Acir, E. (2019). Video games, virtual reality and augmented reality applications in tourism promotion and marketing. *9th advances in hospitality and tourism marketing and management conference proceedings*.

Choe, Y., Stienmetz, J. L., & Fesenmaier, D. R. (2017). Measuring Destination Marketing: Comparing Four Models of Advertising Conversion. *Journal of Travel Research*, *56*(2), 143–157. doi:10.1177/0047287516639161

Dubois, L. E., & Gibbs, C. (2018). Video game–induced tourism: A new frontier for destination marketers. *Tourism Review*, *73*(2), 186–198. doi:10.1108/TR-07-2017-0115

Krishna, A., Cian, L., & Sokolova, T. (2016). The power of sensory marketing in advertising. *Current Opinion in Psychology*, *10*, 142–147. doi:10.1016/j.copsyc.2016.01.007

Li, S., Walters, G., Packer, J., & Scott, N. (2017). A Comparative Analysis of Self-Report and Psychophysiological Measures of Emotion in the Context of Tourism Advertising. *Journal of Travel Research*, *0*(0), 1–15.

Marasco, A., Buonincontri, P., van Niekerk, M., Orlowski, M., & Okumus, F. (2018). Exploring the role of next-generation virtual technologies in destination marketing. *Journal of Destination Marketing & Management*, *9*, 138–148. doi:10.1016/j.jdmm.2017.12.002

Martí-Parreño, J., Bermejo-Berros, J., & Aldás-Manzano, J. (2016). the Effect of Brand Familiarity and Repetition of Product Placement in Video Games. *Journal of Interactive Marketing*, *38*, 725–735.

Skavronskaya, L., Scott, N., Moyle, B., Le, D., Hadinejad, A., Zhang, R., & Shakeela, A. (2017). Cognitive psychology and tourism research: State of the art. *Tourism Review*, *72*(2), 221–237. doi:10.1108/TR-03-2017-0041

KEY TERMS AND DEFINITIONS

Destination Image: Set of attributes, held by the consumer, regarding a destination.

Escapism: Mental escape, through imagination, from unpleasant or boring aspects of everyday life.

Imagery: Experience that allows the individual to visually perceive an object or scene without the object or scene being available to the senses.

Imagination: Ability to visualize images and create new ideas, in the absence of external stimuli.

Product Placement: Reference or placement, with promotional intent, of specific brands or products in works such as video games, movies or television programs.

Tourism Marketing: Marketing strategy, in the field of tourism, through which tourists are attracted to a specific location.

Video Game: Electronic entertainment game where there is player interaction with a visual interface.

Chapter 19
A Content Analysis on Child-Friendly Hotels as an Emerging Concept in Tourism Marketing

Erdem Akkan
https://orcid.org/0000-0001-5684-5676
Mersin University, Turkey

Sezen Bozyiğit
https://orcid.org/0000-0002-0951-3588
Tarsus University, Turkey

ABSTRACT

Children have been an interesting topic in the marketing literature both as consumers and influencers. Some studies claim that children might have an influence for buying decision of household products and services, even if for those which are not directly targeting them. This study is based on web information of 89 hotels, which were categorized as child-friendly on third-party hotel review websites. Through a measurement scale, adapted from similar several studies, the child-friendly characteristics of sample hotels were explored. Measurement scale consists of 35 items and five dimensions. According to the findings, the availability of these dimensions are ranked as (1) entertainment-related services, (2) food and beverage services, (3) room-related services or facilities, (4) other services, and (5) baby-related services in a descending order. It is believed that findings of this study will contribute to related literature and will give practical information to tourism decision-makers.

INTRODUCTION

Family purchase decision-making is one of the most popular subjects in consumer behaviour literature. In this process, husband, wife and children play an active role. It is suggested that children's influence on family purchases consistently accounts for a greater dollar amount than children's direct spending (Bakir, Rose, & Shoham, 2006). However, early studies on family decision-making seem to cover the

DOI: 10.4018/978-1-7998-3156-3.ch019

influence of the husband and wife dyad, rather than the influence of children (Chen, Lehto, Behnke, & Tang, 2016). From the tourism product perspective, it is suggested that parents are not deaf to what their children say related to household purchases. Furthermore, the satisfaction of children is highly rated by parents and if the child does not wish to go to a site, the likelihood of getting satisfied children is poor (Gram, 2007).

The interest in child issues in marketing and tourism literature has been rising for 50 years. Berey and Pollay (1968) referred to three main reasons: (1) the size of the child market is rapidly growing, (2) obviously children influence the family's decision-making and (3) adult consumer behaviour is the direct antecedent of child consumer behaviour. Similarly, McNeal (1992; as cited in Caruana & Vassallo, 2003, p. 61) claimed that children's influence on family purchases is increasing due to four reasons: (1) recently, parents have fewer children, thereby the influence of each child increases, (2) there has been a general increase in one-parent families that has resulted in children doing their own shopping, (3) there has been an increase in working women and delayed childbearing, as a result of which mothers tend to have more money to spend on their children, (4) out of necessity, working couples encourage more household participation and self-reliance. Flurry (2007) discovered that, children's influence is now extending past those products considered traditionally child-oriented to include family-oriented products such as furnishing, automobiles and electronics.

It is reported that children are often involved in choices for family activities, such as vacations, and minimally involved in decisions for major household durables (Swinyard & Sim, 1987). Ceylan (2019) also stated that children are the most effective family members for the holiday destination choice. According to her, regardless of their age, parents consider their children's preferences first. Hotel choices may differ according to demographic characteristics and interests of consumers. Especially, families with children want to choose hotels that are addressing both themselves and their children. These hotels are expected to meet basic needs of both parents and children, be safe for children, and entertain children with those hotel facilities. This ensures that both parents and children will have exciting experiences on the holiday. Being aware of these expectations, many hotel managements promote their services according to these wishes to gain competitive advantage by increasing the demand for them. Jelínková, Tučková, and Jurigová (2017) indicated, that it is necessary for tourism professionals, who are targeting families with children, to establish hotels prepared to meet special needs of children and to provide all kinds of opportunities for both children and their parents.

For the hotel management, it is necessary, but not limited, to have the features to meet expectations of families with children. It is also necessary to make the target market aware of these features. For this purpose, business professionals tend to use various promotion channels. Websites are one of these channels. Websites are critical for hotels, especially for communication, access to tools (i.e., databases, indices, etc.), promotion and marketing reasons (Poock & Lefond, 2001, p. 16). Because prospective customers generally do not have the chance to go and see the hotels on-site, they prefer to gather information about the hotels mostly via the Internet. They may get information from the official website of the hotels or through tourism agencies' websites. Thus, the more information and images of hotels are on websites, the more insight customers will have about the hotels. According to that information, customers can make comparisons and they can decide quickly.

Although it is quite important for families with children to choose a proper hotel for their children, according to literature, limited studies have been found on what expectations families have and which data a "child-friendly hotel" concept should include (Emir & Pekyaman, 2010; Özel, 2015; Aslan, Tosun & Kurtuluş, 2019). Two of the studies (Emir & Pekyaman, 2010; Aslan, Tosun & Kurtuluş, 2019) examined the concept of child-friendly hotels from the customer perspective. Furthermore, in none of those studies, child-friendly hotel dimensions are compared in terms of availabilities. Thus, the purpose of this study is to identify on which child-friendly items hotels focus more and on which they focus less by examining child-friendly hotel web sites. It is predicted that findings of this study will contribute to the growing child-friendly hotel literature and will be beneficial for tourism decision-makers, who are targeting families with children.

BACKGROUND

The literature shows us that in terms of consumption, children significantly influence the family decision-making process. This concept will be discussed in the following part. To clarify it more efficiently, we will describe the theory with related examples under the following three subheadings; influence of children on consumption decisions, product type matters and demographics make a difference.

Influence of Children on Consumption Decisions

According to consumer research over the 40 years, it is established that 'purchase decisions within the family are not always the outcome of individual choices, but rather, family members influence each other' (Hamilton & Catterall, 2006, p. 1032, as cited in Kerrane, Hogg, & Bettany, 2012, p. 811). Solomon (2018) indicated, marketers need to figure out who makes the buying decisions in a family because this information tells them who to target and whether they need to reach both spouses to influence a choice. As children play an active and dominant role in family decision-making, they also become a major source about consumption (Dholakia, 1984). Because of this, children's influence on family decision-making related to purchase is an interesting and popular subject in consumer behavior literature. As a family member, children, who are involved in the recognition or product information search stage, tend to have the greatest influence on the purchasing decision (Beneke, Silverstone, Woods & Schneider, 2011). However, Belch, Ceresino and Belch (1985) claimed that, the influence of children on the family decision-making process is receiving only limited research attention, whereas a great deal of attention has been devoted to children as consumers. Children are influencing the lives of their family at a very broad spectrum of purchases ranging from consumables to services for big ticket items (Aggarwal & Khurana, 2016).

Ward and Wackman (1972) found that, children purchase influence attempts may decrease somewhat with age, depending on the type of product, but mothers' yielding to requests increases with age, probably reflecting a perceived increased competence of older children in making judgements about purchase decisions. Swinyard and Sim (1987) examined the influence of children on a range of products in four decision stages –problem recognition, information search, final decision, and actual purchase. According to their findings, children were found to participate more in decisions about products for which they have apparent self-interest.

Product Type Matters

In one of the early studies related with influence of children on family decision-making, Belch and others (1985) indicated that, the child shows the highest amount of influence on decisions regarding breakfast cereal, and to a lesser degree related to family vacations, and only very little influence is demonstrated on any other products. However, Götze, Prange, & Uhrovska (2009) have found that, children's social power may be perceived much higher for other categories than food because parents are not in the same knowledgeable position as children (e.g. concerning computer games) and perceive children to have the right to influence due to their quasi-expert status. In terms of value, Foxman, Tansuhaj, and Ekstrom (1989) discovered that, children tend to have more to say in the purchase of products that are less expensive and for their own use. Another study also supports this idea by suggesting that the more expensive a product is, the lower is the child's influence (Götze *et al.*, 2009).

For family vacations in particular, the child's greatest influence occurs in the initiation stage and destination choice (Belch *et al.*, 1985). A national survey in the US found out that adolescents influenced their families' purchasing decisions for a variety of products, including vacations and travel (60%), magazine subscriptions (55%) and stereo equipment (37%) (Foxman *et al.*, 1989). According to a more recent survey, one in three millennial parents in the US (34 percent) allow children to make the final decision on where to go on vacation (HomeAway, 2015). Additionally, Yung and Lattimore (2018, p. 64) assert that

Children play an important role in the family travel decision-making process. Whether directly or indirectly, they exert an increasing amount of influence on not only where the family decides to travel to, but also what activities they take part in when they are there.

Demographics Make a Difference

Demographic factors play a major role in children's power to influence their parents (Götze *et al.*, 2009). For instance, the influence of children on their parents' decision-making has been found to increase with age (Beneke *et al.*, 2011). In parallel to this assertion, children in Turkey with the age above 13 have active participation in the holiday purchase decision-making process, while the influence of other age groups (0-5 and 6-12) of children on this process seems limited. Additionally, children older than 13 have a significant effect on the family holiday location decision (Aymankuy & Ceylan, 2013). Similarly, older children were found to participate more in the decision-making processes than younger children, particularly for "non-child" products (Swinyard & Sim, 1987). According to a study that explores children's influence on high-technology products purchasing decisions, it has been found that male children have a stronger influence during the search and decision stages, whereas female children play a more active role during the initiation[1] phase (Beneke *et al.*, 2011). Another study with German, Russian and Turkish families indicated that, German families were generally child-centred, whereas Russian families were parent-centred. Additionally, the education level and child-centeredness related to holiday decisions were negatively, whereas parents' age and child-centeredness related to holiday decisions were positively correlated. In other words, less educated and older parents listen more to their children in terms of holiday decisions (Çakıcı & İyitoğlu, 2010). Another study with Israeli parents revealed that, pluralistic parents allowed their children the highest levels of influence, while protective parents allowed their children the least in the process of family decision-making (Bakir *et al.*, 2006). Seaton and Tagg (1995) carried out a multinational study with families from Italy, Belgium, the UK and France. They found that children from the UK were least likely, and children from Italy were most likely involved in vacation decisions compared to the others. They also claimed that, involving children in the vacation decision-making process improves the possibility of an optimum outcome. Fodness (1992) found that, in families with children, wives were more likely to make individual decisions and less likely to do so in families without children. He also discovered that wives were more likely willing than husbands to conduct the pre-vacation information search.

Characteristics of families may differ in terms of child influence on the consumption choices. After a comprehensive literature review, Kerrane *et al.* (2012) identified some variables, which moderate the amount of influence that a child has within the family decision-making process: (1) family communication pattern, (2) parental socialization style, (3) family size and (4) family type. They suggest that these mentioned variables might moderate the amount of influence a child has within the family. Some evidence also tells us that parental employment status (single versus dual income) appears to affect teen's influences for some categories (stereo phones) but not for others (TVs, VCRs, furniture) (Götze *et al.*, 2009). Additionally, Beneke *et al.* (2011) refer to the social class in terms of child influence. They assert that children who are members of the middle class and higher income families have greater influence on purchasing decisions. However, Chavda, Haley, and Dunn (2005) mention social class in terms of pester power of children. They assert that pester power influences poorer families more than middle-class families, owing to the former not being able to afford expensive (advertised) products. Similarly, Wiman (1983 as cited in Lawlor & Prothero, 2011, p. 565) has found that, children from middle-class backgrounds with well-educated parents tended to be more critical to advertising and also less likely to ask their parents for an advertised product. In one case from Turkey, after a processed meat product advertisement was broadcasted on TV, which depicted sujuk[2] very tasty, many TV audience complained about it and they demanded from the authorities to ban this advertisement. To explore the effectiveness of that controversial ad, a research was conducted and it turned out that those complaints mainly came from poor families who were worried that their children keep pestering them to purchase that sujuk which they could not afford. With suggestion of The Advertising Self-Regulatory Board[3] (RÖK) and the manufacturer's intent, the broadcast of the commercial was shifted after 10:00 pm in the schedule, when children were probably asleep (Hürriyet, 2003).

One reason of the increased children's influence on family consumption choices might be the increasing number of separated parents. Wang and Schofer (2018) indicated that divorce rates had increased dramatically across the globe representing one of the major societal shifts of the late twentieth century. They also claimed that, ideas of individual freedom, consent, and gender equality may influence those shifts. According to official statistics, the global crude divorce rate[4] seems to have doubled from 1960 to 2017. It was 1.00 in 1960 and 2.12 in 2017 (Unifieldlawyers, 2017). In the EU-28, the crude divorce rate was 1.6 in 1990 and it raised to 1.9 in 2015 (Eurostats, 2015a). In Turkey, the number of divorced couples increased by 1.8% compared to the year 2016. The crude divorce rate was calculated 1.6 for the year 2017 (TUIK, 2018a). Even though, a 1.6‰ divorce rate in Turkey seems low compared with western countries, it has been rising significantly. Supporting that hypothesis, it was 1.41 in 2001 (Başkaya & Ünal, 2017) and 0.46 in 1990 (Eyce, 2002). As a result, we will probably see more divorced parents with children in the future in Turkey. In addition, it would not be wrong to say that those single parents' yielding to their children's requests will be much more. Beneke *et al.* (2011) indicate that adolescents in single-parent families appear to have greater influence than adolescents in step and intact families. However, Eckel (1999, as cited in Flurry, 2007, p. 329) claimed that, divorced single parents often suffer a financial burden of a sudden decrease in resources and are thereby interested in cost-effective products, which provide security and time savings.

CHILD-FRIENDLY HOTEL

Roy (2011) indicated that, the hospitality industry is confronted more and more with a growing competition on a global level, a fast advance in technology and greater expectations from the clients. He also claimed that most hotel companies have improved operations and quality. It is pointed out that factors influencing competitiveness in the hospitality industry are as follows: (1) interaction abilities, which are understood as superior abilities, (2) learning organization, which means turning the hotel into an organization that learns, (3) human resources, the people which represent the central point in all aspects of activity in the hospitality industry, from creation and design to the development and delivery of services and (4) technology and information technology, which represent an important strategic asset in obtaining competiveness (Roy, 2011; Maria, Madalina, Catalina & Diana, 2008). Growing competition in the hospitality industry urges industry professionals to seek emerging and profitable markets. Families with children might be one of these markets.

Some demographic features of Turkey make the country attractive for the child-friendly hotel concept. Firstly, Turkey is known as a country with a high children population. According to official stats, for the year 2017, 28.3% of the country's population are children, which are approximately 23 million (TUIK, 2018b). Secondly, for the year 2015, Turkey ranks as top #1 with a 2.14 fertility rate, among 37 European Countries, 28 of which are EU members (Eurostats, 2015b). The average ratio of 0-14 years old children in the total population is about 17.4% in the EU (Eurostats, 2016), whereas it is 23.6% in Turkey (TUIK, 2018c). These data show us that, Turkish households have, and will probably have, more children compared to European countries. From this point of view, it is believed that Turkish children might significantly affect their parents purchase decisions, especially for tourism destinations. There are many various studies considering children's influences on tourism/travel purchases (Fodness, 1992; Thornton, Shaw & Williams, 1997; Gram, 2007; Khoo-Lattimore, 2015; Chen *et al.* 2016; Cicero & Teichert, 2017), whereas for the Turkish tourism market it is very limited (Çakıcı & İyitoğlu, 2010; Emir & Pekyaman, 2010; Aymankuy & Ceylan, 2013). The potentials of the market whet the hotel professionals' appetite, as they want to make more profits. The new emerging child-friendly hotel concept can be seen as a result of this.

As we mentioned previously, literature shows us that children are not only important for the markets mainly directed to them such as snacks or toys, they are also important for such markets targeted to households such as cars, furniture or leisure services. As Schänzel and Yeoman (2015) assert, capturing the diverse needs of children is one of the essential factors for the future of family tourism. From this perspective, child-friendly hotels can be seen as a concept for attracting families with children. It is not easy to say that there is a consensus to define a child-friendly hotel. This concept is mostly created by the hospitality industry, thus there seem to be very limited academic discussions has been made on this topic. A child-friendly hotel can be simply defined as a hotel, which offers additional services intended for children's and parents' special needs. Those needs vary from parents to parents and child to child. This concept has become more important recently in the industry, as one professional in the Turkish hotel industry says:

The child factor in tourism has become more important in the last ten years; changing concepts, the increasing number of hotels and competition force hotels to generate additional services. The child-friendly hotel concept is the result of these services. (Hürriyet, 2010)

As the professional's indication shows us, the industry sees child-friendly concept as a justification for the expansion of their services.

There are only few studies interested in child-friendly hotels in Turkey. A summary of these studies is presented in Table 1. As Özel (2015) indicated, tourism marketing literature lacks of comprehensive and up-to-date studies, which focus on practices of marketing to children in hotel establishments. In none of those studies, child-friendly dimensions are compared in terms of availability. It is predicted that our findings, especially the comparison of the availability of child-friendly characteristics which is given in Table 8, will contribute to literature and will be beneficial for tourism decision-makers. Thanks to our findings, we hope that tourism decision-makers can easily see in which dimensions child-friendly hotels perform well and in which they do not, thus it will become clear where to interfere in the first place.

Table 1. Summary of Some Studies on Child-Friendly Hotels

Authors	Research methodology	Sample	Major findings
Emir and Pekyaman, 2011	Quantitative (survey)	514 hotel customers who have 5-6 years old children	Families concern of safety issues of a hotel (e.g. safe playing areas, providing health service) much
Özel, 2015	Descriptive analysis	77 child-friendly hotels	Child-friendliness characteristic of a hotel can be categorized under six titles: (1) room options for children, (2) food and beverage options for children, (3) activities for children, (5) price options (6) children's health and safety
Aslan, Tosun and Kurtuluş, 2019	Content analysis	152 customer complaints belong to 45 ultra-all-inclusive child-friendly hotels	Customer complaints are generally divided under five categories: (1) general hygiene, (2) health services, (3) child-specific services, (4) general problems of the hotel and (5) employee behavior towards customers

METHOD

Research Sample and Data Collection Method

The universe of research consists of facilities described as "child-friendly hotels" in Turkey. There is no clear or legal description on the child-friendly hotel concept. Because the Ministry of Culture and Tourism of the Republic of Turkey has not determined a standard, this title is given by tourism agencies. In this research, child-friendly hotels have been found from websites of tourism agencies such as Etstur.com, Jollytur.com, Touristica.com.tr, Otelpuan.com.tr, Tatil.com, Tatilbudur.com and Tatilsepeti.com. It has been observed that some of the agencies describe hotels as child-friendly which have only children's pools and water slides. However, considering that these two characteristics are not enough, these hotels are excluded from the scope of the research. As a result of reviewing tourism agencies' websites, 89 child-friendly hotels have been found. Due to this accessible number, the research has been made in the universe without sampling.

The research has been designed qualitatively and secondary data have been used. The data collection method of the research consists of a document review. The document review includes analysis of written materials containing information on facts and events to be researched (Yıldırım & Şimşek, 2013). The reviewed data may be written texts, images or videos. Written texts may be reviewed, or the data may be gathered on the Internet. Baş and Akturan (2017) stated that document analysis can be done through web pages, this method has advantages such as flexibility of data sources, accessing visual and written content and providing the opportunity to sample from a wide geographical region. Additionally, this method provides advantages such as cost and time savings, ease of data collection, avoiding participant reactivity, accuracy and validity (Bowen, 2009).

Considering few studies about which criteria child-friendly hotels should include and the components specified by tourism agencies, 35 items have been determined (Emir & Pekyaman, 2010; Güven, Beydilli, Ceylan, & Bütün, 2017). To summarize the data more easily, these 35 items were grouped into 5 dimensions named as: *baby-related services*, *room-related services or facilities*, *food & beverage services*, *entertainment-related services* and *other services*.

Primarily, a checklist has been created based on these items. Then, the hotels' own websites have been reviewed. Firstly, items are evaluated as available or non-available, then available items of hotels specified on their websites are marked, and items non-available on websites of related hotels have been reviewed on the websites of tourism agencies additionally. For the distinction of where the data were collected, the data gathered from websites of hotels and tourism agencies have been marked by using different coloured pens on the checklist. While reviewing hotels, the key point has been to find the required item on the website of the hotel or tourism agency. The item related to the factors offered by the hotels but not specified on their or the agencies' websites has been considered non-available. The data collection process was between the 21st June to the 4th of July, 2018. All data related to hotels have been analysed using a computer spreadsheet program.

For the data analysis, the content analysis method has been used. The content analysis is an analysis method to attain concepts and relations to clarify the data collected and enable valid deduction (Krippendorff, 2018). In this method, because the purpose is to give meaning to the data collected, data have been categorized by coding. As the data have been categorized, the researcher may make better interpretations of the data. Prasad (2008, p. 2) stated

Broadly content analysis may be seen as a method where the content of the message forms the basis for drawing inferences and conclusions about the content and further, content analysis falls in the interface of observation and document analysis.

Validity and Reliability of the Research

Every research should meet validity and reliability conditions. While validity is about whether the matter to be evaluated is evaluated or not, reliability is about the consistence and stability of the evaluation. Reliability should include stability, repeatability and accuracy (Krippendorff 1980). Ensuring reliability of qualitative research studies is not as easy as it is for quantitative research studies. While validity of quantitative research studies is ensured with a computer program, validity of qualitative research studies depends on the researcher. Reliability of content analysis may be ensured in two ways; researchers and time (Potter & Donnerstein, 1999). In terms of intercoder reliability, validity is related to different researchers doing the same analysis in line with the coding conventions. Reliability in terms of time, on the other hand, is related to one researcher coding the communication content in line with the same coding conventions at two different times. If the repeated evaluations in the research lead to similar conclusions or results, reliability may be attained. In this research, reliability has been ensured with intercoder reliability which is the leading method. The two researchers made the coding by obtaining the data related to child-friendly hotels from the same websites in line with the same criteria. Because the results are similar, it is considered that the research is reliable.

In order to ensure validity, it is required that the data collection method, the research design, sample and the results are in accordance with the research question and purposes (Leung, 2015). Making a wrong choice about one of these may have a negative effect on the validity of the research. Yıldırım & Şimşek (2013) stated that, validity is evaluated in two phases; internal validity and external validity. Internal validity is about researchers' being consistent in data collection, data analysis and interpretation processes, and their explanation on how they ensure this consistency. External validity is about generalizability; it is related to researchers' informing the reader in detail about all phases of the research. Before the field work, items for child-friendly hotels have been determined by reviewing literature and two leading academicians' approvals of these items have been obtained.

FINDINGS

In this part, the information gathered through content analysis will be given. First, general information about the sample hotels, then five dimensions of child-friendly hotels (baby-related services, room-related services or facilities, food & beverage services, entertainment-related services and other services) will be given in separate tables with frequencies and percentages. Table 2 shows general information of child-friendly hotels.

Table 2. General information of child-friendly hotels

Region	n	%
Belek	26	29.2
Kemer	16	18
Side	13	14.6
Bodrum	8	9
Lara	8	9
Fethiye	5	5.6
Alanya	4	4.5
Aydın	2	2.2
Marmaris	2	2.2
İzmir	2	2.2
Dalaman	1	1.1
Kazdagları	1	1.1
Afyon	1	1.1
Total	89	100

Star	n	%
5 Star	85	95.5
4 Star	4	4.5
Total	*89*	*100*

No extra charge for two children	n	%
Yes	13	14.6
No	76	85.4
Total	*89*	*100*

Accommodation Type	n	%
Ultra-all-inclusive	76	85.4
All-inclusive	9	10.1
Half pension	2	2.2
Bed and breakfast	2	2.2
Total	*89*	*100*

Weighted information source[5]	n	%
Agent website	55	61.8
Hotel website	30	33.7
Equivalent	4	4.5
Total	*89*	*100*

According to Table 2, related to the region information, the largest number of child-friendly hotels are located in Belek, followed by hotels located in Kemer and Side, respectively. 95.5% of child-friendly hotels are 5-star, 85.4% of them offer ultra-all-inclusive accommodation, 85.4% of the sample hotels request extra charge for two children. In terms of the weighted information source, 61.8% of the information is acquired mainly from agent websites and 33.7% of the information is acquired mainly from hotel websites. Table 3 shows baby-related services of sample hotels.

Table 3. Baby-related services

Item	Available (%)	Non-available (%)	Total (%)
Solid Baby Food/Mama	62 (69.7)	27 (30.3)	89 (100)
Baby Care Unit in Toilets	61 (68.5)	28 (31.5)	89 (100)
Bottle Sterilization Machine	53 (59.6)	36 (40.4)	89 (100)
Stroller	37 (41.6)	52 (58.4)	89 (100)
Baby Sleeping Room	25 (28.1)	64 (71.9)	89 (100)
Waterproof Baby Diapers	7 (7.9)	82 (92.1)	89 (100)
Total	245 (45.9)	289 (54.1)	534 (100)

According to Table 3, 69.7% of child-friendly hotels offer solid baby food/mama, 68.5% of child-friendly of them offer baby care unit in toilets, 59.6% of them offer bottle sterilization machines, 41.6% of them offer strollers, 28.1% of them offer baby sleeping rooms and 7.9% of them offer waterproof baby diapers. The total availability of baby-related services on hotel or agents' websites is 45.9%. Table 4 shows room-related services or facilities of sample hotels.

Table 4. Room-related services or facilities

Item	Available (%)	Non-available (%)	Total (%)
Family Room	88 (98.9)	1 (1.1)	89 (100)
Children's Bed	79 (88.8)	10 (11.2)	89 (100)
Water Heater	73 (82)	16 (18)	89 (100)
Baby Tub	61 (68.5)	28 (31.5)	89 (100)
Room Cleaning Twice A Day (Upon Request)	47 (52.8)	42 (47.2)	89 (100)
Bottle Heater	24 (27)	65 (73)	89 (100)
Baby Radio	27 (30.3)	62 (69.7)	89 (100)
Milk/Wafer In Minibar	9 (10.1)	80 (89.9)	89 (100)
Humidifier	6 (6.7)	83 (93.3)	89 (100)
Furniture For Children	0 (0)	89 (100)	89 (100)
Total	414 (46.5)	476 (53.5)	890 (100)

According to Table 4, 98.9% of sample hotels have family rooms, 88.8% of them have children's beds, 82% of them have water heaters, 68.5% of them baby tubes, 52.8% of them offer room cleaning twice a day, 27% of them have bottle heaters, 30.3% of them have baby radios, 10.1% of them have milk/wafer in minibars, 6.7% of them have humidifiers. None of the sample hotels have furniture for children in the rooms. The total availability of room-related services or facilities on hotel or agents' websites is 46.5%. Table 5 shows food & beverage services of sample hotels.

Table 5. Food & beverage services

Item	Available (%)	Non-available (%)	Total (%)
Snacks	83 (93.3)	6 (6.7)	89 (100)
Kids' Menu	82 (92.1)	7 (7.9)	89 (100)
Patisserie	81 (91)	8 (9)	89 (100)
Ice Cream	79 (88.8)	10 (11.2)	89 (100)
Baby Feeding Chairs	57 (64)	32 (36)	89 (100)
Special Diet For Children	8 (9)	81 (91)	89 (100)
Total	390 (73)	144 (27)	534 (100)

According to Table 5, 93.3% of child-friendly hotels have snacks, 92.1% of them have kids' menus, 91% of them have patisseries, 88.8% of them have ice cream and 64% of them have baby feeding chairs, whereas only 9% of them offer special diet for children. The total availability of food & beverage services on hotel or agent websites is 73%. Table 6 shows entertainment-related services of sample hotels.

Table 6. Entertainment-related services

Item	Available (%)	Non-available (%)	Total (%)
Kids' Animation	89 (100)	0 (0)	89 (100)
Playground	89 (100)	0 (0)	89 (100)
Pool For Children	87 (97.8)	2 (2.2)	89 (100)
Mini Club	86 (96.6)	3 (3.4)	89 (100)
Free Wi-Fi	83 (93.3)	6 (6.7)	89 (100)
Water Slide	81 (91)	8 (9)	89 (100)
Age Group Distinction	54 (60.7)	35 (39.3)	89 (100)
Kids' Cinema	36 (40.4)	53 (59.6)	89 (100)
Funfair	12 (13.5)	77 (86.5)	89 (100)
Total	617 (77)	184 (23)	801 (100)

According to Table 6, all the child-friendly hotels have kids' animation and playground for kids. 97.8% of them have pools for children, 96.6% have mini clubs, 93.3% of them have free Wi-Fi, 91% of them have water slides, and 60.7% of them offer age group distinction for mini clubs. According to Table 5, 40.4% of sample hotels have kids' cinemas and 13.5% of them have funfairs. The total availability of entertainment-related services on hotel or agent websites is 77%. Table 7 shows others services of sample hotels.

Table 7. Other services

Item	Available (%)	Non-available (%)	Total (%)
Health Service	83 (93.3)	6 (6.7)	89 (100)
Babysitting Services	76 (85.4)	13 (14.6)	89 (100)
Check-in Area for Children	5 (5.4)	84 (94.4)	89 (100)
Children's Corner at the Reception	3 (3.4)	86 (96.6)	89 (100)
Total	167 (46.1)	189 (53.9)	356 (100)

According to Table 7, 93.3% of child-friendly hotels have health services, 85.4% of them have babysitting services, whereas only 5.4% of them have check-in areas for children and 3.4% of them have children's corners at the reception. The total availability of entertainment-related services on hotel or agent websites is 46%. Table 8 summarizes all child-friendly service dimensions on hotel or agent websites.

Table 8. Summary of child-friendly service dimensions

Service dimension	Item#	Available (%)	Non-available (%)	Total (%)
Entertainment-Related	9	77	23	100
Food & Beverage	6	73	27	100
Room-Related	10	46.5	43.5	100
Other	4	46.1	53.9	100
Baby-Related	6	45.9	54.1	100
Total	35			

According to Table 8, the availability of entertainment-related services is 77%, food & beverage services is 73%, room-related services or facilities is 46.5%, other services is 46.1% and baby related services is 45.9%. It can be seen that child-friendly hotels are focusing mainly to entertainment-related and food & beverage services. In order to gain competitive advantage and get differentiation, we also recommend them to offer such services few hotels able to serve, such as furniture for children or children's corner at the reception.

CONCLUSION AND RECOMMENDATIONS

This study aims to depict some characteristics of child-friendly hotels in Turkey. According to the findings, hotels are mainly 5-star hotels and mostly located in Belek, Kemer and Side and most of them offer ultra-all-inclusive accommodation. The information about child-friendly characteristics of hotels is more detailed explained on the agents' websites than the hotels' official websites. In other words, hotels willingly give most of their web promotion activities to their agents. The hotels might reduce marketing costs and increase efficiency by doing that, but they have to check the accuracy of the information and make sure that it is updated regularly.

Child-friendly hotels showed good performance concerning entertainment-related services. Every hotel in the sample has kids' animation and playgrounds for children. Similarly, almost all of them have mini clubs, most of them have free Wi-Fi and water slides. Even if almost all of them have mini clubs, only about sixty percent of the sample hotels have age group distinctions in their mini clubs. In other words, about forty percent of child-friendly hotels offer the same entertainment tools for all age groups. It is thought that this might not be sustainable for hotels, because children tend to play and enjoy with their peers, and as Özel (2015) indicated, children who are satisfied with the services of mini clubs will probably prefer to stay in the same hotel as a customer. However, hotels should depict mini clubs with its distinct features as Seraphin and Yallop (2019) indicate, mini clubs are generally copied from other resorts and they no longer provide a competitive advantage.

Child-friendly hotels have also performed well related to food & beverage services for children. Most of them have snacks, kids' menus, patisseries and ice cream. More than half of the sample hotels offer baby feeding chairs, whereas only nine percent of child-friendly hotels offer special diet for children. Considering that all these kinds of food contain a lot of sugar and calories, and children usually eat much more at any 5-star hotel than they eat at home, we think that hotels should be more responsible and more hotels should offer diet food for children. It's not just a marketing strategy, it is crucial for children's health, because as a recent study has found, the prevalence of child obesity is ten times higher today that it was 40 years ago worldwide (Hürriyet, 2017).

According to room-related services or facilities, it is obvious that almost all child-friendly hotels offer family rooms. Most of them offer children's beds and water heaters. More than half of the sample hotels offer baby tubes and room cleaning twice a day. About a quarter of child-friendly hotels offer bottle heaters, only ten percent of them offer milk/wafer in mini bars and almost seven percent of them offer humidifiers. None of the sample hotels have furniture for children. As Emir and Pekyaman (2010) have found, the safety of a child is one of the top priorities related to families' vacation decisions. However according to our results, only a few of the sample hotels offer humidifiers which make it easier for sick babies to breathe, and as viruses grow in cold and dry air, humidifiers can prevent children from getting sick in the first place (Agadoni, 2015). Furthermore, considering that most of the sample hotels are located in Antalya, where the average temperature is above 25 degrees centigrade in the tourism season, for families with infants, humidifier requests may increase depending on the air conditioner usages there.

For baby-related services, more than half of the sample hotels offer solid baby food/mama, baby care units in toilets and bottle sterilization machines. About forty percent of the sample hotels offer strollers, twenty-eight percent of them have baby sleeping rooms and only about eight percent of them offer waterproof baby diapers. Solid baby food/mama is essential for babies' or toddlers' daily intake; however, about thirty percent of the child-friendly hotels do not offer this kind of food. Furthermore, while sleeping is also essential for children's growth, about seventy percent of the sample hotels do not have baby sleeping room services either. We think that hotels, that are promising safe, peaceful and entertaining vacation experiences to families with children, should give more emphasis on these kinds of services.

For other services, most of the child-friendly hotels offer health services and most of them offer babysitting services, however only about five percent of them offer check-in areas for children and only about three percent of them offer children's corners at the reception. It is interesting to find out that most of the hotels give families babysitting services, whereas only about twenty-eight percent of them have baby sleeping rooms. In our opinion, hotels should consider babysitting services like playing partners or guardians. Besides, almost all of the hotels demand extra charge for health and babysitting services.

Based on the findings of the study, it is possible to make some recommendations. Firstly, measuring guests' perceptions for child-friendly services given in the hotel would be beneficial. Through this, which child-friendly items are needed most or the absence of which ones are tolerable will easily come out. For instance, only thirteen percent of the sample hotels have the option of "no extra charge for two children". For families with children this option is very important, and for the hotel business, we think that letting families free of charge for at most two children can help to create competitive advantage in the industry. Secondly, hotels or agents should design the marketing campaign for the family as a whole. This campaign might involve visuals that promise entertainment and fun for children, but it also should promise that, during the stay, parents would be safe and peaceful regarding their children's all kind of needs. Thirdly, hotels consider entertainment-related and food & beverage services more than the other dimensions. From this point of view, hotels should focus on some other child-friendly items, because according to findings they have almost forgotten some services such as waterproof baby diapers or furniture for children. Last but not least, because the environment, in which the hotels are operating, is continuously evolving, decision-makers should improve hotels' learning organization capabilities. For instance, they should follow up the changes in demographics, which prove to affect child influence, and technology improvements such as mobile marketing possibilities or augmented reality. As Bayraktaroglu and Kutanis (2003, p. 153) claim

learning organization mentality can be seen a key for success in developing and maintaining a competitive advantage for the Turkish Tourism Industry.

Discussion

Based on comprehensive literature review and empirical findings, we think that any child-friendly hotel should define main demographics of its target group before designing promotion strategies. This is especially important because we already know that children's demographic features such as their age (Ward & Wackman, 1972, Aymankuy & Ceylan, 2013) and gender (Beneke *et al.*, 2011), and demographics features of parents such as age and education level (Çakıcı & İyitoğlu, 2010) and family type (Beneke *et al.*, 2011) might affect the influence of purchase decisions. By doing this, any child-friendly hotel might adapt its promotion messages according to its customers. Because older children are suggested to exert greater influence within families in relation to younger siblings (Kerrane *et al.*, 2012), using child-friendly dimensions in hotels' communication messages, especially those for older children such as funfair or kid's cinema, which seem to be comparatively low according to our findings, would be effective for convincing prospective families as customers. For those families, online communication messages should have convincing images portraying that the hotel promises child-friendly attributes, because we already know that children are often regarded as internet mavens and they are especially instrumental role players in the search and decision phases (Beneke *et al.*, 2011). Child-friendly characteristics of a hotel is especially important for low-budget hotels because empirical evidence supports that influence of children will be more when the product (or service) is less expensive (Foxman *et al.*, 1989, Götze *et al.*, 2009).

Unfortunately, child-friendly hotel features vary from hotel to hotel, so there is no clear and common definition of a child-friendly hotel, yet. When any customer makes a hotel reservation just because it is advertised as child-friendly, a dissatisfaction might occur. It is recommended that, especially from the practitioner's side, a child-friendly hotel certification would help to figure out these problems. By doing this, the features of child-friendliness are going to be identified and updated regularly, therefore customers with children will be certain that they are going to get satisfying service in terms of child-friendliness. Similarly with halal hotels, which are organized according to Islamic principles such as halal food, alcohol free services, prayer rooms etc., a child-friendly hotel certificate might contribute value and help to increase customer trust. This certificate can be put into action by either an international (such as International Hotel & Restaurant Association- IH&Ra) or a national organization (such as Turkish Hoteliers Federation- TÜROFED) under the supervision of local and national authorities such as municipalities and the Ministry of Tourism.

FUTURE RESEARCH DIRECTIONS

In the future, studies may contribute to develop a reliable and universal child-friendly hotel evaluation scale, which includes all the dimensions of the services given in the hotel for children and their parents. Additionally, rather than merely on the websites, future studies may also focus on exploring child-friendly characteristics of hotels on different platforms of the Internet, such as mobile applications or social media platforms of the resorts. Exploring some performance measures of child-friendly characteristics on the web such as content, speed, resolution or interactivity is also recommended. Comparing those findings among different cultures, for instance between child-centred or parent-centred cultures would be an interesting topic for further research.

LIMITATIONS

There are some limitations in this study. Firstly, because there are no reliable criteria, sample hotels are chosen through third party agents' websites, advertised as "child-friendly". In other words, we cannot say that all child-friendly hotels in Turkey are included in our sample. Secondly, findings of child-friendly characteristics of the sample hotels are based on the data given on hotels' or agents' websites. Therefore, any dimension which seem low on the findings does not simply mean that the hotel's performance on that dimension is really low. For instance, although a hotel provides babysitting services currently, they might have forgotten to announce that information on the Internet. Thirdly, the current data on the Internet is continuously evolving and updating. Therefore, the findings of this study only represents sample due to the information put on the related websites and only by the time that the data were collected.

ACKNOWLEDGMENT

This academic work was linguistically supported by the Mersin Technology Transfer Office Academic Writing Center of Mersin University.

REFERENCES

Agadoni, L. (2015). *Finding the best humidifier for baby*. Retrieved from https://www.care.com/c/stories/4527/finding-the-best-humidifier-for-baby/

Aggarwal, V. S., & Khurana, S. (2016). Influence of children in family buying process. *BVIMSR's Journal of Management Research*, *8*(2), 146–156.

Aslan, S., Tosun, N., & Kurtuluş, S. A. (2019). A content analysis for complaints related to child-friendly hotels [in Turkish]. *The Journal of Business Science*, *7*(2), 451–475.

Aymankuy, Y., & Ceylan, U. (2013). The role of children on families' decision taking for the purchase of tourism products (a research on native tourists). [in Turkish]. *Electronic Journal of Social Sciences*, *12*(45), 105–122.

Bakir, A., Rose, G. M., & Shoham, A. (2006). Family communication patterns: Mothers' and fathers' communication style and children's perceived influence in family decision making. *Journal of International Consumer Marketing*, *19*(2), 75–95. doi:10.1300/J046v19n02_05

Baş, T., & Akturan, U. (2017). *Computer aided qualitative research methods in social sciences*. Seçkin Publishing. [in Turkish].

Başkaya, Z., & Ünal, A. (2017). Temporal change of marriage and divorce rates at provincial levels in Turkey (2001-2005). [in Turkish]. *The Journal of International Social Research*, *10*(53), 338–358.

Bayraktaroglu, S., & Kutanis, R. O. (2003). Transforming hotels into learning organizations: A new strategy for going global. *Tourism Management*, *24*(2), 149–154. doi:10.1016/S0261-5177(02)00061-4

Belch, G. E., Belch, M. A., & Ceresino, G. (1985). Parental and teenage child influences in family decision making. *Journal of Business Research*, *13*(2), 163–176. doi:10.1016/0148-2963(85)90038-4

Beneke, J., Silverstone, G., Woods, A., & Schneider, G. (2011). The influence of the youth on their parents' purchasing decisions of high-technology products. *African Journal of Business Management*, *5*(10), 3807-3812.

Berey, L. A., & Pollay, R. W. (1968). The influencing role of the child in family decision making. *Journal of Marketing Research*, *5*(1), 70–72. doi:10.1177/002224376800500109

Bowen, G. A. (2009). Document analysis as a qualitative research method. *Qualitative Research Journal*, *9*(2), 27–40. doi:10.3316/QRJ0902027

Çakıcı, C., & İyitoğlu, V. (2010). Market segmentation of the families coming to Antalya by the effect of children on vacation decisions. [in Turkish]. *Marketing and Marketing Research Journal*, *8*, 1–22.

Caruana, A., & Vassallo, R. (2003). Children's perception of their influence over purchases: The role of parental communication patterns. *Journal of Consumer Marketing*, *20*(1), 55–66. doi:10.1108/07363760310456955

Ceylan, E. N. (2019). Child friendly tourism establishment: The Land of Legends [in Turkish]. *Journal of Recreation and Tourism Research*, *6*(3), 263–284. doi:10.31771/jrtr.2019.30

Chavda, H., Haley, M., & Dunn, C. (2005). Adolescents' influence on family decision-making. *Young Consumers*, *6*(3), 68–78. doi:10.1108/17473610510701223

Chen, Y. S., Lehto, X., Behnke, C., & Tang, C. H. (2016). Investigating children's role in family dining-out choices: Evidence from a casual dining restaurant. *Journal of Hospitality Marketing & Management*, *25*(6), 706–725. doi:10.1080/19368623.2016.1077368

Cicero, L., & Teichert, T. (2017). Children's influence in museum visits: Antecedents and consequences. *Museum Management and Curatorship*, *33*(2), 146–157. doi:10.1080/09647775.2017.1420485

Dholakia, R. R. (1984). Intergeneration differences in consumer behavior: Some evidence from a developing country. *Journal of Business Research*, *12*(1), 19–34. doi:10.1016/0148-2963(84)90035-3

Emir, O., & Pekyaman, A. (2010). Important factors for families with children in hotel establishment selection: An application in Afyonkarahisar. [in Turkish]. *Management & Economics: Journal of Celal Bayar University Faculty of Economics and Administrative Sciences*, *17*(2), 159–181.

Etstur. (2020). *ETSTUR travel agency official website*. Retrieved from https://www.etstur.com/

Eurostats. (2015a). *Marriage and divorce statistics*. Retrieved from https://ec.europa.eu/eurostat/statistics-explained/index.php/Marriage_and_divorce_statistics

Eurostats. (2015b). *Total fertility rate, 1960–2015*. Retrieved from https://ec.europa.eu/eurostat/statistics-explained/index.php?title=File:Total_fertility_rate,_1960%E2%80%932015_(live_births_per_woman)_YB17.png

Eurostats. (2016). *Being young in Europe today - demographic trends*. Retrieved from https://ec.europa.eu/eurostat/statistics-explained/index.php/Being_young_in_Europe_today_-_demographic_trends

Eyce, B. (2002). Divorce by demographic characteristics in Turkey. [in Turkish]. *Selçuk University Journal of Faculty of Letters*, *14*, 81–98.

Flurry, L. A. (2007). Children's influence in family decision-making: Examining the impact of the changing American family. *Journal of Business Research*, *60*(4), 322–330. doi:10.1016/j.jbusres.2006.09.029

Fodness, D. (1992). The impact of family life cycle on the vacation decision-making process. *Journal of Travel Research*, *31*(2), 8–13. doi:10.1177/004728759203100202

Foxman, E. R., Tansuhaj, P. S., & Ekstrom, K. M. (1989). Family members' perceptions of adolescents' influence in family decision making. *The Journal of Consumer Research*, *15*(4), 482–491. doi:10.1086/209187

Götze, E., Prange, C., & Uhrovska, I. (2009). Children's impact on innovation decision making. *European Journal of Marketing*, *43*(1/2), 264–295. doi:10.1108/03090560910923328

Gram, M. (2007). Children as co-decision makers in the family? The case of family holidays. *Young Consumers*, *8*(1), 19–28. doi:10.1108/17473610710733749

Güven, O. Z., Beydilli, E. T., Ceylan, U., & Bütün, S. (2017). A study on child-friendly hotel concept. In Proceedings of 3th Tourism Council (pp. 632-641). Republic of Turkey Ministry of Culture and Tourism. [in Turkish].

HomeAway. (2015). *New Survey Reveals the Strong Influence of Kids on Family Travel Plans*. Retrieved from: https://www.homeaway.com/info/media-center/press-releases/2015/new-survey-reveals-the-strong-influence-of-kids-on-family-travel

Hot to make sujuk? (2020). Retrieved from: https://italianbarrel.com/how-to-make-sujuk/

Hürriyet. (2003). *Sujuk commercial has shifted to late night in order children not to gulp*. Retrieved from: https://www.hurriyet.com.tr/ekonomi/cocuklar-yutkunmasin-diye-sucuk-reklami-geceye-kaydi-124884 [in Turkish].

Hürriyet. (2010). *Child-friendly hotels*. Retrieved from https://www.hurriyet.com.tr/ekonomi/cocuk-dostu-oteller-14051915 [in Turkish].

Hürriyet. (2017). *Children are becoming 'quickly obese'*. Retrieved from https://www.hurriyet.com.tr/dunya/dunya-genelinde-cocuklar-hizla-obezlesiyor-40606622 [in Turkish].

Jelínková, D., Tučková, Z., & Jurigová, Z. (2017). Market segment of families with children: A new stimulus for modern Czech spa industry? *Journal of International Studies*, *10*(2), 158–169. doi:10.14254/2071-8330.2017/10-2/12

Jollytur. (2020). *Jollytur travel agency official website*. Retrieved from https://www.jollytur.com/

Kerrane, B., Hogg, M. K., & Bettany, S. M. (2012). Children's influence strategies in practice: Exploring the co-constructed nature of the child influence process in family consumption. *Journal of Marketing Management*, *28*(7-8), 809–835. doi:10.1080/0267257X.2012.698633

Khoo-Lattimore, C., Prayag, G., & Cheah, B. L. (2015). Kids on board: Exploring the choice process and vacation needs of Asian parents with young children in resort hotels. *Journal of Hospitality Marketing & Management*, *24*(5), 511–531. doi:10.1080/19368623.2014.914862

Krippendorff, K. (1980). Validity in content analysis. In E. Mochmann (Ed.), *Computerstrategien für die Kommunkationsanalyse* (pp. 69–112). Campus. Retrieved from https://pdfs.semanticscholar.org/8a09/38e9ee2d55ee64c47e4b4be3f6177c58f20c.pdf

Krippendorff, K. (2018). Content Analysis: An Introduction to its Methodology (4th ed.). SAGE Publications.

Lawlor, M. A., & Prothero, A. (2011). Pester power–A battle of wills between children and their parents. *Journal of Marketing Management*, *27*(5-6), 561–581. doi:10.1080/0267257X.2010.495281

Leung, L. (2015). Validity, reliability and generalizability in qualitative research. *Journal of Family Medicine and Primary Care*, *4*(3), 324–327. doi:10.4103/2249-4863.161306 PMID:26288766

Maria, I., Madalina, T., Catalina, B., & Diana, I. (2008). The factors of competitiveness in the hospitality industry and the competitive strategy of firms. *Annals of the University of Oradea. Economic Science Series*, *17*(2), 213–218.

Otelpuan. (2020). *Hotels evaluation website*. Retrieved from https://www.otelpuan.com/tr

Özel, Ç. H. (2015). Marketing to children in tourism industry: descriptive analysis of kid-friendly hotels' practices in Turkey. In A. Correia, J. Gnoth, M. Kozak, & A. Fyall (Eds.), *Marketing Places and Spaces* (pp. 193–208). Emerald Group Publishing Limited. doi:10.1108/S1871-317320150000010014

Poock, M. C., & Lefond, D. (2001). How college-bound prospects perceive university web sites: Findings, implications and turning browsers into applicants. *College and University*, *77*(1), 15–21.

Potter, J. W., & Donnerstein, D. L. (1999). Rethinking validity and reliability in content analysis. *Journal of Applied Communication Research*, *27*(3), 258–284. doi:10.1080/00909889909365539

Prasad, B. D. (2008). Content analysis: A method in social science research. In D. K. Lal Das & V. Bhaskaran (Eds.), *Research Methods for Social Work* (pp. 1–20). Rawat Publications. Retrieved from http://www.css.ac.in/download/deviprasad/content%20analysis.%20a%20method%20of%20social%20science%20research.pdf

Roy, S. (2011). Competitiveness in service sector: A case of hotel industry in India. *Global Business Review*, *12*(1), 51–69. doi:10.1177/097215091001200104

Schänzel, H. A., & Yeoman, I. (2015). Trends in family tourism. *Journal of Tourism Futures*, *1*(2), 141–147. doi:10.1108/JTF-12-2014-0006

Seaton, A. V., & Tagg, D. S. (1995). The family vacation in Europe: Paedonomic aspects of choices and satisfactions. *Journal of Travel & Tourism Marketing*, *4*(1), 1–21. doi:10.1300/J073v04n01_01

Seraphin, H., & Yallop, A. (2019). Proposed framework for the management of resorts Mini Clubs: An ambidextrous approach. *Leisure Studies*, *38*(4), 535–547. doi:10.1080/02614367.2019.1581249

Solomon, M. R. (2018). *Consumer behavior: Buying, having, and being* (12th ed.). Pearson Education Limited.

Swinyard, W. R., & Peng Sim, C. (1987). Perception of children's influence on family decision processes. *Journal of Consumer Marketing*, *4*(1), 25–38. doi:10.1108/eb008186

Tatilbudur. (2020). *Tatilbudur travel agency official website*. Retrieved from https://www.tatilbudur.com/

Tatilcom. (2020). *Tatil.com travel agency official website*. https://www.tatil.com/

Tatilsepeti. (2020). *Tatilsepeti travel agency official website*. Retrieved from https://www.tatilsepeti.com/

The Advertising Self-Regulatory Board (RÖK). (2010). Retrieved from https://www.rok.org.tr/en/rok-platform/

Thornton, P. R., Shaw, G., & Williams, A. M. (1997). Tourist group holiday decision-making and behavior: The influence of children. *Tourism Management*, *18*(5), 287–297. doi:10.1016/S0261-5177(97)00017-4

Tourstica. (2020). *Touristica travel agency official website*. https://www.touristica.com.tr/

Turkish Statistical Institute (TUIK). (2018a). *Marriage and divorce statistics*. Retrieved from http://www.tuik.gov.tr/PdfGetir.do?id=27593

Turkish Statistical Institute (TUIK). (2018b). *Statistics on children*. Retrieved from http://www.tuik.gov.tr/PreHaberBultenleri.do?id=27596

Turkish Statistical Institute (TUIK). (2018c). *Basic statistics*. Retrieved from http://www.tuik.gov.tr/PreIstatistikTablo.do?istab_id=1588

Unifieldlawyers. (2017). *The love of divorce*. Retrieved from https://www.unifiedlawyers.com.au/blog/global-divorce-rates-statistics/

United Nations. (2020). *Crude Divorce Rate*. Retrieved from https://www.un.org/en/development/desa/population/publications/dataset/marriage/crude-divorces.shtml

Wang, C. T. L., & Schofer, E. (2018). Coming out of the penumbras: World culture and cross-national variation in divorce rates. *Social Forces*, *97*(2), 675–704. doi:10.1093foy070

Ward, S., & Wackman, D. B. (1972). Children's purchase influence attempts and parental yielding. *Journal of Marketing Research*, *9*(3), 316–319. doi:10.1177/002224377200900312

Yıldırım, A., & Şimşek, H. (2013). Qualitative research methods in social sciences (9th ed.). Seçkin Publication. [in Turkish].

Yung, R., & Lattimore, C. K. (2018). Same, same, but different: The influence of children in Asian family travel. In E. Yang & C. Khoo-Lattimore (Eds.), Asian Cultures and Contemporary Tourism. Springer Publication.

ADDITIONAL READING

Carr, N. (2006). A comparison of adolescents' and parents' holiday motivations and desires. *Tourism and Hospitality Research*, *6*(2), 129–142. doi:10.1057/palgrave.thr.6040051

Carr, N. (2011). *Children's and families' holiday experience*. Routledge. doi:10.4324/9780203832615

Davis, H. L. (1976). Decision making within the household. *The Journal of Consumer Research*, *2*(4), 241–260. doi:10.1086/208639

Dunne, M. (1999). The role and influence of children in family holiday decision making. *International Journal of Advertising & Marketing to Children*, *1*(3), 181–191. doi:10.1108/eb027612

Hamilton, K., & Catterall, M. (2006). Consuming love in poor families: Children's influence on consumption decisions. *Journal of Marketing Management*, *22*(9-10), 1031–1052. doi:10.1362/026725706778935655

Lorestani, E. A., & Khalili, M. (2019). Using gamified packaging to increase child influence on family consumption of healthy foods. *Journal of Packaging Technology and Research*, (9) 1-11.

McNeal, J. U. (1992). *Kids as customers: A handbook of marketing to children*. Lexington books.

Wiman, A. R. (1983). Parental influence and children's responses to television advertising. *Journal of Advertising*, *12*(1), 12–18. doi:10.1080/00913367.1983.10672825

KEY TERMS AND DEFINITIONS

Baby-Related Services: One of the five service dimensions of a child-friendly hotel offers. These services include stroller, waterproof baby diapers, etc., and they are especially for families with babies.

Child-Friendly Hotel: A hotel, which offers additional services, intended for children's and parents' special needs.

Family Decision Making: The decision-making process of the household in which husband, wife and children play an active role.

Food and Beverage Services: One of the five service dimensions of a child-friendly hotel offers. These services include kids' menu, ice cream etc., and they are designed especially for children's food and beverage interests.

Influence of Children: The influence of children on family decision-making process especially for household consumption.

Marketing to Children: Marketing activities that are designed for children's special needs, such as healthful and easy to use products, simple promotion messages.

Other Services: One of the five service dimensions of a child-friendly hotel offers. These services include health service, babysitting service, check-in area for children and children's corner at the reception.

Room-Related Services or Facilities: One of the five service dimensions of a child-friendly hotel offers. These services include baby tub, humidifier, etc., and they are usually offered in the hotel room.

ENDNOTES

1. Unlike other studies, this study categorizes purchasing process within three stages as initiation, search and decision.
2. Also known as sucuk, is spicy, dry sausage that's usually made with beef, garlic, sumac, and red pepper. https://italianbarrel.com/how-to-make-sujuk/
3. A non-governmental organization, which aims a proactive and continuous effort for ads to be legal, decent, honest and truthful with the participation of the advertisers – advertising agencies – media trio. It has been founded by the members of Advertisers Association, the Turkish Association of Advertising Agencies and media. https://www.rok.org.tr/en/rok-platform/
4. The crude divorce rate is the annual number of divorces per 1,000 population. https://www.un.org/en/development/desa/population/publications/dataset/marriage/crude-divorces.shtml
5. Indicates that which source contains more information in terms of the 35 child-friendly items.

Compilation of References

Aa (van der), B., Groote, P. & Hulgen, P. (2005). World heritage as NIMBY? The case of the Dutch part of the Wadden Sea. In D. Harrison, & M. Hichcock (Eds.), *The Politics of World Heritage, Negotiating Tourism and Conservation* (pp. 11-22). Clevedon: Channel View Publications.

Aaker, D. A. (1996). Measuring Brand Equity Across Products and Markets. *California Management Review, 38*(3), 102-120.

Aaker, D. (1996). Measuring Brand Equity Across Products and Markets. *California Management Review, 38*(3), 102–120. doi:10.2307/41165845

Aaker, D. A. (1991). *Managing brand equity*. Free Press.

Aaker, J. L. (1999). The Malleable Self: The Role of Self-Expression in Persuasion. *JMR, Journal of Marketing Research, 36*(February), 45–57. doi:10.1177/002224379903600104

Aalbers, M. B. (2019). Introduction To The Forum: From Third To Fifth-Wave Gentrification. *Tijdschrift voor Economische en Sociale Geografie, 110*(1), 1–11. doi:10.1111/tesg.12332

Aas, C., Ladkin, A., & Fletcher, J. (2005). Stakeholder collaboration and heritage management. *Annals of Tourism Research, 32*(1), 28–48. doi:10.1016/j.annals.2004.04.005

Aboim, S. (2013). *A Sexualidade dos Portugueses*. Lisbon: Fundação Francisco Manuel dos Santos.

Adams, K. M. (2004). The genesis of touristic imagery : Politics and poetics in the creation of a remote Indonesian island destination. *Tourist Studies, 4*(2), 115–135. doi:10.1177/1468797604054378

Addis, M., & Holbrook, M. B. (2001). On the conceptual link between mass customisation and experiential consumption: An explosion of subjectivity. *Journal of Consumer Behaviour, 1*(1), 50–66. doi:10.1002/cb.53

Addison, L. (1996). An approach to community-based tourism planning in the Baffin region, Canada's Far North – A retrospective. In L. C. Harrion & W. Husbands (Eds.), *Practicing Responsible Tourism* (pp. 296–329). John Wiley and Sons.

Adhikari, A., & Bhattacharya, S. (2016). Appraisal of literature on customer experience in tourism sector: Review and framework. *Current Issues in Tourism, 19*(4), 296–321. doi:10.1080/13683500.2015.1082538

Adie, B. (2017). Franchising our heritage: The UNESCO World Heritage brand. *Tourism Management Perspectives, 24*, 48–53. doi:10.1016/j.tmp.2017.07.002

Adventure. (2017). In *Cambridge Dictionary online*. Retrieved from https://dictionary.cambridge.org/dictionary/english/adventure

Agadoni, L. (2015). *Finding the best humidifier for baby*. Retrieved from https://www.care.com/c/stories/4527/finding-the-best-humidifier-for-baby/

Agapito, D., Almeida, H., & Campos, A. C. (2016). Sensations, perception and co-creation in rural tourism experiences. In E. Kastenholz, M. J. Carneiro, C. Eusébio, & E. Figueiredo (Eds.), *Meeting Challenges for Rural Tourism through Co-Creation of Sustainable Tourist Experiences* (pp. 39–60). Cambridge Scholars Publishing.

Agapito, D., & Chan, C. S. (2019). A multisensory approach to responsible management in community-based tourism: A case study in Hong Kong. *Journal of Tourism Quarterly*, *1*(1), 1–13.

Agapito, D., Mendes, J., & Valle, P. (2013). Conceptualizing the sensory dimension of tourist experiences. *Journal of Destination Marketing & Management*, *2*(2), 62–73. doi:10.1016/j.jdmm.2013.03.001

Agapito, D., Mendes, J., Valle, P., & Almeida, H. (2014). A contribute to sensory marketing in rural tourist experiences. *PASOS Revista de Turismo y Patrimonio Cultural*, *12*(3), 611–621. doi:10.25145/j.pasos.2014.12.045

Agapito, D., Oom do Valle, P., & Costa Mendes, J. (2013). The Cognitive-Affective-Conative Model of Destination Image: A Confirmatory Analysis. *Journal of Travel & Tourism Marketing*, *30*(5), 471–481. doi:10.1080/10548408.2013.803393

Agapito, D., Valle, P., & Mendes, J. (2014). The sensory dimension of tourist experience: Capturing meaningful sensory-informed themes in South Portugal. *Tourism Management*, *42*, 224–237. doi:10.1016/j.tourman.2013.11.011

Agarez, R. (2013). Regional identity for the leisure of travellers: Early tourism infrastructure in the Algarve (Portugal), 1940-1965. *Journal of Architecture (London)*, *18*(5), 721–743. doi:10.1080/13602365.2013.838982

Aggarwal, V. S., & Khurana, S. (2016). Influence of children in family buying process. *BVIMSR's Journal of Management Research*, *8*(2), 146–156.

Agha, N., & Taks, M. (2015). A theoretical comparison of the economic impact of large and small events. *International Journal of Sport Finance*, *10*(3), 199–216.

Agustín, L. (2010). The (crying) need for different kinds of research. In M. Ditmore, A. Levy, & A. Willman (Eds.), *Sex Work Matters: Exploring Money, Power, and Intimacy in the Sex Industry* (pp. 23–27). Zed Books.

Agustí, P. (2018). Characterizing the location of tourist images in cities. Differences in user-generated images (Instagram), official tourist brochures and travel guides. *Annals of Tourism Research*, *73*, 103–115. doi:10.1016/j.annals.2018.09.001

Ahmad, Y. (2006). The scope and definitions of heritage: From tangible to intangible. *International Journal of Heritage Studies*, *12*(3), 292–300. doi:10.1080/13527250600604639

Ahmed, Z. U. (1991). The inñuence of the components of a state's tourist image on product positioning strategy. *Tourism Management*, *12*(4), 331–340. doi:10.1016/0261-5177(91)90045-U

Akis, S., Peristianis, N., & Warner, J. (1996). Residents attitudes to tourism development: The case of Cyprus. *Tourism Management*, *17*(7), 481–404. doi:10.1016/S0261-5177(96)00066-0

Alagöz, S. B., & Ekici, N. (2014). Experiential Marketing and Vacation Experience: The Sample of Turkish Airlines. *Procedia: Social and Behavioral Sciences*, *150*, 500–510. doi:10.1016/j.sbspro.2014.09.065

Alba, J. W., & Williams, E. F. (2013). Pleasure principles: A review of research on hedonic consumption. *Journal of Consumer Psychology*, *23*(1), 2–18. doi:10.1016/j.jcps.2012.07.003

Albrechts, L., & Denayer, W. (2001). Communicative planning, emancipatory politics and postmodernism. In R. Paddison (Ed.), *Handbook of urban studies* (pp. 369–384). SAGE Publications Ltd. doi:10.4135/9781848608375.n24

Alcañiz, E. B., García, I. S., & Blas, S. S. (2009). The functional-psychological continuum in the cognitive image of a destination: A confirmatory analysis. *Tourism Management*, *30*(5), 715–723. doi:10.1016/j.tourman.2008.10.020

Aleshinloye, K. D., Fu, X., Ribeiro, M. A., Woosnam, K. M., & Tasci, A. D. (2020). The influence of place attachment on social distance: Examining mediating effects of emotional solidarity and the moderating role of interaction. *Journal of Travel Research, ournal of Travel Research, 59*(5), 828-849.

Alexandris, K., & Kaplanidou, K. (2014). Marketing sport event tourism: Sport tourist behaviors and destination provisions. *Sport Marketing Quarterly*, *23*(3), 125–126.

Algarve Events. (2020). *Wedding planner*. Retrieved from: https://algarveevents.com/weddingplanner.php

Algarve Fun. (2017a). *Algarve Stag Dos*. Retrieved from: https://www.algarvefun.com/algarve-stag-dos

Algarve Fun. (2017b). *Best resorts & towns for stag dos and parties in the Algarve*. Retrieved from: https://www.algarvefun.com/algarve-stag-dos/best-resorts-towns-stag-dos-parties-algarve/

Algarve Hen Parties. (n.d.). Retrieved from: https://www.algarvefun.com/algarve-hen-parties

Algarve Weddings. (2016). Retrieved from: https://algarve-weddings.co.uk/

Algarve, P. R. O. T. (2004). Regional Spatial Planning Plan for the Algarve.: Vol. II. *Characterization and Diagnosis*. Algarve Regional Commission for Coordination and Development.

Alhemoud, A. M., & Armstrong, E. G. (1996). Image of Tourism Attractions in Kuwait. *Journal of Travel Research*, *34*(4), 76–80. doi:10.1177/004728759603400413

Ali, F., Hussain, K., & Ragavan, N. A. (2014). Memorable Customer Experience: Examining the Effects of Customers Experience on Memories and Loyalty in Malaysian Resort Hotels. *Procedia: Social and Behavioral Sciences*, *144*, 273–279. doi:10.1016/j.sbspro.2014.07.296

Ali, F., Ryu, K., & Hussain, K. (2016). Influence of Experiences on Memories, Satisfaction and Behavioral Intentions: A Study of Creative Tourism. *Journal of Travel & Tourism Marketing*, *33*(1), 85–100. doi:10.1080/10548408.2015.1038418

Aljandali, A. (2016). *Quantitative Analysis and IBM® SPSS® Statistics: A Guide for Business and Finance*. Springer. doi:10.1007/978-3-319-45528-0

Alkire, S., & Aguilar, G. R. (2011). Least developed countries and the multidimensional poverty index (MPI). Poverty & Human Development Initiative, University of Oxford.

Allen, J. J., & Anderson, C. A. (2018). Satisfaction and frustration of basic psychological needs in the real world and in video games predict internet gaming disorder scores and well-being. *Computers in Human Behavior*, *84*, 220–229. doi:10.1016/j.chb.2018.02.034

Allen, L. R., Long, P. T., Perdue, R. R., & Kieselbach, S. (1988). The impact of tourism development on resident's perception of community life. *Journal of Travel Research*, *27*(1), 16–21. doi:10.1177/004728758802700104

Almeida-García, F., Peláez-Fernández, M. Á., Balbuena-Vazquez, A., & Cortes-Macias, R. (2016). Residents' perceptions of tourism development in Benalmádena (Spain). *Tourism Management*, *54*, 259–274. doi:10.1016/j.tourman.2015.11.007

Altman, D. (2001). *Global Sex*. University of Chicago Press. doi:10.7208/chicago/9780226016047.001.0001

Alvarez, M., de Burgos, J., & Céspedes, J. (2001a). Grupos de interés y estratégias medioambientales de los hoteles españoles. *Tribuna de Economía*, *790*, 163–181.

Alvarez, M., de Burgos, J., & Céspedes, J. (2001b). An analysis of environmental management, organizational context and performance of Spanish hotels. *Omega, Elsevier, 29*(6), 457–471. doi:10.1016/S0305-0483(01)00033-0

Alves, H., Ferreira, J. J., & Fernandes, C. I. (2016). Customer's operant resources effects on co-creation activities. *Journal of Innovation and Knowledge, 1*(2), 69–80. Advance online publication. doi:10.1016/j.jik.2016.03.001

Amadeus Traveller Trend Observatory. (2015). *Future Traveller Tribes 2030, Understanding Tomorrow's Traveller.* Retrieved from: https://amadeus.com/documents/en/retail-travel-agencies/research-report/amadeus-future-traveller-tribes-2030-report.pdf

Andajani, E. (2015). Understanding Customer Experience Management in Retailing. *Procedia: Social and Behavioral Sciences, 211*(September), 629–633. doi:10.1016/j.sbspro.2015.11.082

Andereck, K. L., & Vogt, C. A. (2000). The relationship between residents' attitudes toward tourism and tourism development options. *Journal of Travel Research, 39*(1), 27–36. doi:10.1177/004728750003900104

Andereck, K., & Nyaupane, G. (2011). Exploring the Nature of Tourism and Quality of Life Perceptions among Residents. *Journal of Travel Research, 50*(3), 248–260. doi:10.1177/0047287510362918

Andereck, K., Valentine, K., Knopff, R., & Vogot, C. (2005). Resident's perceptions of community tourism impacts. *Annals of Tourism Research, 32*(4), 1056–1076. doi:10.1016/j.annals.2005.03.001

Andrades, L., & Dimanche, F. (2014). Co-creation of experience value: A tourist behaviour approach. In M. Chen & J. Uysal (Eds.), *Creating experience value in tourism* (pp. 95–112). CABI. doi:10.1079/9781780643489.0095

Andraz, J. M., Silva, J. A., & Viegas, C. M. (2012). The position of tourism activity in the Algarve's economic structure: An integrated region-specific analysis. *Revista Portuguesa de Estudos Regionais, 29*(1), 15–26.

Andresen, L., Boud, D., & Cohen, R. (1999). Experience – Based Learning. In G. Foley (Ed.), *Understanding Adult Education and Training* (pp. 225–239). Allen & Unwin.

Andrews, F. M., & Withey, S. B. (2012). *Social indicators of well-being: Americans' perceptions of life quality.* Plenum Press.

Andriotis, K. (2005). Community Groups' Perceptions of and Preferences for Tourism Development: Evidence from Crete. *Journal of Hospitality & Tourism Research (Washington, D.C.), 29*(1), 67–90. doi:10.1177/1096348004268196

Andriotis, K., & Vaughan, D. (2003). Urban residents' attitudes toward tourism development: The case of Crete. *Journal of Travel Research, 42*(2), 172–185. doi:10.1177/0047287503257488

Antunes, A.M., & Azevedo, A. (1988). *Arquitectura Popular em Portugal, 3rd Volume.* Lisbon: Associação dos Arquitectos Portugueses.

Ap, J. (1992). Residents' Perceptions on Tourism Impacts. *Annals of Travels Research, 19*(4), 665–690. doi:10.1016/0160-7383(92)90060-3

Aragón, J. (1998). Strategic proactivity and firm approach to the natural environment. *Academy of Management Journal, 41*(5), 558–567. doi:10.2307/256942

Aragón, J., Hurtado, N., Sharma, S., & García, V. (2008). Environmental Strategy and Performance in Small Firms: A Resource-Based Perspective. *Journal of Environmental Management, 86*(1), 88–103. doi:10.1016/j.jenvman.2006.11.022 PMID:17239519

Archer, B., & Fletcher, J. (1996). The economic impact of tourism in the Seychelles. *Annals of Tourism Research, 23*(1), 32–47. doi:10.1016/0160-7383(95)00041-0

Aref, F., & Redzuan, M. (2010). Community leaders' perceptions towards socio-cultural impacts of tourism on local communities. *Journal of Human Ecology (Delhi, India)*, *29*(2), 87–90. doi:10.1080/09709274.2010.11906251

Aref, F., Redzuan, M., & Emby, Z. (2009). Assessing community leadership factor in community capacity building in tourism development: A Case Study of Shiraz, Iran. *Journal of Human Ecology (Delhi, India)*, *28*(3), 171–176. doi:10.1080/09709274.2009.11906235

Arnold, M., & Reynolds, K. (2003). Hedonic Shopping Motivations. *Journal of Retailing*, *79*(2), 77–95. doi:10.1016/S0022-4359(03)00007-1

Arnould, E. J., & Price, L. L. (1993). River Magic: Extraordinary Experience and the Extended Service Encounter. *The Journal of Consumer Research*, *20*(1), 24. doi:10.1086/209331

Arnstein, S. R. (1969). A ladder of citizen participation. *Journal of the American Institute of Planners*, *35*(4), 216–224. doi:10.1080/01944366908977225

Arruda, A. M. (1999). Algarve no quadro geocultural do Mediterrâneo Antigo. In *Mª da Graça, Marques (Coord.), O Algarve da Antiguidade aos nossos dias (elementos para a sua história)*. Edições Colibri.

Ashworth, G. J., & Voogd, H. (1990). *Selling the city: Marketing approaches in public sector urban planning*. Belhaven Press.

Aslan, S., Tosun, N., & Kurtuluş, S. A. (2019). A content analysis for complaints related to child-friendly hotels [in Turkish]. *The Journal of Business Science*, *7*(2), 451–475.

Assael, H. (1984). *Consumer Behavior and Marketing Action*. Pub. Co.

Assaker, G., & Hallak, R. (2013). Moderating Effects of Tourists' Novelty-Seeking Tendencies on Destination Image, Visitor Satisfaction, and Short- and Long-Term Revisit Intentions. *Journal of Travel Research*, *52*(5), 600–613. doi:10.1177/0047287513478497

Assaker, G., Vinzi, V. E., & O'Connor, P. (2011). Examining the effect of novelty seeking, satisfaction, and destination image on tourists' return pattern: A two factor, non-linear latent growth model. *Tourism Management*, *32*(4), 890–901. doi:10.1016/j.tourman.2010.08.004

Atkinson, R., & Bridge, G. (2010). Globalisation and the New Urban Colonialism. In J. Brown-Saracino (Ed.), *The Gentrification Debates* (pp. 51–61). Routledge.

Augé, M. (1995). *Non-Places: Introduction to an Anthropology of Supermodernity*. Verso.

Auger, D. (2014). The visitor profile at a major sporting event: the 42nd Final of the Quebec Games (Jeux du Québec) in l'Assomption, Quebec, Canada. *Society & Leisure / Loisir & Société, 37*(1), 151-162.

Australian government. (2019). *Understanding World Heritage: what is outstanding universal value?* https://www.environment.gov.au/heritage/about/world-heritage/outstanding-universal-value

Auzins, A., & Vanags, J. (2012). Analytical assessment of sustainable development concept. *Geomatics*, *8*, 49–53.

Axelsen, M., & Swan, T. (2010). Designing Festival Experiences to Influence Visitor Perceptions: The Case of a Wine and Food Festival. *Journal of Travel Research*, *49*(4), 436–450. doi:10.1177/0047287509346796

Ayazlar, G., & Arslan, R. (2017). A Cross-cultural Investigation of Tourists' Memorable Experiences Between Two Nationalities. *Almatourism - Journal of Tourism. Culture and Territorial Development*, *15*(15), 136–151. doi:10.6092/issn.2036-5195/6558

Aymankuy, Y., & Ceylan, U. (2013). The role of children on families' decision taking for the purchase of tourism products (a research on native tourists). [in Turkish]. *Electronic Journal of Social Sciences, 12*(45), 105–122.

Ayob, N., Wahid, N., & Omar, A. (2011). A measurement model of visitor's event experience within festivals and special events. In *Proceeding of the 5th International Conference of the Asian Academy of Applied Business*, (vol. 9, pp.1-10). Phhom Penh, Cambodia.

Ayuso, S. (2003). *Gestión sostenible en la industria turitica: retórica y práctica en el sector hotelero español* (Unpublished doctoral dissertation). Universitat Autònoma de Barcelona, Barcelona, Spain.

Babin, L. A., & Burns, A. C. (1997). Effects of Print Ad Pictures and Copy Containing Instructions to Imagine on Mental Imagery That Mediates Attitudes. *Journal of Advertising, 26*(3), 33–44. doi:10.1080/00913367.1997.10673527

Bachleitner, R., & Zins, A. H. (1999). Cultural tourism in rural communities: The residents' perspective. *Journal of Business Research, 44*(3), 199–209. doi:10.1016/S0148-2963(97)00201-4

Baker, D., & Crompton, J. (2000). Quality, satisfaction and behavioral intentions. *Annals of Tourism Research, 27*(3), 785–804. doi:10.1016/S0160-7383(99)00108-5

Bakir, A., Rose, G. M., & Shoham, A. (2006). Family communication patterns: Mothers' and fathers' communication style and children's perceived influence in family decision making. *Journal of International Consumer Marketing, 19*(2), 75–95. doi:10.1300/J046v19n02_05

Balbuena, A. (2014). *La actitud del residente ante los impactos del turismo: El caso de Benalmádena* (Doctoral dissertation). Retrieved from RIUMA, Repositorio Institucional de la Universidad de Málaga: (https://riuma.uma.es/xmlui/handle/10630/7492)

Balbuena, A., Barrera, D., & Hernández, M. (2017). El impacto socio-cultural del turismo en Benalmádena. *International Journal of Scientific Management and Tourism, 3*(1), 31–47.

Ballagas, R., Kuntze, A., & Walz, S. P. (2008). Gaming tourism: Lessons from evaluating REXplorer, a pervasive game for tourists. Lecture Notes in Computer Science, 5013, 244–261.

Ballantyne, D., Williams, J., & Aitken, R. (2011). Introduction to Service-Dominant Logic: From propositions to practice. *Industrial Marketing Management, 40*(2), 179–180. doi:10.1016/j.indmarman.2010.06.025

Baloglu, S. (1997). The relationship between destination images and sociodemographic and trip characteristics of international travellers. *Journal of Vacation Marketing, 3*(3), 221–233. doi:10.1177/135676679700300304

Baloglu, S., & Brinberg, D. (1997). Affective images of tourism destinations. *Journal of Travel Research, 35*(4), 11–15. doi:10.1177/004728759703500402

Baloglu, S., & Love, C. (2005). A Cognitive–Affective Positioning Analysis of Convention Cities: An Extension of the Circumplex Model of Affect. *Tourism Analysis, 9*(4), 299–308. doi:10.3727/108354205789807210

Baloglu, S., & Mangaloglu, M. (2001). Tourism destination images of Turkey, Egypt, Greece, and Italy as perceived by US-based tour operators and travel agents. *Tourism Management, 22*(1), 1–9. doi:10.1016/S0261-5177(00)00030-3

Baloglu, S., & McCleary, K. W. (1999). A model of destination image formation. *Annals of Tourism Research, 26*(4), 868–897. doi:10.1016/S0160-7383(99)00030-4

Bandhari, K. (2014). *Tourism and National Identity: Heritage and Nationhood in Scotland. Series Tourism and Cultural Change: 39.* Channel View Publications. doi:10.21832/9781845414498

Bandyopadhyay, R., & Morais, D. (2005). Representative dissonance: India's self and western image. *Annals of Tourism Research, 32*(4), 1006–1021. doi:10.1016/j.annals.2005.02.002

Bandyopadhyay, R., & Nascimento, K. (2010). "Where fantasy becomes reality": How tourism forces made Brazil a sexual playground. *Journal of Sustainable Tourism, 18*(8), 933–949. doi:10.1080/09669582.2010.497220

Bansal, P., & Roth, K. (2000). Why companies go green: A model of ecological responsiveness. *Academy of Management Journal, 43*, 717–736. doi:10.2307/1556363

Baptista, L.V., Nofre, J., & Jorge, M.R. (2018). Mobilidade, Cidade e Turismo: pistas para analisar as transformações em curso no centro histórico de Lisboa. *Revista de Sociologia da Faculdade de Letras da Universidade do Porto*, 14-32.

Barajas, A., Coates, D., & Sánchez-Fernandez, P. (2016). Beyond retrospective assessment. Sport event economic impact studies as a management tool for informing event organization. *European Research on Management and Business Economics, 22*(3), 124–130. doi:10.1016/j.iedee.2015.05.001

Barata, F. T., Molinari, F., Marsh, J., & Cabeça, S. M. (Eds.). (2017). *Creative Innovation and Related Living Lab Experiences: A Mediterranean Model.* UNESCO Chair.

Bardin, L. (1997). *Análise de Conteúdo.* Lisbon: Edições 70.

Barnes, S., Mattsson, J., & Sørensen, F. (2014). Destination brand experience and visitor behavior: Testing a scale in the tourism context. *Annals of Tourism Research, 48*, 121–139. doi:10.1016/j.annals.2014.06.002

Barney, J. (1991). Firm resources and sustained competitive advantage. *Journal of Management, 17*(1), 99–120. doi:10.1177/014920639101700108

Baron, R., & Kenny, D. (1986). The moderator-mediator variable distinction in social psychological research: Conceptual, strategic, and statistical consideration. *Journal of Personality and Social Psychology, 51*(6), 1173–1182. doi:10.1037/0022-3514.51.6.1173 PMID:3806354

Barrett, L. (2017). *How emotions are made: The secret life the brain.* Houghton Mifflin Harcourt Publishing Company.

Barrett, L. F. (2006). Are Emotions Natural Kinds? *Perspectives on Psychological Science, 1*(1), 28–58. doi:10.1111/j.1745-6916.2006.00003.x PMID:26151184

Barrett, L. F., Mesquita, B., & Gendron, M. (2011). Context in emotion perception. *Current Directions in Psychological Science, 20*(5), 286–290. doi:10.1177/0963721411422522

Barroso, C. C., Armario, E. M., & Ruiz, D. M. (2007). The influence of market heterogeneity on the relationship between a destination's image and tourists' future behaviour. *Tourism Management, 28*(1), 175–187. doi:10.1016/j.tourman.2005.11.013

Başkaya, Z., & Ünal, A. (2017). Temporal change of marriage and divorce rates at provincial levels in Turkey (2001-2005). [in Turkish]. *The Journal of International Social Research, 10*(53), 338–358.

BASOC, Diana-Gabriela (2015). A Theoretical Approach to Experiences in Marketing–Mastered through Experience Marketing. Marketing From Information to Decision, (8), 110-118.

Baş, T., & Akturan, U. (2017). *Computer aided qualitative research methods in social sciences.* Seçkin Publishing. [in Turkish].

Batra, R., & Stayman, D. M. (1990). The Role of Mood in Advertising Effectiveness. *The Journal of Consumer Research, 17*(2), 203. doi:10.1086/208550

Baudrillard, J. (2001). Simulacra and Simulations. In M. Porter (Ed.), *Jean Baudrillard – Selected Writings* (pp. 166–184). Polity Press.

Bauman, Z. (1998). *Globalization: The human consequences*. Polity Press.

Bauman, Z. (2000). *Liquid Modernity*. Polity Press.

Baxendale, S., Macdonald, E., & Wilson, H. (2015). The Impact of Different Touchpoints on Brand Consideration. *Journal of Retailing*, *91*(2), 235–253. doi:10.1016/j.jretai.2014.12.008

Bayraktaroglu, S., & Kutanis, R. O. (2003). Transforming hotels into learning organizations: A new strategy for going global. *Tourism Management*, *24*(2), 149–154. doi:10.1016/S0261-5177(02)00061-4

Beck, U. (2008). *Qué es la Globalización?* Barcelona: Paidós.

Beerli, A., & Martin, J. D. (2004a). Factors influencing destination image. *Annals of Tourism Research*, *31*(3), 657–681. doi:10.1016/j.annals.2004.01.010

Beerli, A., & Martin, J. D. (2004b). Tourists' characteristics and the perceived image of tourist destinations: A quantitative analysis – a case study of Lanzarote, Spain. *Tourism Management*, *25*(5), 623–636. doi:10.1016/j.tourman.2003.06.004

Beerli, A., Meneses, G. D., & Gil, S. M. (2007). Self-Congruity and Destination Choice. *Annals of Tourism Research*, *34*(3), 571–587. doi:10.1016/j.annals.2007.01.005

Behm-Morawitz, E. (2017). Examining the intersection of race and gender in video game advertising. *Journal of Marketing Communications*, *23*(3), 220–239. doi:10.1080/13527266.2014.914562

Beinisch, Y., & Paunov, C. (2005). *Broadband Content: The online computer and video game industry*. OECD Report.

Belch, G. E., Belch, M. A., & Ceresino, G. (1985). Parental and teenage child influences in family decision making. *Journal of Business Research*, *13*(2), 163–176. doi:10.1016/0148-2963(85)90038-4

Belisle, F., & Hoy, D. (1980). The Perceived Impact of Tourism by Residents: A Case Study in Santa Marta, Colombia. *Annals of Tourism Research*, *7*(1), 83–101. doi:10.1016/S0160-7383(80)80008-9

Beneke, J., Silverstone, G., Woods, A., & Schneider, G. (2011). The influence of the youth on their parents' purchasing decisions of high-technology products. *African Journal of Business Management*, *5*(10), 3807-3812.

Benjamin, W. (1969). Paris: Capital of the Nineteenth Century. *Perspecta*, *12*, 163–172.

Benson, M., & O'Reilly, K. (2009). Migration and the search for a better way of life: A critical exploration of lifestyle migration. *The Sociological Review*, *57*(4), 608–625. doi:10.1111/j.1467-954X.2009.01864.x

Berdychevsky, L., Gibson, H., & Poria, Y. (2013). Women's sexual behavior in tourism: Loosening the bridle. *Annals of Tourism Research*, *42*, 65–86. doi:10.1016/j.annals.2013.01.006

Berdychevsky, L., Poria, Y., & Uriely, N. (2013). Sexual behaviour in women's tourist experiences: Motivations, behaviors, and meanings. *Tourism Management*, *35*, 144–155. doi:10.1016/j.tourman.2012.06.011

Berey, L. A., & Pollay, R. W. (1968). The influencing role of the child in family decision making. *Journal of Marketing Research*, *5*(1), 70–72. doi:10.1177/002224376800500109

Berg, B. (2009). *Qualitative Research Methods for the Social Sciences*. Allyn & Bacon.

Berlin Declaration. (1997) The Berlin Declaration on Biological Diversity and Sustainable Tourism. *International Conference of Environment Ministers on Biodiversity and Tourism*. Retrieved from: http://www.univeur.org/cuebc/downloads/PDF%20carte/06.%20Berlin%20declaration.PDF

Berndt, A. (2010, Apr.). Diegetic Music: New Interactive Experiences. *Game Sound Technology and Player Interaction*, 60–77.

Berry, L., Carbone, L., & Haeckel, S. (2002). Managing the total customer experience. *MIT Sloan Management Review*, *43*(3), 85–89. doi:10.1371/journal.pone.0015090

Bertella, G. (2013, Fall). Photography and Northern Lights Tourism in Tromsø, Norway. *The Northern Review*, *37*, 167–186.

Bertella, G. (2014). The co-creation of animal-based tourism experience. *Tourism Recreation Research*, *39*(1), 115–125. doi:10.1080/02508281.2014.11081330

Besculides, A., Lee, M., & McCormick, P. (2002). Residents' perceptions of the cultural benefits of tourism. *Annals of Tourism Research*, *29*(2), 303–319. doi:10.1016/S0160-7383(01)00066-4

Bigné, J. E., Sanchez, M. I., & Sanchez, J. (2001). Tourism image, evaluation variables and after purchase behaviour: Inter-relationship. *Tourism Management*, *22*(6), 607–616. doi:10.1016/S0261-5177(01)00035-8

Billig, M. (2006). Is my home my castle? Place attachment, risk perception, and religious faith. *Environment and Behavior*, *38*(2), 248–265. doi:10.1177/0013916505277608

Bimonte, S., & Faralla, V. (2014). Happiness and nature-based vacations. *Annals of Tourism Research*, *46*, 176–178. doi:10.1016/j.annals.2014.02.002

Binkhorst, E. (2008). Turismo de cocreación, valor añadido en escenarios turísticos. *Journal of Tourism Research / Revista de Investigación en Turismo, 1*(1), 40-48.

Binkhorst, E., & Dekker, T. D. (2009). Agenda for Co-Creation Tourism Experience Research. *Journal of Hospitality Marketing & Management*, *18*(2-3), 2–3, 311–327. doi:10.1080/19368620802594193

Biscaia, R., Correia, A., Rosado, A. F., Ross, S. D., & Maroco, J. (2013). Sport sponsorship: The relationship between team loyalty, sponsorship awareness, attitude toward the sponsor, and purchase intentions. *Journal of Sport Management*, *27*(4), 288–302. doi:10.1123/jsm.27.4.288

Blackstock, K. (2005). A critical look at community-based tourism. *Community Development Journal: An International Forum*, *40*(1), 39–49. doi:10.1093/cdj/bsi005

Blake, M. K. (2017). Building an unjust foodscape: Shifting governance regimes, urban place making and the making of Chinese food as ordinary in Hong Kong. *The International Journal of Justice and Sustainability*, *23*(11), 1047–1062.

Blanchard, O., & Buchs, A. (2015). Clarifying Sustainable Development Concepts Through Role-Play. *Simulation & Gaming*, *46*(6), 697–712. doi:10.1177/1046878114564508

Blau, P. M. (2009). *Social Exchange Theory. Exchange and power in social life*. John Wiley and Sons.

Blessi, G. T., Grossi, E., Sacco, P. L., Pieretti, G., & Ferilli, G. (2014). Cultural Participation, Relational Goods and Individual Subjective Well-Being: Some Empirical Evidence. *Review of Economics & Finance*, *4*(8), 33–46.

Boiral, O. (2006). Global Warming: Should companies adopt proactive strategy? *Long Range Planning*, *39*(3), 315–330. doi:10.1016/j.lrp.2006.07.002

Boisen, M., Terlouw, K., & van Gorp, B. (2011). The selective nature of place branding and the layering of spatial identities. *Journal of Place Management and Development*, *4*(2), 135–147. doi:10.1108/17538331111153151

Boksberger, P., Dolnicar, S., Laesser, C., & Randle, M. (2011). Self-congruity theory: To what extent does it hold in tourism? *Journal of Travel Research*, *50*(4), 454–456. doi:10.1177/0047287510368164

Boley, B. B., McGehee, N. G., Perdue, R. R., & Long, P. (2014). Empowerment and resident attitudes toward tourism: Strengthening the theoretical foundation through a Weberian lens. *Annals of Tourism Research*, *49*, 33–50. doi:10.1016/j.annals.2014.08.005

Boley, B. B., Stzelecka, M., & Woosnam, K. M. (2018). Resident perceptions of the Economic Benefits of Tourism: Toward a Common Measure. *Journal of Hospitality & Tourism Research (Washington, D.C.)*, *42*(8), 1295–1314. doi:10.1177/1096348018759056

Bolton, R. (1992). 'Place prosperity vs people prosperity'revisited: An old issue with a new angle. *Urban Studies (Edinburgh, Scotland)*, *29*(2), 185–203. doi:10.1080/00420989220080261

Bonança, J. (2014). *O Milagre do 25 de Abril: Transformação de Olhão*. Arandis.

Bone, P. F., & Ellen, P. S. (1992). The Generation and Consequences of Communication-Evoked Imagery. *The Journal of Consumer Research*, *19*(1), 93–104. doi:10.1086/209289

Bonn, M., Joseph-Mathews, S., Dai, M., Hayes, S., & Cave, J. (2007). Heritage/Cultural Attraction Atmospherics: Creating the Right Environment for the Heritage/Cultural Visitor. *Journal of Travel Research*, *45*(3), 345–354. doi:10.1177/0047287506295947

Boo, S., Busser, J., & Baloglu, S. (2009). A model of customer-based brand equity and its application to multiple destinations. *Tourism Management*, *30*(2), 219–231. doi:10.1016/j.tourman.2008.06.003

Bosangit, C., Hibbert, S., & McCabe, S. (2015). "If I was going to die I should at least be having fun": Travel blogs, meaning and tourist experience. *Annals of Tourism Research*, *55*, 1–14. doi:10.1016/j.annals.2015.08.001

Bosnjak, M., Sirgy, M. J., Hellriegel, S., & Maurer, O. (2011). Postvisit destination loyalty judgments: Developing and testing a comprehensive congruity model. *Journal of Travel Research*, *50*(5), 496–508. doi:10.1177/0047287510379159

Bosque, I., & Martín, H. (2008). Tourist satisfaction: A cognitive-affective model. *Annals of Tourism Research*, *35*(2), 551–573. doi:10.1016/j.annals.2008.02.006

Boswijk, A., Thijssen, T., & Peelen, E. (2006). A New Perspective on the Experience Economy - Meaningful Experiences. The European Centre for the Experience Economy, The Netherlands, 2, 233–247.

Boud, D., Cohen, R., & Walker, D. (1993). *Using Experience for Learning*. Open University Press.

Bourdieu, P. (1996). *Distinction: A Social Critique of the Judgement of Taste* (8th ed.). Harvard University Press.

Bourdieu, P. (2001). *Masculine Domination*. Stanford University Press.

Bowen, G. A. (2009). Document analysis as a qualitative research method. *Qualitative Research Journal*, *9*(2), 27–40. doi:10.3316/QRJ0902027

Boyle, T. I. (2003). *The Impact of Adventure-Based Training on Team Cohesion and Psychological Development in Elite Sporting Teams*. Parkland, FL: Dissertation.com.

Bozon, M. (2004). *Sociologia da Sexualidade*. Editora FGV.

Bozon, M. (2015). Sexualité et genre. In J. Laufer, C. Marry, & M. Maruany (Eds.), *Masculin Féminin: Questions Pour les Sciences de l'Homme* (pp. 169–186). PUF.

Brakus, J. J., Schmitt, B. H., & Zarantonello, L. (2009). Brand Experience: What Is It? How Is It Measured? Does It Affect Loyalty? *Journal of Marketing, 73*(3), 52–68. doi:10.1509/jmkg.73.3.052

Bramwell, B., & Rawding, L. (1996). Tourism marketing images of industrial cities. *Annals of Tourism Research, 23*(1), 201–221. doi:10.1016/0160-7383(95)00061-5

Brandão, R., Lins, P., Lima, F., Arruda, A., Neves, A., & Correia, W. (2014). Design e inovação tecnológica na indústria de video games: Nintendo, um estudo de caso. *SBC - Proceedings of the SBGames*, 1101–1110.

Brandão, V. G. J. (2012). *O Compromisso Marítimo de Olhão (1765-1873)* (Unpublished master's dissertation). University of Algarve.

Brand, R., & Gaffikin, F. (2007). Collaborative planning in an uncollaborative world. *Planning Theory, 6*(3), 282–313. doi:10.1177/1473095207082036

Braun, V., & Clarke, V. (2006). Using thematic analysis in psychology. *Qualitative Research in Psychology, 3*(2), 77–101. doi:10.1191/1478088706qp063oa

Brave, S., Nass, C., & Hutchinson, K. (2005). Computers that care: Investigating the effects of orientation of emotion exhibited by an embodied computer agent. *International Journal of Human-Computer Studies, 62*(2), 161–178. doi:10.1016/j.ijhcs.2004.11.002

Breakey, N. M. (2012). Study in of World Heritage visitors: The case of the remote Riversleigh Fossil Site. *Visitor Studies, 15*(1), 82–97. doi:10.1080/10645578.2012.660845

Brents, B., Jackson, C., & Hausbeck, K. (2010). *The Sate of Sex: Tourism, Sex, and Sin in the New American Heartland.* Routledge.

Brida, J., Riaño, E., & Zapata, S. (2012). Percepciones de los residentes acerca de los impactos del turismo de crucero en la comunidad: Un análisis factorial y de clúster. *Cuadernos de Turismo, 29*, 79–107.

Briedenhann, J. (2011). Economic and tourism expectations of the 2010 FIFA World Cup. A resident perspective. *Journal of Sport & Tourism, 16*(1), 5–32. doi:10.1080/14775085.2011.568085

Brislin, R. W. (1970). Back-Translation for Cross-Cultural Research. *Journal of Cross-Cultural Psychology, 1*(3), 185–216. doi:10.1177/135910457000100301

Brito, S., Oliveira, A., Nobre, L., Macedo, Á., & Marchetti, R. (2017). Brand experience dimensions in the Brazilian context. *Race, 16*(4), 121–146. doi:10.18593/race.v16i1.10527

Brülde, B. (2007). Happiness and the good life. Introduction and conceptual framework. *Journal of Happiness Studies, 8*(1), 1–14. doi:10.100710902-006-9002-9

Brunt, P., & Courtney, P. (1999). Host perceptions of sociocultural impacts. *Annals of Tourism Research, 26*(3), 493-515.

Brunt, P., & Courtney, P. (1999). Host perceptions of sociocultural impacts. *Annals of Tourism Research, 26*(3), 493–515. doi:10.1016/S0160-7383(99)00003-1

Brunt, P., Horner, S., & Semley, N. (2017). *Research Methods in Tourism, Hospitality and Events Management.* Sage.

Bryman, A., & Bell, E. (2011). *Business Research Methods* (3rd ed.). Oxford University Press.

Buda, D., D'Hauteserre, A., & Johnston, L. (2014). Feeling and tourism studies. *Annals of Tourism Research, 46*, 102–114. doi:10.1016/j.annals.2014.03.005

Buffa, F. (2015). Young tourists and sustainability. Profiles, attitudes, and implications for destination strategies. *Sustainability, 7*(10), 14042–14062. doi:10.3390u71014042

Bujosa, A., & Rosselló, J. (2007). Modelling environmental attitudes toward tourism. *Tourism Management, 26*(3), 688–695. doi:10.1016/j.tourman.2006.04.004

Buning, R. J., Cole, Z. D., & McNamee, J. B. (2016). Visitor expenditure within a mountain bike event portfolio: Determinants, outcomes, and variations. *Journal of Sport & Tourism, 20*(2), 103–122. doi:10.1080/14775085.2016.1239547

Buonincontri, P., Morvillo, A., Okumus, F., & van Niekerk, M. (2017). Managing the experience co-creation process in tourism destinations: Empirical findings from Naples. *Tourism Management, 62*, 264–277. doi:10.1016/j.tourman.2017.04.014

Burns, A. C., Biswas, A., & Babin, L. A. (1993). The Operation of Visual Imagery as a Mediator of Advertising Effects. *Journal of Advertising, 22*(2), 71–85. doi:10.1080/00913367.1993.10673405

Butler, F. (n.d.). *Best honeymoon places in the Algarve*. Retrieved from: https://www.myguidealgarve.com/lists/best-honeymoon-places-in-the-algarve

Butler, J. (1999). *Gender Trouble: Feminism and Subversion of Identity*. Routledge.

Butler, J. (2011). *Bodies That Matter: On the Discursive Limits of Sex*. Routledge. doi:10.4324/9780203828274

Butler, R. W. (1980). The concept of a tourist area cycle of evolution: Implications for management of resources. *The Canadian Geographer. Geographe Canadien, 24*(1), 5–12. doi:10.1111/j.1541-0064.1980.tb00970.x

Butler, T. (2007). For gentrification? *Environment & Planning, 39*(1), 162–181. doi:10.1068/a38472

Butler, T., & Lees, L. (2006). Super-Gentrification in Barnsbury, London: Globalization and Gentrifying Global Elites at the Neighbourhood Level. *Transactions of the Institute of British Geographers, 31*(4), 467–487. doi:10.1111/j.1475-5661.2006.00220.x

Byrd, E. T., Bosley, H. E., & Dronberger, M. G. (2009). Comparisons of stakeholder perceptions of tourism impacts in rural eastern North Carolina. *Tourism Management, 30*(5), 693–703. doi:10.1016/j.tourman.2008.10.021

Cabeça, S. M., Gonçalves, A. R., Marques, J. F., & Tavares, M. (Eds.). (2019b). *Book of Abstracts of the 3rd CREATOUR International Conference and Creative Tourism Showcase: Creative Tourism Dynamics: Connecting Travellers, Communities, Cultures, and Places*. Faro: Universidade do Algarve.

Cabeça, S.M., Gonçalves, A.R., Marques, J.F., & Tavares, M. (2019a). Mapping Intangibilities in Creative Tourism Territories through Tangible Objects: a methodological approach for developing creative tourism offers. *Tourism & Management Studies, 15*(SI), 42-49.

Cabeça, S. M. (2016). *Estrutura e processo de formação das Forma Cultural: o Caso do Cante Alentejano*. UNESCO Chair.

Cabecinhas, R., & Amâncio, L. (2004). Estereótipos sociais e assimetria simbólica: três estudos com jovens angolanos e portugueses. *Proceedings of the VIII Luso-Afro Brazilian Congress of Social Sciences*.

Cabezas, A. (2004). Between love and money: Sex, tourism, and citizenship in Cuba and the Dominican Republic. *Signs (Chicago, Ill.), 29*(4), 987–1015. doi:10.1086/382627

Cabezas, A. (2009). *Economies of Desire: Sex and Tourism in Cuba and the Dominican Republic*. Temple University Press.

Cabiddu, F., Lui, T., & Piccoli, G. (2013). Managing Value Co-Creation In The Tourism Industry. *Annals of Tourism Research*, *42*, 86–107. doi:10.1016/j.annals.2013.01.001

Cai, L. A., Wu, B., & Bai, B. (2004). Destination image and loyalty. *Tourism Review International*, *7*(3–4), 153–162.

Cai, L., & Hayes, A. F. (2008). A new test of linear hypotheses in OLS regression under heteroscedasticity of unknown form. *Journal of Educational and Behavioral Statistics*, *33*(1), 21–40. doi:10.3102/1076998607302628

Çakıcı, C., & İyitoğlu, V. (2010). Market segmentation of the families coming to Antalya by the effect of children on vacation decisions. [in Turkish]. *Marketing and Marketing Research Journal*, *8*, 1–22.

Calantone, R. J., Di Benedetto, C. A., Hakam, A., & Bojanic, D. C. (1989). Multiple Multinational Tourism Positioning Using Correspondence analysis. *Journal of Travel Research*, *28*(2), 25–32. doi:10.1177/004728758902800207

Calleja, G. (2013). Narrative Involvement in Digital Games. *Foundations of Digital Games.*

Calleja, G. (2010). Digital games and escapism. *Games and Culture*, *5*(4), 335–353. doi:10.1177/1555412009360412

Campbell, S., Perkins, A., & Mohammed, P. (1999). Come to Jamaica and feel all right: tourism and the sex trade. In K. Kempadoo (Ed.), *Sun, Sex and Gold: Tourism and Sex Work in the Caribbean* (pp. 125–156). Rowman & Littlefield Publishers.

Campos, A. C., Mendes, J., do Valle, P. O., & Scott, N. (2016). Co-Creation Experiences: Attention and Memorability. *Journal of Travel & Tourism Marketing*, *33*(9), 1309–1336. doi:10.1080/10548408.2015.1118424

Campos, A. C., Mendes, J., do Valle, P. O., & Scott, N. (2018). Co-creation of tourist experiences: A literature review. *Current Issues in Tourism*, *21*(4), 369–400. doi:10.1080/13683500.2015.1081158

Campos, A., Mendes, J., Valle, P., & Scott, N. (2017). Co-creating animal-based tourist experiences: Attention, involvement and memorability. *Tourism Management*, *63*, 100–114. doi:10.1016/j.tourman.2017.06.001

Canalis, X. (2016, Mar.). Certifican la muerte del Turismo de Experiencias. *Hosteltur, Innovación Turística.*

Cannon, T. F., & Ford, J. (2002). Relationship of demographic and trip characteristics to visitor spending: An analysis of sports travel visitors across time. *Tourism Economics*, *8*(3), 263–271. doi:10.5367/000000002101298106

Cansel, A., Ekiz, E., & Bavik, A. (2009). Sex tourism in Northern Cyprus: Investigating the current situation. *Tourism Analysis*, *14*(5), 677–689. doi:10.3727/108354209X12597959359374

Cantalice, T. (2009). Feminismo, mercado de sexo e turismo: Reflexões sobre as múltiplas faces e interpretações do sexo mercantil. *Bagoas - Estudos Gays. Género e Sexualidades*, *2*(3), 145–178.

Carbone, L. P. (1998). Total customer experience drives value. *Management Review*, *87*(7), 62.

Carbone, L., & Haeckel, S. (1994). Engineering Customer Experiences. *Marketing Management*, *3*(3), 8–19.

Carmines, E., & Zeller, R. (1979). *Reliability and Validity Assessment*. Sage Publications. doi:10.4135/9781412985642

Carmona, E., Céspedes, J., & de Burgos, J. (2004). Environmental Strategies in Spanish Hotels: Contextual factors and performance. *Service Industries Journal*, *24*(3), 101–130. doi:10.1080/0264206042000247786

Carneiro, M. J., Breda, Z., & Cordeiro, C. (2016). Sports tourism development and destination sustainability: The case of the coastal area of the Aveiro region, Portugal. *Journal of Sport & Tourism*, *20*(3-4), 305–334. doi:10.1080/14775085.2016.1220863

Carneiro, M. J., & Eusébio, C. (2012). Segmentation of the tourism market using the impact of tourism on quality of life. *Tourism & Management Studies*, *7*, 91–100.

Carr, N., & Poria, Y. (2010). Introduction: provision, marketing and consumption of sex and the sexual in the leisure and tourism environment. In N. Carr & Y. Poria (Eds.), *Sex and the Sexual During People's Leisure and Tourism Experiences* (pp. 1–16). Cambridge Scholars Publishing.

Carter, S., & Clift, S. (2000). Tourism, international travel and sex: themes and research. In S. Clift & S. Carter (Eds.), *Tourism and Sex: Culture, Commerce and Coercion* (pp. 1–19). Pinter.

Cartier, C. (1999). The state, property development and symbolic landscape in high-rise Hong Kong. *Landscape Research*, *24*(2), 185–208. doi:10.1080/01426399908706558

Carù, A., Caru, A., & Cova, B. (Eds.). (2007). *Consuming experience*. Routledge; doi:10.4324/9780203390498

Carù, A., & Cova, B. (2003). Revisiting Consumption Experience: A More Humble but Complete View of the Concept. *Marketing Theory*, *3*(2), 267–286. doi:10.1177/14705931030032004

Caruana, A., & Vassallo, R. (2003). Children's perception of their influence over purchases: The role of parental communication patterns. *Journal of Consumer Marketing*, *20*(1), 55–66. doi:10.1108/07363760310456955

Castillo, R. (2014). Feeling at home in the "Chocolate City": An exploration of place-making practices and structures of belonging amongst Africans in Guangzhou. *Inter-Asia Cultural Studies*, *15*(2), 235–257. doi:10.1080/14649373.2014.911513

Cater, E. (1991). Sustainable tourism in the Third World: problems and prospects. *Sustainable tourism in the Third World: problems and prospects.*

Cauberghe, V., & De Pelsmacker, P. (2010). Advergames: The Impact of Brand Prominence and Game Repetition on Brand Responses. *Journal of Advertising*, *39*(1), 5–18. doi:10.2753/JOA0091-3367390101

Caust, J., & Vecco, M. (2017). Is UNESCO World Heritage recognition a blessing or burden? Evidence from developing Asian countries. *Journal of Cultural Heritage*, *27*, 1–9. doi:10.1016/j.culher.2017.02.004

Cavagnaro, E., Staffieri, S., & Postma, A. (2018). Understanding millennials' tourism experience: Values and meaning to travel as a key for identifying target clusters for youth (sustainable) tourism. *Journal of Tourism Futures*, *4*(1), 31–42. doi:10.1108/JTF-12-2017-0058

Cavus, S., & Tanrisevdi, A. (2003). Resident's Attitudes Towards Tourism Development: A Case Study in Kusadasi, Turkey. *Tourism Analysis*, *7*(3), 259–269. doi:10.3727/108354203108750102

Centre for Reviews and Dissemination, University of York (CDR) Systematic Reviews: CDR's Guidance for undertaking reviews in health care. (n.d.). Downloaded on 20/03/2020 Available at: https://www.york.ac.uk/media/crd/Systematic_Reviews.pdf

Céspedes, J., & de Burgos, J. (2004). Un análisis de las dimensiones de la gestión ambiental en los servicios hoteleros. *Dirección y Organización*, *30*, 5–15.

Ceylan, E. N. (2019). Child friendly tourism establishment: The Land of Legends [in Turkish]. *Journal of Recreation and Tourism Research*, *6*(3), 263–284. doi:10.31771/jrtr.2019.30

Chahal, H., Kaur, G., & Rani, A. (2016). Exploring the Dimensions of Customer Experience and its Impact on Word-of-Mouths: A Study of Credit Cards. *Journal of Services Research*, *15*(2), 1–33. https://search.proquest.com/abicomplete/docview/1878857032/8DF011A210874C67PQ/4?accountid=132208

Chalip, L. (2004). Beyond impact: A general model for sport event leverage. In B. Ritchie & D. Adair (Eds.), *Sport tourism: Interrelationships, impacts and issues* (pp. 226–252). Channel View.

Chalip, L., & McGuirty, J. (2004). Bundling sport events with the host destination. *Journal of Sport & Tourism*, *9*(3), 267–282. doi:10.1080/1477508042000320241

Chan, P. (2016). *Place-making – harbourfront for Hong Kong people*. Retrieved September 1, 2019, from https://www.devb.gov.hk/en/home/Blog_Archives1/index_id_180.html

Chan, C.-S., & Peters, M. (2020). Investigating brand strategy gap in Hong Kong: First impressions and unique attractions. *Asian Geographer*, 1–22. doi:10.1080/10225706.2020.1729211

Chandralal, L., & Valenzuela, F.-R. (2013). Exploring Memorable Tourism Experiences: Antecedents and Behavioural Outcomes. *Journal of Economics. Business and Management*, *1*(2), 177–181. doi:10.7763/JOEBM.2013.V1.38

Chandralal, L., & Valenzuela, F.-R. (2015). Memorable Tourism Experiences: Scale Development. *Contemporary Management Research*, *11*(3), 291–310. doi:10.7903/cmr.13822

Chaney, I., Hosany, S., Wu, M. S. S., Chen, C. H. S., & Nguyen, B. (2018). Size does matter: Effects of in-game advertising stimuli on brand recall and brand recognition. *Computers in Human Behavior*, *86*, 311–318. doi:10.1016/j.chb.2018.05.007

Chang, S. (2018). Experience economy in hospitality and tourism: Gain and loss values for service and experience. *Tourism Management*, *64*, 55–63. doi:10.1016/j.tourman.2017.08.004

Chang, S.-H., & Lin, R. (2015). Building a Total Customer Experience Model: Applications for the Travel Experiences in Taiwan's Creative Life Industry. *Journal of Travel & Tourism Marketing*, *32*(4), 438–453. doi:10.1080/10548408.2014.908158

Chang, Y., Yan, J., Zhang, J., & Luo, J. (2010). Online In-Game Advertising Effect: Examining the Influence of a Match Between Games and Advertising. *Journal of Interactive Advertising*, *11*(1), 63–73. doi:10.1080/15252019.2010.10722178

Chan, W., To, C., & Chu, W. (2016). Desire for experiential travel, avoidance of rituality and social esteem: An empirical study of consumer response to tourism innovation. *Journal of Innovation & Knowledge*, *1*(1), 24–35. doi:10.1016/j.jik.2016.02.001

Chavda, H., Haley, M., & Dunn, C. (2005). Adolescents' influence on family decision-making. *Young Consumers*, *6*(3), 68–78. doi:10.1108/17473610510701223

Chen, C. F., & Chen, P. C. (2010). Resident attitudes toward heritage tourism development. *Tourism Geographies*, *12*(4), 525–545. doi:10.1080/14616688.2010.516398

Chen, C. F., & Tsai, D. (2007). How destination image and evaluative factors affect behavioral intentions? *Tourism Management*, *28*(4), 1115–1122. doi:10.1016/j.tourman.2006.07.007

Chen, C. Y., & Chang, S. L. (2019). Moderating effects of information-oriented versus escapism-oriented motivations on the relationship between psychological well-being and problematic use of video game live-streaming services. *Journal of Behavioral Addictions*, *8*(3), 564–573. doi:10.1556/2006.8.2019.34 PMID:31328955

Chen, H., & Haley, E. (2014). Product placement in social games: Consumer experiences in China. *Journal of Advertising*, *43*(3), 286–295. doi:10.1080/00913367.2013.858086

Chen, H., & Rahman, I. (2018). Cultural tourism: An analysis of engagement, cultural contact, memorable tourism experience and destination loyalty. *Tourism Management Perspectives*, *26*, 153–163. doi:10.1016/j.tmp.2017.10.006

Chen, J. S. (2000). Cross-cultural differences in travel information acquisition among tourist from three Pacific-rim countries. *Journal of Hospitality & Tourism Research (Washington, D.C.)*, *24*(2), 239–251. doi:10.1177/109634800002400207

Chen, J. S., & Hsu, C. H. C. (2001). Developing and validating a riverboat gaming impact scale. *Annals of Tourism Research*, *28*(2), 459–476. doi:10.1016/S0160-7383(00)00059-1

Chen, J., & Gursoy, D. (2001). An investigation of tourists' destination loyalty and preferences. *Journal of Contemporary Hospitality Management*, *13*(2), 79–85. doi:10.1108/09596110110381870

Chen, L., Scott, N., & Benckendorff, P. (2017). Mindful tourist experiences: A Buddhist perspective. *Annals of Tourism Research*, *64*, 1–12. doi:10.1016/j.annals.2017.01.013

Chen, N., Ji, S., & Funk, D. C. (2015). An extended study on destination image decay of sport tourists over time. *Journal of Destination Marketing & Management*, *2*(4), 241–252. doi:10.1016/j.jdmm.2013.11.001

Chen, S., & Raab, C. (2012). Predicting resident intentions to support community tourism: Toward an integration of two theories. *Journal of Hospitality Marketing & Management*, *21*(3), 270–294. doi:10.1080/19368623.2011.584268

Chen, Y. S., Lehto, X., Behnke, C., & Tang, C. H. (2016). Investigating children's role in family dining-out choices: Evidence from a casual dining restaurant. *Journal of Hospitality Marketing & Management*, *25*(6), 706–725. doi:10.1080/19368623.2016.1077368

Cherifi, B., Smith, A., Maitland, R., & Stevenson, N. (2014). Destination images of non-visitors. *Annals of Tourism Research*, *49*, 190–202. doi:10.1016/j.annals.2014.09.008

Cheung, P. T. Y. (2011). Civic engagement in the policy process in Hong Kong: Change and continuity. *Public Administration and Development*, *31*(2), 113–121. doi:10.1002/pad.597

Chew, E. Y. T., & Jahari, S. A. (2014). Destination image as a mediator between perceived risks and revisit intention: A case of post-disaster Japan. *Tourism Management*, *40*, 382–393. doi:10.1016/j.tourman.2013.07.008

Chi, C. G.-Q., Cai, R., & Li, Y. (2017). Factors influencing residents' subjective well-being at World Heritage Sites. *Tourism Management*, *63*, 209–222. doi:10.1016/j.tourman.2017.06.019

Chi, C., & Qu, H. (2008). Examining the structural relationships of destination image, tourist satisfaction and destination loyalty: An integrated approach. *Tourism Management*, *29*(4), 624–636. doi:10.1016/j.tourman.2007.06.007

Chin, W. (1998). The Partial Least Square Approach to Structural Equation Modeling. In G. A. Marcoulides (Ed.), *Modern Methods for Business Research* (pp. 295–336). Erlbaum Associates.

Choi W.M., Chan, A. & Wu, J. (1999). A qualitative and quantitative assessment of Hong Kong's image as a tourist destination. *Tourism Management, 20*, 361-365.

Choi, D., & Kim, J. (2004). Why People Continue to Play Online Games : In Search of Online Contents. *Cyberpsychology & Behavior*, *7*(1), 11–25. doi:10.1089/109493104322820066 PMID:15006164

Choi, H., & Murray, I. (2010). Resident attitudes toward sustainable community tourism. *Journal of Sustainable Tourism*, *18*(4), 575–594. doi:10.1080/09669580903524852

Choi, H.-S. C., & Sirakaya, E. (2005). Measuring residents' attitude toward sustainable tourism: Development of sustainable tourism attitude scale. *Journal of Travel Research*, *43*(4), 380–394. doi:10.1177/0047287505274651

Choi, H., & Sirakaya, E. (2006). Sustainability indicators for managing community tourism. *Tourism Management*, *27*(6), 1274–1289. doi:10.1016/j.tourman.2005.05.018

Chon, K.-S. (1992). Self-Image/Destination Image Congruity. *Annals of Tourism Research*, *19*(2), 360–376. doi:10.1016/0160-7383(92)90090-C

Chung, C. Y., & Ngai, P. (2007). Neoliberalization and privatisation in Hong Kong after the 1997 financial crisis. *China Review*, *7*(2), 65–92.

Churchill, G. A. Jr. (1979). A Paradigm for Developing Better Measures of Marketing Constructs. *JMR, Journal of Marketing Research*, *16*(1), 538–562. doi:10.1177/002224377901600110

Cicero, L., & Teichert, T. (2017). Children's influence in museum visits: Antecedents and consequences. *Museum Management and Curatorship*, *33*(2), 146–157. doi:10.1080/09647775.2017.1420485

Clark, E. (2005). The order and simplicity of gentrification – a political challenge. In R. Atkinson & G. Bridge (Eds.), *Gentrification in a Global Context: The new urban colonialism* (pp. 261–269). Routledge. doi:10.4324/9780203392089_chapter_16

Claver, E., Molina, J., Pereira, J., & López, M. (2007). Environmental strategies and their impact in hotel performance. *Journal of Sustainable Tourism*, *15*(6), 663–679. doi:10.2167/jost640.0

Clift, S., & Carter, S. (Eds.). (2000). Tourism and Sex: Culture, Commerce and Coercion. London: Pinter.

Clokes, P., & Jenkins, H. C. (1989). Cracking the Canyon with the Awesome Foursome: Representations of Adventure Tourism in New Zealand. *Environment and Planning. D, Society & Space*, *16*(2), 185–218. doi:10.1068/d160185

Cohen, E. (1971). Arab boys and tourist girls in a mixed Jewish-Arab community. *International Journal of Comparative Sociology*, *12*(4), 217–233. doi:10.1177/002071527101200401

Cohen, E. (1979). A Phenomenology of Tourist Experiences. *Sociology*, *13*(2), 179–201. doi:10.1177/003803857901300203

Cohen, E. (1982). Thai girls and farang men: The edge of ambiguity. *Annals of Tourism Research*, *9*(3), 403–428. doi:10.1016/0160-7383(82)90021-4

Cohen, E. (1988). Authenticity and commoditization in tourism. *Annals of Tourism Research*, *15*(3), 371–386. doi:10.1016/0160-7383(88)90028-X

Cohen, E. (1988). Tourism and AIDS in Thailand. *Annals of Tourism Research*, *15*(4), 467–486. doi:10.1016/0160-7383(88)90044-8

Cohen, E., & Ben-Nun, L. (2009). The Important Dimensions of Wine Tourism Experience from Potential Visitors' Perception. *Tourism and Hospitality Research*, *9*(1), 20–31. doi:10.1057/thr.2008.42

Cohen, E., & Cohen, S. A. (2012). Current sociological theories and issues in tourism. *Annals of Tourism Research*, *39*(4), 2177–2202. doi:10.1016/j.annals.2012.07.009

Cohen, S. A., Duncan, T., & Thulemark, M. (2015). Lifestyle Mobilities: The Crossroads of Travel, Leisure and Migration. *Mobilities*, *10*(1), 155–172. doi:10.1080/17450101.2013.826481

Cohen, S. A., Prayag, G., & Moital, M. (2014). Consumer behaviour in tourism: Concepts, influences and opportunities. *Current Issues in Tourism*, *17*(10), 872–909. doi:10.1080/13683500.2013.850064

Colbert, F. (2003). Entrepreneurship and Leadership in Marketing the Arts. *International Journal of Arts Management*, *6*(1), 30–39.

Collins, K. (2009). Generating Meaningful Sound: Quantifying the Affective Attributes of Sound Effects for Real-Time Sound Synthesis in Audio-Visual Media. *AES 35th International Conference,* 11–16.

Collison, F. M., & Spears, D. L. (2010). Marketing cultural and heritage tourism: The Marshall Islands. *International Journal of Culture, Tourism and Hospitality Research*, *4*(2), 130–142. doi:10.1108/17506181011045208

Colombo, P., & Gold, P. (2004). Multiple memory systems. *Neurobiology of Learning and Memory*, *82*(3), 169–170. doi:10.1016/j.nlm.2004.07.008 PMID:15464401

Community Project Workshop (CPW). (2013). *The renewal and development of Kwun Tong industrial district study*. Hong Kong: Kwun Tong District Council; Kwun Tong Development and Renewal Task Force.

Connell, D. (1997). Participatory development: An approach sensitive to class and gender. *Development in Practice*, *7*(3), 248–259. doi:10.1080/09614529754486 PMID:12348314

Connell, R. W. (1987). *Gender and Power*. Stanford University Press.

Constantin, J. A., & Lusch, R. F. (1994). *Understanding resource management*. The Planning Forum.

Cooper, C., Booth, A., Varley-Campbell, J., Britten, N., & Garside, R. (2018). Defining the process to literature searching in systematic reviews: A literature review of guidance and supporting studies. *BMC Medical Research Methodology*, *18*(1), 85. doi:10.118612874-018-0545-3 PMID:30107788

Coordination Commission of Development for the Region of the Algarve [CCDR Algarve]. (2015). *RIS3 - Algarve 2014-2020 Estratégia regional de investigação e inovação para a especialização inteligente*. CCDR.

Correia, A., Oliveira, C., & Pereira, R. (2017). From Emotions to Place Attachment. In A. Correia, M. Kozak, J. Gnoth, & A. Fyall (Eds.), *Co-Creation and Well-Being in Tourism* (pp. 163–177). Springer International Publishing. doi:10.1007/978-3-319-44108-5_13

Coshall, J. T. (2000). Measurement of Tourists' Images: The Repertory Grid Approach. *Journal of Travel Research*, *39*(1), 85–89. doi:10.1177/004728750003900111

Costanza, R., Fisher, B., Ali, S., Beer, C., Bond, L., Boumans, R., Danigelis, N. L., Dickinson, J., Elliott, C., Farley, J., Gayer, D. E., Glenn, L. M. D., Hudspeth, T., Mahoney, D., McCahill, L., McIntosh, B., Reed, B., Rizvi, S. A. T., Rizzo, D. M., ... Snapp, R. (2007). Quality of life: An approach integrating opportunities, human needs, and subjective well-being. *Ecological Economics*, *61*(2-3), 267–276. doi:10.1016/j.ecolecon.2006.02.023

Cottrell, S. P., Vaske, J. J., Shen, F., & Ritter, P. (2007). Resident perceptions of sustainable tourism in Chongdugou, China. *Society & Natural Resources*, *20*(6), 511–525. doi:10.1080/08941920701337986

Covarrubias, R., Vargas, A., & Rodríguez, M. (2010). Satisfacción de residentes con el desarrollo de la actividad turística en los pueblos mágicos de México: Un indicador de competitividad. *Gestión Turística*, *14*, 33–54.

CPW - Community Project Workshop. (2013). *The renewal and development of Kwun Tong industrial district study*. Hong Kong: Kwun Tong District Council; Kwun Tong Development and Renewal Task Force.

Crandall, R. W., & Sidak, J. G. (2006). *Video Games: Serious Business for America's Economy*. Academic Press.

Crawford, M., & Popp, D. (2003). Sexual double standards: A review and methodological critique of two decades of research. *Journal of Sex Research*, *40*(1), 13–26. doi:10.1080/00224490309552163 PMID:12806528

Cresswell, T. (2006). *On The Move – Mobility in the Modern Western World*. Routledge.

Creswell, J. (2015). *Educational Research: Planning, Conducting, and Evaluating Quantitative and Qualitative Research* (5th ed.). Pearson Education, Inc.

Creswell, J. W. (1998). *Qualitative Inquiry and Research Design*. Sage Publications.

Crick, M. (1989). Representations of international tourism in the social sciences: Sun, sex, sights, savings, and servility. *Annual Review of Anthropology, 18*(1), 307–344. doi:10.1146/annurev.an.18.100189.001515

Crivelli, C., & Fridlund, A. J. (2018). Facial Displays Are Tools for Social Influence. *Trends in Cognitive Sciences, 22*(5), 388–399. doi:10.1016/j.tics.2018.02.006 PMID:29544997

Crivelli, C., Jarillo, S., Russell, J. A., & Fernández-Dols, J. M. (2016). Reading emotions from faces in two indigenous societies. *Journal of Experimental Psychology. General, 145*(7), 830–843. doi:10.1037/xge0000172 PMID:27100308

Crivelli, C., Russell, J. A., Jarillo, S., & Fernández-Dols, J. M. (2016). The fear gasping face as a threat display in a Melanesian society. *Proceedings of the National Academy of Sciences of the United States of America, 113*(44), 12403–12407. doi:10.1073/pnas.1611622113 PMID:27791137

Crivelli, C., Russell, J. A., Jarillo, S., & Fernández-Dols, J. M. (2017). Recognizing spontaneous facial expressions of emotion in a small-scale society of papua New Guinea. *Emotion (Washington, D.C.), 17*(2), 337–347. doi:10.1037/emo0000236 PMID:27736108

Croes, R., & Semrad, K. J. (2015). The relevance of cultural tourism as the next frontier for small island destinations. *Journal of Hospitality & Tourism Research (Washington, D.C.), 39*(4), 469–491. doi:10.1177/1096348013491599

Crompton, J. L. (2006). Economic impact studies: Instruments for political shenanigans. *Journal of Travel Research, 45*(1), 67–82. doi:10.1177/0047287506288870

Cronbach, L. (1951). Coefficient alfa and the internal structure of tests. *Psychometrika, 16*(3), 297–334. doi:10.1007/BF02310555

Csikszentmihalyi, M. (1990). Flow - The Psychology of optimal experience. Academy of Management Review. New York: HarperCollins Publishers Inc. doi:10.5465/AMR.1991.4279513

Csikszentmihalyi, M. (1997). Finding Flow - The Psychology of Engagement with Everyday Life. New York: Basic Books.

Csikszentmihalyi, M. (2004). *Flow: the secret to happiness*. Retrieved from TED talks Web site: www.ted.com/talks/mihaly_csikszentmihalyi_on_flow/transcript#t-1119507

Csikszentmihalyi, M. (2014). Flow and the Foundations of Positive Psychology. Dordrecht: Springer. doi:10.1007/978-94-017-9088-8

Csikszentmihalyi, M. (1990). *Flow: The Psychology of Optimal Experience*. Harper Perennial.

Csikszentmihalyi, M., & Csikszentmihalyi, I. S. (1992). *Optimal experience: Psychological studies of flow in consciousness*. Cambridge University Press.

Csikszentmihalyi, M., & LeFevre, J. (1989). Optimal Experience in Work and Leisure. *Journal of Personality and Social Psychology, 56*(5), 815–822. doi:10.1037/0022-3514.56.5.815 PMID:2724069

Culture2030Goal. (2019). *Culture in the Implementation of the 2030 Agenda*. Author.

Cutler, S. Q., & Carmichael, B. A. (2010). The dimensions of the tourist experience. In M. Morgan, P. Lugosi, & B. Ritchie (Eds.), *The Tourism and Leisure Experience: Consumer and Managerial Perspectives* (pp. 3–26). Channel View Publications. doi:10.21832/9781845411503-004

D'Angella, F., & Go, F. M. (2009). Tale of two cities' collaborative tourism marketing: Towards a theory of destination stakeholder assessment. *Tourism Management, 30*(3), 429–440. doi:10.1016/j.tourman.2008.07.012

Dahles, H. (1998). Redefining Amsterdam as a tourist destination. *Annals of Tourism Research*, *25*(1), 55–69. doi:10.1016/S0160-7383(97)00061-3

Dahles, H., & Bras, K. (1999). Entrepreneurs in romance: Tourism in Indonesia. *Annals of Tourism Research*, *26*(2), 267–293. doi:10.1016/S0160-7383(98)00098-X

Daj, A. (2013). Economic and Legal Aspects of Introducing Novel Ict Instruments: Integrating Sound Into Social Media Marketing - From Audio Branding To Soundscaping. *Bulletin of the Transilvania University of Brasov. Series V, Economic Sciences*, *6*(2), 15–24.

Dalgleish, T. (2004). The emotional brain. *Nature Reviews. Neuroscience*, *5*(7), 583–589. doi:10.1038/nrn1432 PMID:15208700

Damásio, A. (2010). *Self Comes to Mind: Constructing the Conscious Brain*. Random House.

Dann, G. (1977). Anomie, ego-enhancement and tourism. *Annals of Tourism Research*, *4*(4), 184–194. doi:10.1016/0160-7383(77)90037-8

Dann, G. M. S. (1996). Tourists' Images of a Destination - an alternative analysis. *Journal of Travel & Tourism Marketing*, *5*(1/2), 41–55. doi:10.1300/J073v05n01_04

Dann, G., & Jacobsen, J. (2003). Tourism smellscapes. *Tourism Geographies*, *5*(1), 3–25. doi:10.1080/1461668032000034033

Darbellay, F., & Stock, M. (2012). Tourism as complex interdisciplinary research object. *Annals of Tourism Research*, *39*(1), 441–458. doi:10.1016/j.annals.2011.07.002

Dardis, F. E., Schmierbach, M., & Limperos, A. M. (2012a). The Impact of Game Customization and Control Mechanisms on Recall of Integral and Peripheral Brand Placements in Video games. *Journal of Interactive Advertising*, *12*(2), 1–12. doi:10.1080/15252019.2012.10722192

Darr, K. (2011). Introduction to management and leadership concepts, principles, and practices. In R. E. Burke & L. H. Friedman (Eds.), Essentials of management and leadership in public health (p. 7). London: Jones & Bartlett Learning International.

Davidson, J., & Milligan, C. (2004). Embodying Emotion Sensing Space: Introducing Emotional Geographies. *Social & Cultural Geography*, *5*(4), 523–532. doi:10.1080/1464936042000317677

Davis, D. L., & Cosenza, R. M. (1994). *Business Research for Decision Making* (3rd ed.). Wadsworth.

Davis, D., Allen, J., & Cosenza, R. M. (1988). Segmenting local residents by their attitudes, interests, and opinions toward tourism. *Journal of Travel Research*, *27*(2), 2–8. doi:10.1177/004728758802700201

Davison, G. (2008). Heritage: from patrimony to pastiche. In G. Fairclough, R., Harrison, J., Jameson, Jr., & J. Schofield (Eds.), The Heritage Reader, (pp.31-41). Routledge.

Davis, P. (1999). *Ecomuseums: a sense of place*. Leicester University Press.

De Albuquerque, K. (1998). Sex, beach boys and female tourists in the Caribbean. In B. M. Dank & R. Refinetti (Eds.), *Sexuality and Culture* (pp. 87–111). New Brunswinck.

de Burgos, J., & Céspedes, J. (2001). Environmental Performance as an Operations Objective. *International Journal of Operations & Production Management*, *21*(12), 1553–1572. doi:10.1108/01443570110410900

De Lacy, T., Battig, M., Moore, S., & Noakes, S. (2002). *Public / private partnership for sustainable tourism delivering a sustainability strategy for tourism destinations*. Asia Pacific Economic Cooperation.

de Rojas, C., & Camarero, C. (2008). Visitors' experience, mood and satisfaction in a heritage context: Evidence from an interpretation centre. *Tourism Management, 29*(3), 525–537. doi:10.1016/j.tourman.2007.06.004

Deak, A. (2011). Brain and emotion: Cognitive neuroscience of emotions. *Review of Psychology, 18*(2), 71–80.

Deccio, C., & Baloglu, S. (2002). Garfield county resident perceptions of the 2002 Winter Olympics: The spillover effects. *Journal of Travel Research, 41*(1), 45–56. doi:10.1177/0047287502041001006

Deleuze, J., Maurage, P., Schimmenti, A., Nuyens, F., Melzer, A., & Billieux, J. (2019). Escaping reality through video games is linked to an implicit preference for virtual over real-life stimuli. *Journal of Affective Disorders, 245*, 1024–1031.

Demarco, D. (2016). Sustainable Urban Development Perspectives in the Era of Tourism Experience. *Procedia: Social and Behavioral Sciences, 223*, 335–341. doi:10.1016/j.sbspro.2016.05.236

DeMatos, N. (2015). *The impacts of tourism experiences in the destination image. A marketing perspective* (Unpublished doctoral dissertation). University of Algarve.

Demeter, T., & Bratucu, G. (2014). Typologies of youth tourism. *Bulletin of the Transilvania University of Brasov. Series V, Economic Sciences, 7*(56), 115–122.

Development Bureau. (2011). *Urban renewal strategy*. Hong Kong Special Administrative Region Government.

Devine, A., Boyd, S., & Boyle, E. (2010). Unravelling the complexities of inter-organisational relationships within the sports tourism policy arena. *Journal of Policy Research in Tourism, Leisure & Events, 2*(2), 93–112. doi:10.1080/19407963.2010.482269

Devine, A., Boyd, S., & Boyle, E. (2011). Towards a theory of collaborative advantage for the sports tourism policy arena. *International Journal of Public Sector Management, 24*(1), 23–41. doi:10.1108/09513551111099208

Dhebar, A. (2013). Toward a compelling customer touchpoint architecture. *Business Horizons, 56*(2), 199–205. doi:10.1016/j.bushor.2012.11.004

Dholakia, R. R. (1984). Intergeneration differences in consumer behavior: Some evidence from a developing country. *Journal of Business Research, 12*(1), 19–34. doi:10.1016/0148-2963(84)90035-3

Dias, J., Correia, A., & Cascais, T. (2017). Traits in Tourists' Experiences: Senses, Emotions and Memories. In A. Correia, M. Kozak, J. Gnoth, & A. Fyall (Eds.), *Co-Creation and Well-Being in Tourism* (pp. 179–194). Springer International Publishing. doi:10.1007/978-3-319-44108-5_14

Díaz Armas, R., & Gutiérrez Taño, D. (2010). La actitud del residente en el destino turístico de Tenerife: Evaluación y tendencia. *Pasos. Revista de Turismo y Patrimonio Cultural, 8*(4), 431–444. doi:10.25145/j.pasos.2010.08.039

Diedering, M., & Kwiatkowski, G. (2015). Economic impact of events and festivals on host regions. Methods in practice. *Polish Journal of Sport and Tourism, 22*(4), 247–252. doi:10.1515/pjst-2015-0033

Diedrich, A., & García-Buades, E. (2009). Local perceptions of tourism as indicators of destination decline. *Tourism Management, 30*(4), 512–521. doi:10.1016/j.tourman.2008.10.009

Diener, E. (2009). The science of well-being: The collected works of Ed Diener. New York: Springer.

Diener, E., Emmons, R. A., Larsen, R. J., & Griffin, S. (1985). The satisfaction with life scale. *Journal of Personality Assessment, 49*(1), 71–75. doi:10.120715327752jpa4901_13 PMID:16367493

Ding, C., & Tseng, T. (2015). On the relationships among brand experience, hedonic emotions, and brand equity. *European Journal of Marketing, 49*(7/8), 994–1015. doi:10.1108/EJM-04-2013-0200

Dionysopoulou, P., & Mylonakis, J. (2013). Youth tourists' profile and their travel choices as influenced by social media networks. *European Journal of Hospitality and Tourism Research*, *1*(3), 22–35.

Dirsehan, T., & Çelik, M. (2011). Profiling online consumers according to their experiences with a special focus on social dimension. *Procedia: Social and Behavioral Sciences*, *24*, 401–412. doi:10.1016/j.sbspro.2011.09.040

Ditoiu, M.-C., & Căruntu, A.-L. (2014). Sensory Experiences Regarding Five-dimensional Brand Destination. *Procedia: Social and Behavioral Sciences*, *109*, 301–306. doi:10.1016/j.sbspro.2013.12.461

Dixon, A., Backman, S., Backman, K., & Norman, W. (2012). Expenditure-based segmentation of sport tourists. *Journal of Sport & Tourism*, *17*(1), 5–21. doi:10.1080/14775085.2011.635017

Dogan, H. Z. (1989). Forms of adjustment: Sociocultural impacts of tourism. *Annals of Tourism Research*, *16*(2), 216–236. doi:10.1016/0160-7383(89)90069-8

Dolnicar, S., & Grün, B. (2013). Validly Measuring Destination Image in Survey Studies. *Journal of Travel Research*, *52*(1), 3–13. doi:10.1177/0047287512457267

Donnan, H., & Magowan, F. (2010). *The Anthropology of Sex*. Berg.

Dormann, C. (2001). Seducing consumers, evaluating emotions. *Joint Proceedings of IHM-HCI 2001*, *2*, 10–14. Retrieved from papers2://publication/uuid/C5C62B4E-D574-4063-9AEB-76E7C1C01AA5

Doxey, G. V. (1975). A causation theory of visitor-resident irritants: methodology and research inferences. In *The Travel Research Association 6th Annual Conference Proceedings* (pp. 195-198). San Diego, CA: Academic Press.

Doxey, G. V. (1976). When enough is enough: the native are restless in old Niagara. *Heritage Canada, 2*(2), 26-27.

Driver, T. F. (1991). *The magic of ritual: Our need for liberating rites that transform our lives and our communities*. Harper Collins.

Duarte, P. (2012). Tourist experience: Contemporary perspectives. *Journal of Heritage Tourism*, *7*(2), 187–188. doi:10.1080/1743873X.2011.638459

Duarte, P., Folgado-Fernández, J. A., & Hernández-Mogollón, J. M. (2018). Measurement of the impact of music festivals on destination image: The case of a Womad festival. *Event Management*, *22*(4), 517–526. doi:10.3727/152599518X15264726192451

Dubois, L. E., & Gibbs, C. (2018). Video game–induced tourism: A new frontier for destination marketers. *Tourism Review*, *73*(2), 186–198. doi:10.1108/TR-07-2017-0115

Dubois, L. E., Griffin, T., Gibbs, C., & Guttentag, D. (2020). The impact of video games on destination image. *Current Issues in Tourism*, 1–13. doi:10.1080/13683500.2020.1724082

Duerden, M. D., Ward, P. J., & Freeman, P. A. (2015). Conceptualizing Structured Experiences. *Journal of Leisure Research*, *47*(5), 601–620. doi:10.18666/jlr-2015-v47-i5-6096

Duncan, E., Fanderl, H., Maechler, N., & Neher, K. (2016). *Customer experience: Creating value through transforming customer journeys* (Vol. 1). McKinsey & Company., doi:10.1057/9780230291775

Durkheim, E. (1914). Le dualisme de la nature humaine et ses conditions sociales. *Scientia*, *15*, 206–221.

Durr, E., & Jaffe, R. (2012). Theorizing Slum Tourism: Performing, Negotiating and Transforming Inequality. *Revista Europea de Estudios Latinoamericanos y del Caribe*, *93*(93), 113–123. doi:10.18352/erlacs.8367

Duxbury, N., & Campbell, H. (2009). Developing and Revitalizing Rural Communities through Arts and Creativity: A Literature Review. *Creative City Network of Canada*. Retrieved September 1, 2019, from https://www.creativecity.ca/database/files/library/rural_communities_arts_2009.pdf

Duxbury, N., & Jeannotte, S. M. (2010). Culture, Sustainability, and Communities: Exploring the Myths. *Oficina do CES*, *353*, 1–26.

Duxbury, N., & Richards, G. (2019). *A Research Agenda for Creative Tourism*. Edward Elgar. doi:10.4337/9781788110723

Dyer, P., Gursoy, D., Sharma, B., & Carter, J. (2007). Structural modeling of resident perceptions of tourism and associated development on the Sunshine Coast, Australia. *Tourism Management*, *28*(2), 409–422. doi:10.1016/j.tourman.2006.04.002

Eagly, A. H., & Chaiken, S. (1993). *The Psychology of Attitudes*. Harcourt, Brace, & Jovanovich.

Echtner, C. M., & Ritchie, J. R. B. (1991). The meaning and measurement of destination image. *Journal of Tourism Studies*, *2*(2), 2–12.

Echtner, C. M., & Ritchie, J. R. B. (1993). The measurement of destinations image: An empirical assessment. *Journal of Travel Research*, *31*(4), 3–13. doi:10.1177/004728759303100402

Ekinci, Y., Sirakaya-Turk, E., & Preciado, S. (2013). Symbolic consumption of tourism destination brands. *Journal of Business Research*, *66*(6), 711–718. doi:10.1016/j.jbusres.2011.09.008

Ekman, P. (1992). An argument for basic emotions. *Cognition and Emotion*, *6*(3-4), 169–200. doi:10.1080/02699939208411068

Ekman, P. (1993). Facial expression and emotion. *The American Psychologist*, *48*(4), 384–392. doi:10.1037/0003-066X.48.4.384 PMID:8512154

Ekman, P. (2016). What Scientists Who Study Emotion Agree About. *Perspectives on Psychological Science*, *11*(1), 31–34. doi:10.1177/1745691615596992 PMID:26817724

Ekman, P., Friesen, W. V., O'Sullivan, M., Chan, A., Diacoyanni-Tarlatzis, I., Heider, K., Krause, R., LeCompte, W. A., Pitcairn, T., Ricci-Bitti, P. E., Scherer, K., Tomita, M., & Tzavaras, A. (1987). Universals and Cultural Differences in the Judgments of Facial Expressions of Emotion. *Journal of Personality and Social Psychology*, *53*(4), 712–717. doi:10.1037/0022-3514.53.4.712 PMID:3681648

Ek, R., Larsen, J., Hornskov, S. B., & Mansfeldt, O. K. (2008). A dynamic framework of tourist experiences: Space-time and performances in the experience economy. *Scandinavian Journal of Hospitality and Tourism*, *8*(2), 122–140. doi:10.1080/15022250802110091

Elias, N., & Scotson, J. L. (1994). *The Established and the Outsiders: A Sociological Enquiry into Community Problems*. Sage Publications.

Ellis, A. (2003). *Sex Without Guilt in the 21st Century*. Barricade Books.

Ellis, G. D., Jamal, T., & Jiang, J. (2015). Fast-Thinking and Slow-Thinking: A Process Approach to Understand Situated Tourist Experiences. *Tourism Travel and Research Association: Advancing Tourism Research Globally*, *36*, 1–14.

Emir, O., & Pekyaman, A. (2010). Important factors for families with children in hotel establishment selection: An application in Afyonkarahisar. [in Turkish]. *Management & Economics: Journal of Celal Bayar University Faculty of Economics and Administrative Sciences*, *17*(2), 159–181.

Engl, S., & Nacke, L. E. (2013). Contextual influences on mobile player experience - A game user experience model. *Entertainment Computing*, *4*(1), 83–91. doi:10.1016/j.entcom.2012.06.001

Erdem, T., & Swait, J. (1998). Brand equity as a signalling phenomenon. *Journal of Consumer Psychology*, *7*(2), 131–157. doi:10.120715327663jcp0702_02

Erdem, T., & Swait, J. (2004). Brand credibility, brand consideration, and choice. *The Journal of Consumer Research*, *31*(1), 191–198. doi:10.1086/383434

Eshliki, S. A., & Kaboudi, M. (2017). Perception of community in tourism impacts and their participation in tourism planning: Ramsar, Iran. *Journal of Asian Behavioural Studies*, *2*(4), 59–69. doi:10.21834/jabs.v2i4.207

Esman, M. R. (1984). Tourism as ethnic preservation: The Cajuns of Louisiana. *Annals of Tourism Research*, *11*(3), 451–467. doi:10.1016/0160-7383(84)90031-8

Espelt, N. G., & Benito, J. A. D. (2005). The social construction of the image of Girona: A methodological approach. *Tourism Management*, *26*(5), 777–785. doi:10.1016/j.tourman.2004.04.004

Esteves, P., Slongo, L., Esteves, C., & Barcelos, R. (2013). As emoções dos consumidores da terceira idade no processo de escolha de destinos de viagens. *Revista de Administração Da UFSM*, *6*(3), 561–580. doi:10.5902/198346595581

Etstur. (2020). *ETSTUR travel agency official website*. Retrieved from https://www.etstur.com/

European Travel Commission. (2017). *European Tourism Trends & Prospects Q4/2017, ETC*. Retrieved from: https://etc-corporate.org/uploads/reports/ETC-Quarterly-Report-Q4-2017_Final-Public.pdf

Eurostats. (2015a). *Marriage and divorce statistics*. Retrieved from https://ec.europa.eu/eurostat/statistics-explained/index.php/Marriage_and_divorce_statistics

Eurostats. (2015b). *Total fertility rate, 1960–2015*. Retrieved from https://ec.europa.eu/eurostat/statistics-explained/index.php?title=File:Total_fertility_rate,_1960%E2%80%932015_(live_births_per_woman)_YB17.png

Eurostats. (2016). *Being young in Europe today - demographic trends*. Retrieved from https://ec.europa.eu/eurostat/statistics-explained/index.php/Being_young_in_Europe_today_-_demographic_trends

Eyce, B. (2002). Divorce by demographic characteristics in Turkey. [in Turkish]. *Selçuk University Journal of Faculty of Letters*, *14*, 81–98.

Fabre, D. (2001). L'histoire a changé de lieux. In *A. Bensa & D. Fabre (dirs.), Une histoire à soi. Figurations du passé et localités*. Éd Maison des Sciences de l'Homme. doi:10.4000/books.editionsmsh.2952

Fakeye, P. C., & Crompton, J. L. (1991). Image Differences between Prospective, First-Time, and Repeat Visitors to the Lower Rio Grande Valley. *Journal of Travel Research*, *30*(2), 10–16. doi:10.1177/004728759103000202

Falk, R., & Miller, N. (1992). *A primer soft modeling*. The University of Akron Press.

Farahani, B. M., & Sukmajati, D. (2011). Youth travelers as a growing market in urban tourism; potencies and challenges. In *Proceedings Environmental Talk*. Toward A Better Green Living Conference.

Farquhar, P. H. (1989). Managing brand equity. *Marketing Research*, *1*, 24–33.

Farvid, P., Braun, V., & Rowney, C. (2017). 'No girl wants to be called a slut!': Women, heterosexual casual sex and the sexual double standard. *Journal of Gender Studies*, *26*(5), 544–560. doi:10.1080/09589236.2016.1150818

Fave, A., & Bassi, M. (2016). *Flow Experience: Empirical Research and Applications*. doi:10.1007/978-3-319-28634-1_1

Featherstone, M. (1987). Lifestyle and Consumer Culture. *Theory, Culture & Society*, *4*(1), 55–70. doi:10.1177/026327687004001003

Femenia-Serra, F., Neuhofer, B., & Ivars-Baidal, J. A. (2019). Towards a conceptualization of smart tourists and their role within the smart destination scenario. *Service Industries Journal, 39*(2), 109–133. doi:10.1080/02642069.2018.1508458

Fernandes, T., & Cruz, M. (2016). Dimensions and outcomes of experience quality in tourism: The case of Port wine cellars. *Journal of Retailing and Consumer Services, 31*, 371–379. doi:10.1016/j.jretconser.2016.05.002

Ferreira, A. M., & Martins, A. I. (Eds.). (2005). *O Evento FCNC 2005 e o Turismo*. Faro, Portugal: Centro de Estudos da ESGHT, Universidade do Algarve.

Ferreira, P. M. (2010). A actividade sexual: frequência, regularidade e inactividade. In *P. M. Ferreira & M. V. Cabral (Orgs.), Sexualidades em Portugal: Comportamentos e Riscos* (pp. 19–54). Bizâncio.

Ferrer-I-Carbonell, A. (2005). Income and well-being: An empirical analysis of the comparison income effect. *Journal of Public Economics, 89*(5-6), 997–1019. doi:10.1016/j.jpubeco.2004.06.003

Fesenmaier, D. F., & Xiang, Z. (2017). *Design Science in Tourism: Foundations of Destinations Management*. Springer. doi:10.1007/978-3-319-42773-7

Figuerola, M., Fernández, J., Marzal, F., Gómez, D., & Martín, C. (2015). *Medición del impacto del turismo en la Economía, Empleo y el Medio Ambiente en el ámbito local*. Fundación Escuela Organización Industrial.

Finn, M., Elliott-White, M., & Walton, M. (2000). *Tourism and leisure research methods: Data collection, analysis, and interpretation*. Longman.

Fishbein, M., & Ajzen, I. (1974). Attitudes towards objects as predictors of single and multiple behavioral Criteria. *Psychological Review, 81*(1), 59–74. doi:10.1037/h0035872

Flick, U. (2014). *An Introduction to Qualitative Research* (5th ed.). Sage.

Florida, R. (2002). *The rise of the creative class, and how it's transforming work, leisure community and everyday life*. Basic Books.

Flurry, L. A. (2007). Children's influence in family decision-making: Examining the impact of the changing American family. *Journal of Business Research, 60*(4), 322–330. doi:10.1016/j.jbusres.2006.09.029

Fodness, D. (1992). The impact of family life cycle on the vacation decision-making process. *Journal of Travel Research, 31*(2), 8–13. doi:10.1177/004728759203100202

Foris, D. (2014). Study Regarding a New Dimension of Tourism Phenomenon – The Political-Administrative Dimension. *Procedia: Social and Behavioral Sciences, 143*, 1078–1082. doi:10.1016/j.sbspro.2014.07.559

Fornell, C., & Larcker, D. (1981). Evaluating structural equation models with unobservable variables and measurement error. *JMR, Journal of Marketing Research, 18*(1), 39–50. doi:10.1177/002224378101800104

Foucault, M. (1977). *Discipline and Punish: The Birth of the Prison*. A. Lang.

Foulds, A. (2014). *Buying a colonial dream: the role of lifestyle migrants in the gentrification of the historic center of Granada, Nicaragua* (Unpublished master's dissertation). University of Kentucky.

Foxman, E. R., Tansuhaj, P. S., & Ekstrom, K. M. (1989). Family members' perceptions of adolescents' influence in family decision making. *The Journal of Consumer Research, 15*(4), 482–491. doi:10.1086/209187

Frank, K. (2007). Playcouples in paradise: touristic sexuality and lifestyle travel. In M. Padilla, J. Hirsh, M. Laboy, R. Sember, & R. Parker (Eds.), *Love and Globalization: Transformations of Intimacy in the Contemporary World* (pp. 163–185). Vanderbilt University Press.

Franklin, A. (2003). *Tourism: An Introduction.* Sage.

Fredline, E. (2002). *Social impacts of tourism on the Gold Coast.* Altona: Common Ground Publishing.

Fredline, E., & Faulkner, B. (2000). Host community reactions: A cluster analysis. *Annals of Tourism Research, 27*(3), 763–784. doi:10.1016/S0160-7383(99)00103-6

Frey, B. S., Pamino, P., & Steiner, L. (2013). Explaining the Word Heritage List: An Empirical Study. *International Review of Economics, 60*(1), 1–19. doi:10.100712232-013-0174-4

Frias, D. M., Rodriguez, M. A., Castaneda, A., Sabiote, C. M., & Buhalis, D. (2012). The Formation of a Tourist Destination's Image via Information Sources: The Moderating Effect of Culture. *International Journal of Tourism Research, 14*(5), 437–450. doi:10.1002/jtr.870

Friedman, H. S., & Riggio, R. E. (1981). Effect of individual differences in nonverbal expressiveness on transmission of emotion. *Journal of Nonverbal Behavior, 6*(2), 96–104. doi:10.1007/BF00987285

Friedman, J. (2010). Place and place-making in cities: A global perspective. *Planning Theory & Practice, 11*(2), 149–165. doi:10.1080/14649351003759573

Frohlick, S. (2010). The sex of tourism? Bodies under suspicion in paradise. In J. Scott & T. Selwyn (Eds.), *Thinking Through Tourism* (pp. 51–70). Berg.

Frome, J. (2019). Interactive Works and Gameplay Emotions. *Games and Culture, 14*(7–8), 856–874. doi:10.1177/1555412019847907

Fuchs, M., & Pikkemaat, B. (2019). Destination Image Analysis - A Cross-Cultural Segmentation Approach. Rearchgate.net/publication/330841068_Destination_Image_Analysis-A_Cross-Cultural_Segmentation_Approach

Fu, H., Ye, B. H., & Xiang, J. (2016). Reality TV, audience travel intentions, and destination image. *Tourism Management, 55*, 37–48. doi:10.1016/j.tourman.2016.01.009

Gagnon, J., & Simon, W. (1974). *Sexual Conduct. The Social Sources of Human Sexuality.* Aldine Publishing.

Galani-Moutaf, V. (2013). Rural space (re) produced – Practices, performances and visions: A case study from an Aegean island. *Journal of Rural Studies, 32*, 103–113. doi:10.1016/j.jrurstud.2013.04.007

Gallarza, M. G., Saura, I. G., & Garcí, A. H. C. (2002). Destination image: Towards a conceptual framework. *Annals of Tourism Research, 29*(1), 56–78. doi:10.1016/S0160-7383(01)00031-7

Gammon, S., Ramshaw, G., & Wright, R. (2017). Theory in sport tourism: Some critical reflections. *Journal of Sport & Tourism, 21*(2), 69–74. doi:10.1080/14775085.2017.1319515

García, C. (1999). *Las Peregrinaciones Jacobeas.* San Pablo.

García, F. A., Vázquez, A. B., & Macías, R. C. (2015). Resident's attitudes toward the impacts of tourism. *Tourism Management Perspectives, 13*(January), 33–40. doi:10.1016/j.tmp.2014.11.002

Garrod, B. (2009). Understanding the relationship between tourism destination imagery and tourist photography. *Journal of Travel Research, 47*(3), 346–358. doi:10.1177/0047287508322785

Garrod, B., Wornell, R., & Youell, R. (2006). Re-conceptualising rural resources as countryside capital: The case of rural tourism. *Journal of Rural Studies, 22*(1), 117–128. doi:10.1016/j.jrurstud.2005.08.001

Gartner, W. (1993). Image formation process. In M. Uysal & D. Fesenmaier (Eds.), *Communication and channel systems in tourism marketing* (pp. 191–215). The Haworth Press.

Gartner, W. C. (1989). Tourism Image: Attribute Measurement of State Tourism Products Using Multidimensional Scaling Techniques. *Journal of Travel Research, 28*(2), 16–20. doi:10.1177/004728758902800205

Gartner, W. C. (1994). Image formation process. *Journal of Travel & Tourism Marketing, 2*(2-3), 191–216. doi:10.1300/J073v02n02_12

Gefen & Straub. (2003). Managing User Trust in B2C e-Services. *E-Service Journal, 2*(2), 7.

Gelter, H. (2010). Total Experience Management – a conceptual model for transformational experiences within tourism. In *The Nordic Conference on Experience 2008. Research, Education and Practice in Media* (pp. 46–78). Vaasa: Medusa Group of Vaasa Consortium of Higher Education.

Gentile, C., Spiller, N., & Noci, G. (2007). How to Sustain the Customer Experience. An Overview of Experience Components that Co-create Value With the Customer. *European Management Journal, 25*(5), 395–410. doi:10.1016/j.emj.2007.08.005

Gent, W. V., & Boterman, W. (2018). Gentrification of the Changing State. *Tijdschrift voor Economische en Sociale Geografie, 110*(1), 35–46. doi:10.1111/tesg.12331

George, R., & Swart, K. (2012). International tourists' perceptions of crime-risk and their future travel intentions during the 2010 FIFA World Cup in South Africa. *Journal of Sport & Tourism, 17*(3), 201–223. doi:10.1080/14775085.2012.734060

Gerbing, D. W., & Anderson, J. C. (1988). An Updated Paradigm for Scale Development Incorporating Unidimensionality and Its Assessment. *JMR, Journal of Marketing Research, 25*(2), 186–192. doi:10.1177/002224378802500207

Gerring, J. (1997). Ideology: A Definitional Analysis. *Political Research Quarterly, 50*(4), 957–994. doi:10.1177/106591299705000412

Gertner, D. (2011). Unfolding and configuring two decades of research and publications on place marketing and place branding. *Place Branding and Public Diplomacy, 7*(2), 91–106. doi:10.1057/pb.2011.7

Getz, D. (2008). Event Tourism: Definition, evolution, and research. *Tourism Management, 29*(3), 403–428. doi:10.1016/j.tourman.2007.07.017

Getz, D. (2014). Foreword. In L. Sharples, P. Crowther, D. May, & C. Orefice (Eds.), *Strategic Event Creation* (p. vi). Goodfellow Publishers Limited.

Getz, D., & Jamal, T. B. (1994). The environment–community symbiosis: A case of collaborative tourism planning. *Journal of Sustainable Tourism, 2*(3), 152–173. doi:10.1080/09669589409510692

Getz, D., & Page, S. J. (2016). Progress and prospects for event tourism research. *Tourism Management, 52*, 593–631. doi:10.1016/j.tourman.2015.03.007

Geus, S., Richards, G., & Toepel, V. (2016). Conceptualisation and Operationalisation of Event and Festival Experiences: Creation of an Event Experience Scale. *Scandinavian Journal of Hospitality and Tourism, 16*(3), 274–296. doi:10.1080/15022250.2015.1101933

Gibson, H. J., Kaplanidou, K., & Kang, S. J. (2012). Small-scale event sport tourism: A case study in sustainable tourism. *Sport Management Review, 15*(2), 160–170. doi:10.1016/j.smr.2011.08.013

Giddens, A. (1991). *Modernity and Self-Identity: Self and Society in the Late Modern Age.* Polity Press.

Giddens, A. (1992). *The Transformation of Intimacy: Sexuality, Love & Eroticism in Modern Societies.* Polity Press.

Giddens, A. (2000). *As Consequências da Modernidade.* Celta.

Giese, J. L., & Cote, J. A. (2000). Defining Consumer satisfaction. *Academy of Marketing Science Review*, *2000*(1), 1–24.

Gilovich, T., Kumar, A., & Jampol, L. (2015). A wonderful life: Experiential consumption and the pursuit of happiness. *Journal of Consumer Psychology*, *25*(1), 152–165. doi:10.1016/j.jcps.2014.08.004

Gisolf, M. (2013). Liminality and tourism. *Tourism Theories*. Retrieved from: http://www. tourismtheories.org/?p=700

Gisolf, M. (2017). *Tourists' Profiles and Lifestyles. Tourism Theories. Tourist Profiles and Sustainability*. https://www. tourismtheories.org/?p=999

Glaser, B., & Strauss, A. (1967). *The Discovery of Grounded Theory: Strategies for Qualitative Research*. Aldine.

Glass, Z. (2007). The Effectiveness of Product Placement in Video Games. *Journal of Interactive Advertising*, *8*(1), 1–27. doi:10.1080/15252019.2007.10722134

Goffman, E. (1959). *The Presentation of Self in Everyday Life*. Anchor Books.

Goggin, C., Please, P., Ridges, M., Booth, C., Simpson, G., Green, R., & Leys, J. (2017). Connecting with Country in Mungo National Park, Australia: A case study to measure the emotional dimension of experience and place attachment. *Local Environment*, *22*(10), 1217–1236. doi:10.1080/13549839.2017.1334142

Gohary, A., Pourazizi, L., Madani, F., & Chan, E. Y. (2020). Examining Iranian tourists' memorable experiences on destination satisfaction and behavioral intentions. *Current Issues in Tourism*, *23*(2), 131–136. doi:10.1080/13683500.2018.1560397

Golafshani, N. (2003). Understanding reliability and validity in qualitative research. *Qualitative Report*, *8*, 597–606.

Gold, P. (2004). Coordination of multiple memory systems. *Neurobiology of Learning and Memory*, *82*(3), 230–242. doi:10.1016/j.nlm.2004.07.003 PMID:15464406

Gómez-Bantel, A. (2015). Football clubs as symbols of regional identities. *Soccer and Society*, *17*(5), 692–702. doi:10.1080/14660970.2015.1100430

Gonçalves, A. R. (2008). As Comunidades Criativas, o Turismo e a Cultura. *Revista Dos Algarves*, *17*, 10–18.

González-García, R. J., Parra, D., Añó, V., & Calabuig, F. (2016). Percepción de los residentes sobre el impacto del Mundobasket 2014 en Gran Canaria y apoyo a la celebración de eventos deportivos. *Revista Iberoamericana de Psicología del Ejercicio y el Deporte*, *11*(2), 279–288.

González, J., & González, O. (2005). An Analysis of the relationship between environmental motivations and ISO 14001 certification. *British Journal of Management*, *16*(2), 133–148. doi:10.1111/j.1467-8551.2005.00436.x

González-Pérez, J. M. (2019). The dispute over touristic cities. Tourism gentrification in the historic Centre of Palma (Majorca, Spain). *Tourism Geographies*. Advance online publication. doi:10.1080/14616688.2019.1586986

Goossens, C. (2000). Tourism information and pleasure motivation. *Annals of Tourism Research*, *27*(2), 301–321. doi:10.1016/S0160-7383(99)00067-5

Gotham, K. F. (2005). Tourism Gentrification: The Case of New Orlean's Vieux Carre (French Quarter). *Urban Studies (Edinburgh, Scotland)*, *42*(7), 1099–1121. doi:10.1080/00420980500120881

Götze, E., Prange, C., & Uhrovska, I. (2009). Children's impact on innovation decision making. *European Journal of Marketing*, *43*(1/2), 264–295. doi:10.1108/03090560910923328

Govers, R., Go, F. M., & Kumar, K. (2007). Virtual destination image a new measurement approach. *Annals of Tourism Research*, *34*(4), 977–997. doi:10.1016/j.annals.2007.06.001

Graburn, N. (1977). Tourism: the sacred journey. In V. Smith (Ed.), *Hosts and Guests: The Anthropology of Tourism* (pp. 17–31). University of Pensilvania Press.

Graham, C., & Pettinato, S. (2001). Happiness, markets, and democracy: Latin America in comparative perspective. *Journal of Happiness Studies*, *2*(3), 237–268. doi:10.1023/A:1011860027447

Gram, M. (2007). Children as co-decision makers in the family? The case of family holidays. *Young Consumers*, *8*(1), 19–28. doi:10.1108/17473610710733749

Gratton, C. (2005). *The economic impact of sport tourism at major events*. Elsevier Butterworth-Heinemann.

Gratton, C., Shibli, S., & Coleman, R. (2006). The economic impact of major sports events: A review of ten events in the UK. *Socialist Review*, *54*(2), 41–58. doi:10.1111/j.1467-954X.2006.00652.x

Gray, B. (1985). Conditions facilitating interorganisational collaboration. *Human Relations*, *38*(10), 911–936. doi:10.1177/001872678503801001

Green, C. S., & Bavelier, D. (2006). Effect of action video games on the spatial distribution of visuospatial attention. *Journal of Experimental Psychology. Human Perception and Performance*, *32*(6), 1465–1478. doi:10.1037/0096-1523.32.6.1465 PMID:17154785

Greene, H. (2011). Freshmen Marketing: A First-Year Experience with Experiential Learning. *Marketing Education Review*, *21*(1), 79–87. doi:10.2753/MER1052-8008210111

Grenhas, M. C. (2013). *Motivação, Satisfação e Fidelização dos Visitantes do Festival Mediterrâneo de Loulé* (Unpublished master's thesis). University of Algarve, Faro, Portugal.

Grigorovici, D. M., & Constantin, C. D. (2004). Experiencing Interactive Advertising beyond Rich Media: Impacts of Ad Type and Presence on Brand Effectiveness in 3D Gaming Immersive Virtual Environments. *Journal of Interactive Advertising, 4*(3).

Grönroos, C., & Voima, P. (2013). Critical service logic: Making sense of value creation and co-creation. *Journal of the Academy of Marketing Science*, *41*(2), 133–150. doi:10.100711747-012-0308-3

Gross, M. L. (2010). Advergames and the effects of game-product congruity. *Computers in Human Behavior*, *26*(6), 1259–1265. doi:10.1016/j.chb.2010.03.034

Guerra, I. (2010). *Pesquisa Qualitativa e Análise de Conteúdo: Sentidos e Formas de Uso*. Princípia Editora.

Guerreiro, J. A., & Marques, J. F. (2017). Visita Guiada à Fábrica de Antiguidades Sociologia, Turismo e Autenticidade. *Anais Brasileiros de Estudos Turísticos*, *1*(7), 8–22. doi:10.34019/2238-2925.2017.v7.3160

Guerreiro, M., Valle, P., & Mendes, J. (2011). Allgarve events: Implications for the Algarve Image. *Tourism (Zagreb)*, *59*(2), 183–202.

Gu, H., & Ryan, C. (2008). Place attachment, identity and community impacts of tourism—The case of a Beijing hutong. *Tourism Management*, *29*(4), 637–647. doi:10.1016/j.tourman.2007.06.006

Gunn, C. A., & Var, T. (2002). Tourism planning basics, concepts and cases (4th ed.). New York: Routledge, Taylor & Francis Group.

Gunn, C. A. (1972). *Vacationscape. Designing tourist regions*. Taylor & Francis.

Gunn, C. A. (1988). *Vacationscape: Designing tourist regions*. Van Nostrand Reinhold.

Guo, A., Guo, Y., Li, H., & Sun, X. (2017a). Relationship between perceived tourism impacts and perceived quality of life of community residents in tourist destinations. *World Regional Studies*, *26*, 115–127.

Gupta, P. B., & Lord, K. R. (1998). Product Placement in Movies: The Effect of Prominence and Mode on Audience Recall. *Journal of Current Issues and Research in Advertising*, *20*(1), 47–59. doi:10.1080/10641734.1998.10505076

Gursoy, D. S., Chen, J., & Chi, C. (2014). Theoretical examination of destination loyalty formation. *International Journal of Contemporary Hospitality Management*, *26*(5), 809–827. doi:10.1108/IJCHM-12-2013-0539

Gursoy, D., Chi, C. G., & Dyer, P. (2009). An examination of locals' attitudes. *Annals of Tourism Research*, *36*(4), 723–726. doi:10.1016/j.annals.2009.06.003

Gursoy, D., Chi, C. G., & Dyer, P. (2010). Locals' attitudes toward mass and alternative tourism: The case of sunshine coast, Australia. *Journal of Travel Research*, *49*(3), 381–394. doi:10.1177/0047287509346853

Gursoy, D., Jurowski, C., & Uysal, M. (2002). Resident attitudes. A structural modeling approach. *Annals of Tourism Research*, *29*(1), 79–105. doi:10.1016/S0160-7383(01)00028-7

Gursoy, D., & Kendall, K. W. (2006). Hosting mega events: Modeling locals' support. *Annals of Tourism Research*, *33*(3), 603–623. doi:10.1016/j.annals.2006.01.005

Gursoy, D., Ouyang, Z., Nunkoo, B., & Wei, W. (2019). Residents' impact perceptions of and attitudes towards tourism development: A meta-analysis. *Journal of Hospitality Marketing & Management*, *28*(3), 306–333. doi:10.1080/19368623.2018.1516589

Gursoy, D., & Rutherford, D. G. (2004). Host attitudes toward tourism: An improved structural model. *Annals of Tourism Research*, *31*(3), 495–516. doi:10.1016/j.annals.2003.08.008

Gustafson, P. (2009). Your Home in Spain: Residential Strategies in International Retirement Migration. In M. Benson & K. O'Reilly (Eds.), *Lifestyle Migrations: Expectations, Aspirations and Experiences* (pp. 69–86). Ashgate.

Gutiérrez, D. (2010). *Las actitudes de los residentes ante el turismo* (Doctoral dissertation). Retrieved from RIULL, Repositorio Institucional Universidad de La Laguna: (https://riull.ull.es/xmlui/handle/915/2117)

Güven, O. Z., Beydilli, E. T., Ceylan, U., & Bütün, S. (2017). A study on child-friendly hotel concept. In Proceedings of 3th Tourism Council (pp. 632-641). Republic of Turkey Ministry of Culture and Tourism. [in Turkish].

Guzel, O. (2014). The Dimensions of Tour Experience, Emotional Arousal, and Post-experience Behaviors: A Research on Pamukkale in Turkey. *Procedia: Social and Behavioral Sciences*, *150*, 521–530. doi:10.1016/j.sbspro.2014.09.069

Hackworth, J. (2018). Gentrification as a Politico-economic Window: Reflections on the Changing State of Gentrification. *Tijdschrift voor Economische en Sociale Geografie*, *110*(1), 47–53. doi:10.1111/tesg.12330

Hackworth, J., & Smith, N. (2001). The Changing State of Gentrification. *Tijdschrift voor Economische en Sociale Geografie*, *92*(4), 464–477. doi:10.1111/1467-9663.00172

Hair, J., Anderson, R., Tatham, R., & Black, W. (2010). *Multivariate Data Analysis*. Prentice Hall- Pearson.

Hair, J., Hult, G., Ringle, C., & Sarstedt, M. (2017). *A Primer on Partial Least Square Structural Equation Modeling (PLS-SEM)*. Sage.

Haley, A. J., Snaith, T., & Miller, G. (2005). The social impacts of tourism a case study of Bath, UK. *Annals of Tourism Research*, *32*(3), 647–668. doi:10.1016/j.annals.2004.10.009

Halim, A., & Mokhtar, A. (2016). Creating Memorable Visitor Experiences by Assessing the Satisfaction Level and Behavioural Consequences of Attendees. *Procedia Economics and Finance*, *37*, 1–6. doi:10.1016/S2212-5671(16)30084-3

Hall, C. (2004). *Tourism and mobility*. Presented at the CAUTHE 2004: Creating Tourism Knowledge. Retrieved from: https://ourarchive.otago.ac.nz/bitstream/handle/ 10523/738/Mobility_CAUTHE.pdf?sequence=2&isAllowed=y

Hall, C. (1992). Sex tourism in Southeast Asia. In D. Harrison (Ed.), *Tourism and the Less Developed Countries* (pp. 65–74). Belhaven Press.

Hall, C. (2008). *Tourism Planning: Policies, Processes and Relationships*. Pearson Education.

Hall, C. M., & Lew, A. A. (1998). *Sustainable Tourism: A Geographical Perspective*. Addison Wesley Longman Limited.

Hall, C. M., & Valentin, A. (2005). Content analysis. In B. W. Ritchie, P. Burns, & C. Palmer (Eds.), *Tourism research methods: Integrating theory with practice*. Cabi Publishing. doi:10.1079/9780851999968.0191

Halpenny, E., Kulczycki, C., & Moghimehfar, F. (2016). Factors effecting destination and event loyalty: Examining the sustainability of a recurrent small-scale running event at Banff National Park. *Journal of Sport & Tourism*, *20*(3-4), 233–262. doi:10.1080/14775085.2016.1218787

Hammond, N., & Kingston, S. (2014). Experiencing stigma as sex work researchers in professional and personal lives. *Sexualities*, *17*(3), 329–347. doi:10.1177/1363460713516333

Han, J., & Patterson, I. (2007). An Analysis of the Influence That Leisure Experiences Have on a Person's Mood State, Health and Wellbeing. *Annals of Leisure Research*, *10*(3–4), 328–351. doi:10.1080/11745398.2007.9686770

Haobin Ye, B., Qiu Zhang, H., Huawen Shen, J., & Goh, C. (2014). Does social identity affect residents' attitude toward tourism development? An evidence from the relaxation of the Individual Visit Scheme. *International Journal of Contemporary Hospitality Management*, *26*(6), 907–929. doi:10.1108/IJCHM-01-2013-0041

Hardy, A., Beeton, R., & Pearson, L. (2002). Sustainable tourism: An overview of the concept and its position in relation to conceptualisations of tourism. *Journal of Sustainable Tourism*, *10*(6), 475–496. doi:10.1080/09669580208667183

Hareli, S., Shomrat, N., & Hess, U. (2009). Emotional Versus Neutral Expressions and Perceptions of Social Dominance and Submissiveness. *Emotion (Washington, D.C.)*, *9*(3), 378–384. doi:10.1037/a0015958 PMID:19485615

Harrill, R. (2004). Residents' attitudes toward tourism development: A literature review with implications for tourism planning. *Journal of Planning Literature*, *18*(3), 251–266. doi:10.1177/0885412203260306

Hart, A., Kerrigan, F., & vom Lehn, D. (2016). Experiencing film: Subjective personal introspection and popular film consumption. *International Journal of Research in Marketing*, *33*(2), 375–391. doi:10.1016/j.ijresmar.2015.08.005

Ha, S., Huang, R., & Park, J. S. (2019). Persuasive brand messages in social media: A mental imagery processing perspective. *Journal of Retailing and Consumer Services*, *48*(February), 41–49. doi:10.1016/j.jretconser.2019.01.006

Hasani, A., Moghavvemi, S., & Hamzah, A. (2016). The impact of emotional solidarity on residents' attitude and tourism development. *PLoS One*, *11*(6), 157–624. doi:10.1371/journal.pone.0157624 PMID:27341569

Hassan, D., & Connor, S. O. (2009). The socio-economic impact of the FIA World Rally Championship 2007. *Sport in Society*, *12*(6), 709–724. doi:10.1080/17430430902944134

Hassanein, K., & Head, M. (2007). Manipulating perceived social presence through the web interface and its impact on attitude towards online shopping. *International Journal of Human-Computer Studies*, *65*(8), 689–708. doi:10.1016/j.ijhcs.2006.11.018

Hatfield, E., Cacioppo, J. T., & Rapson, R. L. (1993). Emotional contagion. *Current Directions in Psychological Science*, *2*(3), 96–100. doi:10.1111/1467-8721.ep10770953

Haywood, K. M. (1988). Responsible and responsive tourism planning in the community. *Tourism Management*, *9*(2), 105–108. doi:10.1016/0261-5177(88)90020-9

Henriques, C., Guerreiro, M., & Mendes, J. (2014). *Turismo Cultural no Algarve*. RTA/UALG.

Henriques, I., & Sadorsky, P. (1996). The determinants of an environmental responsive firm: An empirical approach. *Journal of Environmental Economics and Management*, *30*(3), 381–395. doi:10.1006/jeem.1996.0026

Henriques, I., & Sadorsky, P. (1999). The relationship between environmental commitment and managerial perceptions of stakeholder importance. *Academy of Management Journal*, *42*, 87–99. doi:10.2307/256876

Henseler, J., Ringle, C. M., & Sarstedt, M. (2015). A new criterion for assessing discriminant validity in variance-based structural equation modeling. *Journal of the Academy of Marketing Science*, *43*(1), 115–135. doi:10.100711747-014-0403-8

Hernández-Lobato, L., Solis-Radilla, M. G., Moliner-Tena, M. G., & Sanchez-Garcia, J. (2006). Tourism Destination Image, Satisfaction and Loyalty: A Study in Ixtapa-Zihuatanejo, Mexico. *Tourism Geographies*, *8*(4), 343–358. doi:10.1080/14616680600922039

Herold, E., Garcia, R., & DeMoya, T. (2001). Female tourists and beach boys: Romance or sex tourism? *Annals of Tourism Research*, *28*(4), 978–997. doi:10.1016/S0160-7383(01)00003-2

Herrera, L. M., Smith, N., & Vera, M. A. M. (2013). Gentrification, Displacement, and Tourism in Santa Cruz De Tenerife. *Urban Geography*, *28*(3), 276–298. doi:10.2747/0272-3638.28.3.276

Hess, U., & Blairy, S. (2001). Facial mimicry and emotional contagion to dynamic emotional facial expressions and their influence on decoding accuracy. *International Journal of Psychophysiology*, *40*(2), 129–141. doi:10.1016/S0167-8760(00)00161-6 PMID:11165351

Hillary, R. (2003). Environmental management systems and the smaller enterprise. *Journal of Cleaner Production*, *12*(6), 561–569. doi:10.1016/j.jclepro.2003.08.006

Hobbs, J., Pattalung, P., & Chandler, R. (2011). Advertising Phuket's nightlife on the internet: A case study of double binds and hegemonic masculinity in sex tourism. *Journal of Social Issues in Southeast Asia*, *26*(1), 80–104. doi:10.1355j26-1e PMID:21919275

Hobson, J., & Dietrich, U. (1994). Tourism, health and quality of life: Challenging the responsibility of using the traditional tenets of sun, sea, sand and sex in tourism marketing. *Journal of Travel & Tourism Marketing*, *3*(4), 21–38. doi:10.1300/J073v03n04_02

Hoeckner, B., Wyatt, E. W., Decety, J., & Nusbaum, H. (2011). Film Music Influences How Viewers Relate to Movie Characters. *Psychology of Aesthetics, Creativity, and the Arts*, *5*(2), 146–153. doi:10.1037/a0021544

Hoffman, A. (1999). Institutional Evolution and Change: Environmentalism and the U.S. Chemical Industry. *Academy of Management Journal*, *42*(4), 351–372. doi:10.2307/257008

Hoijer, B. (2011). Social Representation Theory: A New Theory for Media Research. *Nordicom Review*, *32*(2), 3–16. doi:10.1515/nor-2017-0109

Holbrook, M. B. (2000). The Millennial Consumer in the Texts of Our Times: Experience and Entertainment. *Journal of Macromarketing*, *20*(2), 178–192. doi:10.1177/0276146700202008

Holbrook, M. B., Chestnut, R. W., Oliva, T. A., & Greenleaf, E. A. (1984). Play as a Consumption Experience: The Roles of Emotions, Performance, and Personality in the Enjoyment of Games. *The Journal of Consumer Research, 11*(2), 728–739. doi:10.1086/209009

Holbrook, M. B., & Hirschman, E. C. (1982a). The Experiential Aspects of Consumption: Consumer Fantasies, Feelings, and Fun. *The Journal of Consumer Research, 9*(2), 132. doi:10.1086/208906

Homans, G. C. (1961). Social behaviour: Its elementary forms. New York: Whitestable Litho, Straker Brothers, Ltd. Harcourt, Brace & Worl, Inc.

HomeAway. (2015). *New Survey Reveals the Strong Influence of Kids on Family Travel Plans*. Retrieved from: https://www.homeaway.com/info/media-center/press-releases/2015/new-survey-reveals-the-strong-influence-of-kids-on-family-travel

Hone, K. (2006). Empathic agents to reduce user frustration: The effects of varying agent characteristics. *Interacting with Computers, 18*(2), 227–245. doi:10.1016/j.intcom.2005.05.003

Horn, C., & Simmons, D. (2002). Community adaptation to tourism: Comparisons between Rotorua and Kaikoura, New Zealand. *Tourism Management, 23*(2), 133–143. doi:10.1016/S0261-5177(01)00049-8

Hosany, S., & Gilbert, D. (2009, July 1). Dimensions of Tourists' Emotional Experiences Towards Hedonic Holiday Destinations. Available at SSRN: https://ssrn.com/abstract=1871768

Hosany, S., Ekinci, Y., & Uysal, M. (2006). Destination image and destination personality: An application of branding theories to tourism places. *Journal of Business Research, 59*(5), 638–642. doi:10.1016/j.jbusres.2006.01.001

Hosany, S., & Gilbert, D. (2010). Measuring Tourists' Emotional Experiences toward Hedonic Holiday Destinations. *Journal of Travel Research, 49*(4), 513–526. doi:10.1177/0047287509349267

Hosany, S., & Prayag, G. (2013). Patterns of tourists' emotional responses, satisfaction, and intention to recommend. *Journal of Business Research, 66*(6), 730–737. doi:10.1016/j.jbusres.2011.09.011

Hosany, S., Prayag, G., Deesilatham, S., Caušević, S., & Odeh, K. (2015). Measuring Tourists' Emotional Experiences: Further Validation of the Destination Emotion Scale. *Journal of Travel Research, 54*(4), 482–495. doi:10.1177/0047287514522878

Hosany, S., Prayag, G., Van Der Veen, R., Huang, S., & Deesilatham, S. (2017). Mediating Effects of Place Attachment and Satisfaction on the Relationship between Tourists' Emotions and Intention to Recommend. *Journal of Travel Research, 56*(8), 1079–1093. doi:10.1177/0047287516678088

Hosany, S., & Witham, M. (2010). Dimensions of Cruisers' Experiences, Satisfaction, and Intention to Recommend. *Journal of Travel Research, 49*(3), 351–364. doi:10.1177/0047287509346859

Hot to make sujuk? (2020). Retrieved from: https://italianbarrel.com/how-to-make-sujuk/

Hou, J., & Rios, M. (2003). Community-driven place making. *Journal of Architectural Education, 57*(1), 19–27. doi:10.1162/104648803322336557

Hou, J., Zhao, X., & Zheng, J. (2019). The impact of consistency between the emotional feature of advertising music and brand personality on brand experience. *Journal of Management Analytics, 6*(3), 250–268. doi:10.1080/23270012.2019.1613684

Hsee, C. K., Hatfield, E., Carlson, J. G., & Chemtob, C. (1990). The effect of power on susceptibility to emotion contagion. *Cognition and Emotion, 4*(4), 327–340. doi:10.1080/02699939008408081

Hsieh, S., & Chang, A. (2016). The Psychological Mechanism of Brand Co-creation Engagement. *Journal of Interactive Marketing, 33*, 13–26. doi:10.1016/j.intmar.2015.10.001

Hsu, C. L., & Lu, H. P. (2004). Why do people play on-line games? An extended TAM with social influences and flow experience. *Information & Management*, *41*(7), 853–868. doi:10.1016/j.im.2003.08.014

Huang, J., & Hsu, C. H. C. (2010). The Impact of Customer-to-Customer Interaction on Cruise Experience and Vacation Satisfaction. *Journal of Travel Research*, *49*(1), 79–92. doi:10.1177/0047287509336466

Huang, Y. C., Backman, S. J., Backman, K. F., & Moore, D. W. (2013). Exploring user acceptance of 3D virtual worlds in travel and tourism marketing. *Tourism Management*, *36*, 490–501. doi:10.1016/j.tourman.2012.09.009

Hudson, S., & Ritchie, J. R. B. (2009). Branding a memorable destination experience. The case of 'Brand Canada.'. *International Journal of Tourism Research*, *11*(2), 217–228. doi:10.1002/jtr.720

Hughes, N. (2018). 'Tourists go home': Anti-tourism industry protest in Barcelona. *Social Movement Studies*, *17*(4), 471–477. doi:10.1080/14742837.2018.1468244

Hultén, B. (2011). Sensory marketing: The multi-sensory brand-experience concept. *European Business Review*, *23*(3), 256–273. doi:10.1108/09555341111130245

Hultman, J., & Hall, C. M. (2012). Tourism place-making: Governance of locality in Sweden. *Annals of Tourism Research*, *39*(2), 547–570. doi:10.1016/j.annals.2011.07.001

Human, L. (2012). Adventure-Based Experiences during Professional Training in Psychology: A Follow-up Study. *South African Journal of Psychology. Suid-Afrikaanse Tydskrif vir Sielkunde*, *42*(4), 586–597. doi:10.1177/008124631204200412

Hung, K., & Petrick, J. F. (2011). The role of self- and functional congruity in cruising intentions. *Journal of Travel Research*, *50*(1), 100–112. doi:10.1177/0047287509355321

Hung, K., & Petrick, J. F. (2012). Testing the effects of congruity, travel constraints, and self-efðcacy on travel intentions: An alternative decision-making model. *Tourism Management*, *33*(4), 855–867. doi:10.1016/j.tourman.2011.09.007

Hung, K., Sirakaya, E., & Ingram, L. J. (2011). Testing the efficacy of an integrative model for community participation. *Journal of Travel Research*, *50*(3), 276–288. doi:10.1177/0047287510362781

Hunt, C., & Auster, E. (1990). Proactive environmental management: Avoiding the toxic trap. *Sloan Management Review*, *31*(2), 7–18.

Hunt, C., & Stronza, A. (2014). Stage-based tourism models and resident attitudes towards tourism in an emerging destination in the developing world. *Journal of Sustainable Tourism*, *22*(2), 279–298. doi:10.1080/09669582.2013.815761

Hunter, W. C., & Suh, Y.-K. (2007). Multimethod research on destination image perception: Jeju standing stones. *Tourism Management*, *28*(1), 130–139. doi:10.1016/j.tourman.2005.06.013

Hunt, J. D. (1971). *Image: A Factor in Tourism*. Colorado State University Press.

Hunt, J. D. (1975). Image as a factor in tourism development. *Journal of Travel Research*, *13*(3), 1–7. doi:10.1177/004728757501300301

Hürriyet. (2003). *Sujuk commercial has shifted to late night in order children not to gulp*. Retrieved from: https://www.hurriyet.com.tr/ekonomi/cocuklar-yutkunmasin-diye-sucuk-reklami-geceye-kaydi-124884 [in Turkish].

Hürriyet. (2010). *Child-friendly hotels*. Retrieved from https://www.hurriyet.com.tr/ekonomi/cocuk-dostu-oteller-14051915 [in Turkish].

Hürriyet. (2017). *Children are becoming 'quickly obese'*. Retrieved from https://www.hurriyet.com.tr/dunya/dunya-genelinde-cocuklar-hizla-obezlesiyor-40606622 [in Turkish].

Hwang, S.-N., Lee, C., & Chen, H.-J. (2005). The relationship among tourists' involvement, place attachment and interpretation satisfaction in Taiwan's national parks. *Tourism Management*, *26*(2), 143–156. doi:10.1016/j.tourman.2003.11.006

Hyde, K. F. (2008). Independent Traveler Decision-Making. In A. Woodside (Ed.), *Advances in Culture, Tourism and Hospitality Research* (Vol. 2, pp. 43–151). Emerald Group Publishing Limited., doi:10.1016/S1871-3173(08)02003-X

Hyde, K. F., & Lawson, R. (2003). The nature of independent travel. *Journal of Travel Research*, *42*(1), 13–23. doi:10.1177/0047287503253944

Idziak, W., Majewski, J., & Zmyślony, P. (2015). Community participation in sustainable rural tourism experience creation: A long-term appraisal and lessons from a thematic villages project in Poland. *Journal of Sustainable Tourism*, *23*(8-9), 8–9, 1341–1362. doi:10.1080/09669582.2015.1019513

Ilincic, M. (2014). Benefits of Creative Tourism - The Tourist Perspective. In G. Richards & A.P. Russo (Eds.), Alternative and Creative Tourism. Barcelona: ATLAS - Association for tourism and Leisure Education.

IMPACTUR. (2019). *Indicadores CIITT/UALG IMPACTUR*. http://www.ciitt.ualg.pt/impactur/dt_anuais.asp?nut=17

INE. (2014). *Tourism Statistics 2013*. Statistics Portugal.

INE. (2015). *Anuário Estatístico da Região do Algarve 2014*. Instituto Nacional de Estatística.

INE. (2018a). *Statistical Yearbook of Algarve Region 2017*. Statistics Portugal.

INE. (2018b). *Tourism Statistics 2017*. Statistics Portugal.

INE. (2019). *Statistical Yearbook of Algarve Region 2018*. Statistics Portugal.

INE. (2020). *Number of Beverage Establishments in the Algarve 2018: Database of Statistics Portugal*. Retrieved from: https://www.ine.pt/xportal/xmain?xpid=INE&xpgid=ine_indicadores&userLoadSave=Load&userTableOrder=10765&tipoSeleccao=1&contexto=pq&selTab=tab1&submitLoad=true

Ishikawa, N., & Fukushige, M. (2007). Impacts of tourism and fiscal expenditure to remote islands: The case of the Amami islands in Japan. *Applied Economics Letters*, *14*(9), 661–666. doi:10.1080/13504850500461423

Ismail, A. R. (2011). Experience Marketing: An Empirical Investigation. *Journal of Relationship Marketing*, *10*(3), 167–201. doi:10.1080/15332667.2011.599703

Ismail, A. R., Melewar, T., Lim, L., & Woodside, A. (2011). Customer experiences with brands: Literature review and research directions. *The Marketing Review*, *11*(3), 205–225. doi:10.1362/146934711X589435

Ivanovic, M., & Saayman, M. (2015). Authentic economy shaping transmodern tourism experience. *African Journal for Physical, Health Education. Recreation and Dance*, (Supplement), 24–36.

Izard, C. E. (1994). *Innate and universal facial expressions: Evidence from developmental and cross-cultural research. Psychological Bulletin*. American Psychological Association.

Jackson, S., & Scott, S. (2002). *Gender: A Sociological Reader*. Routledge.

Jackson, S., & Scott, S. (2004). Sexual antinomies in late modernity. *Sexualities*, *7*(2), 233–248. doi:10.1177/1363460704042166

Jafari, J. (1987). Tourism models: The sociocultural aspects. *Tourism Management*, *8*(2), 151–159. doi:10.1016/0261-5177(87)90023-9

Jalis, M. H., Che, D., & Markwell, K. (2014). Utilising Local Cuisine to Market Malaysia as a Tourist Destination. *Procedia: Social and Behavioral Sciences*, *144*, 102–110. doi:10.1016/j.sbspro.2014.07.278

Janoschka, M., Sequera, J., & Salinas, L. (2014). Gentrification in Spain and Latin America: A Critical Dialogue. *International Journal of Urban and Regional Research*, *38*(4), 1234–1265. doi:10.1111/1468-2427.12030

Jeffreys, S. (2003). Sex tourism: Do women do it too? *Leisure Studies*, *22*(3), 223–238. doi:10.1080/026143603200075452

Jelínková, D., Tučková, Z., & Jurigová, Z. (2017). Market segment of families with children: A new stimulus for modern Czech spa industry? *Journal of International Studies*, *10*(2), 158–169. doi:10.14254/2071-8330.2017/10-2/12

Jenkins, O. H. (1999). Understanding and measuring tourist destination images. *International Journal of Tourism Research*, *1*(1), 1–15. doi:10.1002/(SICI)1522-1970(199901/02)1:1<1::AID-JTR143>3.0.CO;2-L

Jennings, G. (2005). Interviewing: a focus on qualitative techniques. In B. W. Ritchie, P. Burns, & C. Palmer (Eds.), *Tourism Research Methods: Integrating Theory with Practice* (pp. 99–117). CABI Publishing. doi:10.1079/9780851999968.0099

Jennings, G. (2005). Perspectives on quality tourism experiences: An introduction. In G. Jennings & N. P. Nickerson (Eds.), *Quality Tourism Experiences* (pp. 1–21). Elsevier Ltd., doi:10.4324/9780080455778

Jensen, R. (1999). *The dream society: how the coming shift from information to imagination will transform your business*. McGraw-Hill.

Jensen, R. (2001). *The Dream Society: How the Coming Shift from Information to Imagination will Transform Your Business* (2nd ed.). McGraw-Hill.

Jeong, E. J., & Biocca, F. A. (2012). Are there optimal levels of arousal to memory? Effects of arousal, centrality, and familiarity on brand memory in video games. *Computers in Human Behavior*, *28*(2), 285–291. doi:10.1016/j.chb.2011.09.011

Jepson, D., & Sharpley, R. (2015). More than sense of place? Exploring the emotional dimension of rural tourism experiences. *Journal of Sustainable Tourism*, *23*(8–9), 1157–1178. doi:10.1080/09669582.2014.953543

Jiménez-Naranjo, H. V., Coca-Pérez, J. L., Gutiérrez-Fernández, M., & Sánchez-Escobedo, M. C. (2016). Cost-benefit analysis of sport events: The case of World Paddle Tour. *European Research on Management and Business Economics*, *22*(3), 131–138. doi:10.1016/j.iedee.2015.04.001

Jimura, T. (2011). The impact of world heritage site designation on local communities - A case study of Ogimachi, Shirakawa–mura, Japan. *Tourism Management*, *32*(2), 288–296. doi:10.1016/j.tourman.2010.02.005

Johanson, M. M., & Woods, R. H. (2008). Recognizing the Emotional Element in Service Excellence. *Cornell Hospitality Quarterly*, *49*(3), 310–316. doi:10.1177/1938965508316267

Johnson, J., Snepenger, D., & Akis, S. (1994). Residents' perceptions of tourism development. *Annals of Tourism Research*, *21*(3), 629–642. doi:10.1016/0160-7383(94)90124-4

Johnston, L., & Longhurst, R. (2010). *Space, Place and Sex: Geographies of Sexualities*. Rowman & Littlefield Publishers.

Jollytur. (2020). *Jollytur travel agency official website*. Retrieved from https://www.jollytur.com/

Joo, D., & Woosnam, K. M. (2019). Measuring tourists' emotional solidarity with one another—A modification of the Emotional Solidarity Scale. *Journal of Travel Research*. doi:10.1177/0047287519878503

Joppe, M. (1996). Sustainable community tourism development revisited. *Tourism Management*, *17*(7), 475–479. doi:10.1016/S0261-5177(96)00065-9

Jordan, F., & Aitchison, C. (2008). Tourism and the sexualisation of the gaze: Solo female tourists' experiences of gendered power, surveillance and embodiment. *Leisure Studies*, *27*(3), 329–349. doi:10.1080/02614360802125080

Josiam, B., Hobson, J., Dietrich, U., & Smeaton, G. (1998). An analysis of the sexual, alcohol and drug related behavioural patterns of students on spring break. *Tourism Management, 19*(6), 501–513. doi:10.1016/S0261-5177(98)00052-1

Josiassen, A., Assaf, A. G., Woo, L., & Kock, F. (2016). The imagery–image duality model—An integrative review and advocating for improved delimitation of concepts. *Journal of Travel Research, 55*(6), 789–803. doi:10.1177/0047287515583358

Jurosky, C., & Gursoy, D. (2004). Distance effects on resident's attitudes toward tourism. *Annals of Tourism Research, 31*(2), 296–304. doi:10.1016/j.annals.2003.12.005

Jurowski, C., Uysal, M., & Williams, D. R. (1997). A theoretical analysis of host community resident reactions to tourism. *Journal of Travel Research, 36*(2), 3–11. doi:10.1177/004728759703600202

Juslin, P. N., & Laukka, P. (2004). Expression, Perception, and Induction of Musical Emotions: A Review and a Questionnaire Study of Everyday Listening. *Journal of New Music Research, 33*(3), 217–238. doi:10.1080/0929821042000317813

Kalish, R., & Kimmel, M. (2011). Hooking up: Hot hetero sex or the new numb normative? *Australian Feminist Studies, 26*(67), 137–151. doi:10.1080/08164649.2011.546333

Kaplanidou, K., Kerwin, S., & Karadakis, K. (2013). Understanding sport event success: Exploring perceptions of sport event consumers and event providers. *Journal of Sport & Tourism, 18*(3), 137–159. doi:10.1080/14775085.2013.861358

Kapoor, A. (2015). *Consumer Experiences and Emotion Management* (N. Malhotra, Ed.). Business Expert Press.

Karuizawa Tourist Association. (2018) *Events*. Retrieved from: https://karuizawa-kankokyokai.jp/event/ (in Japanese)

Karuizawa Town. (2017). *Genetic Strategy for Regional Revitalisation in Karuizawa Town*. Retrieved from: https://www.town.karuizawa.lg.jp/www/contents/1454307204126/index.html

Kashdan, T. B., Biswas-Diener, R., & King, L. A. (2008). Reconsidering happiness: The costs of distinguishing between hedonics and eudaimonia. *The Journal of Positive Psychology, 3*(4), 219–233. doi:10.1080/17439760802303044

Kastenholz, E., Carneiro, M. J. A., & Marques, C. P. (2012b). Marketing the rural tourism experience. In R. H. Tsiotsou & R. E. Goldsmith (Eds.), *Strategic marketing in tourism services* (pp. 247–264). Emerald.

Kastenholz, E., Carneiro, M. J. A., Marques, C. P., & Lima, J. (2012a). Understanding and managing the rural tourism experience – The case of a historical village in Portugal. *Tourism Management Perspectives, 4*, 207–214. doi:10.1016/j.tmp.2012.08.009

Kastenholz, E., Carneiro, M., Marques, C., & Loureiro, S. (2017). The dimensions of rural tourism experience: Impacts on arousal, memory, and satisfaction. *Journal of Travel & Tourism Marketing, 35*(2), 1–13. doi:10.1080/10548408.2017.1350617

Katz, E., & Foulkes, D. (1962). On the Use of the Mass Media as "Escape": Clarification of a Concept. *Public Opinion Quarterly, 26*(3), 377–388. doi:10.1086/267111

Kavaratzis, M., & Hatch, M. J. (2013). The dynamics of place brands: An identity-based approach to place branding theory. *Marketing Theory, 13*(1), 69–86. doi:10.1177/1470593112467268

Kayat, K., Sharif, N. M., & Karnchanan, P. (2013). Individual and collective impacts and residents' perceptions of tourism. *Tourism Geographies, 15*(4), 640–653. doi:10.1080/14616688.2012.762690

Kee, T. (2017). Innovation in sustainable development: Sustainable place-making strategies in Hong Kong. *The 19th International Conference on Innovation and Sustainable Development Conference Proceedings.*

Keitumetse, S. O. (2011). Sustainable development and cultural heritage management in Botswana: Towards sustainable communities. *Sustainable Development*, *19*(1), 49–59. doi:10.1002d.419

Keller, K. L. (1993). Conceptualizing, measuring, and managing customer-based brand equity. *Journal of Marketing*, *57*(1), 1–22. doi:10.1177/002224299305700101

Kelley, K., Harrolle, M., & Casper, J. (2014). Estimating consumer spending on tickets, merchandise, and food and beverage: A case study of a NHL Team. *Journal of Sport Management*, *28*(3), 253–265. doi:10.1123/jsm.2012-0275

Kempadoo, K. (Ed.). (1999). *Sun, Sex and Gold: Tourism and Sex Work in the Caribbean*. Rowman & Littlefield Publishers.

Kendig, L. H., & Keast, B. C. (2010). *Community character: Principles for design and planning*. Island Press.

Kerrane, B., Hogg, M. K., & Bettany, S. M. (2012). Children's influence strategies in practice: Exploring the co-constructed nature of the child influence process in family consumption. *Journal of Marketing Management*, *28*(7-8), 809–835. doi:10.1080/0267257X.2012.698633

Kessler, G. (2013). Collaborative language learning in co-constructed participatory culture. *CALICO Journal*, *30*(3), 307–322. doi:10.11139/cj.30.3.307-322

Khan, I., & Rahman, Z. (2015). A review and future directions of brand experience research. *International Strategic Management Review*, *3*(1), 1–14. doi:10.1016/j.ism.2015.09.003

Khizindar, T. M. (2012). Effects of tourism on residents' quality of life in Saudi Arabia: An empirical study. *Journal of Hospitality Marketing & Management*, *21*(6), 617–637. doi:10.1080/19368623.2012.627226

Khoo-Lattimore, C., Prayag, G., & Cheah, B. L. (2015). Kids on board: Exploring the choice process and vacation needs of Asian parents with young children in resort hotels. *Journal of Hospitality Marketing & Management*, *24*(5), 511–531. doi:10.1080/19368623.2014.914862

Kiili, K. (2005). Digital game-based learning: Towards an experiential gaming model. *The Internet and Higher Education*, *8*(1), 13–2. doi:10.1016/j.iheduc.2004.12.001

Killion, G. (1992). *Understanding tourism*. Study Guide.

Kim, H., Gursoy, D., & Lee, S. (2006). The impact of the 2002 World Cup on South Korea: Comparisons of pre- and post-games. *Tourism Management*, *27*(1), 86–96. doi:10.1016/j.tourman.2004.07.010

Kim, H., & Richardson, S. L. (2003). Motion picture impacts on destination images. *Annals of Tourism Research*, *30*(1), 216–237. doi:10.1016/S0160-7383(02)00062-2

Kim, H., & Stepchenkova, S. (2015). Effect of tourist photographs on attitudes towards destination: Manifest and latent content. *Tourism Management*, *49*, 29–41. doi:10.1016/j.tourman.2015.02.004

Kimiecik, J. C., & Harris, A. T. (1996). What Is Enjoyment? A Conceptual/Definitional Analysis with Implications for Sport and Exercise Psychology. *Journal of Sport & Exercise Psychology*, *18*(3), 247–263. doi:10.1123/jsep.18.3.247

Kim, J. H. (2014). The antecedents of memorable tourism experiences: The development of a scale to measure the destination attributes associated with memorable experiences. *Tourism Management*, *44*, 34–45. doi:10.1016/j.tourman.2014.02.007

Kim, J. H., & Ritchie, J. R. B. (2014). Cross-Cultural Validation of a Memorable Tourism Experience Scale (MTES). *Journal of Travel Research*, *53*(3), 323–335. doi:10.1177/0047287513496468

Kim, J. H., Ritchie, J. R. R., & McCormick, B. (2012). Development of a scale to measure memorable tourism experiences. *Journal of Travel Research*, *51*(1), 12–25. doi:10.1177/0047287510385467

Kim, J., & Fesenmaier, D. (2015). Designing Tourism Places: Understanding the Tourism Experience through our Senses. In *Tourism Travel and Research Association - Turning Insights into Actions, the Crucial Role of Tourism Research* (p. 5). Travel and Tourism Research Association. Retrieved from https://scholarworks.umass.edu/ttra/ttra2015/Academic_Papers_Oral/19

Kim, J.-H. (2010). Determining the factors affecting the memorable nature of travel experiences. *Journal of Travel & Tourism Marketing*, *27*(8), 780–796. doi:10.1080/10548408.2010.526897

Kim, J.-H. (2018). The Impact of Memorable Tourism Experiences on Loyalty Behaviors: The Mediating Effects of Destination Image and Satisfaction. *Journal of Travel Research*, *57*(7), 856–870. doi:10.1177/0047287517721369

Kim, J.-H., & Jang, S. (2016). Memory Retrieval of Cultural Event Experiences: Examining Internal and External Influences. *Journal of Travel Research*, *55*(3), 322–339. doi:10.1177/0047287514553058

Kim, J.-H., & Youn, H. (2017). How to Design and Deliver Stories about Tourism Destinations. *Journal of Travel Research*, *56*(6), 808–820. doi:10.1177/0047287516666720

Kim, K., Uysal, M., & Sirgy, M. J. (2013, June). 92013). How does tourism in a community impact the quality of life of community residents? *Tourism Management*, *36*, 527–540. doi:10.1016/j.tourman.2012.09.005

Kim, S. (2012). Audience involvement and film tourism experiences: Emotional places, emotional experiences. *Tourism Management*, *33*(2), 387–396. doi:10.1016/j.tourman.2011.04.008

Kim, S. B., Kim, D. Y., & Bolls, P. (2014). Tourist mental-imagery processing: Attention and arousal. *Annals of Tourism Research*, *45*, 63–76. doi:10.1016/j.annals.2013.12.005

Kim, S. S., McKercher, B., & Lee, H. (2009). Tracking tourism destination image perception. *Annals of Tourism Research*, *36*(4), 715–718. doi:10.1016/j.annals.2009.04.007

Kim, S., Lee, K. Y., Shin, S., & Yang, S. (2017). Effects of tourism information quality in social media on destination image formation: The case of Sina Weibo. *Information & Management*, *54*(6), 687–702. doi:10.1016/j.im.2017.02.009

Kim, S., & Lehto, X. Y. (2013). Projected and perceived destination brand personalities: The case of South Korea. *Journal of Travel Research*, *52*(1), 117–130. doi:10.1177/0047287512457259

King, B., Pizam, A., & Milman, A. (1993). Social impacts of tourism: Host perceptions. *Annals of Tourism Research*, *20*(4), 650–665. doi:10.1016/0160-7383(93)90089-L

King, C., Chen, N., & Funk, D. C. (2015). Exploring destination image decay: A study of sport tourists' destination image change after event participation. *Journal of Hospitality & Tourism Research (Washington, D.C.)*, *39*(1), 3–31. doi:10.1177/1096348012461547

King, N., & Horrocks, C. (2010). *Interviews in Qualitative Research*. Sage Publications.

Kinsey, A. (1998). *Sexual Behavior in the Human Female*. Indiana University Press. (Original work published 1953)

Kirchherr, J., & Charles, K. (2018). Enhancing the sample diversity of snowball samples: Recommendations from a research project on anti-dam movements in Southeast Asia. *PLoS One*, *13*(8), e0201710. doi:10.1371/journal.pone.0201710 PMID:30133457

Kirillova, K., Lehto, X., & Cai, L. (2017). What triggers transformative tourism experiences? *Tourism Recreation Research*, *42*(4), 498–511. doi:10.1080/02508281.2017.1342349

Kirstin, H., & Christoph, B. (2010). Image ðt between sport events and their hosting destinations from an active sport tourist perspective and its impact on future behaviour. *Journal of Sport & Tourism*, *15*(3), 215–237. doi:10.1080/14775085.2010.513147

Kislali, H., Kavaratzis, M., & Saren, M. (2019). Destination image formation: Towards a holistic approach. *International Journal of Tourism Research*, 1–11.

Kitnuntaviwat, V., & Tang, J. C. (2008). Residents' attitudes, perception and support for sustainable tourism development. *Tourism and Hospitality Planning & Development, 5*(1), 45-60.

Klein, J., Moon, Y., & Picard, R. W. (2002). This computer responds to user frustration: Theory, design, and results. *Interacting with Computers*, *14*(2), 119–140. doi:10.1016/S0953-5438(01)00053-4

Klimmt, C. (2008). Escapism. In The International Encyclopedia of Communication. Academic Press.

Klimmt, C., Hartmann, T., & Frey, A. (2007). Effectance and control as determinants of video game enjoyment. *Cyberpsychology & Behavior*, *10*(6), 845–847. doi:10.1089/cpb.2007.9942 PMID:18085976

Kline, R. B. (2011). *Principles and practice of structural equation modeling*. Guilford Press.

Kneafsey, M. (1998). Tourism and Place Identity: A case-study in rural Ireland. *Irish Geography*, *31*(2), 111–123. doi:10.1080/00750779809478623

Knowles, C., & Harper, D. (2009). *Hong Kong: Migrant lives, landscapes, and journeys*. University of Chicago Press. doi:10.7208/chicago/9780226448589.001.0001

Koc, E. (2013). Inversionary and liminoidal consumption: Gluttony on holidays and obesity. *Journal of Travel & Tourism Marketing*, *30*(8), 825–838. doi:10.1080/10548408.2013.835669

Koch, S., & Artmayr, P. (2019). Stability and development of user innovation strategies for video game producers. *European Journal of Innovation Management*, *ahead-of-print*(ahead-of-print). Advance online publication. doi:10.1108/EJIM-05-2019-0117

Ko, D., & Stewart, W. (2002). A structural equation model of resident attitudes for tourism development. *Tourism Management*, *23*(5), 521–530. doi:10.1016/S0261-5177(02)00006-7

Kokosalakis, C., Bagnall, G., Selby, M., & Burns, S. (2006). Place image and urban regeneration in Liverpool. *International Journal of Consumer Studies*, *30*(4), 389–397. doi:10.1111/j.1470-6431.2006.00520.x

Konecnik, M., & Gartner, W. (2007). Customer-based brand equity for a destination. *Annals of Tourism Research*, *34*(2), 400–421. doi:10.1016/j.annals.2006.10.005

Korça, P. (1996). Resident attitudes toward tourism impacts. *Annals of Tourism Research*, *23*(3), 695–697. doi:10.1016/0160-7383(95)00092-5

Korea Convention. (2019). *Top reasons for choosing*. Retrieved from http://www.koreaconvention.org/eng/info/reason.kc

Korea Tourism Organization. (2017). *2017 hankook ganging trend bogosu*. Retrieved from http://kto.visitkorea.or.kr/kor/notice/data/report/org/board/view.kto?id=430336&isNotice=false&instanceId=127&rnum=12

Korea Tourism Organization. (2018). *Sukjaro bonen Hankook ganging* Retrieved from http://kto.visitkorea.or.kr/kor/notice/data/report/org/board/view.kto?id=431195&isNotice=false&instanceId=127&rnum=5

Kotler, P. (1973). Atmospherics as a Marketing Tool. *Journal of Retailing*, *49*(4), 48–64.

Kramer, A. D. I., Guillory, J. E., & Hancock, J. T. (2014). Experimental evidence of massivescale emotional contagion through social networks. *Proceedings of the National Academy of Sciences of the United States of America, 111*(29), 8788–8790. doi:10.1073/pnas.1320040111 PMID:24889601

Kreziak, D., & Frochot, I. (2011). Co-construction de l'expérience touristique: Les stratégies des touristes en stations de sport d'hiver [Co-construction of the tourism experience: Tourists strategies in ski resorts]. *Décisions Marketing, 64*, 23–33. doi:10.7193/DM.064.23.33

Krippendorff, K. (2018). Content Analysis: An Introduction to its Methodology (4th ed.). SAGE Publications.

Krippendorff, K. (1980). Validity in content analysis. In E. Mochmann (Ed.), *Computerstrategien für die Kommunkationsanalyse* (pp. 69–112). Campus. Retrieved from https://pdfs.semanticscholar.org/8a09/38e9ee2d55ee64c47e4b4be3f6177c58f20c.pdf

Krippendorff, K., & Bock, M. (2009). *The content analysis reader*. Sage.

Krishna, A. (2012). An integrative review of sensory marketing: Engaging the senses to affect perception, judgment and behavior. *Journal of Consumer Psychology, 22*(3), 332–351. Advance online publication. doi:10.1016/j.jcps.2011.08.003

Krishna, A., & Schwarz, N. (2014). Sensory marketing, embodiment, and grounded cognition: A review and introduction. *Journal of Consumer Psychology, 24*(2), 159–168. doi:10.1016/j.jcps.2013.12.006

Krumhansl, C. L. (1997). An exploratory study of musical emotions and psychophysiology. *Canadian Journal of Experimental Psychology/Revue Canadienne de Psychologie Expérimentale, 51*(4), 336–353.

Kurtz, J., & Welch, C. (2016). The Emotion of Happiness. In H. L. Meiselman (Ed.), *Emotions Measurement*. Elsevier Ltd. doi:10.1016/B978-0-08-100508-8.00020-5

Kvasova, O. (2011). Socio-demographic determinants of eco-friendly tourist attitudes and behaviour. *Tourism Today (Nicosia), 11*, 73–95.

Kwiatkowski, G. (2016). Composition of event attendees: A comparison of three small-scale sporting events. *International Journal of Sport Finance, 11*(2), 163–180.

Kwon, J., & Vogt, C. (2010). Identifying the role of cognitive, affective and behavioural components in understanding residents' attitudes toward place marketing. *Journal of Travel Research, 49*(4), 423–435. doi:10.1177/0047287509346857

Kwortnik, R. J. Jr, & Ross, W. T. Jr. (2007). The role of positive emotions in experiential decisions. *International Journal of Research in Marketing, 24*(4), 324–335. doi:10.1016/j.ijresmar.2007.09.002

Kyngäs, H., Mikkonen, K., & Kääriäinen, M. (2020). The Application of Content Analysis in Nursing Science Research. In H. Kyngäs, K. Mikkonen, & M. Kääriäinen (Eds.), *The Application of Content Analysis in Nursing Science Research* (pp. 13–21). Springer. doi:10.1007/978-3-030-30199-6_1

Labadi, S. (2007). Representations of the nation and cultural diversity in discourses on World Heritage. *Journal of Social Archaeology, 7*(2), 147–170. https://www.academia.edu/516837/Representations_of_the_nation_and_cultural_diversity_in_discourses_on_World_Heritage

Ladhari, R., Souiden, N., & Ladhari, I. (2011). Determinants of loyalty and recommendation: The role of perceived service quality, emotional satisfaction and image. *Journal of Financial Services Marketing, 16*(2), 111–124. doi:10.1057/fsm.2011.10

Lai, K., & Li, Y. (2012). Core-periphery structure of destination image concept, evidence and implication. *Annals of Tourism Research, 39*(3), 1359–1379. doi:10.1016/j.annals.2012.02.008

Lai, K., & Li, Y. (2015). Tourism destination image: Conceptual problems and definitional solutions. *Journal of Travel Research, 55*(8), 1065–1080. doi:10.1177/0047287515619693

Landau, S., & Everitt, B. (2003). *A Handbook of Statistical Analysis using SPSS*. CRC Press. doi:10.1201/9780203009765

Landorf, C. (2009). Managing for sustainable tourism: A review of six cultural world heritage sites. *Journal of Sustainable Tourism, 17*(1), 53–70. doi:10.1080/09669580802159719

Landry, C. (2000). *The creative city: A toolkit for urban innovation*. Comedia.

Landsheer, J., & Boeije, H. (2010). In search of content validity: Facet analysis as a qualitative method to improve questionnaire design. *Quality & Quantity, 44*(1), 59–69. doi:10.100711135-008-9179-6

Lane, B. (1994). Sustainable rural tourism strategies: A tool for development and conservation. *Journal of Sustainable Tourism, 2*(1-2), 102–111. doi:10.1080/09669589409510687

Lankford, S. V., & Howard, D. R. (1994). Developing tourism impact attitude scale. *Annals of Tourism Research, 21*(1), 121-139.

Lankford, S., & Howard, D. (1994). Developing a tourism impact attitudes scales. *Annals of Tourism Research, 21*(1), 121–139. doi:10.1016/0160-7383(94)90008-6

Larsen, G., & George, V. (2004). *The Social Construction of Destination Image – A New Zealand Film Example*. Working Papers, 4/01.February.

Larsen, S. (2007). Aspects of a Psychology of the Tourist Experience. *Scandinavian Journal of Hospitality and Tourism, 7*(1), 7–18. doi:10.1080/15022250701226014

Látková, P., & Vogt, C. A. (2012). Residents' attitudes toward existing and future tourism development in rural communities. *Journal of Travel Research, 51*(1), 50–67. doi:10.1177/0047287510394193

Latour, B. (1987). *Science in Action: How to follow scientists and engineers through society*. Harvard University Press.

Lawlor, M. A., & Prothero, A. (2011). Pester power–A battle of wills between children and their parents. *Journal of Marketing Management, 27*(5-6), 561–581. doi:10.1080/0267257X.2010.495281

Law, S., & Braun-LaTour, K. A. (2000). I'll have what she's having: Gauging the impact of product placements on viewers. *Psychology and Marketing, 17*(12), 1059–1075. doi:10.1002/1520-6793(200012)17:12<1059::AID-MAR3>3.0.CO;2-V

Lawson, F., & Baud-Bovy, M. (1977). *Tourism and Recreational Development*. Architectural Press.

Lawton, L. J., & Weaver, D. B. (2015). Using residents' perceptions research to inform planning and management for sustainable tourism: A study of the Gold Coast Schoolies Week, a contentious tourism event. *Journal of Sustainable Tourism, 23*(5), 660–682. doi:10.1080/09669582.2014.991398

Lazzeroni, M., Bellini, N., Cortesi, G., & Loffredo, A. (2013). The Territorial Approach to Cultural Economy: New Opportunities for the Development of Small Towns. *European Planning Studies, 21*(4), 4. doi:10.1080/09654313.2012.722920

LCPED - Legislative Council Panel on Economic Development. (2018). *2018 Policy Address Policy Initiatives of Commerce, Industry and Tourism Branch, Commerce and Economic Development Bureau*. Retrieved September 1, 2019, from https://www.legco.gov.hk/yr18-19/english/panels/edev/papers/edevcb4-10-2-e.pdf

LDWHP. (n.e.). *World Heritage Status. Is there opportunity for economic gain. Research and Analysis of the Socio Economic Impact Potential of UNESCO World Heritage Site Status*. RACES, RC &TBR Rebanks Consulting Ltd and Trends Business Research Ltd. Research and Creative Enterprise Services (RACES). http://icomos.fa.utl.pt/documentos/2009/WHSTheEconomicGainFinalReport.pdf

Leal, A., & Roldán, J. (2001). Validación de un modelo de implantación de gestión de calidad total y su efecto en los resultados empresariales: un estudio causal predictivo mediante análisis PLS (Partial Least Square). In *Proceedings of the X International Conference of the European Association of Management and Business Economics* (pp. 541-555). Reggio Calabria: AEDEM.

LeDoux, J. (2012). Rethinking the Emotional Brain. *Neuron*, *73*(4), 653–676. doi:10.1016/j.neuron.2012.02.004 PMID:22365542

Lee, C. K., Kang, S. K., Long, P., & Reisinger, Y. (2010). Residents' perceptions of casino impacts: A comparative study. *Tourism Management*, *31*(2), 189–201. doi:10.1016/j.tourman.2009.02.011

Lee, C. K., Lee, Y. K., & Lee, B. K. (2005). Korea's destination image formed by the 2002 World Cup. *Annals of Tourism Research*, *32*(4), 839–858. doi:10.1016/j.annals.2004.11.006

Lee, M., & Faber, R. J. (2007). Effects of Product Placement in On-Line Games on Brand Memory: A Perspective of the Limited-Capacity Model of Attention. *Journal of Advertising*, *36*(4), 75–90. doi:10.2753/JOA0091-3367360406

Lee, S., Heere, B., & Chung, K. (2013). Which Senses Matter More? The Impact of Our Senses on Team Identity and Team Loyalty. *Sport Marketing Quarterly*, *22*(4), 203–213. http://search.ebscohost.com/login.aspx?direct=true&db=bth&AN=93709410&lang=pt-br&site=ehost-live

Lees, L., Slater, T., & Wyly, E. (2008). *Gentrification*. Routledge.

Lee, T. H. (2013). Influence analysis of community resident support for sustainable tourism development'. *Tourism Management*, *34*, 37–46. doi:10.1016/j.tourman.2012.03.007

Lee, T. H., & Chang, P.-S. (2017). Examining the relationships among festivalscape, experiences, and identity: Evidence from two Taiwanese aboriginal festivals. *Leisure Studies*, *36*(4), 453–467. doi:10.1080/02614367.2016.1190857

Lee, T. H., & Crompton, J. (1992). Measuring Novelty Seeking in Tourism. *Annals of Tourism Research*, *19*(4), 732–751. doi:10.1016/0160-7383(92)90064-V

Lee, W., & Gretzel, U. (2012). Designing persuasive destination websites: A mental imagery processing perspective. *Tourism Management*, *33*(5), 1270–1280. doi:10.1016/j.tourman.2011.10.012

Lee, W., Gretzel, U., & Law, R. (2010). Quasi-Trial Experiences through Sensory Information on Destination Web Sites. *Journal of Travel Research*, *49*(3), 310–322. doi:10.1177/0047287509346991

Lee, Y. H., & Lim, E. A. I. C. (2010). When good cheer goes unrequited: How emotional receptivity affects evaluation of expressed emotion. *JMR, Journal of Marketing Research*, *47*(6), 1151–1161. doi:10.1509/jmkr.47.6.1151

Legardinier, C., & Villette, S. (2007). O turismo sexual. In C. Ockrent & S. Treiner (Eds.), *O Livro Negro da Condição das Mulheres* (pp. 490–496). Difel.

Leheny, D. (1995). A political economy of Asian sex tourism. *Annals of Tourism Research*, *22*(2), 367–384. doi:10.1016/0160-7383(94)00082-4

Lehu, J.-M. (2007). *Branded entertainment: Product placement & brand strategy in the entertainment business*. Kogan Page Publishers.

Lemon, K., & Verhoef, P. (2016). Understanding Customer Experience Throughout the Customer Journey. *Journal of Marketing*, *80*(6), 69–96. doi:10.1509/jm.15.0420

Lempert, K., & Phelps, E. (2016). Affect in Economic Decision Making. In L. F. Barrett, M. Lewis, & J. M. Haviland-Jones (Eds.), *Handbook of Emotions* (4th ed., Vol. 1, pp. 98–112). The Guilford Press.

Lems, A. (2016). Placing displacement: Place-making in a world of movement. *Ethnos*, *81*(2), 315–337. doi:10.1080/00141844.2014.931328

Lenzerini, F. (2011, February). Intangible Cultural Heritage: The Living Culture of Peoples. *European Journal of International Law*, *22*(1), 101–120. doi:10.1093/ejil/chr006

Lepp, A. (2007). Residents' attitudes towards tourism in Bigodi village, Uganda. *Tourism Management*, *28*(3), 876–885. doi:10.1016/j.tourman.2006.03.004

Lepp, A. (2008). Attitudes towards initial tourism development in a community with no prior tourism experience: The Case of Bigodi, Uganda. *Journal of Sustainable Tourism*, *16*(1), 5–22. doi:10.2167/jost630.0

Leung, L. (2015). Validity, reliability and generalizability in qualitative research. *Journal of Family Medicine and Primary Care*, *4*(3), 324–327. doi:10.4103/2249-4863.161306 PMID:26288766

Lew, A. A. (2012). Geography and the marketing of tourism destinations. In J. Wilson (Ed.), *The Routledge handbook of tourism geographies* (pp. 181–186). Routledge.

Lew, A. A. (2017). Tourism planning and place making: Place-making or placemaking? *Tourism Geographies*, *19*(3), 448–466. doi:10.1080/14616688.2017.1282007

Lewis, B., & Porter, L. (2010). In-game advertising effects: Examining player perceptions of advertising schema congruity in a massively multiplayer online role-playing game. *Journal of Interactive Advertising*, *10*(2), 46–60. doi:10.1080/15252019.2010.10722169

Liang, Z. X., & Hui, T.-K. (2016). Residents' quality of life and attitudes toward tourism development in China. *Tourism Management*, *57*, 56–67. doi:10.1016/j.tourman.2016.05.001

Light, D. (2009). *The Senses: The Human Body, How It Works*. Chelsea House Publishers.

Li, H., Edwards, S. M., & Lee, J.-H. (2002). Measuring the Intrusiveness fo Advertisements: Scale Development and Validation. *Journal of Advertising*, *31*(2), 37–48. doi:10.1080/00913367.2002.10673665

Likert, R. (1974). The Method of Constructing an Attitude Scale. In G. M. Maranell (Ed.), *Scalling: a sourcebook for behavioural scientists* (pp. 90–96). Transaction Publishers.

Liljander, V., & Strandvik, T. (1997). Emotions in service satisfaction. *International Journal of Service Industry Management*, *8*(2), 148–169. doi:10.1108/09564239710166272

Li, M., Cai, L. A., Lehto, X. Y., & Huang, J. (2010). A missing link in understanding revisit intention—The role of motivation and image. *Journal of Travel & Tourism Marketing*, *27*(4), 335–348. doi:10.1080/10548408.2010.481559

Lindberg, K., & Johnson, R. L. (1997). Modeling resident attitudes toward tourism. *Annals of Tourism Research*, *24*(2), 402–424. doi:10.1016/S0160-7383(97)80009-6

Lindsay, P., & Norman, D. (1977). *Human information processing: an introduction to Psychology*. Academic Press.

Lin, Z., Chen, Y., & Filieri, R. (2017). Resident-tourist value co-creation: The role of residents' perceived tourism impacts and life satisfaction. *Tourism Management*, *61*, 436–442. doi:10.1016/j.tourman.2017.02.013

Lita, R. P., & Ma'ruf. (2015). Relationship model among sport event image, destination image, and tourist satisfaction of Tour de Singkarak in West Sumatera. *Journal of Economics, Business, and Accountancy Ventura, 18*(1), 91-102.

Litvin, S. W., & Goh, H. K. (2002). Research Note: Self-Image Congruity: A Valid Tourism Theory? *Tourism Management, 23*(1), 81–83. doi:10.1016/S0261-5177(01)00065-6

Liu, Y., & Cheng, J. (2016). Place Identity: How Tourism Changes Our Destination. *International Journal of Psychological Studies, 8*(2).

Liu, J. C., & Var, T. (1986). Resident attitudes toward tourism impacts in Hawai. *Annals of Tourism Research, 13*(2), 193–214. doi:10.1016/0160-7383(86)90037-X

Liu, J., Sheldon, P., & Var, T. (1987). Resident Perceptions of The Environment Impacts of Tourism of Tourism. *Annals of Tourism Research, 14*(1), 17–37. doi:10.1016/0160-7383(87)90045-4

Li, Y. (2000). Geographical consciousness and tourism experience. *Annals of Tourism Research, 27*(4), 863–883. doi:10.1016/S0160-7383(99)00112-7

Lodrà-Riera, I., Martínez-Ruiz, M. P., Jiménez-Zarco, A. I., & Izquierdo-Yusta, A. (2015). A multidimensional analysis of the information sources construct and its relevance for destination image formation. *Tourism Management, 48*, 319–328. doi:10.1016/j.tourman.2014.11.012

Long, J. A. (2007). Subjects, respondents, participants and collaborators. (2007). In J. A. Long (Ed.), *Researching leisure, sport, and tourism* (pp. 34–46). SAGE Publications Ltd. doi:10.4135/9780857024619

Long, P. T., Perdue, R. R., & Allen, L. (1990). Rural resident tourism perceptions and attitudes by community level of tourism. *Journal of Travel Research, 28*(3), 3–9. doi:10.1177/004728759002800301

Lopes, A. C., Rodrigues, E. B., & Vera-Cruz, R. (2019). Tourism Gentrification. *IOP Conf. Series: Materials Science and Engineering*, 471.

López, G., & Palomino, B. (2008). El turismo como actividad emergente para las comunidades y pueblos indígenas. In A. Castellanos-Guerrero & J. A. Machuca (Eds.), *Turismo, identidades y exclusión* (pp. 31–50). Universidad Autónoma Metropolitana y Casa Juan Pablos.

Lorentzen, A. (2009). Cities in the experience economy. *European Planning Studies, 17*(6), 829–845. doi:10.1080/09654310902793986

Lourenço, R. (2008). *Turismo de prática desportiva. Estudo de caso: Os percursos pedestres e os termalistas clássicos em Monfortinho* [Tourism of sports practice. Case study: The pedestrian and classical thermal trails in Monfortinho] (Unpublished master's thesis). University of Beira Interior, Covilhã, Portugal.

Lucas, K., & Sherry, J. L. (2004). Sex differences in video game play: A communication-based explanation. *Communication Research, 31*(5), 499–523. doi:10.1177/0093650204267930

Lu, L., Chi, C. G., & Liu, Y. (2015). Authenticity, involvement, and image: Evaluating tourist experiences at historic districts. *Tourism Management, 50*, 85–96. doi:10.1016/j.tourman.2015.01.026

Lull, R. B., Gibson, B., Cruz, C., & Bushman, B. J. (2016, May). Killing Characters in Video Games Kills Memory for In-Game Ads. *Psychology of Popular Media Culture*.

Lull, R. B., Gibson, B., Cruz, C., & Bushman, B. J. (2018). Killing characters in video games kills memory for in-game ads. *Psychology of Popular Media Culture, 7*(1), 87–97. doi:10.1037/ppm0000108

Lundqvist, L.-O., Carlsson, F., Hilmersson, P., & Juslin, P. N. (2009). Emotional responses to music: Experience, expression, and physiology. *Psychology of Music, 37*(1), 61–90. doi:10.1177/0305735607086048

Lusch, R. F., Vargo, S. L., & O'Brien, M. (2007). Competing through service: Insights from service-dominant logic. *Journal of Retailing, 83*(1), 5–18. doi:10.1016/j.jretai.2006.10.002

Lutz, K. A., & Lutz, R. J. (1978). Imagery-Eliciting Strategies: Review and Implications of Research. *Advances in Consumer Research. Association for Consumer Research (U. S.), 5*, 611–620.

Lynch, K. (1960). *The image of the city*. MIT Press.

MacCannell, D. (1973). Staged Authenticity: Arrangements of Social Space in Tourist Settings. *American Journal of Sociology, 79*(3), 589–603. doi:10.1086/225585

MacInnis, D. J., & Price, L. L. (1987). The Role of Imagery in Information Processing: Review and Extensions. *The Journal of Consumer Research, 13*(4), 473–491. doi:10.1086/209082

MacInnis, D. J., & Price, L. L. (1990). An Exploratory Study of the Effects of Imagery Processing and Consumer Experience on Expectations and Satisfaction. *Advances in Consumer Research. Association for Consumer Research (U. S.), 17*, 41–47.

MacKay, K. J., & Fesenmaier, D. R. (1997). Pictorial element of destination in image formation. *Annals of Tourism Research, 21*(3), 537–565. doi:10.1016/S0160-7383(97)00011-X

MacKay, K. J., & Fesenmaier, D. R. (2000). An Exploration of Cross-Cultural Destination Image Assessment. *Journal of Travel Research, 38*(4), 417–423. doi:10.1177/004728750003800411

MacKay, K., & Campbell, J. (2004). An examination of residents' support for hunting as a tourism product. *Tourism Management, 25*(4), 443–452. doi:10.1016/S0261-5177(03)00127-4

Madhavaram, S., & Hunt, S. D. (2008). The Service-Dominant Logic and a Hierarchy of Operant Resources: Developing Masterful Operant Resources and Implications for Marketing Strategy. *Journal of the Academy of Marketing Science, 36*(1), 67–82. doi:10.100711747-007-0063-z

Madrigal, R. (1993). A tale of tourism in tow cities. *Annals of Tourism Research, 22*(2), 336–353. doi:10.1016/0160-7383(93)90059-C

Madrigal, R. (1995). Residents' perceptions and the role of government. *Annals of Tourism Research, 22*(1), 86–102. doi:10.1016/0160-7383(94)00070-9

Main, K., & Sandoval, G. F. (2015). Placemaking in a translocal receiving community: The relevance of place to identity and agency. *Urban Studies (Edinburgh, Scotland), 52*(1), 71–86. doi:10.1177/0042098014522720

Maitland, C., Granich, J., Braham, R., Thornton, A., Teal, R., Stratton, G., & Rosenberg, M. (2018). Measuring the capacity of active video games for social interaction: The Social Interaction Potential Assessment tool. *Computers in Human Behavior, 87*, 308–316.

Ma, J., Gao, J., Scott, N., & Ding, P. (2013). Customer delight derived from theme park experiences: The antecedents of delight based on cognitive appraisal theory. *Annals of Tourism Research, 42*(1), 359–381. doi:10.1016/j.annals.2013.02.018

Ma, L., & Lew, A. A. (2012). Historical and geographical context in festival tourism development. *Journal of Heritage Tourism, 7*(1), 13–31. doi:10.1080/1743873X.2011.611595

Manente, M., & Minghetti, V. (2006). Destination management organizations and actors. In D. Buhalis & C. Costa (Eds.), *Tourism business frontiers: Consumers, products and industry* (pp. 228–237). Elsevier Butterworth Heinemann. doi:10.1016/B978-0-7506-6377-9.50032-8

Manhas, P. S., & Ramjit. (2013). Customer Experience and Its Relative Influence on Satisfaction and Behavioural Intention in Hospitality and Tourism Industry. *South Asian Journal of Tourism and Heritage*, *6*(1), 53–68.

Mannell, R. C., & Iso-Ahola, S. E. (1987). Psychological nature of leisure and tourism experience. *Annals of Tourism Research*, *14*(3), 314–331. doi:10.1016/0160-7383(87)90105-8

Mano, H., & Oliver, R. (1993). Assessing the dimensionality and structure of the consumption experience: Evaluation, feeling, and satisfaction. *The Journal of Consumer Research*, *20*(3), 451. doi:10.1086/209361

Maria, I., Madalina, T., Catalina, B., & Diana, I. (2008). The factors of competitiveness in the hospitality industry and the competitive strategy of firms. *Annals of the University of Oradea. Economic Science Series*, *17*(2), 213–218.

Marine-Roig, E., & Ferrer-Rosell, B. (2018). Measuring the gap between projected and perceived destination images of Catalonia using compositional analysis. *Tourism Management*, *68*, 236–249. doi:10.1016/j.tourman.2018.03.020

Marks, L. J., & Olson, J. C. (2001). Toward a cognitive structure conceptualization of product familiarity. *Advances in Consumer Research. Association for Consumer Research (U. S.)*, *8*, 145–150.

Markus, H., & Kunda, Z. (1986). Stability and malleability of the self-concept. *Journal of Personality and Social Psychology*, *51*(4), 858–866. doi:10.1037/0022-3514.51.4.858 PMID:3783430

Marôco, J. (2014). *Análise estatística com o SPSS statistics* [Statistical analysis with SPSS statistics]. ReportNumber.

Marques, J. F. (2016). Sun, sand, sea and sex, tourism. In J. Jafari & X. Honggen (Eds.), *Encyclopedia of Tourism*. Springer. doi:10.1007/978-3-319-01384-8_536

Marques, J. F., & Lança, M. (2016). Para além da vida quotidiana. Amor e sexualidade em contexto turístico: Resultados preliminares de uma pesquisa exploratória no Algarve (Portugal). *Anais Brasileiros de Estudos Turísticos*, *6*(2), 7–22.

Marsh, A. A., Ambady, N., & Kleck, R. E. (2005). The effects of fear and anger facial expressions on approach- and avoidance-related behaviors. *Emotion (Washington, D.C.)*, *5*(1), 119–124. doi:10.1037/1528-3542.5.1.119 PMID:15755225

Martín de la Rosa, B. (2009). *Turismo en ecosistemas insulares. Antropología en el paraíso. Colección PASOS Edita, 3*. Asociación Canaria de Antropología.

Martin, D., & Woodside, A. (2008). Grounded Theory of International Tourism Behavior. *Journal of Travel & Tourism Marketing*, *24*(4), 245–258. doi:10.1080/10548400802156695

Martín-Duque, C. (2017). Los impactos del turismo en el camino de Santiago Francés: Una aproximación cualitativa. *Methaodos. Revista de Ciencias Sociales*, *5*(1), 62–73.

Martínez, M., & Fierro, E. (2018). Aplicación de la técnica PLS-SEM en la gestión del conocimiento: Un enfoque práctico. *Revista Iberoamericana para la Investigación y Desarrollo Educativo*, *8*(16), 1–35.

Martin, H. S., & del Bosque, I. A. R. (2008). Exploring the cognitive-affective nature of destination image and the role of psychological factors in its formation. *Tourism Management*, *29*(2), 263–277. doi:10.1016/j.tourman.2007.03.012

Martín-Santana, J. D., Beerli-Palacio, A., & Nazzareno, P. A. (2017). Antecedents and consequences of destination image gap. *Annals of Tourism Research*, *62*, 13–25. doi:10.1016/j.annals.2016.11.001

Martins, J., Gonçalves, R., Branco, F., Barbosa, L., Melo, M., & Bessa, M. (2017). A multisensory virtual experience model for thematic tourism: A Port wine tourism application proposal. *Journal of Destination Marketing & Management*, *6*(2), 103–109. doi:10.1016/j.jdmm.2017.02.002

Martins, L., Varela, M. L. R., Fernandes, N. O., Carmo–Silva, S., & Machado, J. (2020). Literature review on autonomous production control methods. *Enterprise Information Systems*, *00*(00), 1–13. doi:10.1080/17517575.2020.1731611

Martí-parreño, J., Bermejo-berros, J., & Aldás-manzano, J. (2017). Product Placement in Video Games: The Effect of Brand Familiarity and Repetition on Consumers ' Memory. *Journal of Interactive Marketing*, *38*, 55–63. doi:10.1016/j.intmar.2016.12.001

Maruyama, N. U., Woosnam, K. M., & Boley, B. B. (2017). 'Residents' attitudes toward ethnic neighborhood tourism (ENT): Perspectives of ethnicity and empowerment'. *Tourism Geographies*, *19*(2), 265–286. doi:10.1080/14616688.2016.1258432

Marx, G., & Gilon, C. (2012). The Molecular Basis of Memory. *ACS Chemical Neuroscience*, *3*(8), 633–642. doi:10.1021/cn300097b PMID:23050060

Marx, K. (1982). *Capital* (3rd ed., Vol. 1). Penguin Books.

Masa'deh, R. E., Nasseef, M. A., Alshayeb, H., Ojilat, J., & Alshafiee, M. (2017). The Effect of Sport Tourism Management on Support for Tourism Development. *Journal of Management and Strategy*, *8*(3), 20–34. doi:10.5430/jms.v8n3p20

Mason, P. (2014). *Researching tourism, leisure and hospitality for your dissertation*. Oxford: Goodfellow Publishers.

Mason, P. (2003). *Tourism Impacts, Planning and Management*. Butterworth-Heinemann.

Mason, P., & Cheyne, J. (2000). Residents' attitudes to proposed tourism development. *Annals of Tourism Research*, *27*(2), 391–411. doi:10.1016/S0160-7383(99)00084-5

Massey, D. (2005). *For space*. Sage.

Mathieson, A., & Wall, G. (1982). *Tourism, economic, physical and social impacts*. Longman.

Mathisen, L. (2013). Staging natural environments: A performance perspective. *Advances in Hospitality and Leisure*, *9*, 163–183. doi:10.1108/S1745-3542(2013)0000009012

Matos, N. M. S. (2014). *The Impact of Tourism Experiences in the destination image. A Marketing Perspective* (Unpublished doctoral dissertation). University of Algarve, Faro.

Matos, N., Mendes, J., & Pinto, P. (2015). The Role of Imagery and Experiences in the Construction of a Tourism Destination Image. *Journal of Spatial and Organizational Dynamics*, *3*(2), 65–84.

Mau, G., Silberer, G., & Constien, C. (2008). Communicating brands playfully: Effects of in-game advertising for familiar and unfamiliar brands. *International Journal of Advertising*, *27*(5), 827–851. doi:10.2501/S0265048708080293

Mazumdar, S., & Mazumdar, S. (2008). Religious placemaking and community building in Diaspora. *Environment and Behavior*, *41*(3), 1–31.

Mccabe, S., & Johnson, S. (2013). The happiness factor in tourism: Subjective well-being and social tourism. *Annals of Tourism Research*, *41*, 42–65. doi:10.1016/j.annals.2012.12.001

McCartney, G., Butler, R., & Bennett, M. (2008). A Strategic Use of the Communication Mix in the Destination Image-Formation Process. *Journal of Travel Research*, *47*(2), 183–196. doi:10.1177/0047287508321201

Mccombes, L., Vanclay, F., & Evers, Y. (2015). Putting social impact assessment to the test as a method for implementing responsible tourism practice. *Environmental Impact Assessment Review*, *55*, 156–168. doi:10.1016/j.eiar.2015.07.002

McCool, S. F., & Martin, S. R. (1994). Community attachment and attitudes toward tourism development. *Journal of Travel Research*, *32*(3), 29–34. doi:10.1177/004728759403200305

McCracken, G. (1988). *The Long Interview*. Sage. doi:10.4135/9781412986229

McGehee, N. G., & Andereck, K. L. (2004). Factors predicting rural residents' support of tourism. *Journal of Travel Research*, *43*(2), 131–140. doi:10.1177/0047287504268234

Mcgillivray, M. (2007). Human well-being: Issues, concepts and measures. In M. Mcgillivray (Ed.), *Human Well-Being* (pp. 1–22). Palgrave Macmillan. doi:10.1057/9780230625600_1

McIntosh, A. J., & Siggs, A. (2005). An exploration of the experiential nature of Boutique accommodation. *Journal of Travel Research*, *44*(1), 74–81. doi:10.1177/0047287505276593

McIntyre, N. (2013). Mobilities, Lifestyles and Imagined Worlds: Exploring the Terrain of Lifestyle Migration. In T. Duncan, S. A. Cohen, & M. Thulemark (Eds.), *Lifestyle Mobilities: Intersections of Travel, Leisure and Migration* (pp. 193–207). Ashgate.

McKean, P. (1978). Towards a theoretical analysis of tourism: economic dualism and cultural involution in Bali. In V. Smith (Ed.), *Hosts and guests. The anthropology of tourism* (pp. 119–138). University of Pennsylvania Press.

McKercher, B., & Bauer, T. (2003). Conceptual framework of the nexus between tourism, romance and sex. In T. Bauer & B. McKercher (Eds.), *Sex and Tourism. Journeys of Romance, Love and Lust* (pp. 3–18). The Haworth Hospitality Press.

Meacci, L., & Liberatore, G. (2015). Towards a Senses-Based Model for Experimental Tourism: The YOUTOOSCANY.COM Case. *Proceedings 1st International Conference on Experiential Tourism*, (10), 9–11.

Megeirhi, H. A., Woosnam, K. M., Ribeiro, M. A., Ramkissoon, H., & Denley, T. J. (2020). Employing a value-belief-norm framework to gauge Carthage residents' intentions to support sustainable cultural heritage tourism. *Journal of Sustainable Tourism*, *28*(9), 1351–1370. doi:10.1080/09669582.2020.1738444

Meho, L. I. (2006). E-Mail Interviewing in Qualitative Research: A Methodological Discussion. *Journal of the American Society for Information Science and Technology*, *57*(10), 1284–1295. doi:10.1002/asi.20416

Mendes, J., Guerreiro, M., & Matos, N. (2016). Monitoring the Quality of Tourism Experience. *Proceedings of the Asia Tourism Forum 2016 - the 12Th Biennial Conference of Hospitality and Tourism Industry in Asia, 19*(1), 294–304. 10.2991/atf-16.2016.44

Mendes, L. M. (2017). Gentrificação turística em Lisboa: Neoliberalismo, financeirização e urbanismo austeritário em tempos de pós-crise capitalista 2008-2009. *Cadernos Metrópole*, *19*(39), 479–512. doi:10.1590/2236-9996.2017-3906

Mendoza, M., & Motenrrubio, J. (2012). Actitud de la comunidad residente en Acapulco hacia los Spring Breakers y su comportamiento. *Revista Análisis Turístico*, *12*(1), 27–38.

Meng, F., Li, X., & Uysal, M. (2010). Tourism Development and Regional Quality of Life: The Case of China: 旅游发展和区域生活质量: 中国案例研究. *Journal of China Tourism Research*, *6*(2), 164–182. doi:10.1080/19388160.2010.481602

Mercille, J. (2005). Media effects on image – The case of Tibet. *Annals of Tourism Research*, *32*(4), 1039–1055. doi:10.1016/j.annals.2005.02.001

Merritt, J. (1998). EM into SME won't go? Attitudes, awareness and practices in the London Borough of Croydon. *Business Strategy and the Environment*, *7*(2), 90–100. doi:10.1002/(SICI)1099-0836(199805)7:2<90::AID-BSE141>3.0.CO;2-F

Mesquita, B. (2010, Nov.). Emoting: A contextualized process. *The Mind in Context*, 83–104.

Michel, F. (2009). *Planeta Sexo. Turismos Sexuais, Mercantilização e Desumanização dos Corpos*. Campo das Letras.

Miettinen, S. (2005). Creative tourism as tool for local empowerment. In G. Richards & J. Wilson (Eds.), From Cultural Tourism to Creative Tourism. Part 4 (Changing Experiences - the development of Creative Tourism). ATLAS.

Milano, C., Cheer, J., & Novelli, M. (2019). *Overtourism: Excesses, discontents and measures in travel and tourism*. Cabi Publishing. doi:10.1079/9781786399823.0000

Milhausen, R., & Herold, E. (1999). Does the sexual double standard still exist? Perceptions of university women. *Journal of Sex Research*, *36*(4), 361–368. doi:10.1080/00224499909552008

Miller, D. W., & Stoica, M. (2003a). Comparing the effects of a photograph versus artistic renditions of a beach scene in a direct-response print ad for a Caribbean resort island: A mental imagery perspective. *Journal of Vacation Marketing*, *10*(1), 11–21. doi:10.1177/135676670301000102

Milman, A., & Pizam, A. (1995). The Role of Awareness and Familiarity with a Destination: The Central Florida Case. *Journal of Travel Research*, *33*(3), 21–27. doi:10.1177/004728759503300304

Ministry of Land, Infrastructure and Transport. (2018). *The White Paper of Tourism*. Retrieved from: http://www.mlit.go.jp/statistics/file000008.html (in Japanese)

Miniterio das Finanças da República Democrática de East-Timor (MF-RDTL). (2019). *State budget*. East-Timor.

Minkiewicz, J., Evans, J., & Bridson, K. (2013). How do consumers co-create their experiences? An exploration in the heritage sector. *Journal of Marketing Management*, •••, 1–30. doi:10.1080/ 0267257X.2013.800899

Miró, S. V. (2011). Producing a "Successful City": Neoliberal Urbanism and Gentrification in the Tourist City – The Case of Palma (Majorca). *Urban Studies Research*, *2011*. Advance online publication. doi:10.1155/2011/989676

Miskinyte, E., & Smirnova, E. (2011). *Effective governance in tourism destination development. Federal Agency for Tourism, Russian Federation & World Tourism Organization*. UNWTO.

Mizumori, S., Yeshenko, O., Gill, K., & Davis, D. (2004). Parallel processing across neural systems: Implications for a multiple memory system hypothesis. *Neurobiology of Learning and Memory*, *82*(3), 278–298. doi:10.1016/j.nlm.2004.07.007 PMID:15464410

Moghavvemi, S., Woosnam, K. M., Paramanathan, T., Musa, G., & Hamzah, A. (2017). The effect of residents' personality, emotional solidarity, and community commitment on support for tourism development. *Tourism Management*, *63*, 242–254. doi:10.1016/j.tourman.2017.06.021

Moher, D., Liberati, A., Tetzlaff, J., Altman, D. G., & the The PRISMA Group. (2009). Preferred reporting items for systematic reviews and meta-analyses: the PRISMA Statement. *PloS Medicine, 6*. Retrieved from: http://www. Plosmedicine.org/article/fetchObject.action?uri=info

Molesworth, M. (2006). Real brands in imaginary worlds: Investigating players' experiences of brand placement in digital games. *Journal of Consumer Behaviour*, *5*(4), 355–366. doi:10.1002/cb.186

Monterrrubio, C., Coredero, J., & García, M. (2011). Percepciones de la comunidad local infantil en Huatulco México. Un acercamiento al estudio del cambio sociocultural asociado al turismo. *El Periplo Sustentable*, *20*, 149–185.

Monterrubio, J. (2008). Comunidades receptoras y percepciones: Un estudio sobre turismo y sexualidad. *Teoría y Praxis*, *5*(5), 145–160. doi:10.22403/UQROOMX/TYP05/11

Monterrubio, J. (2018). *Impactos socioculturales del turismo. Dimensiones culturales, teóricas y empíricas*. Triullas.

Monterrubio, J. C., Mendoza, M. M., Fernández, M. J., & Gullete, G. S. (2011). Turismo y cambios sociales. Estudio cualitativo sobre percepciones comunitarias en Bahías de Huatulco, México. *Cuadernos de Turismo*, *28*, 171–189.

Montoya, O. (2007). Aplicación del análisis factorial a la investigación de mercados: Caso de estudio. *Sciences et Techniques (Paris)*, *XII*(35), 281–286.

Moreno-Gil, S., & Martín-Santana, J. (2015). Understanding the image of self-Contained and serviced apartments: The Case of sun and Beach destinations. *Journal of Hospitality & Tourism Research (Washington, D.C.)*, *39*(3), 373–400. doi:10.1177/1096348013491594

Morgan, M. (2006). Making space for experiences. *Journal of Retail & Leisure Property*, *5*(4), 305–313. doi:10.1057/palgrave.rlp.5100034

Morgan, M., Elbe, J., & Curiel, J. de E. (2009). Has the experience economy arrived? The views of destination managers in three visitor-dependent areas. *International Journal of Tourism Research*, *11*(2), 201–216. doi:10.1002/jtr.719

Morgan, M., Lugosi, P., & Ritchie, J. B. (Eds.). (2010). *The tourism and leisure experience: Consumer and managerial perspectives* (Vol. 44). Channel View Publications. doi:10.21832/9781845411503

Moscardo, G. (1996). Mindful Visitors. *Annals of Tourism Research*, *23*(2), 376–397. doi:10.1016/0160-7383(95)00068-2

Moscardo, G., & Beckendorff, P. D. (2009). Mythbusting: Generation Y and Travel. In P. Benckendorff, G. Moscardo, & D. Pendergast (Eds.), *Tourism and Generation Y* (pp. 16–26). CABI Publishing. doi:10.1079/9781845936013.0016

Mossberg, L. (2007). A Marketing Approach to the Tourist Experience. *Scandinavian Journal of Hospitality and Tourism*, *7*(1), 59–74. doi:10.1080/15022250701231915

Mossberg, L. (2008). Extraordinary Experiences through Storytelling. *Scandinavian Journal of Hospitality and Tourism*, *8*(3), 195–210. doi:10.1080/15022250802532443

Moutinho, L. (1987). Consumer Behaviour in Tourism. *European Journal of Marketing*, *21*(10), 5–44. doi:10.1108/EUM0000000004718

Mules, T., & Dwyer, L. (2005). Public sector support for sport tourism events: The role of cost-benefit analysis. *Sport in Society*, *8*(2), 338–355. doi:10.1080/17430430500087864

Murillo, J., Garcés, C., & Rivera, P. (2007). What prevents firms from advancing in their environmental strategy? *International Advances in Economic Research*, *13*(1), 35–46. doi:10.100711294-006-9059-6

Murillo, J., Garcés, C., & Rivera, P. (2008). Estrategia medioambiental y expectativas de ventajas competitivas. *Cuadernos de Estudios Empresariales*, *18*, 9–31.

Murphy, P. E. (1985). *Tourism: A community approach*. Methuen.

Murphy, P., Pritchard, M. P., & Smith, B. (2000). The destination product and its impact on traveller perceptions. *Tourism Management*, *21*(1), 43–52. doi:10.1016/S0261-5177(99)00080-1

Murtas, D., & Davis, P. (2009). The role of the Ecomuseo dei terrazzamenti e della vite, (Cortemilia, Italy) in community development. *Museum & Society*, *7*(3), 150–186.

Myrttinen, H. (2009). East-Timor: A relapsing "success" story. *Taiwan Journal of Democracy*, *5*(1), 219–239.

Nadel, L., Hupbach, A., Gomez, R., & Newman-Smith, K. (2012). Memory formation, consolidation and transformation. *Neuroscience and Biobehavioral Reviews*, *36*(7), 1640–1645. doi:10.1016/j.neubiorev.2012.03.001 PMID:22465050

Naderifar, M., Goli, H., & Ghaljaie, F. (2017). Snowball Sampling: A Purposeful Method of Sampling in Qualitative Research. *Strides in Development of Medical Education*, *14*(3). Advance online publication. doi:10.5812dme.67670

Nadkarni, N. M., Hasbach, P. H., Thys, T., Crockett, E. G., & Schnacker, L. (2017). Impacts of nature imagery on people in severely nature-deprived environments. *Frontiers in Ecology and the Environment*, *15*(7), 395–403. doi:10.1002/fee.1518

Nagano Prefectural Government. (2017). *The Results of Survey on the Total Number of Foreign Lodgers in 2016* (in Japanese). Retrieved from: https://www.pref.nagano.lg.jp/kankoki/sangyo/kanko/toukei/gaikokujin.html

Nasution, R. A., Sembada, A. Y., Miliani, L., Resti, N. D., & Prawono, D. A. (2014). The Customer Experience Framework as Baseline for Strategy and Implementation in Services Marketing. *Procedia: Social and Behavioral Sciences*, *148*, 254–261. doi:10.1016/j.sbspro.2014.07.041

Natali, L., Handa, S., Peterman, A., Seidenfeld, D., & Tembo, G. (2018). Does money buy happiness? Evidence from an unconditional cash transfer in Zambia. *SSM - Population Health*, *4*, 225–235. doi:10.1016/j.ssmph.2018.02.002 PMID:29854906

National Immigration Directorate (NID). (2018). *Estatistica dos visitantes e pedidos de visto em Timor-Leste*. East-Timor.

National Statistics Directorate (NSD) & United Nations Population Fund (UNFPA). (2010). *Highlights of the 2010 census main results in East-Timor*. Dili, East-Timor: Authors.

National Statistics Directorate (NSD). (2015). *East-Timor population and housing census 2015*. NSD.

National Tourism Directorate (NTD). (2018). *Estatistica dos visitantes e turistas em Timor-Leste*. NTD.

Nawijn, J., & Mitas, O. (2012). Resident attitudes to tourism and their effect on subjective well-being: The case of Palma de Mallorca. *Journal of Travel Research*, *51*(5), 531–541. doi:10.1177/0047287511426482

Neal, J. D., Sirgy, M. J., & Uysal, M. (1999). The role of satisfaction with leisure travel/tourism services and experience in satisfaction with leisure life and overall life. *Journal of Business Research*, *44*(3), 153–163. doi:10.1016/S0148-2963(97)00197-5

Neal, J. D., Uysal, M., & Sirgy, M. J. (2007). The effect of tourism services on travelers' quality of life. *Journal of Travel Research*, *46*(2), 154–163. doi:10.1177/0047287507303977

Nelson, M. R. (2002). Recall of Brand Placements in Computer / Video Games. *Journal of Advertising Research*, *42*(2), 80–92. doi:10.2501/JAR-42-2-80-92

Nelson, M. R. (2005). Exploring Consumer Response to "Advergaming.". In C. Haugtvedt, K. Machleit, & R. Yalch (Eds.), *Online Consumer Psychology. Understanding and Influencing Consumer Behavior in the Virtual World* (pp. 167–223). Lawrence Erlbaum Associates, Inc.

Nelson, M. R., Keum, H., & Yaros, R. (2004). Advertainment or Adcreep: Game players' attitudes toward advertising and product placements in computer games. *Journal of Interactive Advertising*, *5*(1), 3–21. doi:10.1080/15252019.2004.10722090

Nepal, S. (2008). Residents' Attitudes to Tourism in Central British Columbia, Canada. *Tourism Geographies*, *10*(1), 42–65. doi:10.1080/14616680701825123

Netimperative - latest digital marketing news. (2013). *The MyCapeTownHoliday.com*. Retrieved October 27, 2019, from http://www.netimperative.com/2013/02/video-viral-of-the-week-facebook-users-get-virtual-holiday-in-cape-town/

Neuhofer, B., Buhalis, D., & Ladkin, A. (2012). Conceptualising technology enhanced destination experiences. *Journal of Destination Marketing and Management*, *1*(1–2), 36–46. doi:.jdmm.2012.08.001 doi:10.1016/j

Neumann, R., & Strack, F. (2000). "Mood contagion": The automatic transfer of mood between persons. *Journal of Personality and Social Psychology*, *79*(2), 211–223. doi:10.1037/0022-3514.79.2.211 PMID:10948975

Newsome, D., Moore, S. A., & Dowling, R. K. (2013). *Natural Area Tourism: Ecology, Impacts and Management* (2nd ed.). Channel View Publications.

Ng, M. N. (2009). *Pilgrimages: Memories of Colonial Macau and Hong Kong*. Hong Kong University Press.

Ng, W., & Diener, E. (2014). What matters to the rich and the poor? Subjective well-being, financial satisfaction, and postmaterialist needs across the world. *Journal of Personality and Social Psychology*, *107*(2), 326–338. doi:10.1037/a0036856 PMID:25090131

Nicholas, L. N., Thapa, B., & Ko, Y. J. (2009). Residents' perspectives of a World Heritage Site: The Pitons Management Area, St. Lucia. *Annals of Tourism Research*, *36*(3), 390–412. doi:10.1016/j.annals.2009.03.005

Nickerson, N. (2005). Some Reflections on Quality Tourism Experiences. In G. Jennings & N. P. Nickerson (Eds.), *Quality Tourism Experiences* (pp. 227–236). Elsevier Ltd.

Nicola, E. S. S. (2011). The relevance of visitors' nation brand embeddedness and personality congruence for nation brand identiðcation, visit intentions and advocacy. *Tourism Management*, *32*(6), 1282–1289. doi:10.1016/j.tourman.2010.12.004

Nicolson, P., & Burr, J. (2003). What is 'normal' about women's (hetero)sexual desire and orgasm? A report of an in-depth interview study. *Social Science & Medicine*, *57*(9), 1735–1745. doi:10.1016/S0277-9536(03)00012-1 PMID:12948581

Nispen, T. Van, Huiberts, S., Donders, S., & Koch, S. (2011). *The Nln-Player : a System for Nonlinear Music in Games, 2009–2012*. Academic Press.

Nobre, A. (2008). *História Breve da Vila de Olhão da Restauração*. Olhão: APOS.

Norman, D. A., & Verganti, R. (2014). Incremental and radical innovation: Design research vs. technology and meaning change. *Design Issues*, *30*(1), 78–96. doi:10.1162/DESI_a_00250

North, A. C., Hargreaves, D. J., & McKendrick, J. (1999). The influence of in-store music on wine selections. *The Journal of Applied Psychology*, *84*(2), 271–276. doi:10.1037/0021-9010.84.2.271

North, A. C., Sheridan, L. P., & Areni, C. S. (2016). Music Congruity Effects on Product Memory, Perception, and Choice. *Journal of Retailing*, *92*(1), 83–95. doi:10.1016/j.jretai.2015.06.001

Northouse, P. G. (2013). Leadership: theory and practice (6th ed.). Thousand Oaks, CA: Sage Publications, Inc.

Numally, J. (1978). *Psychometric Theory*. MacGraw-Hill.

Nunes, P. (2010). *Desporto, turismo e ambiente: O turismo de Natureza como pólo de atração turística complementar ao produto Sol & mar na sub-região do Litoral Alentejano* [Sports, tourism and environment: Nature tourism as a tourist attraction pole complementary to the Sun & beach product in the Alentejo Litoral sub-region] (Unpublished doctoral dissertation). University of Lisbon, Lisbon.

Nunkoo, R., & Gursoy, D. (2012). Resident's Support for Tourism an Identity Perspective. *Annals of Tourism Research*, *39*(1), 243–268. doi:10.1016/j.annals.2011.05.006

Nunkoo, R., & Ramkissoon, H. (2011). Developing a community support model for tourism. *Annals of Tourism Research*, *38*(3), 964–988. doi:10.1016/j.annals.2011.01.017

Nunkoo, R., & Ramkissoon, H. (2011). Residents' satisfaction with community attributes and support for tourism. *Journal of Hospitality & Tourism Research (Washington, D.C.)*, *35*(2), 171–190. doi:10.1177/1096348010384600

Nunkoo, R., & Ramkissoon, H. (2012). Power, trust, social exchange and community support. *Annals of Tourism Research*, *39*(2), 997–1023. doi:10.1016/j.annals.2011.11.017

Nunkoo, R., Smith, S. L., & Ramkissoon, H. (2013). Residents' attitudes to tourism: A longitudinal study of 140 articles from 1984 to 2010. *Journal of Sustainable Tourism*, *21*(1), 5–25. doi:10.1080/09669582.2012.673621

Nunkoo, R., & So, K. K. F. (2016). Residents' support for tourism: Testing alternative structural models. *Journal of Travel Research*, *55*(7), 847–861. doi:10.1177/0047287515592972

Nunnally, J. C. (1978). Assessment of reliability. In *Psychometric theory* (2nd ed.). McGraw-Hill.

Nunnally, J. C. (1978). *Psychometric theory* (2nd ed.). McGraw-Hill.

Nzama, A. (2008). Socio-cultural impacts of tourism on the rural areas within the World Heritage Sites–The case of KwaZulu-Natal, South Africa. *South Asian Journal of Tourism and Heritage*, *1*, 1–8.

O'Connell-Davidson, J. (2001). The sex tourist, the expatriate, his ex-wife and her 'other: The politics of loss, difference and desire. *Sexualities*, *4*(1), 5–24. doi:10.1177/136346001004001001

O'dell, T., & Billing, P. (2005). *Experiencescapes: Tourism, culture and economy*. Business School Press DK.

O'Halloran, R. M. (1992). Tourism management profiles: Implications for tourism education. *Hospitality Review*, *10*(1).

O'Leary-Kelly, S. W., & Vokurka, R. J. (1998). The Empirical Assessment of Construct Validity. *Journal of Operations Management*, *16*(4), 387–405. doi:10.1016/S0272-6963(98)00020-5

OECD. (2019). *South Korea*. Reatrived from https://data.oecd.org/korea.htm#profile-economy

Oh, H., Fiore, M., & Jeoung, M. (2007). Measuring Experience Economy Concepts: Tourism Applications. *Journal of Travel Research*, *46*(2), 119–132. doi:10.1177/0047287507304039

Okazaki, E. (2008). A Community-Based Tourism Model: Its Conception and Use. *Journal of Sustainable Tourism*, *16*(5), 511–529. doi:10.1080/09669580802159594

Olya, H., & Gavilyan, Y. (2016). Configurational Models to Predict Residents' Support for Tourism Development. *Journal of Travel Research*, *56*(7), 893–912. doi:10.1177/0047287516667850

Oppermann, M. (1999). Sex tourism. *Annals of Tourism Research*, *26*(2), 251–266. doi:10.1016/S0160-7383(98)00081-4

Oppermann, M. (Ed.). (1998). *Sex Tourism and Prostitution: Aspects of Leisure, Recreation and Work*. Cognizant Communication Corporation.

Ortiz, Y., Martínez, E., Rogel, R., & Rochel, M. (2016). Los stakeholders de la industria hotelera: Una clasificación a partir de sus intereses ambientales. *Universidad & Empresa*, *18*(30), 97–120. doi:10.12804/rev.univ.empresa.30.2016.05

Otelpuan. (2020). *Hotels evaluation website*. Retrieved from https://www.otelpuan.com/tr

Othman, S., Nishimura, Y., & Kubota, A. (2013). Memory association in place making: A review. *Procedia: Social and Behavioral Sciences*, *85*, 554–563. doi:10.1016/j.sbspro.2013.08.384

Otto, J. E., & Ritchie, J. R. B. (1996). The service experience in tourism. *Tourism Management*, *17*(3), 165–174. doi:10.1016/0261-5177(96)00003-9

Overy, K., & Molnar-Szakacs, I. (2009). Being together in time: Musical experience and the mirror neuron system. *Music Perception, 26*(5), 489–504. doi:10.1525/mp.2009.26.5.489

Oyster. (2020). *Most romantic hotels in the Algarve*, Portugal. Retrieved from: https://www.oyster.com/algarve/hotels/roundups/most-romantic-hotels-in-algarve-faro-district-portugal/

Özel, Ç. H. (2015). Marketing to children in tourism industry: descriptive analysis of kid-friendly hotels' practices in Turkey. In A. Correia, J. Gnoth, M. Kozak, & A. Fyall (Eds.), *Marketing Places and Spaces* (pp. 193–208). Emerald Group Publishing Limited. doi:10.1108/S1871-317320150000010014

Page, S. J., & Hall, C. M. (2014). *The Geography of Tourism and Recreation: Environment, Place and Space*. Routledge.

Paivio, A. (1971). Imagery and Verbal Processes. Holt, Rinehart & Winston.

Paivio, A. (1969). Mental imagery in associative learning and memory. *Psychological Review, 76*(3), 241–263. doi:10.1037/h0027272

Papadimitriou, D., Apostolopoulou, A., & Kaplanidou, K. (2015). Destination personality, affective image, and behavioral intentions in domestic urban tourism. *Journal of Travel Research, 54*(3), 302–315. doi:10.1177/0047287513516389

Papadimitriou, D., Kaplanidou, K., & Apostolopoulou, A. (2018). Destination Image Components and Word-of-Mouth Intentions in Urban Tourism: A Multigroup Approach. *Journal of Hospitality & Tourism Research (Washington, D.C.), 42*(4), 503–527. doi:10.1177/1096348015584443

Park, M., & Yoo, J. (2020). Effects of perceived interactivity of augmented reality on consumer responses: A mental imagery perspective. *Journal of Retailing and Consumer Services, 52.*

Park, D. B., Nunkoo, R., & Yoon, Y. S. (2015). Rural residents' attitudes to tourism and the moderating effects of social capital. *Tourism Geographies, 17*(1), 112–133. doi:10.1080/14616688.2014.959993

Park, K., Lee, J., & Lee, T. (2017). Residents' attitudes toward future tourism development in terms of community well-being and attachment. *Asia Pacific Journal of Tourism Research, 22*(2), 160–172. doi:10.1080/10941665.2016.1208669

Park, R., Burgess, E., & McKenzie, R. D. (1967). *The City. Suggestions for Investigation of Human Behavior in the Urban Environment*. The University of Chicago Press. (Original work published 1925)

Park, S., & Santos, C. (2017). Exploring the Tourist Experience. A Sequential Approach. *Journal of Travel Research, 56*(1), 16–27. doi:10.1177/0047287515624017

Patch, J., & Brenner, N. (2007). Gentrification. In G. Ritzer (Ed.), *The Blackwell Encyclopedia of Sociology* (pp. 1917–1920). Blackwell Publishing. doi:10.1002/9781405165518.wbeosg035

Patton, M. (1990). *Qualitative Evaluation and Research Methods* (2nd ed.). Sage.

Patwardhan, V., Ribeiro, M. A., Payini, V., Woosnam, K. M., Mallya, J., & Gopalakrishnan, P. (2020). Visitors' Place Attachment and Destination Loyalty: Examining the Roles of Emotional Solidarity and Perceived Safety. *Journal of Travel Research, 59*(1), 3–21. doi:10.1177/0047287518824157

Pavli, A., Silvestros, C., Patrinos, S., & Maltezou, H. C. (2015). Vaccination and malaria prophylaxis among Greek international travelers to Asian destinations. *Journal of Infection and Public Health, 8*(1), 47–54. doi:10.1016/j.jiph.2014.07.002 PMID:25129447

Pawaskar, P., & Goel, M. (2014). A Conceptual Model: Multisensory Marketing and Destination Branding. *Procedia Economics and Finance, 11*(11), 255–267. doi:10.1016/S2212-5671(14)00194-4

Pearce, P. L., & Moscardo, G. M. (1986). The Concept of Authenticity in Tourist Experiences. *The Australian and New Zealand Journal of Sociology*, *22*(1), 121–132. doi:10.1177/144078338602200107

Pearson, J., Naselaris, T., Holmes, E. A., & Kosslyn, S. M. (2015). Mental Imagery: Functional Mechanisms and Clinical Applications. *Trends in Cognitive Sciences*, *19*(10), 590–602. doi:10.1016/j.tics.2015.08.003 PMID:26412097

Peirce, J., Martin, D. G., & Murphy, J. T. (2011). Relational place-making: The networked politics of place. *Transactions of the Institute of British Geographers*, *36*(1), 54–70. doi:10.1111/j.1475-5661.2010.00411.x

Peñaloza, L., & Venkatesh, A. (2006). Further evolving the new dominant logic of marketing: From services to the social construction of markets. *Marketing Theory*, *6*(3), 299–316. doi:10.1177/1470593106066789

Pendergast, D. (2009). Getting to know the Y generation. In Tourism and generation Y (pp. 1–15). Wallingford, UK: CABI Publishing. doi:10.1079/9781845936013.0001

PENT. (2012). *National Strategic Tourism Plan: Horizon 2013-2015*. Ministry of Economy and Employment.

Pentus, K., Mehine, T., & Kuusik, A. (2014). Considering Emotions in Product Package Design through Combining Conjoint Analysis with Psycho Physiological Measurements. *Procedia: Social and Behavioral Sciences*, *148*, 280–290. doi:10.1016/j.sbspro.2014.07.044

Perdue, R. R., Long, P. T., & Allen, L. (1987). Rural resident tourism perceptions and attitudes. *Annals of Tourism Research*, *14*(3), 420–429. doi:10.1016/0160-7383(87)90112-5

Perdue, R. R., Long, P. T., & Allen, L. (1990). Resident support for tourism development. *Annals of Tourism Research*, *17*(4), 586–599. doi:10.1016/0160-7383(90)90029-Q

Pereira, E., Mascarenhas, M., Flores, A., Chalip, L., & Pires, G. (2020). Strategic leveraging: Evidences of small-scale sport events. *International Journal of Event and Festival Management*, *11*(1), 69–88. doi:10.1108/IJEFM-07-2018-0046

Pereira, E., Mascarenhas, M., Flores, A., & Pires, G. (2015). Nautical small-scale sports events portfolio: A strategic leveraging approach. *European Sport Management Quarterly*, *1*(15), 27–47. doi:10.1080/16184742.2015.1007883

Perkins, H., & Thorns, D. (2001). Reflections on Urry's tourist gaze in the context of contemporary experience in the antipodes. *International Sociology*, *16*, 185–2004. doi:10.1177/0268580901016002004

Perritano, J. (2011). *Science of Emotions*. Marshall Cavendish Benchmark.

Peters, S., & Leshner, G. (2013). Get in the Game: The Effects of Game-Product Congruity and Product Placement Proximity on Game Players' Processing of Brands Embedded in Advergames. *Journal of Advertising*, *42*(2–3), 113–130. doi:10.1080/00913367.2013.774584

Petrick, J. (2004). Are loyal visitors desired visitors? *Tourism Management*, *25*(4), 463–470. doi:10.1016/S0261-5177(03)00116-X

Petroshius, S. M., & Crocker, K. E. (1989). An empirical analysis of spokesperson characteristics on advertisement and product evaluations. *Journal of the Academy of Marketing Science*, *17*(3), 217–225. doi:10.1007/BF02729813

Pezzi, E., & Vianna, S. (2015). Tourist Experience and Experiential Tourism: A study about dimensions of memorable experience. *Turismo Em Análise*, *26*(1), 165–187. doi:10.11606/issn.1984-4867.v26i1p165-187

Phau, I., Shanka, T., & Dhayan, N. (2010). Destination image and choice intention of university student travellers to Mauritius. *International Journal of Contemporary Hospitality Management*, *22*(5), 758–764. doi:10.1108/09596111011053846

Phelps, A. (1986). Holiday destination image: The problem of assessment. *Tourism Management, 7*(3), 168–180. doi:10.1016/0261-5177(86)90003-8

Phelps, E. A. (2004). Human emotion and memory: Interactions of the amygdala and hippocampal complex. *Current Opinion in Neurobiology, 14*(2), 198–202. doi:10.1016/j.conb.2004.03.015 PMID:15082325

Picazo, P., & Moreno-Gil, S. (2019). Analysis of the projected image of tourism destinations on photographs: A literature review to prepare for the future. *Journal of Vacation Marketing, 25*(1), 3–24. doi:10.1177/1356766717736350

Pickering, C., & Byrne, J. (2014). The benefits of publishing systematic quantitative literature reviews for PhD candidates and other early-career researchers. *Higher Education Research & Development, 33*(3), 534–548. doi:10.1080/07294360.2013.841651

Pike, S. (2002). Destination image analysis – a review of 142 papers from 1973 to 2000. *Tourism Management, 23*(5), 541–549. doi:10.1016/S0261-5177(02)00005-5

Pike, S. (2009). Destination brand positions of a competitive set of near-home destinations. *Tourism Management, 30*(6), 857–866. doi:10.1016/j.tourman.2008.12.007

Pike, S., & Ryan, C. (2004). Destination positioning analysis through a comparison of cognitive, affective, and conative perceptions. *Journal of Travel Research, 42*(4), 333–342. doi:10.1177/0047287504263029

Pine, B. J., & Gilmore, J. H. (1999). The Experience Economy: Work is Theatre and Every Business a Stage. Harvard Business School Press.

Pine, B. J., & Gilmore, J. H. (1998). Welcome to the experience economy. *Harvard Business Review, 76*(4), 97–105. PMID:10181589

Pine, J., & Gilmore, J. (1998). Welcome to the Experience Economy. *Harvard Business Review, 76*(4), 97–105. PMID:10181589

Pine, J., & Gilmore, J. (1999). *The Experience Economy: Work is Theater & Every Business a Stage*. Harvard Business School Press.

Pine, J., & Gilmore, J. H. (1999). *The experience economy: Work is theatre & every business a stage*. Harvard Business School Press.

Ping, R. A. Jr. (2004). On Assuring Valid Measures for Theoretical Models Using Survey Data. *Journal of Business Research, 57*(2), 125–141. doi:10.1016/S0148-2963(01)00297-1

Pinkster, F. M., & Boterman, W. R. (2017). When the spell is broken: Gentrification, urban tourism and privileged discontent in the Amsterdam canal district. *Cultural Geographies, 24*(3), 457–472. doi:10.1177/1474474017706176 PMID:29278248

Piscitelli, A. (2004). On «gringos» and «natives»: Gender and sexuality in the context of international sex tourism in Fortaleza, Brazil. *Vibrant, 1*(1/2), 87–114.

Ponce, M. D. (2004). Percepción del modelo turístico de sol y playa: El caso del mar menor. *Papeles de Geografía, 39*, 173–186.

Poock, M. C., & Lefond, D. (2001). How college-bound prospects perceive university web sites: Findings, implications and turning browsers into applicants. *College and University, 77*(1), 15–21.

Poria, Y., & Carr, N. (2010). What have we done and where should we go? In N. Carr & Y. Poria (Eds.), *Sex and the Sexual During People's Leisure and Tourism Experiences* (pp. 181–196). Cambridge Scholars Publishing.

Poria, Y., Reichel, A., & Cohen, R. (2013). Tourists perceptions of World Heritage Site and its designation. *Tourism Management, 35*, 272–274. doi:10.1016/j.tourman.2012.02.011

Porras, N., Plaza, M. A., & Vargas, A. (2019). Quality of Life and Perception of the Effects of Tourism: A Contingent Approach. In A. M. Campón, J. M. Hernández, & J. A. Folgado (Eds.), *Best Practices in Hospitality and Tourism Marketing and Management* (pp. 109–132). Springer. doi:10.1007/978-3-319-91692-7_6

Post, J., & Altman, B. (1994). Managing the Environmental Change Process: Barriers and Opportunities. *Journal of Organizational Change Management, 7*(4), 64–81. doi:10.1108/09534819410061388

Potter, J. W., & Donnerstein, D. L. (1999). Rethinking validity and reliability in content analysis. *Journal of Applied Communication Research, 27*(3), 258–284. doi:10.1080/00909889909365539

PPS (Project for Public Spaces). (2007). *What Is Placemaking?* Retrieved September 1, 2019, from https://www.pps.org/article/what-is-placemaking

PPS (Project for Public Spaces). (2013). Stronger citizens, stronger cities: Changing governance through a focus on place. *Project for Public Spaces website*. Retrieved September 1, 2019, from http://www.pps.org/blog/stronger-citizens-stronger-cities-changing-governance-through-a-focus-on-place/

Prahalad, C. K., & Ramaswamy, V. (2004). Co-creation experiences: The next practice in value creation. *Journal of Interactive Marketing, 18*(3), 5–14. doi:10.1002/dir.20015

Prahalad, C. K., & Ramaswamy, V. (2004). *The future of competition: Co-creating unique value with customers*. Harvard Business School Press.

Prasad, B. D. (2008). Content analysis: A method in social science research. In D. K. Lal Das & V. Bhaskaran (Eds.), *Research Methods for Social Work* (pp. 1–20). Rawat Publications. Retrieved from http://www.css.ac.in/download/deviprasad/content%20analysis.%20a%20method%20of%20social%20science%20research.pdf

Pratt, S. (2015). The economic impact of tourism in SIDS. *Annals of Tourism Research, 52*, 148–160. doi:10.1016/j.annals.2015.03.005

Prayag, G. (2009). Tourists' Evaluations of Destination Image, Satisfaction, and Future Behavioral Intentions-The Case of Mauritius. *Journal of Travel & Tourism Marketing, 26*(8), 836–853. doi:10.1080/10548400903358729

Prayag, G., Dookhony-Ramphul, K., & Maryeven, M. (2010). Hotel development and tourism impacts in Mauritius: Hoteliers' perspectives on sustainable tourism. *Development Southern Africa, 27*(5), 697–712. doi:10.1080/0376835X.2010.522832

Prayag, G., Hosany, S., Muskat, B., & Del Chiappa, G. (2017). Understanding the Relationships between Tourists' Emotional Experiences, Perceived Overall Image, Satisfaction, and Intention to Recommend. *Journal of Travel Research, 56*(1), 41–54. doi:10.1177/0047287515620567

Prayag, G., Hosany, S., Nunkoo, R., & Alders, T. (2013). London residents' support for the 2012 Olympic Games: The mediating effect of overall attitude. *Tourism Management, 36*, 629–640. doi:10.1016/j.tourman.2012.08.003

Prayag, G., Hosany, S., & Odeh, K. (2013). The role of tourists' emotional experiences and satisfaction in understanding behavioral intentions. *Journal of Destination Marketing & Management, 2*(2), 118–127. doi:10.1016/j.jdmm.2013.05.001

Prayag, G., & Ryan, C. (2011). The relationship between the 'push' and 'pull' factors of a tourist destination: The role of nationality – an analytical qualitative research approach. *Current Issues in Tourism, 14*(2), 121–143. doi:10.1080/13683501003623802

Prayag, G., & Ryan, C. (2012). Antecedents of Tourists' Loyalty to Mauritius: The Role and Influence of Destination Image, Place Attachment, Personal Involvement, and Satisfaction. *Journal of Travel Research, 51*(3), 342–356. doi:10.1177/0047287511410321

Prebensen, N. K., Campos, A. C., & Blomstervik, I. (2018). Facilitating for Co-creation in experiential consumption through perceived behaviour control and self-efficacy. *Tourism & Management Studies International Conference 2018.*

Prebensen, N. K. (2017). Staging for value co-creation in nature-based experiences: the case of a surfing course at Surfer's Paradise, Australia. In N. K. Prebensen, J. S. Chen, & M. S. Uysal (Eds.), *Co-creation in tourism Experiences* (pp. 36–49). Routledge. doi:10.4324/9781315645919-4

Prebensen, N. K., & Campos, A. C. (2017). Operant resources in experiential consumption: self-efficacy (SE) versus perceived behavioral control (PBC). *7th ATMC Conference Advances in Tourism Marketing Conferences: The Art of Living Together.*

Prebensen, N. K., & Foss, L. (2011). Coping and co-creating in tourist experiences. *International Journal of Tourism Research, 13*(1), 54–67. doi:10.1002/jtr.799

Prebensen, N. K., Kim, H., & Uysal, M. (2015). Cocreation as moderator between the experience value and satisfaction relationship. *Journal of Travel Research.* Advance online publication. doi:10.1177/0047287515583359

Prebensen, N. K., Woo, E., Chen, J. S., & Uysal, M. (2013). Motivation and involvement antecedents of perceived value of the destination experience. *Journal of Travel Research, 52*(2), 253–264. doi:10.1177/0047287512461181

Prebensen, N., Vitterso, J., & Dahl, T. (2013). Value Co-creation Significance of Tourist Resources. *Annals of Tourism Research, 42*, 240–261. doi:10.1016/j.annals.2013.01.012

Prebensen, N., & Xie, J. (2017). Efficacy of co-creation and mastering on perceived value and satisfaction in tourists' consumption. *Tourism Management, 60*, 166–176. doi:10.1016/j.tourman.2016.12.001

Prentice, R. (1993). Community-driven tourism planning and residents' preferences. *Tourism Management, 14*(3), 218–227. doi:10.1016/0261-5177(93)90023-E

Presenza, A., Del Chiappa, G., & Sheehan, L. (2013). Residents' engagement and local tourism governance in maturing beach destinations. Evidence from an Italian case study. *Journal of Destination Marketing & Management, 2*(1), 22–30. doi:10.1016/j.jdmm.2013.01.001

Preuss, H., Seguin, B., & O'Reilly, N. (2007). Profiling major sport event visitors: The 2002 Commonwealth Games. *Journal of Sport & Tourism, 12*(1), 5–23. doi:10.1080/14775080701496719

Priest, S., & Bunting, C. (1993). Changes in Perceived Risk and Competence During Whitewater Canoeing. *Journal of Applied Recreation Research, 18*, 265–280.

Proença, S., & Soukiazis, E. (2008). Tourism as an economic growth factor: A case study for Southern E,uropean countries. *Tourism Economics, 14*(4), 791–806. doi:10.5367/000000008786440175

Pruitt, D., & LaFont, S. (1995). For love and money: Romance tourism in Jamaica. *Annals of Tourism Research, 22*(2), 422–440. doi:10.1016/0160-7383(94)00084-0

Prytherch, D. L., & Maiques, J. V. B. (2009). City profile: Valencia. *Cities (London, England), 26*(2), 103–115. doi:10.1016/j.cities.2008.11.004

Pyke, S., Hartwell, H., Blake, A., & Hemingway, A. (2016). Exploring well-being as a tourism product resource. *Tourism Management, 55*, 94–105. doi:10.1016/j.tourman.2016.02.004

Qiu, J. P., & Zou, F. (2004). Research about Content Analysis Method. *Journal of Library Science in China*, *2*, 14–19.

Quadri-Felitti, D., & Fiore, A. (2013). Destination loyalty: Effects of wine tourists' experiences, memories, and satisfaction on intentions. *Tourism and Hospitality Research*, *13*(1), 47–62. doi:10.1177/1467358413510017

Quan, S., & Wang, N. (2004). Towards a structural model of the tourist experience: An illustration from food experiences in tourism. *Tourism Management*, *25*(3), 297–305. doi:10.1016/S0261-5177(03)00130-4

Quest, M. (1990). *Howarth Book of Tourism*. Macmillan. doi:10.1007/978-1-349-11687-4

Qu, H., Kim, L. H., & Im, H. H. (2011). A model of destination branding: Integrating the concepts of the branding and destination image. *Tourism Management*, *32*(3), 465–476. doi:10.1016/j.tourman.2010.03.014

Quintal, G., Paipe, G., Felipe, J. L., & Carvalho, M. J. (2016). Strategic tool to estimate the consumption patterns of non-residents spectators at sporting events (Liga NOS): Adaptation to Portuguese reality. *Brazilian Business Review*, *15*(2), 102–120. doi:10.15728/edicaoesp.2016.6

Rada, V., & Martín, V. (2014). Random route and quota sampling: Do they offer any advantage over probably sampling methods? *Open Journal of Statistics*, *4*(05), 391–401. doi:10.4236/ojs.2014.45038

Radzuan, I. S. M., Fukami, N., & Ahmad, Y. (2015). Incentives for the conservation of traditional settlements: Residents' perception in Ainokura and Kawagoe, Japan'. *Journal of Tourism and Cultural Change*, *13*(4), 301–329. doi:10.1080/14766825.2014.952302

Rahmann, H., & Rahmann, M. (1992). *The Neurobiological Basis of Memory and Behavior*. Springer. doi:10.1007/978-1-4612-2772-4

Rais, N., Musa, R., & Muda, M. (2016). Reconceptualisation of Customer Experience Quality (CXQ) Measurement Scale. *Procedia Economics and Finance*, *37*(16), 299–303. doi:10.1016/S2212-5671(16)30128-9

Ramaswamy, V., & Gouillart, F. (2010). *The power of co-creation*. Free Press.

Ramaswamy, V., & Ozcan, K. (2018). What is co-creation? An interactional creation framework and its implications for value creation. *Journal of Business Research*, *84*, 196–205. doi:10.1016/j.jbusres.2017.11.027

Ramírez, P., Mariano, A., & Salazar, E. (2014). Propuesta metodológica para aplicar modelo de ecuaciones estructurales con PLS: El caso del uso de bases de datos científicas en estudiantes universitarios. *Revista ADMpg Gestão Estratégica*, *7*(2), 133–139.

Ramón-Cardona, J. (2014). Tipos de oferta turística y actitudes de los residentes: El caso de Ibiza. *Revista Brasileira de Pesquisa em Turismo*, *8*(1), 3–22. doi:10.7784/rbtur.v8i1.704

Ramón-Cardona, J. (2015). Comparativa entre islas de la valoración de distintos tipos de oferta turística por parte de los residentes: El caso de las Islas Baleares (España). *Turismo e Sociedade*, *8*(2), 255–275.

Rasoolimanesh, S. M., Ringle, C. M., Jaafar, M., & Ramayah, T. (2017). Urban vs. rural destinations: Residents' perceptions, community participation and support for tourism development. *Tourism Management*, *60*(June), 147–158. doi:10.1016/j.tourman.2016.11.019

Ratten, V. (2016a). Sport innovation management: Towards a research agenda. *Innovation (North Sydney, N.S.W.)*, *18*(3), 238–250. doi:10.1080/14479338.2016.1244471

Ratten, V. (2016b). The dynamics of sport marketing: Suggestions for marketing intelligence and planning. *Marketing Intelligence & Planning*, *34*(2), 162–168. doi:10.1108/MIP-07-2015-0131

Redondo, I. (2012). The effectiveness of casual advergames on adolescents' brand attitudes. *European Journal of Marketing*, *46*(11), 1671–1688. doi:10.1108/03090561211260031

Regensburg Experience. (2019). Retrieved October 27, 2019, from https://rex.julienbiere.de/

Região de Turismo do Algarve. (2014). *Plano de Marketing Estratégico para o Turismo do Algarve 2015-2018*. Retrieved from https://www.turismodeportugal.pt/SiteCollectionDocuments/estrategia/Estrategias-Regionais-Algarve/Plano-Marketing-Estrategico-Turismo-Algarve-2015-2018.pdf

Regional Board of Statistics of Madeira. (2018). *Estatísticas do Turismo da Região Autónoma da Madeira. Resultados Provisórios* [Tourism Statistics of the Autonomous Region of Madeira. Interim Results]. Author.

Reichel, A., Fuchs, G., & Uriely, N. (2009). Israeli backpackers. The Role of Destination Choice. *Annals of Tourism Research*, *36*(2), 222–246. doi:10.1016/j.annals.2008.11.002

Reisenzein, R., Studtmann, M., & Horstmann, G. (2013). Coherence between emotion and facial expression: Evidence from laboratory experiments. *Emotion Review*, *5*(1), 16–23. doi:10.1177/1754073912457228

Ren, H. (2010). *Neoliberalism and culture in China and Hong Kong: The countdown of time*. Routledge. doi:10.4324/9780203844939

República Democrática de Timor Leste (RDTL) Ministerio das Finanças (MF). (2019). *Orçamento geral do estado 2019*. Dili, East-Timor: Author.

Rérat, P., Söderström, O., & Piguet, E. (2009). New Forms of Gentrification: Issues and debates. *Population Space and Place*, *16*(5), 335–343. doi:10.1002/psp.585

Reynolds, W. H. (1965). The role of the consumer in image building. *California Management Review*, *7*(Spring), 69–76. doi:10.2307/41165634

Ribeiro, F., & Sacramento, O. (2006). A ilusão da conquista: Sexo, amor e interesse entre *gringos* e *garotas* em Natal (Brasil). *Cronos (Valencia)*, *7*(1), 161–172.

Ribeiro, M. A., Pinto, P., Silva, J. A., & Woosnam, K. M. (2017). Residents' attitudes and the adoption of pro-tourism behaviours: The case of developing island countries. *Tourism Management*, *61*, 523–537. doi:10.1016/j.tourman.2017.03.004

Ribeiro, M. A., Pinto, P., Silva, J. A., & Woosnam, K. M. (2018). Examining the predictive validity of SUS-TAS with maximum parsimony in developing island countries. *Journal of Sustainable Tourism*, *26*(3), 379–398. doi:10.1080/09669582.2017.1355918

Ribeiro, M. A., Valle, P. O., & Silva, J. A. (2013). Residents' Attitudes towards Tourism Development in Cape Verde Islands. *Tourism Geographies*, *15*(4), 654–679. doi:10.1080/14616688.2013.769022

Ribeiro, M. A., Woosnam, K. M., Pinto, P., & Silva, J. A. (2018). Tourists' destination loyalty through emotional solidarity with residents: An integrative moderated mediation model. *Journal of Travel Research*, *57*(3), 279–295. doi:10.1177/0047287517699089

Ribeiro, M., Silva, M. C., Schouten, J., Ribeiro, F., & Sacramento, O. (2007). *Vidas na Raia: Prostituição Feminina em Regiões de Fronteira*. Edições Afrontamento.

Ribes-Giner, G., Perello-Marín, M., & Díaz, O. (2016). Co-creation Impacts on Student Behavior. *Procedia: Social and Behavioral Sciences*, *228*, 72–77. doi:10.1016/j.sbspro.2016.07.011

Richards, G., & Hall, D. (2002). The Community: A sustainable concept in tourism development? In G. Richards & D. Hall (Eds.), Tourism and sustainable community development (2nd ed., pp. 1-13). London: Routledge, Taylor & Francis Group.

Richards, G. (2011). Creativity and tourism. The state of the art. *Annals of Tourism Research, 38*(4), 1225–1253. doi:10.1016/j.annals.2011.07.008

Richards, G. (2014). Creativity and tourism in the city. *Current Issues in Tourism, 17*(2), 119–144. doi:10.1080/13683500.2013.783794

Richards, G. (2015). Eventfulness and the quality of life. *Tourism Today (Nicosia), 14*, 23–36.

Richards, G. (2015). The new global nomads: Youth travel in a globalizing world. *Tourism Recreation Research, 40*(3), 340–352. doi:10.1080/02508281.2015.1075724

Richards, G. (2016). The challenge of creative tourism. *Ethnologies (Québec), 38*, 1–2, 31–42.

Richards, G. (2017). Measuring event experiences: An international view. In G. Richards & A. R. Lanuza (Eds.), *Experiencias turisticas de festivals y eventos* (pp. 11–27). Colección Pasos.

Richards, G. (2019). Event experience research directions. In J. Ambrecht, E. Lumdberg, & T. Andersson (Eds.), *A research agenda for event management* (pp. 79–93). Edward Elgar Publishing Limited. doi:10.4337/9781788114363.00015

Richards, G., & King, B. (2003). International Youth Travel. *Travel & Tourism Analyst, 6*(December).

Richards, G., & Palmer, R. (2010). *Eventful Cities: Cultural Management and Urban Revitalisation.* Elsevier.

Richards, G., & Raymond, C. (2000). Creative Tourism. *ATLAS News, 23*, 16–20.

Richards, G., & Wilson, J. (2003). *New horizons in independent youth and student travel. A report to the international student travel confederation (ISTC) and the association of tourism and leisure education (ATLAS).* International Student Travel Confederation.

Richards, G., & Wilson, J. (2005). Changing Places - the spatial challenge of creativity. In G. Richards & J. Wilson (Eds.), *From Cultural Tourism to Creative Tourism. Part 3 - Changing Places - the spatial challenge of creativity* (pp. 7–8). ATLAS.

Richards, G., & Wilson, J. (2006). Developing creativity in tourist experiences: A solution to the serial reproduction of culture? *Tourism Management, 27*(6), 1209–1223. doi:10.1016/j.tourman.2005.06.002

Rickly-boyd, J. M. (2010). The Tourist Narrative. *Tourist Studies, 9*(3), 259–280. doi:10.1177/1468797610382701

Rickly-Boyd, J. M. (2012). Authenticity & aura. A benjaminian approach to tourism. *Annals of Tourism Research, 39*(1), 269–289. doi:10.1016/j.annals.2011.05.003

Ridderstaat, J., Croes, R., & Nijkamp, P. (2016). A two-way causal chain between tourism development and quality of life in a small island destination: An empirical analysis. *Journal of Sustainable Tourism, 24*(10), 1461–1479. doi:10.1080/09669582.2015.1122016

Rieber, L. P. (1996). Seriously Considering Play: Designing Interactive Learning Environments Based on the Blending of Microworlds, Simulations, and Games. *Educational Technology Research and Development, 44*(2), 43–58. doi:10.1007/BF02300540

Riegelsberger, J. (2003). Interpersonal cues and consumer trust in e-commerce. *Conference on Human Factors in Computing Systems - Proceedings*, 674–675. 10.1145/765891.765923

Riegelsberger, J., Angela, S. M., & McCarthy, J. D. (2003). Shiny happy people building trust?: photos on e-commerce websites and consumer trust. *Proceedings of the Conference on Human Factors in Computing Systems*, 121–128. 10.1145/642611.642634

Ritchie, B. W., & Inkari, M. (2006). Host community attitudes toward tourism and cultural tourism development: The case of the Lewes District, Southern England. *International Journal of Tourism Research*, *8*(1), 27–44. doi:10.1002/jtr.545

Ritchie, J., & Spencer, L. (1994). Qualitative Analysis for Applied Policy Research. In A. Bryman & R. G. Burgess (Eds.), *Analyzing Qualitative Data* (pp. 173–194). Routledge. doi:10.4324/9780203413081_chapter_9

Ritzer, G. (2011). *Globalization: The essentials*. Wiley-Blackwell.

Roberts, L., & Hall, D. (2001). *Rural tourism and recreation: principles to practice*. CABI Publishing. doi:10.1079/9780851995403.0000

Rodrigues, F., Moreira, J., & Vitorino, L. (Eds.). (2013). Comportamento do Consumidor: Quando a Neurociência, a Psicologia, a Economia e o Marketing se encontram! Viseu: Psicosoma Editora.

Rodrigues, A., Correia, A., Kozak, M., & Tuohino, A. M. (2015). Lake-destination image attributes: content analysis of text and pictures. In A. Correia, J. Gnoth, M. Kozak, & A. Fyall (Eds.), *Marketing Places and Spaces (Advances in Culture, Tourism and Hospitality Research)* (Vol. 10, pp. 293–314). Emerald Group Publishing Limited. doi:10.1108/S1871-317320150000010022

Rofe, M. W. (2003). 'I Want to be Global': Theorising the Gentrifying Class as an Emergent Élite Global Community. *Urban Studies (Edinburgh, Scotland)*, *40*(12), 2511–2526. doi:10.1080/0042098032000136183

Roldán, J., & Sánchez, M. (2012). Variance-Based Structural Equation Modeling: Guidelines for Using Partial Least Square in Information System Research. In M. Mora, O. Gelman, A. L. Steenkamp, & M. Raisinghani (Eds.), *Research Methodologies, Innovations and Philosophies in Software Systems Engineering and Information Systems* (pp. 193–221). IGI Global. doi:10.4018/978-1-4666-0179-6.ch010

Rollero, C., & Piccoli, N. (2010). Place attachment, identification, and environment perception: An empirical study. *Journal of Environmental Psychology*, *30*(2), 198–205. doi:10.1016/j.jenvp.2009.12.003

Roome, N. (1992). Developing environmental management strategies. *Business Strategy and the Environment*, *1*(1), 11–24. doi:10.1002/bse.3280010104

Rosenberg, M. (1979). *Conceiving the Self*. Basic Books.

Rose-Redwood, R., & Alderman, D. (2011). Critical interventions in political toponymy. *ACME: An International E-Journal for Critical Geographies*, *10*(1), 1–6.

Rossiter, J. R., & Percy, L. (1980). Attitude change through visual imagery in advertising. *Journal of Advertising*, *9*(2), 10–16. doi:10.1080/00913367.1980.10673313

Rossler, M. (2006). World Heritage cultural landscapes. *Landscape Research*, *31*(4), 333–353. doi:10.1080/01426390601004210

Royo, M., & Ruiz, M. (2009). Actitud del residente hacia el turismo y el visitante: Factores determinantes en el turismo y excursionismo rural-cultural. *Cuadernos de Turismo*, *23*, 214–236.

Roy, S. (2011). Competitiveness in service sector: A case of hotel industry in India. *Global Business Review*, *12*(1), 51–69. doi:10.1177/097215091001200104

Russell, J. A. (1980). A circumplex model of affect. *Journal of Personality and Social Psychology*, *39*(6), 1161–1178. doi:10.1037/h0077714

Russell, J. A. (2003). Core Affect and the Psychological Construction of Emotion. *Psychological Review*, *110*(1), 145–172. doi:10.1037/0033-295X.110.1.145 PMID:12529060

Ryan, C. (2000). Sex tourism: paradigms of confusion? In S. Clift & S. Carter (Eds.), *Tourism and Sex: Culture, Commerce and Coercion* (pp. 23–40). Cassel.

Ryan, C. (2002). Equity, management, power sharing and sustainability – issues of the 'new tourism'. *Tourism Management*, *23*(1), 17–26. doi:10.1016/S0261-5177(01)00064-4

Ryan, C., & Hall, M. (2001). *Sex Tourism: Marginal People and Liminalities*. Routledge.

Ryan, C., & Kinder, R. (1996). Sex, tourism and sex tourism: Fulfilling similar needs? *Tourism Management*, *17*(7), 507–518. doi:10.1016/S0261-5177(96)00068-4

Ryan, J., & Silvanto, S. (2010). World Heritage Sites: The Purposes and Politics of Destination Branding. *Journal of Travel & Tourism Marketing*, *27*(5), 533–545. doi:10.1080/10548408.2010.499064

Saayman, M., & Saayman, A. (2014). Appraisal of measuring economic impact of sport events. *S.A. Journal for Research in Sport Physical Education and Recreation*, *36*(3), 151–181.

Sai Kung District Council. (2011). *Yim Tin Tsai Village and St. Joseph's Chapel*. Retrieved April 20, 2018, from http://travelinsaikung.org.hk/en/Routes/view.aspx?menu=20&Nid=59

Salgado-Barandela, J., Barajas, A., & Sánchez-Fernández, P. (2017). Economic impact of sport: Topic of growing interest for the scientific literature. *Revista Internacional de Medicina y Ciencias de la Actividad Física y del Deporte*, *17*(68), 729–755.

Salmela, M. (2014). *True Emorions* (R. Ellis & P. Zachar, Eds.). John Benjamins Publishing Company.

Salo, J., & Karjaluoto, H. (2007). A conceptual model of trust in the online environment. *Online Information Review*, *31*(5), 604–621. doi:10.1108/14684520710832324

Samadhi, T. N. (2001). The urban design of a Balinese town: Placemaking issues in the Balinese urban setting. *Habitat International*, *25*(4), 559–575. doi:10.1016/S0197-3975(01)00024-8

Sameer, H., & Martin, D. (2012). Self-image congruence in consumer behavior. *Journal of Business Research*, *65*(5), 685–691. doi:10.1016/j.jbusres.2011.03.015

Sánchez-Fernández, P., & Rodríguez-Campo, M. L. (2012). Sports events as strategic tool of citymarketing: The case of basketball Galicia. *Revista Intercontinental de Gestão Desportiva*, *2*(1), 16–29.

Sanchez-Taylor, J. (2000). Tourism and "embodied" commodities: Sex tourism in the Caribbean. In S. Clift & S. Carter (Eds.), *Tourism and Sex: Culture, Commerce and Coercion* (pp. 41–53). Pinter.

Sanchez-Taylor, J. (2001). Dollars are a girl's best friend? Female tourists' sexual behaviour in the Caribbean. *Sociology*, *35*(3), 749–764. doi:10.1177/S0038038501000384

Sanchez-Taylor, J. (2006). Female sex tourism. A contradiction in terms? *Feminist Review*, *83*(1), 42–59. doi:10.1057/palgrave.fr.9400280

Santa Cruz, F., & López-Guzmán, T. (2017). Culture, Tourism and World Heritage Sites. *Tourism Management Perspectives*, *24*, 111–116. doi:10.1016/j.tmp.2017.08.004

Santana, R. (2011). *Estudio sociodemográfico de los impactos del turismo para el desarrollo local del municipio Morón* (Unpublished dissertation). University of Ciego de Ávila, Ciego de Ávila, Cuba.

Santana, A. (1997). *Antropología y turismo. ¿Nuevas hordas, viejas culturas?* Ariel.

Sanz, V. A., Moreno, F. C., & Camacho, D. P. (2012). Social impact of a major athletic event: The Formula 1 Grand Prix of Europe. *Cultura, Ciencia y Deporte*, *7*(19), 53–65. doi:10.12800/ccd.v7i19.23

São José, J. M., & Amado, C. A. F. (2016). On studying ageism in long-term care: A systematic review of the literature. *International Psychogeriatrics*, 1–15. PMID:27852342

Sarafa, J., & Brito-Henriques, E. (2017). O Hotel Santiago de Alfama Como Alegoria da Política Recente de Regeneração Urbana. *Finisterra*, *52*(106), 65–84.

Sassen, S. (2007). *A Sociology of Globalization*. W. W. Norton & Company.

Saunders, M., Lewis, P., & Thornhill, A. (2012). *Research methods for business students* (6th ed.). Pearson.

Saveriades, A. (2000). Establishing the social tourism carrying capacity for the tourist resorts of the east coast of the Republic of Cyprus. *Tourism Management*, *21*(2), 147–156. doi:10.1016/S0261-5177(99)00044-8

Schänzel, H. A., & Yeoman, I. (2015). Trends in family tourism. *Journal of Tourism Futures*, *1*(2), 141–147. doi:10.1108/JTF-12-2014-0006

Schenk, C. T., & Holman, R. H. (1980). A Sociological Approach to Brand Choice: The Concept of Situational Self Image. *Advances in Consumer Research. Association for Consumer Research (U. S.)*, *7*, 610–614.

Schenk, C. T., & Holman, R. H. (1980). A sociological approach to brand choice: The concept of situational self-image. *Advances in Consumer Research. Association for Consumer Research (U. S.)*, *7*, 610–614.

Scherer, K., & Zentner, M. (2001). Emotional effects of music: Production rules. *Music and Emotion: Theory and Research*, 361–392.

Schmitt, B. (1999). Experiential Marketing. *Journal of Marketing Management*, *15*(1–3), 53–67. doi:10.1362/026725799784870496

Schmitt, B. (2010). Experience marketing: Concepts, frameworks and Consumer Insights. *Foundations and Trends in Marketing*, *5*(2), 55–112. doi:10.1561/1700000027

Schmitt, B., Brakus, J., & Zarantonello, L. (2015). From experiential psychology to consumer experience. *Journal of Consumer Psychology*, *25*(1), 166–171. doi:10.1016/j.jcps.2014.09.001

Schneider, L.-P., & Cornwell, B. T. (2005). Cashing in on crashes via brand placement in computer games. *International Journal of Advertising*, *24*(3), 321–343. doi:10.1080/02650487.2005.11072928

Schnitzer, M., Schlemmer, P., & Kristiansen, E. (2017). Youth multi-sport events in Austria: Tourism strategy or just a coincidence? *Journal of Sport & Tourism*, *21*(3), 179–199. doi:10.1080/14775085.2017.1300102

Schroeder, K. (2015). Cultural values and sustainable tourism governance in Bhutan. *Sustainability*, *7*(12), 16616–16630. doi:10.3390/su71215837

Scott, N., Laws, E., Richins, H., & Agrusa, J. (2011). *Tourist destination governance practice, Theory and Issues*. CAB International - MPG Books Group.

Seaton, A. V., & Tagg, D. S. (1995). The family vacation in Europe: Paedonomic aspects of choices and satisfactions. *Journal of Travel & Tourism Marketing*, *4*(1), 1–21. doi:10.1300/J073v04n01_01

Sebele, L. S. (2010). Community-based tourism ventures, benefits and challenges: Khama Rhino Sanctuary Trust, Central District, Botswana. *Tourism Management*, *31*(2), 136–146. doi:10.1016/j.tourman.2009.01.005

Secretario Estado Formação Profissional e Emprego (SEFOPE) e Direção Naciona da Estatistica (DNE). (2018). *East-Timor labour force survey 2018*. Author.

Seetanah, B. (2011). Assessing the dynamic economic impact of tourism for island economies. *Annals of Tourism Research, 38*(1), 291–308. doi:10.1016/j.annals.2010.08.009

Selänniemi, T. (2003). On holiday in the liminoid playground: place, time, and self in tourism. In T. Bauer & B. McKercher (Eds.), *Sex and Tourism. Journeys of Romance, Love and Lust* (pp. 19–31). The Haworth Hospitality Press.

Sepehr, S., & Head, M. (2018). Understanding the role of competition in video gameplay satisfaction. *Information & Management, 55*(4), 407–421. doi:10.1016/j.im.2017.09.007

Seraphin, H., & Yallop, A. (2019). Proposed framework for the management of resorts Mini Clubs: An ambidextrous approach. *Leisure Studies, 38*(4), 535–547. doi:10.1080/02614367.2019.1581249

Serazio, M. (2017). Branding politics: Emotion, authenticity, and the marketing culture of American political communication. *Journal of Consumer Culture, 17*(2), 225–241. doi:10.1177/1469540515586868

Servidio, R., & Ruffolo, I. (2016). Exploring the relationship between emotions and memorable tourism experiences through narratives. *Tourism Management Perspectives, 20*, 151–160. doi:10.1016/j.tmp.2016.07.010

Sfandla, C., & Björk, P. (2012). Tourism experience network: Co-creation of experiences in interactive processes. *International Journal of Tourism Research*. Advance online publication. doi:10.1002/jtr1892

Shani, A., Chen, P.-J., Wang, Y., & Hua, N. (2010). Testing the Impact of a Promotional Video on Destination Image Change: Application of China as a Tourism Destination. *International Journal of Tourism Research, 12*, 116–133.

Shankar, R. S. (2018). Destination Personality and Destination Image: A Literature Review. *The IUP Journal of Brand Management, 15*(4).

Sharpley, R. (2000). Tourism and sustainable development: Exploring the theoretical divide'. *Journal of Sustainable Tourism, 8*(1), 1–19. doi:10.1080/09669580008667346

Sharpley, R. (2014). Host perceptions of tourism: A review of the research. *Tourism Management, 42*(June), 37–49. doi:10.1016/j.tourman.2013.10.007

Sharpley, R., & Roberts, L. (2004). Rural Tourism – 10 Years On. *International Journal of Tourism Research, 6*(3), 119–124. doi:10.1002/jtr.478

Shaw, C. (2005). *Revolutionize Your Customer Experience* (1st ed.). Palgrave Macmillan. doi:10.1057/9780230513457

Sheldon, P. J., & Abenoja, T. (2001). Resident attitudes in a mature destination: The case of Waikiki. *Tourism Management, 22*(5), 435–443. doi:10.1016/S0261-5177(01)00009-7

Sheller, M. (2011). Mobility. *Sociopedia.isa*. Doi:10.1177/205684601163

Sheller, M. (2014). The new mobilities paradigm for a live sociology. *Current Sociology Review, 62*(6), 1–23. doi:10.1177/0011392114533211

Sheller, M., & Urry, J. (2006). The New Mobilities Paradigm. *Environment & Planning A, 38*(2), 207–226. doi:10.1068/a37268

Sherry, J. L. (2004). Flow and media enjoyment. *Communication Theory, 14*(4), 328–347. doi:10.1111/j.1468-2885.2004.tb00318.x

Shields, R. (1990). The 'system of pleasure': Liminality and the carnivalesque at Brighton. *Theory, Culture & Society*, *7*(1), 39–72. doi:10.1177/026327690007001002

Shoval, N., Schvimer, Y., & Tamir, M. (2018). Real-Time Measurement of Tourists' Objective and Subjective Emotions in Time and Space. *Journal of Travel Research*, *57*(1), 3–16. doi:10.1177/0047287517691155

Shucksmith, M. (2018). Re-imagining the rural: From rural idyll to Good Countryside. *Journal of Rural Studies*, *59*, 163–172. doi:10.1016/j.jrurstud.2016.07.019

Siemens, J. C., Smith, S., & Fisher, D. (2015). Investigating the Effects of Active Control on Brand Recall Within In-Game Advertising. *Journal of Interactive Advertising*, *15*(1), 43–53. doi:10.1080/15252019.2015.1021432

Sigler, T., & Wachsmuth, D. (2016). Transnational gentrification: Globalisation and neighbourhood change in Panama's Casco Antiguo. *Urban Studies (Edinburgh, Scotland)*, *53*(4), 705–722. doi:10.1177/0042098014568070

Silicon Republic. (2011). *IrelandTown*. Retrieved October 27, 2019, from https://www.siliconrepublic.com/play/tourism-ireland-launches-facebook-game-ireland-town

Silva, J. A. (2019, Jan. 3). O turismo no Algarve. Desafios em tempos de incerteza – Parte I. *Sulinformação*.

Silva, A., & Blanchette, T. (2005). "Nossa Senhora da Help": Sexo, turismo e deslocamento transnacional em Copacabana. *Cadernos Pagu*, *25*(25), 249–280. doi:10.1590/S0104-83332005000200010

Simon, J. P. (2018). Triggering the emergence of digital ecosystems: The role of mobile and video games in emerging economies. *Digital Policy. Regulation & Governance*, *20*(5), 449–478. doi:10.1108/DPRG-03-2018-0008

Simpson, M. C. (2008). Community Benefit Tourism Initiatives - A conceptual oxymoron? *Tourism Management*, *29*(1), 1–18. doi:10.1016/j.tourman.2007.06.005

Sinclair-Maragh, G. (2017). Demographic analysis of residents' support for tourism development in Jamaica. *Journal of Destination Marketing & Management*, *6*(1), 5–12. doi:10.1016/j.jdmm.2016.03.005

Singer, J., & Pope, K. (2012). *The Power of Human Imagination: New Methods in Psychotherapy*. Springer Science & Business Media.

Singh, J., & Hart, S. (2007). Sex workers and cultural policy: Mapping the issues and actors in Thailand. *The Review of Policy Research*, *24*(2), 155–173. doi:10.1111/j.1541-1338.2007.00274.x

Sirakaya, E., Ekinci, Y., & Kaya, A. G. (2008). An examination of the validity of SUS-TAS in cross-cultures. *Journal of Travel Research*, *46*(4), 414–421. doi:10.1177/0047287507308328

Sirakaya, E., Ingram, L., & Harrill, R. (2009). Resident typologies within the integrative paradigm of sustaincentric tourism development. *Tourism Analysis*, *13*(5), 531–544. doi:10.3727/108354208788160405

Sirakaya, E., Teye, V., & Sönmez, S. (2002). Understanding residents' support for tourism development in the Central Region of Ghana. *Journal of Travel Research*, *41*(8), 57–67. doi:10.1177/004728750204100109

Sirakaya, E., & Woodside, A. G. (2005). Building and Testing Theories of Decision Making by Travellers. *Tourism Management*, *26*(6), 815–832. doi:10.1016/j.tourman.2004.05.004

Sirgy, M. J. (2010). Toward a quality-of-life theory of leisure travel satisfaction. *Journal of Travel Research*, *49*(2), 246–260. doi:10.1177/0047287509337416

Sirgy, M. J., & Su, C. (2000). Destination Image, Self-Congruity and Travel Behaviour: Toward an Integrative Model. *Journal of Travel Research*, *38*(2), 340–352. doi:10.1177/004728750003800402

Skavronskaya, L., Scott, N., Moyle, B., Le, D., Hadinejad, A., Zhang, R., Gardiner, S., Coghlan, A., & Shakeela, A. (2017). Cognitive psychology and tourism research: State of the art. *Tourism Review*, *72*(2), 221–237. doi:10.1108/TR-03-2017-0041

Skidmore, W. (1975). *Theoritical thinking in sociology*. Cambrige University Press.

Skoll, G. R., & Korstanje, M. (2014). Urban heritage, gentrification, and tourism in Reverwest and El Abasto. *Journal of Heritage Tourism*, *9*(4), 349–359. doi:10.1080/1743873X.2014.890624

Smith, M. (2003). Holistic holidays: Tourism and the reconciliation of body, mind and spirit. *Tourism Recreation Research*, *28*(1), 103–108. doi:10.1080/02508281.2003.11081392

Smith, N. (2002). New Globalism, New Urbanism: Gentrification as Global Urban Strategy. *Antipode*, *34*(3), 434–457. doi:10.1111/1467-8330.00249

Smith, N. (2010). A Short History of Gentrification. In J. Brown-Saracino (Ed.), *The Gentrification Debates* (pp. 31–36). Routledge.

Solomon, M. R. (2018). *Consumer behavior: Buying, having, and being* (12th ed.). Pearson Education Limited.

Sönmez, S., Apostolopoulos, Y., Theocharous, A., & Massengale, K. (2013). Bar crawls, foam parties, and clubbing networks: Mapping the risk environment of a Mediterranean resort. *Tourism Management Perspectives*, *8*, 49–59. doi:10.1016/j.tmp.2013.05.002

Souiden, N., Ladhari, R., & Chiadmi, N. E. (2017). Destination personality and destination image. *Journal of Hospitality and Tourism Management*, *32*, 54–70. doi:10.1016/j.jhtm.2017.04.003

Stabler, M. J. (1992). Modelling the Tourism Industry: A New Approach. In M. T. Sinclair & M. I. Stabler (Eds.), *the Tourism Industry: An International Analysis* (pp. 15–43). CAB.

Starcevic, S., & Konjikusic, S. (2018), Why Millenials as digital travelers transformed marketing strategy in tourism industry. *International Thematic Monograph Tourism in Function of Development of the Republic of Serbia - Tourism in the Era of Digital Transformation*. University of Kragu. Available at SSRN: https://ssrn.com/abstract=3280320

Stasiak, A. (2013). Tourist Product in Experience Economy. *Tourism (Zagreb)*, *23*(1), 27–35. doi:10.1016/j.tourman.2015.08.005

Statista.com. (2019). *Video game market value worldwide from 2012 to 2021*. Retrieved November 19, 2019, from https://www.statista.com/statistics/292056/video-game-market-value-worldwide/

Stebbins, R. (1997). Lifestyle as a Generic Concept in Ethnographic Research. *Quality & Quantity*, *31*(4), 347–360. doi:10.1023/A:1004285831689

Stein, A., & Ramaseshan, B. (2016). Towards the identification of customer experience touch point elements. *Journal of Retailing and Consumer Services*, *30*, 8–19. doi:10.1016/j.jretconser.2015.12.001

Sthapit, E. (2017). Exploring tourists' memorable food experiences: A study of visitors to Santa's official hometown. *Anatolia*, *28*(3), 404–421. doi:10.1080/13032917.2017.1328607

Sthapit, E., & Coudounaris, D. (2018). Memorable tourism experiences: Antecedents and outcomes. *Scandinavian Journal of Hospitality and Tourism*, *18*(1), 72–94. doi:10.1080/15022250.2017.1287003

Stins, J. F., Roelofs, K., Villan, J., Kooijman, K., Hagenaars, M. A., & Beek, P. J. (2011). Walk to me when I smile, step back when I'm angry: Emotional faces modulate whole-body approach-avoidance behaviors. *Experimental Brain Research*, *212*(4), 603–611. doi:10.100700221-011-2767-z PMID:21698468

Stone, L. (1989). Cultural Crossroads of Community Participation in Development: A Case from Nepal. *Human Organization, 48*(3), 206–213. doi:10.17730/humo.48.3.d436524749371702

Storz, C., Riboldazzi, F., & John, M. (2015). Mobility and innovation: A cross-country comparison in the video games industry. *Research Policy, 44*(1), 121–137. doi:10.1016/j.respol.2014.07.015

Stylidis, D., Biran, A., Sit, J., & Szivas, E.M. (2014). Residents' support for tourism development: The role of residents' place image and perceived tourism impacts. *Tourism Management, 45*, 260-274.

Stylidis, D. (2018). Residents' place image: A cluster analysis and its links to place attachment and support for tourism. *Journal of Sustainable Tourism, 26*(6), 1007–1026. doi:10.1080/09669582.2018.1435668

Stylidis, D., Biran, A., Sit, J., & Szivas, E. M. (2014). Residents' support for tourism development: The role of residents' place image and perceived tourism impacts. *Tourism Management, 45*, 260–274. doi:10.1016/j.tourman.2014.05.006

Su, X. (2018). Yim Tin Tsai village in Hong Kong: a model for effective conservation. *South China Morning Post.* Retrieved April 20, 2018, from https://www.scmp.com/photos/hong-kong/2161879/yim-tin-tsai-village-hong-kong-model-effective-conservation

Suess, C. (2014). *Factors Influencing the Perceived Impacts of Medical Tourism Development on Quality of Life* (Doctoral dissertation). Retrieve from UNLV Theses, Dissertations, Professional Papers, and Capstones: https://digitalscholarship.unlv.edu/thesesdissertations/2302/

Su, H. J., Huang, Y. A., Brodowsky, G., & Kim, H. J. (2011). The impact of product placement on TV-induced tourism: Korean TV dramas and Taiwanese viewers. *Tourism Management, 32*(4), 805–814. doi:10.1016/j.tourman.2010.06.020

Suh, E., Diener, E., Oishi, S., & Triandis, H. C. (1998). The shifting basis of life satisfaction judgments across cultures: Emotions versus norms. *Journal of Personality and Social Psychology, 74*(2), 482–493. doi:10.1037/0022-3514.74.2.482

Su, M. M., Wall, G., & Xu, K. (2016). Heritage tourism and livelihood sustainability of a resettled rural community: Mount Sanqingshan World Heritage Site, China. *Journal of Sustainable Tourism, 24*(5), 735–757. doi:10.1080/09669582.2015.1085868

Sun, M., Ryan, C., & Pan, S. (2015). Using Chinese Travel Blogs to Examine Perceived Destination Image: The Case of New Zealand. *Journal of Travel Research, 54*(4), 543–555. doi:10.1177/0047287514522882

Su, X., & Teo, P. (2008). Tourism politics in Lijiang, China: An analysis of state and local interactions in tourism development. *Tourism Geographies, 10*(2), 150–168. doi:10.1080/14616680801999992

Sweetser, P., & Wyeth, P. (2005). GameFlow: A model for evaluating player enjoyment in games. *Computers in Entertainment, 3*(3), 1–23. doi:10.1145/1077246.1077253

Swinyard, W. R., & Peng Sim, C. (1987). Perception of children's influence on family decision processes. *Journal of Consumer Marketing, 4*(1), 25–38. doi:10.1108/eb008186

Taghipour, A. (2017). Product Placement in Mobile Games and Thai Students Attitude Toward the Brand Recalling. *PEOPLE: International Journal of Social Sciences, 3*(2), 2279–2294. doi:10.20319/pijss.2017.32.22792294

Takahashi, N. (2013). *A Study on the Environmental Change and Conservation of the Summer Retreat Areas in Karuizawa* (in Japanese). Retrieved from: http://nc.heritage.tsukuba.ac.jp/?action=common_download_main&upload_id=578

Tan, S. K., Kung, S. F., & Luh, D. B. (2013). A Model of Creative Experience in Creative Tourism. *Annals of Tourism Research, 41*, 153–174. doi:10.1016/j.annals.2012.12.002

Tao, S.-P. (2014). Experiential marketing and marketing experience: An empirical study of the influence of Summer Boot Camp on military propensity. *Military Psychology*, *26*(5–6), 422–433. doi:10.1037/mil0000059

Tapia, S. A. (2018). Youth tourism? Traveling in the biographies of young people of popular neighborhoods of Buenos Aires. *Revista Brasileira de Pesquisa em Turismo*, *12*(3), 104–123. doi:10.7784/rbtur.v12i3.1508

Tarnowski, P., Kołodziej, M., Majkowski, A., & Rak, R. J. (2017). Emotion recognition using facial expressions. *Procedia Computer Science*, *108*, 1175–1184. doi:10.1016/j.procs.2017.05.025

Tasci, A. D. A., & Gartner, W. C. (2007). Destination Image and Its Functional Relationships. *Journal of Travel Research*, *45*(4), 413–425.

Tasci, A. D. A., Gartner, W. C., & Cavusgil, S. T. (2007). Conceptualization and operationalization of destination image. *Journal of Hospitality & Tourism Research (Washington, D.C.)*, *31*(2), 194–223. doi:10.1177/1096348006297290

Tasci, A. D. A., Gartner, W. C., & Cavusgil, S. T. (2007). Measurement of destination brand bias using a quasi-experimental design. *Tourism Management*, *28*(6), 1529–1540. doi:10.1016/j.tourman.2007.02.009

Tasci, A., & Gartner, W. (2007). Destination image and its functional relationships. *Journal of Travel Research*, *45*(4), 413–425. doi:10.1177/0047287507299569

Tatilbudur. (2020). *Tatilbudur travel agency official website*. Retrieved from https://www.tatilbudur.com/

Tatilcom. (2020). *Tatil.com travel agency official website*. https://www.tatil.com/

Tatilsepeti. (2020). *Tatilsepeti travel agency official website*. Retrieved from https://www.tatilsepeti.com/

Taylor, J. G., Gomes, R. A., Rasmussen, T., Ross, M., Rosser, A., Sandbu, M., & van Staveren, T. (2011). East-Timor human development report 2011. United Nations Development Programme.

Taylor, K., & Lennon, J. L. (2011). Cultural landscapes: A bridge between culture and nature? *International Journal of Heritage Studies*, *17*(6), 537–554. doi:10.1080/13527258.2011.618246

Taylor, S. J., Bogdan, R., & DeVault, M. (2015). *Introduction to qualitative research methods: A guidebook and resource*. John Wiley & Sons.

Terlutter, R., & Capella, M. L. (2013). The Gamification of Advertising: Analysis and Research Directions of In-Game Advertising, Advergames, and Advertising in Social Network Games. *Journal of Advertising*, *42*(2–3), 95–112. doi:10.1080/00913367.2013.774610

Tessitore, T., & Geuens, M. (2013). PP for "product placement" or "puzzled public"? The effectiveness of symbols as warnings of product placement and the moderating role of brand recall. *International Journal of Advertising*, *32*(3), 419–442. doi:10.2501/IJA-32-3-419-442

Tessitore, T., Pandelaere, M., & Van Kerckhove, A. (2014). The Amazing Race to India: Prominence in reality television affects destination image and travel intentions. *Tourism Management*, *42*, 3–12. doi:10.1016/j.tourman.2013.10.001

Teye, V., Sirakaya, E., & Sönmez, S. (2002). Resident's attitudes toward tourism development. *Annals of Tourism Research*, *29*(3), 668–688. doi:10.1016/S0160-7383(01)00074-3

The Advertising Self-Regulatory Board (RÖK). (2010). Retrieved from https://www.rok.org.tr/en/rok-platform/

The cultural program 365 Algarve. (2016). Cultural Program.

The United Nations Environment Programme and World Tourism Organization. (2005). *Making Tourism More Sustainable*. Retrieved from: http://www.unep.fr/shared/publications/pdf/dtix0592xpa-tourismpolicyen.pdf

The United Nations World Tourism Organisation. (2017). *UNWTO Tourism Highlights 2017 Edition*. Retrieved from: https://www.e-unwto.org/doi/pdf/10.18111/9789284419029

The United Nations World Tourism Organisation. (2018). *UNWTO World Tourism Barometer*. Retrieved from: https://cf.cdn.unwto.org/sites/all/files/pdf/unwto_barom18_01_january_excerpt_hr.pdf

The World Commission on Environment and Development. (1987). *Our Common Future*. Retrieved from: http://www.un-documents.net/our-common-future.pdf

Thomas, M. (2005). 'What happens in Tenerife stays in Tenerife: Understanding women's sexual behaviour on holiday. *Culture, Health & Sexuality*, *7*(6), 571–584. doi:10.1080/13691050500256807 PMID:16864223

Thomas, N. J., Harvey, D. C., & Hawkins, H. (2013). Crafting the Region: Creative Industries and Practices of Regional Space. *Regional Studies*, *47*(1), 75–88. doi:10.1080/00343404.2012.709931

Thompson, A., Martin, A., Gee, S., & Geurin, A. (2016). Fans' perceptions of professional tennis events' social media presence. *Communication & Sport*, *5*(5), 579–603. doi:10.1177/2167479516650442

Thomson, A., Schlenker, K., & Schulenkorf, N. (2013). Conceptualizing sport event legacy. *Event Management*, *17*(2), 111–122. doi:10.3727/152599513X13668224082260

Thomson, M., MacInnis, D. J., & Park, C. W. (2005). The ties that bind: Measuring the strength of consumers' emotional attachments to brands. *Journal of Consumer Psychology*, *15*(1), 77–91. doi:10.120715327663jcp1501_10

Thornton, P. R., Shaw, G., & Williams, A. M. (1997). Tourist group holiday decision-making and behavior: The influence of children. *Tourism Management*, *18*(5), 287–297. doi:10.1016/S0261-5177(97)00017-4

Thurlow, C., & Jaworski, A. (2014). 'Two hundred ninety-four': Remediation and multimodal performance in tourist placemaking. *Journal of Sociolinguistics*, *18*(4), 459–494. doi:10.1111/josl.12090

Timothy, D. J. (2011). *Cultural Heritage and tourism: An Introduction*. Chanel View Publications.

Tinsley, R., & Lynch, P. (2001). Small tourism business networks and destination development. *International Journal of Hospitality Management*, *20*(4), 367–378. doi:10.1016/S0278-4319(01)00024-X

Tissot, S. (2015). *Good Neighbors: Gentrifying Diversity in Boston's South End*. Verso.

Tiwsakul, R., Hackley, C., & Szmigin, I. (2005). Explicit, non-integrated product placement in British television programmes. *International Journal of Advertising*, *24*(1), 95–111. doi:10.1080/02650487.2005.11072906

Todd, R., Talmi, D., Schmitz, T., Susskind, J., & Anderson, A. (2012). Psychophysical and neural evidence for emotion enhanced perceptual vividness. *The Journal of Neuroscience: The Official Journal of the Society for Neuroscience*, *32*(33), 11201–11212. doi:10.1523/JNEUROSCI.0155-12.2012 PMID:22895705

Tolman, D., Hirschman, C., & Impett, E. (2005). There is more to the story: The pace of qualitative research on female adolescent sexuality in policy making. *Sexuality Research & Social Policy*, *2*(4), 4–17. doi:10.1525rsp.2005.2.4.4

Torkington, K. (2012). Place and Lifestyle Migration: The Discursive Construction of 'Glocal' Place-Identity. *Mobilities*, *7*(1), 71–92. doi:10.1080/17450101.2012.631812

Tosun, C. (1999). Towards a Typology of Community Participation in the Tourism Development Process. *Anatolia*, *10*(2), 113–134. doi:10.1080/13032917.1999.9686975

Tosun, C. (2000). Limits to community participation in the tourism development process in developing countries. *Tourism Management*, *21*(6), 613–633. doi:10.1016/S0261-5177(00)00009-1

Tosun, C. (2002). Host perceptions of impacts: A comparative tourism study. *Annals of Tourism Research, 29*(1), 231–253. doi:10.1016/S0160-7383(01)00039-1

Tourism Authority of Thailand. (2011). *Smile Land.* Retrieved October 27, 2019, from http://www.smilelandgame.com

Tourstica. (2020). *Touristica travel agency official website.* https://www.touristica.com.tr/

Tovar, C., & Lockwood, M. (2008). Social impacts of tourism: An Australian regional case study. *International Journal of Tourism Research, 10*(4), 365–378. doi:10.1002/jtr.667

TP. (2017). *Estratégia Turismo 2027.* https://estrategia.turismodeportugal.pt/

TP/Governo de Portugal. (2015). *Turismo 2020: Plano de Ação para o Desenvolvimento do Turismo em Portugal.* http://www.turismodeportugal.pt/Portugu%C3%AAs/turismodeportugal/apresentacoes/Documents/Turismo-2020-Sintra.pdf

TP/UALG. (2017). *Perfil do turista que visita o Algarve (2016).* https://issuu.com/turismo_algarve/docs/perfil_do_turista_2016_relatorio_fi

Trauer, B., & Ryan, C. (2005). Destination image, romance and place – an application of intimacy theory in tourism. *Tourism Management, 26*(4), 481–491. doi:10.1016/j.tourman.2004.02.014

Trudeau, D. (2006). Politics of belonging in the construction of landscapes: Place-making, boundary-drawing and exclusion. *Cultural Geographies, 13*(3), 421–443. doi:10.1191/1474474006eu366oa

Truong, T. (1990). *Sex, Money and Morality: Prostitution and Tourism in Southeast Asia.* Zed Books.

Tsai, C.-T. (2016). Memorable Tourist Experiences and Place Attachment When Consuming Local Food. *International Journal of Tourism Research, 18*(6), 536–548. doi:10.1002/jtr.2070

Tsaur, S.-H., Chiu, Y.-T., & Wang, C.-H. (2007). The Visitors Behavioral Consequences of Experiential Marketing. *Journal of Travel & Tourism Marketing, 21*(1), 47–64. doi:10.1300/J073v21n01_04

Tsutsui, K., Sakuma, Y., & Kasami, K. (2015). Regional Regeneration by Business in Rural Community of In-migrants from Urban Areas' [in Japanese]. *Journal of Association of Rural Planning, 34*(1), 45–50. doi:10.2750/arp.34.45

Tung, V. W. S., Lin, P., Qiu Zhang, H., & Zhao, A. (2017). A framework of memory management and tourism experiences. *Journal of Travel & Tourism Marketing, 34*(7), 853–866. doi:10.1080/10548408.2016.1260521

Tung, V. W. S., & Ritchie, J. R. B. R. B. (2011). Exploring the essence of memorable tourism experiences. *Annals of Tourism Research, 38*(4), 1367–1386. doi:10.1016/j.annals.2011.03.009

Tuohino, A., & Pitkänen, K. (2004). The transformation of a neutral lake landscape into a meaningful experience-interpreting tourist photos. *Journal of Tourism and Cultural Change, 2*(2), 77–93. doi:10.1080/14766820408668170

Turismo de Portugal. (2017). *Relatório de Sustentabilidade.* Retrieved from https://travelbi.turismodeportugal.pt/pt-pt/Documents/Sustentabilidade/Estrat%C3%A9gia%20de%20Sustentabilidade/relatorio-sustentabilidade-2017.pdf

Turkish Statistical Institute (TUIK). (2018a). *Marriage and divorce statistics.* Retrieved from http://www.tuik.gov.tr/PdfGetir.do?id=27593

Turkish Statistical Institute (TUIK). (2018b). *Statistics on children.* Retrieved from http://www.tuik.gov.tr/PreHaberBultenleri.do?id=27596

Turkish Statistical Institute (TUIK). (2018c). *Basic statistics.* Retrieved from http://www.tuik.gov.tr/PreIstatistikTablo.do?istab_id=1588

Turner, V. (1974a). *Dramas, Fields & Metaphors. Symbolic Action in Human Societies*. Cornell University Press.

Turner, V. (1974b). *O Processo Ritual. Estrutura e Anti-Estrutura*. Vozes.

Tussyadiah, I. (2014). Toward a Theoretical Foundation for Experience Design in Tourism. *Journal of Travel Research, 53*(5), 543–564. doi:10.1177/0047287513513172

Tussyadiah, I. P., & Fesenmaier, D. R. (2009). Mediating Tourist Experiences. *Annals of Tourism Research, 36*(1), 24–40. doi:10.1016/j.annals.2008.10.001

Twining-Ward, L., & Butler, R. (2002). Implementing STD on a small island: Development and use of sustainable tourism development indicators in Samoa. *Journal of Sustainable Tourism, 10*(5), 363–387. doi:10.1080/09669580208667174

Tynan, C., & McKechnie, S. (2009). Experience marketing: A review and reassessment. *Journal of Marketing Management, 25*(5–6), 501–517. doi:10.1362/026725709X461821

Uchida, Y. (1989). 'Regarding the establishment of images of 'luxurious summer resort/ cottage destination' in Karuizawa' (in Japanese), *A review of Geography. Ser. A, 62*(7), 495–512.

Um, S., & Crompton, J. L. (1987). Measuring resident's attachment levels in a host community. *Journal of Travel Research, 26*(1), 27–29. doi:10.1177/004728758702600105

UNESCO Creative Network. (2006). *Towards Sustainable Strategies for Creative Tourism*. Discussion Report of the Planning Meeting for 2008 International Conference on Creative Tourism, Santa Fe, NM. Retrieved April 20, 2019, from https://unesdoc.unesco.org/ark:/48223/pf0000159811

UNESCO. (2019). *World Heritage List*. https://whc.unesco.org/en/list/

UNESCO. (2019a). *Browse the Lists of Intangible Cultural Heritage and the Register of good safeguarding practices*. UNESCO.

UNESCO/ICCROM/ICOMOS/IUCN. (2016). *Gestão do patrimônio mundial cultural* [International Council on Monuments and Sites, International Union for Conservation of Nature, International Centre for the Study of the Preservation and Restoration of Cultural Property]. https://unesdoc.unesco.org/ark:/48223/pf0000244283

Unifieldlawyers. (2017). *The love of divorce*. Retrieved from https://www.unifiedlawyers.com.au/blog/global-divorce-rates-statistics/

United Nations Educational, Scientific and Cultural Organisation. (2006). *Towards Sustainable Strategies for Creative Tourism. Discussion Report of the Planning Meeting for 2008 International Conference on Creative Tourism, Santa Fe, New Mexico, U.S.A., October 25–27, 2006*. UNESCO Creative Cities Network. Retrieved from: https://unesdoc.unesco.org/ark:/48223/pf0000159811

United Nations. (2020). *Crude Divorce Rate*. Retrieved from https://www.un.org/en/development/desa/population/publications/dataset/marriage/crude-divorces.shtml

UNWTO (United Nations World Tourism Organization), Government of East-Timor, & UNDP (United Nations Development Programme). (2007). Sustainable tourism sector development and institutional strengthening project Volume 1 - Main report. Authors.

UNWTO. (2018). *Tourism Highlights 2018 Edition International Tourism Trends 2017*. UNWTO.

UNWTO. (2019). *Baseline Report on the Integration of Sustainable Consumption and Production Patterns into Tourism Policies*. https://www.e-unwto.org/doi/book/10.18111/9789284420605

UNWTO. (2019). *International Tourism Highlights 2019 Edition*. https://www.e-unwto.org/doi/pdf/10.18111/9789284421152

UNWTO. (2019). *International Tourism Highlights International tourism continues to outpace the global economy*. UNWTO.

UNWTO/UN. (2017). *Tourism and the Sustainable Development Goals – Journey to 2030 – Highlights*. https://www.e-unwto.org/doi/book/10.18111/9789284419340

UNWTO/UNESCO. (2019). *4th World Conference on Tourism and Culture: Investing in future generations*. https://www.unwto.org/4th-unwtounesco-world-conference-tourism-and-culture-investing-future-generations

Urry, J. (2002). Mobility and Proximity. *Sociology*, *36*(2), 255–274. doi:10.1177/0038038502036002002

Uysal, M., & Sirgy, M. J. (2019). Quality-of-life indicators as performance measures. *Annals of Tourism Research*, *76*, 291–300. doi:10.1016/j.annals.2018.12.016

Valle, P. O. D., Silva, J. a., Mendes, J., & Guerreiro, M. (2006). Tourist satisfaction and destination loyalty intention : A structural and categorical analysis. *International Journal of Business Science and Applied Management*, *1*(1), 25–44. https://business-and-management.org/library/2006/1_1--25-44--Oom_do_Valle,Silva,Mendes,Guerreiro.pdf

Valle, P., Guerreiro, M. M., Mendes, J., & Silva, J. (2011). Tourist satisfaction and destination loyalty intention: A structural and categorical analysis. *International Journal of Business Science and Applied Management*, 25–44.

Van Gennep, A. (1960). *The Rites of Passage. A Classical Study of Cultural Celebrations*. University of Chicago Press. (Original work published 1909)

Van Wart, M. (2003). Public-sector leadership theory: An assessment. *Public Administration Review*, *63*(2), 214–228. doi:10.1111/1540-6210.00281

Vargas, A., Porras, N., Plaza, M. A., & Riquel, F. (2008). Tourism and wine-marking industry. Perspective and proposals for development in the Spanish province of Huelva. In *Proceedings of the lASK International Conference, Advance in Tourism Research* (pp. 33-43). Aveiro: IASK.

Vargas, A., Plaza, M. A., & Porras, N. (2007). Desarrollo del turismo y percepción de la comunidad local: Factores determinantes de su actitud hacia un mayor desarrollo turístico. In *Proceedings of the XXI National Conference of AEDEM* (vol. 1, pp. 24-34). Madrid: AEDEM.

Vargas, A., Plaza, M. A., & Porras, N. (2009). Understanding Residents' Attitudes toward the Development of Industrial Tourism in a Former Mining Community. *Journal of Travel Research*, *47*(3), 373–387. doi:10.1177/0047287508322783

Vargas, A., Plaza, M. A., & Porras, N. (2011). Explaining resident's attitudes to tourism. Is it a universal model possible? *Annals of Tourism Research*, *38*(2), 460–480. doi:10.1016/j.annals.2010.10.004

Vargas, A., & Riquel, F. (2012). Influence of the institutional context on the performance of golf courses, considering the natural environment. *Environmental Engineering and Management Journal*, *11*(11), 2001–2012. doi:10.30638/eemj.2012.249

Vargas, A., Valle, P., Mendes, J. C., & Silva, J. A. (2015). Residents' attitude and level of destination development: An international comparison. *Tourism Management*, *48*(June), 199–210. doi:10.1016/j.tourman.2014.11.005

Vargas-Sánchez, A., Porras-Bueno, N., & Plaza-Mejía, M. de L. Á. (2009). Residents' attitudes towards tourism development in the Spanish province of Huelva. Narodnostopanski Arhiv, International edition, University of Huelva, Spain, 31-63.

Vargas-Sánchez, A., Porras-Bueno, N., & de los Ángeles Plaza-Mejía, M. (2014). Residents' attitude to tourism and seasonality. *Journal of Travel Research*, *53*(5), 581–596. doi:10.1177/0047287513506295

Vargo, S. L., & Lusch, R. F. (2004). Evolving to a new dominant logic for marketing. *Journal of Marketing*, *68*(1), 1–17. doi:10.1509/jmkg.68.1.1.24036

Vargo, S. L., & Lusch, R. F. (2006). Service-dominant logic: Reactions, reflections and refinements. *Marketing Theory*, *68*(3), 281–288.

Vargo, S. L., & Lusch, R. F. (2008). Service-dominant logic: Continuing the evolution. *Journal of the Academy of Marketing Science*, *36*(1), 1–10. doi:10.100711747-007-0069-6

Vargo, S. L., & Lusch, R. F. (2016). Institutions and axioms: An extension and update of service-dominant logic. *Journal of the Academy of Marketing Science*, *44*(1), 5–23. doi:10.100711747-015-0456-3

Vargo, S. L., & Lusch, R. F. (2017). Service-dominant logic 2025. *International Journal of Research in Marketing*, *34*(1), 46–67. Advance online publication. doi:10.1016/j.ijresmar.2016.11.001

Vargo, S. L., Maglio, P. P., & Akaka, M. A. (2008). On value and value co-creation: A service system and service logic perspective. *European Management Journal*, *26*(3), 145–152.

Var, T., Kendall, K. W., & Tarakcioglu, E. (1985). Resident attitudes towards tourists in a Turkish resort town. *Annals of Tourism Research*, *12*(4), 652–658. doi:10.1016/0160-7383(85)90086-6

Vashisht, D. (2019). Effect of interactivity and congruence on brand advocacy and brand acceptance. *Arts and the Market,* (ahead-of-print).

Vashisht, D., & Royne, M. B. (2016). Advergame speed influence and brand recall: The moderating effects of brand placement strength and gamers' persuasion knowledge. *Computers in Human Behavior*, *63*, 162–169. doi:10.1016/j.chb.2016.05.022

Vastag, G., Kerekes, S., & Rondinelli, D. (1996). Evaluation of Corporate Environmental Management Approaches: A Framework and Application. *International Journal of Production Economics*, *43*(2-3), 193–211. doi:10.1016/0925-5273(96)00040-0

Veal, A. J. (2011). *Research Methods for Leisure and Tourism* (4th ed.). London: Financial Times/Prentice Hall.

Veal, A. J. (1993). The concept of lifestyle: A review. *Leisure Studies*, *12*(4), 233–252. doi:10.1080/02614369300390231

Vela, M., & Molina, M. (2009). Actitud del residente hacia el turismo y el visitante: Factores determinantes en el turismo y excursionismo Rural-Cultural. *Cuadernos de Turismo*, *23*, 217–236.

Verberckmoes, S., Poels, K., Dens, N., Herrewijn, L., & De Pelsmacker, P. (2016). When and why is perceived congruity important for in-game advertising in fantasy games? *Computers in Human Behavior*, *64*, S29–S38. doi:10.1016/j.chb.2016.07.062

Verhoef, P., Lemon, K., Parasuraman, A., Roggeveen, A., Tsiros, M., & Schlesinger, L. (2009). Customer Experience Creation: Determinants, Dynamics and Management Strategies. *Journal of Retailing*, *85*(1), 31–41. doi:10.1016/j.jretai.2008.11.001

Vicente, J., Molina, V., & Ruiz, I. (2008). *La Gestión Medioambiental en las Empresas de Economía Social.* Paper presented at VII Colóquio Ibérico de Estudos Rurais, Coimbra, Portugal.

Viegas, M. (2011). Good Environmental Practices in the Hospitality Sector. *Journal of Tourism and Sustainability*, *I*(2), 5–13. Retrieved December 18, 2019, from https://sapientia.ualg.pt/bitstream/10400.1/13382/3/goodpractices.pdf

Villani, D., Carissoli, C., Triberti, S., Marchetti, A., Gilli, G., & Riva, G. (2018). Video games for Emotion Regulation: A Systematic Review. *Games for Health Journal*, *7*(2), 85–99. doi:10.1089/g4h.2017.0108 PMID:29424555

Vincent, V., & Thompson, W. (2002). Assessing Community Support and Sustainability for Ecotourism Development. *Journal of Travel Research, 41*(2), 153–160. doi:10.1177/004728702237415

Visit Korea. (2019). *100 Must-Visit Tourist spots in Korea.* Retrieved from http://english.visitkorea.or.kr/enu/TRV/TV_ENG_Must100.jsp#

Vittersø, J., Vorkinn, M., Vistad, O. I., & Vaagland, J. (2000). Tourist experiences and attractions. *Annals of Tourism Research, 27*(2), 432–450. doi:10.1016/S0160-7383(99)00087-0

Vo Thanh, T., & Kirova, V. (2018). Wine tourism experience: A netnography study. *Journal of Business Research, 83*, 30–37. doi:10.1016/j.jbusres.2017.10.008

Volo, S. (2009). Experience : A Tourist Based Approach Conceptualizing Experience : A Tourist. *Journal of Hospitality Marketing & Management, 18*(April), 111–126. doi:10.1080/19368620802590134

Volo, S. (2010). Bloggers' reported tourist experiences: Their utility as a tourism data source and their effect on prospective tourists. *Journal of Vacation Marketing, 16*(4), 297–311. doi:10.1177/1356766710380884

Vong, M. (2014). *Tourism in Timor-Leste: Contributions to a sustainable development* (Unpublished doctoral dissertation). University of Algarve.

Vong, M., & do Valle, P. O. (2013). A opinião das empresas turísticas sobre o turismo em East-Timor. In J. A. Silva & F. Anjos (Eds.), Revista internacional em língua portuguesa - Turismo (III Série, pp. 215-227). Associação das Universidades de Língua Portuguesa, Lisboa: Clássica, Artes Gráficas S.A.

Vong, M., do Valle, P. O., & Silva, J. A. (2014). Turismo em Timor-Leste: Presente e futuro. In C. Costa, F. Brandão, R. Costa, & Z. Breda (Eds.), *Turismo nos países lusófonos: Conhecimento, estratégia e territótios* (Vol. I, pp. 399–410). Escolar Editora.

Vong, M., Pinto, P., & Silva, J. A. (2017). Diaspora tourism: The case of East-Timor. *Tourism (Zagreb), 65*(2), 218–233.

Vukic, M., Kuzmanovic, M., & Stankovic, M. K. (2014). Understanding the Heterogeneity of Generation Y's Preferences for Travelling: A Conjoint Analysis Approach. *International Journal of Tourism Research.* Advance online publication. doi:10.1002/jtr.2015

Wagner, U. (1977). Out of time and place: Mass tourism and charter trips. *Ethnos, 42*(1-2), 38–52. doi:10.1080/00141844.1977.9981130

Wallerstein, I. (1999). Análise dos Sistemas Mundiais. In A. Giddens & J. Turner (Eds.), *Teoria Social Hoje* (pp. 447–470). UNESP.

Walters, G., Sparks, B., & Herington, C. (2007). The effectiveness of print advertising stimuli in evoking elaborate consumption visions for potential travelers. *Journal of Travel Research, 46*(1), 24–34. doi:10.1177/0047287507302376

Wan, C. T. (2004). *From salt pan to resort plan: heritage conservation for the Island of Yim Tin Tsai, Sai Kung* (Unpublished master dissertation). The University of Hong Kong, Hong Kong.

Wang, L., Tang, H., Liu, D., & Xing, C. (2017). Study on Customer Experience Dimension Based on Online Shopping reviews of MI ' s Products. In *MATEC Web of Conferences* (Vol. 100, pp. 1–7). Academic Press.

Wang, C. T. L., & Schofer, E. (2018). Coming out of the penumbras: World culture and cross-national variation in divorce rates. *Social Forces, 97*(2), 675–704. doi:10.1093foy070

Wang, N. (1999). Rethinking authenticity in tourism experience. *Annals of Tourism Research, 26*(2), 349–370. doi:10.1016/S0160-7383(98)00103-0

Wang, W. (2019). The effect of early-life outdoor experiences on residents' attitudes towards sustainable tourism within an urban context. *Journal of Outdoor Recreation and Tourism*, *25*, 1–9. doi:10.1016/j.jort.2018.10.002

Wang, Y. A., & Pfister, R. E. (2008). Residents' attitudes toward tourism and perceived personal benefits in a rural community. *Journal of Travel Research*, *47*(1), 84–93.

Wang, Z., Mao, H., Li, Y. J., & Liu, F. (2017). Smile big or not? Effects of smile intensity on perceptions of warmth and competence. *The Journal of Consumer Research*, *43*(5), 787–805.

Ward, C., & Bernom, T. (2011). Beyond social exchange theory: Attitudes toward tourists. *Annals of Tourism Research*, *38*(4), 1556–1569. doi:10.1016/j.annals.2011.02.005

Ward, S., & Wackman, D. B. (1972). Children's purchase influence attempts and parental yielding. *Journal of Marketing Research*, *9*(3), 316–319. doi:10.1177/002224377200900312

Weaver, A. (2007). Product Placement and Tourism-Oriented Environments: An Exploratory Introduction. *International Journal of Tourism Research*, *284*(4), 275–284. doi:10.1002/jtr.612

Weber, M. (1963). *Economy and Society: An outline of interpretive Sociology*. University of California Press.

Weed, M., & Bull, C. (2009). *Sports tourism: participants, policy and providers* (2nd ed.). Elsevier.

Weis, D. (2002). The need to integrate sexual theory and research. In M. Wiederman & B. Whitley (Eds.), *Handbook for Conducting Research on Human Sexuality* (pp. 7–24). Lawrence Erlbaum Associates.

Weitman, S. (1998). On the elementary forms of the socioerotic life. *Theory, Culture & Society*, *15*(3-4), 71–110. doi:10.1177/0263276498015003005

Wild, B., Erb, M., & Bartels, M. (2001). Are emotions contagious? Evoked emotions while viewing emotionally expressive faces: Quality, quantity, time course and gender differences. *Psychiatry Research*, *102*(2), 109–124. doi:10.1016/S0165-1781(01)00225-6 PMID:11408051

Williams, D. A. H., Cowling, P. I., & Murphy, D. T. (2019). Emotional Congruence in Video Game Audio. Encyclopedia of Computer Graphics and Games, 1–3.

Williams, M. (2018). *Hong Kong's only working salt pans and a Robinson Crusoe moment just a boat ride away from Sai Kung*. Retrieved September 1, 2019, from https://www.scmp.com/lifestyle/travel-leisure/article/2130134/hong-kongs-only-working-salt-pans-and-robinson-crusoe

Williams, A. M., & Patterson, G. (1998). 'An Empire Lost but a Province Gained': A Cohort Analysis of British International Retirement in the Algarve. *International Journal of Population Geography*, *4*(2), 135–155. doi:10.1002/(SICI)1099-1220(199806)4:2<135::AID-IJPG99>3.0.CO;2-6 PMID:12348626

Williams, J., & Chinn, S. J. (2010). Meeting relationship marketing goals through social media: A conceptual model for sport marketers. *International Journal of Sport Communication*, *3*(4), 422–437. doi:10.1123/ijsc.3.4.422

Wilson, E. (1975). *Sociobiology: The New Synthesis*. Harvard University Press.

Wilson, E. (1978). *On Human Nature*. Harvard University Press.

Wolak, R., Kalafatis, S., & Harris, P. (1998). An investigation into four characteristics of services. *Journal of Empirical Generalisations in Marketing Science*, *3*, 22–41.

Woo, E., Kim, H., & Uysal, M. (2015). Life satisfaction and support for tourism development. *Annals of Tourism Research*, *50*, 84–97. doi:10.1016/j.annals.2014.11.001

Woosnam, K. M. (2011). Testing a model of Durkheim's theory of emotional solidarity among residents of a tourism community. *Journal of Travel Research, 50*(5), 546–558. doi:10.1177/0047287510379163

Woosnam, K. M. (2012). Using emotional solidarity to explain residents' attitudes about tourism and tourism development. *Journal of Travel Research, 51*(3), 315–327. doi:10.1177/0047287511410351

Woosnam, K. M., Aleshinloye, K. D., & Maruyama, N. (2016). Solidarity at the Osun Osogbo Sacred Grove—A UNESCO world heritage site. *Tourism Planning & Development, 13*(3), 274–291. doi:10.1080/21568316.2015.1104380

Woosnam, K. M., Aleshinloye, K. D., Ribeiro, M. A., Stylidis, D., Jiang, J., & Erul, E. (2018). Social determinants of place attachment at a World Heritage Site. *Tourism Management, 67*, 139–146. doi:10.1016/j.tourman.2018.01.012

Woosnam, K. M., Aleshinloye, K. D., Van Winkle, C. M., & Qian, W. (2014). Applying and expanding the theoretical framework of emotional solidarity in a festival context. *Event Management, 18*(2), 141–151. doi:10.3727/15259951 4X13947236947428

Woosnam, K. M., Maruyama, N., Boley, B. B., & Erul, E. (2018). Stereotypes and perceived solidarity in ethnic enclave tourism. *Journal of Tourism and Cultural Change, 16*(2), 138–154. doi:10.1080/14766825.2016.1242596

Woosnam, K. M., & Norman, W. C. (2010). Measuring residents' emotional solidarity with tourists: Scale development of Durkheim's theoretical constructs. *Journal of Travel Research, 49*(3), 365–380. doi:10.1177/0047287509346858

Woosnam, K. M., Norman, W. C., & Ying, T. (2009). Exploring the theoretical framework of emotional solidarity between residents and tourists. *Journal of Travel Research, 48*(2), 245–258. doi:10.1177/0047287509332334

Woosnam, K., Maruyama, N. U., Ribeiro, M. A., & Joo, D. (2019). Explaining minority residents' attitudes of ethnic enclave tourism from general perceptions of tourism impacts. *Journal of Tourism and Cultural Change, 17*(4), 467–484. doi:10.1080/14766825.2019.1601407

World Golf Awards. (2018). Retrieved from: https://worldgolfawards.com/winners/2018

World Tourism Organisation and World Youth Student & Educational Travel Confederation (UNWTO and WYSE Travel Confederation). (2008). *Youth travel matters: understanding the global phenomenon of youth travel.* WTO.

World Tourism Organization and World Youth Student & Educational Travel Confederation (UNWTO and WYSE Travel Confederation). (2011). *The power of youth travel.* WTO.

World Travel & Tourism Council. (1997). *Agenda 21 for the travel and tourism industry: towards environmentally sustainable development.* World Travel & Tourism Council.

World Travel Awards. (2017). Retrieved from: https://www.worldtravelawards.com/award-europes-leading-beach-destination

Wortham-Galvin, B. D. (2008). Mythologies of placemaking. *Places, 20*(1), 32–39.

WTA. (2018). *Site World Travel Awards.* https://www.worldtravelawards.com/

WTO (World Tourism Organization). (2004). Indicators of sustainable development for tourism destinations: A guidebook. Madrid, Spain: World Tourism Organization.

WTO (World Tourism Organization). (2009). Handbook on tourism destinations branding with an introduction by Simon Anholt. Madrid, Spain: The World Tourism Organization and the European Travel Commission.

WTTC. (2019). *Travel & Tourism Economic Impact 2019 – World.* https://www.wttc.org/-/media/files/reports/economic-impact-research/regions-2019/world2019.pdf

WTTC. (2019a). *Travel & Tourism Economic Impact 2019 – Portugal.* https://www.wttc.org/-/media/files/reports/economic-impact-research

Wyckoff, M. A., Neumann, B., Pape, G., & Schindler, K. (2015). *Placemaking as and economic development tool: A placemaking guide.* Land Policy Institute, Michigan State University.

Xiong, J., Hashim, N. H., & Murphy, J. (2015). Multisensory image as a component of destination image. *Tourism Management Perspectives, 14*, 34–41. doi:10.1016/j.tmp.2015.03.001

Xu, F., Tian, F., Buhalis, D., & Weber, J. (2013). Marketing tourism via electronic games: Understanding the motivation of tourist players. *5th International Conference on Games and Virtual Worlds for Serious Applications, VS-GAMES 2013.* 10.1109/VS-GAMES.2013.6624235

Xu, J., Broekens, J., Hindriks, K., & Neerincx, M. A. (2015). Mood contagion of robot body language in human robot interaction. *Autonomous Agents and Multi-Agent Systems, 29*(6), 1216–1248. doi:10.100710458-015-9307-3

Yamane, T. (1967). *Statistics: An introductory analysis* (2nd ed.). Harper and Row.

Yang, C. H., Lin, H.-L., & Han, C. C. (2010). Analysis of international tourists arrivals in China: The role of world heritage sites. *Tourism Management, 316*(6), 827–837. doi:10.1016/j.tourman.2009.08.008 PMID:32287732

Yang, M., Roskos-Ewoldsen, D., Dinu, L., & Arpan, L. (2006). The Effectiveness of "in-Game" Advertising: Comparing College Students' Explicit and Implicit Memory for Brand Names. *Journal of Advertising, 35*(4), 143–152. doi:10.2753/JOA0091-3367350410

Yan, L., Xu, J., & Zhou, Y. (2018). 'Residents' attitudes toward prostitution in Macau'. *Journal of Sustainable Tourism, 26*(2), 205–220. doi:10.1080/09669582.2017.1338293

Yau, E. (2016). Abandoned Hong Kong island gets new life as heritage site and ecotourism destination. *South China Morning Post.* Retrieved September 1, 2019, from https://www.scmp.com/lifestyle/travel-leisure/article/1908076/abandoned-island-sai-kung-comes-back-life-heritage-site-and

Ye, Y., Tussyadiah, I. P., & Fesenmaier, D. R. (2009). *Phenomenological Capture of Tourist Experience.* Paper presented at Tourism Experiences: Meanings, Motivations, Behaviours Conference, Preston, UK.

Yıldırım, A., & Şimşek, H. (2013). Qualitative research methods in social sciences (9th ed.). Seçkin Publication. [in Turkish].

Yin, R. (2014). *Case study research.* Sage Publications.

Yoo, C., & MacInnis, D. (2005). The brand attitude formation process of emotional and informational ads. *Journal of Business Research, 58*(10), 1397–1406. doi:10.1016/j.jbusres.2005.03.011

Yoon, Y., Gursoy, D., & Chen, J. S. (2001). Validating a tourism development theory with structural equation modeling. *Tourism Management, 22*(4), 363–372. doi:10.1016/S0261-5177(00)00062-5

Yoon, Y., Kim, S., & Kim, S. (2015). Successful and unsuccessful film tourism destinations: From the perspective of Korean local residents, Perceptions of film tourism impacts. *Tourism Analysis, 20*(3), 297–311. doi:10.3727/108354215X14356694891852

Young, B. (1983). Touristization of traditional Maltese fishing-farming villages: A general model. *Tourism Management, 4*(1), 35–41. doi:10.1016/0261-5177(83)90048-1

Yousaf, A., Amin, I., & Santos, J. A. C. (2018). Tourists' motivations to travel: A theoretical perspective on the existing literature. *Tourism and Hospitality Management, 24*(1), 197–211. doi:10.20867/thm.24.1.8

Yu, C., Chancellor, H., & Cole, S. (2011). Measuring Residents' Attitudes toward Sustainable Tourism: A Reexamination of the Sustainable Tourism Attitude Scale. *Journal of Travel Research*, *50*(1), 57–63. doi:10.1177/0047287509353189

Yu, C.-P., Charles Chancellor, H., & Tian Cole, S. (2011). Examining the effects of tourism impacts on resident quality of life: Evidence from rural midwestern communities in USA. *International Journal of Tourism Sciences*, *11*(2), 161–186. doi:10.1080/15980634.2011.11434643

Yuksel, A., Yuksel, F., & Bilim, Y. (2010). Destination attachment: Effects on customer satisfaction and cognitive, affective and conative loyalty. *Tourism Management*, *31*(2), 274–284. doi:10.1016/j.tourman.2009.03.007

Yung, R., & Lattimore, C. K. (2018). Same, same, but different: The influence of children in Asian family travel. In E. Yang & C. Khoo-Lattimore (Eds.), Asian Cultures and Contemporary Tourism. Springer Publication.

Yung, R., & Khoo-Lattimore, C. (2019). New realities: A systematic literature review on virtual reality and augmented reality in tourism research. *Current Issues in Tourism*, *22*(17), 2056–2081. doi:10.1080/13683500.2017.1417359

Yürük, P., Akyol, A., & Şimşek, G. G. (2017). Analyzing the effects of social impacts of events on satisfaction and loyalty. *Tourism Management*, *60*(June), 367–378. doi:10.1016/j.tourman.2016.12.016

Yustisiana, R. A. (2018). Elements of Nature As the Object of Escapism in Maya Angelou'S "Woman Work.". *Journal of Language and Literature*, *18*(02), 158–167. doi:10.24071/joll.2018.180207

Zatori, A. (2013). *Tourism Experience Creation From a Business Perspective*. Corvinus University of Budapest. Retrieved from http://phd.lib.uni-corvinus.hu/801/7/Zatori_Anita_den.pdf

Zatori, A., Smith, M. K., & Puczko, L. (2018). Experience-involvement, memorability and authenticity: The service provider's effect on tourist experience. *Tourism Management*, *67*, 111–126. doi:10.1016/j.tourman.2017.12.013

Zhang, H., Wu, Y., & Buhalis, D. (2018). A model of perceived image, memorable tourism experiences and revisit intention. *Journal of Destination Marketing and Management*, *8*, 326–336. doi:10.1016/j.jdmm.2017.06.004

Zhang, H., Fu, X., Cai, L. A., & Lu, L. (2014). Destination image and tourist loyalty: A meta-analysis. *Tourism Management*, *40*, 213–223. doi:10.1016/j.tourman.2013.06.006

Zhang, H., & Xu, H. (2019). A structural model of liminal experience in tourism. *Tourism Management*, *71*, 84–98. doi:10.1016/j.tourman.2018.09.015

Zhang, J., & Fu, X. (2015). The Influence of Background Music of Video Games on Immersion. *Journal of Psychology & Psychotherapy*, *05*(04).

Zhang, J., Inbakaran, R., & Jackson, M. S. (2006). Understanding community attitudes towards tourism and host-guest interaction in the urban-rural border region. *Tourism Geographies*, *8*(2), 182–204. doi:10.1080/14616680600585455

Ziakas, V., & Costa, C. A. (2011). Event portfolio and multi-purpose development: Establishing the conceptual grounds. *Sport Management Review*, *14*(4), 409–423. doi:10.1016/j.smr.2010.09.003

Zilahy, G. (2004). Organizational factors determining the implementation of cleaner production measures in the corporate sector. *Journal of Cleaner Production*, *12*(4), 311–319. doi:10.1016/S0959-6526(03)00016-7

About the Contributors

Patrícia Pinto holds a PhD in Quantitative Methods Applied to Economics and Management (QMAEM) from the University of Algarve. She is Associate Professor at the Faculty of Economics, University of Algarve, Portugal, where she got the Aggregation title. She is Director of the PhD Program in QMAEM and Coordinator of the Research Centre in Tourism, Sustainability and Well-being (CinTurs). Her research interests include applications of statistical methods in tourism and marketing. She has published in influential journals in the fields of Tourism Economics and Management.

Manuela Guerreiro is an Auxiliary Professor at University of Algarve (Portugal). PhD in Economic and Management Sciences, Msc in Cultural Management, and graduated in Communication Sciences. Director of the Master in Marketing Management. Co-coordinator of the Research Centre CinTurs. Research interests: Marketing and tourism, especially in brand management and place branding, consumer behavior, e-Marketing and social media, tourism experiences, arts, and events Marketing. Researcher at CinTurs.

* * *

Dora Agapito holds a PhD in Tourism, an MA in Marketing and a BSc in Communication Sciences. She is an Assistant Professor at the University of Algarve (Faculty of Economics) and integrated researcher at the Research Centre for Tourism, Sustainability and Well-being (Cinturs), Portugal. Previously, she was a Senior Lecturer in Tourism and Events Management at the University of Lincoln, UK. Her research focuses on responsible tourism management, sensory-informed tourist experiences, experience design, destination management, sustainable tourism experiences, consumer psychology and social marketing.

Erdem Akkan is currently a full-time faculty member as an assistant professor in the Maritime Faculty, Department of Maritime Business Administration at Mersin University. He received his Ph.D. degree in marketing from Çukurova University, M.A. degree in marketing from Anadolu University and B.A. degree in the department of business administration from Anadolu University. He currently lectures principles of marketing, production management, maritime English and entrepreneurship courses. His research interests include; sustainable logistics, corporate social responsibility, digital marketing and marketing ethics.

Sofia Almeida holds a PhD in Tourism and a degree in Business Communication. She is Assistant Professor at the Faculty of Tourism and Hospitality (FTH), Universidade Europeia. Her research focuses on the areas of hotel management, networks, coopetition, hotel marketing consortia, hotel branding and digital marketing. Sofia has been publishing in hospitality studies with several of her colleagues, as in Cooperation in the Hospitality of the Center: Myth or Reality? with C. Costa and J. M. Simões (Edições Actual, Almedina, 2018) and The network cooperation as a lever factor in Hospitality (Journal of Tourism and Development, 2015). She is a research member of the Centre for Geographical Studies, IGOT, Universidade Lisboa.

Mário Ascenção is a Principal Lecturer at Haaga-Helia UAS, a Certified Experience Expert by the Lapland Centre of Expertise for the Experience Industry (now House of Lapland), an Accredited Service Design Master Trainer by the Service Design Network (SDN), and a Certified Facilitator of LEGO® SERIOUS PLAY® method by the Association of Master Trainers. He is also a Certified Hospitality Educator (CHE®) and a Certified Guest Service Professional (CGSP®) by the American Hotel & Lodging Educational Institute.

Sezen Bozyiğit is currently a full-time faculty member as an associate professor in the Applied Technology and Management School, Department of International Trade and Logistics at Tarsus University. She received her Ph.D. degree in marketing from Çukurova University; MBA degree in Business Administration from University of Cag and her B.A. degree in the department of public administration from University of Ankara. She currently lectures consumer behavior, international marketing, principles of marketing, research methods, corporate social responsibility and transportation management courses. Her research interests include; logistics, digital marketing, consumer socialization process and consumer behavior.

Sónia Cabeça, PhD, Sociologist, with thesis on fanaticism, and cultural forms. Integrated researcher at the Centre for Spatial and Organizational Dynamics (CIEO) and researcher at the UNESCO chair in Intangible Heritage and Traditional Know-How: Linking Heritage. Specialist in Ethnossociology and Intangible Heritage, devoted several years to the study of Cante Alentejano, a traditional chant from Alentejo, and its bearers, having written the doctoral thesis "Structure and Formation Process of Cultural Forms: the case of Cante Alentejano" (2016). Is currently post doctorate in the project CREATOUR, working on creative tourism, cultural mapping, cultural heritage, and sustainability.

Ana Cláudia Campos holds a PhD in Tourism, a MA in Tourism Management, and a BA in Philosophy. Currently she is Assistant Professor at the Faculty of Tourism and Hospitality, European University of Lisbon, Portugal. Her research interests are in tourism marketing, tourism experience, tourist psychology and co-creation. She is a member of the Research Centre for Spatial and Organizational Dynamics (CIEO), University of Algarve. Ana has published several articles and book chapters with fellow authors J. Mendes, P. Valle and N. Scott, such as Co-creation of tourist experiences: A literature review in Current Issues in Tourism (2018) and Co-creating animal-based tourist experiences: attention, involvement and memorability in Tourism Management (2017). More recent publications include joint work with J. Ramos on creative tourism and salt tourism, e.g. Tourism in a Salt Pan: Does Creativity Matter? (IGI Global, 2019). She also contributed to The Encyclopedia of Sustainable Tourism by editors C. Cater, B. Garrod, & T. Low (CABI, 2016).

Fatima Lampreia Carvalho, PhD in Government (Essex University, UK), MA in Political Sciences (IUPERJ) and MA in Theology (ACU) works at Project TurExperience – "Tourist experiences' impacts on the destination image: searching for new opportunities to the Algarve ", financed by the European Fund for Regional Development through the Operational Program CRESC Algarve 2020, at the Faculty of Economics (CIEO-CinTurs), University of Algarve. Fatima is research associate of CinTurs with current research interest in destination image, overtourism and tourism sustainability and qualitative methods.

Chung-Shing Chan is a Research Assistant Professor at the Department of Geography and Resource Management, The Chinese University of Hong Kong. His doctoral research investigates the potential of green resources for city branding in Hong Kong. His teaching and research interests include place branding and marketing, sustainable tourism and eco-tourism.

Chen Cui holds a master degree in International Tourism Management in the School of Hospitality and Tourism Management and the University of Surrey, UK. Her research interest focuses on host community support for tourism development, quality-of-life and well-being.

Fábia Pires Esteves is a PhD candidate in Management and a MSc in Service Management from the Faculty of Economics, University of Porto (Portugal). She is a member of the Artificial Intelligence and Decision Support group of the internationally oriented multidisciplinary Associate Laboratory INESC TEC, and also works as a manager of cultural, environmental and social innovation projects. Her current research interests include consumer behaviour, relational marketing, digital marketing based on video games, and consumer experiences in tourism.

Alexandra Gonçalves was the Regional Director of Culture of the Algarve, and councilor of the Municipality of Faro – areas of Culture, Social Action, Health, Advertising, Economy and Tourism, from 2009 to 2013. PhD in Tourism by the University of Évora and Master in Management and Administration of Cultural Heritage, by the Faculty of Human and Social Sciences of University of Algarve and University of Paris-8, she is currently the Head Director of the School of Management, Hospitality and Tourism from the University of Algarve. She has several articles and research published in tourism, cultural heritage and events. She collaborates in several research projects in these areas, highlighting the CREATOUR project, where she is the lead researcher for the regional team of the University of Algarve.

Jorge André Guerreiro holds a degree and MA in Sociology from the Faculty of Economics of the University of Algarve, having written his dissertation on lifestyle mobilities and authenticity in tourism. Currently is a PhD student of sociology, specializing in tourism gentrification and anti-tourism social movements in Portugal. Guerreiro has published in fields such as philosophy of science, poverty and social exclusion, regional economy and in sociology of tourism.

Cláudia Henriques has a PhD in Economics (Algarve University, 2002), thesis under the subject of "Management and Planning of Sustainable Tourism in cities and historic city quarters. The case of Lisbon". She is lecturer of Tourism Economics, Cultural and Creative Sector and Tourism Planning and Managing at Algarve University (ESGHT_UALG) from 1993 until today. Her Research interests are in Cultural Tourism, Tourism Planning and Management; Cultural and Tourism Policies and Cultural heritage (Historic Centres). She is member of CINTurs (Research Centre for Tourism, Sustainability and Well-being - Centro de Investigação em Turismo, Sustenta-bilidade e Bem-estar - CinTurs).

Yo Han Kim holds a Master Degree in International Tourism Management at the University of Surrey, UK. His research interests concern sustainable tourism development, residents support for tourism development and residents' perceptions of social-cultural impacts of tourism within local communities.

Milene Lança holds a Degree and a Master in Sociology assigned by the University of Algarve (Portugal). She is currently PhD student in Tourism at the Faculty of Economics of this University. She received a doctoral scholarship from the Portuguese Foundation for Science and Technology and she is researcher at the Research Centre for Tourism, Sustainability and Well-being (CinTurs). Her PhD thesis addresses the relationship between sexuality and tourism, which are her main scientific interests.

João Filipe Marques holds a PhD in Sociology by the École des Hautes Études de Paris and a BA and a MA in Anthropology by the New University of Lisbon. He is Assistant Professor at the Faculty of Economics of the University of Algarve and Integrated Researcher at the Research Center for Tourism Sustainability and Well-being (CinTurs – UAlg). Currently he is the Director of the Sociology Master Program. He has been teaching Social Sciences and Sociological Theories since the 1990's and is main field of research has been the Sociology of Racism and Ethnicity. Recently is research interests shifted towards the sociology of tourism, leisure and travel and he has been teaching Sociology of Tourism in the Sociology Programs of the University of Algarve.

Margarida Mascarenhas, PhD in Sport Sciences, is Professor in the scientific area of Sport Management at the University of Lisbon and Integrated Researcher at the Research Center for Tourism, Sustainability and Well-being at the University of Algarve. Her research interests are located in the field of sports management, namely sports finance and economics, sports strategy, sports sustainable development, sports environmental sustainability, sports events and sports tourism.

Nelson de Matos is an adjunct professor at the University of the Algarve, with experience in the hospitality and marketing industry. He holds a PhD in Tourism from the same university. His interests are tourism experience and consumer behaviour.

Júlio da Costa Mendes is a PhD holder in Management, specializing in strategy and organizational behavior, and a MA holder in Economic and Management Sciences. Assistant Professor and a member of the Research Centre on Spatial and Organizational Dynamics, he is also the Director of the Master Course in Tourism Organizations Management and a member of the coordinating body of the PhD Program in Tourism. His current research interests include excellence and integrated quality management in tourism destinations, tourism experience, events management, tourism destinations image, and branding and customer satisfaction. He has published both in national and international journals.

Mayu Moritomo holds a Master Degree in International Tourism Development at the University of Surrey, UK. Her research interests concern sustainable tourism development and residents' perceptions of social-cultural impacts of tourism within local communities. Currently, she works for KSC Co LTD

Consulting company as an engineer with specialisation in Tourism, Public Transport and Urban Planning in Tokyo, Japan.

Dália Paulo, Museologist and Cultural Management. Municipal Director at Loulé's Municipality. President of the Board Acesso Cultura (2019/2021). Master in Art History (2007), University of Algarve. Post-graduated in Roman Arqcheology (1999), University of Coimbra. Degree in History –Archaeology (1996), University of Coimbra. She holds the PhD curricular part in Socialmuseology at the University Lusófona de Humanidades e Tecnologias, Lisbon. Invited expert at the University of Algarve in different Masters and faculties.

Rui Pedro is a PhD candidate at the University of Algarve.

Elsa Pereira, PhD in Sport Sciences, is professor at University of Algarve and Integrated Researcher at the Research Centre for Tourism, Sustainability and Well-being (CinTurs), University of Algarve. Her research interests are situated in the field of Sport Sciences, namely sport management, sport strategy and sustainable development, sports events, sport tourism and nautical tourism.

Célia M. Q. Ramos graduated in Computer Engineering from the University of Coimbra, obtained her Master in Electrical and Computers Engineering from the Higher Technical Institute, Lisbon University, and the PhD in Econometrics in the University of the Algarve (UALG), Faculty of Economics, Portugal. She is Adjunct Professor at School for Management, Hospitality and Tourism, also in the UALG, where she lectures computer science. Areas of research and special interest include conception and development of information systems, tourism information systems, big data, e-tourism, econometric modeling and panel-data models. She is member of CinTurs.

Fernando Bessa Ribeiro holds a PhD in Social Sciences assigned by the University of Trás-os-Montes and Alto Douro (Portugal). He is currently Associate Professor of the Sociology Department at the University of Minho and Integrated Researcher at CICS.NOVA.UMinho. Capitalism and development, gender and sexuality, tourism, and HIV/AIDS are his main research topics, on which he has published several books and articles in national and international journals. More recently, he is also working on urban issues, on a project financed by the Portuguese Foundation for Science and Technology.

Manuel Alector Ribeiro, Ph.D., is a Lecturer in Tourism Management at the University of Surrey, UK. His research focuses on applied quantitative methods and modelling of community support for sustainable tourism development, consumer behaviour, gender, and entrepreneurship in tourism and related services. His work is published in leading international journals in the field of tourism and he is currently the editor of the Methodology section of The Service Industries Journal.

Francisco Riquel is an associate lecturer of business administration at the University of Huelva (Spain). Active member of its research group on "Innovation and Development Strategies in Tourism Firms" (GEIDETUR). His research is mainly focused on the study of golf tourism and environmental management.

Bernadete Dias Sequeira, PhD in Sociology (University of Évora), Msc in Organization and Information Systems (Department of Management, University of Évora), and graduated in Sociology (University of Évora). Assistant Professor at the Faculty of Economics, University of Algarve (Portugal). She is member of the Interdisciplinary Centre of Social Sciences – CICS.NOVA. Current research interests includes Sociology of Organizations, Sociology of knowledge, knowledge Management, Human Resource Management and Tourism.

João Albino Silva is Full Professor at the Faculty of Economics, University of the Algarve (Portugal) and Member of the Research Center for Tourism, Sustainability and Well-Being (2011). He is the director of the PhD Program in Tourism, University of the Algarve, since 2007, and acts as a member of International and National Editorial Boards of scientific journals in tourism, management and economics. He has published in influential journals in the fields of Tourism Economics and Management.

Ivo Sousa-Ferreira holds a Master in Mathematics from the University of Madeira. Currently, he is a PhD candidate in Statistics and Operational Research at the Faculty of Science of the University of Lisbon and a researcher member of the Center of Statistics and Applications. During the period of 2016-2018, he was lecturer at the Faculty of Exact Science and Engineering of the University of Madeira, where taught in the areas of Statistics and Probability.

Mirian Nogueira Tavares is Associate Professor at the University of Algarve. With academic studies in Communication Sciences, Semiotics and Cultural Studies (Ph.D. in Communication and Contemporary Culture, from the Federal University of Bahia), she has developed research work and theoretical production in fields related to Cinema, Literature and other Arts, as well as artistic and aesthetic film studies. As a lecturer at the University of Algarve, she has participated in the development of the Visual Arts degree, the Master's programmes in Communication, Culture and Arts and Cultural Management and the PhD programmes in Communication, Culture and Arts and Digital Media and Arts. She is the current Coordinator of CIAC - http://ciac.pt/index.php/pt/ (Arts and communication Research Centre), funded by FCT. In the past, she was assistant to the State Secretary for Culture of the State of Sergipe, in Brazil, where she coordinated several cultural activities.

Isabel Teotónio has Master in Probability and Statistics, Faculty of Sciences, University of Lisbon, Lisbon, Portugal. Lecturer at Algarve University, Faro, Portugal.

Alfonso Vargas-Sánchez is full professor of Strategic Management at the Spanish University of Huelva. Active member of its research group on "Innovation and Development Strategies in Tourist Firms" (GEIDETUR). Author of a number of papers published in top journals, particularly in the field of tourism management. Prof. Vargas-Sánchez is the editor of "Enlightening Tourism. A Pathmaking Journal" (indexed in Scopus), since its launch in 2011.

Margarida Viegas is an Adjunct Professor of Applied Quantitative Methods at School of Management, Hospitality and Tourism of the University of the Algarve. She holds an undergraduate degree in Decisional Systems Engineering from the ISMA-COCITE, a postgraduate degree in Financial Management from the University of Algarve and a postgraduate degree in Strategic Management and E-Business from the University of Huelva. She holds a Masters degree in Statistics and Information Management from the ISEGI-New University of Lisbon. She holds a European PhD in Economics and Management from the University of Huelva. Her research interests include sustainable tourism and environmental management. She has participated in several national and international conferences and published in national and international refereed journals.

Manuel Vong holds a PhD in Tourism from the University of Algarve (Portugal). He is Senior Lecture at the School of Tourism and Hospitality Management, Dili Institute of Technology (DIT), Timor-Leste, and Member of the Centre of Applied Research and Policy Studies (CARPS), Dili Institute of Technology (DIT). He is the Director of the Graduate School, Master of Business Administration (MBA), and Coordinator of Tourism Management Program, Dili Institute of Technology. He acts as a member of Timor-Leste Journal of Business and Management (TLJBM) in the fields of Tourism, Hospitality and Management. He has published in influential journals in the field of Tourism.

Magda Wikesjö, MSc in Marketing Management (Faculty of Economics, University of Algarve), and graduated in Landscape Architecture (University of Algarve). Lecturer at the Faculty of Economics, University of Algarve (Portugal). She is currently preparing her PhD in Economic and Management Sciences at University of Algarve. Current research interests includes, Consumer Behaviour, Cultural Event Experience, Eventscapes, Marketing Management and Tourism.

Kyle Maurice Woosnam, Ph.D., is an Associate Professor in Parks, Recreation and Tourism Management in the Warnell School of Forestry and Natural Resources at the University of Georgia, USA. Kyle's research interests concern social-cultural and economic impacts of tourism, resident-tourist interactions within tourist destinations, and sustainable tourism development and planning. Over the course of the last 15 years, he has undertaken numerous research projects focusing on community-level tourism impacts, in more than 25 countries.

Index

F

G

H

I

L

M

N

O

P

Q

R

S

Ensure Quality Research is Introduced to the Academic Community

Become an IGI Global Reviewer for Authored Book Projects

The overall success of an authored book project is dependent on quality and timely reviews.

In this competitive age of scholarly publishing, constructive and timely feedback significantly expedites the turnaround time of manuscripts from submission to acceptance, allowing the publication and discovery of forward-thinking research at a much more expeditious rate. Several IGI Global authored book projects are currently seeking highly-qualified experts in the field to fill vacancies on their respective editorial review boards:

Applications and Inquiries may be sent to:
development@igi-global.com

Applicants must have a doctorate (or an equivalent degree) as well as publishing and reviewing experience. Reviewers are asked to complete the open-ended evaluation questions with as much detail as possible in a timely, collegial, and constructive manner. All reviewers' tenures run for one-year terms on the editorial review boards and are expected to complete at least three reviews per term. Upon successful completion of this term, reviewers can be considered for an additional term.

If you have a colleague that may be interested in this opportunity,
we encourage you to share this information with them.

www.ingramcontent.com/pod-product-compliance
Lightning Source LLC
LaVergne TN
LVHW080407011026
849161LV00009B/2361
* 9 7 8 1 7 9 9 8 3 1 5 6 3 *